# 10<sup>TH</sup> ESAFORM CONFERENCE ON MATERIAL FORMING

To learn more about AIP Conference Proceedings, including the
Conference Proceedings Series, please visit the webpage
**http://proceedings.aip.org/proceedings**

# 10TH ESAFORM CONFERENCE ON MATERIAL FORMING

Zaragoza, Spain    18 – 20 April 2007

**PART A**

*EDITORS*

## Elías Cueto
*Aragon Institute of Engineering Research (I3A)*
*University of Zaragoza*
*Zaragoza, Spain*

## Francisco Chinesta
*Laboratoire de Mécanique des Systèmes et Procédés*
*UMR CNRS-ENSAM-Polytech'Orleans*
*Paris, France*

*All papers have been peer-reviewed*

**SPONSORING ORGANIZATIONS**
European Scientific Association for Material Forming (ESAFORM)
European Community on Computational Methods in Applied Sciences (ECCOMAS)
Spanish Ministry of Education and Science
Regional Government of Aragón
University of Zaragoza

Melville, New York, 2007
**AIP CONFERENCE PROCEEDINGS ■ VOLUME 907**

PHYS

**Editors:**

Elías Cueto
Aragon Institute of Engineering Research (I3A)
University of Zaragoza
Betancourt Building, Maria de Luna, s.n.
E-50018 Zaragoza
Spain

E-mail: ecueto@unizar.es

Francisco Chinesta
Laboratoire de Mécanique des Systémes et Procédés
UMR 8106 CNRS-ENSA-ESEM
151 Bd de l'Hôpital
F-75013 Paris
France

E-mail: francisco.chinesta@paris.ensam.fr

L.C. Catalog Card No. 2007902208
ISBN 978-0-7354-0414-4
ISSN 0094-243X
Printed in the United States of America

cm 6/27/07

TA 401
.3
E83
2007
v. A
PHYS

# PART A

## 1. INVERSE ANALYSIS OPTIMIZATION AND STOCHASTIC APPROACHES
### *(E. Massoni and T. Van Den Boogaard)*

## 2. MULTISCALE APPROACHES
### *(A. M. Habraken and S. Bouvier)*

*The italicized names below each section title represent the coordinators of each mini-symposium.

### 3. SHEET METAL FORMING
*(P. Picart, T. Welo, and T. Meinders)*

vii

### 4. ANISOTROPY AND FORMABILITY
*(D. Banabic, F. Barlat, O. Cazacu, and T. Kuwabara)*

## 5. HYDROFORMING
### (J. C. Gelin)

## 6. FORGING AND ROLLING
### *(J. L. Chenot and J. Kusiak)*

## 7. EXTRUSION AND DRAWING
### *(S. Støren and J. Huetink)*

## 8. MICROFORMING AND NANOSTRUCTURED MATERIALS
### (U. Engel and A. Rosochowski)

## 9. MACHINING AND CUTTING
### (F. Micari and Ph. Lorong)

xiii

# PART B

## 10. STRUCTURES AND PROPERTIES OF POLYMERS
### *(J. M. Haudin and E. Mitsoulis)*

# 11. PROCESSING OF POLYMERS

*(J. F. Agassant, G. Menary, F. Schmid, J. Covas, and F. Dupret)*

## 12. COMPOSITES
### (Ph. Boisse, R. Akkerman, J. Cao, A. Long, and S. Lomov)

### 13. SEMI-SOLID PROCESSES
*(H. Atkinson, A. Rassili, and G. Hirt)*

### 14. HEAT TRANSFER MODELING
*(F. Schmidt, Y. Le Maoult, and K. Mocellin)*

## 15. SUPERPLASTIC FORMING OF ALUMINIUM AND MAGNESIUM ALLOYS
### (G. Bernhart and V. Berdin)

## 16. NEW AND ADVANCED NUMERICAL STRATEGIES
## IN FORMING PROCESSES SIMULATION
### *(F. Chinesta and E. Cueto)*

# $10^{th}$ ESAFORM Conference on Material Forming. Zaragoza, Spain

E. Cueto* and F. Chinesta[†]

*I3A. University of Zaragoza, Spain
[†]ENSAM, Paris, France

## PREFACE

The series of ESAFORM Conferences on Material Forming has arrived to its $10^{th}$ anniversary. Since the first conference, held in Sophia-Antipolis, France, in 1998, ESAFORM conferences have continued growing continuously. The need for production efficiency and quality in industries requires using the very best approaches to modelize and simulate the forming processes, and the ESAFORM conferences have been witnesses of this fact.

In this year's conference, more than 250 papers have been accepted. In addition, it features for the first time an ECCOMAS thematic conference devoted to "new and advanced numerical strategies in forming simulation", which has been traditionally one of the mini-symposia of the conference.

On behalf of the organizing team, we would like to thank all the symposia coordinators for their excellent (and hard) work in reviewing the manuscripts and to wish all the participants in the conference the most pleasant stay at Zaragoza, a city with a long history.

*Elías Cueto*
*Francisco Chinesta*
*Zaragoza and Paris, February 2007*

## Organizing Committee

Elías Cueto. I3A. University of Zaragoza
Francisco Chinesta. ENSAM Paris
Iciar Alfaro. I3A. University of Zaragoza
David González. I3A. University of Zaragoza
David Bel. I3A. University of Zaragoza
Maurice Touratier. ENSAM Paris
Manuel Doblaré. I3A. University of Zaragoza

## Symposia Organizers

E. Massoni
T. van den Boogaard
A.M. Habraken
S. Bouvier
P. Picard
T. Welo
T. Meinders
D. Banabic
F. Barlat
O. Cazacu
T. Kuwabara
J.C. Gelin
J.L. Chenot
J. Kusiak
S. Støren
J. Huetink
U. Engel
A. Rosochowski
F. Micari
Ph. Lorong
J.M. Haudin

E. Mitsoulis
J.F. Agassant
G. Menary
F. Schmid
J. Covas
F. Dupret
Ph. Boisse
R. Akkerman
J. Cao
A. Long
S. Lomov
H. Atkinson
A. Rassili
G. Hirt
F. Schmidt
Y. Le Maoult
K. Mocellin
G. Bernhart
V. Berdin
F. Chinesta
E. Cueto

Dr. Valey BERDIN
Mr. R.K. ISLAMGALIEV
Prof. Miroslav PLANCAK
Prof. Elías CUETO
Dr. Roger ANDERSSON
Prof. Kjell MATTIASSON
Dr. Jean-Marie M DREZET
Prof. R. AKKERMAN
Prof. Marc GEERS
Dr. Timo MEINDERS
Dr. A.H. VAN DEN BOOGAARD
Prof. Andrew LONG
Mr. Qin YI
Dr. Frédéric BARLAT
Prof. J. CAO
Prof. O. CAZACU
Prof. Wojciech Z. MISIOLEK
Dr. R. SHIVPURI

Prof. Robert H.WAGONER
Dr. Jeong-Whan YOON
Prof. K. CHUNG
Prof. T. KUWABARA
Prof. K. NARASHIMHAM
Prof. Dr. ir. A. Herman TEKKAYA
Prof. G. ALMARAZ DOMINGUEZ
Prof. Stefania BRUSCHI
Prof. Bruno BUCHMAYR
Prof. Julie CHEN
Prof. D. S. COMSA
Prof. Eduardo DVORKIN
Prof. Gang FANG
Prof. Sergio R. IDELSOHN
Prof. Antti S. KORHONEN
Prof. Veli-Tapani KUOKKALA
Prof. Marion MERKLEIN

# 1 – INVERSE ANALYSIS OPTIMIZATION AND STOCHASTIC APPROACHES

*(E. Massoni and T. van den Boogaard)*

# Modelling, screening, and solving of optimisation problems: Application to industrial metal forming processes

M.H.A. Bonte, A.H. van den Boogaard and E. Veldman

*University of Twente, P.O. Box 217, 7500 AE Enschede, The Netherlands*

**Abstract.** Coupling Finite Element (FEM) simulations to mathematical optimisation techniques provides a high potential to improve industrial metal forming processes. In order to optimise these processes, all kind of optimisation problems need to be mathematically modelled and subsequently solved using an appropriate optimisation algorithm. Although the modelling part greatly determines the final outcome of optimisation, the main focus in most publications until now was on the solving part of mathematical optimisation, i.e. algorithm development. Modelling is generally performed in an arbitrary way.

In this paper, we propose an optimisation strategy for metal forming processes using FEM. It consists of three stages: a structured methodology for modelling optimisation problems, screening for design variable reduction, and a generally applicable optimisation algorithm. The strategy is applied to solve manufacturing problems for an industrial deep drawing process.

**Keywords:** optimisation modelling, screening, optimisation algorithm, metal forming, FEM

## INTRODUCTION

Product improvement and cost reduction have always been important goals in the metal forming industry. The rise of Finite Element simulations for metal forming processes has contributed to these goals in a major way. More recently, coupling FEM simulations to mathematical optimisation techniques has shown the potential to make a further giant contribution to product improvement and cost reduction.

Mathematical optimisation consists of the modelling and solving of optimisation problems. Much research on the optimisation of metal forming processes has been published during the last couple of years, see e.g. [1]. Most of this research focussed on the solving part of optimisation, i.e. the development of a specific algorithm and its application to a specific optimisation problem for a specific metal forming process. Although the modelling of the optimisation problem determines the final outcome of optimisation, it is often done quite arbitrarily. To our opinion, much more attention should be paid to the proper modelling of optimisation problems in metal forming, rather than implementing another slightly more efficient optimisation algorithm.

In this paper, we propose a generally applicable optimisation strategy which makes use of FEM simulations of metal forming processes. The strategy includes both a structured methodology for modelling and an algorithm for solving optimisation problems. The optimisation strategy is applied to the optimisation of an automotive deep drawing process.

CP907, *10th ESAFORM Conference on Material Forming*, edited by E. Cueto and F. Chinesta
© 2007 American Institute of Physics 978-0-7354-0414-4/07/$23.00

# THE OPTIMISATION STRATEGY FOR METAL FORMING PROCESSES

The proposed optimisation strategy consists of three stages: modelling, screening and solving.

## Modelling

The first stage is to model the optimisation problem, i.e. defining objective function, constraints and design variables. We propose a structured 7 step methodology for modelling optimisation problems in metal forming [2]:

1. Determine the appropriate optimisation stage
2. Select only the necessary responses
3. Select one response as objective function, the others as implicit constraints
4. Quantify the objective function and implicit constraints
5. Select possible design variables
6. Define the ranges on the design variables
7. Identify explicit constraints

Being based on the generally applicable Product Development Cycle [2, 3], this modelling methodology can be applied to any metal forming process, product and problem. In the end, it yields a specific mathematical optimisation model, which can subsequently be solved using a suitable optimisation algorithm. The 7 step methodology is further demonstrated in the next section when it is applied to an industrial deep drawing process.

## Screening

If many design variables are still present in the modelled optimisation problem, screening techniques can be applied to reduce the number of design variables. We propose to apply a Resolution III fractional factorial DOE strategy [5] for screening. Resolution III designs allow for independently estimating the linear effects of the design variables on the responses (objective function and constraints). After having run the corresponding FEM simulations, the linear effects can be estimated by applying statistical techniques such as ANalysis Of VAriance (ANOVA) [5]. The amount and direction of the effect of each variable on each response can be nicely displayed in Pareto and Effect plots. Examples of Pareto and Effect plots are presented in the Figures 1(a) and (b), respectively.

Using these techniques, the variables with the largest effects may be kept in the optimisation model whereas the variables having less effect may be omitted. In such a way, the amount of design variables may be significantly decreased while maintaining control over objective function and constraints during optimisation.

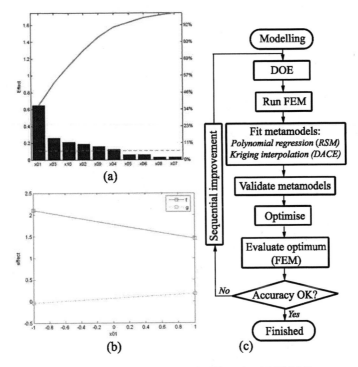

**FIGURE 1.** (a) Pareto plot; (b) Effect plot; (c) SAO [4]

## Solving

The final stage of the optimisation strategy is to solve the optimisation problem by a suitable algorithm. We propose the Sequential Approximate Optimisation (SAO) algorithm presented in several publications, see e.g. [4]. An overview of the algorithm is presented in Figure 1. It comprises a spacefilling Latin Hypercubes Design Of Experiments (DOE) strategy, RSM and Kriging metamodelling and validation techniques, and a multistart SQP algorithm for optimising the metamodels. The algorithm allows for sequential improvement of the accuracy.

## APPLICATION TO DEEP DRAWING

The optimisation strategy presented in the previous section is applied to an industrial metal forming process. The automotive part concerned is shown in Figure 2(a). It is manufactured from 3 mm thick steel sheet in 19 stages, of which 10 are deep drawing operations.

In an early stage, cracks were encountered during or after the deep drawing process ("reference process"). The location of the cracks is indicated in Figure 2(a). To reduce

**(a)**        **(b)**

**FIGURE 2.** The automotive part: (a) Part and crack location; (b) FEM simulation

the scrap rate, some stages of the deep drawing process have been altered resulting in a significant decrease of the defect rate ("modified process").

To decrease material and control costs, the optimisation strategy proposed in the previous section is applied to this production problem. Two optimisation cases will be treated: (i) optimisation of the reference process; and (ii) optimisation of the modified process.

## Modelling

Steps 1, 2, 3 and 4 of the 7 step methodology for modelling the optimisation problem yielded the following response model:

$$\min f = \frac{d_{f1}}{d_{f1ref}} + \frac{d_{f2}}{d_{f2ref}} \tag{1}$$
$$s.t.\, g = d_g \leq 0$$

$f$ is the objective function that aims to reduce the crack occurrence. Two causes of the cracks have been determined [6]: (i) large deformations in the critical region; and (ii) a thickness concentration in the critical region. One can observe the thickness concentration in the FEM calculation presented in Figure 2(b). AutoForm has been adopted as FEM code. Both causes are visualised in the Forming Limit Diagram (FLD) in Figure 3(b): $d_{f1}$ depicts the definition of the amount of deformation in the critical region, $d_{f2}$ the distance to the line of equal thickness which represents the thickness in the critical region. $d_{f1ref}$ and $d_{f2ref}$ are the objective function values corresponding to the reference process. As an implicit constraint, necking is not allowed to occur. $d_g$ is the distance to the material dependent Forming Limit Curve (FLC), which is a measure

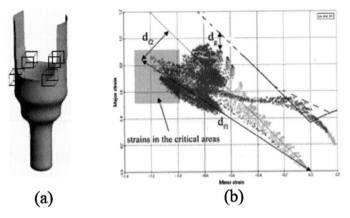

(a)                                    (b)

**FIGURE 3.**  (a) Definition of critical region; (b) Definition of responses

for necking. This response is not bounded to the critical region: necking is not allowed at any location throughout the automotive part.

Steps 5, 6 and 7 of the modelling methodology assisted in defining 10 design variables for this application: the blank size and thickness, punch and die radii and blank holder forces of several deep drawing operations, and coefficients of friction.

## Screening

Ten design variables is, however, still quite many to optimise using the SAO algorithm. Therefore, a $2_{III}^{(10-6)}$ design has been generated and the corresponding 16 FEM calculations have been conducted. ANOVA has been applied and the Pareto and Effect plots of $f$ and $g$ have been generated: Figure 1(a) showed the Pareto plot of $f$, Effect plots of $f$ and $g$ were displayed in Figure 1(b). It can be seen that the blank size is the most dominant variable. The influence of the other variables has been attributed to numerical noise. Hence, the blank size was the only variable taken into account for optimisation. The other 9 variables were kept at their original settings.

## Solving

The one variable optimisation problem has subsequently been solved using SAO. Table 1 presents the results, which were obtained after running 14 FEM calculations. Applying the optimisation strategy significantly reduced the objective function value of the reference process. The modified process still performs somewhat better. Table 1 also presents the results of optimising the modified process [6]. Note the further, significant improvement of the objective function, which suggests cracks can be reduced further using the proposed optimisation strategy.

7

**TABLE 1.** Optimisation results

| | Reference process | Modified process | Optimised reference process | Optimised modified process |
|---|---|---|---|---|
| Blank size | 204 | 204 | 209.9 | – |
| f | 2 | 1.4823 | 1.6732 | 0.9097 |
| g | -0.0171 | -0.0459 | -0.0049 | -0.0048 |

# CONCLUSIONS

A generally applicable optimisation strategy for metal forming processes using FEM has been proposed. It contains a structured methodology for modelling optimisation problems, screening techniques for design variable reduction, and a generally applicable algorithm for solving the optimisation problem. The strategy has been successfully applied to an industrial deep drawing process. The strategy has been and is being applied to other industrial metal forming processes, products and optimisation problems as well, which indicates its general applicability.

# ACKNOWLEDGMENTS

This research has been carried out in the framework of the project "Optimisation of Forming Processes MC1.03162". This project is part of the research programme of the Netherlands Institute for Metals Research (NIMR). The industrial partners co-operating in this project are gratefully acknowledged for their useful contributions to this research.

# REFERENCES

1. O. Schenk, and M. Hillmann, *Computers and Structures* **82**, 1695–1705 (2004).
2. M. Bonte, A. van den Boogaard, and J. Huétink, *Submitted to: Structural and Multidisciplinary Optimization* (2006).
3. K. Yang, and B. El-Haik, *Design For Six Sigma; A roadmap for Product Development*, McGraw-Hill, Inc., New York, USA, 2003, ISBN 0-07-141208-5.
4. M. Bonte, T. Do, L. Fourment, A. van den Boogaard, J. Huétink, and A. Habbal, "A comparison of optimisation algorithms for metal forming processes," in *Proceedings of ESAFORM*, Glasgow, UK, 2006, pp. 883–886.
5. R. Myers, and D. Montgomery, *Response Surface Methodology: Process and Product Optimization Using Designed Experiments*, John Wiley and Sons, Inc., New York, USA, 2002, 2nd edn., ISBN 0-471-41255-4.
6. E. Veldman, *Using optimisation techniques to solve a production problem*, CTW.06/TM-5544, University of Twente, Enschede, The Netherlands (2006).

# The robust optimisation of metal forming processes

M.H.A. Bonte, A.H. van den Boogaard and R. van Ravenswaaij

*University of Twente, P.O. Box 217, 7500 AE Enschede, The Netherlands*

**Abstract.** Robustness, reliability, optimisation and Finite Element simulations are of major importance to improve product quality and reduce costs in the metal forming industry. In this paper, we review several possibilities for combining these techniques and propose a robust optimisation strategy for metal forming processes. The importance of including robustness during optimisation is demonstrated by applying the robust optimisation strategy to an analytical test function: for constrained cases, deterministic optimisation will yield a scrap rate of about 50% whereas the robust counterpart reduced this to the required $3\sigma$ reliability level.

**Keywords:** optimisation, metal forming, FEM, robustness, reliability

## INTRODUCTION

Product improvement and cost saving have always been important goals in the metal forming industry. One way of achieving these two goals is optimising towards robust metal forming processes. A robust metal forming process will yield metal products at a more constant quality level. Hence, it will (i) improve the product' quality; and (ii) save costs because the number of non-feasible products (scrap) is decreased. Generally, optimisation strategies only include deterministic control variables. To assess the robustness of a metal forming process, the noise variables (e.g. material variation) need to be taken into account during optimisation.

In [1], we presented three ways to optimise towards robust metal forming processes using time consuming Finite Element simulations of these processes: deterministic optimisation, robust optimisation and reliability based optimisation. An example of deterministic optimisation to yield a robust deep drawing process has been included in the same paper. In this paper, we continue on robust optimisation techniques. We review several possibilities for optimising towards robust metal forming processes and present a robust optimisation strategy that includes both design and noise variables into optimisation. The robust optimisation strategy is demonstrated and compared to deterministic optimisation by application to an analytical test function.

## POSSIBILITIES FOR ROBUST OPTIMISATION

From the three possibilities published in [1], we propose to continue with robust optimisation techniques. This is based on the following considerations:

- Deterministic optimisation does not take into account process robustness and reliability during optimisation;

CP907, *10th ESAFORM Conference on Material Forming*, edited by E. Cueto and F. Chinesta
© 2007 American Institute of Physics 978-0-7354-0414-4/07/$23.00

- Reliability based optimisation algorithms are generally very time consuming;
- Robust optimisation takes into account process robustness and even reliability if a response distribution is assumed (e.g. a normal distribution);
- Robust optimisation is relatively efficient.

Two ways for robust optimisation are the Taguchi and Dual Response Surface Methods [2]. *Taguchi methods* are based on crossed orthogonal array Design Of Experiments (DOE) plans. After having run the physical or in our case numerical experiments, response measurements can be analysed based on ANalysis Of VAriance (ANOVA) and Signal-to-Noise ratios (S/N-ratios). Taguchi methods possess several severe disadvantages [3, 4]: (i) Use of the S/N-ratio implies the mean and variance of a response distribution are confounded; (ii) crossed orthogonal arrays lack flexibility and efficiency; (iii) Taguchi methods do not allow for sequential experimentation/optimisation.

Alternatively one can employ *Dual Response Surface Methods (DRSM)*: one Response Surface Model (RSM) is fitted for the mean and one for the variance of a response [2]. The most straight-forward way of robust optimisation using DRSM is *Direct Variance Modelling*: for each of the DOE points in the control variable space, one can perform an orthogonal array in the noise variable space to assess the probability distribution of the response for those control variable settings. Basically, this resembles Taguchi methods, but overcomes the three disadvantages mentioned above. Direct Variance Modelling is, however, very time consuming since noise variable assessment requires performing several FEM calculations for each control variable setting.

A much more efficient way of robust optimisation using DRSM is fitting one *single RSM metamodel* in both the control and noise design variable space, e.g. the following RSM metamodel which is quadratic in the design variable space and linear + interaction in the noise variable space:

$$\hat{y}(\mathbf{x},\mathbf{z}) = \beta_0 + \mathbf{x}^T\beta + \mathbf{x}^T\mathbf{B}\mathbf{x} + \mathbf{z}^T\gamma + \mathbf{x}^T\Delta\mathbf{z} + \varepsilon \tag{1}$$

where $\hat{y}$ is a single metamodel of a response dependent on the control variables $\mathbf{x}$ and noise variables $\mathbf{z}$. $\beta_0$, $\beta$, $\mathbf{B}$, $\gamma$ and $\Delta$ denote the fitted regression coefficients and $\varepsilon$ is the random error term. From Equation 1, one can analytically determine two RSM metamodels for mean and variance [2]:

$$\mu_y = E[\hat{y}(\mathbf{x},\mathbf{z})] = \beta_0 + \mathbf{x}^T\beta + \mathbf{x}^T\mathbf{B}\mathbf{x} \tag{2}$$
$$\sigma_y^2 = \text{var}[\hat{y}(\mathbf{x},\mathbf{z})] = \sigma_z^2(\gamma^T + \mathbf{x}^T\Delta)(\gamma + \Delta^T\mathbf{x}) + \sigma^2$$

with $\mu_y$ and $\sigma_y^2$ the metamodels for mean and variance of the response. Overcoming the disadvantages of the Taguchi method and being more efficient than Direct Variance Modelling, we propose a robust optimisation strategy based on fitting single response surfaces in the combined control-noise variable space.

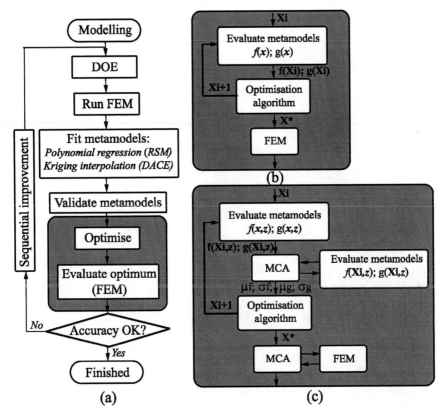

**FIGURE 1.** (a) Flow chart of the optimisation strategy; (b) Deterministic; (c) Robust

# A ROBUST OPTIMISATION STRATEGY FOR METAL FORMING PROCESSES

The proposed robust optimisation strategy is an extension of a deterministic optimisation strategy for metal forming processes presented in amongst others [5]. A flowchart of the optimisation strategy is presented in Figure 1(a). The robust optimisation strategy differs from the deterministic strategy in the modelling, optimisation and evaluation parts.

Concerning the modelling, noise variables are included in addition to deterministic control variables. For the noise variables, a normal distribution is assumed. For each response (objective function or constraint), one now obtains a response distribution ($\mu_y$ and $\sigma_y$) instead of a response value $y$. As objective function $f$ one can optimise $\mu_f$, $\sigma_f$ or a weighted sum $\mu_f \pm w\sigma_f$. If $\mu_f$ or $\sigma_f$ are optimised, it is advised to include the weighted sum as a constraint: this takes into account process reliability in the optimisation problem. Also other constraints $g$ are taken into account as a weighted sum $\mu_g \pm w\sigma_g$.

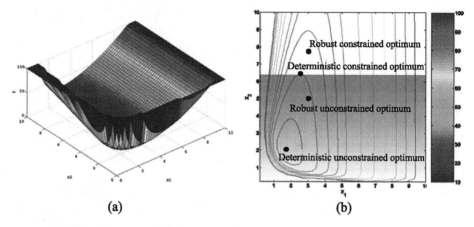

**FIGURE 2.** (a) Analytical test function; (b) Contour plot including optima

Figures 1(b) and (c) compare the differences in the optimisation algorithms and optimum evaluation for the deterministic and robust optimisation strategies. The difference in optimisation is the determination of the separate metamodels for $\mu_y$ and $\sigma_y$. This is done by Equation 2 when RSM is used as metamodelling technique; when Kriging is employed instead of RSM, an analytical derivation of $\mu_y$ and $\sigma_y$ is not possible. In this case we run a Monte Carlo Analysis (MCA) on the fitted metamodel as shown in Figure 1(c). The difference in the evaluation of the optimum $\mathbf{X}^*$ is that, in the deterministic case, this can be done by running one final FEM calculation. In case the robustness and reliability need to be assessed after optimisation, it is necessary to run an MCA using FEM calculations, which is quite time consuming.

## APPLICATION TO AN ANALYTICAL TEST FUNCTION

The robust optimisation strategy will now be applied to the analytical test function presented in Figure 2(a). Figure 2(b) presents the contour of this objective function as well as a constraint. The constrained deterministic optimisation problem is:

$$\min f = 12 + x_1^2 + \frac{1 + x_2^2}{x_1^2} + \frac{x_1^2 x_2^2 + 100}{(x_1 x_2)^4}; \text{ s.t. } g = 6.5 - x_2 \leq 0; \ 0.1 \leq x_1, x_2 \leq 10 \quad (3)$$

For the unconstrained deterministic optimisation model, the constraint $g$ is simply omitted. Both the unconstrained and constrained deterministic optima are presented in Figure 2(b).

The robust optimisation problem is modelled as follows:

$$\min \mu_f; \text{ s.t. } \mu_f + 3\sigma_f \leq 50; \ \mu_g + 3\sigma_g \leq 0; \ 1 \leq x_1, x_2 \sim N(\mu, 0.4) \leq 10 \quad (4)$$

12

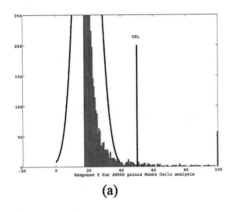 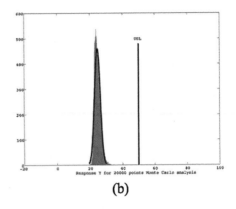

| (a) | (b) |

**FIGURE 3.** Response distributions: (a) Deterministic unconstrained optimum; (b) Robust unconstrained optimum

Again the unconstrained ($g$ omitted) and the constrained problem have been optimised, this time using the robust optimisation strategy. 100 function evaluations are run for each optimisation. Both corresponding optima are again displayed in Figure 2(b).

After optimisation, the reliability of all optima has been evaluated using an MCA of 20000 function evaluations. Figure 3 compares the results of deterministic and robust unconstrained optimisation. The scrap rate has been reduced from 0.92% for the deterministic optimum to $\ll 0.005\%$ for the robust optimum. The improvement of the robust optimisation strategy w.r.t. the deterministic one is even much more dramatic in constrained cases as depicted in Figure 4. For the deterministic optimum, the scrap rate due to violation of the constraint $g$ is 50.3% (Figure 4(b)). For the robust optimum, Figure 4(d) shows that the scrap rate has been reduced to 0.1%, which nicely corresponds to the $3\sigma$ reliability level modelled in Equation 4.

## CONCLUSIONS

Robustness, reliability, optimisation and Finite Element simulations are of major importance to improve product quality and reduce costs in the metal forming industry. In this paper, we proposed a robust optimisation strategy for metal forming processes. In addition to deterministic control variables, the strategy explicitly takes into account noise variables such as material variation and optimises probability distributions of objective function and constraints in order to achieve a robust and reliable metal forming process. The importance of including robustness during optimisation has been demonstrated by applying the robust optimisation strategy to an analytical test function: for constrained cases, deterministic optimisation will yield a scrap rate of about 50% whereas the robust optimisation strategy reduced this scrap rate to the demanded $3\sigma$ reliability level.

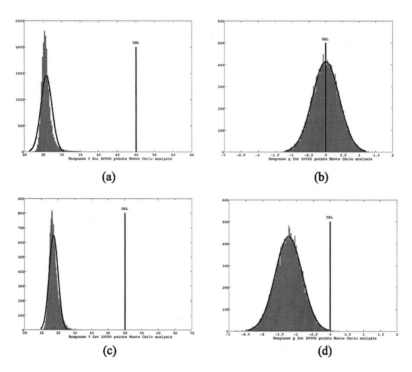

(a)                                 (b)

(c)                                 (d)

**FIGURE 4.** (a) Objective function distribution deterministic constrained optimum; (b) Constraint distribution deterministic constrained optimum; (c) Objective function distribution robust constrained optimum; (d) Constraint distribution robust constrained optimum

## ACKNOWLEDGMENTS

This research has been carried out in the framework of the project "Optimisation of Forming Processes MC1.03162". This project is part of the research programme of the Netherlands Institute for Metals Research (NIMR). The industrial partners co-operating in this project are gratefully acknowledged for their useful contributions to this research.

## REFERENCES

1. M. Bonte, A. van den Boogaard, and B. Carleer, "Optimising towards robust metal forming processes," in *Proceedings of ESAFORM*, Glasgow, UK, 2006, pp. 47–50.
2. R. Myers, and D. Montgomery, *Response Surface Methodology: Process and Product Optimization Using Designed Experiments,* John Wiley and Sons, Inc., New York, USA, 2002, 2nd edn., ISBN 0-471-41255-4.
3. R. Myers, A. Khuri, and G. Vining, *The American Statistician* **46**, 131–139 (1992).
4. V. Nair, *Technometrics* **34**, 127–161 (1992).
5. M. Bonte, A. van den Boogaard, and J. Huétink, *Submitted to: Structural and Multidisciplinary Optimization* (2006).

# A Practical Approach To Preform Design For Different Materials

Otto Harrer*, Christof Sommitsch*,**, Guntram Rüf*
and Bruno Buchmayr*

*Chair of Metal forming, University of Leoben
Franz Josefstr. 18, A-8700 Leoben, AUSTRIA

* **Christian Doppler Laboratory for Materials Modelling and Simulation, University of Leoben
Franz Josefstr. 18, A-8700 Leoben, AUSTRIA

**Abstract.** To forge an H-shaped cross section, various preform designs have been tested for steel 42CrMo4, aluminum 7075 and nickel base alloy 80 A (Bohler L306). The influence of different boundary conditions like temperature and friction on the preform and hence on the forming process have been investigated by means of two dimensional finite element analyses. Furthermore, the influence of the preform on the microstructure was computed and the structural damage evolution in the forged parts depending on the preform design has been considered for alloy 80 A.

**Keywords:** Preform design, Material flow, Finite element analysis, Microstructure, Damage.
**PACS:** 01.10.Fv

## INTRODUCTION

The primary objective of forging is to induce a desired change of shape on a piece of metal. However, in modern manufacturing technology and product application a proper shape of the workpiece is not sufficient. Usage properties play an outstanding role too. These desired material state and geometry of the final product depend on several process parameters such as die surfaces, die lubrication conditions, preform, material properties of the initial workpiece and the microstructure evolution during the forming process.

The final product with a desired material state and geometry must be achieved economically. That is why an overall process design involves the entire design of the sequence of operations, the die shapes for each stage, and the initial preform shape.

In former years die and perform design followed experimental guidelines [1, 2]. Nowadays finite element methods are state of the art. For example die and preform design problems were systematically studied by Kobayashi, who introduced the so called backward tracing technique, in which he retraced the loading path in the actual forming process from a given final configuration [3]. Terms like upper bound elemental technique and modified upper bound elemental technique are very well known in preform optimization [4].

CP907, 10ʰ ESAFORM Conference on Material Forming, edited by E. Cueto and F. Chinesta
© 2007 American Institute of Physics 978-0-7354-0414-4/07/$23.00

Preform design problems can be formulated as optimization problems, which can be solved by employing a sequential search method starting from a reference solution [5]. Herein, Badrinarayanan presents a methodology for designing the die and preform for metal forming operations by posing them as optimization problems for minimizing certain error norms [6]. Bonte compares various optimization algorithms for metal forming by application to forging [7]. In other publications not only the desired shape of the finished product is regarded from the mathematical point of view but also the influence of the deformation history on both the local microstructure and the mechanical properties are highlighted (see e.g. [8]).

As different materials demand different preforms according to their peculiar properties an ideal preform does not exist. This will be shown by simulating the forging process of an H-shaped cross section using the finite element method. The analysed materials are aluminum 7075, 42CrMo4 and nickel base alloy 80 A. For the latter, a damage model has been implemented in the computations in order to predict any critical areas with possible cracks in the forged part. Furthermore, a grain structure model is used to consider recrystallization during hot forging.

## PROCEDURAL METHOD

Various two dimensional finite element analyses have been performed to find a convenient preform to forge the desired part. Figure 1 shows the axisymmetric part as well as the half of the simulated preform and die, respectively. The preform has been divided constructively into three parts with a volume fraction of $V_{1pf}/V_{1ff} = 0.76$, $V_{2pf}/V_{2ff} = 1.53$ and $V_{3pf}/V_{3ff} = 0.85$ (pf: preform, ff: final form). With the preform which has been found for steel, the authors investigated if it is possible to forge the same part with the same preform but different materials like aluminum and nickel base alloys. Thereby a special attention has been turned to the filling of the die and the occurring stresses and strains, respectively.

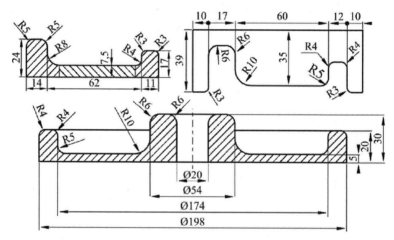

**FIGURE 1.** Axisymmetric forged workpiece (bottom) and sectional drawing of preform and die (top).

# FINITE ELEMENT MODEL

The simulation was done by means of the finite element programme DEFORM™. The model consists only of the rigid die and the visco-plastic material. The forging machine was a hydraulical press. As the part is symmetric only one quarter has been modelled. A Coulomb friction model and heat transfer between die and workpiece has been used. Table 1 shows the input data for all three materials.

TABLE 1. Input data for the FEM-simulation.

| Material | 42CrMo4 | Aluminum 7075 | ALLOY 80 A |
|---|---|---|---|
| Elements | 5000 bricks | 5000 bricks | 5000 bricks |
| Friction | Coulomb 0.35 | Coulomb 0.10 | Coulomb 0.40 |
| Heat transfer, W/m²K | 9000 | 9000 | 9000 |
| Initial temperature, °C | 1100 | 500 | 1010 |

# GRAIN STRUCTURE AND DAMAGE MODEL

The softening process in materials with relative low stacking fault energy like nickel-based superalloys, is mainly governed by recrystallization. After reaching a critical strain during deformation, the nucleation and growth of grains start. Inserting the time for 50% dynamic recrystallisation $t_{0.5}$ into the Avrami type equation for the recrystallized fraction $X = 1 - \exp\left[-\ln(2)\left(t/t_{0.5}\right)^k\right]$ delivers the dynamically recrystallized (DRX) fraction $X_{dyn}$

$$X_{dyn} = 1 - \exp\left\{-0.693\left[\frac{t\,Z^b}{B^k}\exp\left(-\frac{Q_{def}}{RT}\right)\right]^k\right\} \tag{1}$$

with $Q_{def}$ as the activation energy for deformation and $b$, $B$ as well as $k$ as material parameters. For the investigated Alloy 80 A, two temperature regimes with the critical temperature of 1020°C have to be stated. The separation in two temperature regimes is necessary in order to account for the precipitation of carbides and the γ'-phase $Ni_3$(Al,Ti) in the lower temperature regime [9].

Since the introduction of the finite elements method in the hot bulk forming also damage criteria were defined. In the model of effective stresses the stressed material is divided into representative volume elements. With the occurrence of a damage $D$ it is assumed that by formation of pores only the fraction $(1-D)$ of the section of a volume element carries the applied loads. All by ductile failure affected parameters are accordingly treated as effective values. For the effective stress tensor $\tilde{\tilde{\sigma}}$ thus follows $\tilde{\tilde{\sigma}} = \sigma /(1-D)$, where $\sigma$ is the Cauchy stress tensor. This is valid for tensile and also for compressive stresses, if the microcracks and microcavities remain open. For certain materials and certain conditions of loading, the defects may remain open. This is often the case for very brittle materials. If the defects close completely in compression, the area which effectively carries the load equals the initial undamaged area. To define an effective area in compression, a crack closure parameter $h$ was defined [10] that depends a priori upon the material and the loading. The law of evolution of damage derives from the potential of dissipation $\Psi$, which is a scalar convex function of the state variables in case of isotropic plasticity and isotropic damage

$$\dot{D} = -\frac{\partial \Psi}{\partial Y} = \left(\frac{Y}{S_0}\right)^{s_0} \dot{\varepsilon}_{eq} \qquad (2),$$

where $S_0$ and $s_0$ are material and temperature dependent and $\dot{\varepsilon}_{eq}$ is the equivalent true strain rate. The damage strain energy release rate $Y$ corresponds to the variation of internal energy density due to damage growth at constant stress and is given by [11] and [12]

$$Y = \frac{1+v}{2E}\left[\frac{\langle\bar{\sigma}\rangle:\langle\bar{\sigma}\rangle}{(1-D)^2} + \frac{h\langle-\bar{\sigma}\rangle:\langle-\bar{\sigma}\rangle}{(1-hD)^2}\right] - \frac{v}{E}\left[\left(\frac{tr(\langle\bar{\sigma}\rangle)}{(1-D)}\right)^2 + h\left(\frac{tr(\langle-\bar{\sigma}\rangle)}{(1-hD)}\right)^2\right] \qquad (3),$$

where $E$ is the elastic modulus, $v$ is the Poisson's ratio, $tr(\bar{x})$ denotes the trace of a tensor and $\langle x\rangle$ is the Macauley bracket. For the implementation of the crack closure parameter $h$ in Eq. 3, tensile and compressive stresses have to be distinguished in a multi-axially stressed state, thus the stress tensor is split into a positive and a negative part, related to the signs of the principal stresses $\sigma_i$. For the prediction of the material parameter, tensile tests of Alloy 80 A were carried out for different temperatures in the range of 900°C - 1050°C (for details see [11]).

Both microstructure and damage models, described above, were implemented into the finite element programme. The flow potential after von Mises was modified in order to describe the damaged material behaviour, i.e. to reduce the flow stress $k_f = \sigma_{eq}/(1-D)$, where $\sigma_{eq}$ is the equivalent von Mises stress. The evolution of damage as a function of the dynamically recrystallized fraction was calculated by

$$D_i = D_{i-1} + \frac{\dot{D}\Delta t}{D_c}(1-f) \qquad (4),$$

where $i$ demarks the time step, $\Delta t$ the time increment and $D_c$ is the rupture criterion. Therefore rupture is assumed if $D_i$ equals $1$. If a fully recrystallized structure is reached, i.e. $f=1$, the progress of materials damage stops.

## RESULTS

Figure 2 and Fig. 3 show the tangential stress and the effective strain for the steel 42CrMo4 and for aluminum 7075, respectively. Since the tangential tensile stress is not maximal at the end of the forming process it is displayed when the die is going to be filled. One can see that for both materials, steel and aluminum, the tensile stresses are rather high at the top of the rib as well as at the corner radii. We assume that in these areas the likelihood of crack formation is very high, which sometimes really comes true in practice.

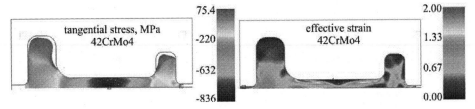

**FIGURE 2.** Tangential stress and effective strain for 42CrMo4.

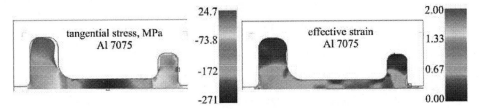

**FIGURE 3.** Tangential stress and effective strain for aluminum 7075.

As the friction between die and workpiece differs with various materials, the material flow also changes. The steel fills the die completely and moreover, material loss into the flash can be observed. Although the volume of the preform is only 1.02 times the volume of the finished product, it seems to be too much. However, forging a part of aluminum with the same preform, the die is not filled completely. It seems that the material has difficulties to rise into the rib and therefore flows in the outer diameter of the flash. That is why the partial volumes (rib, web) and the particular dimensions (web thickness, width of the web, height of the rib, fillet radii, corner radii) have to be optimized for both steel and aluminum. Thereby material losses shall be minimized.

Figure 4 shows the damage parameter $D$ and the recrystallized fraction $f$ for the alloy 80 A. A damage parameter of 1 means that a macrocrack will occur and the corresponding element will be deleted by the program. This phenomenon can be observed especially in the ribs and corners of the forging. In comparison to Fig. 2 and Fig. 3, high damage occurs in areas with high tensile stresses. However, some occurring cracks could be reweld if the surface is free from oxidation. But cracks during forging also indicate that the preform must be corrected. With the occurrence of cracks the computation was stopped and hence the die was not completely filled although showing much material in the flash.

According to the material law the recrystallized fraction is high where the accumulated strains are high, of course (see also Fig. 2 and Fig. 3). In the rib the material only is forced to rise but is not deformed. That is why there is no recrystallization, which can lead to a coarse grain structure. This fact must not be neglected with respect to the required usage properties.

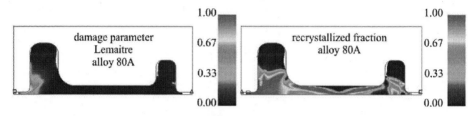

**FIGURE 4.** Damage distribution and recrystallized fraction for the alloy 80 A.

# CONCLUSIONS

Two dimensional axisymmetric finite element analyses with respect to preform optimization have been performed for the materials steel 42CrMo4, aluminum 7075 and nickel base alloy 80 A. It has been demonstrated that no ideal preform exists and each of the above mentioned materials demands its peculiar preform according to its special forming characteristic. Thereby the interface between die and material and hence friction has a significant influence on the die filling. Furthermore, a damage and grain structure model has been implemented for the alloy 80 A. It shows that cracks can occur due to an inadequate preform and thus high local tensile stresses. For the alloy 80 A the recrystallization during forging also has been considered. This helps to detect areas with coarse grains and to avoid resulting shortcomings a priori.

# REFERENCES

1. K. Lange, Closed Die Forging in Steel, Springer-Verlag, Berlin, (1958).
2  R. C. Jones, Drop Forging Die Design, The Association of Engineering and Shibbuilding Draftmen, Richmond, VA, (1965).
3. S. Kobayashi, S. Oh, and T. Altan, Metal Forming and the Finite-Element Method, Oxford University Press, New York, (1989).
4. C. S. Han, R. V. Grandhi, and R. Srinivasan, Optimum design of forging die shapes using nonlinear finite element analysis, AIAA Journal 31(4) (1993) pp. 774-781.
5. N. Zabaras and S. Badrinarayanan, Inverse problems and techniques in metal forming processes, in: N. Zabaras et al., eds., Inverse Problems in Engineering: Theory and Practice, ASME, New York, (1993) pp. 65-76.
6. S. Badrinarayanan, Preform and die design problems in metal forming, A Dissertation Presented to the Faculty of the Graduate School of Cornell University in Partial Fulfilment of the Requirements for the Degree of Doctor of Philosophy, (1997).
7. M. Bonte, ESAFORM conference, April 28, (2006), Glasgow, UK.
8. H. Grass, C. Krempaszky and E. Werner, 3-D FEM-simulation of hot forming processes for the production of a connecting rod, Computational Materials Science 36 (2006), pp. 480–489.
9. C. Sommitsch and W. Mitter, Acta Mater. 54 (2006), pp. 357.
10. J. Lemaitre, A Course on Damage Mechanics, (1996), Springer Verlag, Berlin.
11. C. Sommitsch and G. Rüf, Ductile fracture analysis with damage models of effective stresses at high temperatures, Proc. 8th Internat. Conf. on Technology of Plasticity, October 9-13, (2005), Verona, Italy, P.F. Bariani (Ed.), Edizione Progetto, Padova, ISBN 88-87331-74-X (CD-ROM).
12. F. Andrade Pires, J. Cesar de Sa, L. Costa Sousa and R. Natal Jorge, Numerical modelling of ductile plastic damage in bulk metal forming, Int. J. Mech. Sci. 45 (2003) pp. 273-294.

# Meta-Model Based Optimisation Algorithms for Robust Optimization of 3D Forging Sequences

Lionel Fourment

*CEMEF, Ecole des Mines de Paris, BP 207, 06 904 Sophia Antipolis Cedex, France*

**Abstract.** In order to handle costly and complex 3D metal forming optimization problems, we develop a new optimization algorithm that allows finding satisfactory solutions within less than 50 iterations (/function evaluation) in the presence of local extrema. It is based on the sequential approximation of the problem objective function by the Meshless Finite Difference Method (MFDM). This changing meta-model allows taking into account the gradient information, if available, or not. It can be easily extended to take into account uncertainties on the optimization parameters. This new algorithm is first evaluated on analytic functions, before being applied to a 3D forging benchmark, the preform tool shape optimization that allows minimizing the potential of fold formation during the two-stepped forging sequence.

**Keywords:** Optimization Algorithm, Response Surface, Uncertainties, Design of Experiments, Sequential Approximation, Shape Optimization, Tool Design, Forging
**PACS:** 81.05.Zx, 46.15.Cc, 07.05.Tp, 07.05.Fb, 47.54.Jk, 81.05.Bx

## INTRODUCTION

A key issue in metal forming optimization is the computational time required for a single evaluation of the considered 3D design. It usually requires between several hours on a single computer machine to several days on a parallel computer. Therefore, it is reasonable to look for optimization algorithms that can find satisfactory solutions within less than 50 process simulations. They should also be able to escape local extrema by investigating the full space of parameters. These constraints speak in favor of surrogate models that can be based on interpolation methods like Polynomial, Krigging [1-3], Moving Least Square [4], or else Meshless Finite Difference Method (MFDM) [5, 6], which is used here. They can be utilized either inside expensive global algorithms like Evolution Strategies [1] or Genetic Algorithms [5], or to build Successive Sequential Approximations [3, 7], which is the approach selected here.

This paper has been written in the frame of a meta-model based on the MFDM. It allows using both non gradient and gradient interpolations. It makes it possible to introduce uncertainties on the optimization parameters. The presented algorithm is applied first to analytical optimizations tests and then to a benchmark forging problem.

CP907, *10th ESAFORM Conference on Material Forming*, edited by E. Cueto and F. Chinesta
© 2007 American Institute of Physics 978-0-7354-0414-4/07/$23.00

# BENCHMARK OPTIMIZATION PROBLEM

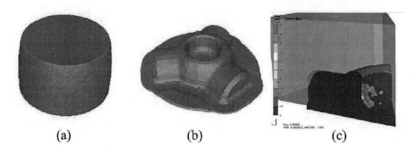

(a)          (b)          (c)

**FIGURE 1.** Initial billet (a) and forged spindle (b) at the end of the second operation. Formation of a lap in the colored zone (c) without an adequate preform.

In the two-stepped sequence to forge a spindle out of a cylinder (see Figure 1), the preforming operation is axisymmetric, while the second and finishing one is closed-die forging with flash utilized. With the initial design of perform tool, simple flat dies, the material fold over during the second operation, as shown in Figure 1-c. This surface defect is quantified by an unusual increase of the equivalent strain rate $\dot{\bar{\varepsilon}}$ on the free surface with respect to a reference value $\dot{\bar{\varepsilon}}_{ref}$, so that the following objective function is utilized:

$$f(p) = \int_{t=0}^{t=t_{end}} \frac{1}{\left|\partial\Omega_{free}^t\right|} \left( \int_{\partial\Omega_{free}^t} \left( \frac{\dot{\bar{\varepsilon}}}{\dot{\bar{\varepsilon}}_{ref}} \right)^{\alpha} ds \right)^{\frac{1}{\alpha}} dt \qquad (1)$$

where $p$ represent the shape parameters, $\partial\Omega_{free}^t$ the free surface at time $t$, and $\alpha$ a function parameter that it taken equal to 10 in the present applications. With the initial design that produces a fold, $f(p) = 10.49$. The axisymmetric shape of the preforming tool is parameterized with a Bspline curve (see Figure 2) with 2 active parameters in the considered applications. The computational cost of a single evaluation of the objective function requires about 3 hours on a personal computer.

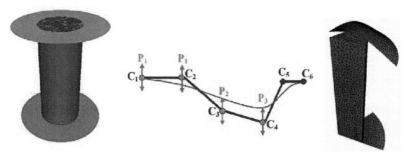

**FIGURE 2.** Initial design of the preforming operation, parameterization of the axisymmetric tool surface with Bspline functions, and resulting tool.

# META-MODEL

## Meta-Model Using The Gradient Information

The present meta-model is based on the Meshless Finite Difference Method (MFDM) [6]. For any point $i$ located at $x_i$, the approximation $\tilde{f}_i$ of $f(x_i)$ is a function of the $f_v = f(x_v)$ and $\nabla f_v$ values at the $n_v$ master points $v$. $\tilde{f}_i$ minimizes the residue (3) of the first Taylor series expansions (2) at $x_v$:

$$\forall v = 1, n_v, \tilde{f}_i = f_v + \nabla f_v (x_i - x_v) + O\left(\|x_i - x_v\|^2\right) \tag{2}$$

$$\tilde{f}_i = \underset{\tilde{f}_i'}{MIN}\, \pi_1\left(\tilde{f}_i'\right) \quad \text{with:} \quad \pi_1\left(\tilde{f}_i\right) = \sum_{v=1, n_v} \frac{\left(\tilde{f}_i - f_v - \nabla f_v (x_i - x_v)\right)^2}{\|x_i - x_v\|^4} \tag{3}$$

After minimization: $\quad \tilde{f}_i = \left(\sum_{v=1, n_v} \frac{1}{\|x_i - x_v\|^4}\right)^{-1} \sum_{v=1, n_v} \frac{\tilde{f}_i - f_v - \nabla f_v (x_i - x_v)}{\|x_i - x_v\|^4} \tag{4}$

## Meta-Model Without The Gradient Information

When the gradient of $(f_v)_{v=1, n_v}$ is not known, the Taylor series expansion is written in the inverse way, at $x_i$:

$$\forall v = 1, n_v, f_v = \tilde{f}_i + \nabla \tilde{f}_i (x_v - x_i) + O\left(\|x_v - x_i\|^2\right) \tag{5}$$

$$\left(\tilde{f}_i, \nabla \tilde{f}_i\right) = \underset{(\tilde{f}_i', \nabla \tilde{f}_i')}{MIN}\, \pi_0\left(\tilde{f}_i', \nabla \tilde{f}_i'\right) \quad \text{with:} \quad \pi_0\left(\tilde{f}_i, \nabla \tilde{f}_i\right) = \sum_{v=1, n_v} \frac{\left(\tilde{f}_i + \nabla \tilde{f}_i (x_v - x_i) - f_v\right)^2}{\|x_v - x_i\|^4} \tag{6}$$

$\left(\tilde{f}_i, \nabla \tilde{f}_i\right)$ is then the solution of the linear system: $\quad A\begin{pmatrix} \tilde{f}_i \\ \nabla \tilde{f}_i \end{pmatrix} = b\left((f_v)_{v=1, n_v}\right) \tag{7}$

## Approximation Error

An estimation of the local interpolation error $\Delta \tilde{f}_i$ of $\tilde{f}_i$ at $x_i$ can be computed by:

$$\Delta \tilde{f}_i = \sqrt{\left(\sum_{v=1, n_v} \frac{1}{\|x_i - x_v\|^4}\right)^{-1} \pi\left(\tilde{f}_i\right)} \tag{8}$$

$\Delta \tilde{f}_i$ allows computing a lower bound of $f(x_i)$, which is equal to $\tilde{f}_i - \Delta \tilde{f}_i$. A more global interpolation error $E$ is also computed over the entire parameter space $\Omega$:

$$E = \int_\Omega \left(\tilde{f}(x) - f(x)\right)^2 dx \tag{9}$$

By substituting $\tilde{f}(x)$ by (4), or by the value resulting from the resolution of (7), a first order approximation of $E\left((x_v)_{v=1,n_v}\right)$ is given by:

$$E\left((x_v)_{v=1,n_v}\right) = \int_\Omega \left(\sum_{v=1,n_v} \frac{1}{\|x-x_v\|^4}\right)^{-1} \left(\sum_{v=1,n_v} \frac{1}{\|x-x_v\|^2}\right) dx \qquad (10)$$

an *a priori* error estimation of the meta-model provided by the points $(x_v)_{v=1,n_v}$.

## Design Of Experiments (DOE)

The minimization of $E\left((x_v)_{v=1,n_v}\right)$ with respect to the positions of the $n_v$ master points $(x_v)_{v=1,n_v}$ provides the best initial design of experiments (DOE) for the considered meta-model, in a fully consistent way. If there already exist $n_u$ master points $(x_u)_{u=1,n_u}$, the a priori interpolation error $E\left((x_u)_{u=1,n_u},(x_v)_{v=1,n_v}\right)$ resulting from the addition of the new $n_v$ master points $(x_v)_{v=1,n_v}$ can be written as:

$$E\left((x_u)_u,(x_v)_v\right) = \int_\Omega \frac{\displaystyle\sum_{u=1,n_u} \|x-x_u\|^{-2} + \sum_{v=1,n_v} \|x-x_v\|^{-2}}{\displaystyle\sum_{u=1,n_u} \|x-x_u\|^{-4} + \sum_{v=1,n_v} \|x-x_v\|^{-4}} dx \qquad (11)$$

The minimization of $E\left((x_u)_{u=1,n_u},(x_v)_{v=1,n_v}\right)$ with respect to $(x_v)_{v=1,n_v}$ provides an optimal way to enrich the existing design of experiments based on $(x_u)_{u=1,n_u}$.

## SEQUENTIAL APPROXIMATION ALGORITHMS (SAA)

Contrary to most Response Surface Algorithms, the proposed Sequential Approximation Algorithm (SAA) starts with a rather coarse DOE and tries to continuously improve the current meta-model. It follows three successive steps. I) First, the **initialization** is the DOE. It consists in finding the best $n_v$ master points that minimize $E\left((x_v)_{v=1,n_v}\right)$ and then in evaluating $f_v = f(x_v)$.

II) The second step is a **sequential improvement** of the meta-model that uses two procedures. 1) The point that minimizes $\tilde{f} - \Delta\tilde{f}$ is added; it is regarded to be the point that has the best potential to minimize $f$. It so allows investigating the parameter space where the error is large and the approximated function is small. 2) If this point has already been found in a previous iteration, the meta-model is enriched by DOE: the point $x_v$ that minimizes $E\left((x_u)_{u=1,n_u},x_v\right)$ is added. In practice, it is advisable that $x_v$ has a good potential to minimize $f$, and so satisfies

24

$\tilde{f}(x_v) - \Delta \tilde{f}(x_v) \le f_{min}$, where $f_{min}$ is the calculated minimal value of $f$. III) Finally, the third step is an **exploitation** of the obtained meta-model: the objective function is evaluated at the point that minimizes $\tilde{f}$. In order to carry on the sequential improvement of the meta-model, the previous two procedures are kept. However, in the DOE one, $x_v$ has to belong to $\Omega_{zoom}$, the hyper-sphere centered in $x_{min}$ (such that $f(x_{min}) = f_{min}$) with a radius such that $(d+1)$ master points are also included in $\Omega_{zoom}$. In the presented applications, 5 points are generated during the first step, 35 during the second one, and 10 during the last one.

## Sequential Approximation Algorithms With Uncertainties (SSA-U)

A meta-model allows taking into account uncertainties on the optimization parameters at a very low computational cost, whereas other type of uncertainties on material, friction or process data, for instance, would require constructing a higher dimension and more expensive meta-model. The uncertainties on the shape parameters are supposed to follow a uniform law of $\|\Delta x\|$ range. They are taken into account by replacing $\tilde{f}(x)$ by $\overline{\tilde{f}}(x)$ (12) at any stage of the optimization procedure, so that the sequential improvement of the meta-model goes together with an improvement of the factoring in uncertainties.

$$\forall x \in \Omega, \quad \overline{\tilde{f}}(x) = \underset{\overline{x} \in \upsilon(x)}{MIN} \tilde{f}(\overline{x}) \; ; \; \upsilon(x) = \left\{ x, \forall i = 1, d, x_i \in \left[ x_i - \|\Delta x\|, x_i + \|\Delta x\| \right] \right\} \qquad (12)$$

## APPLICATIONS

Table 1 presents the results obtained with different versions of SAA for different analytical function: camel-back, Rosenbrock, Rastrigin, Grienwanck, and a semi-analytical function, which is the meta-model constructed from a very large number of simulations of the benchmark forging problem. SSA0 is a variant of SAA where the initial DOE is sequentially built, adding master points one by one. In SAA0-H, the gradient is utilized to enhance the meta-model, which is confirmed by the results of Table 1, where SAA0-H is better than SAA0. However, a better initialization with a global DEO (SAA) provides even better results.

For the benchmark problem, the new SAA is compared to the Meta-model based Evolution Strategy (MES) developed by Emmerich et al. [1], which has shown quite efficient for this problem where gradient algorithm are trapped into local minima and do not suggest solutions that remove the folding defect. SAA provides a better and significantly different solution (see Table 2).

For this problem, the objective function can be very sensitive to small variations of the preform shape, so the inevitable uncertainties on the parameter values (the perform tool shape cannot be exactly machined according to optimization recommendations) have to be taken into account. They are taken equal to 1mm for shape parameters ranging between -10mm and +20mm. It is noticeable that SAA-U allows finding not only a more robust solution but also a globally better one (see Table 2).

TABLE 1. Results of the various SSA strategies for some analytical and semi-analytical functions.

| | Camel-back | Rosenbrock | Rastrigin | Grienwank | Semi-analytical |
|---|---|---|---|---|---|
| SAA0-H | -0.795 | 0.215 | -199.960 | -3.996 | 4.233 |
| p1 | 0.172 | 0.897 | 0.500 | -0.057 | -6.667 |
| p2 | -0.862 | 0.759 | 0.499 | -0.038 | 9.697 |
| SAA0 | -0.847 | 0.643 | -199.980 | -2.773 | 4.622 |
| p1 | -0.088 | 0.418 | 0.499 | -6.724 | -1.724 |
| p2 | -0.607 | 0.129 | 0.500 | -0.517 | 14.828 |
| SAA | -0.951 | 0.046 | -199.890 | -4.000 | 4.228 |
| best at iter. | 49 | 47 | 18 | 26 | 49 |
| p1 | 0.120 | 1.009 | 0.498 | 0.002 | -6.711 |
| p2 | -0.610 | 1.040 | 0.499 | 0.006 | 9.722 |
| Exact | -1.0316 | 0.000 | -200.000 | -4.000 | - |
| p1 | -0,0898 / 0,0898 | 1.000 | 0.500 | 0.000 | - |
| p2 | 0,7126 / -0,7126 | 1.000 | 0.500 | 0.000 | - |

TABLE 2. Results of the different tested algorithms for the benchmark spindle optimization problem.

| Algorithm | Best Value | best at iter. | uncertainties | p1 | p2 | Fold |
|---|---|---|---|---|---|---|
| MES | 8.140 | 37 | - | -7.870 | -10.000 | removed |
| SAA | 8.091 | 51 | - | -9.676 | -9.527 | removed |
| SAA-U | 7.877 | 48 | 1.0 | -8.294 | -8.094 | removed |

# CONCLUSIONS

The newly developed SAA provides better results that the robust MES. Higher order approximations using gradient information allows improving the algorithm efficiency, but not significantly enough: a better initialization with better design of experiments is more effective. Taking into account uncertainties on shape parameters allows escaping local minima and finding a better and more robust solution.

# REFERENCES

1 M. Emmerich, A. Giotis, M. Özdemir, T. Bäck, K. Giannakoglou, Metamodel-assisted evolution strategies, in: Anonymous Int. Conference on parallel problem solving from nature Springer, Berlin, GERMANY, 2002.
2 M. Bonte, Van den Boogaard, A.H., J. Huetink, A metamodel based optimisation algorithm for metal forming processes. Int J Numer Methods Eng in prints (2006)
3 D. Buche, N. Schraudolph, P. Koumoutsakos, Accelerating Evolutionary Algorithms with Gaussian Process Fitness Function Models, IEEE Transactions on Systems, Man and Cybernetics 35 (2005) 183-194.
4 H. Naceur, S. Ben-Elechi, C. Knopf-Lenoir, J. L. Batoz, Response SurfaceMethodology for the Design of SheetMetalForming Parameters to Control Springback Effects usingthe Inverse Approach, in: Anonymous 2004.
5 L. Fourment, T. T. Do, A. Habbal, M. Bouzaïane, Gradient, non-gradient and hybrid algorithms for optimizing 2D and 3D forging sequences, in: D. Banabic Eds.8th International ESAFORM Conference on Material Forming 2005.
6 T. Liszka, J. Orkisz, The finite difference method at arbitrary irregular grids and its application in applied mechanics, Comp.and Struct 11 (1980) 83-95.
7 M. H. A. Bonte, L. Fourment, Van den Boogaard, A.H., J. Huetink, Optimisation of metal forming processes using Finite Element simulations - A Sequential Approximate Optimisation algorithm and its comparison to other algorithms by application to forging Structural and Multidisciplinary Optimization (in preparation)

# Research on Softening of A95456 Alloy Deformed Under Elevated Temperatures

Pavel A. Petrov* and Victor I. Perfilov*

*Department of Autobody building and metal forming, Moscow State Technical University "MAMI",
107023, B.Semenovskaya str., 38, GPS, Moscow, Russia

**Abstract.** The present paper describes the results of the research on the softening of aluminium alloy A95456 deformed at elevated temperatures. The investigations were carried out within the temperature range of 310-450 °C and strain rate of 0.01-0.4 $s^{-1}$. The strain rate was either constant or variable in performed experiments. In case of variable strain rate two different schemes were observed. Firstly, the deformation of alloy A95456 was performed at constant die velocity and so the strain rate increased monotonically. Secondly, the die velocity was changed suddenly during the deformation of A95456 alloy. In turn, it caused the sudden strain rate change. To describe the softening behaviour of A95456 alloy several equations were investigated. The accuracy of each equation was estimated. Some practical recommendations for use of those equations were given.

**Keywords:** flow stress, aluminum alloy, alloy A95456, isothermal deformation, elevated temperatures, hardening, softening, hot deformation
**PACS:** 81.40.-z, 81.40.Lm, 81.70.-q.

## INTRODUCTION

To chose the regime of hot massive forming processes of metals or alloys it is necessary to take into account a deformable material behavior for definite temperature or/and strain rate interval. In turn, it requires creation of an accurate mathematic model of flow stress, which would describe the behavior of ferrous or non ferrous material behavior during hot deformation.

A great variety of flow stress models [1, 2, 3, etc.] are available for estimating the behavior of deformable material at high temperatures. But their efficiency is limited by the definite strain-rate range or they can be applied for definite conditions of deformation. The rheological models based on phenomenological approach [4, 5, etc.] are more appropriate for the description of material behavior during hot deformation especially in wide range of temperatures as well as strain-rates. Such models are extremely flexible to the conditions of deformation as well as any changes in macro- and microstructure of a material under study. It means that they could take into account as much physical phenomena as possible, for instance dynamic and static softening, transient processes, etc.

In general, softening is a physical effect which goes with the hot deformation of ferrous or non-ferrous materials and is opposed to the hardening effect of a deformable

CP907, 10th ESAFORM Conference on Material Forming, edited by E. Cueto and F. Chinesta
© 2007 American Institute of Physics 978-0-7354-0414-4/07/$23.00

material. It results in a deformable material structure evolution and implies a combination of such processes as recrystallization, grains recovery, cell formation, annihilation, etc.

Depending on the conditions of deformation at least two types of softening can be marked out. These are dynamic softening and static one. The term "dynamic" means that an investigated process happens during deformation while the term "static" characterizes a process which occurs after deformation of a material, for instance after the interruption of deformation or after unloading of deformed material or during post deformation heating or annealing, etc.

To sum up, the present paper implies the investigation of static softening of aluminium alloy A95456 under elevated temperatures and with strain rate range of 0.01-0.4 $s^{-1}$. The research on softening of alloy A95456 was carried out experimentally and theoretically as well.

## EXPERIMETAL PROCEDURE

The experimental investigation was carried out with the help of electromechanical universal testing machine (nominal load = 50 kN). The cylindrical samples were cut from a bar of aluminium alloy A95456. The chemical composition of that alloy is given in Table 1.

TABLE 1. Chemical composition.

| Element | Al | Cu | Mg | Mn | Fe | Si | Zn | Ti |
|---------|------|------|------|------|------|------|------|------|
| % | base | 0.04 | 6.80 | 0.53 | 0.22 | 0.16 | 0.20 | 0.10 |

The sizes of the samples were as follows: diameter = 10 mm; height = 10 mm. The samples were heated to temperatures of 310°C, 400°C, 430°C in the electric furnace that the testing machine is equipped. Deformation of the heated samples was carried out on flat dies inside the furnace. Samples were compressed up to about 60% of the initial height of sample without lubrication. Then the further displacement of testing machine ram was stopped. The design of machine allowed us to keep the ram position almost invariable within time period of 0-200s after the interruption of a sample compression. It means that the distance between upper and lower dies was constant for that period of time and so the increment of deformation per time unit was equal to zero. Two parameter were measured, namely deformation load and time. The value of the first parameter was measured because after the ram was stopped the decrease in load occurred due to static softening, i.e. recrystallization, recovery, etc. During compression of samples the strain-rate was either constant or variable. The investigated interval of strain-rate is 0.01–0.4 $s^{-1}$. The detailed description of conditions for each test is given in Table 2.

In accordance with the test trial 4 (see Table 2), there were two stages of deformation. At the first stage the ram velocity was constant and equal to 0.1 mm/s (6 mm/min) and its displacement was about 0.5 mm. At the second stage of test the ram velocity was 4.0 mm/s (240 mm/min) and the displacement of ram was equal to 5.5 mm. So after compression each sample had about 4 mm in height.

**TABLE 2.** Description of tests conditions.

| Test trial | Temperature of sample heating, °C | Strain-rate $\dot{\varepsilon}_i$, s$^{-1}$ | Time of post deformation interruption, s | Notes |
|---|---|---|---|---|
| 1 | 430 | 0.1 | 180.12 | $\dot{\varepsilon}_i$ =const=0.1 s$^{-1}$ |
| 2 | 400 | 0.4 | 180.18 | $\dot{\varepsilon}_i$ =const=0.4 s$^{-1}$ |
| 3 | 310 | 0.4 | 180.11 | $\dot{\varepsilon}_i$ =const=0.4 s$^{-1}$ |
| 4 | 430 | - | 180.13 | $\dot{\varepsilon}_i$ = variable, $V_{ram}$= variable |

## RESULTS

Figures 1a and 1b illustrate the flow curves for alloy A95456 while the softening curves are shown in the Figure 2.

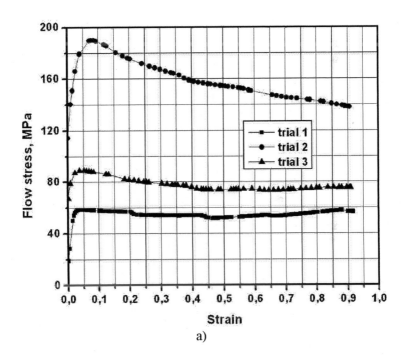

a)

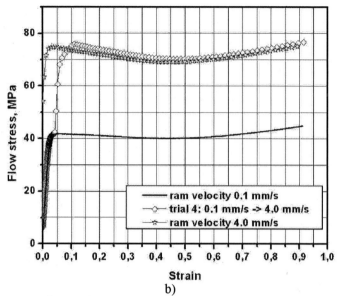

FIGURE 1. Flow stress-strain curve for A95456 alloy

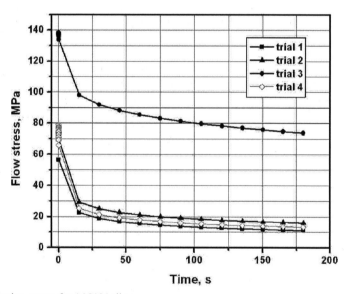

FIGURE 2. Softening curves for A95456 alloy

The analysis of experimental data (see figure 2) provides us the advance in deformable material behaviour within strain range 0-0.9. The hardening effect in A95456 alloy is significant when the strain value is less than 0.05-0.1. The increase in strain value occurs decreasing the flow stress value due to the great effect of softening on material flow.

Sudden change in ram velocity from 0.1 mm/s to 4 mm/s (see figure 1b, trial 4) tends the increase in flow stress value up the value which corresponds to the ram velocity of 4.0 mm/s.

## DISCUSSION

Softening curves presented in the Figure 3 can be divided into two zones. The first zone corresponds to the time period of 0-15s while the second one is within time range of 15-180s. The first zone characterizes the dynamic softening. It occurs due to the inertia of the ram when it was stopped. The second part of a softening curve (see Figure 3) is referred to the static softening of alloy under study. During that stage the ram velocity is equal to zero.

To describe the effect of static softening two models were used. The first model (see equation (1)) is Maxwell's equation which is regarded as a classical equation of softening. The second model (see equation (2)) is an equation which is widely used in theory of creep [4]. Both models are given below:

$$\sigma_i = \sigma_o^i e^{-t/\tau_o} ,\qquad (1)$$

$$\sigma_i(t) = \sigma_o + \left(\sigma_o^i - \sigma_o\right)e^{-gt}, \qquad (2)$$

where $\tau_0$ = softening time; $\sigma_o^i$ = initial flow stress corresponds to time moment of t=0; $\sigma_o^i$ = initial flow stress corresponds to time moment of t=0; $\sigma_o$ and g = coefficients depends on deformation conditions, kind of material and microstructure evolution.

The unknown coefficients which these equations include can be determined with the help of any standard optimization method. In the present paper Levenberg-Marquardt technique was used for that.

Figure 3 illustrates the comparison between experimental data and calculated one in terms of softening. The values of identified coefficients are given in Table 3.

TABLE 3. Values of unknown coefficients in equation (1) and (2)

| Test trial | Equation (1) | | Equation (2) | | |
|---|---|---|---|---|---|
| | $\tau_0$, s | $\sigma_o^i$, MPa | $\sigma_o$, MPa | $\sigma_o^i$, MPa | g, s$^{-1}$ |
| 1 | 142.86 | 25 | 11 | 28 | 0.023 |
| 2 | 142.86 | 40 | 16 | 34 | 0.019 |
| 3 | 285.71 | 110 | 74 | 105 | 0.016 |
| 4 | 142.86 | 30 | 13 | 31 | 0.022 |

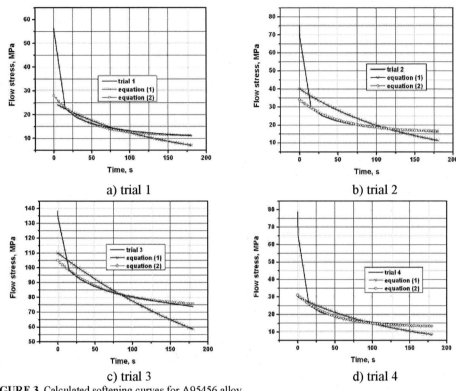

a) trial 1

b) trial 2

c) trial 3

d) trial 4

**FIGURE 3.** Calculated softening curves for A95456 alloy

## CONCLUSIONS

Figure 3 shows that the equation (2) is in the best agreement with experimental data within the second zone of softening curves of A95456 alloy (see Figure 2). The obtained data can be used for description of softening effect of aluminium alloys like A95456. Moreover, it can be applied for numerical simulation of metal forging processes under elevated temperatures when the effect of softening is more significant in comparison with hardening one.

## REFERENCES

1. Z. Gronostajski, *Journal of Materials Processing Technology* **106**, 40-44 (2000).
2. O.M. Smirnov, *Metal Forming Under Superplastic Conditions*, Moscow: Mashinostroenie, 1979.
3. M. Pietrzyk and R. Kuziak. "Development of the Constitutive Law for Microalloyed Steels Deformed in the Two-Phase Range of Temperatures" in *Proc. 10th International Conference on Metal Forming*, edited by J.Kusiak, P.Hartley, etc. Poland, Krakow, 19-22 September, 2004, pp.465-47
4. P.I.Poluxin, G.Ya.Gun, A.M.Galkin, et al. *Flow Stress of Ferrous and Nonferrous Materials*, Moscow: Metallurgiya, 1983.

# Behaviour model identification based on inverse modeling and using Optical Full Field Measurements (OFFM): application on rubber and steel

V. Velay, L. Robert, F. Schmidt, S. Hmida and T. Vallet

*Research Centre on Tools Materials and Processes (CROMeP)*
*Ecole des mines d'Albi-Carmaux, 81013 ALBI Cedex 9, France*

**Abstract.** Biaxial properties of materials (polymer or steel) used in many industrial processes are often difficult to measure. However, these properties are useful for the numerical simulations of plastic-processing operations like blow moulding or thermoforming for polymers and superplastic forming or single point incremental forming for steels. Today, Optical Full Field Measurements (OFFM) are promising tools for experimental analysis of materials. Indeed, they are able to provide a very large amount of data (displacement or strain) spatially distributed. In this paper, a mixed numerical and experimental investigation is proposed in order to identify multi-axial constitutive behaviour models. The procedure is applied on two different materials commonly used in forming processes: polymer (rubber in this first approach) and steel. Experimental tests are performed on various rubber and steel structural specimens (notched and open-hole plate samples) in order to generate heterogeneous displacement field. Two different behaviour models are considered. On the one hand, a Money-Rivlin hyperelastic law is investigated to describe the high levels of strain induced in tensile test performed on a rubber open-hole specimen. On the other hand, Ramberg-Osgood law allows to reproduce elasto-plastic behaviour of steel on a specimen that induces heterogeneous strain fields. Each parameter identification is based on a same Finite Element Model Updated (FEMU) procedure which consists in comparing results provided by the numerical simulation (ABAQUS$^{TM}$) with full field measurements obtained by the DISC (Digital Image Stereo-Correlation) technique (Vic-3D$^{®}$).

**Keywords:** Behaviour identification, Inverse problem, Full field measurements, Stereo-correlation
**PACS:** 62.20.-x

## INTRODUCTION

The aim of this investigation is to develop an original methodology which will be able to determine multiaxial material properties. The approach can be applied to identify behaviour models of steels or polymers commonly used in many industrial forming processes like blow moulding or single point incremental forming. The paper is based on a mixed experimental and numerical approach using Optical Full Field Measurements (OFFM) and Finite Element model results. Today, OFFM methods are more and more applied in experimental mechanics, they allow to measure kinematic fields (displacement or strain) and provide extended possibilities to identify multiaxial behaviour models. Several recent investigations using different experimental full field methods and various identification procedures have been developed [1, 2]. In this work, a Finite Element Model Updated (FEMU) method is considered [3, 4, 5]. It consists in comparing experimental measurements obtained by DISC (Digital Image Stereo-Correlation) technique

*CP907, 10th ESAFORM Conference on Material Forming,* edited by E. Cueto and F. Chinesta
© 2007 American Institute of Physics 978-0-7354-0414-4/07/$23.00

(Vic-3D®) with numerical simulation (ABAQUS™). Two different kind of tensile tests are investigated on two structural specimens and two materials (rubber and steel).

## EXPERIMENTAL INVESTIGATION

Both of tensile tests performed on rubber and steel have been carried out with a servo-electric testing machine and controller connected to a computer. It allows to measure the displacement rate and the global strength induced into the specimen. A stereo-rig composed of two 8-bit Qimaging Qicam digital cameras with CCD resolution of $1360 \times 1036$ pixels are located in front of the testing machine in order to measure the displacement field during the tests by DISC technique [6] using the Vic-3D® software. In each case, the investigation area only considers a part of the structural specimen and can not include the whole sample. This experimental method allows to provide displacement or strain field components induced within the investigation area by the global loading. Each displacement or strain map corresponds to a time increment of the tensile test. Figure 1a presents the global tensile curve obtained for the steel specimen. For the identification process, three images will be selected, corresponding to three time increments (100s, 200s and 300s) as indicated in figure 1a. These time increments characterize the plastic behaviour of the material. Indeed, global plasticity increases with time (figure 1a) but local plasticity also occurs and depends on the investigation area. For instance, plasticity will be the most important within the notch whatever the time increment. All these considerations allow to consider a very large amount of plastic range even if only three time increments will be used into the analysis. Similar assumption will be done for the rubber specimen. As an illustration, figure 1b shows axial and shear strain components within the steel notched specimen at a time of 300s.

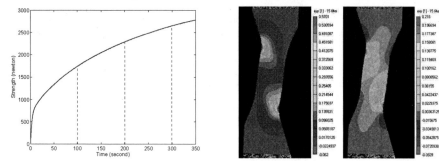

**FIGURE 1.** a- Tensile test curve performed on the notched steel specimen; b- Axial strain field (left) and shear strain field (right) measured by DISC technique

## IDENTIFICATION METHODOLOGY

### Strategy

Identification strategy can be divided into several stages. First, some preliminary works have to be performed on the experimental files in order to keep only the good matched points and to avoid mismatched informations. Then, locations where exper-imental data are measured have to be projected onto the finite element mesh. Figure

2a presents a comparison between nodes considered in experimental and numerical meshes, for the rubber open-hole specimen. Only the investigation area is modelized in the numerical simulation. Experimental and calculated displacements will be considered to evaluate the cost function. Moreover, experimental displacements measured at the upper and lower parts of the investigation area will be introduced into the simulation to define the boundary conditions. Thus, rigid body motions will be taken into account in the simulation. Finite element simulation will be performed with ABAQUS$^{TM}$, a script allows to extract node displacements and global strength necessary to the analysis. The main program automatically runs the finite element calculation, extracts calculated and experimental data for the cost function evaluation and updates behaviour model parameters to be identified. Cost function minimization is performed with a modified Nelder-Mead Simplex algorithm implemented in Matlab$^{®}$. This robust method attempts to minimize a scalar-valued nonlinear function of several variables using only function values, without any derivative information. In this investigation, this method is suitable as the cost function can not be explicitly formulated. The cost function $J(P)$ depends on experimental and calculated displacements and strength. Its formulation includes a spatial (nodes) and time informations (images), as:

$$J(P) = \beta \cdot \frac{||u^{exp} - u^{sim}||}{||u^{exp}||} + (1 - \beta) \cdot \frac{||F^{exp} - F^{sim}||}{||F^{exp}||}$$

where: 
$$||u^{exp} - u^{sim}|| = \sqrt{\sum_{k,i=1}^{N_n,N_i} \left( \left( u_1^{sim} - u_1^{exp} \right)_{k,i}^2 + \left( u_2^{sim} - u_2^{exp} \right)_{k,i}^2 \right)}$$

and: 
$$||F^{exp} - F^{sim}|| = \sqrt{\sum_{i=1}^{N_i} \left( \left( F^{sim} - F^{exp} \right)_i^2 \right)}$$

$N_n$ is the number of nodes considered for the analysis, $N_i$ the number of images obtained at various time increments, $u_\alpha, \alpha = 1, 2$ the displacement components, $F$ the global strength in the loading direction, $P$ the parameter vector to be identified and $\beta$ a weight parameter.

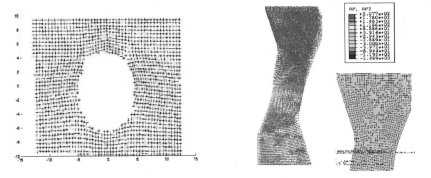

**FIGURE 2.** a- Experimental data projection onto the finite element mesh; b- Displacement magnitude field induced in the steel notched specimen (left) and reaction forces induced by the boundary conditions (right)

Figure 2b shows the calculated displacement magnitude field corresponding to the first image (t=100s). Applying the experimental displacement field as a boundary conditions (upper and lower parts of the investigation area), rigid body displacement fields are naturally taken into account into the calculation. Figure 2b illustrates reaction forces induced by the previous applied boundary conditions. They are considered for the cost function evaluation.

## Behaviour modeling

Two relevant laws are carried out in order to investigate rubber and steel behaviour. Indeed, a Ramberg-Osgood law is considered to describe steel behaviour whereas a Mooney-Rivlin hyperelastic law is used to reproduce behaviour of rubber.

### *Steel*

A Ramberg-Osgood model is selected to reproduce the steel behaviour. This law is available into ABAQUS$^{\mathrm{TM}}$ software. Basic one-dimensional model is:

$$E\varepsilon = E(\varepsilon_e + \varepsilon_p) = \sigma + \frac{E}{K}\left(\frac{|\sigma|}{K}\right)^{\frac{1}{M}-1}\sigma \quad \text{where } E \text{ is the Young Modulus, } \varepsilon \text{ the total}$$

strain, $\varepsilon_e$ and $\varepsilon_p$ the elastic and plastic parts, $\sigma$ the stress induced, $K$ and $M$ the parameters of the non linear term defining the hardening. Multiaxial formulation includes Hooke linear elastic relation used to generalize the first term of the previous equation and a nonlinear term to define plasticity through the use of the Mises stress potential and associated flow law.

$$E\underline{\underline{\varepsilon}} = (1+v)\underline{\underline{\sigma}}' - (1-2v)p\underline{\underline{I}} + \frac{3}{2}\frac{E}{K}\left(\frac{\sigma_{eq}}{K}\right)^{\frac{1}{M}}\underline{\underline{\sigma}}' \quad \text{where Elastic properties } E \text{ and } v \text{ are}$$

assumed to be known and only both parameters of the nonlinear term $K$ and $M$ will be identified. In order to assess the material parameters of the forward problem to be identified, various direct simulations have been performed in both cases (steel and rubber). Figure 3a presents the evolution of the cost function versus the two parameters $K$ and $M$ of the Ramberg-Osgood law. A global minimum seems to be reached around a value of 5 for $M$ and 500 MPa for $K$.

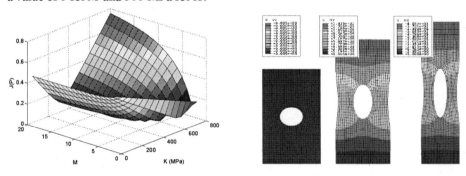

**FIGURE 3.** a- Evolution of the cost function (Ramberg-Osgood law); b- Axial displacement fields provided by a multi-stage numerical simulation performed on rubber open-hole specimen)

### *Rubber*

Mooney-Rivlin hyperelastic model is known to provide a good description of rubber behaviour. Moreover, it only requires two material parameters $C_{01}$ and $C_{10}$ to be identified. Energy function is formulated as $W = C_{10}(I_1 - 3) + C_{01}(I_2 - 3)$, where $I_1$ and $I_2$

are the first and the second invariant of the left Cauchy-Green tensor, respectively. A multi-stage identification procedure is performed on a open-hole specimen. Figure 3b presents direct finite element simulations at different time increments.

## Results and discussion

For both cases (steel and rubber), preliminary classical tensile tests performed on flat specimens allows to assess a first set of material parameters and to initialize the identification procedure.

### Steel

Several minimizations are performed in order to investigate the influence of the weight parameter $\beta$. A value of 0.9 confers a more important weight on the displacement field. On the contrary, a value of 0.2 allows to maximize the effect of the force induced. Figure 4a gives the evolution of the Ramberg-Osgood parameters versus the number of iterations and figure 4b the evolution of the cost function, both for a weight parameter equals to 0.9. A good correlation on both displacement fields and reaction forces is obtained for this value of $\beta$.

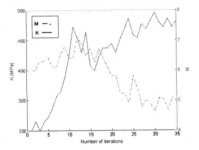 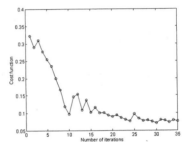

**FIGURE 4.** a- Evolution of the Ramberg-Osgood parameters $K$ and $M$ over the number of iterations for a weight parameter of 0.9; b- Evolution of the cost function over the number of iterations for a weight parameter of 0.9 and the Ramberg-Osgood law

Figure 5a presents respectively the axial (left) and shear (right) strain components for a time of 300s. Results can be compared with those provided by DISC technique (Figure 1b). They are in a good agreement with experiment. A weight parameter of 0.2 induces calculated reaction forces closer to those provided by measurements. However, calculated displacement field will be less accurate than in the previous case. Thus, a good compromise should be studied in details into the choice of the weight parameter. Young modulus $E$ and Poisson ratio $v$ are assumed to be known and fixed ($E = 166000$ MPa and $v = 0.34$). Optimized parameters provided by the analysis are $K = 480$ MPa and $M = 4.92$.

### Rubber

In this case, DISC technique is more difficult to carry out due to the very large displacement induced in tensile tests. Actually, experimental results allow to assess the heterogeneous deformation within the specimen. But these measurements can not yet be used to determine behaviour parameters by the inverse approach. Figure 5b illustrates a comparison between measured and simulated shear strain field obtained with coefficients identified in uniaxial test conditions ($C_{10} = 0.136$MPa and $C_{01} = 0.097$MPa). Strain concentrations can be predicted by optical measurements as it can be seen on the left half of the specimen.

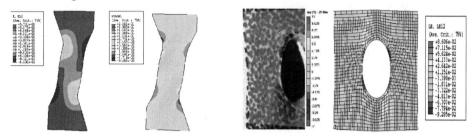

**FIGURE 5.**  a- Axial strain field (left) and shear strain field (right) calculated by FE Simulation; b-Comparison between experimental (left) and calculated (right) shear strain fields performed on open-hole rubber specimen

## CONCLUSIONS

In this investigation, an efficient methodology was carried out in order to identify multiaxial behaviour laws within various test conditions and different materials. It is based on a forward problem using optical full field measurements and finite element simulations. For that purpose, a Finite Element Model Updated method was developed. A cost function was formulated in term of displacement fields and reaction forces and minimized by a modified Nelder-Mead Simplex algorithm. Two different applications were considered. On the one hand, identification of elastoplastic behaviour law of steel was successfully performed. Strain component fields provided by numerical simulation are in a good agreement with those obtained from an experimental point of view. On the other hand, the inverse approach was applied on a hyperelastic behaviour of rubber. In this case, some experimental problems occur. Test conditions have to be improved in order to be able to perform a complete multiaxial identification.

## REFERENCES

1. M. Meuwissen, *An inverse method for the mechanical characterization of metals*, Ph.D. thesis, Eindhoven university of technology (1998).
2. M. Bonnet, M. Grédiac, F. Hild, S. Pagano, and F. Pierron, *Mécanique et Industries* **11**, 297–305 (2003).
3. H. Haddadi, S. Belhabib, M. Gaspérini, and P. Vacher, "Identification of the parameters of Swift law using strain field measurements," in 8*th* *European Mechanics of Materials Conference*, Conference on Material and structural identification from full-field measurements, Cachan, France, 2005.
4. M. Giton, A.-S. Bretelle, and P. Ienny, *Strain* **42**, 291–297 (2006).
5. D. Lecompte, A. Smits, H. Sol, J. Vantomme, and D. V. Hemelrijck, "Elastic orthotropic parameter identification by inverse modeling of biaxial tests using Digital Image Correlation," in 8*th* *European Mechanics of Materials Conference*, Conference on Material and structural identification from full-field measurements, Cachan, France, 2005.
6. J. Helm, S. McNeill, and M. Sutton, *Optical Engineering* **35**, 1911–1920 (1996).

# Numerical Predictions on the Final Properties of Metal Injection Moulded Components after Sintering Process

J. Song[*,†], T. Barriere[*], B. Liu[†], J.C. Gelin[*]

*Femto-ST Institute/LMA, ENSMM, 26 Rue de l'Epitaphe, 25000 Besancon, France
†Department of Applied Mechanics, Southwest Jiaotong University, 610031 Chengdu, China

**Abstract.** A macroscopic model based on a viscoplastic constitutive law is presented to describe the sintering process of metallic powder components obtained by injection moulding. The model parameters are identified by the gravitational beam-bending tests in sintering and the sintering experiments in dilatometer. The finite element simulations are carried out to predict the shrinkage, density and strength after sintering. The simulation results have been compared to the experimental ones, and a good agreement has been obtained.

**Keywords:** Metal injection moulding; Sintering; Stainless steel; Numerical simulation
**PACS:** 81.20.Ev; 81.20.Hy

## INTRODUCTION

Metal injection moulding (MIM) is relatively a new forming technology in powder metallurgy industries, which is especially efficient and beneficial for manufacturing small and intricate metallic components. It includes the four basic steps of mixing the powders and binders, injection moulding, debinding and sintering. Due to the large shrinkage arising in the sintering stage, the numerical prediction of the final properties is very important for the design of the injection mould and the optimization of the process parameters. In the study, the numerical simulations are employed to compute the final shrinkage, density and strength of the sintered components in 316L stainless steel powders. The experimental data are used for the identification of the parameters entering in the viscoplastic constitutive law and verify the predicted results issued from simulations.

## MACROSCOPIC MODEL FOR SINTERING

A macroscopic model based on the continuum mechanics can be used to predict the shrinkages and distortions of the components during and at the end of the sintering stage. The green part after injection moulding and debinding is regarded as a compressible porous material part. The deformation of the sintered body is governed

CP907, 10th ESAFORM Conference on Material Forming, edited by E. Cueto and F. Chinesta
© 2007 American Institute of Physics 978-0-7354-0414-4/07/$23.00

by the mass, momentum and energy conservation equations [1]. The important issue in the model is to determine the constitute law of the part during the sintering process.

## Viscoplastic Constitutive Law for Sintering

Under high temperature sintering, the densification of the polycrystalline materials is governed by diffusion processes. The resulting macroscopic behavior associated can be regarded as the creep deformation [2]. The deformation of the sintered body is rate dependent, including the shrinkage and distortion. A viscoplastic constitutive law as formulated in continuum mechanics can be used to describe this process. It can be expressed as below:

$$\dot{\varepsilon}_{vp} = \frac{\sigma'}{2G_p} + \frac{\sigma_m - \sigma_s}{3K_p} \mathbf{I} \tag{1}$$

where $\dot{\varepsilon}_{vp}$ is the viscoplastic strain rate, $\boldsymbol{\sigma}$ is the Cauchy stress tensor, $\boldsymbol{\sigma}'$ the deviatoric stress tensor, $\sigma_m = \mathrm{tr}(\boldsymbol{\sigma})/3$ is the hydrostatic stress, $\mathbf{I}$ is a second order identity tensor, $G_p$ and $K_p$ are the shear and bulk viscosity moduli of the porous material, $\sigma_s$ is the sintering stress that drives the densification process. The elastic-viscous analogy is used to determine the viscosity moduli [3]:

$$G_p = \frac{\eta_z}{2(1+\nu_{vp})}, K_p = \frac{\eta_z}{3(1-2\nu_{vp})}, \nu_{vp} \approx \frac{1}{2}\sqrt{\frac{\rho}{3-2\rho}} \tag{2}$$

where $\eta_z$ is the uniaxial viscosity, $\nu_{vp}$ is the viscous Poisson's ratio, and $\rho$ is the relative density of the porous materials, that is governed by the mass conservation equation $\dot{\rho} = -\rho \cdot \mathrm{tr}(\dot{\varepsilon}_{vp})$. $\eta_z$ and $\sigma_s$ are two parameters entering in the constitutive law that should be properly determined for the further numerical simulations.

## Determination of Uniaxial Viscosity

For the sintering of 316L stainless steel powder MIM components, grain boundary diffusion is the main densification mechanism. Based on the Coble's creep model and the mechanics of porous materials, the uniaxial viscosity is expressed as [4]:

$$\eta_z = \frac{kTG^3\rho^2}{\Omega D_{b0} \exp(-Q_b/RT)} \tag{3}$$

where $T$ is the absolute temperature, $G$ is the grain size, $k$ is a constant, $\Omega$ is the atomic volume, $D_{b0}$ is the coefficient of grain boundary diffusion, $R$ is the gas constant, and $Q_b$ is the activation energy for grain boundary diffusion mechanism.
The following equation is chosen to describe the behaviour of grain growth during the sintering of the 316L stainless steel powders [5]:

$$\frac{dG}{dt} = \frac{B \exp(-Q_G / RT)}{G} \qquad (4)$$

where $Q_G$ is the activation energy for grain growth and $C$ is a material coefficient. For 316L stainless steel, when the temperature is less than 1200 °C, $Q_G = 315.8$ kJ/mol, otherwise $Q_G = 50$ kJ/mol. Koseski $et$ $al$ investigated the changes of grain size occurring during sintering for 316L stainless steel powders [6]. In their experiments, the material characteristics of gas-atomized powders are closed to the feedstock used in our experiments. With these experimental data, the determined value of the material constant $B$ in Equation (4) is equal to 0.98 $(\mu m)^2/s$.

## Determination of Sintering Stress

The expression of sintering stress proposed by Olevsky has been widely used for numerical simulation of the stainless steel powders [1, 5], as following:

$$\sigma_s = \frac{C\rho^2}{r} \qquad (5)$$

where $r$ is the radius of the particle, $C$ is the material constant. This expression is then used in our work.

## Model for Evaluating the Strength after Sintering

For the porous materials after sintering, the following expressions are proposed to determine the yield strength and ultimate tensile strength [7]:

$$\sigma_y = \sigma_y^0 \frac{1-\theta}{K_c}, \sigma_{UTS} = \sigma_{UTS}^0 \frac{1-\theta}{1+\alpha_s(K_c-1)\theta} \qquad (6)$$

where the superscripts 0 indicates the strength of the wrought material, and the subscript y and UTS indicate the yield strength and ultimate tensile strength, $\theta = 1 - \rho$ is the porosity factor, $K_c$ is the strength concentration factor, $\alpha_s$ is a constant.

The following empirical expression is proposed to determine the strength concentration factor [7]:

$$K_c = \exp(0.0946(\ln(\frac{X}{8D}))^2 + 0.1746\ln(\frac{X}{8D}) + 0.4576) \qquad (7)$$

where $X$ is the diameter of the neck between the particles, $D$ is the diameter of the particle. $X/D$ is often used to evaluate the sintering bonding effects. Skorohod proposed the following expression to determine $X/D$ in sintering [8]:

$$\left(\frac{X}{D}\right)^2 = 1 - \left(\frac{\theta}{\theta_0}\right)^{\frac{4}{3}} \qquad (8)$$

where $\theta_0$ is the porosity of the green part before sintering.

## IDENTIFICATION OF THE MATERIAL PARAMETERS

The gravitational beam-bending tests in sintering have been carried out in our lab to identify the parameters related to uniaxial viscosity [9]. In the identification algorithm, the activation energy for grain boundary diffusion $Q_b$ in Equation (4) is chosen as 167 kJ/mol [2]. An optimization method is used to determine the material constant $A = k / \Omega D_{b0}$. Based on the determined uniaxial viscosity, the shrinkage curves in sintering obtained by experiments in dilatometer have been used to determine the sintering stress parameter $C$ entering in Equation (5) [4, 9]. The entire sintering process is divided into three stages. The identified parameters for the thermal heating cycle to 1360 °C at 8 °C/min and holding for 1 h are presented in Table 1.

**TABLE 1.** Identified parameters in the sintering model for 316L stainless steel powder.

| Sintering Stages | Parameter A ( Pa·s/(m3·K)) | Parameter C (N/m) |
|---|---|---|
| $0.000 < \rho \leqslant 0.645$ | $5.19 \times 10^{15}$ | 0.696 |
| $0.645 < \rho \leqslant 0.930$ | $1.97 \times 10^{15}$ | 6.454 |
| $0.930 < \rho \leqslant 1.000$ | $16.1 \times 10^{15}$ | 1.640 |

The calculated uniaxial viscosity and shrinkage based on the presented model and the related identified parameters are shown in Figure 1.

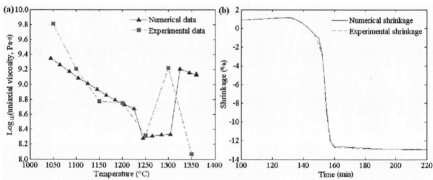

**FIGURE 1.** The uniaxial viscosity and shrinkage obtained by the proposed model with the identified parameters and the experiments: (a) uniaxial viscosity, (b) shrinkage.

## NUMERICAL SIMULATIONS

The fully coupled thermal-stress analysis solver provided by Abaqus® finite element software is used to carry out the numerical simulations. The sintering model described in Equations (1)-(8) and the identified parameters are implemented through the user subroutine UMAT. The gravity, inhomogeneous green density and friction are considered in the simulation. The Coulomb friction law is used in the simulation. The frictional coefficient between the part and the alumina support is set to be 0.5. The initial density contours obtained by in-house bi-phasic injection simulation software [10] was imported into the sintering simulation by nodal interpolation.

## Evolution of the Density and Shrinkage during Sintering

A simulation example for sintering of a part with a wheel shape (external diameter 50 mm, thickness 3.3 mm) has been conducted. The initial density of the green part and the final density after sintering obtained by simulations are shown in Figure 2.

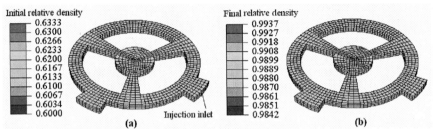

**FIGURE 2.** Relative densities of the part resulting from simulations: (a) before sintering, (b) after sintering.

Due to the factors as friction, inhomogeneous green density and gravity, it represents the uneven shrinkages in the final sintered part. The shrinkage in radial direction is less than that in thickness direction, as shown in Figure 3. The dimensional shrinkages in dimension of the parts obtained by experiments are presented in Figure 4. The measured average shrinkages in radial and thickness directions are -13.94% and -16.53% respectively

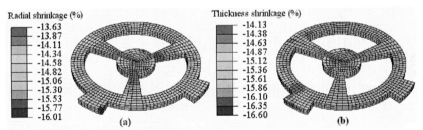

**FIGURE 3.** The uneven shrinkages obtained through simulation, due to friction, inhomogeneous green density, gravity and sintering: (a) in radial direction, (b) in thickness direction.

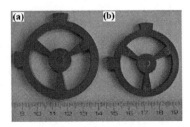

**FIGURE 4.** Shrinkage of the part in experiments: (a) before sintering, (b) after sintering.

## Prediction for Strength of the Sintered Parts

The tensile test specimens sintered up to various peak temperatures are used to evaluate their strength after the sintering through experiments [9]. Simultaneously, the strengths are also predicted by numerical simulation. The strengths obtained by tensile tests and simulations are presented in Figure 5, and one can notice that the agreement is rather good.

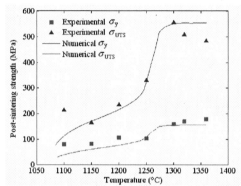

**FIGURE 5.** Numerical prediction and the experimental data for yield strength and ultimate tensile strength of the 316L stainless steel sintered parts.

## CONCLUSIONS

A macroscopic model for the complete sintering stage has been proposed for predicting the final shrinkages and strength of the MIM components. The parameters in the employed viscoplastic constitutive law have been identified through an inverse approach by the gravitational beam-bending tests and sintering experiments in dilatometer, resulting in results obtained through simulation more accurate. The simulation results are in good agreement with the experimental data obtained in the lab. It proves well that the physical model is valid for sintering simulation, and the proposed methods to identify the parameters are reliable.

## REFERENCES

1. E. A. Olevsky, *Mat. Sci. Eng. R* **23**, 41-100(1998).
2. R. M. German, *Sintering Theory and Practices,* New York: John Wiley, 1996.
3. R. K. Bordia and G. W. Scherer, *Acta Metall.* **36**, 2393-2397(1988).
4. J. Song, J.C. Gelin, T. Barriere and B. Liu, *J. Mater. Process. Technol.* **177**, 352-355(2006).
5. R. Zhang, "Numerical simulation of solid-state sintering of metal powder compact dominated by grain boundary diffusion ", Ph.D. Thesis, The Pennsylvania State University, 2005.
6. P. Suri, R. P. Koseski and R. M. German, *Mat. Sci. Eng. A* **402**, 341-348(2005).
7. P. Suri, D. F. Heaney and R. M. German, *J. Mater. Sci.* **38**, 4875-4881(2003).
8. E. A. Olevsky, Shoales E. A. and R. M. German, *Mater. Res. Bull.***36**, 449-459(2001).
9. J.C. Gelin, T. Barriere, J. Song and B. Liu, "Experimental investigations and numerical modeling of sintering process for 316L stainless steel MIM components ", in *PIM 2006 Conference*, March 2006.
10. T. Barriere, J.C. Gelin and B. Liu, *Powder Metall.***44**, 228-234(2001).

# 2 – MULTISCALE APPROACHES

## (A. M. Habraken and S. Bouvier)

# A Multiscale Model Based On Intragranular Microstructure – Prediction Of Dislocation Patterns At The Microscopic Scale

Gérald Franz [1], Farid Abed-Meraim [1], Tarak Ben Zineb [2],
Xavier Lemoine [3], Marcel Berveiller [1]

*1 LPMM, UMR CNRS 7554, ENSAM 4 rue Augustin Fresnel, 57078 Metz Cedex 3, France*
*2 LEMTA, UMR CNRS 7563, Nancy Universités 2 rue Jean Lamour,*
*54519 Vandœuvre-lès-Nancy, France*
*3 Centre Automobile Produit, Arcelor Research S.A., voie Romaine B.P. 30320,*
*57283 Maizières-lès-Metz, France*

**Abstract.** A large strain elastic-plastic single crystal constitutive law, based on dislocation annihilation and storage, is implemented in a new self-consistent scheme, leading to a multiscale model which achieves, for each grain, the calculation of plastic slip activity, with help of regularized formulation drawn from visco-plasticity, and dislocation microstructure evolution. This paper focuses on the relationship between the deformation history of a BCC grain and induced microstructure during monotonic and two-stage strain paths.

**Keywords:** Crystal plasticity, Microstructure, Dislocations, Complex strain paths.
**PACS:** 61.72.Ff, 61.72.-y, 62.20.Fe, 83.10.Gr, 83.60.-a

## INTRODUCTION

The evolution of the plastic anisotropy of polycrystals during plastic flow can be attributed to several sources taking place at different scales. These include at the mesoscopic scale the slip processes and consequently, the texture and internal stresses development, and at the microscopic scale the development of intragranular dislocations patterns. During sheet metal forming processes, strain-path changes often occur in the material and consequently, some macroscopic effects appear due to the induced plastic anisotropy during the previous deformation. These softening/hardening effects must be correctly predicted because they can significantly influence the strain distribution and may lead to flow localization, shear bands and even material failure. The physical cause of these effects can be associated to the intragranular microstructural evolution. This implies that an accurate description of the dislocations patterning during monotonic or complex strain-paths is needed to lead to a relevant constitutive model. In this paper, a crystal plasticity model coupled with a description of the microstructural evolution at the grain scale is presented. Then the capability of the model to predict at a microscopic scale the evolution of dislocation sheets during changing strain paths will be shown on a simple shear loading and on a reverse test.

*CP907, 10th ESAFORM Conference on Material Forming,* edited by E. Cueto and F. Chinesta
© 2007 American Institute of Physics 978-0-7354-0414-4/07/$23.00

# SINGLE CRYSTAL MODEL

The elastic-plastic single crystal constitutive law written within the large strain framework presented here can be found in several works [1-3]. The single crystal behaviour is assumed to be elastic-plastic and the plastic deformation is only due to crystallographic slip. The other plastic deformation modes as twinning or phase transformation are not considered. In BCC metals, 24 independent slip systems are potentially active, i.e. the slip planes {110} and {112} and the slip directions <111>.

The velocity gradient, written below $g$, consists of a symmetric part $d$ corresponding with strain rate of crystalline lattice, and a skew symmetric part $w$ representing the rotation rate:

$$g_{ij} = d_{ij} + w_{ij} \qquad (1)$$

Plastic strain and rotation rates can be expressed by introducing the slip rate $\dot{\gamma}$:

$$d_{ij}^p = d_{ij} - d_{ij}^e = R_{ij}^g \dot{\gamma}^g, \; w_{ij}^p = w_{ij} - w_{ij}^e = S_{ij}^g \dot{\gamma}^g \qquad (2)$$

where $R$ and $S$ are the symmetric and skew symmetric parts of Schmid tensor.

A slip system is active if its resolved shear stress $\tau^g$ achieves a critical value $\tau_c^g$ and if its rate $\dot{\tau}^g$ reaches the critical shear stress rate value $\dot{\tau}_c^g$. Consequently, the considered slip rate is different of zero. That can be written as:

$$\tau^g < \tau_c^g \Rightarrow \dot{\gamma}^g = 0$$
$$\tau^g = \tau_c^g, \; \dot{\tau}^g < \dot{\tau}_c^g \Rightarrow \dot{\gamma}^g = 0 \qquad (3)$$
$$\tau^g = \tau_c^g, \; \dot{\tau}^g = \dot{\tau}_c^g \Rightarrow \dot{\gamma}^g \neq 0$$

To avoid combination analysis and save computing time, the relationship (3) can be expressed by a visco-plastic type regularization, without introducing time dependency. The new formulation is written for a single system g:

$$\dot{\gamma}^g = k^g \dot{\tau}^g, \; k^g = \frac{1}{H^{gg}} \left\{ \frac{1}{2} \left( 1 + \tanh \left[ k_1 \left( \frac{|\tau^g|}{\tau_c^g} - 1 \right) \right] \right) \right\} \left\{ \frac{1}{2} \left( 1 + \tanh\left( k_2 \dot{\tau}^g \tau^g \right) \right) \right\} \qquad (4)$$

where $H^{gg}$ is the self-hardening term.

In large strain framework, the elastic law is written as:

$$\hat{\sigma}_{ij} = C_{ijkl} \left( d_{kl} - d_{kl}^p \right) - \sigma_{ij} d_{kk} \qquad (5)$$

where $C$ is the elasticity tensor and $\hat{\sigma}$ is the Cauchy stress co-rotational derivative. After some mathematical developments, the slip rate expression becomes:

$$\dot{\gamma}^g = \left( \delta_{hg} + k^h R_{ij}^h C_{ijkl} R_{kl}^g \right)^{-1} k^h R_{ij}^h \left( C_{ijkl} - \sigma_{ij} \delta_{kl} \right) d_{kl} \qquad (6)$$

where $\delta$ is the Kronecker symbol.

The behaviour law is described using a tangent operator linking the nominal stress rate $\dot{n}$ with the velocity gradient $g$. The single crystal incremental constitutive law is obtained by:

$$\dot{n}_{ij} = l_{ijkl} g_{kl}, \; l_{ijkl} = \left[ C_{ijkl} - \frac{1}{2} \left( \delta_{ik} \sigma_{lj} + \delta_{il} \sigma_{kj} \right) - \frac{1}{2} \left( \sigma_{ik} \delta_{lj} - \sigma_{il} \delta_{jk} \right) \right] -$$
$$\left[ C_{ijpq} R_{pq}^g + S_{ip}^g \sigma_{pj} - \sigma_{ip} S_{pj}^g \right] \left[ \delta_{hg} + k^h R_{mn}^h C_{mnpq} R_{pq}^g \right]^{-1} k^h R_{mn}^h \left[ C_{mnkl} - \sigma_{mn} \delta_{kl} \right] \qquad (7)$$

# MODELING OF INTRAGRANULAR MECHANISMS

This microscopic model, based on experimental observations on BCC grains, is inspired by the works of Peeters [4]. The hardening is described through several families of dislocation densities and their evolution.

During plastic deformation, an intragranular microstructure develops, consisting of straight planar dislocations walls and of statistically stored dislocations in the cells. This microstructure is characterized by three dislocation densities (Figure 1.). The dislocation cells are represented by a single dislocation density for the whole slip systems. Two different dislocation densities are associated with six dislocations walls families: the density of immobile dislocations $\rho^{wd}$ stored in walls, and the polarity dislocation density $\rho^{wp}$, that is assumed to have a sign.

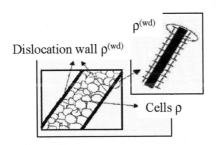

**FIGURE 1.** Schematic representation of the intragranular microstructure: dislocation sheets parallel to {110} planes of the most active slip systems, cells with a more random character [4].

The model constructs at most two families of walls, in agreement with the experimental observations ; the primary family is generated parallel to the {110}-plane of the most active slip system, the second one is constructed in the same way, parallel to the {110}-plane of the second active slip system. The model distinguishes the evolution of walls created by the current slip processes from the evolution of walls formed by the prior slip activity.

The immobile dislocation density of each current wall $i$ can be calculated by:

$$\dot{\rho}_i^{wd} = \frac{1}{b}\left(I^{wd}\sqrt{\rho_i^{wd}} - R^{wd}\rho_i^{wd}\right)\dot{\Gamma}_i \qquad (8)$$

with $b$ the magnitude of the Burgers vector, $\dot{\Gamma}_i$ the total slip rate on the plane of the $i^{th}$ most active slip system and $I^{wd}$ and $R^{wd}$ the immobilization and the recovery coefficients. The last one scales with the annihilation length $y_c$.

The storage and recovery of polarity dislocations for each current wall $i$, can be described by:

$$\dot{\rho}_i^{wp} = \left(sign\left(\Phi_i^{wp}\right)I^{wp}\sqrt{\rho_i^{wd} + \left|\rho_i^{wp}\right|} - R^{wp}\rho_i^{wp}\right)\left|\Phi_i^{wp}\right| \qquad (9)$$

where $\Phi_i^{wp} = \sum_{s=1}^{n}\frac{\dot{\gamma}^s}{b}m^s n_i^w$ is the net flux of dislocations from slip systems non-coplanar of each current wall $i$, the scalar product of the unit slip direction vector $m^s$

49

of the system $s$ with the normal unit vector $n_i^w$ of the wall $i$ being equal to zero for slip activity coplanar with this wall. $I^{wp}$ and $R^{wp}$ are respectively the immobilization and the recovery coefficients of the polarity dislocations.

When a flux $\Phi_i^{wp}$, associated with a family $i$ of current walls, is reversed (for example during reverse tests), the polarity dislocations stuck at the border of theses walls are annihilated:

$$\dot{\rho}_i^{wp} = -R_{rev}\rho_i^{wp}\left|\Phi_i^{wp}\right| \tag{10}$$

where $R_{rev}$ is the recovery coefficient of the polarity dislocations.

A change in deformation path or a rotation of a crystal can lead to the activation of new slip systems. The mobile dislocations form new walls corresponding to the current deformation mode, but also disintegrate the old walls formed by prior slip activity, according to:

$$\dot{\rho}_i^{wd} = -\frac{R_{ncg}}{b}\rho_i^{wd}\dot{\Gamma}_{new}, \quad \dot{\rho}_i^{wp} = -\frac{R_{ncg}}{b}\rho_i^{wp}\dot{\Gamma}_{new} \tag{11}$$

where $\dot{\Gamma}_{new}$ denotes the total slip rate on the two crystallographic planes containing the highest slip activity and $R_{ncg}$ the annihilation coefficient of the latent walls.

During reverse tests, a change in deformation path leads to the activation of the same slip systems but in the opposite sense. The polarity dislocations stuck at the border of the walls can move easily away and are annihilated by dislocations of opposite sign in cells. There is an increase of the annihilation rate for the randomly distributed cells:

$$\dot{\rho} = \frac{1}{b}\left\langle \left(I\sqrt{\rho} - R\rho\right)\sum_{s=1}^n\left|\dot{\gamma}_s\right| - \Psi R_2\rho\frac{\rho_{bausch}}{2\rho_{sat}^{wp}}\sum_{s=1}^n\left|\dot{\gamma}_s\right|\right\rangle \tag{12}$$

with $I$ and $R$ the immobilization and the recovery coefficients of the cells. If no fluxes are reversed $\Psi = 0$, otherwise $\Psi = 1$ with $\rho_{bausch} = \left|\rho_i^{wp}\right|$ if only one flux, corresponding to the family $i$ of wall, is reversed, and $\rho_{bausch} = \sum_{i=1}^2\left|\rho_i^{wp}\right|$ if two fluxes are reversed.

The critical shear stress on slip system g includes several contributions: $\tau_0$ represents all aspects of the microstructure that are not included in the internal variables (e.g. initial grain size), $\tau^{cells}$ depicts isotropic hardening due to the cells, $\tau^w$ introduces the latent hardening of the walls, and takes the contribution of polarity associated to the walls into account. The resultant critical shear stress is given by:

$$\tau_c^g = \tau_0 + (1-f)\tau^{cells} + f\sum_{i=1}^6\tau_{ig}^w \tag{13}$$

with

$$\tau^{cells} = \alpha G b\sqrt{\rho}, \quad \tau_{ig}^w = \alpha G b\left(\left\langle m_s.n_i^w sign\left(\rho_i^{wp}\right)\right\rangle\sqrt{\left|\rho_i^{wp}\right|}\right) + \left|m_s.n_i^w\right|\sqrt{\rho_i^{wd}}\right) \tag{14}$$

where $\alpha$ represents the dislocation interaction parameter, $G$ the shear modulus and $f$ the volume fraction of walls.

50

# RESULTS ON A SINGLE CRYSTAL

The parameters given in the Table 1. were used here. These parameters are obtained by fitting the simulated results with five different experimental tests (two reverse tests and two cross tests with different predeformation and a monotonic shear). The values of the material constants used in the model are shown in Table 2. The initial values of the dislocations densities equal $1.10^9$ m$^{-2}$.

**TABLE 1.** Material parameters [4].

| I | R [m] | I$^{wd}$ | R$^{wd}$ [m] | R$_{ncg}$ [m] | I$^{wp}$ | R$^{wp}$ [m] | R$_{rev}$ [m] | R$_2$ [m] |
|---|---|---|---|---|---|---|---|---|
| $2,2.10^{-2}$ | $8,5.10^{-10}$ | $9,4.10^{-1}$ | $2,6.10^{-8}$ | $2,3.10^{-9}$ | $5,0.10^{-2}$ | $3,8.10^{-9}$ | $1,0.10^{-8}$ | $1,0.10^{-8}$ |

**TABLE 2.** Material constants [4].

| b [m] | G [MPa] | $\alpha$ | f | $\tau_0$ [MPa] |
|---|---|---|---|---|
| $2,48.10^{-10}$ | $8,16.10^4$ | 0,2 | 0,2 | 42,0 |

In Figure 2 (left) a TEM micrograph in a grain of a specimen after a 15% shear test with shear direction (SD) parallel to rolling direction (RD) is depicted. The traces of only one pronounced family of dislocation wall can be observed, parallel to the (101) slip plane. Figure 2 (right) shows a TEM micrograph in a grain of a specimen after a reverse test with SD parallel to RD. The traces of two pronounced families of dislocation walls can be observed, ones are parallel to the (011) slip plane and the others are parallel to the (10-1) slip plane.

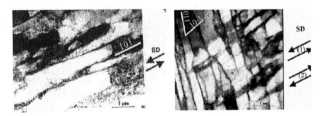

**FIGURE 2.** TEM micrographs in a grain with initial orientation: (-27.2°, 133.4°, 53°) after a shear test of 15% (left) ; (43.3°, 127.2°, -43.2°) after a reverse test of -30%/30% (right) [4].

In Figure 3 and Figure 4, the dislocation density $\rho^{wd}$ stored in the walls (left) and the polarity dislocation density $\rho^{wp}$ associated with the walls (right), are depicted as a function of the shear strain $\gamma$.

For the simple shear mode, there is only one pronounced family of walls, parallel to (101)-plane, predicted by the model, as it is shown on Figure 3 (left). This result is in agreement with the TEM observation. The polarity dislocation density is low (Figure 3 (right)) because there is only one pronounced activated slip system and the wall is generated parallel to the plane of this system.

For the reverse test, the model predicts the traces of both families of walls (Figure 4 (left)). The decrease of the dislocation density $\rho^{wd}$ at the beginning of the second load, not present in the Peeters simulation results [4], is due to the progressive activation of the slip systems. During the prestrain, the polarity along the walls is

51

generated, due to the accumulation of dislocations of opposite sign at either side of the walls. During the reversal of the load, the polarity dislocations are mobilized again and are annihilated with dislocations of opposite sign, leading to a depolarization of the walls. After this transition zone, the walls are polarized again, the polarity dislocations are accumulated at the borders of the current walls. These different steps are depicted on Figure 4 (right).

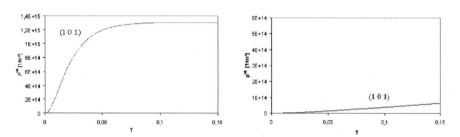

**FIGURE 3.** Evolution of the intensity and polarity of the dislocation walls in a crystal with initial orientation (-27.2°, 133.4°, 53°) during a shear test.

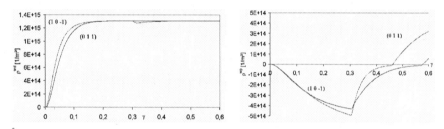

**FIGURE 4.** Evolution of the intensity and polarity of the dislocation walls in a crystal with initial orientation (43.3°, 127.2°, -43.2°) during a reverse test.

## CONCLUSIONS

The present paper shows two examples of intragranular microstructure prediction with the model for a monotonic deformation and two-stage strain paths. Our results are similar to the results obtained by Peeters [4].

The model is validated at the microscopic scale because it is able to reproduce the dislocation microstructure observed in TEM after monotonic deformation, reverse test and cross test (not shown here) for crystal with different initial orientations.

## REFERENCES

1. R. J. Asaro, *Journal Of Applied Mechanics* **50**, pp. 921-934 (1983).
2. D. Pierce, R. J. Asaro and A. Needleman, *Acta Metallurgica* **31-12**, pp. 1951-1976 (1983)
3. T. Iwakuma and S. Nemat-Nasser, "Finite Elastic-Plastic Deformation Of Polycrystalline Metals And Composites", Technical rapport 83-3-51.
4. B. Peeters, "Multiscale Modelling Of The Induced Plastic Anisotropy In IF Steel During Sheet Forming", Ph.D. Thesis, Katholieke Universiteit Leuven, 2002

# Formations of Matter, Light and Sound described with Bernoulli's principle

## L. Strömberg

*Fredrlg 38, se-38237 Nybro, previously Department of Solid Mech, KTH, Sweden*
*e-mail: lena_str@hotmail.com*

**Abstract.** Properties of Bernoulli's principle are investigated. Assuming a velocity field, dynamics of a hurricane are described. In a multi-scale model, formulas for gravitational red-shift are recovered, by regarding the flow of light particles as a mixture with a Newtonian gravitating body. For sound in a mixture, a Doppler formula is deduced. Plastic deformation is considered in conjunction with sound.

**Keywords**: whirlwind, mixture, multiscale model, gravitational red-shift, thermal blue-shift, Acoustic Doppler effect, plasticity

## INTRODUCTION

Bernoulli's principle in a flow states that decreased pressure occurs simultaneously as increased velocity. In the present context, this is used in conjunction with mixture theory to describe flow, light beams and sound. Consider a flow with density $\rho$, pressure p, and velocity v. When incompressible, and no explicit dependence on time, Bernoulli's principle give that $p+\rho v^2$ is constant for different points on a streamline.

For a mixture of two constituents [1], BP for each constituent i, is modified to read, $p_i+\rho_i v_i^2 + m_i = $const, where $m_i$ is the sum of momentum supply.

## BUOYANCY

For subwater streams, water has a velocity, and according to BP the pressure decreases. This results in decreased uplift/buoyancy of a floating object.

## COMMERCIAL PUMPS AND MIXERS

ITT Flygt AB is the manufacturer of pumps and mixers for a variety of applications. Each pump is delivered with its own characteristic flow pressure curve. Also more details on processes, products and components, eg. simulation of flow in the casted pump-wheel, are given in 'Scientific Impeller'.

*CP907, 10th ESAFORM Conference on Material Forming,* edited by E. Cueto and F. Chinesta
© 2007 American Institute of Physics 978-0-7354-0414-4/07/$23.00

Figure 1. Impeller pumps.

## WHIRLWIND OF A HURRICANE

Consider the hurricane as a 'large eddy' in a region, with density $\rho$, pressure p, and velocity $\mathbf{v}$, that fulfil BP. Motion is assumed as a velocity field $\mathbf{v} = \omega \times \mathbf{r} + \mathbf{v}_z$, where $\omega = \omega \mathbf{e}_z$ is angular velocity and $\mathbf{r}$ is radius vector in a circular in-plane motion, and $\mathbf{v}_z$ is upward velocity.

It is known that the whirlwind build up energy at sea, from the hot surface water,(JH). Assuming that the medium of the hurricane fulfils the ideal gas law, then pressure will rise with temperature. In more detailed modeling, the region close to water surface could be considered as a mixture [1], such that a momentum supply enters the equation of motion, and allows increased pressure. Or the increased temperature may be modeled with heat conduction in a boundary condition.

When the whirlwind reaches land, momentum supply vanishes, or heat transfer from ground or air is much less. If pressure drops, velocity will increase, which make a more severe wind, with larger 'strength'. As a measure of strength is used the change in vertical direction $p_{,z}$ [2], the Lie derivative [3]. With the above velocity field in notation $\mathbf{v} = (\omega r, v_z)$ and BP, this will read

$$p_{,z} = -2\mathbf{v} \cdot (r\omega_{,z}, v_{z,z})\rho = -2(\omega_{,z} r^2 \omega + v_z v_{z,z})\rho$$

where $_{,z}$ denotes differentiation with respect to z.

Pressure drop may occur (as a result of change in boundary conditions) entering land from sea, or at spatial extension of the whirlwind.

## LIGHT REGARDED AS A FLOW OF PARTICLES

Next it will be exemplified how BP apply to observations in astrophysics.

### Gravitational red-shift

In a multiscale model, the light-beam is considered a mixture with a gravitating body. Hereby, the Newtonian gravitational potential enters the equation of BP as a momentum supply, given by $\rho GM/r$, where G is the Newton constant, M is mass and r is distance.

In predictions of the Pound-Rebka experiment of red-shift, is used an energy-balance and equivalent relativistic mass, such that $E=mc^2$. A wave-length and frequency f, is associated with the kinetic energy through de Broglie correspondance, as $E=hf$, where h is Plancks constant. Hereby the shift will be $GM/(rc^2)$. In BP, this is obtained with $v=c$, $p=0$, and integrating over the volume.

## Scattering

Light with the lower wavelengths scatters more, since it reflects against both smaller and larger particles, eg. Tyndall effect, blue sky.

Figure 2. Tyndall effect: Bluish light reflected in fog.

If, for a beam of light, internal scattering give a pressure, then according to BP, velocity will be lower for the blue light. This is also the case (Snell's law), as is seen for dispersion in a prism. Scattering from the beam would correspond to friction losses not accounted for in the present version of BP.

*Blue-shift, hypothesis*: Promoted by heat (radiation), flashing light. To model/measure blue-shift, a mixture of emitted and reflected light may be considered.

## DOPPLER SHIFT OF SOUND

In [4], to describe frequency-shift, is assumed a constituent (fluid or gas) with velocity $v_1$ and density $\rho_1$, that transmits an acoustic wave with sound velocity $c_1-v_1$, mixed [1] with an acoustic medium at rest, denoted by 2. Constituting the mixture to be an acoustic medium with sound/wave velocity c (JH), a Doppler effect that depend on ratio of velocity and sound speed, as well as densities of the constituents will read

$$\omega_1^2\rho_1c_1^2\,(1-(v_1/c_1))^2=\omega_2^2\rho_2c^2\,(1-f(\rho_1,\rho_2)(\,v_1/c)^2)-d_t^2p_2 \qquad (1)$$

$\omega_2$ is detected frequency at location2, $\omega_1$ is transmitted frequency at the location1 and f is a function of the densities. For $c_1=c$, the 'average' values $\rho_1=\rho_2$, $f(\rho_1,\rho_2)=1$ (since an integration over volume was presumed), and assuming that 2 fulfils BP, the classical relativistic Doppler formula is recovered. The assumption of sound velocity $c_1-v_1$, is probably valid at a limited distance to the moving sound generator. When the function $f(\rho_1,\rho_2)$ vanishes, which is the case far from 1 or 2, the formula alters, which agrees with that the Doppler effect is more noticable, at close distance.

The formula is of interest for example in the design of damping materials [5] and composites.

*Example 1:* Duck Talk. Speech after inhaling helium will be more high frequent, than if transported in usual air. This is seen from (1) with $v_1=0$, since, at 1, in the body where sound is generated, properties are the same for both media. And at 2, where the sound is heard, for helium $\rho_2$ and c are lower than for air.

Some observations in (1) on measured frequencies and time constants, in conjunction with sound generation in deforming solid media (described as a fluid), will be made in forthcoming subsection.

# SOLIDS

## Bouncing ball

As a generalisation, a BP for a bouncing solid ball is established. When falling, the path of the ball is considered a stream line, and then BP reads $p+\rho v^2+m=$const, where $m=-\rho gy$ (y being a vertical coordinate) is momentum supply from gravity, and 'pressure at depth y' is also $p=-\rho gy$.

At bounce, the ball is deformed, and to model this process, a weak formulation of BP may be considered. This is obtained by integrating with a weight function over the volume of the ball. Hereby is obtained energies, both linear and nonlinear, that describe the impact, dependant of material parameters.

In a first approximation, ball deform as a spring and motion will be of a harmonic oscillator, with the Hamiltonian $H=ax^2+b(d_tx)^2$, where x is the coordinate from center of mass and a,b are material constants. $H(x=0)$ is initial kinetic energy. Invoking other weighting, nonlinearity and dissipation is included as $H=ax^2+b(d_tx)^2-cx(d_tx)^2-d$ where c,d are material constants. For $d=0$, the acceleration may be calculated to read

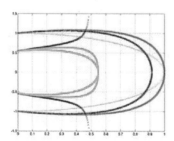

Figure 3. Phase portraits $d_tx$ versus x for [H,a,c]/b=[1,1,0.7],[1,1,0], ,[1,1.2,0.7],[0.3,1,1.6],[0.3,1,1] from right, and diverging [0.3,1,2].

$-(2ax-c(d_tx)^2)/(2(b-cx))$. It is seen that, the larger c, the more rapid bounce, since x increases and $d_tx$ decreases. The dimension of c/b will be 1/Length, a material length scale, and thus the rapid bounce correspond to a smaller length. Phase portraits $d_t x$ versus x, for constant H-values, and material parameters, are seen in Figure 3. Dim(a/b) will be $1/Time^2$, and for larger (a/b) i.e smaller time, maximum abbreviation max(x) will decrease, and the bounce will be almost as rapid.

## Plastic flow

For material descriptions of plastic solids, analogies with BP will be made, at minor level. From averaging over a representative volume; so called homogenisation, the governing equation at continuum level is obtained.

### *Yield criteria*

At continuum level, a Drucker-Prager, or Mohr-Colomb-like yield criterion consider a bound for sum of hydrostatic pressure and von Mises or shear stress. In a formulation of flow at micro-scale, the BP can be assumed. Then the velocity part in BP is associated with shear stress in a homogenisation procedure, to obtain the yield criterion.

In a slip line/shear band, deformation consists of shear strain and some normal deformation. The latter is often given as a jump in the displacement (over a band of finite width). The plastic stiffness properties in the zone are known as cohesion. In [6], a relation between shear and normal deformation is given.

### *Deformation hardening and time constants*

Plasticity or slip and micro-creep often occur at a constant loading, resulting in additional hardening. Comparing with a time dependency, for the larger time constants, the deformation hardening will increase.

### *Sounds at micro-cracking*

Plastic flow may occur in conjunction with small vibrations and wave propagation [7], in shear band. Also actual sound is produced, eg. 'Tin-scream' in Sn, and short-time snaps in wood. In Sn, the sound correspond to a larger time constant, and deformation hardening will probably occur, whereas for a short duration snap, the response is more brittle.

## Doppler shift in materials

In the same manner as in Example 1, some hypothesis of frequencies in plastic flow will be concluded. Assume that a sound in eg. Sn, with frequency $\omega_1$ is generated in a slip line. Then from (1), $\omega_2^2 = \omega_1^2 \rho_1 c_1^2 / (\rho_2 c^2)$, and if sound velocity and density is larger in adjacent more dense medium, $\omega_2$ will be lower.

To consider creep effects, and time constants in a material, the Doppler formula will be expressed in time-domain. So-called eigen-times, are defined as $T_i = 2\pi/\omega_i$. Hereby $T_1^2 = T_2^2 \rho_1 c_1^2 / (\rho_2 c^2)$. If $T_i$:s are identified as time-constants of hardening, then due to deformation hardening at 1, the ability of hardening in adjacent regions not so dense is decreased. This could lead to material degradation such as inter-granular cracking, and brittleness of a hard material. However, if the 'pressure' $p=\rho_2 c^2$, in the mixture is increased, time constant will also be higher.

## ESCAPE VELOCITY AND SILENT HOLE

Let $v_1$ be the escape velocity from a gravitating body with radius r. Then $v_1{}^2=2(\omega r)^2$, where $\omega$ is the angular velocity, or expressed in mass and Newton constant, $v_1{}^2=2GM/r$. Following the derivation of the Schwarzschild singularity and radius of a black hole [3], a similar object for sound, a so-called silent hole, will be discussed. Comparing with (1), there will be a singularity for $r=f2GM/c^2$. Insertion of data [3] give that within this model, when f approaches 1 (f<0.2), a body of approximate size and density of the moon is a silent hole, for sound.

## CONCLUSION

More detailed models of continuum properties, may increase the knowledge about macroscopic and measurable phenomena. For a whirlwind, by calculating the directional-derivative, it was found that the energy released from a decreased pressure is decomposed in increased angular velocity and upward velocity. Alternate behaviour in terms of added viscosity and boundary conditions, is possible.

Within the framework of acoustic, when constituents and mixture fulfil the wave equation, characterised by sound speed, a modified formula for Doppler effect, with a dependence of spatially distributed densities of constituents and mixture, were derived. For plasticity, analogies with flow at microscale were discussed. The Doppler formula was applied to sound in conjunction with slip and micro-cracking.

## ACKNOWLEDGEMENTS

Author is greatly acknowledged to Mr J. Hamberg, in text referred to as (JH), for valuable discussion, during the course of this work.

## REFERENCES

1. Bowen, R.M. *Th of Mixtures*. Continuum Phys (1976) ed. by A.C. Eringen, Vol III, Acad Press.

2. Nygårdshaug, G. *Alle orkaners mor*. P.13, Cappelen (2004), ISBN 820224072

3. Wikipedia, www.wikipedia.org

4. L. Strömberg Continuum Mixture theory as an approach to Fluid-Structure Interaction Part II. Dynamics with relative motions. Proceedings. NSCM 19, 2006, Lund, Sweden. pp. 227-230.

5. A.R. Mackintosh et al. Influence of fillers on the relaxation characteristics and processing of polyurethane elastomers. Proc. Esaform 2006. pp. 13-26.

6. L. Strömberg Solutions to integral equations in nonlocal plasticity. Report 2001.

7. N.S. Ottosen and K. Runesson. (1991) Properties of discontinous bifurcation solutions in elasto-plasticity. *Int J Sol Struct*. 27, pp 401-421.

# Modeling Material Flow Characteristics Over Multiple Recrystallization Cycles

A. D. Foster[1], J. Lin[1], D. C. J. Farrugia[2], and T. A. Dean[1]

[1]*Department of Mechanical and Manufacturing Engineering, University of Birmingham, Edgbaston, Birmingham B15 2TT, UK*
[2]*Corus R, D & T, Swinden Technology Centre, Moorgate, Rotherham S60 2AR, UK*

**Abstract.** The evolution of steel microstructure during multi-pass hot rolling processes may include several recrystallization (RX) stages. Assuming a sufficient dislocation density is present at the exit of the roll, RX will occur during the interpass time between rolling stages. This will result in the material reaching the next rolling stand in either a full recrystallized or part recrystallized state. In this study an equation set is proposed in which complex RX cycles may be modeled, including overlapping cycles. Other microstructure information such as grain size evolution may be mapped.

**Keywords:** Hot Rolling, Recrystallization, Multi-Pass Rolling, Material Evolution
**PACS:** 81.05.-t

## INTRODUCTION

Material flow behavior during hot deformation is often complex. Viscoplasticity, temperature dependence and history dependent hardening and softening mechanisms will all influence the dynamic yield point and thus working forces.

Dislocations are formed within grains to accommodate plastic deformation and will accumulate during hot working. The high energy levels associated with a high dislocation density provide the driving force for recovery and recrystallization (RX) processes [1]. These competing material softening processes annihilate dislocations by two distinct mechanisms: Recovery acts to annihilate pairs of dislocations and form subgrains; and RX forms new grains with low dislocation density, which grow into existing grains from nucleation sites [2, 3].

The multi-pass hot rolling roughing process shown in figure 1 is used to create low carbon steel slab, plate, and bar sections and typically has over 10 passes. Plastic deformation during the first pass creates a high dislocation density. During the interpass, metadynamic RX commences and new grains develop. The fully recrystallized material has a small-grained, wrought structure before entering the next pass. However if the interpass time is short or the driving energy low, material may enter the second pass in a part-recrystallized state. The RX cycle may take place several times during a multi-pass roughing process.

Many material models are available to describe RX and recovery mechanisms [3-6]. Advanced models use constitutive dislocation-based equations capable of mapping

CP907, *10th ESAFORM Conference on Material Forming*, edited by E. Cueto and F. Chinesta
© 2007 American Institute of Physics 978-0-7354-0414-4/07/$23.00

the interaction of dislocation-based hardening, the RX cycle, grain size, recovery processes and the resulting viscoplastic flow behavior. However for multi-pass hot rolling there is a need to simulate multiple RX processes. The work presented here advances a unified, constitutive material model for RX and recovery proposed by Lin et al. [3]. Multiple RX cycles may be modeled and the detail to which grain size and dislocation density are mapped within recrytallized material is enhanced when compared to traditional continuum models.

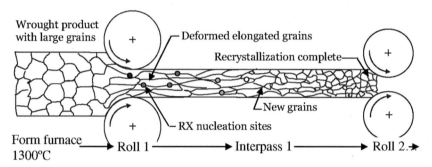

**FIGURE 1.** Grain evolution through the initial hot rolling roughing mills.

## FORMULATION OF MODEL

The core of the unified material model is given in equations (1-6). The material model assumes isotropic behavior and is expressed by: plastic flow rate, $\dot{\varepsilon}_p$ (1); normalized dislocation density, $\dot{\bar{\rho}}$ (2) (the first term describes accumulation, the second recovery); isotropic dislocation hardening, $R$ (3); RX onset, $\dot{X}$ (4); RX fraction, $\dot{S}$ (5); grain size, $\dot{d}$ (6); and flow stress, $\sigma$ (7). The mechanics behind each equation and details of the formulation can be found in the original paper [3]. This set is valid while the RX parameter $S_1$ is zero.

$$\dot{\varepsilon}_p = A_1 \sinh\left[A_2\left(\sigma - R - k\right)\right]_+ d^{-\gamma_1} \tag{1}$$

$$\dot{\bar{\rho}}_0 = \dot{\bar{\rho}}_0^+ - \dot{\bar{\rho}}_0^- \text{ where } \dot{\bar{\rho}}_0^+ = \left(d_0/A_3\right)^{\gamma_2}\left(1 - \bar{\rho}_0\right)\dot{\varepsilon}^{\gamma_3}{}_p \ , \ \dot{\bar{\rho}}_0^- = A_4\bar{\rho}_0^{\gamma_4} \tag{2}$$

$$R = A_5\sqrt{\bar{\rho}} \tag{3}$$

$$\dot{X}_1 = A_6\left(1 - X_1\right)\bar{\rho}_0 \tag{4}$$

$$\dot{S}_1 = A_7\left[X_1\bar{\rho}_0 - \bar{\rho}_c\right]_+\left(1 - S_1\right)^{\gamma_5} \tag{5}$$

$$\dot{d}_0 = A_8 d_0^{-\gamma_6} \tag{6}$$

$$\sigma = E\left(\varepsilon_T - \varepsilon_p\right) \tag{7}$$

The subscript '+' indicates zero or positive outputs only; $A_{1-7}$ & $\gamma_{1-6}$ are material-dependent constants; $k$ is the initial yield stress; $\bar{\rho}_c$ is the critical dislocation density for RX; $E$ is the Young's modulus; $\varepsilon_T$ is the total strain; $d$ is the averaged grain size, equal to $d_0$ here; and $\bar{\rho}$ is the averaged, normalized dislocation density, equal to $\bar{\rho}_0$ here. Equations (1), (2), (4), (5), and (6) are given as history dependent rate equations that need integrating via a suitable numerical technique.

The RX modeling is based on the notion of an RX front, as described by Sandstrom and Lagneborg [6]. A newly nucleated mobile grain boundary migrates into an area of high dislocation density and leaves a recrystallized, low dislocation density matrix in its wake (see figure 2). The difference in dislocation density, and thus stored energy, provide the driving force for RX. The speed of RX will depend on the driving force, the material composition, and the number of RX nucleation points. If the driving force is reduced by recovery mechanisms or a drop in temperature, complete RX may not be achieved.

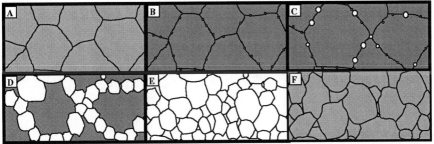

**FIGURE 2.** RX front passing through a deformed material. A) Deformed microstructure RX. B) RX nucleation. C) Newly formed RX grains. D) RX fronts expand into old grains. E) Fully RX material. F) Grain growth and continued deformation.

The subscript '0' given in the equations above denote the material in the region before an RX front passes through. Once $S_1$ is non-zero a new set of equations (subscript '1') is automatically generated to describe the dislocation density, RX onset, and grain size of the new grains. The material is mapped as two parts: a non-RX fraction $(1- S_1)$ and a new RX fraction $S_1$. A new RX onset parameter $X_2$ is generated in preparation for the next RX cycle. The average grain size is given by the grain size within each front multiplied by the relative fraction of that front. The same principal is applied to material hardness $R$ and to the accumulation of new dislocations. For computational efficiency, equations with subscript '0' may be removed when $S_1$ reaches the fully RX state $(S_1=1)$.

# RESULTS

A suitable routine was developed to automatically generate new material equations and allow up to two RX fronts to be active at any one time. Forward Euler integration was used to progress material evolution with time.

To demonstrate the capabilities of this modeling method, the material model has been calibrated using previously published data [3]. The initial grain size was 190µm. A nominal grain size of 10µm was given to new RX grains.

**TABLE 1.** Material constants determined for C-Mn steel at 1373°K.

| Const. | $k$ | $E$ | $\bar{\rho}_c$ | $A_1$ | $A_2$ | $A_3$ | $A_4$ | $A_5$ | $A_6$ | $A_7$ | $A_8$ | $\gamma_1$ | $\gamma_2$ | $\gamma_3$ | $\gamma_4$ | $\gamma_5$ | $\gamma_6$ |
|---|---|---|---|---|---|---|---|---|---|---|---|---|---|---|---|---|---|
| Value | 34.0 | 1.0e3 | 0.04 | 0.68 | 8.8e-2 | 148.4 | 0 | 6 | 26.0 | 10.9 | 48.8 | 983.3 | 0.55 | 1.92 | 1.60 | 0.87 | 1.65 | 1.93 |

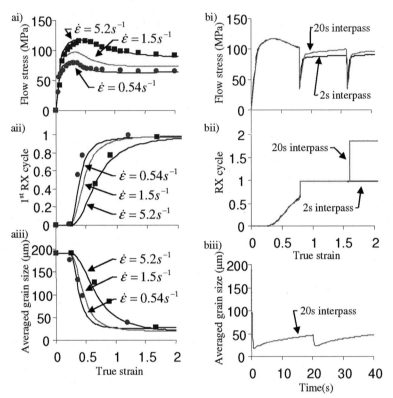

**FIGURE 3.** a) C-Mn steel material fit at 1373°K. Computed (lines) against experiment data (points). b) Simulated 3-part multiple-interrupt test at 1373°K and $\dot{\varepsilon} = 5.2s^{-1}$. Interrupted at ε=0.8 and ε=1.6. i) Flow stress. ii) Recrystallization cycle completeness. iii) Average grain size.

Figure 3a) shows the material model fit against experimentally derived data. The model is capable of accurately predicting flow stress, recrystallized fractions, and the average grain size. The model could also be used to reveal the disparate grain size found in a part recrystallized material. This detail is not available from other models, but is a natural consequence of the way this model has been formulated.

Figure 3b) shows material properties resulting from a simulated 3 stage multiple-interrupt test. In two runs, constant strain rate deformation was halted for approximately 2s (black) and 20s (grey) at a strain of 0.8 and again for the same durations at a strain of 1.6. During the first interrupt pass both materials fully recrystallize. During subsequent deformation the 20s material is harder (higher flow stress). This is due to extended grain growth during the longer 20s interrupt.

During the second interrupt, only the material which experienced a 20s interrupt recrystallized. The 20s interrupt was of sufficient time for the material to completely recrystallize. The smaller grain size of the 2s material during the $2^{nd}$ pass resulted in a lower dislocation density being reached, insufficient for RX to commence.

## CONCLUSIONS

The model developed in this work has the functionality required to predict material microstructure over multiple RX cycles. It has been shown that the period of interrupt will affect the onset of RX during subsequent deformation, in turn affecting the material grain size.

The next stage of the work will be to refine and fit the model to multiple interrupt test data and implement into an FE simulation of multipass rolling. In this way multiple RX cycles and material evolution predictions may be used to predict product grain size and optimize rolling interpass times.

## ACKNOWLEDGMENTS

The authors gratefully acknowledge the finance provided by Corus UK Ltd and the EPSRC.

## REFERENCES

1. R. Sandstrom and R. Lagneborg, *Acta Metallurgica* **23**, 481-488 (1975).
2. T. Sakai and J.J. Jonas, *Acta Metallurgica* **32**, 189-209 (1984).
3. J. Lin, Y. Liu, D.C.J. Fareugia and M. Zhou, *Philosophical Magazine A* **85**, 1967-1987 (2005).
4. S. Serajzadeh, *Materials Letters* **59**, 3319-3324 (2005).
5. E. Nes, *Progress in Materials Science* **41**, 129-193 (1998).
6. R. Sandstrom and R. Lagneborg, *Acta Metallurgica* **23**, 387-398 (1975).

# Different Approach to the Aluminium Oxide Topography Characterisation

Sanja Mahovic Poljacek[*], Miroslav Gojo[*], Pero Raos[†] and Antun Stoic[†]

[*]Faculty of Graphic Arts, University of Zagreb, Getaldiceva 2, 10000 Zagreb, Croatia
[†]Mechanical Engineering Faculty, J.J. Strossmayer University of Osijek,
Trg Ivane Brlić Mažuranić 2, 35000 Slavonski Brod, Croatia

**Abstract.** Different surface topographic techniques are being widely used for quantitative measurements of typical industrial aluminium oxide surfaces. In this research, specific surface of aluminium oxide layer on the offset printing plate has been investigated by using measuring methods which have previously not been used for characterisation of such surfaces. By using two contact instruments and non-contact laser profilometer (LPM) 2D and 3D roughness parameters have been defined. SEM micrographs of the samples were made. Results have shown that aluminium oxide surfaces with the same average roughness value ($R_a$) and mean roughness depth ($R_z$) typically used in the printing plate surface characterisation, have dramatically different surface topographies. According to the type of instrument specific roughness parameters should be used for defining the printing plate surfaces. New surface roughness parameters were defined in order to insure detailed characterisation of the printing plates in graphic reproduction process.

**Keywords:** printing plate, aluminium oxide surface, 2D and 3D roughness parameters.
**PACS:** 81.40.–z, 81.70.-q

## INTRODUCTION

The offset printing technique is based on different physical and chemical properties of printing and nonprinting areas on the aluminium printing plate. The printing areas have explicitly hydrophobic properties and mostly consist of photosensitive organic material. The nonprinting areas have explicitly hydrophilic properties in order to ensure selective adsorption of molecules of different formulation on the printing plate surface [1]. In the printing process nonprinting areas are damped with water whose molecules are of polar character, and the printing areas are covered by ink which contains non-polar molecules of the higher fat acids [2]. The quality level of the imprints mostly depends on the water-ink balance on the printing plate in the printing process.

In the printing plate making process photosensitive layer is exposed to the specific irradiation, due to which it becomes soluble in development solution. The result of a chemical development process in alkaline solution is a printing plate consisting of areas covered with photoactive organic layer (printing areas) and aluminium oxide surfaces (non printing areas). In the printing plate making process quality control of

CP907, 10th ESAFORM Conference on Material Forming, edited by E. Cueto and F. Chinesta
© 2007 American Institute of Physics 978-0-7354-0414-4/07/$23.00

the printing plate surfaces is often monitored by different visual methods [3][4]. These methods are suitable for detailed quality control of the printing surfaces (organic coating) and do not provide information about the quality of nonprinting surfaces (aluminium oxide). The aim of this research was to make a new approach to topography analysis of aluminium oxide surfaces which represent the nonprinting areas in offset printing technique.

## MATERIAL SURFACES AND EXPERIMENTAL METHODS

Aluminium base of a printing plate is produced by electrochemical roughening process of a thin aluminium foil (AA1050) [5][6]. It is made by the process of rolling which causes specific structure of the aluminium in the rolling direction. The thickness of the aluminium foils is typically from 0,15 to 0,51 mm and depends on the size and type of the printing machine. Electrochemical graining process is carried out in order to improve the water adhesion on the aluminium oxide film and to enhance the adhesion of the photosensitive layer. During this process aluminium surface becomes uniformly and finely grained. Size and quality of the grained microstructure will influence the printing performance of the printing plates. Electrochemical roughening process can be carried out in different acids by using alternating current [7]. Oxide film on the roughened surface is produced by anodizing. Various types of surface structures can be obtained under different electrochemical conditions [5]. They differ in shape and size of the porous structure, dimensions of the peaks and valleys, and accordingly in amount of water adsorption abilities, etc. The shape and quality of the microstructure can be monitored by using different microscopic systems and by measuring the surface roughness by contact and noncontact non-destructive methods. The most common technique for surface roughness measuring is the contact stylus profilometry and mostly used parameters for characterising the printing plate surfaces are average surface roughness value (Ra) and mean roughness depth (Rz).

In this paper three different printing plate samples were selected in order to characterise their surfaces. The samples were 0,30 mm thick and were commercially grained and anodized. Commercial aluminium sheets typically have the average surface roughness parameter (Ra) in the range from 0,4-0,5 μm and the mean roughness depth (Rz) typically in the range from 3,0-6,0 μm, measured by mechanical stylus profilometer [5].

**TABLE 1.** Definition of the measured surface roughness parameters.

| Roughness parameter | Description |
| --- | --- |
| $R_a$ | Arithmetic mean deviation of the surface |
| $R_{zDIN}$ | Mean peak-to-valley height in 10 dots |
| $R_y$ | Vertical distance between highest peak-deepest valley |
| $R_q$ | Root-mean-square deviation of the surface |
| $R_p$ | Distance between highest peak and the reference line |
| $R_{pk}$ | Reduced peak high |
| $R_k$ | Core roughness depth |
| $R_{vk}$ | Reduced valley depth |
| $M_{r1}$ | Smallest material ratio of the roughness core profile |

According to the ISO standard 12218:1997 printing plates were exposed and developed in alkaline solution. Selected samples had equal values of the average surface roughness (Ra) measured by contact stylus profilometer. The average surface values were obtained by measuring the samples in the rolling direction (x) and in the direction vertical to the one from the rolling process (y). The optical micrographs were made by JEOL JMS T300 scanning electron microscope. Perthometer S8P Mahr GmbH Göttingen and Taylor-Hobson Surtronic 3+ were used for contact and UBM Microfocus System for noncontact measurements. The defined surface roughness parameters are listed in Table 1 [8][9][10].

## RESULTS OF THE CONTACT MEASUREMENTS

Samples have been selected according to the same values of average surface roughness parameter measured by electronical-mechanical profilometer Perthometer (Ra=0,54 μm) and Surtronic 3+ (Ra=0,32 μm) with diamond stylus of 5 μm radius and stylus angle 90°. All measurements were performed in the same measuring conditions using Gauss filter (DIN 4776) limited wave length λ=0,8 mm with the evaluation length lm=5,6 and 4,0 mm. Surface roughness values were obtained by measuring the samples in the rolling direction (x) and in the direction vertical to the one from the rolling process (y). Obtained results are given in Table 2 and 3.

**TABLE 2.** Surface roughness parameters (Perthometer).

| Parameter | Sample 1-x | Sample 1-y | Sample 2-x | Sample 2-y | Sample 3-x | Sample 3-y |
|---|---|---|---|---|---|---|
| $R_a$ (μm) | 0,54 | 0,54 | 0,54 | 0,54 | 0,54 | 0,54 |
| $R_z$ (μm) | 4,02 | 4,22 | 4,21 | 4,18 | 4,82 | 4,32 |
| $R_y$ (μm) | 5,32 | 5,17 | 5,00 | 5,26 | 5,93 | 5,85 |
| $R_p$ (μm) | 1,76 | 1,74 | 1,87 | 1,71 | 1,98 | 1,87 |
| $R_{pk}$ (μm) | 0,36 | 0,45 | 0,37 | 0,43 | 0,43 | 0,43 |
| $R_k$ (μm) | 1,59 | 1,53 | 1,72 | 1,52 | 1,71 | 1,63 |
| $R_{vk}$ (μm) | 0,96 | 1,05 | 1,02 | 1,03 | 1,13 | 1,07 |
| $M_{r1}$ (%) | 7,17 | 8,07 | 5,47 | 7,80 | 6,70 | 7,23 |
| $M_{r2}$ (%) | 84,73 | 84,17 | 85,30 | 84,83 | 86,93 | 85,70 |

**TABLE 3.** Surface roughness parameters (Surtronic 3+).

| Parameter | Sample 1-x | Sample 1-y | Sample 2-x | Sample 2-y | Sample 3-x | Sample 3-y |
|---|---|---|---|---|---|---|
| $R_a$ (μm) | 0,32 | 0,32 | 0,32 | 0,32 | 0,32 | 0,32 |
| $R_z$ (μm) | 2,50 | 2,85 | 2,45 | 2,45 | 2,30 | 2,67 |
| $R_y$ (μm) | 3,25 | 3,70 | 3,10 | 3,20 | 3,08 | 3,30 |
| $R_p$ (μm) | 0,42 | 0,44 | 0,41 | 0,42 | 0,37 | 0,44 |

## RESULTS OF THE NONCONTACT MEASUREMENTS

Surface roughness parameters measured by UBM noncontact laser profilometer have been obtained by two kinds of measurements: line and area measurements. Line measurements have been taken by using Gauss filter (DIN 4776), limited wave length λ=0,8 mm with the evaluation length lm=4,0 mm. Measurement range was ±50 μm, laser spot size was 1 μm. Each sample was measured with 1000 points/mm. The resolution in the vertical direction was ± 0.1 μm. The laser output was 0.2 mW and the laser wavelength 780 nm. Surface roughness values were obtained by measuring the samples in the rolling direction (x) and in the direction vertical to the one from the rolling process (y) by the stepwise type of measurement (Table 4). Area measurings were made in the same conditions as line measurement with 500 points/mm over 0,6 mm x 1,0 mm square (Table 5).

TABLE 4. Surface roughness parameters (line measurement).

| Parameter | Sample 1-x | Sample 1-y | Sample 2-x | Sample 2-y | Sample 3-x | Sample 3-y |
|---|---|---|---|---|---|---|
| $R_a$ (μm) | 2,49 | 2,50 | 2,58 | 2,58 | 2,64 | 2,61 |
| $R_z$ (μm) | 14,92 | 14,42 | 15,02 | 15,23 | 15,51 | 15,89 |
| $R_y$ (μm) | 17,28 | 15,92 | 16,98 | 17,22 | 17,61 | 15,84 |
| $R_q$ (μm) | 3,00 | 3,02 | 3,16 | 3,00 | 3,20 | 3,12 |
| $R_p$ (μm) | 8,42 | 7,21 | 8,19 | 7,95 | 7,80 | 7,12 |
| $R_{pk}$ (μm) | 1,58 | 1,40 | 1,80 | 1,56 | 1,55 | 1,49 |
| $R_k$ (μm) | 8,57 | 8,64 | 9,33 | 9,25 | 9,26 | 9,33 |
| $R_{vk}$ (μm) | 2,42 | 2,44 | 2,14 | 2,35 | 2,52 | 2,21 |
| $M_{r1}$ (%) | 6,03 | 5,35 | 5,80 | 6,05 | 5,50 | 5,80 |
| $M_{r2}$ (%) | 90,16 | 89,15 | 92,35 | 92,02 | 89,25 | 90,24 |

TABLE 5. Surface roughness parameters (area measurement).

| Parameter | Sample 1 | Sample 2 | Sample 3 | Parameter | Sample 1 | Sample 2 | Sample 3 |
|---|---|---|---|---|---|---|---|
| $R_a$ (μm) | 3,44 | 3,41 | 3,40 | $R_{pk}$ (μm) | 1,72 | 1,68 | 1,78 |
| $R_z$ (μm) | 21,98 | 21,71 | 21,38 | $R_k$ (μm) | 12,04 | 11,94 | 11,71 |
| $R_y$ (μm) | 24,42 | 26,50 | 22,58 | $R_{vk}$ (μm) | 2,48 | 2,42 | 2,53 |
| $R_q$ (μm) | 4,03 | 4,00 | 3,99 | $M_{r1}$ (%) | 3,60 | 3,60 | 3,60 |
| $R_p$ (μm) | 12,51 | 10,78 | 10,30 | $M_{r2}$ (%) | 90,90 | 90,80 | 89,80 |

## DISCUSSION

The aim of this research was the comparison between different surface roughness parameters measured on the aluminium oxide film of the printing plate. Most commonly used roughness parameters for characterisation of this type of surfaces are arithmetical average value of all absolute distances of the roughness profile from the center line (Ra) and the mean roughness depth (Rz). These are frequently used in the electrochemical processing of aluminium foil, as means of controlling the porous oxide microstructure. Considering the different types of instruments used in this research the results have shown acceptable deviations.

Results obtained by two mechanical styluses were in correlation, although the values given by Surtronic were insignificantly smaller. Results obtained with laser profilometer were higher and gave dramatically different information about the same surfaces. It is obvious that measuring results performed by different kinds of instruments are not suitable for quantitative comparison. SEM analysis and roughness profiles were made in order to obtain more detailed information about the surface structure of the aluminium oxide film (Fig. 1).

Fig. 1 represents the SEM images of aluminium oxide film on the printing plate surface. These samples were selected from the printing plate making process according to their same average roughness value. Standardised quality control showed that the printing plate samples were of corresponding level of quality according to the ISO 12218:1997 and ISO 12647-2:2004 standards.

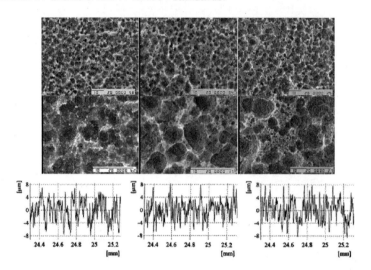

**FIGURE 1.** SEM images and roughness profiles of the samples 1, 2 and 3 (from left).

One can see that microstructure of the samples is specific and differs in the number of depths and pits with the average diameter of 1-10 μm (Fig. 1). Roughness profiles of the samples point out at differences in microstructure as well. Sample 1 in correlation with sample 2 has a structure of higher porosity with deeper and narrower valleys and higher and narrower peaks. Sample 2, according to the roughness profile, has more complex structure with smaller peaks and valley values. Sample 3 has a structure with broader tracks, which is probably in correlation with the surface slicing visible on SEM image.

Measured roughness parameters point out that analysis of printing plate surface can be performed with different stylus methods, but separation of single parameter suitable for giving detailed information about the surface is not possible. The parameter which could be stressed out is material (bearing) ratio (Mr) which is being calculated from the roughness profile. It was observed in similar investigations whose aim was the characterisation of surface topography of different materials [10][11][12]. Sometimes it is used to estimate the percent of the contact area of the material during the exploitation of the surface and can be useful for defining the state of porosity of the surface. According to the results in Tables 2,3,4,5, surface parameter which could be sufficient for characterisation of the printing plate surface has to be selected according to the type of the instrument. Contact stylus methods can give reliable information about surface topography, but they have certain limitations. Contact force of the stylus could damage the surface during the measurement. There could be a problem with measuring deeper and sharper surface features which are narrower than the stylus. This might lead to misinterpretation of the results. Measurements obtained with laser stylus gave similar results, but all the values were significantly higher, most likely because of the optical reflection on sharp edges of the surface. Results of the measurings in the rolling direction (x direction) and in the direction vertical to the one from the rolling process (y direction) did not show any significant differences.

## CONCLUSION

Surface of the aluminium printing plate is completely covered with porous and irregular structures which are necessary to obtain the substantial adhesion of photosensitive layer on the aluminium oxide surface and to improve wettability of nonprinting elements. SEM micrographs and roughness profiles have shown different structures of the porous oxide layer on the foil but they can not be used for obtaining quantitative topography information. On the other hand, todays equipment for quality control of printing plate surfaces provides only visual quality assessment of printing plate surfaces. It is not even possible to select only a single roughness parameter which could provide quantitative topography characterisation of this kind of surfaces. Using the results from this paper and from other investigations still in progress, a few parameters which can show significant correlations have been selected. Besides material ratio ($M_r$) which can be graphically represented on the curve of relative length carrying capacity (Abbot-Firestone curve) [10][12][13], roughness parameters $R_{pk}$, $R_k$ and $R_{vk}$, can be selected as parameters which could show notable differences in surface structures. They are pinpointed because their values are in correlation with SEM images and roughness profiles (Fig. 1).

Nevertheless, comparison of surface roughness measurement and optical analysis by other methods (i.e. fractal geometry) [14] can give more detailed information about surface topography.

## REFERENCES

1. U. Fiebag and C. Savariar-Hauck, U.S. Patent No. 6,649,324 (18 Nov. 2003)
2. P. Aurenty, S. Lemery and A. Gandini, "Dynamic Spreading of Fountain Solution onto Lithographic Anodized Aluminium Oxide", in TAGA Proceedings, Rochester, NY, 1997, pp. 563-576.
3. U. Schmitt, "Korrekter Einsatz der FOGRA-CtP- und der FOGRA-Prozessor-Testform", in FOGRA Symposium: Computer to Plate, edited by Fogra, Erfahrungen für die Produktionspraxis, Fogra, München, 2005, pp.31-38.
4. J. Neues and H.J. Falge, Ugra/FOGRA Reproduction Test Chart 1999, Sonderdruck Nr. 6, Fogra, München, (2000).
5. P.K.F. Limbach, M. P. Amor and J. Ball, U.S. Patent No. 6,524,768 B1 (25 Feb. 2003)
6. R. Hutchinson, *Trans. Inst. Met. Finish.* **79**, 57-59 (2001).
7. C.S. Lin, C.C. Chang and H.M. Fu, *Materials Chemistry and Physics* **68**, 217–224 (2001).
8. M. Wieland, P. Hänggi, W. Hotz, M. Textor, B.A. Keller, N.D. Spencer, *Wear* **237**, 231-252 (2000).
9. K.J. Stout and L. Blund, *Surf. Coat. Technol.* **71**, 69-81 (1995).
10. P. Seitavuopio, "The roughness and imaging characterisation of different pharmaceutical surfaces", Ph.D. Thesis, University of Helsinki, 2006.
11. H.J. Pahk, K. Stout and L. Blunt, *Int J Adv Manuf Technol* **16**, 564-570 (2000).
12. M. Wieland, "Experimental Determination and Quantitative Evaluation of the Surface Composition and Topography of Medical Implant Surfaces and their Influence on Osteoblastic Cell-Surface Interactions", Ph.D. Thesis, Swiss Federal Institute of Technology Zurich, 1999.
13. S.A. Whitehead, A.C. Shearer, D.C. Watts and N.H.F. Wilson, *Dental Materials* **15**, 79–86 (1999).
14. D. Chappard, I. Degasne, G. Huré, E. Legrand, M. Audran, M.F. Baslé, Biomaterials 24, 1399-1407 (2003).

# Lattice Boltzmann Discrete Numerical Simulation of Droplets Coalescence

Bachir Mabrouki [(1)], Kamal Mohammedi [(1)], Idir Belaidi [(1)]
and Djemai Merrouche [(2)]

*(1) Groupe Modélisation en Mécanique et Productique/ LMMC*
*M. Bougara University, Boumerdès 35000 Algeria*
*(2) CRNB,COMENA Ain-Ouessara 17000 Djelfa, Algeria*

**Abstract.** In this paper, a liquid vapor program has been developed and applied for the study of the dynamics of two droplets coalescence. The underlying theoretical model makes it possible to couple the state equation of a non ideal fluid with the pressure tensor at the interface and uses the excess free-energy density formalism. The Maxwell reconstruction procedure is reproduced by our simulation results. The Laplace and Gibbs-Thomson equations are well satisfied for a liquid droplet at equilibrium. This proved the consistency of the model with thermodynamics. As an application, we have demonstrated the ability of LBM to predict complex phenomena within complex geometries in the case of droplets coalescence.

Keywords: Lattice Boltzmann, Coalescence, Discrete Simulation, Droplets.
PACS: 47.55.D, 47.11.Qr,

## INTRODUCTION

The Lattice Boltzmann Method (LBM) is becoming widely used in multiphase multifluid flows simulation for solving the Navier-Stokes equations in a parallel fashion in complex geometries. This particle method is a valuable tool in multi-scale simulation with better performances than the 'continuum' CFD methods. Because the LBM multiphase and multi-component models do not track interfaces, it is easy to simulate fluid flows with complex geometry interfaces. During the last few years, numerous models based on single phase lattice Gas (LG) and lattice Boltzmann (LB) models have been developed. They originated in the FHP two-dimensional LG automaton introduced by Frisch and al. and in FCHC four dimensional LG developed by D'Humières and al. for 3D simulation [6,7,8]. Incompressible Navier-Stokes equations are recovered by both approaches in an asymptotic limit. Some of these single phase methods have been extended to simulate the behavior of two or more immiscible fluids. Immiscible lattice Boltzmann (ILB) multiphase models are developed on the basis of the single phase LB models and immiscible lattice gas (ILG) models; they consist of Boltzmann equations supplemented by perturbation of populations near the interface in order to introduce surface tension; the separation of phases is performed in the same manner as in ILG models.

CP907, *10ʰ ESAFORM Conference on Material Forming*, edited by E. Cueto and F. Chinesta
© 2007 American Institute of Physics 978-0-7354-0414-4/07/$23.00

# LATTICE BOLTZMAN METHOD

In the LBM method, a typical elementary volume of fluid is described as a collection of particles that are represented in terms of a particle velocity distribution function at each point of space. The single particle distribution function, $f_i(x) = f_i(x, e_i)$, defined for each lattice vector $\vec{e}_i$ at each site $x$. Considering a single-time relaxation approximation (BGK), we get for a given $f_i$ the equation:

$$f_i(\vec{x} + \vec{e}_i \Delta t, t + \Delta t) = f_i(\vec{x}, t) - \frac{1}{\tau}(f_i - f_i^0),\tag{1}$$

(with $\Delta t$ :time step, $\tau$ a relaxation parameter, $f_i^0$ : equilibrium distribution function).

For a one component non ideal fluid, the density $\rho$ and the fluid momentum $\rho\vec{u}$ are related to the distribution functions by:

$$\rho = \Sigma_i f_i = \sum_i f_i^0,\tag{2}$$

$$\rho u_\alpha = \sum_i f_i e_{i\alpha} = \sum_i f_i^0 e_{i\alpha}.\tag{3}$$

# FREE ENERGY APPROACH

The choice of $f_i^0$ higher moments must be such that the resulting continuum equations correctly describe the one-component non-ideal fluid hydrodynamics [11]. Defining the second moment as:

$$\sum_i f_i^0 e_{i\alpha} e_{i\beta} = P_{\alpha\beta} + \rho u_\alpha u_\beta,\tag{4}$$

where $\alpha$ and $\beta$ represent a cartesian coordinates

The Van Der Waals fluid for non ideal systems at a fixed temperature has the following free-energy functional within a gradient-squared approximation:

$$\Psi = \int d\vec{r}\left(\psi(T,\rho) + \frac{k}{2}(\nabla\rho)^2\right),\tag{5}$$

The first term is the bulk free-energy density at a temperature T, which is given by:

$$\psi(T,\rho) = \rho KT \ln\left(\frac{\rho}{1-\rho b}\right) - a\rho^2,\tag{6}$$

while the second term gives the free-energy contribution from density gradients in an inhomogeneous system and is related to the surface tension through the coefficient $\kappa$. The pressure tensor is related to the free energy in the usual way:

$$P_{\alpha\beta}(\vec{r}) = P(\vec{r})\delta_{\alpha\beta} + k\frac{\partial\rho}{\partial x_\alpha}\frac{\partial\rho}{\partial x_\beta},\tag{7}$$

with

$$p(\vec{r}) = P_0 - k\rho\nabla^2\rho - \frac{k}{2}\left|\vec{\nabla}\rho\right|^2,\tag{8}$$

where $P_0 = n\psi'(n) - \psi(n)$ is the equation of state of the fluid.

The shear viscosity $\nu$ is given by:

$$\nu = \frac{2\tau-1}{8}(\Delta t)c^2.\tag{9}$$

where $c$ is the sound velocity.

71

# FLUID PROPERTIES

## *Equilibrium densities and Interfacial tension*

At a fixed temperature $T < T_c$ the two phases can coexist at a single pressure value only, $P_0(T)$, which is determined by the so-called Maxwell's equal area reconstruction and satisfies the following equation [16]:

$$\mu(\rho_1,T)=\mu(\rho_2,T)=\mu_{eq}(T),$$ (10)

Where $\rho_1$ and $\rho_2$ are the densities of the vapor (gas) and the liquid, respectively, and $\mu_{eq}(T)$ is the equilibrium chemical potential.

For a flat interface and using the assumption that the temperature $T$ is close to the critical temperature $T_c$, the fluid density across the interface, $\rho(z)$, can be expressed by [15]:

$$\rho(z)=\rho_c+\frac{1}{2}(\rho_2-\rho_1)\tanh(2z/D).$$ (11)

where D is a measure of the interface thickness and $\rho_c$ is a critical density.

The Van Der Waals theory gives the following expression for the interfacial tension at a flat interface [16]:

$$\sigma=k\int_{-\infty}^{+\infty}\left(\frac{\partial\rho}{\partial z}\right)^2(z)dz,$$ (12)

where z is the coordinate perpendicular to the interface.

Applying the approximation done for the density near the critical temperature this expression becomes [15]:

$$\sigma=\frac{2k(\rho_2-\rho_1)^2}{3D},$$ (13)

The capillary pressure is defined by Young-Laplace equation [17]:

$$P_c=P_2-P_1=\frac{\sigma}{R},$$ (14)

and the capillary number is given by [1]: $C_a=\frac{\mu u}{\sigma},$ (15)

According to the Gibbs-Thomson [15, 18] the vapor and liquid pressures are given, respectively, by:

$$P_1=P_0+\frac{\rho_1}{\rho_2-\rho_1}\frac{\sigma}{R},$$ (16)

and

$$P_2=P_0+\left(1+\frac{\rho_1}{\rho_2-\rho_1}\right)\frac{\sigma}{R}.$$ (17)

# RESULTS

In this paper we implemented the lattice Boltzmann model for non-ideal fluids to simulate the coalescence of two droplets. The two steps "stream and collide" algorithm [19] for a hexagonal lattice (D2Q7) is used to simulate lattice Boltzmann equation on 40×40 site lattices. The domain can be decomposed into unit cells of length $L$ and only the content of such unit cell is displayed. The fluid chosen by Swift et al. [11] was selected for our study, which has as coefficients $a=9/49$ and $b=2/21$, corresponding to a critical density $n_c=7/2$ and a critical temperature $T_c=4/7$. Throughout this work, $k=0.01$ and the interface spurious velocity is nearly eliminated by choosing $\tau=\left(1+1/\sqrt{3}\right)/2$ [11]. No slip boundary conditions are imposed on the walls and periodic boundary conditions are imposed on the two domain ends. Before considering our results, we discuss the model validation. For the following calculations unless other specifications, the parameter values are $\rho_1=1.7$, $\rho_2=5.17$, $D=6$ and $\sigma=1.37\times10^{-2}$.

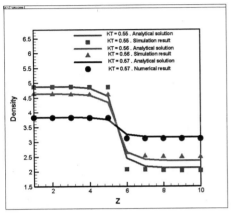

**FIGURE 1.** Coexistence curve (temperature vs. density for the Van der Waals fluid).

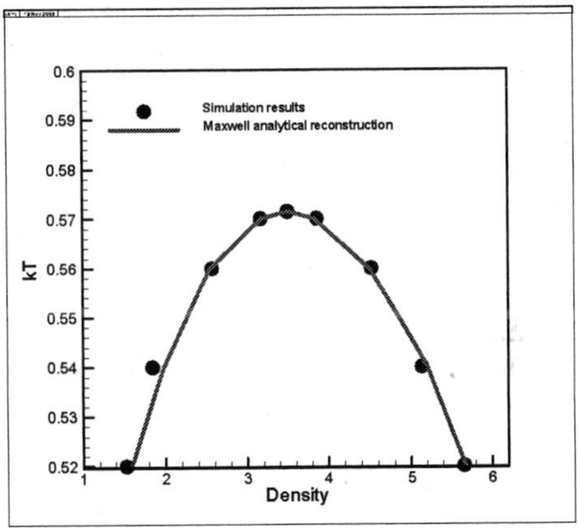

**FIGURE 2.** Equilibrium density profiles normal to flat interface for a Van Der Waals fluid

In Figure 1, a flat interface is equilibrated for $6 \times 10^4$ time steps between the liquid and gas phases for different temperatures and observing the maximum and minimum densities. The model is, therefore, consistent with the Maxwell's equal-area reconstruction procedure for the free energy (11) [11- 12].

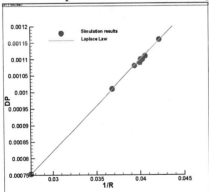
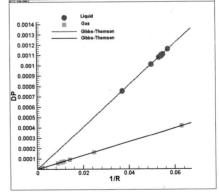

**FIGURE 3.** Laplace Law test for 2D Bubble

**FIGURE 4.** Calculated liquid pressure values vs the Gibbs-Thomson estimates

The calculated density values which have relaxed for $6 \times 10^4$ time steps agree well with the values obtained from Eq. 15 with different temperatures as shown in Figure 2 [11, 12]. The next tests deal with the thermodynamic consistency of the model. The first one associated with interfacial phenomena is to check Laplace's formula by measuring the pressure difference between the inside and the outside of the droplet. The simulated values of $\sigma$ has been compared with theoretical prediction, Eq. 16, and good agreement was reported [2,13,15] as shown in Figure 3. The results of the second test is presented in Figure 4, where the differences $P_1 - P_0$ and $P_2 - P_0$ plotted against the droplet curvature. The agreement of the calculated values with the theoretical ones shown in Figure 4 is satisfactory [15]. The system was equilibrated for $10^4$ time steps for both tests.

The coalescence phenomenon is visualized in Figures 5 where two circular droplets are equilibrated for 15000 time steps.

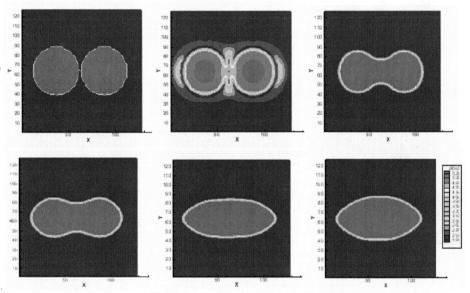

**FIGURE 5.** Coalescence of 2D liquid droplets (time steps: 1, 50, 800, 3000, 20000, 60000)

The thickness of the vapor film initially separating the two droplets is smaller than the interface thickness. The density gradients in the interface region give rise to the development of the so-called Otswald "ripening" phenomenon [15] and lead to complete coalescence into a spherical droplet of volume nearly equal to the sum of the volumes of the original droplets. The spherical form is due to a non Galilean invariant effect [11-12].

## CONCLUSION

In this paper we have developed a lattice Boltzmann program to simulate the dynamics of a non-ideal fluid. Our results are in good agreement with those obtained by other authors [11,12,15]. In particular we have considered non slip boundary conditions at the walls and a periodic boundary condition in the direction parallel to the translating planes. The Maxwell reconstruction procedure is reproduced by our simulation results. The Laplace and Gibbs-Thomson equations are well satisfied for a liquid droplet at equilibrium. This proved the consistency of the model with thermodynamics. Our results confirmed the ability of LBM to predict complex phenomena within complex geometries like the coalescence of two droplets.

## REFERENCES

1. S. Chen, G. Doolean, 1998, "Lattice Boltzmann Method for Fluid Flow," Annu. Rev. Fluid Mech.
2. X. Shan, H.Chen, Lattice Boltzmann Model for Simulating Flows with Multiple Phases and Components, Physical Review E, March 1993, 47 (3), 1815-1819
3. X. Shan, H. Chen, Simulation of Nonideal Gases and Liquid-Gas Phase Transition by Lattice Boltzmann Equation, Physical Review E, April 1994, 49 (4), 2941-2948,
4. X. Shan, G. Doolean, Multicomponent Lattice-Boltzmann Model with Interparticle Interactions, Journal of Statistical Physics, 81 (1/2), 379-393
5. X. Shan, G. Doolean, Diffusion in Multicomponent Lattice Boltzmann Equation Model, Physical Review E, October 1996, 54 (4), 3614-3620
6. U.Frish, B.Hasslacher, Y. Pomeau 1986. Lattice-gas automata for the Navier-Stokes equations. Phys. Rev. Lett. 56: 1505-8
7. U. Frish, D. D'Humières, B. Hasslacher, P. Lallemand, Y. Pomeau, J.P. Rivet, 1987. Lattice Gas Hydrodynamics in Two and Three dimensions. Complex Syst. 1: 649-707
8. G.R. McNamara, G. Zanetti 1988. Use of the Boltzmann Equation to Simulate Lattice-Gas Automata. Phys. Rev. Lett. 61: 2332-35
9. F.J. Higuera, J. Jiménez 1989. Boltzmann Approach to Lattice Gas Simulations. Europhys. Lett. 9:663-68
10. M.R. Swift, S.E Orlandini, W.R. Osborn, J.M. Yeomans, 1996. Lattice Boltzmann Simulation of Liquid-Gas and Binary-Fluid Systems. Phys. Rev. E 54:5041-52
11. M.R. Swift, W.R. Osborn, J..M. Yeomans. 1995. Lattice Boltzmann simulation of nonideal fluids. Phys. Rev. Lett. 75:830-33
12. BT Nadiga, S. Zaleski 1996. Investigations of a two-phase fluid model. Eur. J. Mech. B/Fluids 15:885-96
13. Y. H. Qian, D. D'Humière, P. Lallemand, "Lattice BGK models for Navier-Stokes equation", Europhys. Lett. 17, 479 (1992).
14. A. D. Angelopoulos, V. N. Paunov, V. N. Burganos, A. C. Payatakes, Lattice Boltzmann simulation of non ideal vapor-liquid flow in porous media, March 1998, Physical Review E, Volume 57, Number 3.
15. J.W. Cahn, , J.E Hilliard, Free Energy of a Non uniform System. 1. Interfacial Free Energy. J. Chem. Phys., 28 (2), 258-267 (1958)
16. L.D. Landau, E.M. Lifschitz, Theoretical Physics, v. VI, Fluid Mechanics, Second Edition, 1987, Pergamon.
17. J.S. Hsieh, , 1975, Principles of Thermodynamics, McGraw-Hill
18. N.N. Bogoliobov, Problems of a Dynamical Theory in Statistical Physics, Moscow,1946
19. A.K. Gunstensen, D.H. Rothman, S. Zaleski, G. Zanetti. 1991, Lattice Boltzmann model of immiscible fluids, Phys. Rev. A. 43:4320-27

# Experimental Analysis and Modelling of Fe-Mn-Al-C Duplex Steel Mechanical Behaviour

M. N. Shiekhelsouk[a], V. Favier[a], K. Inal[b], O. Bouaziz[d], M. Cherkaoui[a]

[a]LPMM,ENSAM Metz, 4 rue Augustin Fresnel, Technopôle, 57078 Metz Cedex 3, France
[b]MECASURF, ENSAM Aix, 2 cours des Arts et Métiers, 13617 Aix en Provence, France
[d]ARCELOR RESEARCH, Voie Romaine, BP 30320, F-57283 Maizière les Metz Cedex, France

**Abstract.** A new variety of duplex steels with high content of manganese and aluminum has been elaborated in Arcelor Research. These steels contain two phases: austenite and ferrite combining the best features of austenitic and ferritic steels. In this work, four duplex steels with different chemical composition and phase volume fraction are studied. The evolution of internal stresses for the two phases has been determined by X-ray diffraction during an in situ tensile test. These measurements results were used to determine the mechanical behaviour of the duplex steel using a micromechanical approach by scale transition for tensile tests. Though a good agreement between experiments and simulations is found at the macroscopic level, the calculated internal stresses of the austenitic phase do not match experimental results. These discrepancies are attributed to (i) a bad estimation of the austenite yield stress or (ii) the presence of kinematic hardening in the austenitic phase. A new step is then proposed to test these two hypotheses.

**Keywords:** Duplex, Multi-scale modeling, internal stresses, X-Rays Diffraction (XRD).
**PACS:** 83.60.-a Material behavior

## INTRODUCTION

The development of steels for automotive applications is focused on an increase of strength combined with the improvement of its ductility. A new variety of duplex Fe-Mn-Al-C steels with high content of manganese and aluminum has been elaborated in Arcelor Research. These steels contain two phases: austenite and ferrite combining the best features of austenitic and ferritic steels. The aim of this research is to obtain a better combination in term of mechanical resistance, ductility and ultra lightweight. In this work, four duplex steels with different chemical composition and phase volume fraction are experimentally analyzed and modelled using a micromechanical approach by scale transition for tensile tests.

## MATERIAL

Four duplex alloys classified as (A, B, C, and D) are studied. They differs by the carbon content which offers four volumes fractions of austenite phase (Table 1). These alloys have been hot rolled at high temperature where the dominant phase is the ferrite

CP907, 10[h] ESAFORM Conference on Material Forming, edited by E. Cueto and F. Chinesta
© 2007 American Institute of Physics 978-0-7354-0414-4/07/$23.00

contrary to the classical steels. After rolling, the specimens were slowly cooled down in the air. A treatment of recrystallization has been realized. Metallographic analysis of the four alloys was done on polished and etched specimens according to standard preparation method [1]. The micrographs in figure 1 reveal grains elongated in the rolling direction. Austenitic grains are distributed in the ferritic matrix.

TABLE 1. Carbon content and phases fraction for the 4 alloys.

| Grade | C % | γ % | α % |
|-------|------|-----|-----|
| A | 0.126 | 31 | 69 |
| B | 0.175 | 45 | 55 |
| C | 0.218 | 52 | 48 |
| D | 0.260 | 61 | 39 |

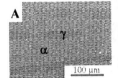 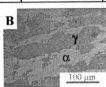 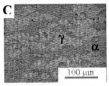 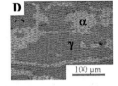

FIGURE 1. Microstructure of the rolling surface for the four duplex steel .

FIGURE 2. Orientation imaging maps of the same area of a specimen B obtained by the EBSD method at LETAM . The colors indicate grains of ferrite (a) and austenite (b) with misorientations < 10°.

The grain size and orientation were determined using the Electron Back Scattering Diffraction (EBSD) technique for the nuance B. The ferritic grains are strongly elongated in the rolling direction and can reach several hundreds of μm with intragranular disorientations < 10° (figure 2a). The same kind of features is found for austenitic grains except that these ones are smaller and less disorientated than the ferritic ones (figure 2b). In addition, one observes the presence of small austenitic grains with size lower than 10 μm strongly disorientated. One notes that the indexing rate of the austenitic phase is low and obstructs an analysis more quantitative of the grains size distribution. This low indexing rate is probably due to the presence of many dislocations in the austenitic grains displaying a strong deformed state.

## STRESSES ANALYSIS BY X-RAYS DIFFRACTION

In this work, the technique of x-rays diffraction is used to determine the internal stresses, for each phase at the initial state and during the deformation, using classical $\sin^2\psi$ method [2]. The details of experimental conditions were presented in previous work [3,4].

# Internal stresses at initial state

For each alloy, the sample surface was prepared by mechanical polishing followed by electropolished treatment to avoid grinding stresses at the surface. The longitudinal stress $\sigma_1$ and transverse stress $\sigma_2$ in the two phases were determined at different depths (every 15 µm) until the measured value does not evolve anymore. The corresponding depth (called dm) and longitudinal stress $\sigma_1$ and transverse stress $\sigma_2$ components for the four grades are listed in table 2. It has been checked that both determined stresses in the two phases satisfy the mixture rule where the macroscopic stress $\Sigma_I$ is closed to zero.

$$\Sigma_I = f^\alpha \sigma_I^\alpha + f^\gamma \sigma_I^\gamma \qquad (1)$$

**TABLE 2.** Longitudinal and transverse residual stress for the four grades in the austenite and the ferrite at dm depth and the corresponding macroscopic stress following equation (1)

| Grades | $f^\gamma$ (%) | $d_m$ (µm) | Longitudinal stress (MPa) | | | Transverse stress (MPa) | | |
|---|---|---|---|---|---|---|---|---|
| | | | $\sigma_1^\alpha$ | $\sigma_1^\gamma$ | $\Sigma_1$ | $\sigma_2^\alpha$ | $\sigma_2^\gamma$ | $\Sigma_2$ |
| A | 31 | 135 | $-55 \pm 20$ | $410 \pm 20$ | 89 | $-50 \pm 15$ | $225 \pm 5$ | 31,8 |
| B | 45 | 90 | $-250 \pm 30$ | $290 \pm 10$ | -7 | $-150 \pm 35$ | $205 \pm 15$ | 9,75 |
| C | 52 | 75 | $-325 \pm 20$ | $205 \pm 10$ | -49 | $-235 \pm 20$ | $125 \pm 15$ | -47,8 |
| D | 61 | 30 | $-350 \pm 30$ | $195 \pm 10$ | -17,5 | $-245 \pm 25$ | $105 \pm 15$ | -31,5 |

We find that the etched depth (dm) varies as a function of austenitic volume fraction. It increases when the volume fraction of the austenite decreases (see table 2). In addition, a gradient of residual stress is found as dm is quite high for a polishing with diamond paste. The residual stress gradient results essentially from thermal stresses induced by the cooling process. The residual stresses in the austenitic phase decrease when its volume fraction increases. Moreover, the ferrite is in a compression state whereas the austenite is in a tension state (see table 2).This residual stress distribution result from both the difference in the thermal expansion coefficient and the specific volume leading to thermal stresses generation during cooling from an elevated temperature (last step of the elaboration process). Indeed, as already mentioned in section 2, the austenite is formed from the ferrite around 1000 °C. Consequently, the created austenite that has a lower specific volume than the ferrite is in a tension state. Moreover, during final cooling after the transformation stage, the tension state is emphasized because the thermal expansion coefficient of the austenite is higher than the one of the ferrite. Although the thermal dilation tensor is isotropic in cubic crystal structure, one can see that the absolute value of the residual stress is almost higher in the longitudinal direction compared to the transverse direction. This is not explained at the present time but could come from anisotropic plastic deformation (in relation with crystallographic texture) occurring during cooling [5].

# Internal stresses during loading

One of our objectives is to analyze the behavior of these duplex by accounting for the evolution of internal stresses in the two phases during the deformation. In this context, a first series of experiments were carried out on the specimen B in order to determine the internal stresses in the two phases at the initial state and their evolution during an in situ tensile test coupled with XRD. The experimental results of this test were presented elsewhere [3]. In order to check the results obtained previously and to improve the reliability of measurements, a second series of experiments was carried out on specimen C. The experimental analysis details were presented in previous work [4]. The stress state evolution of each phase was determined in the elastic domain, during elastic-plastic transition and also in the hardening range (see figure 3). In the elastic regime, the longitudinal stress $\sigma_1$ for the two phases increases linearly with the macroscopic applied stress. The stress in the austenitic phase is higher than the stress in the ferrite. When plastic flow occurs in both phases, there is a redistribution of the internal stresses due to the incompatibility of plastic deformation between the two phases. The average stress on the two phases is almost equal to the macroscopic one. The stress difference between the austenite and the ferrite decreases up to a macroscopic strain of 11% where the stress in the austenite is about 850 MPa whereas the stress in the ferrite is approximately 715 MPa. At higher macroscopic strain, the ferrite starts to soften whereas the austenite continues to harden. At 15% of deformation, the stress is about 935 MPa for the austenite and 655 MPa for the ferrite. These types of measurements enable us to determine the mechanical behaviour of the duplex using a micromechanical model by scale transition and also to predict the influence of the volume fraction of each phase on the macroscopic behaviour.

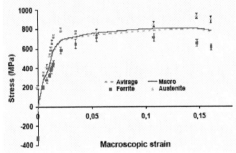

**FIGURE 3.** Longitudinal internal stresses in each phase as a function of the macroscopic strain

## MODELING OF THE DUPLEX STEELS

A scale transition model for tensile tests basing on ISOW model [6] for two phases has been developed to describe the behaviour of the duplex steels. An elastoplastic phenomenologic law of Ludwig type was selected to describe the behaviour of each phase I:

If $\sigma^i \leq \sigma^i_y$ then $\sigma^i = E \, \varepsilon^i$ if not $\sigma^i = \sigma^i_y + K^i \left( \varepsilon^i_{plastique} \right)^{n^i}$ (2)

where $\sigma^i_y$ is the yield stress in MPa, $K^i$ in MPa and $n^i$ are constant material.

The objective is to minimize the number of parameters to be identified by using physical arguments. The macroscopic response of material was given by tensile testes. Considering the chemical composition of the two phases, the yield stress is calculated by empirical laws resulting from many tensile tests carried out on ferrite and austenite alloys [7]. The values of yield stress calculated from the chemical composition for the 4 duplex are given in table 3. The experiments indicate that the work hardening of the ferrite depends little on its chemical composition. Consequently, the work hardening parameters of the ferrite are prescribed identical for the 4 grades and correspond to those classically identified ($K^\alpha = 1350$ MPa and $n^\alpha = 0,3$). The behaviour of the ferritic phase is thus entirely given. Only the work hardening parameters of the austenite $K^\gamma$ and $n^\gamma$ remain to be identified by successive comparisons between calculated and experimental results.

TABLE 3. : Values of the yield stresses and identified work hardening parameters of the austenite

|  |  | A | B | C | D |
|---|---|---|---|---|---|
| Calculated | $\sigma_y^\gamma$ (MPa) | 270 | 268 | 282 | 283 |
| | $\sigma_y^\alpha$ (MPa) | 556 | 520 | 520 | 520 |
| Fitted | $K^\gamma$ (MPa) | 1350 | 1350 | 1350 | 1200 |
| | $n^\gamma$ | 0,38 | 0,44 | 0,38 | 0,39 |

The values of identified parameters, listed in Table 3, vary little from one grade to another and it is then possible to determine an average behaviour of the austenite (Figure 4a). The average values obtained are: $\sigma_y^\gamma = 275.75$ (MPa), $K^\gamma = 1312$ (MPa) and $n^\gamma = 0.3975$. To improve the identification procedure reliability, the internal stresses evolution in each phase predicted by the model for the specimen C were confronted to the ones obtained experimentally from an in situ tensile test coupled with XRD (Figure 4b). The stress level in the ferritic phase and its evolution with the macroscopic strain predicted by the model is quite close to experimental measurements. On the other hand, significant variations exist in the case of the austenite. Indeed, the stress level predicted in the austenitic phase is too low up to 5% of deformation (beginning of plasticity). The preceding results indicate that the austenite yield stress at the beginning of the plasticity is not well presented. This difference between the experimental and modelling results can be explained by: (i) an incorrect estimation of $\sigma_y^\gamma$ (calculated by using an empirical law integrating the effect of the alloying elements) (ii) the presence of residual intraphase stresses probably due to hot rolling processes that are not accounted for. These two hypotheses were integrated successively in the model leading to an increase in the yield stress of the austenite at the beginning of plasticity. The first hypothesis (i) requires to modify the yield stress value of the austenite leading to a new identification of the other parameters ($K^\gamma$ and $n^\gamma$). The new set of parameters is: $\sigma_y^\gamma = 700$ (MPa), $K^\gamma = 550$ (MPa) and $n^\gamma = 0.5$. The strong value of the austenite yield stress found by the identification procedure can be experimentally justified from the observation by EBSD (see section 2). Indeed, the micrographs show the presence of small grains with grain size lower than 10µm and also the presence of many dislocations at the initial state in the phase austenitic. To perform the second hypothesis (ii), the introduction of a kinematical component $X_0$ in our model presenting the intergranulair internal stresses through the phase is required. Then, the constitutive equation for tensile loading for each phase is written as:

$$\sigma - X_0 = \sigma_y + K\left(\varepsilon^p\right)^n \tag{3}$$

It is thus necessary to identify the new parameter $X_0$ material. Only the parameter $n^\gamma$ was modified ($n^\gamma$=0,7). The values of kinematical component are $X_0^\alpha$ =0 MPa for the ferrite and $X_0^\gamma$ = 350MPa for the austenite. These values indicate the absence and the presence of intergranular stresses in the ferritic and austenitic phase respectively. This result can appear surprising although the properties of these phases are rather close. However, to better understand the mechanical interactions between the two phases and to test these two hypotheses, additional investigations are required by in situ compression testes coupled with XRD. These measurements constitute real indications for the construction of a self-consistent type model which could be validated on several scales.

(a) 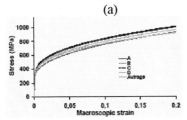  (b)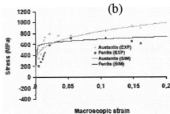

**FIGURE 4.** (a) Simulated austenite response in each grade, ( b) *Simulated and experimental Internal stresses evolution*

## ACKNOWLEDGMENTS

We would like to thank the LETAM for EBSD measurements and more particularly David Barbier. We also thank the ENSAM in Metz for their support and more particularly Denis Bouscaud for help with Proto iXRD system.

## REFERENCES

1. A.O. Benscoter, Metallography and Microstructures, Metals Handbook, vol. 9, ninth ed., ASM International, Materials Park, OH, 1985,pp. 165–196.
2. E. Macherauch, P. Muller: Rev.Appl.Phys Vol. 13, 1961, pp..305-312.
3. M. N. Shiekhelsouk , V. Favier, K. Inal, S. Allain, O. Bouaziz, M. Cherkaoui, Materials Science Forum Vols. 524-525, 2006,pp. 833-838.
4. M. N. Shiekhelsouk, O. Lorrain, V. Favier, K. Inal, S. Migot, O. Bouaziz, M. Cherkaoui, Colloque matériaux 2006 ,13 au 17 novembre 2006, Dijon, France.
5. J.Johansson, M. Odén, X.-H. Zeng, Acta Mater.V.47, No. 9, 1999, pp. 2696–2684.
6. O. Bouaziz, P. Buessler, Rev. Metall. V 1, 2002, pp.71–77.
7. Internal paper Arcelor Research

# A Coupled Mean Field / Gurson-Tvergaard Micromechanical Model For Ductile Fracture In Multiphase Materials With Large Volume Fraction of Voids

Thibaut Van Hoof, Olivier Piérard and Frédéric Lani

*CENAERO - Centre de Recherches en Aéronautique, Avenue Mermoz, 30, B-6041, Gosselies*

**Abstract.** In the framework of the European project PROHIPP (New design and manufacturing processes for high pressure fluid power product – NMP 2-CT-2004-50546), CENAERO develops a library of constitutive models used to predict the mechanical response of a family of cast iron. The present contribution focuses on one particular microstructure, corresponding to a ferrite matrix containing spheroidal graphite and isolated inclusions of pearlite. An incremental mean field homogenisation scheme such as the one developed by Doghri and Ouaar [1] is used. In the present application, the ferrite matrix is described by a Gurson type [2] constitutive law (porous plasticity) while the pearlite inclusions are assumed to obey the classical isotropic $J_2$ plasticity. The predictions of the micromechanical model are compared to the results of Finite Element simulations performed on three-dimensional representative volume elements (RVEs).

**Keywords:** Micromechanics, Mean field modeling, Cast iron, Ductility, Finite Element simulations
**PACS:** 46.35.+z - 46.50.+a - 62.20.Fe - 81.40.Jj - 81.40.Lm

## INTRODUCTION

The PROHIPP project mainly focuses on the development of new design and manufacturing processes for high pressure fluid power products. One of the tasks of this project consists of increasing the lifetime and the load carrying capacity of cast iron hydraulic cylinders both by improving the design and by optimizing the microstructure. However, a real microstructural optimization would require a huge number of experiments in order to cover the whole variety of expected behaviors (see FIGURE 1). Therefore, the PROHIPP project includes an important research effort aiming at predicting numerically the microstructure resulting from the casting process and deriving the mechanical properties of the cast material through the use of appropriate micro-macro models. This contribution presents the modeling approach followed for the second of the five microstructures: a ferrite matrix containing spherical nodules of graphite and isolated inclusions of pearlite.

CP907, *10th ESAFORM Conference on Material Forming,* edited by E. Cueto and F. Chinesta
© 2007 American Institute of Physics 978-0-7354-0414-4/07/$23.00

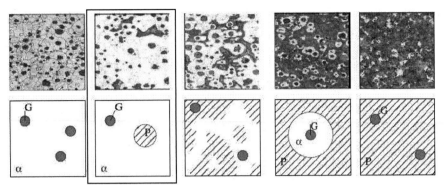

**FIGURE 1.** Identification of the five main categories of microstructure found in as-cast iron. The diameter of the graphite nodules is about 10µm. This contribution focuses on the second of the five microstructures: a ferrite matrix containing spherical nodules of graphite and isolated inclusions of pearlite (dark islands).

## MODEL DESCRIPTION

The starting point of the micromechanical model is the incremental mean-field (IMF) approach described by Doghri and Ouaar [1]. The IMF model used in the present contribution is based on the solution of Eshelby [3] for a single ellipsoidal inclusion embedded in an infinite matrix subjected to a far field loading and on the mean field approach by Mori and Tanaka [4] for a non-dilute concentration of inclusions. The incremental formulation allows for imposing non-monotonic loading conditions and is thus appropriate for future use as constitutive law in a FE code.

In the present study, the two-phase *ferrite + spheroidal graphite* system, seen as the matrix, is given a Gurson-Tvergaard constitutive law. Even though void nucleation is implemented in the model, it is not considered in the present system. Since the ferrite-graphite interface is not coherent, the graphite nodules can be considered as already nucleated voids (the situation where compressive stresses load the graphite nodules is not considered in this contribution), though this assumption is open to discussion. The Gurson-Tvergaard porous plasticity model is basically a yield surface of which the form depends on the accumulated plastic strain and the volume fraction of voids in the material. This particular yield surface writes:

$$\Phi \equiv \frac{\sigma_e^2}{\sigma_y^2} + 2q_1 f \cosh\left(\frac{q_2}{2}\frac{\sigma_{kk}}{\sigma_y}\right) - 1 - \left(q_1 f\right)^2 = 0 \tag{1}$$

where $\sigma_e$ is the Von Mises stress in the matrix, $\sigma_y$ is the matrix yield stress, $\sigma_{kk}$ is 3 times the hydrostatic stress, $f$ is the volume fraction of voids and $q_1$ and $q_2$ are fitting parameters usually taken as 1.5 and 1. The model was implemented following the generalized mid point procedure proposed by Zhiliang Zhang in his thesis [5], based on the decomposition proposed by Aravas [6].

The *ferrite + spheroidal graphite* matrix contains *pearlite* islands modeled as spherical elastic-plastic inclusions. Note that the voids are *not* modeled as a phase of the mean-field model.

The undamaged constituents obey the classical isotropic $J_2$ plasticity with a Swift hardening law:

$$\sigma_e = \sigma_y \left(1 + h \varepsilon_e^{pl}\right)^n \qquad (2)$$

where $h$ is the Swift hardening coefficient, $\varepsilon_e^{pl}$ is the accumulated plastic strain and $n$ is the Swift hardening exponent.

In this contribution the phases have fictitious materials parameters. Indeed, the properties of the constituents should be measured in-situ. This has not been done yet. The parameters are listed in TABLE 1.

TABLE 1. Parameters of the undamaged constituents

| Phase | Young Modulus | Poisson Ratio | Yield stress $\sigma_y$ | Hardening coefficient $h$ | Hardening Exponent $n$ |
|---|---|---|---|---|---|
| Ferrite | 210 GPa | 0.3 | 295 MPa | 137.5 | 0.17476 |
| Virtual pearlite | 210 GPa | 0.3 | 885 MPa | 137.5 | 0.17476 |

## MODEL VALIDATION

The model is assessed through the following validation procedure:
1. Validation of the implementation of the Gurson-Tvergaard model by comparison to results from the literature and similar implementations available in commercial FE software
2. Validation of the implementation of the IMF model by comparison to results from the literature and similar implementations available in commercial software
3. Comparison of the results obtained with the Gurson-Tvergaard model with FE simulations on 3D RVEs
4. Comparison of the results obtained with the IMF model with FE simulations on 3D RVEs
5. Comparison of the results obtained with the coupled IMF/Gurson-Tvergaard model with FE simulations on 3D RVEs.

Steps 1 to 4 led to the following conclusions:
1. & 2. The implementations of the GT and IMF models are correct.
3. The GT model gives accurate predictions of the overall stress-strain curve and of the evolution of the porosity for materials containing up to ~10% of voids.
4. The IMF gives accurate predictions of the overall stress-strain curves for materials containing up to 15% of hard inclusions. It significantly underestimates the stress level in the inclusions compared to the FE simulations. The yield ratio (i.e. $\sigma_{Y\text{-}Pearlite} / \sigma_{Y\text{-}Ferrite}$) has a strong impact on the quality of the IMF prediction of the stresses in the inclusions.

Step 5 is presented hereafter.

# The coupled IMF/GT model Vs. FE simulations on 3D RVEs

In order to evaluate the ability of the IMF/GT model to describe the *ferrite + graphite + pearlite* system, its predictions are compared with the results of FE simulations performed on 3D RVEs. The boundary conditions, the unstructured mesh and the distribution of voids and inclusions are tri-periodical. The microstructure is meshed with $2^{nd}$ order tetrahedra. Uniaxial tension tests are simulated using the general purpose commercial finite element software Abaqus v6.6.1.

At this stage, two types of RVEs can be considered: i) a 3D RVE consisting of damaged ferrite containing pearlite inclusions. The damaged ferrite is modeled with the GT model. Let us nickname it the GP-RVE, ii) a 3D RVE consisting of un-damaged ferrite containing *voids and pearlite* inclusions. Let us nickname it the FVP-RVE. These RVEs are illustrated in FIGURE 2.

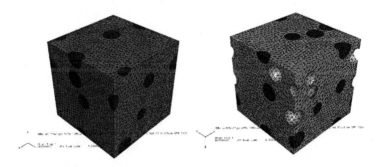

**FIGURE 2**. On the left, a 3D RVE of damaged ferrite modeled with a GT model containing pearlite inclusions (the GP-RVE). On the right, a 3D RVE of undamaged ferrite containing isolated voids and pearlite inclusions (the FVP-RVE).

The overall stress-strain curves obtained with the coupled IMF/GT and with the two FE models described here above are shown in FIGURE 3.

The agreement between the predictions of the IMF/GT model and the results of the simulations performed on the FVP-RVE is satisfactory. The IMF/GT model only slightly underestimates the overall effective stress but correctly predicts the overall hardening rate.

While the prediction MT/GT model and the results of simulations performed on the GP-RVE look closer in terms of overall stress-level, they clearly depart for what concerns the overall hardening rate. This difference in hardening rate is due to localization of plastic strain and consequently of damage at the interface between the damaged ferrite and the pearlite inclusions (Note that the FE simulations performed on the GP-RVE usually crash due to this severe localization).

This void localization occurs due to the fact that each gauss point in the matrix actually hides a RVE corresponding to the GT model. It means that in this case, the voids are several orders of magnitude smaller than the actual voids / spheroidal graphite inclusions. This situation does not correspond to the physical reality of cast iron where the inclusions of pearlite and the voids have about the same size. Nevertheless this second population of voids might be of utmost importance at larger deformations.

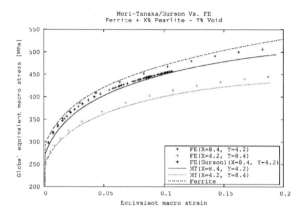

**FIGURE 3**. Overall stress-strain curves obtained with the IMF/GT (lines) and FE (+) models. Results in medium and light grey extending to a macro strain of about 0.18-0.2 come from the IMF/GT model and from FE simulations performed on the FVP-RVE, they correspond to the following pairs of (volume fraction of pearlite, initial volume fraction of voids): (8.4%, 4.2%) and (4.2%, 8.4%). The dark grey crosses (+) up to a macro strain of about 0.11 give the results of the FE simulation performed on the GP-RVE with an initial volume fraction of voids of 8.4% and a volume fraction of pearlite of 4.2%.

These observations are confirmed by the predictions of the phases stresses shown in FIGURE 4. The limitations of the IMF scheme for high yield ratios are observed, i.e. it significantly underestimates the stress level in the hard inclusions.

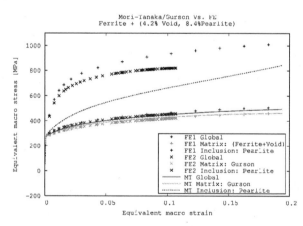

**FIGURE 4**. Average Von Mises stress in the phases predicted by the IMF/GT model (lines), resulting from FE simulations performed on the GP-RVE (+) and resulting from FE simulations performed on the FVP-RVE(x) with a volume fraction of pearlite of 8.4% and a volume fraction of voids of 4.2%. The inclusions feel the highest stress in the three models.

FIGURE 5 shows the increase of the volume fraction of voids as a function of the overall strain. Results obtained with the coupled IMF/GT model are close to FE estimations for small strains. The agreement would be better at larger strains if the yield ratio was lower (<2.5).

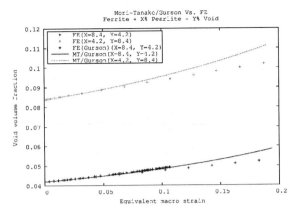

**FIGURE 5.** Evolution of the volume fraction of voids as a function of the overall equivalent strain obtained with the IMF/GT (lines) and FE (+) models. Results reaching macro strains of about 0.18-0.2 come from the IMF/GT model and from FE simulations performed on the FVP-RVE, they correspond to the following pairs of (volume fraction of pearlite, initial volume fraction of voids): (8.4%, 4.2%) and (4.2%, 8.4%) . The dark grey crosses (+) stopping at a macro strain of about 0.11 give the results of the FE simulation performed on the GP-RVE with an initial volume fraction of voids of 4.2% and a volume fraction of pearlite of 4.2%.

## CONCLUSION

From these observations, the following conclusion can be drawn: the coupled IMF/GT model is promising for predicting correctly the overall and per-phase mechanical responses of elasto-plastic systems containing voids and hard spherical inclusions when:
- the volume fraction of inclusions is low
- the yield ratio is not too high
- plastic and damage localization are not too important, i.e. when the hard inclusions and the voids are about of the same size.

The use of the IMF/GT model is thus appropriate for the prediction of the mechanical response of the examined spheroidal cast iron microstructure.

## ACKNOWLEDGMENTS

This research is supported by the European Commission under the 6th Framework Programme through Contract nr NMP 2-CT-2004-50546. The authors thank Prof. T. Pardoen and I. Doghri for all the fruitful discussions.

## REFERENCES

1. I. Doghri and A. Ouaar, *Int. J. of Sol. and Struct.* **40**, 1681-1712 (2003).
2. A.L. Gurson, *J. Eng. Mater. Technol.* **99**, 2-16 (1977).
3. J.D. Eshelby, *Proc. Roy. Soc.* **A241**, 376-396 (1057).
4. T. Mori and K. Tanaka, *Acta Met.* **21**, 571-574 (1973)
5. Z. Zhang, PhD thesis, Finland: Lappeenranta University of Technology, 1994.
6. N. Aravas, *Int. J. Numer. Meth. Eng.* **24**, 1395-1416 (1987)

# Assessment of Convex Plastic Potentials Derived from Crystallographic Textures

## A. Van Bael[1,2] and P. Van Houtte[1]

[1] Dept. MTM, Katholieke Universiteit Leuven, Kasteelpark Arenberg 44, B-3001 Heverlee, Belgium
[2] Dept. IWT, Katholieke Hogeschool Limburg, Campus Diepenbeek, Agoralaan Gebouw B, bus 3, B-3590 Diepenbeek, Belgium

**Abstract.** The paper is concerned with a method to derive analytical plastic potentials in strain rate space, the parameters of which are identified from the crystallographic texture and the Taylor polycrystal plasticity model. Such potentials are especially suited to account for texture anisotropy in finite element (FE) simulations at an engineering length scale. The potentials strictly guarantee the convexity of the corresponding yield locus in stress space, which is a critical requirement for the numerical stability of the FE simulations. The method is assessed here for a number of industrial steel sheets and aluminium alloy sheets. Particular attention is given to the effects of the modifications needed to ensure convexity on the parameters and on the predicted $r$-values. The results are compared to both experimental values and the ones obtained directly from the Taylor model.

**Keywords:** anisotropy, yield loci, Taylor model, finite-element method.
**PACS:** 62.20.Fe, 68.55.Jk, 81.05.Bx, 81.40.Lm, 83.10.Ff, 83.10.Gr, 83.60.-a, 83.60.La, 83.80.Ab.

## INTRODUCTION

Finite-element (FE) models for metal forming simulations require accurate constitutive models that take the effects of texture, microstructure and substructure into account [1]. The present paper is concerned with texture anisotropy. For applications at an engineering length-scale it is proposed to work with an analytical constitutive model, the parameters of which are identified using results of a crystal plasticity model and a homogenization procedure. The model is of the phenomenological type and it contains equations that mimic the material behaviour that would result from direct calls to the crystal plasticity model using the Taylor homogenisation assumption. Using the terminology explained in [1], it is a hierarchical model at the macro-scale, describing the behaviour of a polycrystal with typically 1000 grains or more. The approach is based on the concept of plastic potentials in strain rate space [2], and particular mathematical expressions of 6th rank strictly guarantee the convexity of the corresponding yield loci in stress space [3]. The model has been implemented in implicit elastic-plastic FE codes [4], and initial results have also been presented for the implementation in explicit FE codes [5]. It is the purpose of the present paper to evaluate the plastic potentials for various industrial materials, with special attention to the effects of the iterative procedure to ensure convexity.

CP907, *10th ESAFORM Conference on Material Forming*, edited by E. Cueto and F. Chinesta
© 2007 American Institute of Physics 978-0-7354-0414-4/07/$23.00

# CONVEX POTENTIALS AND IDENTIFICATION PROCEDURE

The plastic behaviour of polycrystalline materials can be described with dual plastic potentials [2]. One is the yield locus in stress space, and the other, in strain rate space, is the function $\Psi(\mathbf{D})$ representing the plastic work dissipated per unit volume for a given macroscopic strain rate tensor $\mathbf{D}$. For materials with negligible strain-rate sensitivity, the plastic flow stress $\mathbf{S}$ (which is deviatoric for plastically incompressible materials) can be calculated as

$$\mathbf{S} = \frac{\partial \Psi}{\partial \mathbf{D}} \qquad (1)$$

The following analytical expression was proposed [3], using a vector representation for $\mathbf{D}$ with 5 components $D_p$ for incompressible materials:

$$\psi(\mathbf{D}) = \tau^c \sqrt[6]{\alpha'_{pqrstu} D_p D_q D_r D_s D_t D_u} \quad \text{with} \quad 1 \le p \le q \le r \le s \le t \le u \le 5 \qquad (2)$$

$\tau^c$ is a scaling factor (with the dimensions of a stress) to take strain hardening into account, and $\alpha'_{pqrstu}$ are 210 dimensionless parameters that determine the shape of the corresponding yield locus.

Using the Taylor theory and assuming that all critical resolved shear stresses in all grains have the same value $\tau^c$, the rate of work per unit volume can be written as:

$$\Psi(\mathbf{D}) = \tau^c D_{vM} \, \overline{M} \, (\mathbf{a}) \qquad (3)$$

with

$$D_{vM} = (2/3)^{1/2} D \quad \text{and} \quad D = (\mathbf{D}:\mathbf{D})^{1/2} \qquad (4)$$

$D_{vM}$ is the von Mises equivalent strain rate and $\overline{M}$ the average Taylor factor which depends on the plastic strain rate mode $\mathbf{a}$:

$$\mathbf{a} = \mathbf{D} / D \qquad (5)$$

For a given texture, a three-step procedure allows to calculate the parameters $\alpha'_{pqrstu}$ in Equation (2). First, the average Taylor factor $\overline{M}$ is evaluated very efficiently for a large number (around 100000) of strain rate modes using the series expansion method [6]. Then, the parameters $\alpha'_{pqrstu}$ are identified by least squares fitting of the function $\overline{M}(\mathbf{a})$ derived from equations (2-5) to these $\overline{M}$ values. Finally, the coefficients may need to be slightly modified by an additional iterative algorithm in order to ensure that the associated yield locus is strictly convex everywhere in stress space [3,4]. The eigenvalues of a 35x35 matrix $\alpha_{PQ}$, obtained from the parameters by an additional index contraction, should be positive or zero to guarantee convexity, and possibly negative eigenvalues are iteratively modified towards zero.

# MATERIAL DATA

Four industrial sheet metals with thickness around 1.0 mm are considered: low-carbon steel (0.03 wt% C), an interstitial-free (IF) steel (0.003 wt% C), a solution strengthened aluminium alloy AA5182 and a precipitation hardened aluminium alloy AA6016-T4 (aged at room temperature). All sheets are cold rolled and annealed, the steel sheets also being skin-passed. Tensile tests with $r$-value determination have been performed between 0° and 90° to the rolling direction, every 22.5° for the steels and every 15° for the aluminium alloys. Texture measurements have been carried out by X-ray diffraction on flat specimens at half sheet thickness. The reflection method was used, measuring four incomplete pole figures [7]. The orientation distribution function (ODF) was calculated using the series expansion method [8]. ODF-sections of interest are shown in Fig. 1: $\varphi_2 = 45°$ for the steels, and $\varphi_2 = 90°$ for the aluminium alloys. The texture of the low-carbon steel is relatively weak ($f_{max} = 4.4$), with a $\gamma$-fibre, an $\alpha$-fibre with decreasing intensities towards rotated cube and rotated Goss, and a weak Goss-component. The texture of the IF-steel consists of a very strong $\gamma$-fibre ($f_{max} = 12.9$) with spread towards a partial $\alpha$-fibre and a maximum on {111}<112>. The AA5182 alloy has a relatively weak texture ($f_{max} = 4.1$), showing an ND-rotated cube fibre with a peak at {001} <310>, and furthermore weak Goss {011} <100> and {011} <122> components. The aluminium alloy AA6016-T4 consists of a very strong cube texture ($f_{max} = 12.7$) and a weak {011} <111> component.

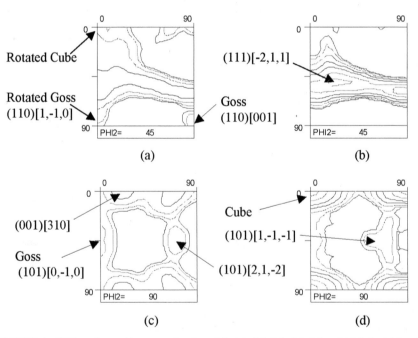

**FIGURE 1.** ODF-sections with contour levels at 1.0, 1.3, 2.0, 3.2, 5.0, 8.0 and 11.0 for (a) low-C steel, (b) IF-steel, (c) AA5182 and (d) AA6016-T4.

# RESULTS AND DISCUSSION

## The Convexity Procedure Quantitatively Assessed

The previously described three-step procedure has been applied to obtain the plastic potentials of $6^{th}$ rank for the four materials, assuming $\{110\}+\{112\} <111>$ slip for the steels and $\{111\}<110>$ for the aluminium alloys. In order to quantify the modifications during the third step, a normalised difference is calculated as:

$$\frac{\sum_i (\alpha_i - \beta_i)^2}{\sum_i \alpha_i^2}. \qquad (6)$$

Here $\alpha_i$ are the 210 coefficients obtained directly from the least squares fitting, and $\beta_i$ the modified ones after the convexity procedure. Modifications prove to be necessary for the four materials. The evolution of the most negative eigenvalue of the 35x35 matrix $\alpha_{PQ}$ is shown in Fig. 2. It can be noticed that a larger number of iteration steps is required for the aluminium alloys. Also, these modifications result in larger normalised differences for the aluminium alloys (0.152 and 0.165 for AA5182 and AA6016-T4, respectively) than for the steels (0.017 and 0.025 for low-C and IF, respectively). Evidently, the larger values for the aluminium alloys result from the lower number of available slip systems. Indeed, (erroneously) using the bcc slip systems $\{110\}+\{112\} <111>$ for the aluminium textures gives lower normalised differences (0.005 and 0.008 for AA5182 and AA6016-T4, respectively), and using fcc slip systems $\{111\} <110>$ for the steel textures leads to larger values (0.134 and 0.119 for low-C and IF, respectively).

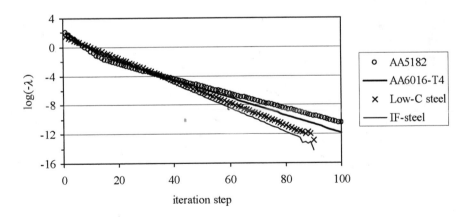

**FIGURE 2.** Evolution of the absolute value of the most negative eigenvalue $\lambda$ of the 35x35 matrix $\alpha_{PQ}$ during the iterative procedure to ensure convexity for $6^{th}$ rank plastic potentials, using a base-10 logarithmic scale for $-\lambda$.

# *R*-value Validations

The variations of experimental and modelled *r*-values with the angle of the tensile specimen to the rolling direction are shown in Fig. 3. Modelled values are obtained from the Taylor model directly and from the 6th order plastic potentials, both prior to and after the convexity procedure. The curves permit to evaluate the predictive quality of the Taylor model on the one hand, and the capability of the plastic potentials to mimic this model on the other. The global shapes of the experimental variations are predicted by the Taylor model for all considered materials. The predictions oscillate around the experimental values for the steels with a maximum difference of 0.25, but always underestimate those for the aluminium alloys by an amount of 0.05 up to 0.25. In terms of *q*-values (defined as the ratio between the plastic strains in the width and the tensile direction and given by $q = r/(1+r)$), the maximum differences are 0.05 and 0.12 for the steels and aluminium alloys, respectively. Furthermore, the curves reveal that the mimicking capacity of the plastic potentials is the best for the steels, where a maximum difference in *r*-values of 0.05 is observed. For the AA5182 alloy, the potential resulting from the least squares fitting leads to *r*-values that are consistently lower (up to 0.1) than the Taylor predictions, whereas an accurate fit is obtained for the AA6016-T4 aluminium alloy (at most 0.05 difference). The convexity procedure turns out to increase the *r*-values consistently for both aluminium alloys, on average by a value of 0.2, thereby artificially bringing them close to the experimental results.

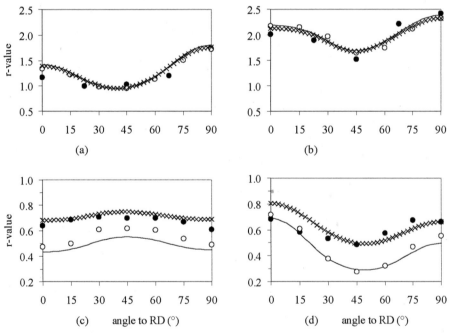

**FIGURE 3.** Variations of *r*-values vs. angle to rolling direction for (a) low-carbon steel, (b) IF-steel, (c) AA5182 and (d) AA6016-T4, obtained from experiments (?), directly from Taylor model (?), from 6th order potentials before (- ) and after the convexity procedure (×).

# CONCLUSIONS AND PERSPECTIVES

Analytical potentials in strain rate space, based on the Taylor plasticity model for polycrystalline textured materials, have been evaluated for industrial sheet materials. These potentials have been designed for implementation into finite-element models of forming processes at an engineering scale, where the guaranteed convexity of the yield locus in stress space is essential for numerical stability of the simulations. Two aspects have been studied: the quality of the underlying Taylor model for the prediction of $r$-values, and the capability of the analytical approximation to mimic this model. The Taylor model is confirmed to be more accurate for low-carbon steels than for aluminium alloys. Furthermore, the procedure to enforce convexity artificially improves the predicted $r$-values for aluminium alloys. The larger impact of the convexity procedure for these materials is related to the smaller number of available slip systems. The present study indicates that further improvements would be welcome for the accuracy of the underlying plasticity model and for the mimicking quality of the analytical approximations.

Plasticity models are now available that have proven to be better than the Taylor model for deformation texture predictions [9], but the quality of these models for the mechanical response during plastic deformation still needs to be investigated. Furthermore, there is a need for more general plastic potential functions that would also be suited for materials with a-symmetrical deformation mechanisms, such as titanium alloys with mechanical twinning systems. Indeed, the current $6^{th}$ order potential assumes identical behaviour upon reversal of the macroscopic deformation. Finally, the identification of the parameters currently requires the application of the plasticity model for a very large number ($\pm$ 100000) of strain rate modes. Whereas this has been accomplished in a very efficient way for the Taylor model, future potential functions should be designed to allow for a determination of the parameters on the basis of a smaller number of strain rate modes.

# ACKNOWLEDGMENTS

The authors gratefully acknowledge the financial support by the Federal Government of Belgium (Belgian Science Policy Contract P5/08).

# REFERENCES

1. P. Van Houtte, A. Van Bael and M. Seefeldt, *Materials Science Forum* **539-543**, 3454-3459 (2007).
2. P. Van Houtte, *Int. J. Plasticity* **10**, 719-748 (1994).
3. P. Van Houtte and A. Van Bael, *Int. J. Plasticity* **20**, 1505-1524 (2004).
4. A. Van Bael, S. He, S. Li and P. Van Houtte, "The simulation of cold metal deformation using convex plastic potentials calculated from texture data", VII International Conf. on Computational Plasticity (COMPLAS 2003), E. Oñate and D.R.J. Owen (Eds), CIMNE, Barcelona, 2003.
5. S. Ristic, S. He, A. Van Bael, P. Van Houtte, *Materials Science Forum* **495-497**, 1535-1540 (2005).
6. P. Van Houtte, *Int. J. Plasticity* **17**, 807-818 (2001).
7. P. Van Houtte, *Textures and Microstructures* **6**, 137-162 (1984).
8. H.J. Bunge, "Texture analysis in material science", Butterworths London (1982).
9. P. Van Houtte, S. Li, M. Seefeldt and L. Delannay, *Int. J. Plasticty* **21**, 589-624 (2005).

# Strategy of Material Parameters Identification for Non Linear Mechanical Behavior: Sensitivity of FE Computation

S. Bouvier[+], L. Alves[*] and A.M. Habraken[++]

[+]LPMTM-CNRS, UPR9001, University Paris 13, 99 av. J-B. Clément, 93430 Villetaneuse, France
[*]Dept. of Mech. Engineering, University of Minho, Campus de Azurèm, 4800-058 Guimarães, Portugal
[++]ArGEnCo Dept., University of Liege, Chemin des Chevreuils 1, 4000 Liege, Belgium

**Abstract.** The purpose of the present work is to analyze several aspects related to the connection between the constitutive models, their identification and the FEM predictions. Several issues are addressed: the experimental data base that should be used in the identification procedure, the choice of the mechanical tests involved (monotonous and/or non-proportional loading, homogeneous or heterogeneous tests...), the identification strategies (direct or inverse FE optimization, simultaneous or sequential material parameters identification...). Besides its obvious interest, such study aim to find a good balance between the number and the type of relevant involved mechanical tests in material behavior characterization. This is an important issue for industrial applications.

**Keywords:** Non-Linear Behavior, Anisotropic material, Constitutive laws, Sensitivity analysis, Identification.
**PACS:** Replace this text with PACS numbers; choose from this list: http://www.aip.org/pacs/index.html

## INTRODUCTION

In the context of numerical accuracy requirement, the hardening laws and yield criteria used to describe the material behavior play a very significant role. In recent years, intensive efforts have been done in order to develop new constitutive models that allow a more accurate description of the mechanical behavior of metal sheets. However, one should emphasize the fact that the question of the suitable way for material parameters identification is not completely solved so far. The purpose of the present work is to analyze several issues related to the connection between the constitutive models, their identification and the FEM predictions. Several practical applications are presented and discussed where we mainly focus on the effect of the identification strategies, on the numerical simulations results. The sensitivity of the FE results to the modeling description was discussed in [Bouvier et al., 2006]. The first problem compares different strategies of yield locus identification and their effect on ears prediction in deep drawing. The second investigated problem analyses two identification approaches: (i) direct identification (*i.e.* local volume element computation) developed in University Paris 13, (ii) inverse identification using FE

CP907, *10th ESAFORM Conference on Material Forming*, edited by E. Cueto and F. Chinesta
© 2007 American Institute of Physics 978-0-7354-0414-4/07/$23.00

computation, developed in University of Liege. The third problem deals with the sensitivity of FE computation to the type of mechanical data involved in the identification process.

## YIELD SURFACE IDENTIFICATION

The initial and the induced anisotropy in materials are described through the shape, the size and the position of the yield locus. For well-annealed materials, it is rather admitted that their crystallographic textures are responsible for their initial anisotropy. The latter define the shape of the yield surface. Therefore, sequential strategy based on the identification of the yield surface material parameter using only the initial anisotropy (*i.e.* crystallographic texture, in-plane strain and/or stress anisotropy) seems to be quite acceptable. The hardening material parameters are identified on the stress-strain curves with fixed yield surface parameters. In this context, several approaches can be compared. An example is presented in Figure 1(a) in case of [Hill, 1948] criterion. The labels indicated on the figure mean:

(1)    <u>Taylor-Bishop-Hill model (TBH)</u>: predicted $r(\alpha)$[1] using TBH model and experimental crystallographic texture.

(2)    <u>Experimental data</u>: measured $r(\alpha)$ using uniaxial tensile test along different orientations with respect to the rolling direction.

(3)    <u>[Hill, 1948] using mechanical data</u>: identification of [Hill, 1948] parameters {F, G, H and N} using the measured $r(\alpha)$.

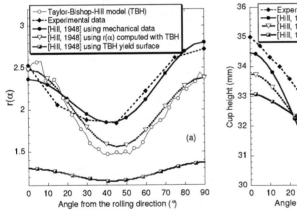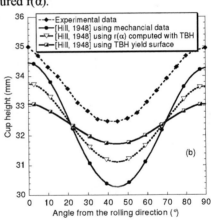

**FIGURE 1.** (a) Different strategies in the identification of the material parameters for the quadratic [Hill, 1948] yield criterion. (b) Ears prediction using the previous identified material parameters in deep drawing of a cylindrical cup.

(4)    <u>[Hill, 1948] using $r(\alpha)$ computed with TBH</u>: identification of [Hill, 1948] parameters {F, G, H, L, M, N} using TBH predicted $r(\alpha)$.

---

[1] Hill coefficient of anisotropy.

(5)     [Hill, 1948] using TBH yield surface: identification of [Hill, 1948] parameters {F, G, H, L, M, N} using best fit of [Hill, 1948] yield locus to the TBH one.

As general comments: (i) there are some discrepancies in the predicted r(α) using the TBH model and the mechanical data; (ii) the predicted r(α) using the last strategy (*i.e.* yield surface fitting) is relatively far from the others. In this situation, the identified material parameters describe the whole anisotropy of the material. On the contrary, the identified material parameters using the mechanical data reduce the material anisotropy to the in-plane one.

The identified parameters are used in the simulation of a deep drawing of a cylindrical cup (Figure 1(b)). The results clearly show a significant sensitivity of the FE predictions to the strategy of identification. An improvement of the predicted solution is obtained when the whole material anisotropy is described.

## SEQUENTIAL VS SIMULTANEOUS MATERIAL PARAMETERS IDENTIFICATION

The material parameters identification of constitutive laws requires the use of mechanical tests suitable for the behavior under investigation. Such mechanical tests provide the global response (*e.g.* force-displacement) of the material. In order to be used in the parameters identification, this response should be converted in a local one (*i.e.* stress-strain curve). This can be simply done for homogeneous mechanical test. However, this condition may not be satisfied even for some classical mechanical tests (*e.g.* the simple shear test). Keeping in mind the boundary effects and the accuracy of the experimental measurement, the assumption of strain field homogeneity is commonly accepted. But, it can be also partly inspected using the full-field measurement techniques. Therefore, two identification methods can be adopted. The first one is based on a homogeneous interpretation of the mechanical test. This method requires an analytical computation of the stress state for a given strain state and vice-versa. For some specific stress and strain states, this method was implemented in the SiDoLo software [Haddadi *et al.*, 2006]. The second method doesn't require the homogeneity of the strain field. It uses an inverse identification procedure through FE computation. An example was proposed by [Flores *et al.*, 2007] in Lagamine FE code.

In all cases, the parameters identification procedure is based on a minimization of a cost function using least squares estimation. Such cost function measures the agreement between experimental and simulated data. Different material parameters strategies can be considered, as simultaneous (yield locus and hardening laws) or sequential material parameters identification (yield locus then hardening laws). An example is proposed in Figure 2.

The stress-strain curves for uniaxial tensile test and simple shear test along the rolling direction obtained from different identification strategies are presented. The material work-hardening is described using Teodosiu-Hu model [Haddadi *et al.*, 2006] coupled to von Mises or [Hill, 1948] criterion. Figure 2(a) presents results using inverse FE-identification. The labels indicated on the figure mean:

(1) [Hill, 1948] using $r(\alpha)$: sequential identification strategy is adopted. First the material parameters {F, G, H and N} for [Hill, 1948] are determined using the Hill coefficients of anisotropy $r_0$, $r_{45}$ and $r_{90}$ and $\sigma_0^2$. Then, the material parameters of the hardening laws are determined with fixed {F, G, H and N}.

(2) [Hill, 1948] using $\sigma(\alpha)$ and N not fitted: sequential identification strategy is adopted. First the material parameters {F, G, H and N} for [Hill, 1948] are determined using the initial yield stresses for different mechanical tests. Then, the material parameters of the hardening laws are determined with fixed {F, G, H and N}.

(3) [Hill, 1948] using $\sigma(\alpha)$ and N fitted: simultaneous identification strategy is adopted. First the material parameters {F, G, H and N} for [Hill, 1948] are determined using the initial yield stresses for different mechanical tests. Then, the material parameters of the hardening laws are determined with fixed {F, G, H} only.

According to [Flores *et al.*, 2007], the best fitting is obtained when simultaneous identification is adopted (Figure 2(a)). This means that the change of the yield surface during the hardening is taken into account. The identification strategy in this context proposes average values of the material parameters that describe such change.

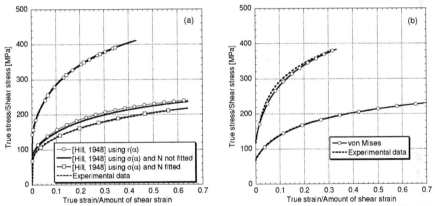

**FIGURE 2.** Teodosiu-Hu hardening laws identification using von Mises or [Hill, 1948] yield loci for an IF mild steel. (a) Different identification strategies proposed by [Flores *et al.*, 2006], and (b) by [Bouvier *et al.*, 2003].

However for this material, a micromechanical computation shows that the contribution of the texture (*i.e.* the geometrical hardening) to the macroscopic hardening is rather small compared to the evolution of the microstructure (*i.e.* density of dislocations and their patterning…), even when significant texture evolution takes place (*e.g.* the simple shear test). The identification result of Figure 2(b) is obtained using direct optimization with SiDoLo software, of the hardening parameters using an isotropic von Mises yield surface for the same material of Figure 2(a).

These two strategies of identification lead to the same behavior description for the uniaxial tensile test and the simple shear test. However, the in-plane description of the

---

[2] The initial yield stress of uniaxial tensile test along the rolling direction

strain anisotropy is very different (Figure 3). This leads to a significant effect on FE simulations as discussed in the previous section.

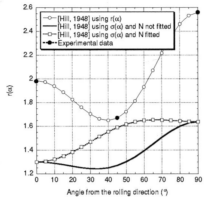

**FIGURE 3.** In-plane description of the strain anisotropy obtained with the identified parameters of [Hill, 1948] yield locus after [Flores *et al.*, 2006].

## SELECTION OF THE EXPERIMENTAL DATA FOR THE PARAMETERS IDENTIFICATION

It is an important issue that the stress and/or the strain states involved in the identification strategy include the domain of intended applications.

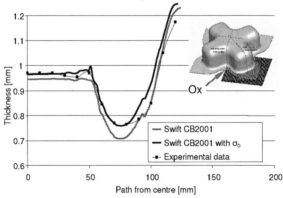

**FIGURE 4.** Thickness distribution along the rolling direction Ox for an aluminum alloy AA5182-O deformed up to 60mm depth.

However, taking into account such considerations is not always feasible due to the difficulty in carrying out the mechanical test that reproduces the desired stress or strain fields. The cross tool test (Figure 4) was specially designed in order to impose different strain paths to the material. It is a suitable tool for the investigation of the capabilities of constitutive models. However, the question of the identification strategy should be carefully taken under consideration here. In the example of Figure 4, the behavior is described by the isotropic Swift law coupled to the non quadratic criterion recently proposed by [Cazacu and Barlat, 2001]. Two identification strategies are performed using or not the data from the equibiaxial test. A significant sensitivity of the FE results in term of thickness prediction is observed using the two identification strategies. This can be explained by the presence of the biaxial stress state in this forming process.

# CONCLUSION

In this paper, we investigate the problem of material parameter identification of inelastic constitutive laws in connection with FE predictions. It is worth noting that the identification procedure of constitutive laws is far from being trivial and may have a large effect on FE predictions. The following conclusion can be drawn:

(i)     The question of non uniqueness solution of the identification problem may be partly solved knowing the physical significance of the material parameters involved in the constitutive laws. As an example, the sequential identification strategy discussed in the paper is based on the assumption of weak contribution of the texture evolution on the hardening. In such situation, the resort to micromechanical models may bring some clarification.

(ii)    Another encounter problem in material parameter characterization concerns the absence of the experimental data. This situation occurs when the conventional mechanical tests are not possible to be performed (*e.g.* characterization of through thickness anisotropy for sheet material). The example discussed in section 2 (*i.e.* deep drawing of a cylindrical cup) clearly show that the generally adopted through-thickness isotropic behavior assumption deteriorates the FE prediction.

(iii)   The identification strategy can not be completely disconnected from the simulated forming process. Indeed, a very accurate material behavior can be obtained for the experimental database involved in the identification process. However, such specific stress and/or strain states can be far from the ones involved in the simulated forming process, leading to bad prediction. An example is given with the biaxial stress state in the cross tool simulation.

Other applications are also under investigation (full-field measurement input in the identification strategy, kinemetic hardening characterization and springback sensitivity...).

# REFERENCES

1.  S. Bouvier, C. Teodosiu, H. Haddadi, V. Tabacaru (2003), J. Physique IV 105, pp.215-222 (2003).
2.  S. Bouvier, J.L. Alves, M. Oliveira, L.F. Menezes and C. Teodosiu, in 12th International Symposium Plasticity-2006, edited by A. Khan, 2006, Halifax (Canada).
3.  O. Cazacu and F. Barlat, Math. Mech. Solids 6, pp. 613-630 (2001).
4.  P. Flores, L. Duchêne, C. Bouffioux, T. Lelotte, C. Henrard, N. Pernin, A. Van Bael, S. He, J. Duflou and A.M. Habraken, Int. J. Plasticity 23, 420-449 (2007).
5.  H. Haddadi, S. Bouvier, M. Banu, C. Maie rand C. Teodosiu, Int. J. Plasticity 22, 2226-2271 (2006).

# Phenomenological Analysis of the Kinematic Hardening of HSLA and IF Steels Using Reverse Simple Shear Tests

A. Aouafi[a,b], S. Bouvier[a], M. Gaspérini[a], X. Lemoine[b]
and O. Bouaziz[b]

[a] *LPMTM-CNRS, UPR 9001, Université Paris 13, 99 av JB Clément, Villetaneuse, 93430 France*
[b] *Centre Automobile Produit, Arcelor Research S.A., voie Romaine B.P. 30320, 57283 Maizières-lès-Metz, France*

**Abstract.** Reverse simple shear tests are used to analyse the Bauschinger effect and the evolution of the kinematic hardening for a wide range of equivalent von Mises strain [0.025 - 0.3]. This work is carried out on two high strength low-alloyed steels. In order to investigate the effect of the precipitates on the macroscopic behaviour, a ferritic mild steel is used as a reference. Different phenomenological descriptions of the back-stress tensor are examined in order to analyse their ability to describe the experimental behaviour.

**Keywords:** Kinematic hardening, Simple shear test, high strength low-alloyed steels.

## INTRODUCTION

Most of the commercial alloys contain second phase particles (*e.g.* hard phase, precipitates…) to achieve the desired mechanical properties. Those particles serve as barriers for the dislocations motion and contribute to the overall strengthening of the materials. Nevertheless in the same time, such particles introduce sources of internal stresses which may significantly change the macroscopic behaviour during reverse loading. This phenomenon is classically known as the Bauschinger effect. An accurate description of this effect is a key point in order to obtain reliable FE simulations in forming processes. In this context, the experimental characterization of the Bauschinger effect for a wide range of strain becomes crucial. The work proposed here aims to investigate the consequence of the presence of the precipitates on the evolution of the Bauschinger effect for two grades of low-alloyed steels. As a reference, the behaviour of single phase interstitial free ferritic steel, is first reviewed. For the specific case of sheet materials, the uniaxial tensile test is not suitable due to the occurrence of a localized necking and buckling. The simple shear test is known to be more appropriate [1]. In this work, this mechanical test is used under monotonic and reverse loading with different amounts of shear strain in the forward direction. A comprehensive analysis of the macroscopic behaviour is carried out (work-hardening rate evolution, analysis of the transitory work-hardening regime…) in order to have some insights on the kinematic hardening evolution as a function of the amount of the

CP907, *10th ESAFORM Conference on Material Forming*, edited by E. Cueto and F. Chinesta
© 2007 American Institute of Physics 978-0-7354-0414-4/07/$23.00

plastic strain. Preliminary phenomenological models of the kinematic hardening are examined in order to assess their ability to describe the experimental behaviour.

## MATERIALS AND EXPERIMENTAL TESTS

Three different steels are studied in this work: an Interstitial Free mild steel (IF-steel) of 1 mm thickness and two high strength low-alloyed steels with different volume fraction of NbC precipitates, namely H360 and H280, provided by Arcelor as cold-rolled sheets of 1mm and 1.7mm thickness, respectively. The simple shear specimens are rectangular specimens of 30x18 mm². The gauge area is limited to the central area of 30x2 mm². All the tests are performed along the rolling direction on the as-received materials. Monotonic tests are conducted up to 80pct amount of shear strain. For the reverse simple shear tests, different amounts of forward shear strain, typically 4, 7, 10, 20, 25, 30 and 50pct, are used in order to characterize the evolution of the kinematic hardening as a function of the amount of preshear strain.

## RESULTS

Inspection of strain vs stress curves obtained by monotonic and Bauschinger tests reveals a typical decrease of the flow stress at the beginning of the reverse loading (i.e. the Bauschinger effect), but also an occurrence of a transitory work-hardening regime after a certain amount of forward shear strain (typically > 10pct). In case of IF-steel, microstructural investigations [2, 3, 4] reveal that such macroscopic behaviour can be associated to a gradual dissolution of predeformed dislocations walls followed by development of new ones with opposite polarity.

The transitory work-hardening regime observed in the three steels is first analysed using two strain parameters $\gamma 1$ and $\gamma 2$, which are determined as shown in Figure 1(a):
  – $\gamma 1$ indicates the beginning of the transitory work-hardening regime
  – $\gamma 2$ indicates the resumption of the work-hardening.
The results are reported in Figure 1(b) for different values of amount of forward shear strain. The occurrence of the transitory work-hardening regime is delayed for the low-alloyed steels. Preliminary microstructural investigations (Figure 2) are consistent with this macroscopic behaviour. Indeed, the precipitates act as barriers to the dislocations motions, leading to somehow more statistical distribution of dislocations between the dislocations boundaries in the HSLA compared to the IF steel. Hence, at the early stages of reverse loading, some annihilation of such trapped dislocations could occur, which may explain some delay (or absence) in partial dissolution of the preformed dislocation sheets.

It can be noticed that the behaviour of the two low-alloyed steels is different for $\gamma <$ 20pct: only H280 steel exhibits transitory regime for $\gamma$=0.1 and 0.2pct, as in IF steel, which is consistent with its lower volume fraction of the precipitates. Finally, the size of the transitory work-hardening regime gradually increases with the amount of forward preshear strain and tends to reach some stagnation for $\gamma > 30$pct. (in relation with the degree of patterning of the dislocations).

101

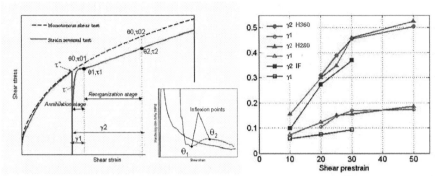

**FIGURE 1-** (a) Schematic definition of γ1 and γ2 parameters (b) Evolution of the transitory work-hardening parameters γ1 and γ2 with shear prestrain

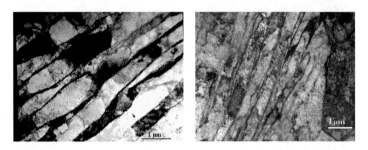

**FIGURE 2-** TEM investigation after simple shear test up to 30pct amount of shear strain. (a) IF-steel (b) H360.

## Kinematic Hardening Evolution

To quantify the change of the flow stresses upon stress reversal, different phenomenological parameters can be used [5, 6]. Such parameters describe the decrease of the flow stress in the reversed direction as a function of the amount of forward strain. In the present study, the parameter X which classically describes the back stress is used. In the case of the simple shear test, it can be determined as:

$$X = \frac{\tau^+ - \tau^-}{2},$$

where $\tau^+$ denotes the flow stress in the forward straining and $\tau^-$ is the absolute value of the reversed yield stress. The latter is determined with an offset of 0.1pct of von Mises equivalent plastic strain (see below). Figure 3 shows that the Bauschinger effect increases with the increase of the amount of forward shear strain, and also with the volume fraction of the precipitates. The relative back-stress ratio $X/\tau^+$ values are relatively high, but are consistent with other results on similar materials [5] as shown on Figure 4. As reverse shear testing permit larger prestrain values than tension-compression tests, it can be concluded from Figure 3 that no saturation tendency occurs for X values before 30pct of shear strain, which was not obvious from Figure 4 only.

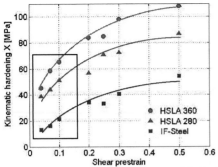

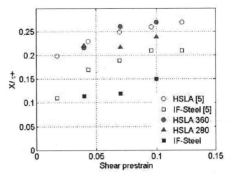

**FIGURE 3-** Experimental evolution of the back stress X for different amount of forward shear strain.

**FIGURE 4-** Experimental evolution of X / τ⁺ for forward shear strain γ values lower than 15 pct. Data from [5] are introduced for comparison.

## Influence of the Offset Value

The quantitative characterisation of the Bauschinger effect using X parameter needs an arbitrary choice of a plastic strain offset after the strain reversal. The influence of this offset on the experimental determination of the back stress X values is reported in Figure 5. The X values are significantly different (up to 25pct variation) when the offset value is switched from 0.2pct (conventional offset for yield stress) to 0.1pct, whereas this effect is strongly reduced when it is switched from 0.1 to 0.05pct, justifying the value 0.1pct used in this work.

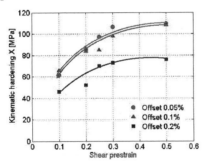

**FIGURE 5-** Influence of the offset on the determination of the X value in the case of H360 steel.

## MODELLING

From a macroscopic point of view, four phenomenological hardening models within the framework of standard plasticity using von Mises yield criterion are considered in order to compare their ability to describe the experimental behaviour.

Model 1 (*4 material parameters*): Isotropic hardening using Voce law and non linear kinematic hardening (ECNL) using Armstrong-Frederick law;

Model 2 (*5 material parameters*): Isotropic hardening using Voce law and both Prager's linear (ECL) and non linear (ECNL) kinematic hardening;

Model 3 (*4 material parameters*): The evolution law for the isotropic hardening R is coupled to the non linear kinematic hardening (ECNL);

$$dR = C_R \left( \frac{R_{sat}}{1 + \|d\mathbf{X}\| / \|\mathbf{X}\|} - R \right) d\bar{\varepsilon}^p, \quad d\mathbf{X} = KNd\varepsilon^p - \beta \mathbf{X} d\bar{\varepsilon}^p$$

Model 4 (*9 material parameters*): Teodosiu-Hu Model [7];

The material parameters involved in the hardening models are identified using the monotonic simple shear test and Bauschinger tests up to 7, 10 and 30pct amount of forward shear strain. The remaining reverse tests (4, 20 and 50pct) are used in order to assess the capability of the models. The comparison between the experimental and simulated stress-strain curves is reported in Figure 6. As expected, due to the number of material parameters, a good agreement on the whole curves is obtained using the Teodosiu-Hu model. However, considering only the back stress values X, all the models underestimate the experimental values, as illustrated by Figure 7.

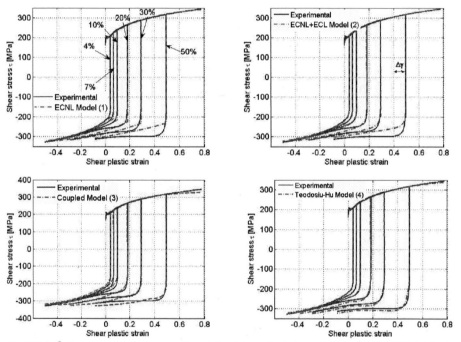

**FIGURE 6** - Comparison between experimental and theoretical stress-strain curves for HSLA 360 steel

Focussing on the curves after strain reversal (the "elbow" zone), the following error parameter is used to quantify the discrepancy between the different models:

$$\text{Error parameter} = \frac{1}{N} \sum_{i=1}^{N} \left( \frac{\tau_i^{exp} - \tau_i^{num}}{\tau_i^{num}} \right)^2,$$

where N is the number of points in the elbow zone of the curves, corresponding to the strain interval $\Delta\gamma$ defined by a 10pct decrease after the strain reversal. The results are reported in Figure 7. One can notice that for small prestrains, the model (3) accurately describes the elbow zone. However, for larger prestrains leading to the transitory work-hardening regime, a more sophisticated description of the kinematic hardening is required.

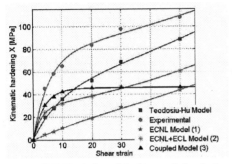

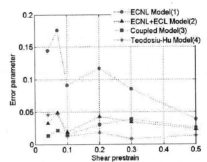

**FIGURE 7**- Evolution of the kinematic hardening for different models

**FIGURE 8**-Evolution of normalized error between experimental and theoretical curves

## CONCLUSION

Thanks to reverse shear tests, phenomenological analysis of kinematic hardening of HSLA steels was performed. Compared with IF steel, HSLA steels present more pronounced Bauschinger effect, and due to the presence of precipitates the transitory hardening regime is delayed.

A simply coupled hardening model can describe the "elbow" zone of the curves for small prestrain, but more complex models are needed to describe larger prestrain effects. In order to have a physically-based coupled model, further work is in progress based on the work of Bouaziz [8] to analyse the relations between the mechanical behaviour and the microstructure ingredients.

## ACKNOWLEDGMENTS

The authors are grateful to Arcelor Research SA for the financial support of this work.

## REFERENCES

1.  S. Bouvier, H. Haddadi, P. Levée, C. Teodosiu, Mater. Proc. Techno. 174, pp.115, 2006.
2.  E. V. Nesterova, V. Richard, T. Chauveau, LPMTM Internal Report (University Paris 13) France, 1999.
3.  E. V. Nesterova, B. Bacroix and C. Teodosiu, Metal. Mater. Trans. A, Vol. 32A, pp.2527, 2001.
4.  E. F. Rauch, Mater. Sci. Eng. A234, pp.653, 1997.
5.  K. Han, C. J. Van Tyne and B. S. Levy, Metal. Mater. Trans. Vol. 36A, pp. 2379, 2005.
6.  M. Choteau, P. Quaegebeur and S. Degallaix, Mech. Mater. 37, pp. 1143, 2005.
7.  C. Teodosiu, Z. Hu, In: Cartensen, J.V., Leffers, T., Lorentzen, T., Pedersen, O.B., Sørensen, B.F., Winther, G. (Eds.), Proc. Risø Int. Symp. Mat. Sci., (Denmark), pp.149, 1998.
8.  O. Bouaziz and G. Dirras, MATERIAUX 2006, Dijon, France, 2006.

# Characterisation for numerical study of mechanical behaviour of cermet Mo-TiC

Cédat Denis [a], Rey Colette [a], Clavel Michel [a], Schmitt J.H. [a], Le Flem Marion [b], Allemand Alexandre [c]

[a] MSSMAT ,Ecole Centrale·Paris, 92295 Chatenay Malabry, France

[b] CEA DEN/DMN/SRMA, 91191 Gif-sur-Yvette, France

[c] CEA DRT/LITEN/LTMEx, 91191 Gif-sur-Yvette, France

**Abstract.** The aim of this study is to characterise the mechanical behaviour of composite material: CERamic-METal. This material is composed of molybdenum-titanium carbide.A numerical mesoscopic approach is built, which consists in creating a 3D- finite element layer by integrating the results produced by the EBSD acquisition performed on the undeformed material. The parameters used for this simulation were identified using a polycrystalline model and an inverse method from experimental results.

**Keywords:** Mesoscopic approach, Cermet, 3D-Finite element
**PACS:** 60

## INTRODUCTION:

Titanium carbide (TiC) matrix cermets were widely studied in 1980's [1]. Because of their high hardness, thermal stability and excellent wear resistance, those cermets have been successfully used for cutting tools in finishing work of steel and iron. However, here the major constituents of this cermet consist of fine molybdenum particles, which are soft and ductile, and the minor constituent of titanium carbide which is hard and brittle. Both hard and metal phases are, to some extent, modified by, respectively, refractory carbides and other iron group transition metals in order to achieve certain aimed end properties. In other words, the materials science of a such compound, must include the scientific concepts related to both ceramic and metal phases.

In order to understand the role of material initial microstructure and their actives system of deformation on macro-zones formation, a simulation of compressive test has been undertaken, allowing to follow the evolution of the crystallographic texture and morphology for different deformation rates.To describe the microstructure close to the real one, a layer of 130 prismatic grains has been undertaken, then on an actual 3D

CP907, *10th ESAFORM Conference on Material Forming,* edited by E. Cueto and F. Chinesta
© 2007 American Institute of Physics 978-0-7354-0414-4/07/$23.00

aggregate constituted of about 1500 grains. In both cases, the initial orientations and positions of the grains in the aggregate were determined experimentally by EBSD analysis.

## MATERIAL:

Molybdenum - titanium carbide cermet were prepared according to Hot Isostatic Pressure (HIP) facilities. Firstly, the cermet analysed was manufactured by mixing appropriate amounts of raw material powders in alcoholic synthesis route with zirconium balls. A green sample was cold isostatic pressed during 1 minute. After outgassing at 600°C in an high-vaccum during 12 hours, a hot isostatic pressure is applied at 1600°C during 2 hours under 160MPa.

The microstructure of the sintered TiC/Mo cermet (d > 0,95) can be seen in figure 1 which shows a scanning electron micrograph. This micrograph was obtained using backscattered electrons, an imaging technique which makes full use of atomic number contrast between the different constituents. It can be seen that there are both bright and dark cores, surrounded by grey rim.

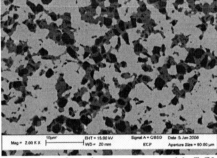

**FIGURE 1.** *SEM picture showed microstructure in Mo-TiC25%at. cermet*

So the bright cores contain heavier atoms, in this case Mo, and the darker are TiC. Thus, the hard particles (TiC) can be seen to have what is referred to as a core/shell or core/rim structure with molybdenum as binder phase.

In order to confirm the grain size, the specimens structure was examined by transmission electron microscopy (TEM) using a JEOL, the accelerating voltage being 120KV. As shown on figure 2, the grain size of TiC is close to 1µm whereas the grain size of Mo is about 2-3µm.

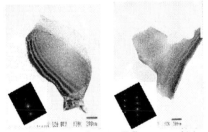

**FIGURE 2.** *TEM image showing TiC grain and Mo grain*

TEM observations were carried out to highlight the core/shell structure of the cermet. Thus, the shell phase (Mo,Ti)C, already observed figure 1, can be identified at the grain boundary between TiC and Mo grains on the figure 3. Grain boundaries between molybdenum grains are particle free.

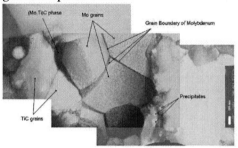

**FIGURE 3.** *Transmission electron micrograph picture showing core/shell structure in a Mo-TiC25%at cermet.*

## CRYSTAL PLASTICITY MODELLING:

In order to understand the microstructure evolutions of our material, a polycrystalline model ([2] and [3]) was used to simulate compressive test. The crystalline plasticity constitutive law uses a large deformation framework based on the multiplicative decomposition of the deformation gradient into elastic and plastic parts.. The model has been implemented in the finite element code ABAQUS and accounts for crystallographic glide and lattice rotation. At room temperature, for bcc structure (as molybdenum), slip occurs on the 24 slip systems {110}<111> and {112}<111>. For fcc structure (as titanium carbide), slip is considered to occur on the {111}<110>.

The resolved shear stress $\tau^s$, the critical shear stress $\tau^s_c$ and the slip rate $\dot{\gamma}^s$ are related by a viscoplastic power law with $\dot{\gamma}_0$ as a reference slip rate.

$$\frac{\dot{\gamma}^s}{\dot{\gamma}_0} = \left(\frac{\tau^s}{\tau^s_c}\right)^{1/m} \text{sgn}(\tau^s) \quad \text{if } |\tau^s| \geq \tau^s_c \text{ and } \dot{\gamma}^s = 0 \text{ otherwise} \qquad \text{Eq.1}$$

$$\text{With} \quad \tau^s_c = \tau_0 + \mu b \sqrt{\sum_u a^{su} \rho^u} \qquad \text{Eq.2}$$

$\rho^u$ is the dislocation density on system (u) and $a^{su}$ is an hardening matrix which coefficients depend on the interactions between dislocation families (s) and (u). The coefficient 1/m corresponds to the velocity exponent, $\mu$ and $b$ are the respective shear modulus and Burgers vector magnitude.

The hardening terms contained in $h^{su}$ do not remain constant during deformation, but depend on internal variables (dislocation densities on all slip systems) in order to take into account the physical aspects of plasticity [4]. On a particular slip system (s), the evolution of the dislocation density $\rho^s$ is governed by a production term based on Orowan's relationship and by an annihilation term which takes into account the dynamic recovery during deformation. Thus, it may be shown ([5] and [6]) that :

$$h^{su} = \frac{\mu}{2} \frac{a^{su}}{\sqrt{\sum_t a^{st}\rho^t}} \left( \frac{1}{\Lambda^u} - g_c \rho^u \right)$$

Eq.3

$$\Lambda^u = \frac{\sqrt{\sum_{t\neq s} \rho^t}}{K} + \frac{1}{D}$$

Eq.4

$g_c$ is proportional to the annihilation distance of dislocation dipoles. $\Lambda^u$ representing the mean free path of the mobile dislocations gliding on the system u. K is a material parameter, and D is the mean polycrystal grain size.

$a^{su}$ is a hardening matrix which coefficients depend on the interactions between dislocation families s and u. The physical meaning of the whole parameters set is detailed in [2].

## AGGREGATES CHARACTERISATION, MESHING:

The finite element mesh was constituted by 8 nodes cubic elements (reduced integration), obtained directly from EBSD acquisitions, performed on the sintered material before forging with a FEG-SEM (LEO 1530) using a 0,3 µm step. By this method, the special grain arrangements and the crystallographic orientations (local texture) were taken into account. But the average grain size (3µm or less) makes this task difficult and the different layers are unlinked by the grain. So, the pattern dimensions are equal to the analyzed one. Its thickness is set to 0,3 µm in order to obtain original cubic elements.

FIGURE 4. *(a) SEM picture (b) initial inverse pole figure map showing the position of the normal axis in the stereographic triangles (c) Aggregate picture*

The analysed size of the aggregate were $20x20x3\mu m^3$ (ten layers) and a cubic mesh of 0,3µm. As example, the mesh of aggregate is given on Fig. 4. The advantage of such aggregate, with actual local orientations and actual morphology, is that the interaction between grains is taken into account.

## RESULTS:

The parameters used for this simulation were identified using a polycrystalline model and an inverse method from experimental results. We observe that each parameter must be identified in these three phases.

TEM investigations were also performed on sample in order to identify the dislocation density before any deformation. The dislocations in TiC phase, shown in figure 5, have a <110> Burgers vector.

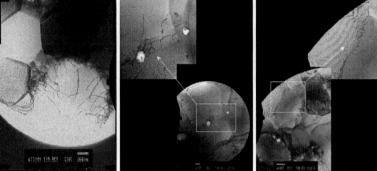

**FIGURE 5.** *Transmission electron micrograph showing (a) dislocations in TiC, (b) Hexagonal reordered of dislocations in TiC grain and (c) dislocations network in molybdenum*

These results indicate that the hard phase is subjected to local stress concentrations which are sufficient to generate dislocations in titanium carbide grains (figure 5 (a)). Nevertheless, molybdenum phase doesn't present any dislocation and that is due to the recrystallisation during forming (figure 5 (b) and (c)).

Table 1 and 2 review the parameters identified during this study for molybdenum and titanium carbide respectively.

*Table 1. Behaviour law parameters for molybdenum at room temperature*

| elasticity | | viscoplasticity | | | | | | | | | |
|---|---|---|---|---|---|---|---|---|---|---|---|
| $E$ [GPa] | $\mu$ [GPa] | $a^{ss}$ | $a^{su}$ | $K$ | $\tau_0$ [MPa] | $g_c$ [nm] | $n$ | $\dot{\gamma}_0$ [$s^{-1}$] | $\rho_0[m^{-2}]$ | $D$ [$\mu m$] | $b$ [nm] |
| 330 | 127 | 0.215 | 0.25 | 41 | 185 | 8 | 100 | $10^{-10}$ | $10^5$ | 3 | 0.251 |

*Table 2. Behaviour law parameters for titanium carbide at room temperature*

| elasticity | | viscoplasticity | | | | | | | | | |
|---|---|---|---|---|---|---|---|---|---|---|---|
| $E$ [GPa] | $\mu$ [GPa] | $a^{ss}$ | $a^{su}$ | $K$ | $\tau_0$ [MPa] | $g_c$ [nm] | $n$ | $\dot{\gamma}_0$ [$s^{-1}$] | $\rho_0[m^{-2}]$ | $D$ [$\mu m$] | $b$ [nm] |
| 440 | 183 | 0.215 | 0.25 | 41 | 1320 | 8 | 100 | $10^{-12}$ | $10^9$ | 2 | 0.251 |

$\tau_0$ and $\rho_0$ are the respective initial shear stress and dislocation density on every slip system $s$. $E$ is the Young modulus.

TEM observations were carried out to highlight the core/shell structure of the cermet. Thus, the shell phase (Mo,Ti)C, observed figure 1, is localized at the grain boundary between TiC and Mo grains. In the Mo-TiC cermets, the core (TiC) of each hard particle is thought, essentially, to be what remains of the original raw material; the shell has the same crystal structure as the core but contains additional elements, primarily Mo, arising from contact with the binder phase. For compositional analysis of this phase, X-ray detection (EDX) and wavelength dispersive spectrometers (WDS), outfitting the electron microprobe, was used.

It would prove that (Mo,Ti)C phase presents a composition close to the solid solubility of molybdenum in the titanium carbide: $TiC-Mo_{10-15at.\%}$.

Moreover, nanoindentation experiments have been made to obtain the Young modulus (350 GPa). From a mixture law applied on Poisson's ratios, we are able to find the shear modulus of this phase.

## CONCLUSION:

The simulation part undertake research at different scale levels in predicting material macroscopic behaviour, by taking into account underlying mechanisms at microscopic levels. The model is implemented in ABAQUS finite element code, using the framework of large transformation, with small elastic distortion but large lattice rotation. The local cleavage and shear stresses are computed at the grain scale and at different temperatures from crystalline aggregates simulations generated with EBSD mappings.

A multiscale approaches is needed to implement this model. Indeed, the model parameters have been identified with an inverse method from mechanical tests, carried out at different temperatures and strain rates, and from TEM observations.

In order to make simulation on real microstructure, a work of characterisation on the phase $TiC-Mo_{10-15at.\%}$ have been made.

## REFERENCES

1. P. Ettmayer and H. Kolaska, 12th Plansee Seminar '89, Vol.2 (H.Bildstein and H.M. Ortner,Eds), pp771-801
2. T. Hoc, C. Rey and J. Raphanel, *Acta mater.* 49, 2001, 1835-1846.
3. P. Erieau, C. Rey, Modelling of deformation and rotation bands and of deformation induced by grain boundaries in IF Steel aggregate during large plain strain, IJP (in press).
4. P. Franciosi: Acta metal. **33** (1985), p.1601
5. D.Pierce, R.J. Asaro and A. Needleman. Material rate dependence and localized deformation in crystalline solids, overview 32, vol 31, No12 pp1951-1976,1983
6. C. Teodosiu, J.L. Raphanel, and L. Tabourot, in *Large Plastic Deformation. Proceeding of the seminar MECAMAT'91*, edited by C. Teodosiu, J.L. Raphanel, and F. Sidoroff, (Fontainebleau, 1991), p. 153

# Description of microstructural intragranular heterogeneities in a Ti-IF steel using a micromechanical approach

A. Wauthier*,†, R. Brenner*, H. Réglé†,* and B. Bacroix*

*Laboratoire des Propriétés Mécaniques et Thermodynamiques des Matériaux,
Université Paris Nord, avenue Jean-Baptiste Clément, 93430 Villetaneuse
†Arcelor Research, Voie romaine – BP30320, 57283 Maizières-lès-Metz

**Abstract.** A classical problem in metallurgical research is to control the recrystallisation texture which forms during the last annealing process and which determine the mechanical behaviour of the final products. It is now widely admitted that the local deformed state and the substructural heterogeneities within the polycrystal are key parameters to understand the recrystallisation mechanisms.

In this work, we present a micromechanical approach based on the use of the affine extension of the self-consistent scheme for viscoplastic behaviours and a phenomenological description of dislocation patterning using a hardening model recently developed for two-stage strain paths. These two ingredients allow to compare the model with experimental crystallographic texture after rolling as well as experimental observations of the intragranular substructure using orientation imaging by Electron Back-Scattered Diffaction. It is shown that the rolling texture is correctly simulated and successful predictions of the orientation of dislocation sheets are obtained.

## 1. INTRODUCTION

To create the desirable crystallographic textures that improve the formability properties of low carbon steels, a high degree of deformation by cold-rolling is needed. Since it is evident that this deformation substructure plays an important role in the nucleation process during recrystallisation (see e. g. [1]), it is important to characterize deformation heterogeneities and to explore the possibilities to link them with a micromechanical model describing the deformation process. We present here a micromechanical approach based on the use of the affine extension of the self-consistent scheme. A local hardening law based on dislocation patterning is used to describe the evolution of the dislocation density within the walls. In this paper, we report results obtained in the characterization of the deformation traces often observed in several grains after etching in a cold-rolled IF steel. The deformation traces are first analysed from EBSD scans in relation with their crystallographic orientation and then compared to cell-block boundaries (CBBs) predicted by the micromechanical model.

CP907, 10th ESAFORM Conference on Material Forming, edited by E. Cueto and F. Chinesta

## 2. EXPERIMENTAL PROCEDURE

### 2.1. Material studied

The material studied is a Ti-IF steel, issued from a previous hot rolling, which presents an average grain size of $20\mu m$ and a low crystallographic texture (Figure 1). It has been deformed by cold-rolling with different thickness reductions (15%, 30%, 40% and 50%) in a laboratory rolling mill. The crystallographic texture of the material after cold-rolling presents classically two fibers : $\alpha$ ($\langle 110 \rangle$ direction aligned with the rolling direction) and $\gamma$ ($\{111\}$ pole aligned with the normal direction of the sheet). The maximum value of the orientation distribution function (ODF) is located in the $\alpha$ fiber.

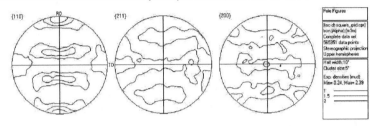

**FIGURE 1.** Pole figures of the Ti-IF steel in the as-received state (hot band)

### 2.2. Experimental analysis of the substructure

The dislocations tend to form structures during plastic deformation. It is well established that the morphology of the structure depends on numerous parameters : the number of active slip systems, the grain size, the crystalline structure ... In our case, a cellular structure with dislocation sheets parallel to the crystallographic planes of highest plastic slip activity is expected [2]. The development of such structures, implying significant intragranular misorientations, can be studied using (i) the orientation imaging technique based on electron back-scattered diffraction (EBSD) within a scanning electron microscope (SEM) and (ii) "direct" observations of the local strain field caused by dislocations by transmission electronic microscopy (TEM).

EBSD measurements were carried out on a Cambridge S360 with an operating voltage of 25kV and a probe current of about 3nA. All observations were made on the section normal to the transverse direction (TD) with a step size of 200nm. The samples were mechanically polished and then electro-polished to remove any work-hardened surface layer with a solution made up of 5% of perchloric acid and 95% of ethanol during 10s at 38V. The inclination of the traces along the rolling direction is regarded versus their orientation or, more exactly, the orientation just near that trace inside the considered grain. For example, in figure 2 we can see two areas (delimited by white ellipsoids) inside the same grain with several parallel traces. In this case only one orientation and one inclination for this set of traces are noted.

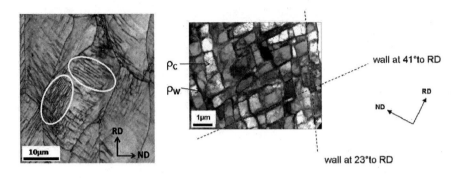

**FIGURE 2.** Image quality EBSD map after 30% of thickness reduction (step size of 0.1 μm) and TEM micrograph in bright field showing dislocation microstructure after 40% of thickness reduction

Their main inclination can be compared to the dislocation walls observed with TEM in these materials. For instance, for a sample cold-rolled up to 40% thickness reduction, the mean trace inclination along RD is about 38° (average value obtained for a set of 20 grains) and in TEM we generally observed 2 dislocation walls.

## 3. MICROMECHANICAL MODELLING

### 3.1. Transition-scale scheme

To predict the local and overall response of the Ti-IF steel, we adopt a micromechanical approach which allows to derive the mechanical behaviour and the microstructural evolution (especially the development of crystallographic texture) of the polycrystal from the behaviour of the constituent grains. In this framework, a polycrystal is viewed as a composite material with a large number of mechanical phases and a granular topology. A phase includes all the grains with the same crystalline orientation. The homogenisation procedure delivers an estimation of the macroscopic response and a statistical description of the stress and strain fields within each phase (average, standard deviation etc ...) by making use of Eshelby's inclusion formalism. For polycrystalline microstructures, it is known that the most relevant model is the self-consistent scheme which assumes that each phase plays the same role from a morphological point of view (concept of perfect disorder [3]).

For plastic behaviours, and more generally for any nonlinear constitutive laws, an additional choice has to be made regarding the way the nonlinear response can be obtained from a classical linear homogenisation model. Here, we adopt the so-called affine extension of the self-consistent scheme. It has been shown [4, 5] that this affine approach leads to a more realistic description of the (visco)plastic behaviour than the usual secant model of Hill and Hutchinson [6, 7]. Besides, it presents interesting connections with some more advanced variational formulations [8].

## 3.2. Constitutive viscoplastic flow rule

It is assumed that the deformation occurs by viscoplastic glide on the following slip systems : $\langle 111 \rangle \{1\bar{1}0\}$ and $\langle 111 \rangle \{11\bar{2}\}$. On a slip system $k$, the flow rule reads

$$\dot{\gamma}_k = \dot{\gamma}_0 \left| \frac{\tau_k}{\tau_0^k} \right|^{n-1} \frac{\tau_k}{\tau_0}, \tag{1}$$

with $\tau_k$ the resolved shear stress, $\dot{\gamma}_k$ the slip rate and $\tau_0^k$ the reference (critical)) shear stress. To describe the plastic behaviour of steel at room temperature, we consider a nonlinear exponent $n = 40$ and a reference slip rate $\dot{\gamma}_0 = 10^{-11}\text{s}^{-1}$ [9]. Finally, it is necessary to specify the value of the reference shear stress $\tau_0^k$ and its evolution with the deformation.

## 3.3. Local hardening law based on dislocation patterning

To take into account phenomenologically the dislocation patterning, Peeters *et al.* [10] proposed a local hardening law which makes use of dislocation densities. Compared to previous approaches like the one developed by Tabourot *et al.* [11], this model supposes that cell-block boundaries form during deformation. It is assumed that these dislocation sheets form parallel to $(110)$ crystallographic planes. To describe this structure, use is made of different variables : the dislocation density $\rho_c$ within the cells, the dislocation density $\rho_w$ within the walls (dislocation sheets) and a "polarized" dislocation density $\rho_p$. Each of these variables has its own evolution equation which follows the general form $\dot{\rho} = (A\sqrt{\rho} - B\rho)|\dot{\gamma}|$. This leads to a stationary state with a saturation of the dislocation densities. For more details, the reader is referred to [10].

The following features can be pointed out :

- At each time step, the model assumes that cell-block boundaries form on the two $(110)$ planes of highest slip activity.
- During a strain-path change, progressive destruction of old CBBs occurs. This feature must be considered even in the present case of a radial and monotonic loading because of the crystalline rotation which implies a continuous strain-path change for each orientation. Consequently, it is worth noting that more than two CBBs can exist within a crystalline orientation.
- The CBBs are associated with latent hardening since they act as obstacles for the mobile dislocations on subsequently activated slip systems.
- The reference shear stress $\tau_0$ is obtained by a rule of mixture within each crystalline orientation according to the "composite" model of Mughrabi [12]. It reads

$$\tau_0 = (1 - f_w)\tau^C + f_w \sum_{i=1}^{6} \tau_i^{CBB} \tag{2}$$

with $f_w$ the volumic fraction of CBB, $\tau^C$ the reference shear stress within the cells and $\tau_i^{CBB}$ the reference shear stress for each potential cell-block boundary.

The parameters identified in [10] have been used. They lead to a saturating dislocation density of $1.3 \, 10^{-15} \text{m}^{-2}$ within the CBBs. A nice feature of this model is that it allows to link intragranular crystalline misorientations with a specific dislocation density $\rho_w$.

## 4. RESULTS AND DISCUSSION

The model has been used to simulate (i) the overall response, including the crystallographic texture evolution and (ii) the dislocation patterning at the intragranular level. Simulations are performed up to 50% of thickness reduction. Concerning the microstructural description of the initial state of the polycrystal, it must be noted that the crystallographic texture of the hot-band (Figure 1) is used and that a random distribution of equiaxed grain is assumed.

The model is able to capture the intergranular heterogeneity within the polycristal. Indeed, it predicts a higher mean deformation for the grains belonging to the $\alpha$ fiber compared to the mean deformation within the $\gamma$ fiber. Concerning the overall crystallographic texture, a rather good description of the main texture components is obtained (Figure 3). Especially, compared to Taylor-like models, the $\gamma$ fiber is correctly predicted. Nevertheless, it can be observed that the intensity of this fiber is too high with respect to the experimental texture.

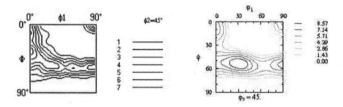

**FIGURE 3.** Experimental versus simulated ODF obtained with the affine self-consistent model (50% thickness reduction)

To compare experimental traces with CBBs predicted by the model, we considered different initial crystallographic textures : isotropic texture described by a set of 2016 orientations following the vector method and two sets of orientations randomly taken from an EBSD scan of the as-received state (respectively 1000 and 3000 points). For the experimental crystallographic texture, it has been checked that the ODF obtained with the two sets is comparable to the one obtained with the entire EBSD scan.

From EBSD observations, we know the orientation of the grain ($g_{exp}$) near each experimental trace and for each strain level observed. From the modeling, for each deformation step, we have a new set of orientations and we extract the ones at less than $5°$ from $g_{exp}$. Then, we have access to the dislocation density and the inclination along the rolling direction for each dislocation walls for a set of orientations near experimental ones. We then compare the inclinations of the two predicted walls with maximum dislocation densities, for all orientations at less than $5°$ from $g_{exp}$, with the inclination of the experimental trace. It is worth noting that the model often predicts only two dislocations walls with a high dislocation density for a given crystalline orientation.

Besides, we mention that within the set of crystalline orientations predicted by the micromechanical model, it is possible to have grains with close crystalline orientations but different deformation histories because of the texture evolution and work-hardening.

Our analysis shows that there is a correct agreement between experimental traces and predicted CBBs. Besides, the best agreement is obtained when considering the real initial texture represented by 3000 orientations (Table 1). Interestingly, it has to be noted that the majority of experimental traces that do not correspond to the CBBs predicted by the model belong to grains with a high orientation gradient. This observation might be related with the fact that the self-consistent model only consider average orientations per phase. It could also be necessary to consider the formation of CBBs on $(112)$ planes in the hardening model.

TABLE 1. Agreement between experimental traces and predicted CBBs.

| Isotropic | Exp. ODF (1000 points) | Exp. ODF (3000 points) |
|-----------|------------------------|------------------------|
| 60% ± 12% | 57% ± 9% | 66% ± 7% |

## 5. CONCLUSIONS

By comparing experimental with simulation results we show that the evolution of the deformation substructure of this Ti-IF steel is quite well reproduced by the present model with a good prediction of dislocations sheets up to 50% cold-rolling. Since the deformation texture is also well estimated, we now aim at coupling these results with a recrystallisation model to predict the kinetic of subsequent static recrystallisation and the final crystallographic texture. For that goal, the possibility of considering specific dislocations densities, linked to interior cells and cell-block boundaries, is very attractive to describe phenomenologically the so-called grain fragmentation and its influence on recrystallisation.

## REFERENCES

1.  M. R. Barnett, *ISIJ Int.* **38**, 78–85 (1998).
2.  B. Peeters, B. Bacroix, C. Teodosiu, P. van Houtte, and E. Aernoudt, *Acta Mat.* **49**, 1621–1632 (2001).
3.  E. Kröner, *J. Mech. Phys. Solids* **25**, 137–156 (1977).
4.  R. Masson, and A. Zaoui, *J. Mech. Phys. Solids* **47**, 1543–1568 (1999).
5.  R. Brenner, O. Castelnau, and L. Badea, *Proc. R. Soc. Lond.* **A460**, 3589–3612 (2004).
6.  J. W. Hutchinson, *Proc. R. Soc. Lond.* **A319**, 247–272 (1970).
7.  J. W. Hutchinson, *Proc. R. Soc. Lond.* **A348**, 101–127 (1976).
8.  R. Masson, M. Bornert, P. Suquet, and A. Zaoui, *J. Mech. Phys. Solids* **48**, 1203–1226 (2000).
9.  P. Erieau, Ph.D. thesis, Ecole Centrale Paris (2003).
10. B. Peeters, M. Seefeldt, C. Teodosiu, S. R. Kalidindi, P. van Houtte, and E. Aernoudt, *Acta Mat.* **49**, 1607–1619 (2001).
11. L. Tabourot, M. Fivel, and E. Rauch, *Mater. Sci. Engin.* **A234-236**, 639–642 (1997).
12. H. Mughrabi, *Mater. Sci. Engin.* **85**, 15–31 (1987).

# Research Concerning The Mechanical And Structural Properties Of Warm Rolled Construction Carbon Steels

## C. Medrea[1], G. Negrea[2], S. Domsa[2]

[1]*Technological Educational Institute of Piraeus, Depmartment of Physics, Chemistry and Materials Technology, 250 Thivon and P. Ralli Str, 12244, Aigaleo, Athens, Greece*
[2]*Technical University of Cluj-Napoca, Faculty of Materials Science and Engineering, ClujNapoca, Romania*

**Abstract.** Construction carbon steels represent an important steel class due to the large quantity in which it is produced. Generally, these steels are delivered in as-rolled or normalized condition heaving a ferrite-pearlite microstructure. For a given chemical composition, the mechanical characteristics of this microstructure are largely influenced by the grain size. Rolling is the deformation process which is most widely used for grain size refinement. Situated in the intermediate temperature range, warm-rolling presents certain advantages as compared to classical hot- or cold-working processes.
The paper presents a study on the microstructure and mechanical properties of Ck15 carbon steel samples warm-rolled. After deformation, the microstructure was investigated by light microscopy. Hardness measurements were made on the section parallel to the rolling direction. The mechanical properties of the steel after warm-rolling were assessed by tensile and impact tests. Additional information concerning the fracture behavior of warm-rolled samples was obtained by examining the fracture surface by scanning electron microscopy. The microstructure of the steel proved to have good mechanical properties. By considering the technologic and energy aspects, the paper shows that warm-rolling can lead to the improvement of mechanical properties of construction carbon steels.

**Keywords:** Construction carbon steels, warm rolling, microstructure, mechanical properties
PACS: 81.05.Bx,81.20.Hg,89.20.-a

## INTRODUCTION

Construction carbon steels are iron–carbon alloys with generally less than 0.25% C. They are used mainly for welded structures and, considering the large production scale, represent an important class of steels [1]. Construction carbon steels are supplied, generally, in as hot-rolled or normalized state. Due to their poor hardenability, the transformation during cooling from the rolling temperature or from the austenitization temperature for normalizing leads to a microstructure composed of ferrite and pearlite grains. Construction carbon steels have to fulfill a range of requirements regarding the mechanical characteristics. Among these, two appear as essential: the tensile strength and the brittle fracture strength [2]. For a given chemical composition, the mechanical characteristics depend on the microstructure. The mechanical properties of the ferrite – pearlite structure are strongly influenced by the size of the ferrite grains [3]. A range of methods to refine the ferrite – pearlite

CP907, *10th ESAFORM Conference on Material Forming*, edited by E. Cueto and F. Chinesta
© 2007 American Institute of Physics 978-0-7354-0414-4/07/$23.00

structure are applied in the industrial practice: altering of the chemical composition [4], normalizing heat treatment [5], plastic forming by controlled rolling [6] , rapid cooling [7]. The possibilities to alter the chemical composition are very restraint and the high cost of the rapid cooling equipment is limiting the industrial use of this method. The normalizing treatment gives good results only for fully deoxidized steels and increases the price of the products. Semi-deoxidized steels are suitable for controlled rolling but the restrictions it imposes are changing the manufacturing process into a very complex one. Therefore, rolling is the deformation process which is most widely used for grain size refinement The practice of rolling in the upper ferritic region instead of the austenitic region has been termed as warm rolling. Warm rolling causes significant changes on the final microstructure of the product. There have been several researches to study the process. The effect of warm rolling on the structure and properties of a low carbon steel has been investigated by Hawkins [8] and Haldar [9]. The effect of strain on the microstructure and mechanical properties of multi-pass warm caliber rolled low carbon steel have been conducted by Torizuka [10,11]. The modeling of warm rolling of a low carbon steel has been considered by Serajzadeh [12]. Ultra-fine ferrite microstructure in a warm rolled C-Mn steel has been obtained by Santos [13]. A submicron mild steel microstructure has been produced by one simple warm deformation by Liu [14]. The physical metallurgy involved during warm rolling is not yet fully understood and calls for further research activity on this process. The process is used for the manufacturing of small and middle sized parts and for a narrow range of steels. This paper presents the mechanical and structural properties of a warm rolled construction steel Ck15.

## EXPERIMENTAL DETAILS

Parallelepipedic samples with the dimensions of 20x20x200 mm were cut from normalized Ck15 carbon steel bars and subjected to full annealing (austenitizing followed by slow cooling) before rolling. The samples had the following chemical composition: 0.16%C, 0.32%Si, 0.56%Mn, 0.02%P, 0.01%S. Therefore, deformation of samples started from a coarse microstructure. The warm-rolling was performed by using a laboratory duo-reversible rolling mill under the following conditions: strain rate: 2.04 s$^{-1}$, deformation ratio: 36.4% and sample temperature: 700°C at the beginning and 550 °C, respectively, at the end of rolling[15].

After deformation, the microstructure was investigated by light microscopy on a section parallel to the rolling direction. The microstructure of warm-rolled specimens was compared with the microstructure of normalized samples, normalizing being frequently used for microstructure refinement of low carbon steels.

In order to evaluate the uniformity of deformation across the section of warm-rolled specimens, hardness measurements were made on the section parallel to the rolling direction. The results were compared with those determined on a separate group of samples, which were cold-rolled with the same deformation ratio.

The mechanical properties of the steel after warm-rolling and normalizing were assessed by tensile and impact tests. The tensile tests, performed according to standard procedure, allowed for determination of the yield strength, tensile strength, elongation and reduction in area. Additional information concerning the fracture behavior of warm-rolled samples was obtained by examining the fracture surface by scanning electron microscopy. Toughness of warm-rolled and, for comparison, of normalized samples was determined by Charpy impact tests carried out at room temperature on U–notch specimens using a Charpy impact machine with a potential energy of 300 J.

## RESULTS AND DISCUSSION

Usually, the carbon steels are delivered in normalized condition which, for Ck15 consists of fine-grained ferrite–pearlite microstructure (fig. 1.a). The effects of previous plastic deformations are eliminated during normalizing, and the microstructure consists of equiaxed grains with a uniform distribution of pearlite. The samples subjected to warm-rolling display a specific microstructure with pearlite grains distributed in bands (fig. 1.b). The ferrite grains, which are more ductile, are strongly elongated along the rolling direction. The ferrite grains are subjected to strain hardening, and their grain boundaries are difficult to distinguish. The pearlite grains undergo two simultaneous processes during deformation. Firstly, the pearlite grains are divided into subgrain blocks embedded in ferrite and they appear in the microstructure as very fine pearlite grains distributed in bands parallel to the rolling direction. Secondly, inside these small pearlite grains, the cementite lamellae are broken into small fragments which will undergo a spheroidization process and take a globular shape before cooling. Thus, the pearlite grains are just slightly deformed and are broken into smaller fragments that are redistributed along the rolling bands [15].

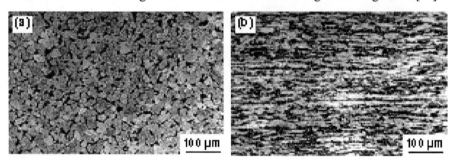

**FIGURE 1.** The microstructure of normalized (a) and warm rolled (b) samples.

Hardness measurements showed a very uniform hardness distribution in the cross–section of warm-rolled samples as opposed to the cold worked samples (Table 1).

**TABLE 1**. Hardness values of warm-rolled and cold-rolled samples: point A – in the middle of the section, point B – at half distance between A and C, point C – near the surface.

| Type of deformation | Hardness, HV20 | | |
|---|---|---|---|
| | A | B | C |
| Warm-rolling | 199 | 200 | 201 |
| Cold-rolling | 214 | 249 | 276 |

This is an indication of a uniform deformation across the section produced by warm-rolling. During warm-rolling, strain hardening and recrystallization processes take place simultaneously. Strain hardening is indicated by the vanishing of the ferrite

120

grain boundaries and by the increase of the hardness from an initial value of 119 HV (in the annealed state) to 201 HV after warm-rolling. Strain hardening is more significant in the case of cold-rolling, from 119 HV to 276 HV. Therefore, the warm-rolled sample is partially strain hardened and partially recrystallized. The incomplete recrystallization consists in a first stage recovery (at lattice level only) of ferritic grains, a process that lowers the internal stresses induced during rolling. Due to this process, warm-rolling allows much higher deformations to be achieved as compared to cold-rolling, without the risk of cracking. If the hardness variation across the section is insignificant after warm-rolling, it reaches a value of 29 % after cold-rolling.

Table 2 shows the mechanical properties of warm-rolled and normalized samples, determined by tensile and impact tests. The results indicate that warm-rolled steel has a higher yield strength and tensile strength compared to normalized steel. However, due to partial strain hardening during warm-rolling, the elongation, reduction in area and toughness are slightly lower. Because the mechanical properties of the steel after warm-rolling and normalizing are comparable, it appears that warm-rolling can be used as an adequate process for achieving the required delivery state properties for construction carbon steels.

**TABLE 2.** Mechanical properties of samples after warm-rolling and in normalized condition.

| Condition | Yield strength, MPa | Tensile strength, MPa | Elongation, % | Reduction in area, % | Toughness, J/cm$^2$ |
|-----------|---------------------|------------------------|----------------|----------------------|---------------------|
| Warm-rolled | 359 | 493 | 24.7 | 59 | 162.5 |
| Normalized | 338 | 471 | 25.5 | 61.5 | 185 |

The examination of the fracture surfaces by scanning electron microscopy after tensile testing showed similar surface morphologies for both, warm-rolled and normalized samples as shown in figure 2.

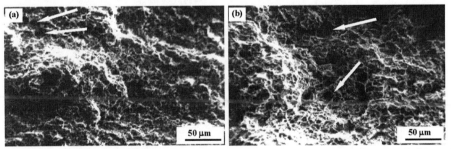

**FIGURE 2.** SEM micrographs showing the morphology of the fracture surfaces of warm-rolled (a) and normalized (b) samples subjected to tensile testing.

The samples display very rough intergranular fracture surfaces with the classical cup–and–cone appearance. The cup dimensions for warm-rolled specimens were considerably smaller, most probably due to strain hardening induced by warm-rolling (figure 2.a). Small cavities can be seen on the central area of the fracture surfaces (indicated by arrows). These cavities were found to be finer and more uniformly distributed on the warm-rolled sample compared to the normalized sample (figure 2.b). At the interface between the straining zone and the pulling-out one, inter-granular straining zones occurred (figure 3.a). According to the crystalline orientation of the grains, these zones show different aspects. In one grain, the straining strip appears in a "painting" shape (figure 3.b,d), while in another it looks like a wall in whose interior parallel straining strips, of variable thickness, occur (figure 3.c).

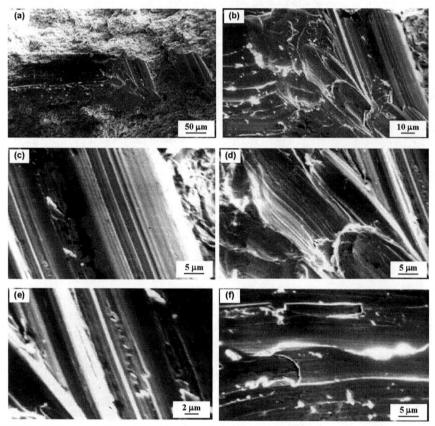

**FIGURE 3.** The morphological aspects of the fracture surface of warm-rolled samples after tensile testing.

For large magnifications, the wall shows fracture crests, as jig-saw teeth, which represents a front of edge dislocations (figure 3.e). The warm-rolled structure is reach in this kind of defects, due to a partial work hardening supported by the material during forming. The existence of the edge dislocations fronts is confirming the theory describing the spherulization of pearlite through the globulization of the cementite in its structure as a result of the motion of dislocations at the rolling temperature.

## CONCLUSIONS

Warm-rolling of Ck15 carbon steel generated a fine microstructure. The two constituents of the steel, ferrite and pearlite, behave differently during rolling. Ferrite grains, are plastically deformed and elongated along the rolling direction. The pearlite grains undergo only a slight plastic deformation. Instead, they are broken during rolling and then gain a globular appearance. After warm-rolling, pearlite grains are very small in size and are distributed in bands parallel to the rolling direction. After warm-rolling, the steel is partially strain hardened and partially recrystallized.

The mechanical properties after warm-rolling are similar to those obtained by normalizing. The warm-rolling process is of practical interest because it has certain economical advantages as compared to both, hot- and cold-working: reduction of material losses (there are no significant oxidation or decarburization processes), superior surface quality and dimensional control, and significant energy savings (lower deformation forces required and normalizing can be eliminated). In addition, warm-rolling can be applied to a broader range of steels, higher deformation can be achieved, the deformation is more uniform in the cross–section of the rolled product and the microstructure is less strain hardened.

## REFERENCES

1. I. Hrisulakis and D. Pandelis, "Hardening methods for metallic materials", in *Science and Technology of Metallic Materials*, edited by Papasotiriu Publishing House, Athens,1996, pp. 427.
2. S. Domsa and M. Bodea,C. Prica,"Design of Construction Steels" in *Design of Materials*, edited by Casa Cartii de Stiinta, Cluj-Napoca, 2005 pp.119-131.
3. A. J.DeAdro, C.L. Garcia and E.J. Palmiere," Heat Treating", in *ASM Handbook*, Vol. 4, 1991, pp. 237-255.
4. G. Vermesan, " Guide for Heat Treating" ,in *Thermochemical Treatments* , edited by Dacia Publishing House, Cluj-Napoca, 1987 , pp. 167-234.
5. T. Dulamita, et. al., "Normalizing", in *Technology of Heat Treatments*, edited by Editura Didactica si Pedagogica, Bucharest, 1982, pp. 108-115.
6. J. K. Mac Donald, "Developments in the production of notch ductile steels", *Journal of Australian Institute of Metals*, No. 5, 1965, pp. 52-58.
7. E. R. Morgan," Improved steels through hot strip mill controlled coding", *Journal of Metals*,August, 1965, pp. 829-835.
8. D. N. Hawkins and A. A. Shuttleworth ,"The effect of warm rolling on the structure and properties of a low carbon steel", *Journal of Mechanical Working Technology*, Vol.2,1979, pp. 333-345.
9. A. Hadar, R.K.Tay, "Microstructural and textural development in a extra low carbon steel during warm rolling".Materials Science and Engineering A, Vol. 391, 2005, pp. 402-407.
10. S. Torizuka, A. Ohmori, S. V. S. Narayama Murty and K. Nagai, "Effect of Strain on the Micrustructure and Mechanical Properties of Multi-pass Warm Caliber Rolled Low Carbon Steel", *Scripta Materialia*, Vol. 54, 2006, pp.563-568.
11. S. Torizuka, E. Muramatsu, S. V. S. Narayama Murty and K. Nagai, "Microstructure Evolution and Strenght-reduction in Area Balance of Ultrafine-grained Steels Produced by Warm Caliber Rolling", *Scripta Materialia*, Vol. 55, 2006, pp.751-754.
12. S. Serajzadeh, "Modeling the Warm Rolling of a Low Carbon Steel", *Materials Science and Engineering A*, Vol.371, 2004, pp.318-323.
13. D. B. Santos, R.K. Bruzszek, P.C.M.Rodriguez and E.V.Pereloma, " Formation of Ultra-fine Ferrite Microstructure in a Warm Rolled and Annealed C-Mn Steel", *Materials Science and Engineering*, Vol. 346,2003, pp.189-195.
14. M. Liu, Bi Shi, H.Cao, X. Cai and H. Song ,"A Submicron Mild Steel Produced By Simple Warm Deformation", *Materials Science and Engineering* , Vol. 360,2003, pp.101-106.
15. C. Medrea-Bichtas,I. Chicinas, S. Domsa, "Study on Warm Rolling of AISI1015 Carbon Steel", International Journal of Materials Research and Advanced Techniques,Vol. 6, 2002, pp554-449.

# 3 – SHEET METAL FORMING

## *(P. Picart, T. Welo, and T. Meinders)*

# Parameters Controlling Dimensional Accuracy of Aluminum Extrusions Formed in Stretch Bending

Henry Ako Baringbing[1], Torgeir Welo[1,2]

[1] *Department of Engineering Design and Materials, Norwegian University of Science and Technology (NTNU), N-7491, Trondheim, Norway*
[2] *Hydro Aluminium Structures Raufoss AS, P.O. Box 15, N-2831, Raufoss, Norway*

**Abstract.**. For stretch formed components used in the automotive industry, such as bumper beams, it is of primary importance to control parameters affecting dimensional accuracy. The variations in geometry and mechanical properties induced in extrusion and stretch forming lead to subsequent dimensional inaccuracy of the final product. In this work, tensile and compression samples were taken at three different positions along AA7108W extruded profiles in order to determine material parameters for a constitutive model particularly suited for strong texture materials. In addition, geometry were measured and analyzed statistically in order to study its impact on local cross sectional distortions (sagging) and springback in stretch bending of a bumper beam. These full scale experiments were combined with analytical and numerical simulations to quantify the impact of each basic parameter on product quality. It is concluded that this methodology provides a means to systematically control the product quality by focusing on reducing the acceptance limits of the main parameters controlling basic mechanisms in stretch forming. Despite the assumptions and simplifications made in order to make the analytical expressions solvable, the approach has proven its capability in establishing accurate closed-form expressions including the main influential parameters.

**Keywords:** Stretch bending, variability, sagging, springback
**PACS:** 40, 80

## INTRODUCTION

Sheet or extruded profiles that can be formed from bending processes depend on many variables. To understand the origin of dimensional variations in selected forming processes is in order to improve production robustness for high volume applications. In most manufacturing companies, it is of great importance that the product produced has high repeatability, e.g. in geometry. Variability in geometry, due to several manufacturing steps, causes distortion from the nominal shape of the product. Aluminium is commonly used for automotive components, e.g. bumpers, engine cradles and space frames. The aluminium profile is produced in several steps, including extrusion, heat treatment, stretching, etc. The geometry of the profile varies due to die deflections and temperatures during extrusion. Bumpers are usually made from hollow square profile, being formed in bending processes. There are several bending processes used in industry, one of them is rotary stretch bending. The main challenge related to bending of extrusion is to control the variability. There are two main factors that influence the final shape of the profile after bending, i.e. springback

CP907, *10th ESAFORM Conference on Material Forming*, edited by E. Cueto and F. Chinesta
© 2007 American Institute of Physics 978-0-7354-0414-4/07/$23.00

and sagging. The aim of this paper is to study the influence of geometry and material properties on different positions to the final shape of the bumper.

## STRETCH BENDING

### Springback

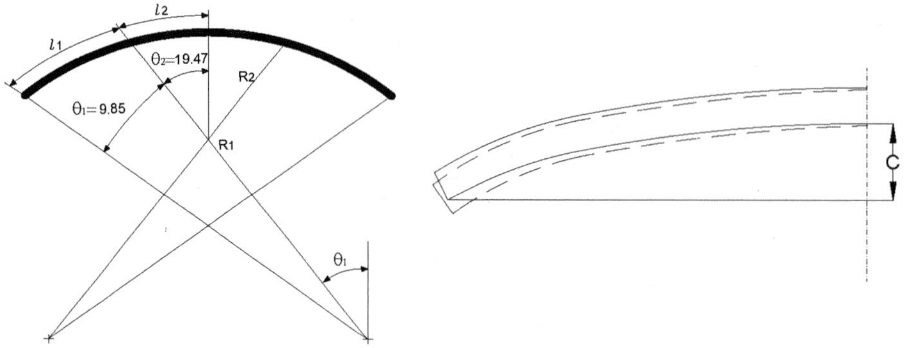

**FIGURE 1.** Dies and Springback Offset

Springback ($\Delta c$) is defined as the vertical offset of the end profile after unloading. Figure 1 shows basic dies sketch and springback offset.

$$\Delta c = \int \kappa^e m_v ds = \frac{2\Delta\sigma}{Eh}\left\{\left[R_1^2\left(\theta_1\sin\theta_1+\cos\theta_1-1\right)\right]+\left[R_2^2\left(\theta_1\sin\theta_2+\cos\theta_2-1\right)\right]\right\} \quad (1)$$

where $E$ is elastic modulus, $h$ is the height of the profile, $R_1 \& R_2$ are radii of curvature; 1600 mm and 450 mm respectively. The strain of the profile at the upper flange and $\Delta\sigma$ can be defined as,

$$\varepsilon = \varepsilon_{min} + \frac{h}{2(R+0.5h)} = \frac{l-l_0}{l} + \frac{h}{2(R+0.5h)} \quad (2)$$

$$\Delta\sigma = \frac{\sigma_p - \sigma_0}{2} = \frac{K\varepsilon_p^n - \sigma_0}{2} = \frac{1}{2}\left(K\left(\frac{l-l_0}{l}+\frac{h}{2(R+0.5h)}\right)^n - \sigma_0\right) \quad (3)$$

where $\sigma_0$ is the yield strength of material, $K$ is strength coefficient, $n$ is strain hardening coefficient, $\varepsilon_p$ is plastic strain, $\sigma_p$ is flow stress which follows Ludwik-Holomon material model; $\sigma = K\varepsilon_p^n$.

### Sagging

Torgeir Welo [1] had developed theoretical model of sagging for pure and stretch bending with deformation theory of plasticity. The maximum sagging depth ($w$) is inferenced by geometry shape and material properties. After bending, the shape of the profile is distorted, see Figure 2.

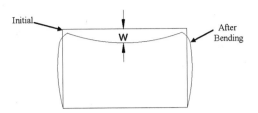

**FIGURE 2.** Sagging on Profile

Sagging depth can be defined as,

$$w = c_1 c_2 c_3 \frac{hb^4}{R^2 t_f^2} \qquad (4)$$

where $w$ is sagging depth, $c_1$, $c_2$, & $c_3$ on equation (5) and (6) are sagging coefficient. Anisotropy value ($\bar{r}$) is given on equation (5).

$$c_1 = \frac{2\bar{r}+1}{\bar{r}+2} \left( \frac{1}{3}\sqrt{2(\bar{r}+2)} \right)^{1-n}, \quad \bar{r} = \frac{r_0 + 2r_{45} + r_{90}}{4} \qquad (5)$$

$$c_2 = \frac{(1-\cos\theta) - \dfrac{w}{2R}}{\dfrac{0.5}{1-0.27\theta^2}\sin^2\theta}, \quad c_3 = \frac{3}{256}\left[ 5 - \frac{4}{\left(\dfrac{t_f}{t_w}\right)^2 \dfrac{1}{2-n}\dfrac{h}{b}+1} \right] \qquad (6)$$

## EXPERIMENTAL SETUP

In rotary stretch bending, the profile (workpiece) is placed on top of the dies and clamped. A press is used to rotate the die about the pivot points, see Figure 3. Stretch bending tests were done by clamping the profile's end to the dies and then applying bending moment from the press. Effect of stretching can be gained because the end profiles clamped and follow dies rotation ($29.3^0$). The tension force and bending moment are applied at the same time. The profile is stretched slightly beyond elastic limit. Tests are designed with dies R450 2% strain along the internal flange. The set up of the dies as follows,

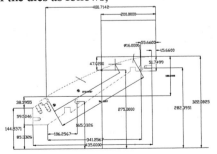

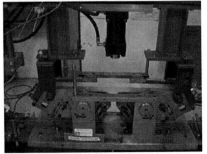

**FIGURE 3.** Dies Geometry and Test Rig

# NUMERICAL SIMULATION PROCESSES

## Material Parameter

Extruded aluminium alloys typically have strong texture. An anisotropic yield criterion YLD2003 is well suited for the present purpose. Material models, which have been implemented as user subroutines in LS-DYNA, was used; Strong Texture Model (STM). The STM material model uses anisotropic yield criterion, isotropic elasticity, isotropic strain hardening, strain-rate hardening, and instability criterion, see Berstad et al. [3]. The yield criterion YLD2003 of Aretz [4] is defined as,

$$2\overline{f} = \left|\sigma_1'\right|^m + \left|\sigma_2'\right|^m + \left|\sigma_1'' - \sigma_2''\right|^m \tag{7}$$

where $m$ is an integer that is proposed to 8 for FCC material. The yield criterion contains eight parameters ($a_1$, $a_2$, ... $a_8$) and $\sigma_j'$ and $\sigma_k''$ are the principal values of two linear transformations of the stress deviator. In the present work, the flow stress is defined with Voce hardening relation,

$$\sigma_y = \sigma_0 + \sum_{i=1}^{2} Q_i \left(1 - \exp\left(-C_i \overline{\varepsilon}\right)\right) \tag{8}$$

where $\sigma_0$ is the yield stress, $Q_i$ and $C_i$ is strain hardening constant, and $\overline{\varepsilon}$ is effective plastic strain. Voce parameter for $\sigma_0$, $Q_i$, $C_i$ ($i$=1, 2) in Eq. (8) were found by curve fitting true stress-strain curves in $0^0$ direction, which is chosen as reference direction.

Samples were taken from two dies and three different positions; back (b), middle (m) and front (f), see Figure 4.

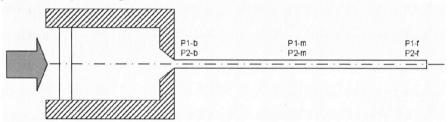

**FIGURE 4.** Extrusion Processes

Table 1 shows the results from material characteristic tests. Ludwik-Holomon material data is used for analytical calculation, and STM material data is used for simulation calculation. Material characteristic test procedures for STM material are given by Reyes [5].

**TABLE 1. Material Parameter**

| Dimension | | Dies 1 | | | Dies 2 | | |
|---|---|---|---|---|---|---|---|
| | | P1-b | P1-m | P2-f | P2-b | P2-m | P2-f |
| 1 Thickness | t | 3,99 | 3,91 | 3,85 | 4,01 | 4,00 | 4,01 |
| 2 Width | w | 40,11 | 40,78 | 40,15 | 40,12 | 40,07 | 40,12 |
| 3 Height | h | 59,95 | 60,42 | 60,10 | 60,10 | 60,12 | 60,10 |
| 4 Mod. Elasticity | E | 70000 | 70000 | 70000 | 70000 | 70000 | 70000 |

| | Ludwik - Halomon Material | | | | | | | |
|---|---|---|---|---|---|---|---|---|
| 5 | Hardening coef. (n) | 0 deg | 0,3371 | 0,3429 | 0,3447 | 0,3469 | 0,3466 | 0,3317 |
| 6 | Strength coef. (K) | 0 deg | 447,28 | 437,13 | 442,84 | 443,09 | 447,29 | 450,14 |
| 7 | Yield strength ($\sigma_0$) | 0 deg | 83,73 | 81,96 | 80,97 | 83,47 | 83,25 | 86,51 |
| | STM Material | | | | | | | |
| 8 | Flow stress ratio | $R_0$ | 1 | 1 | 1 | 1 | 1 | 1 |
| | (FSR) | $R_{45}$ | 0,915 | 0,96 | 0,946 | 0,975 | 0,972 | 0,934 |
| | | $R_{90}$ | 1,06 | 1,143 | 1,099 | 1,11 | 1,086 | 1,061 |
| 9 | Compression ratio | 1/rb | 2,313 | 3,195 | 3,368 | 3,547 | 3,864 | 3,289 |
| 10 | Lankford Parameter | $r_0$ | 0,319 | 0,289 | 0,302 | 0,297 | 0,268 | 0,268 |
| | (Anisotropy coef.) | $r_{45}$ | 1,657 | 1,983 | 1,834 | 1,769 | 1,739 | 1,772 |
| | | $r_{90}$ | 1,194 | 1,067 | 1,008 | 1,124 | 1,11 | 1,123 |
| | | $\bar{r}$ | 1,207 | 1,331 | 1,245 | 1,240 | 1,214 | 1,234 |
| 11 | Hardening | $\sigma_0$ | 58,881 | 77,04 | 74,425 | 77,188 | 53,899 | 78,120 |
| | coefficient | $Q_1$ | 23,009 | 29,62 | 16,225 | 5,201 | 26,874 | 15,977 |
| | | $C_1$ | 2091,538 | 29,99 | 65,908 | 878,259 | 3742,616 | 86,535 |
| | | $Q_2$ | 198,114 | 193,63 | 204,011 | 208,418 | 203,928 | 202,546 |
| | | $C_2$ | 10,642 | 6,65 | 7,899 | 8,267 | 9,362 | 8,647 |
| 12 | YLD 2003 coefficient | $a_1$ | 0,897 | 0,863 | 0,910 | 0,894 | 0,913 | 0,920 |
| | | $a_2$ | 1,029 | 1,004 | 1,059 | 1,045 | 1,080 | 1,080 |
| | | $a_3$ | 0,938 | 0,873 | 0,909 | 0,891 | 0,912 | 0,934 |
| | | $a_4$ | 1,133 | 0,994 | 1,048 | 0,991 | 1,000 | 1,082 |
| | | $a_5$ | 0,898 | 0,906 | 0,909 | 0,907 | 0,894 | 0,887 |
| | | $a_6$ | 0,965 | 0,881 | 0,910 | 0,909 | 0,929 | 0,956 |
| | | $a_7$ | 1,117 | 1,079 | 1,089 | 1,060 | 1,059 | 1,110 |
| | | $a_8$ | 1,089 | 1,108 | 1,052 | 1,066 | 1,040 | 1,044 |

# RESULT

Sagging and springback were measured with optical measurement technique, see [2] for further information. Figure 5 shows springback and sagging results. It shows that springback analytical, simulation and experimental results are in the same tendency; springback at the back profile is a higher than other positions. Springback analytical solutions are still far from experiment results.

Sagging results show the same tendency, except for analytical solutions, where at the middle position; the sagging is less than other positions. Anisotropy factors at the middle position for dies 1 are higher than other position.

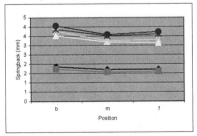 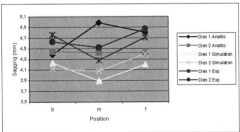

FIGURE 5. Springback and Sagging result

Anisotropy factor contributes to different sagging result. Flow stress ratio is relatively constant in all position, see Figure 6.

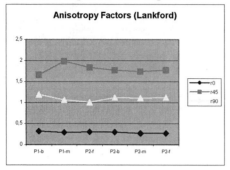
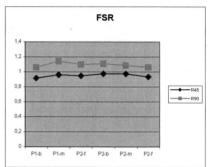

**FIGURE 6.** Anisotropy Factors and FSR (Flow Stress Ratio)

## CONCLUSION

The results, discussed in terms of response surfaces, show that cross sectional (local) dimensions is mainly controlled by geometry parameters and anisotropy of the material, whereas the overall bend geometry is mainly controlled by process and material's yield stress and hardening.

## ACKNOWLEDGEMENT

The financial support of Norwegian University of Science and Technology, Hydro Aluminium and Norwegian Research Council is grateful acknowledged. The assistance from Dr. Ing Odd-Geir Lademo for STM material models is of great value. I wish to thank Odd Perry Søvik for his support in material for stretch bending test.

## REFERENCES

1. T. Welo., Bending of Alumunium Extrusion, 1995, Sintef Report.
2. P. T. Moe, "Optical Measurement Technology For Aluminium Extrusions." In Preceding of the 10th ESAFORM conference 2007, Zaragosa, Spain.
3. T. Berstad, O.-G. Lademo, O.S. Hopperstad, O.S., Weak and Strong Texture Models in LS-Dyna (WTM-2D/3D and STM-2D), 2005, Sintef Report.
4. H. Aretz, "Applications of a new plane stress yield function to orthotropic steel and aluminium sheet metals", Modell. Simulat. Mater. Sci. Eng. 12 (2004) 491–509.
5. Reyes, A., et al, "Modeling of textured aluminium alloys used in a bumper system: Material tests and characterization", Computational Material Science, 2005.

# The Influence of the Welding Line Placement on Springback Effect in Case of a U-shape Part Made from Tailor Welded Stripes

Albut Aurelian

*University of Bacau, 157 Marasesti Street, 600115 Bacau, Romania*

**Abstract.** This paper deals with some numerical simulation and experimental tests related to forming and springback of a U-shaped part manufactured from tailor welded stripes. Final shape of the formed part manufactured by tailor welded stripes is seriously affected by springback effect. The materials tailored in the welded stripes have different springback values. This paper work is trying to prove out the important role of welding line placement has on the springback phenomenon. The influence of the welding line placement on the tailor welded stripes springback is examined using the simulation by finite element method (ABAQUS). Experimental tests have been carried out using different positions of the welding line and maintaining constant all other parameters. The resulted parts were measured using a 3D scanning machine. A comparison between simulation and experiment results is presented in the final section of this paper.

**Keywords:** tailor welded stripes, forming, springback, welding line.
**PACS:** 81.20.Hy, 81.20.Vj

## INTRODUCTION

A tailor welded stripe consists of two sheets that have been welded together in a single plane prior to forming. The sheets joined by welding can be identical, or they can have different thickness, mechanical properties or surface coatings. Various welding processes, i.e. laser welding, mash welding, electron-beam welding or induction welding, can join them. And, the techniques of numerical analysis applicable for sheet metal forming have been considerably developed for the last several years. However, accurate prediction of the springback remains elusive. Many studies presents a wide range of information about the formability and failure patterns of welded stripes. A wide range of information about the formability and failure patterns of tailor welded stripes and the springback of non-welded sheet metal parts has been presented. However, the springback characteristics of tailor-welded stripes have hardly been found. Published results on springback prediction of tailor welded stripes are minimal. The welding line was insignificant influence when is placed perpendicular to the direction of the deformation force.

In case of U-shape forming of tailor welded stripes, the welding line placement has great influence on springback phenomenon, even the welding line is placed perpendicular to the direction of the deformation force.

CP907, *10th ESAFORM Conference on Material Forming*, edited by E. Cueto and F. Chinesta
© 2007 American Institute of Physics 978-0-7354-0414-4/07/$23.00

In this study, the tailor welded stripes (joined together without taking in consideration the mechanical properties of the welding line) with two types of material having the same thickness bud different mechanical properties, are used to investigate springback characteristics in U-shape forming.

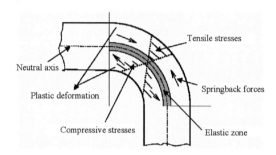

**FIGURE 1.** Bending of sheet metal

Springback (figure 1) is mainly influenced by the punch and die profile radii, initial clearance between punch and die, friction conditions, rolling direction of the materials, blankholder force, material properties (elastic modulus, Poisson's coefficient, constitutive behavior in plastic field) etc. The purpose of this study was to investigate the welding line placement influence on the springback effect of the tailor welded stripes. To achieve this goal, simulation and experimental test were carried with different positions of the welding line.

## EXPERIMENTAL RESEARCHES REGARDING THE WELDING LINE INFLUENCE

The tailor welded stripes used in the experiments were made from FEPO and E220 steel. Strips of 350×30 mm dimensions and 0.7 mm thickness were cut from the metal sheet along to the material rolling direction (fig. 2).

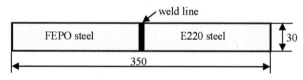

**FIGURE 2.** Tailor welded strips for U-shape forming (unit: mm)

The mechanical properties of FEPO steel and E220 steel determined for 0° material rolling direction are presented in table 1.

TABLE 1. Mechanical properties of the base materials

| Material name | Young modulus MPa | Tensile strength MPa | Uniform Elongation % | Total Elongation % | Plastic strain ratio r | Strain-hardening coefficient n |
|---|---|---|---|---|---|---|
| FEPO | 200 825 | 281 | 17.3 | 28,8 | 1.86 | 0.234 |
| E220 | 204 000 | 348 | 10.2 | 20,4 | 1.42 | 0,190 |

134

# Experimental Layout

The experimental tests were realized using a die for rectangular parts that allowed utilization of different blank holder forces. The device is presented in figure 3. Tool geometry is presented in table 3. The experimental tests have been done with varied welding line placement with respect to the part axis. The blankholder force was maintained constant to 10 kN. The forming force was generated using a mechanical tensile test machine. The profile of the obtained part and the parameters of springback were measured with a numerical controlled scanning machine Roland Model MDX-15 (Fig. 4), and the obtained data was processed in CAD software.

**TABLE 3.** Die geometric parameters

| | |
|---|---|
| Punch geometry (mm) | 78×120 |
| Punch profile radius (mm) | 10 |
| Die opening (mm) | 80 |
| Die profile radius (mm) | 5 |
| Punch stroke (mm) | 50 |

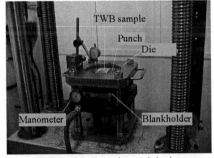

**FIGURE 3.** Experimental device

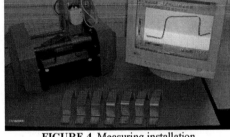

**FIGURE 4.** Measuring installation

Springback parameters that were observed during the tests are presented in figure 5:
- $\theta_1$ – sidewall angle between real profile and theoretical profile;
- $\theta_2$ – flange angle between real profile and theoretical profile;
- $\rho$ – curvature radius of the sidewall.

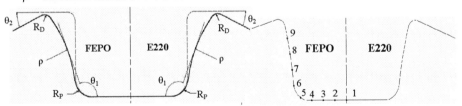

**FIGURE 5.** Geometrical springback parameters    **FIGURE 6.** Welding line position after forming

# Experimental Results

In order to experimentally determine the influence of the welding line placement on springback parameters, the samples have been cut along to the material rolling direction (RD is perpendicular to the weld line). Initially the welding line was placed

135

on the central axis of the part, on the following tests the welding line was moved each time toward to the FEPO steel with 8.7 mm. To minimize the influence of the blank holder force, its value was constant at 10kN. The forming tests have been done with lubrication of the tools and of the tailor welded stripes sample. The values of springback parameters are recorded in table 4.

TABLE 4. Springback parameters

| Welding line displacement [mm] | Zone of the part made from FEPO steel | | | | | | Zone of the part made from E220 steel | | | | | |
|---|---|---|---|---|---|---|---|---|---|---|---|---|
| | Angle $\theta_1$ [grd] | | Angle $\theta_2$ [grd] | | Sidewall Radius [mm] | | Angle $\theta_1$ [grd] | | Angle $\theta_2$ [grd] | | Sidewall Radius [mm] | |
| | Ideal | Measured | Ideal | Measured | Ideal | Measured | Ideal | Measured | Ideal | Measured | Ideal | Measured |
| 0.00 | | 99.14 | | 13.64 | | 203.72 | | 101.31 | | 18.43 | | 103.31 |
| 8.75 | | 99.31 | | 13.67 | | 203.96 | | 101.29 | | 18.41 | | 103.37 |
| 16.5 | | 99.23 | | 13.79 | | 203.86 | | 101.42 | | 18.45 | | 103.39 |
| 26.25 | | 99.25 | | 13.84 | | 203.84 | | 101.45 | | 18.43 | | 103.34 |
| 35.00 | 90 | 100.75 | 0 | 14.31 | ∞ | 191.59 | 90 | 101.35 | 0 | 18.42 | ∞ | 103.32 |
| 43.75 | | 101.45 | | 16.21 | | 182.08 | | 101.37 | | 18.44 | | 103.33 |
| 52.50 | | 102.53 | | 17.62 | | 170.21 | | 101.41 | | 18.41 | | 103.34 |
| 61.25 | | 102.81 | | 18.37 | | 160.79 | | 101.39 | | 18.39 | | 103.36 |
| 70.00 | | 103.12 | | 19.12 | | 151.00 | | 101.42 | | 18.38 | | 103.38 |

# SIMULATION RESEARCHES REGARDING THE WELDING LINE INFLUENCE

The simulation of U-shape part forming was made using finite element method. The objective is to create a model that allows an accurate prediction of springback intensity, stress and strain state at the end of the forming process. The analyzed geometrical parameters are sidewall radius $\rho$ and springback angles $\theta_1$ and $\theta_2$. In order to validate the model, the results of the FE analysis were compared with the experimental results.

## Simulation Methodology

The simulations considered a plane strain state. The material was modelled as elastic-plastic, where elasticity is considered isotropic and plasticity is modelled as anisotropic using Hill quadratic anisotropic yield criterion.

The geometrical model is presented in figure 4. The initial dimensions of the sheet were 350 mm length, 30 mm width and 0.7 mm thick. The sheet was modelled as deformable body with 400 shell elements (S4R) on one row with 5 integration points through the thickness. The tools (punch, die and blankholder) were modelled as analytical rigid because they have the advantage of reduced calculus efforts and a good contact behaviour. Rigid body movements are controlled by reference points.

## Simulation Results

As in the experimental tests, initially the welding line was place exactly on the central axis of the part. After that, the welding line was moved toward the FEPO steel area incrementing each time with 8.75 mm. The last simulation was made with a welding line displacement of 70 mm with respect to the central axis of the part, toward the part area made from FEPO steel, more E220 steel flows into the forming process. The influence of the welding line position on the parameters of springback is illustrated in figure 7.

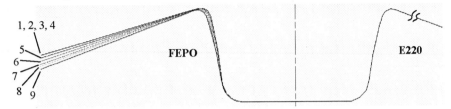

**FIGURE 7.** Influence of the welding line position on springback of tailor welded strips

The variations of springback parameters ($\theta_1$, $\theta_2$, $\rho$) as a function of the welding line position are graphically presented in figures no 8, 9, 10 and recorded in table 5.

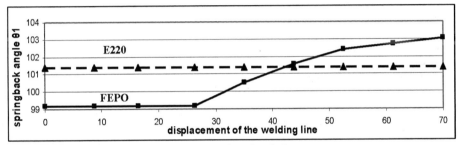

**FIGURE 8.** Variation of angle $\theta_1$

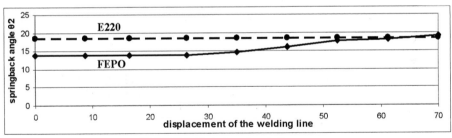

**FIGURE 9.** Variation of angle $\theta_2$

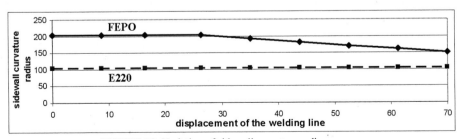

**FIGURE 10.** Variation of sidewall curvature radius $\rho$

TABLE 5. Springback parameters

| Welding line displacement [mm] | Zone of the part made from FEPO steel | | | | | | Zone of the part made from E220 steel | | | | | |
|---|---|---|---|---|---|---|---|---|---|---|---|---|
| | Angle $\theta_1$ [grd] | | Angle $\theta_2$ [grd] | | Sidewall Radius [mm] | | Angle $\theta_1$ [grd] | | Angle $\theta_2$ [grd] | | Sidewall Radius [mm] | |
| | Ideal | Measured | Ideal | Measured | Ideal | Measured | Ideal | Measured | Ideal | Measured | Ideal | Measured |
| 0.00 | | 99.20 | | 13.80 | | 203.93 | | 101.4 | | 18.4 | | 103.34 |
| 8.75 | | 99.20 | | 13.80 | | 203.93 | | 101.4 | | 18.4 | | 103.34 |
| 16.5 | | 99.20 | | 13.80 | | 203.93 | | 101.4 | | 18.4 | | 103.34 |
| 26.25 | | 99.20 | | 13.80 | | 203.93 | | 101.4 | | 18.4 | | 103.34 |
| 35.00 | 90 | 100.51 | 0 | 14.68 | $\infty$ | 191.64 | 90 | 101.4 | 0 | 18.4 | $\infty$ | 103.34 |
| 43.75 | | 101.56 | | 16.15 | | 182.15 | | 101.4 | | 18.4 | | 103.34 |
| 52.50 | | 102.39 | | 17.62 | | 170.17 | | 101.4 | | 18.4 | | 103.34 |
| 61.25 | | 102.72 | | 18.14 | | 160.86 | | 101.4 | | 18.4 | | 103.34 |
| 70.00 | | 103.03 | | 19.06 | | 151.05 | | 101.4 | | 18.4 | | 103.34 |

# CONCLUSIONS

It can be considered that the results generated by the analysis of springback phenomenon using finite element method are sufficiently accurate and can be considered valid (table 6). When properly used, simulation by finite element method can be considered a valuable tool in the study of the influencing factors of the springback phenomenon able to offer accurate data even from the design stage.

TABLE 6. Error between experimental and simulation results for the springback parameters

| Welding line displacement [mm] | Zone of the part made from FEPO steel | | | Zone of the part made from E220 steel | | |
|---|---|---|---|---|---|---|
| | Angle $\theta_1$ [grd] | Angle $\theta_2$ [grd] | Sidewall Radius [mm] | Angle $\theta_1$ [grd] | Angle $\theta_2$ [grd] | Sidewall Radius [mm] |
| 0.00 | -0.06 | -0.16 | -0.21 | -0.09 | 0.03 | -0.03 |
| 8.75 | 0.11 | -0.13 | 0.03 | -0.11 | 0.01 | 0.03 |
| 16.5 | 0.03 | -0.01 | -0.07 | 0.02 | 0.05 | 0.05 |
| 26.25 | 0.05 | 0.04 | -0.09 | 0.05 | 0.03 | 0.00 |
| 35.00 | 0.24 | -0.37 | -0.05 | -0.05 | 0.02 | -0.02 |
| 43.75 | -0.11 | 0.06 | -0.07 | -0.03 | 0.04 | -0.01 |
| 52.50 | 0.14 | 0.00 | 0.04 | 0.01 | 0.01 | 0.00 |
| 61.25 | 0.09 | 0.23 | -0.07 | -0.01 | -0.01 | 0.02 |
| 70.00 | 0.09 | 0.06 | -0.05 | 0.02 | -0.02 | 0.04 |

Analyzing the variations of springback parameters ($\theta_1$, $\theta_2$, $\rho$), obtained experimentally and by simulation, the following observation can be presented:

- positioning of the welding line on the bottom of the part has no influence on the springback parameters;
- placement of the welding line on the punch radius and on the part side wall determine an important grow of the springback parameters in the area of the part made from FEPO steel;
- displacement of the welding line cause insignificant variation of the springback parameters in the area of the part made from E220 steel.

# REFERENCES

1. L. Papeleux, J.P. Ponthot, "Finite element simulation of springback in sheet metal forming", J. of Mat. Proc. Tech., 2002, pp. 125-126.
2. A. Albut, "Springback of Tailor Welded Blaanks", Ph.D. Thesis, Politehnica University Bucharest, 2006.
3. E. Chu, "Springback in plane strain stretch/draw sheet forming", Int. J. Mech. Sci., vol. 36, no. 3, 1995.

# A finite strain isotropic/kinematic hardening model for springback simulation of sheet metals

Ivaylo N. Vladimirov and Stefanie Reese

*Institute of Solid Mechanics, Braunschweig University of Technology,*
*D-38106 Braunschweig, Germany*

**Abstract.** Crucial for the accurate prediction of the blank springback is the use of an appropriate material model, which is capable of modelling the typical cyclic hardening behaviour of metals (e.g. Bauschinger effect, ratchetting). The proposed material model combines both nonlinear isotropic hardening and nonlinear kinematic hardening, and is defined in the finite strain regime. The kinematic hardening component represents a continuum extension of the classsical rheological model of Armstrong-Frederick kinematic hardening. The evolution equations of the model are integrated by a new form of the exponential map algorithm, which preserves the plastic volume and the symmetry of the internal variables. Finally, the applicability of the model for springback prediction has been demonstrated by performing simulations of the draw-bending process.

**Keywords:** finite strains, kinematic and isotropic hardening, exponential map, springback
**PACS:** 46.15.-x; 46.35.+z

## INTRODUCTION

Springback of the blank after unloading and removal from tooling causes major problems in the metal forming industry. It presents difficulties during assembly and complicates the die design process because the required die rework is elaborate and expensive. Therefore, the development of material models enabling the realistic simulation of springback is certainly of primary interest in academic and industrial research.

A number of publications dealing with springback and its modelling have appeared in the literature lately. Some of them (e.g. Chung et al. [2], Wang et al. [14], Yoshida & Uemori [16]) focus on material modelling aspects related to springback, whereas others (e.g. Li et al. [6], Ragai et al. [8]) present experimental investigations. Several other more theoretical papers (e.g. Choi et al. [1], Dettmer & Reese [3], Hakansson et al. [4], Menzel et al. [7], Svendsen et al. [11], Wallin et al. [13]) propose models for kinematic hardening, which is the decisive model component in simulating springback.

In this work, we present an extension to the finite strain kinematic hardening model (see Reese & Vladimirov [10], Vladimirov & Reese [12]) with nonlinear isotropic hardening. The model is derived in a thermomechanically-consistent format. Further features of the model include the multiplicative split of the plastic part of the deformation gradient, and a strain-like internal variable to describe kinematic hardening.

The numerical implementation of the model deals with the integration of the evolution equations for the internal variables. The algorithm used here represents a new form of the exponential map based on an implicit time integration scheme. It automatically fulfills plastic incompressibility in every time step, and has the advantage of retaining the symmetry of the internal variables.

CP907, *10th ESAFORM Conference on Material Forming,* edited by E. Cueto and F. Chinesta

# CONSTITUTIVE MODELLING

To start with, the constitutive model is based on the multiplicative split $\mathbf{F}_p = \mathbf{F}_{p_e}\mathbf{F}_{p_i}$ of the plastic deformation gradient into "elastic" and "inelastic" parts, $\mathbf{F} = \mathbf{F}_e\mathbf{F}_p$ being the classical multiplicative split of $\mathbf{F}$. Based on the principle of material objectivity, the Helmholtz free energy depends on the deformation only through the elastic right Cauchy-Green tensors $\mathbf{C}_e$ and $\mathbf{C}_{p_e}$ and has the form

$$\psi = \psi_e\left(\mathbf{C}_e\right) + \psi_{kin}\left(\mathbf{C}_{p_e}\right) + \psi_{iso}\left(\kappa\right) \tag{1}$$

where the term $\psi_{iso}\left(\kappa\right)$ represents the amount of stored energy due to isotropic hardening, and $\kappa$ is the isotropic hardening variable. Inserting the form of the Helmholtz free energy (1) into the Clausius-Duhem inequality $-\dot\psi + \mathbf{S} \cdot \frac{1}{2}\dot{\mathbf{C}} \geq 0$, and making use of several tensor calculus rules, one obtains the reduced form of the entropy inequality

$$\left(\mathbf{M} - \chi\right) \cdot \mathbf{d}_p + \mathbf{M}_{kin} \cdot \mathbf{d}_{p_i} + R\dot\kappa \geq 0 \tag{2}$$

where $\mathbf{M}$ and $\mathbf{M}_{kin}$ are the so-called Mandel stresses, $\mathbf{d}_p$ and $\mathbf{d}_{p_i}$ are the plastic rate-of-deformation tensor and the inelastic plastic rate-of-deformation tensor, respectively, and $R = -\partial\psi_{iso}/\partial\kappa$ is the stress-like isotropic hardening variable. This inequality is sufficiently satisfied by the evolution equations:

$$\mathbf{d}_p = \dot\lambda\,\frac{\partial\Phi}{\partial\mathbf{M}} = \dot\lambda\,\frac{\mathbf{M}^D - \chi^D}{\|\mathbf{M}^D - \chi^D\|} \tag{3}$$

$$\mathbf{d}_{p_i} = \dot\lambda\,\frac{b}{c}\,\mathbf{M}^D_{kin} \tag{4}$$

$$\dot\kappa = \sqrt{\frac{2}{3}}\dot\lambda, \quad R = -Q(1 - e^{-\beta\kappa}) \tag{5}$$

The evolution equations (3) and (4) have to be transformed to the reference configuration. Finally, the set of constitutive equations of the model is summarized below:

- Stress tensors

$$\mathbf{S} = 2\mathbf{F}_p^{-1}\frac{\partial\psi_e}{\partial\mathbf{C}_e}\mathbf{F}_p^{-T}, \quad \mathbf{X} = 2\mathbf{F}_{p_i}^{-1}\frac{\partial\psi_{kin}}{\partial\mathbf{C}_{p_e}}\mathbf{F}_{p_i}^{-T}, \quad \mathbf{Y} = \mathbf{CS} - \mathbf{C}_p\mathbf{X}, \quad \mathbf{Y}_{kin} = \mathbf{C}_p\mathbf{X} \tag{6}$$

- Evolution equations

$$\dot{\mathbf{C}}_p = 2\dot\lambda\,\frac{\mathbf{Y}^D\mathbf{C}_p}{\sqrt{\mathbf{Y}^D \cdot (\mathbf{Y}^D)^T}}, \quad \dot{\mathbf{C}}_{p_i} = 2\dot\lambda\,\frac{b}{c}\,\mathbf{Y}^D_{kin}\mathbf{C}_{p_i}, \quad \dot\kappa = \sqrt{\frac{2}{3}}\dot\lambda \tag{7}$$

- Yield function

$$\Phi = \sqrt{\mathbf{Y}^D \cdot (\mathbf{Y}^D)^T} - \sqrt{\frac{2}{3}}(\sigma_y - R), \quad R = -Q(1 - e^{-\beta\kappa}) \tag{8}$$

- Kuhn-Tucker conditions

$$\dot{\lambda} \geq 0, \quad \Phi \leq 0, \quad \dot{\lambda}\Phi = 0 \tag{9}$$

Obviously, all constitutive equations in the reference configuration are represented in terms of the arguments $\mathbf{C}, \mathbf{C}_p, \mathbf{C}_{p_i}$ and the plastic multiplier $\dot{\lambda}$. Thus, $\mathbf{C}_p$ and $\mathbf{C}_{p_i}$ play the role of internal variables of the model, describing the evolution of plastic deformation and the evolution of kinematic hardening, respectively.

## NUMERICAL IMPLEMENTATION

It is widely accepted in the literature (see e.g. Weber & Anand [15], Reese & Govindjee [9]) that the exponential map algorithm is a suitable tool for the integration of evolution equations of deviatoric character. Its main advantage is the fact that it exactly preserves plastic incompressibility, i.e., $\det \mathbf{F}_p = 1$.

Consider the plastic flow rule (7), written here in the following more general format

$$\dot{\mathbf{C}}_p = \dot{\lambda}\,\mathbf{f}(\mathbf{C}, \mathbf{C}_p, \mathbf{C}_{p_i}) = \dot{\lambda}\,\mathbf{g}(\mathbf{C}, \mathbf{C}_p, \mathbf{C}_{p_i})\,\mathbf{C}_p \tag{10}$$

where

$$\mathbf{f} = 2\,\frac{\mathbf{Y}^D\,\mathbf{C}_p}{\sqrt{\mathbf{Y}^D \cdot (\mathbf{Y}^D)^T}} = \mathbf{f}\mathbf{C}_p^{-1}\,\mathbf{C}_p = \mathbf{g}\mathbf{C}_p \tag{11}$$

Since $\dot{\mathbf{C}}_p$ is a symmetric tensor, it follows that also $\mathbf{f}$ is symmetric. The same, however, does not hold for $\mathbf{g}$, i.e., $\mathbf{g} \neq \mathbf{g}^T$. If the function $\dot{\lambda}\,\mathbf{g} = \mathbf{g}_0$ were constant, one could solve the differential equation $\dot{\mathbf{C}}_p = \mathbf{g}_0\,\mathbf{C}_p$ analytically. In general, however, $\dot{\lambda}\,\mathbf{g}$ is a nonlinear, time-dependent function, and an analytical solution is not available. Therefore, for an implicit integration scheme $\dot{\lambda}\,\mathbf{g}$ is kept piecewise constant in the time interval $\Delta t = t_{n+1} - t_n$ and approximated by the value at the end of the interval.

Due to the fact that $\mathbf{g}$ is not symmetric, direct use of the exponential map will result in a violation of the symmetry requirements:

$$\mathbf{C}_{p_{n+1}}^T = \mathbf{C}_{p_{n+1}} \neq \exp\left(\Delta t\,\dot{\lambda}_{n+1}\,\mathbf{g}_{n+1}\right)\mathbf{C}_{p_n} \tag{12}$$

The symmetry of $\mathbf{C}_{p_{n+1}}$ could be guaranteed if $\mathbf{g}_{n+1}$ were symmetric and the tensors $\mathbf{g}_{n+1}$ and $\mathbf{C}_{p_n}$ were coaxial. Thus, in order to retain the symmetry of the internal variable, additional assumptions are required.

A solution to preserve the symmetry was suggested by Dettmer & Reese [3], where the form (12) was multiplied with $\mathbf{C}_{p_n}^{-1}\,\mathbf{C}_p$ from the right to give

$$\mathbf{C}_p\mathbf{C}_{p_n}^{-1}\mathbf{C}_p = \exp(\bar{\lambda}\,\mathbf{g})\,\mathbf{C}_p \tag{13}$$

In the above, and also from now on, the index $n+1$ for the quantities evaluated at time $t_{n+1}$, has been omitted. In addition, $\bar{\lambda}$ is defined as $\bar{\lambda} := \dot{\lambda}\,\Delta t$.

This integration rule requires the use of a series representation because the exponent of a non-symmetric tensor has to be computed. If the exponential function had a symmetric tensor as argument, it could be represented by means of a spectral decomposition. One can show that $\exp(\lambda \mathbf{g})$ can be replaced by $\mathbf{U}_p \exp(\lambda \mathbf{U}_p^{-1} \mathbf{f} \mathbf{U}_p^{-1}) \mathbf{U}_p^{-1}$ in Equation (13) to obtain the final form of the integrated plastic flow rule:

$$\mathbf{C}_p \mathbf{C}_{p_n}^{-1} \mathbf{C}_p = \mathbf{U}_p \exp(\bar{\lambda} \mathbf{U}_p^{-1} \mathbf{f} \mathbf{U}_p^{-1}) \mathbf{U}_p$$

$$\Rightarrow \mathbf{C}_{p_n}^{-1} = \mathbf{U}_p^{-1} \exp(\bar{\lambda} \mathbf{U}_p^{-1} \mathbf{f} \mathbf{U}_p^{-1}) \mathbf{U}_p^{-1} \tag{14}$$

The integration of the evolution equation for kinematic hardening is treated analogously. Due to the symmetry of the internal variables, the resulting system of equations consists of 13 nonlinear scalar equations, which have to be solved iteratively by means of e.g. the Newton method. The internal variables that have to be updated during the iteration are $\mathbf{U}_p^{-1}$, $\mathbf{U}_{p_i}^{-1}$ and $\lambda$.

## EXAMPLES

### Gauss point investigations

In this work, a large deformation elastoplastic material model with combined nonlinear kinematic and isotropic hardening has been presented. It should be capable of describing the cyclic hardening behaviour and the Bauschinger effect. The latter means that straining in one direction reduces the yield stress in the opposite direction.

In Fig 1(a), two cycles of uniaxial tension/compression are shown. The material parameters used for the numerical tests read: $\mu = 80000$ MPa, $\Lambda = 119999.67$ MPa, $\sigma_y = 300$ MPa, $c = 1900$ MPa, $b = 8.5$, $Q = 400$ MPa, $\beta = 2.5$. The Bauschinger effect is clearly visible and due to the presence of nonlinear isotropic hardening, the cyclic hardening is also correctly displayed. Figure 1(b) depicts one cycle of simple shear. It should be noted that both for uniaxial tension and for simple shear logarithmic strains of more than 50 % have been computed, clearly showing the applicablity of the material model for large deformation problems.

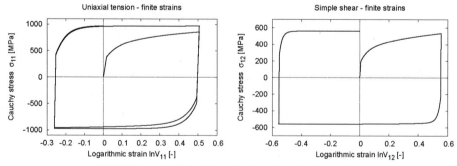

**FIGURE 1.**   (a) Uniaxial tension (2 cycles), (b) simple shear

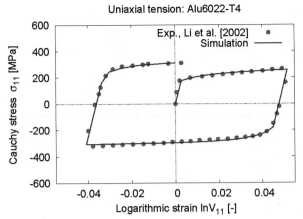

**FIGURE 2.** Aluminium Alu6022-T4: simulation vs. experiment

Next, the model has been validated by fitting its parameters to results from experiments (see Li et al. [6]). Figure 2 depicts the very good agreement between test and simulation data. The obtained set of material parameters is $\mu = 25000$ MPa, $\Lambda = 37500$ MPa, $\sigma_y = 174$ MPa, $c = 1920$ MPa, $b = 51.5$, $Q = 98$ MPa, $\beta = 10.5$, and it has been used for the finite element simulations in the next section.

## FE Examples

In this subsection, the structural behaviour is taken into account. The model has been implemented as a user subroutine (UMAT) in the finite element program ABAQUS and then used in the simulation of draw-bending. The draw-bending test is conducted as follows: a strip of sheet metal is held fixed between two clamp jaws and is being drawn and bent over a cylindrical tool. Afterwards, the strip is unloaded and removed from the machinery to observe its springback.

Figures 3 and 4 show how the simulation of the draw-bending test was done in this study. Fig. 3 (left) depicts the position of the sheet strip, after it has been bent over the cylindrical tool. By applying a tensile force at the right end and prescribing displacement at the left one, the sheet is being drawn over the tool to the end position, illustrated in Fig. 3 (right). Then, loading and boundary conditions are removed and the strip is allowed to "spring back" (Fig. 4). Figures 3 and 4 represent the distribution of the von Mises stress over the volume of the sheet metal strip. The decrease of the stresses during the springback phase is due to the elastic unloading.

The above simulation has been performed by means of ABAQUS/Standard, with a converged mesh of 4800 (length x thickness x breadth = 300 x 4 x 4) C3D8R linear solid elements with reduced integration and hourglass stabilization.

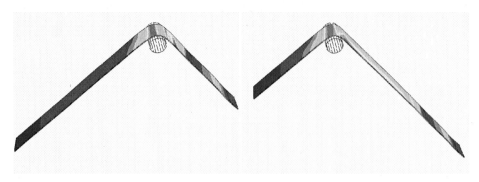

**FIGURE 3.** Draw-bending: states 1 and 2

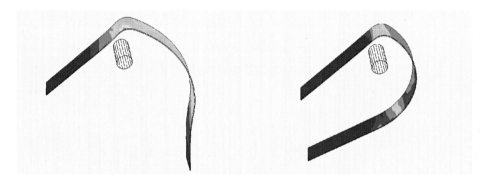

**FIGURE 4.** Draw-bending: states 3 and 4

# REFERENCES

1. Y. Choi, C. S. Han, J. K. Lee, and R. Wagoner, *Int. J. Plast.*, **22**, 1745–1764, (2006).
2. K. Chung, M. G. Lee, D. Kim, C. Kim, M. L. Wenner and F. Barlat, *Int. J. Plast.*, **21**, 861–882, (2005).
3. W. Dettmer and S. Reese, *Comp. Meth. App. Mech. Engrg*, **193**, 87–116, (2004).
4. P. Hakansson, M. Wallin and M. Ristinmaa, *Int. J. Plast.*, **21**, 1435–1460, (2005).
5. Hibbit, Karlsson & Sorensen, ABAQUS/Standard User's Manual, Version 6.3 (2003).
6. K. P. Li, W. D. Carden, and R. H. Wagoner, *Int. J. Mech. Sci.*, **44**, 103–122, (2002).
7. A. Menzel, M. Ekh, K. Runesson and P. Steinmann, *Int. J. Plast.*, **21**, 397–434, (2005).
8. I. Ragai, D. Lazim and J. A. Nemes, *J. Mat. Proc. Techn.*, **166**, 116–127, (2005).
9. S. Reese, and S. Govindjee, *Int. J. Sol. Struct.*, **35**, 3455–3482, (1998).
10. S. Reese, and I. N. Vladimirov, in *Proceedings of the 9th International ESAFORM Conference on Material Forming*, edited by N. Juster, A. Rosochowski, Publishing House Akapit, Krakow, Poland, 2006, pp. 415–418.
11. B. Svendsen, V. Levkovitch, J. Wang, F. Reusch, and S. Reese, *Comp. Struct.*, **84**, 1077–1085, (2006).
12. I. N. Vladimirov, and S. Reese, in *Proceedings of the International Deep Drawing Research Group 2006 Conference*, edited by A. D. Santos, A. Barata da Rocha, Porto, Portugal (2006), pp. 177-182.
13. M. Wallin and M. Ristinmaa, *Int. J. Plast.*, **21**, 2025–2050, (2005).
14. J. Wang, V. Levkovitch, F. Reusch and B. Svendsen, *Arch. Appl. Mech.*, **74**, 890–899, (2005).
15. G. Weber, and L. Anand, *Comp. Meth. App. Mech. Engrg*, **79**, 173–202, (1990).
16. F. Yoshida, and T. Uemori, *Int. J. Mech. Sci.*, **45**, 1687–1702, (2003).

# Experimental and Numerical Investigations in Single Point Incremental Sheet Forming For Micro parts

S. Dejardin, S. Thibaud, J.C. Gelin

*FEMTO-ST Institute, Applied Mechanics Laboratory – ENSMM. 26 rue de l'Epitaphe, 25000 Besançon. France*

**Abstract.** Incremental Sheet Forming processes have been demonstrated in recent studies as a very promising technology to manufacture sheet metal parts by the CNC controlled movement of a simple generative tool. In glance with its various advantages, such process has been introduced as an alternative to reduce costs resulting from stamping technology when small batches or prototypes have to be manufactured. In this paper, an application is carried out accounting flexibility of the process linked to the fact that the punches or dies are avoided. Although the process still needs a further optimization, preliminary results have been obtained through experimental tests to manufacture micro parts. At the same time, a FEM analysis has been carried out in order to get the characteristics of the formed parts.

**Keywords:** Incremental forming, sheet metal, FEM, experimental investigations, micro parts.
**PACS:** 81.20.Hy

## INTRODUCTION

In the last few years, Incremental Sheet Forming has been developed as a very strategic process to produce sheet metal parts. Indeed, due to the low set-up costs, the use of this technology may be interesting when small batches or simple products have to be manufactured [1]. The basic concept consists in fact in avoiding traditional die and in using a simple punch as forming tool for which tooling path is controlled by CNC. Besides the slowness of the process, incremental sheet forming process seems to be an effective response to the increasing flexibility requirement associated to sheet metal forming and offer the possibility to establish a powerful alternative to manufacture parts which cannot be made with conventional processes. This possibility becomes a need in applications where the part has to be unique as in medical field as example                                                                                                        [2].
As far as the process understanding is concerned, several studies were carried out to verify and justify the larger material formability that characterizes incremental forming in comparison with conventional stamping operation [3-4]. On the basis of the positive effect of the extremely localized deformation imposed by a simple hemispherical tool, this paper presents an application of ISF for producing micro sheet metal parts.

On the other hand, it is necessary to point out the main difficulties of ISF associated to the industrial suitability of the process. More in detail, as process mechanics is

CP907, *10ʰ ESAFORM Conference on Material Forming*, edited by E. Cueto and F. Chinesta
© 2007 American Institute of Physics 978-0-7354-0414-4/07/$23.00

mainly characterized by the stretching deformation mode of the sheet metal added to the lack of any dies, a relative sheet thinning and rough dimensional accuracy determine precise limits as the steepness of formed parts.

In this sense, it is very important to increase the knowledge of such a technology through both experimental and numerical investigations. Therefore, this paper summarizes the developments carried out in our lab to set up an experimental equipment to prove that SPIF could be a solution for manufacturing micro parts. This study is also based on an accurate FE analysis of the process able to predict thickness and stress contours during the incremental sheet metal forming.

## EXPERIMENTAL INVESTIGATIONS

In the first step, a set of experiments, characterized through different parts, were performed in order to study the feasibility to manufacture micro parts in using ISF process. Experimental tests were carried out on a 3-axis controlled CNC milling machine. An experimental tooling system has been used to clamp the blank as shown in Figure 1.

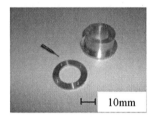 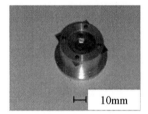

**FIGURE 1.** The experimental tooling system used in ISF

A simple punch with 1mm diameter hemispherical head was used as a forming tool for which the path was specified on the CNC machine through a proper routine. Such path includes both the movement in the horizontal plane (i.e. the x-y displacement plane of the milling machine) and the penetration of the punch in the vertical axis direction. The sheet, with dimensions of 30mm * 30mm and 0.24mm thickness, is supported around its contour with a circular blank holder. The blank material was CuNiP. Thus, as several investigations have shown in recent studies [5], the blank is mainly stretched by the local action of the punch in experimental parts obtained in this paper.

As Ambrogio et al. [6] have shown in previous investigations that the single point incremental sheet forming process mainly depends on geometrical and process conditions. Particularly, the accuracy of the final geometry is mainly influenced by the tool depth step. Indeed, the increase in stretching conditions with the tool path pitch involves the decrease in surface quality. For these reasons, the experimental tests were carried out accordingly to the parameters defined in Table 1.

| TABLE 1. Test configuration. | |
|---|---|
| **Process parameters** | **Value** |
| Tool diameter | 1mm |
| Tool depth step | 0.01mm |
| Feed rate | 300m/min |
| Speed rotation | 250tr/min |

Several shapes related in Figure 2 were chosen to lead experimental tests.

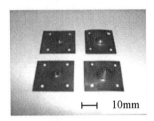
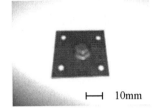

FIGURE 2. Experimental shapes

## NUMERICAL ANALYSIS

In order to simulate the process, a finite element model was implemented in LS-Dyna® FEM simulation code. As incremental forming is a progressive sheet metal forming process characterized by large displacements and localized strains, an explicit solution scheme was adopted, resulting in the choice of LS-Dyna® FEM as the simulation code.

According to experimental tests, the investigated shape used to perform the simulation is a pyramid one with a hexagonal base with 12mm side and 3mm height. The slope of the wall is fixed to 45°, the main data are summarized in Figure 3.

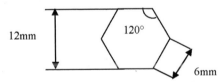

FIGURE 3. Geometry of the considered part

Due to the three-dimensional tool path, a fully three-dimensional analysis is required. As a consequence, shell elements with 4 nodes and 6 degrees of freedom per node and five integration points along the thickness were used. Furthermore, an adaptive mesh refinement was performed in order to reduce element size when the distortion level reached a maximum value. These ingredients allow a proper modeling

147

of the progressive deformation of the sheet by increasing the number of nodes in contact with the tool surface.

The deformed mesh reproducing the investigated shape is reported in figure 4.

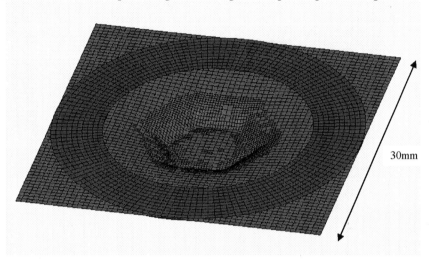

**FIGURE 4.** Deformed mesh corresponding to a vertical punch displacement equal to 2mm

The sheet was initially meshed with about 900 elements. Sheet metal behavior over the yield stress has been accounted by means of a Swift type hardening law.

$$\overline{\sigma} = k.\left(\varepsilon_p + \overline{\varepsilon}\right)^n \tag{1}$$

The tool is considered as a rigid body and the corresponding boundary conditions are related to the path that it should follow during the process.

At the end of simulation, the final profile was numerically revealed and compared with the experimental one (Figure 5).

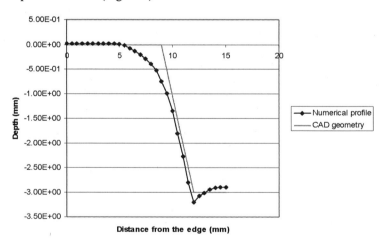

**FIGURE 5.** Comparison between numerical and experimental profiles for the considered part

The blankholder force is modeled thanks to the Ls-Dyna® load_rigid_body card. Several tests have been carried out in order to study the influence of the pressure on the formed parts. In particular, numerical analysis is focused on thickness distribution and geometrical accuracy. Figure 6 shows first results concerning the blankholder load influence on geometrical accuracy. Numerical results highlight that a pressure value higher than 0.5MPa (about 2% of the material yield strength) is enough to clamp the sheet [7]. Nevertheless first results concerning the blankholder load show that there is no particular influence on geometrical accuracy (Figure 6). On the other hand, the higher the blankholder load is, the higher the thinning is (Figure 7).

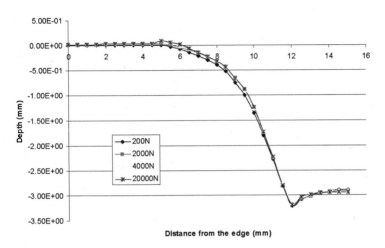

**FIGURE 6.** Thickness profiles comparison between two blankholder loads (F=200N i.e. P=0.5MPa and F=20000N i.e. P=50MPa)

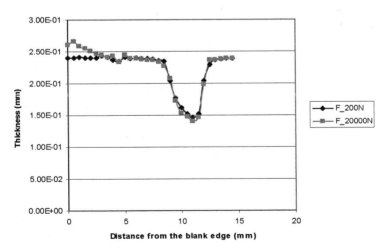

**FIGURE 7.** Thickness comparisons between two blankholder load (F=200N i.e. P=0.5MPa and F=20000N i.e. P=50MPa)

# CONCLUSION

This paper has demonstrated the possibility to get micro sheet metal parts in using the Incremental Sheet Forming process. The reported results show that is possible to get complex shapes with small thickness using such a process, avoiding the risks of failure and necking that classically occur in using standard sheet metal processing. The paper has also investigated on the influence of main process parameters as blankholder load on geometrical accuracy and thickness variation. It has also be demonstrated that the finite element simulation of the process is possible using a commercial transient dynamic explicit code, permitting to get the deformed shape depending on tool path, as well as thickness variations. Further developments in progress are concerned with the possibility to online thickness measurement with the aim to fully control the process.

# ACKNOWLEDGMENTS

This work has been carried out with the financial support from the European Commission under the STREP project SCULPTOR (project no. NMP2-CT-2005-014026).

# REFERENCES

1. J.J. Park, Y.H. Kim, Fundamental studies on the incremental sheet metal forming technique, Journal of Materials Processing Technology 140 (2003) 447-453
2. G. Ambrogio, L. De Napoli, L. Filice, F. Gagliardi, M. Muzzupappa, Application of Incremental Forming process for high customised medical product manufacturing, Journal of materials Processing Technology 162-163 (2005) 156-162
3. H. Iseki, An experimental and theoretical study on a forming limit curve in incremental forming of sheet metal using spherical roller, Proc. of Metal Forming, (2000), pp. 557-562.
4. M.-S. Shim, J.J. Park, The formability of Aluminum sheet in incremental forming, Journal of materials Processing Technology 113 (2001) 654-658
5. L. Filice, L. Fratini, F. Micari, Analysis of Material Formability in Incremental Forming, Proc. of Metal Forming, (2000), pp. 557-562
6. G. Ambrogio, I. Costanino, L. De Napoli, L. Filice, L. Fratini, M. Muzzupappa, Influence of some relevant process parameters on the dimensional accuracy in incremental forming: a numerical and experimental investigation, Journal of materials Processing Technology 153-154, (2004), pp. 501-507
7. G. Ambrogio, L. Filice, F. Gagliardi, F. Micari, Sheet Incremental Forming: A new process configuration allowing a sheet material controlled flow under the blank-holder, ICTP, 2005

# An Advanced Numerical Differentiation Scheme for Plastic Strain-Rate Computation

Holger Aretz

*Hydro Aluminium Deutschland GmbH, Research & Development,*
*Georg-von-Boeselager-Str. 21, D-53117 Bonn, Germany*
*E-mail: holger.aretz@hydro.com*

**Abstract.** Modern yield functions for orthotropic sheet metals are quite complex in a mathematical sense, mainly due to their non-quadratic character and/or the incorporation of eigenvalues of linearly transformed stress tensors (e.g. [5, 6]). In particular, the analytical computation of first and second order yield function gradients, which are, for instance, required in finite element codes, can become a very lengthy task. Thus, the numerical differentiation is a very convenient method to circumvent these difficulties [1, 2]. In the present article an advanced numerical differentation scheme is presented that exploits consequently the homogeneity property of the yield function.

**Keywords:** Metal plasticity, Plastic strain-rate, Numerical differentiation, Yield function
**PACS:** 83.10.-y, 83.10.Bb, 83.10.Ff, 83.50.-v, 83.50.Uv, 83.60.-a, 83.60.La, 83.80.Nb, 83.85.Tz, 89.20.Kk, 02.60.Jh, 02.70.-c, 02.70.Bf, 02.70.Dh, 45.10.-b, 46.05.+b, 46.15.-x, 46.35.+z

## INTRODUCTION

Modern non-quadratic yield functions capable of describing accurately the initial plastic anisotropy of sheet metals require frequently a high implementation effort regarding numerical codes, in particular when the general 3D stress state is considered (e.g. [5, 6]). In particular, the first and second order yield function gradients result often in lengthy expressions. Moreover, natural singularities that appear in these expressions must be identified and require a special mathematical treatment, a solveable but tedious task. As an attractive alternative numerical differentiation can be conveniently used to circumvent all these problems [1, 2]. This has the important advantage that new yield functions can be implemented very rapidly, without worrying about complex expressions for the components of the yield function gradient. In this paper, an advanced numerical differentiation method is proposed that offers a higher accuracy than standard difference schemes.

## THEORETICAL BASIS

### Yield Function and Flow-Rule

The equivalent stress, $\bar{\sigma}(\boldsymbol{\sigma}) \geq 0$, is introduced as a scalar-valued, non-negative, stress-like load measure representative for a multiaxial stress state $\boldsymbol{\sigma}$. Plastic yielding sets in when the equivalent stress equals the (instantaneous) reference yield stress, $Y_{ref}$, of the

CP907, *10th ESAFORM Conference on Material Forming,* edited by E. Cueto and F. Chinesta
© 2007 American Institute of Physics 978-0-7354-0414-4/07/$23.00

considered material. This yield condition reads as

$$\bar{\sigma}(\boldsymbol{\sigma}) = Y_{\text{ref}} \tag{1}$$

The reference yield stress is assumed to be positive (i.e. $Y_{\text{ref}} > 0$) and can, for instance, be determined in uniaxial or equibiaxial tensile test. Alternatively, one may introduce a yield function, $F$, as follows:

$$F(\boldsymbol{\sigma}) = \bar{\sigma}(\boldsymbol{\sigma}) - Y_{\text{ref}} \leq 0 \tag{2}$$

For $F < 0$ the material deforms elastically, for $F = 0$ plastically. $F > 0$ is non-admissible in the rate-independent theory of plasticity. $F = 0$ describes the yield surface.

Reasonably, when the stress state $\boldsymbol{\sigma}$ is scaled by a scalar $\xi$, i.e. $\boldsymbol{\sigma} \to \xi \cdot \boldsymbol{\sigma}$, then the equivalent stress $\bar{\sigma}$ should also be scaled by the same factor. Thus, we demand additionally that

$$\bar{\sigma}(\xi \cdot \boldsymbol{\sigma}) = \xi \cdot \bar{\sigma}(\boldsymbol{\sigma}), \quad \xi \geq 0 \tag{3}$$

Note that $\xi \geq 0$ must be used to maintain the non-negativity of $\bar{\sigma}$. Eqn. (3) states that $\bar{\sigma}$ is a *homogeneous function of degree one* with respect to a non-negative multiplier $\xi$.

The associated flow-rule (see [3]) governs the tensor of plastic strain-rates $\dot{\boldsymbol{\varepsilon}}^P$ given by

$$\dot{\boldsymbol{\varepsilon}}^P = \dot{\bar{\varepsilon}} \cdot \partial\bar{\sigma}/\partial\boldsymbol{\sigma} = \dot{\bar{\varepsilon}} \cdot \partial F/\partial\boldsymbol{\sigma}, \quad F(\boldsymbol{\sigma}) = 0 \tag{4}$$

with the equivalent plastic strain-rate $\dot{\bar{\varepsilon}} \geq 0$. Since $\dot{\bar{\varepsilon}} > 0$ during plastic loading it follows that $\dot{\boldsymbol{\varepsilon}}^P$ and $\partial\bar{\sigma}/\partial\boldsymbol{\sigma}$ are parallel.

## Numerical Differentiation in Scaled Stress Space

In the following the general 3D stress space is considered, but the developed theory can be readily adapted to plane stress or plane strain problems as well. The yield function gradient $\partial\bar{\sigma}/\partial\boldsymbol{\sigma}$ appearing in the associated flow rule is conveniently calculated using numerical differentiation. The numerical differentiation is done in a *scaled* stress space. This is possible due to the homogeneity of the yield function. With the homogeneity property $\bar{\sigma}(\xi\,\boldsymbol{\sigma}) = \xi\,\bar{\sigma}(\boldsymbol{\sigma})$, $\xi \geq 0$, we can write:

$$\frac{\partial\bar{\sigma}(\boldsymbol{\sigma})}{\partial\boldsymbol{\sigma}}\xi = \frac{\partial}{\partial\boldsymbol{\sigma}}(\xi\,\bar{\sigma}(\boldsymbol{\sigma})) = \frac{\partial}{\partial\boldsymbol{\sigma}}(\bar{\sigma}(\xi\boldsymbol{\sigma})) = \frac{\partial\bar{\sigma}(\xi\boldsymbol{\sigma})}{\partial(\xi\boldsymbol{\sigma})} : \frac{\partial(\xi\boldsymbol{\sigma})}{\partial\boldsymbol{\sigma}} = \frac{\partial\bar{\sigma}(\xi\boldsymbol{\sigma})}{\partial(\xi\boldsymbol{\sigma})}\xi \tag{5}$$

Hence

$$\boxed{\partial\bar{\sigma}(\boldsymbol{\sigma})/\partial\boldsymbol{\sigma} = \partial\bar{\sigma}(\boldsymbol{\sigma}')/\partial\boldsymbol{\sigma}'} \tag{6}$$

with $\boldsymbol{\sigma}' \equiv \xi \cdot \boldsymbol{\sigma}$. This means that the yield function gradient appearing in the associated flow-rule can also be calculated in the *scaled* stress space $\boldsymbol{\sigma}'$. In the present work the scaling factor $\xi$ is chosen as $\xi = \{\sigma_{11}^2 + \sigma_{22}^2 + \sigma_{33}^2 + \sigma_{12}^2 + \sigma_{23}^2 + \sigma_{31}^2\}^{-1/2}$, without referring to other possibilities. For numerical differentiation the *forward difference scheme* is applied as follows:

$$\frac{\partial \bar{\sigma}(\sigma')}{\partial \sigma'_{ij}} = \frac{\bar{\sigma}(\sigma'_{ij} + \delta\sigma', \sigma'_{kl}) - \bar{\sigma}(\sigma'_{ij}, \sigma'_{kl})}{\delta\sigma'} \tag{7}$$

with $ij \neq kl$ and $\{i,j,k,l\} \in \{1,2,3\}$. $\delta\sigma' = 1 \cdot 10^{-6}$ is applied. The computation in the scaled stress space is preferred because the choice of an appropriate value for $\delta\sigma'$ is then significantly simplified.

The *second order* yield function gradient $\partial^2 \bar{\sigma}/\partial\sigma^2$ is frequently required in implicit finite element codes and is computed using numerical differentiation as well. The computation is also conducted in the *scaled* stress space. The relationship between the gradient in the scaled and unscaled stress space is given as follows:

$$\frac{\partial^2 \bar{\sigma}(\sigma)}{\partial\sigma^2} = \frac{\partial}{\partial\sigma}\left(\frac{\partial\bar{\sigma}(\sigma)}{\partial\sigma}\right) = \frac{\partial}{\partial\sigma}\left(\frac{\partial\bar{\sigma}(\xi\sigma)}{\partial(\xi\sigma)}\right) = \frac{\partial}{\partial(\xi\sigma)}\left(\frac{\partial\bar{\sigma}(\xi\sigma)}{\partial\sigma}\right)$$
$$= \frac{\partial}{\partial(\xi\sigma)}\left(\frac{\partial\bar{\sigma}(\xi\sigma)}{\partial(\xi\sigma)} : \frac{\partial(\xi\sigma)}{\partial\sigma}\right) \tag{8}$$

Hence

$$\boxed{\partial^2\bar{\sigma}(\sigma)/\partial\sigma^2 = \xi \cdot \partial^2\bar{\sigma}(\sigma')/\partial\sigma'^2} \tag{9}$$

with $\sigma' \equiv \xi \cdot \sigma$. Again, numerical differentiation is used to calculate $\partial^2\bar{\sigma}(\sigma')/\partial\sigma'^2$. The numerical differentiation scheme is as follows. For $i = j$

$$\frac{\partial^2\bar{\sigma}(\sigma')}{\partial\sigma'_{ij}\partial\sigma'_{ij}} = \frac{\bar{\sigma}_+ - 2\bar{\sigma}(\sigma') + \bar{\sigma}_-}{(\delta\sigma')^2} \tag{10}$$

and else

$$\frac{\partial^2\bar{\sigma}(\sigma')}{\partial\sigma'_{ij}\partial\sigma'_{mn}} = \frac{\bar{\sigma}_{++} - \bar{\sigma}_{+-} - \bar{\sigma}_{-+} + \bar{\sigma}_{--}}{(2\delta\sigma')^2} \tag{11}$$

using the abbreviations

$$\bar{\sigma}_+ \equiv \bar{\sigma}(\sigma'_{ij} + \delta\sigma', \sigma'_{kl}), \qquad \bar{\sigma}_- \equiv \bar{\sigma}(\sigma'_{ij} - \delta\sigma', \sigma'_{kl}) \tag{12}$$

with $ij \neq kl$ and

$$\begin{aligned}\bar{\sigma}_{++} &\equiv \bar{\sigma}(\sigma'_{ij} + \delta\sigma', \sigma'_{mn} + \delta\sigma', \sigma'_{kl}), & \bar{\sigma}_{+-} &\equiv \bar{\sigma}(\sigma'_{ij} + \delta\sigma', \sigma'_{mn} - \delta\sigma', \sigma'_{kl}) \\ \bar{\sigma}_{-+} &\equiv \bar{\sigma}(\sigma'_{ij} - \delta\sigma', \sigma'_{mn} + \delta\sigma', \sigma'_{kl}), & \bar{\sigma}_{--} &\equiv \bar{\sigma}(\sigma'_{ij} - \delta\sigma', \sigma'_{mn} - \delta\sigma', \sigma'_{kl})\end{aligned} \tag{13}$$

with $ij, mn \neq kl$. For second order gradients $\delta\sigma' = 1 \cdot 10^{-3}$ is used. Note the symmetry

$$\frac{\partial}{\sigma'_{mn}}\left(\frac{\partial\bar{\sigma}(\sigma')}{\partial\sigma'_{ij}}\right) = \frac{\partial}{\sigma'_{ij}}\left(\frac{\partial\bar{\sigma}(\sigma')}{\partial\sigma'_{mn}}\right) \tag{14}$$

due to Schwarz' theorem. In case of *plastic incompressibility* the identity

$$\partial\bar{\sigma}(\sigma')/\partial\sigma'_{33} = -\partial\bar{\sigma}(\sigma')/\partial\sigma'_{11} - \partial\bar{\sigma}(\sigma')/\partial\sigma'_{22} \tag{15}$$

holds thereby reducing the computational effort. Similar relationships may be found for the second order gradient, but for the sake of brevity this is not discussed here.

## FINITE ELEMENT IMPLEMENTATION

The numerical differentiation scheme outlined above is part of an elastic-plastic user material subroutine linked to the commercial finite element codes ABAQUS/Standard (implicit) and ABAQUS/Explicit (explicit) version 6.5, respectively. The finite element implementation is based on the assumptions of isotropic linear elasticity and the additive decomposition of the strain-rate $\dot{\varepsilon}$ into an elastic and a plastic part according to

$$\dot{\sigma} = C : (\dot{\varepsilon} - \dot{\varepsilon}^p), \qquad \dot{\varepsilon} = \dot{\varepsilon}^e + \dot{\varepsilon}^p \tag{16}$$

Isotropic hardening is used, i.e. the reference yield stress depends on the accumulated equivalent plastic strain according to $Y_{\text{ref}} = Y_{\text{ref}}(\bar{\varepsilon})$ with $\bar{\varepsilon} = \int \dot{\bar{\varepsilon}}\, dt$. The elastic-plastic boundary value problem is solved in an incremental-iterative fashion. Variables related to the previous and the current solution step are denoted by $(\bullet)_n$ and $(\bullet)_{n+1}$, respectively. For computational use, rates are replaced by finite differences, e.g. $\dot{\varepsilon}$ is replaced by $\Delta\varepsilon$. In each integration point of the finite elements the total strain increment $\Delta\varepsilon$ is imposed while the corresponding stress state must be computed. The calculation of the stress response is based on the classical elastic predictor-plastic corrector method (see e.g. [7]). If the elastic trial stress $\sigma^* = \sigma_n + C : \Delta\varepsilon$ leads to $F(\sigma^*) \geq 0$ the stress state is iteratively returned to the yield surface by means of the 'Tangent Cutting Plane' (TCP) algorithm. The implemented TCP algorithm is summarized as follows (for details it is referred to [8, 9]):

1. Initialization: $i = 0$, $\sigma^{(i)}_{n+1} := \sigma^*$, $\bar{\varepsilon}^{(i)}_{n+1} := \bar{\varepsilon}_n$, $\Delta\bar{\varepsilon}^{(i)}_{n+1} := 0$

2. IF $F(\sigma^{(i)}_{n+1}) < 0$ THEN exit

3. WHILE $(|F(\sigma^{(i)}_{n+1}, \bar{\varepsilon}^{(i)}_{n+1})|/Y_{\text{ref}}(\bar{\varepsilon}^{(i)}_{n+1}) > \text{tol})$ DO

$$\delta\bar{\varepsilon} := \left. \frac{F(\sigma, \bar{\varepsilon})}{\dfrac{\partial F}{\partial\sigma} : C : \dfrac{\partial F}{\partial\sigma} - \dfrac{\partial\bar{\sigma}}{\partial\bar{\varepsilon}} + \dfrac{\partial Y_{\text{ref}}}{\partial\bar{\varepsilon}}} \right|_{\sigma^{(i)}_{n+1}, \bar{\varepsilon}^{(i)}_{n+1}}$$

$$\sigma^{(i+1)}_{n+1} := \sigma^{(i)}_{n+1} - \delta\bar{\varepsilon}\, C : \left. \frac{\partial F}{\partial\sigma} \right|_{\sigma^{(i)}_{n+1}}$$

$$\Delta\bar{\varepsilon}^{(i+1)}_{n+1} := \Delta\bar{\varepsilon}^{(i)}_{n+1} + \delta\bar{\varepsilon}$$

$$\bar{\varepsilon}^{(i+1)}_{n+1} := \bar{\varepsilon}^{(i)}_{n+1} + \max(0, \Delta\bar{\varepsilon}^{(i+1)}_{n+1})$$

$$i := i + 1$$

END DO

whereby tol is a user-defined tolerance (here: $\text{tol} = 10^{-6}$). The TCP algorithm converges very quickly in about five iterations, even if small tolerances are demanded. In order to increase the accuracy of the TCP method the imposed total strain increment $\Delta\varepsilon$ is subdivided into a finite number of sub-increments $N_{sub} \geq 1$, i.e. the sub-increment of strain entering the TCP algorithm is given by $\Delta\varepsilon_{sub} = \frac{1}{N_{sub}} \cdot \Delta\varepsilon$ and the procedure is repeated $N_{sub}$ times. $N_{sub}$ is calculated using the following empirical formula:

$$N_{sub} := \max\left(N_0, \text{integer}\left(A \cdot \frac{\bar{\sigma}(\boldsymbol{\sigma}^*)}{Y_{ref}(\bar{\varepsilon}_n)} + 0.5\right)\right) \tag{17}$$

whereby the term 0.5 has been introduced for rounding up. $N_0$ is the minumum number of sub-increments. In the present implementation $N_0 = 5$ is used. The factor $A$ is an empirical constant taken as 2.0 which was found to be suitable for the investigated test cases. It should be mentioned that objectivity (see e.g. [7]) requires that $\boldsymbol{\sigma}_n$ and $\Delta\varepsilon$ are neutralized with respect to rigid-body rotation before entering the stress update, which is achieved using the co-rotational formulation provided by ABAQUS.

## APPLICATION EXAMPLE

The example presented here is a deep drawing process of a square box, taken from the ABAQUS Examples Manual [4]. ABAQUS/Explicit is used. Due to material and geometrical symmetries only 1/4 of the model needs to be discretized by finite elements. The blank is modelled by means of S4R shell elements (four nodes reduced integrated shell element). The von Mises yield function is adopted in this analysis, but in order to demonstrate that the developed theory applies to more complex constitutive models the yield function 'Yld91' according to Barlat [5] is used with anisotropy parameters $a = b = c = f = g = h = 1$ and exponent $m = 2$ thus representing a von Mises yield function. In this way, the user material subroutine can be compared with the standard von Mises implementation of ABAQUS. The computed thickness distributions and the total (elastic + plastic) major strain distributions are displayed in Figs. 1 and 2. One may see that the results are qualitatively and quantitatively almost identical. The vanishing small differences are attributed to the different treatments of the isotropic hardening curve (standard implementation: equidistant piecewise linear interpolated tabular data; user subroutine: analytical expression) and the constant transverse shear stiffness in case of the user material subroutine. In summary, the present example confirms the correctness of the numerical differentation scheme outlined above. It should be mentioned that the explicit solution procedure involves only the first order yield function gradient. Results using the implicit solution procedure (in which the second order yield function gradient appears) are not shown here for the sake of brevity, but it should be mentioned that the numerical differentiation scheme works also excellently in this case.

## CONCLUSIONS

An advanced numerical differentiation scheme for first and second order yield function gradient calculation was proposed that exploits the homogeneity property of the yield function. The method has the following important advantages: (i) it allows a rapid and unified implementation of yield functions, (ii) singularities that naturally appear

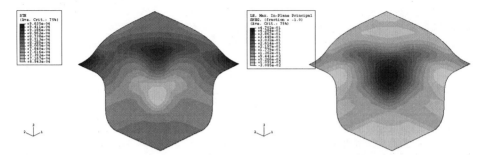

**FIGURE 1.** Results using the standard implementation. Left: Calculated sheet thickness distribution. Right: Calculated total (elastic + plastic) major strain distribution distribution.

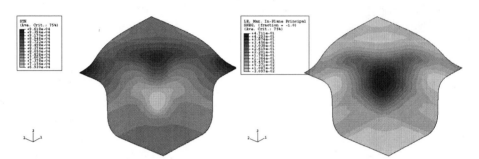

**FIGURE 2.** Results using the user material subroutine. Left: Calculated sheet thickness distribution. Right: Calculated total (elastic + plastic) major strain distribution distribution.

in analytical yield function gradient expressions do not appear. By means of a finite element simulation the capabilities of the method have been demonstrated.

# REFERENCES

1. H. Aretz: *Modellierung des anisotropen Materialverhaltens von Blechen mit Hilfe der Finite-Elemente-Methode*, doctoral thesis, RWTH Aachen University, Shaker-Verlag, 2003
2. H. Aretz: Applications of a new plane stress yield function to orthotropic steel and aluminium sheet alloys, Modelling and Simulation in Materials Science and Engineering 12 (2004) 491-509
3. H. Aretz: A less hypothetical perspective on rate-independent continuum theory of metal plasticity, Mechanics Research Communications 33 (2006) 734-738
4. N.N.: ABAQUS 6.5 Example Problems Manual, ABAQUS Inc., www.abaqus.com
5. F. Barlat, D.J. Lege, J.C. Brem: A six-component yield function for anisotropic materials, International Journal of Plasticity 7 (1991) 693-712
6. F. Barlat, H. Aretz, J.W. Yoon, M.E. Karabin, J.C. Brem, R.E. Dick: Linear transformation-based anisotropic yield functions, International Journal of Plasticity 21 (2005) 1009-1039
7. T. Belytschko, W.K. Liu, B. Moran: *Nonlinear Finite Elements for Continua and Structures*, John Wiley & Sons, 2000
8. M. Ortiz, J.C. Simo: An analysis of a new class of integration algorithms for elastoplastic constitutive relations, International Journal for Numerical Methods in Engineering 23 (1986) 353-366
9. J.C. Simo, T.J.R. Hughes: *Computational Inelasticity*, Springer-Verlag, 1998

# High Velocity Forming of Magnesium and Titanium Sheets

A. Revuelta*, J. Larkiola*, A.S. Korhonen[†] and K. Kanervo[†]

*Technical Research Centre of Finland, P.O. Box 1000, 02044 VTT, Finland
[†]Helsinki University of Technology, P.O. Box 6200, 02015 HUT, Finland

**Abstract.**
Cold forming of magnesium and titanium is difficult due to their hexagonal crystal structure and limited number of available slip systems. However, high velocity deformation can be quite effective in increasing the forming limits. In this study, electromagnetic forming (EMF) of thin AZ31B-O magnesium and CP grade 1 titanium sheets were compared with normal deep drawing. Same dies were used in both forming processes. Finite element (FE) simulations were carried out to improve the EMF process parameters. Constitutive data was determined using Split Hopkinson Pressure Bar tests (SHPB). To study formability, sample sheets were electromagnetically launched to the female die, using a flat spiral electromagnetic coil and aluminum driver sheets. Deep drawing tests were made by a laboratory press-machine.

Results show that high velocity forming processes increase the formability of Magnesium and Titanium sheets although process parameters have to be carefully tuned to obtain good results.

**Keywords:** EMF, Titanium, Magnesium, FEM
**PACS:** 82.20.Wt, 83.10.Gr, 83.50.-v

## INTRODUCTION

High velocity deformation can be effective in increasing the formability of metal sheets. In addition, some other common metal forming problems, such as wrinkling and spring back can be minimized. It is known that increasing the strain rate increases the flow stress required for slip and thus promotes twinning [1]. In this study, the impact forming of AZ31B-O magnesium and CP grade 1 titanium sheets were examined with a special designed test dies, Fig. 1a. To study forming, the material sheets, ranging from 0.2 mm to 0.55 mm, were electromagnetically launched to the female die using a flat coil and an aluminum driver sheet.

The high impact pressures generated when two solid bodies impact at high velocity, can produce significant plastic deformation of the sheet (velocities over 100 m/s can be generated in EMF). This dynamic forming process is more like ironing than conventional sheet metal stretching and this largely compressive state of stress inhibits localization. The failure strains can be increased beyond those obtained in tensile tests and the forming limit diagrams [3].

CP907, 10th ESAFORM Conference on Material Forming, edited by E. Cueto and F. Chinesta
© 2007 American Institute of Physics 978-0-7354-0414-4/07/$23.00

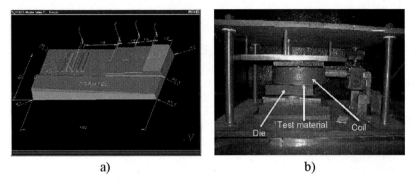

a)                               b)

**FIGURE 1.** a) Tool Dimensions, b) EMF Setup

**FIGURE 2.** Upper and lower tools for deep drawing

## EXPERIMENTAL WORK

### Electromagnetic Forming

The basic experimental setup is shown in Fig. 1b. To implement EMF, a capacitor bank is charged to a high voltage and then the stored charge is discharged through a coil. In this case, maximum discharge voltage attainable was $15kV$. EMF is most easily practiced with materials of high electrical conductivity (resistivity under $15\mu\Omega \cdot cm$). Due to the different nature of the materials tested an auxiliary driver aluminium sheet was used. In addition to EMF experiments, deep drawing tests were made in laboratory press machine. The upper and lower tool can been seen in Figs. 2a and 2b. Same tool (Fig.2b) was also used in EMF-tests.

**TABLE 1.** Tensile and SHPB results summary

| | $R_{p0.2}(N/mm^2)$ | $R_m(N/mm^2)$ | $A_{50,2mm}()$ |
|---|---|---|---|
| Mg (tensile test) | 161 | 242 | 11 |
| Mg (HSB) | 192 | 309 | 12 |
| Ti (Tensile test) | 203 | 325 | 48 |
| Ti (HSB) | 335 | 610 | 29 |

# RESULTS

## Tensile and SHPB Tests

Results of the uniaxial tensile and Split Hopkinson Pressure Bar (SHPB) tests can be seen in Table 1. SHPB, also known as Kolsky bar, technique is used to measure the stress-strain response of materials at high strain rates, typically in the range $10^2-10^4s^{-1}$. Ultimate strength increases with increasing strain rate. In the SHPB experiments strain rate was approximately $1500s^{-1}$.

From this results it's evident that, as expected, strength of both Mg and Ti increases in high strain rates.

## Using FEM Simulation as a Support Design Tool

Proper estimation of the forming forces generated requires a detailed knowledge of the geometry of the coil as well as a simulation of the electromagnetic field generated by such a coil [6]. The FEM package employed didn't allow for such a calculation and detailed information about the coil geometry was also unavailable. Therefore, in order to link the FEM simulations with the experimental results, some tests were designed. In this tests, an open hollow die with the same dimensions of the testing part was used. Sheets of the same thickness and materials were formed using several voltages and the height of the formed sample in the middle point was measured (see Fig.3). This process was then simulated by FEM and the boundary conditions employed were correlated from the experimental results.

Once a relation between boundary conditions for the FE model and discharge voltage of the EMF process was established, it was possible to use the FEM simulations as a design tool for finding appropiate discharge levels (Fig.4b). These simulations are carried out using an explicit approach and due to the high velocities of the EMF process, are very unexpensive from a computational point of view.

## EMF Tests

At the beginning, electromagnetic forming tests were made using one impulse to form a sample sheet. The objective was finding an optimum result by varying the discharge voltage. However, when the discharge was high enough to get a significant

a) 5kV                                  b) 8kV

**FIGURE 3.**   Tests for EMF – Simulation tuning

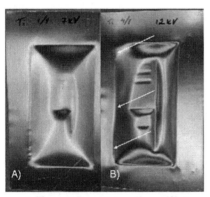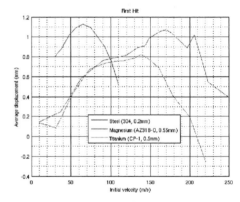

a) Shape after rebounding effect        b) FEM estimation of threshold to avoid reboundin

**FIGURE 4.**

plastic deformation against the tool, it was observed that some areas of the sheet were rebounding back from the tool surface due to the impact. This rebounding can be seen in Fig.4. Also, sides are not anymore straight (arrows). Afterwards, EMF-tests were made by utilizing simulations results where the number of hits and voltage was tuned to get desirable shape, the main idea is reaching the final shape by an incremental forming.

As it can be seen in Fig.4b, simulation results showed that bouncing will occur when the discharge voltage/ sheet velocity is over certain limit. Furthermore, there will always be at least small rebounding and that should be taken into account when optimized process parameters are searched. A typical calculated shape and corresponding experi-

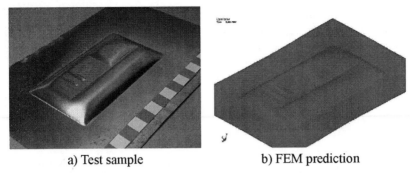

a) Test sample                         b) FEM prediction

**FIGURE 5.**   Comparison between experimental test and FEM prediction for Titanium 0.5mm thickness sheet

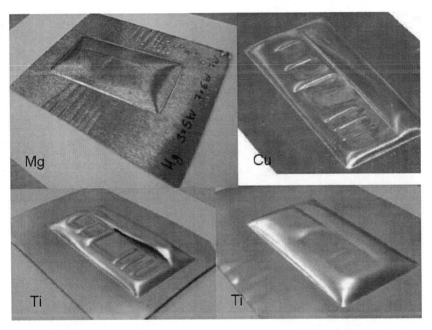

**FIGURE 6.**   Forming examples for Titanium, Magnesium and Copper

mental result can be seen in Fig.5. Material is Titanium and the final calculated shape has good similarity to the EMF-test result.

Some examples of the final forms are seen in Fig.6. For comparative purposes, there is also a sample of oxygen free copper sheet (OFCu). OFCu is a soft material with yield strength of 80 MPa which has good properties for EMF-process.

Deep drawing tests were made with all materials. The radii of the punch and die corners were so small that all test sheets were fractured before total draw height was attained. However, clear differences between tested materials were observed. The effect of high strain rate during forming can be seen crisply for Mg-sheets. In deep drawing process, Mg-sample was ruptured at the very beginning of the draw but in EMF-tests no cracking was attained.

## CONCLUSIONS

At high forming velocities, high strains to failure can be obtained with materials having low quasi-static ductility. In this work the formability of AZ31B-O magnesium and CP grade 1 titanium was tested in two different forming process; deep drawing and electromagnetic forming. In addition, the Hopkinson split bar material test was made to get mechanical properties of Ti and Mg in high forming velocities. Both materials were suitable for EMF-process at least when an aluminum driver sheet was used. From the results of this study we conclude:

- Formability of Magnesium seemed to improve at high strain rates.
- Rebounding effect should be taken into account when the EMF process and tools are on the design stage.
- Although the tools and dies are the same in deep drawing and EMF process, the most severe forming is localized on different areas of metal sheet. Combining EMF and deep drawing in consecutive forming processes, this phenomena could be utilized to increase the overall forming of the process.

## ACKNOWLEDGMENTS

Electromagnetic forming and deep drawing tests were made in The Studio of Stainless Steel in Tornio by Pauli Ylipelto, Heidi Kalliosalo and Rauno Toppila. SHPB tests were made in Tampere University of Technology by Prof. V-T. Kuokkala and his group.

## REFERENCES

1. W.F. Hosford and R.M. Cadell, *Metal Forming; Mechanics and Metallurgy*, 2nd ed., Prentice-Hall, New Jersey, 1991.
2. V.S. Balanethiram and G.S. Daehn, *Enhanced formability of interstitial free iron at high strain rates*, Scripta Mater. 27 (1992), p. 1783.
3. Mala Seth, Vincent J. Vohnout, Glenn S. Daehn, *Formability of steel sheet in high velocity impact*, J. of Mat. Processing Tech., 168 (2005), p. 390 – 400
4. Glenn S. Daehn, M. Altynova, V.S. Balanethiram, G. Fenton, M. Padmanabhan, A. Tamhane and E. Winnard, *High-velocity metal forming – an old technology addresses new problems*, JOM 47 (1995) (7), p. 42.
5. M.C. Noland, *Designing for the High-Velocity Metalworking Processes – Electromagnetic, Electrohydraulic, Explosive and Pneumatic-Mechanical, Design Guide*, Machine Design, 1967, p. 163.
6. J. Unger, M. Stiemer, B. Svendsen, H. Blum, *Multifield modeling of electromagnetic metal forming processes*, J. of Mat. Processing Tech., 177 (2006), p. 270 – 273
7. D.A.Oliveira, M.J.Worswick, M.Finn, D.Newman, *Electromagnetic forming of aluminium alloy sheet: Free-form and cavity fill experiments and model*, J. of Mat. Processing Tech., 170 (2005), p. 350 – 362

# Use of TPIF or SPIF for Prototype Productions: an Actual Case

A. Attanasio [1], E. Ceretti [1], L. Mazzoni[1], C. Giardini [2]

[1] University of Brescia – Dept. of Mech. & Industrial Engineering, via Branze 38, 25123 Brescia, Italy
[2] University of Bergamo – Dept. Of Design & Technology., viale Marconi 5, 24044 Dalmine (BG), Italy

**Abstract.** The present research deals with sheet incremental forming (SIF), a recently developed technique characterized by high flexibility, reduced production times and costs. This is ideal when low volume batches or customized parts or prototypes have to be manufactured. Two different SIF techniques can be identified: Single Point Incremental Forming (SPIF) and Two Points Incremental Forming (TPIF). The main difference between them consists of the presence or not of a die under the sheet. The aim of the present research is to experimentally compare these two different approaches when producing the same part, that is a door handle cavity. In addition, a FE analysis of the process has been developed.

**Keywords:** Forming, Deformation, Simulation
**PACS:** 81.20.Hy, 81.40.Lm, 87.53.Vb

## INTRODUCTION

The sheet metal forming technique called Sheet Incremental Forming (SIF) [1] has been recently developed in order to answer to the market demands in terms of flexibility, development time reduction, cost reduction. In fact, this sheet metal forming method uses as upper die a simple hemispherical tool which locally deforms the sheet. The bottom die is absent (Single Point Incremental Forming SPIF) or made of resin, wood or low carbon steel which assure low cost of the equipment (Two Point Incremental Forming TPIF).

There are relevant differences between these two SIF configurations. Many researchers [1-3] underlined a high flexibility of SPIF, due to the absence of dies, and an increase of the drawing ratio compared with conventional deep drawing. SPIF is indicated for the realization of simple, nearly symmetric shapes with few exceptions. The limit of this technique is the low geometrical and dimensional accuracy. Possible solutions are multistage SPIF or the use of trajectory correction algorithms. These solutions lead to a better dimensional accuracy, but they increase the part production time.

As far as the TPIF is concerned, the conducted researches showed a decrease of sheet formability, but a higher geometrical and dimensional accuracy, also in the case of complex, not axis-symmetric geometries.

This paper aims to better understand the geometrical and dimensional differences between SPIF and TPIF by means of experimental tests and of FE simulations when

CP907, 10th ESAFORM Conference on Material Forming, edited by E. Cueto and F. Chinesta
© 2007 American Institute of Physics 978-0-7354-0414-4/07/$23.00

163

working a car door handle cavity (Figure 1). In particular, the FE results for conventional deep drawing, TPIF and SPIF were compared with the experiments in terms of sheet thickness and dimensional errors so identifying the limits and the advantages of the different techniques.

**FIGURE 1.** The part, the lower die for TPIF and the blankholder.

## EXPERIMENTAL TESTS AND SIMULATIVE MODEL

The SIF experimental tests were realized on a 3-axis CNC milling machine able to move the spherical punch according to a given tool path. As far as TPIF is concerned, a lower die (reproducing the piece to be manufactured) and a blankholder (Figure 1), made of epoxy resin, were used. Since the car door handle cavity is realized on a shaped surface it was necessary to build a shaped die and blankholder (see Figure 1) and to perform a pre-bending of the sheet.

During SPIF process a die with a cavity greater than the part shape was utilized. The sheet was 260 mm x 135 mm x 0.8 mm and made of FeP04 steel. Three tests were carried out for each working process by using the same tool path. The tool path strategy used in the experiments is the result of previous researches [4] and it represents the best identified solution in terms of final part quality (dimensional and geometrical accuracy). In particular, this tool path is a step down path characterized by a series of Z-constant contours with a variable step depth ($\Delta Z$) whose value is determined by the limit value of the *scallop* parameter. In analogy with milling operations, this parameter represents the height of the peaks left on the worked material surface. This means that when working almost flat surfaces the number of contours is increased so achieving a better surface quality. The process variables, together with a sketch showing the meaning of the scallop parameter, are reported in Table 1.

**TABLE 1.** Process Parameters.

| Parameter | Value |
|---|---|
| Tool diameter [mm] | 20 |
| Spindle speed [rpm] | 100 |
| Feed rate [mm/min] | 1200 |
| Maximum step depth [mm] | 0.2 |
| Limit scallop value [mm] | 0.02 |

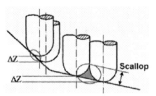

FEM models of the different working techniques were developed with the aim of evaluating the process feasibility and the sheet deformation and thickness. In fact FEM can help in evaluating the process parameter influences on the final part quality (shape, thinning, surface finishing, stress and strain distributions) prior to the experimental phase.

The simulations were conducted using the explicit code PAMSTAMP® considering the sheet as an elastic-plastic element and the die, the punch and the blank-holder as rigid bodies. Friction was modeled according to the Coulomb law with a friction coefficient equal to 0.04 for all the contact surfaces (considering that in the experimental phase the contact area between tool and blank is flooded with lubricant). The first part of the simulation, for all the tested cases, is the sheet pre-bending so reproducing the actual process.

Since PAMSTAMP needs the tool velocity vs. time as input, a suitable SW was developed to calculated them starting from the coordinates of the trajectory defined in the CL file obtained by the CAM module. The punch moves continuously and the working parameters are reported in Table 1. Figure 2 shows the simulation results in terms of stress distributions and sheet thinning for deep drawing, TPIF and SPIF processes.

By comparing the simulative stress and thinning distributions for TPIF and SPIF processes it is evident that no big differences can be identified. Only a small increase of the thinning in TPIF can be found due to the higher deformation caused by the contact between die, sheet and punch. This increase in the deformation resulted in reduced sheet spring back and, as a consequence, in better dimensional accuracy (this also shown in the experiments reported in Figure 3). When the results of TPIF and SPIF are compared with the ones of classical deep drawing differences can be found. In fact, the DD resulting stresses are bigger and the sheet thickness is more uniform and thinning is lower.

## RESULT ANALYSIS

Figure 3 reports the results obtained in terms of dimensional error calculated as the distance between the CAD die geometry and the part profiles (i.e., geometrical and dimensional errors) for both experimental and simulative campaigns measured along two orthogonal directions together with the CAD die profile and the shape of the undeformed blank. It can be noticed how the chosen technique affects the dimensional error. In particular, SPIF, compared to TPIF, is characterized by a higher dimensional error when working steeper surfaces with respect to the initial blank orientation as in points A and B of Figure 3 where the angles between initial sheet and die are higher. This error affects also the zones close to these angles. In the other areas the two techniques demonstrate a similar behavior in terms of dimensional error.

These considerations are valid both for the experimental and the simulative results. In all the tested cases FEM outputs underestimate the dimensional error. This can be due to the fact that in the developed model the sheet elastic recovery due to the springback is not taken into account.

Comparing both the simulative and experimental results of the TPIF and SPIF processes with the simulative ones of the deep drawing it is evident that the dimensional errors characterizing these innovative technologies are higher than in conventional methods. In fact, deep drawing assures an almost constant error of 0.1 mm.

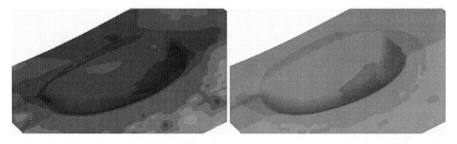

Stress & Thinning distributions for DD

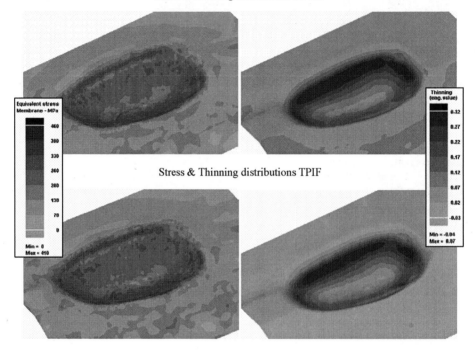

Stress & Thinning distributions TPIF

Stress & Thinning distributions for SPIF

**FIGURE 2.** Simulative stress and thinning distributions for Deep Drawing, TPIF and SPIF processes.

# CONCLUSIONS

The present research aimed to improve the knowledge on sheet incremental forming processes for the two techniques known as TPIF and SPIF. The research consisted of experiments on a car door handle cavity for evaluating geometrical and dimensional errors with respect to the CAD die profile. The same working parameters and the same tool path strategy were adopted.

By comparing the results it can be noticed that TPIF assures the achievement of a better dimensional accuracy (improvement of 25-30% in the best cases) even if this error is not negligible.

In any case TPIF and SPIF techniques demonstrate small differences. This can be due to the fact that for the SPIF tests a shaped blankholder, instead of the conventional holding of the sheet on its edges (flat blankholder), and a partial support die (in SPIF it is absent) were used.

Regarding the part quality, TPIF assures a better surface finishing especially near to the fillet radius around the die cavity.

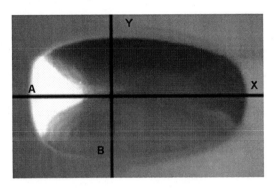

Dimensional error along X plane                    Dimensional error along Y plane

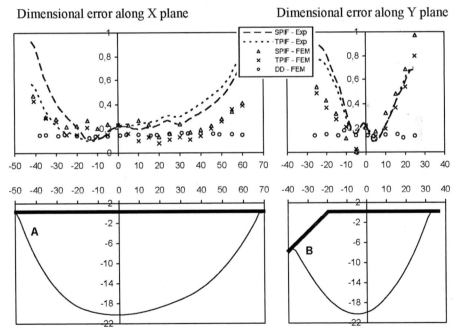

**FIGURE 3.** Measuring directions, experimental and simulative dimensional error for TPIF, SPIF and DD, die geometries and undeformed sheet profiles.

The FEM model realised has shown to be sufficiently reliable in representing the actual process. The accuracy could be improved by simulating the sheet elastic springback too. The limit consists in the high calculation time.

The best results in terms of dimensional error and accuracy were obtained with the deep drawing simulation which represents the working solution for series production.

Of course, the present research demonstrates that, if the market need is for prototypes realised in short time, the sheet incremental forming is capable of realising mechanical components fulfilling these requests at low cost compared with the deep drawing process.

## REFERENCES

1. J. Jeswiet, F. Micari, G. Hirt, A. Bramley, J. Duflou and J. Allwood, "Asymmetric Single Point Incremental Forming of Sheet Metal" in *Annals of CIRP*, Vol. 54/2/2005, pp. 623-649.
2. Q. Qin, E.S. Masuku, A.N. Bramley, A.R. Mileham and G.W. Owen, "Incremental Sheet Forming Simulation and Accuracy" in *Proceedings of the 8th ICTP*, Verona, Italay, 2005, pp.333-334.
3. G. Ambrogio, L. De Napoli, L. Filice, F. Micari and M. Muzzupappa, "Some Considerations on the Precision of Incrementally Formed Double-Curvature Sheet Components" in *Proceedings of the 9th International Conference on Material Forming ESAFORM 2006*, Glasgow, UK, 2006, pp.199-202.
4. A. Attanasio, E. Ceretti and C. Giardini, "Optimization of tool path in two points incremental forming" in *Proceeding of 11th International Conference on Metal forming*, The University of Birmingham, UK, 2006, pp.409-412.

# Numerical Analysis
# Of The Resistance To Pullout Test
# Of Clinched Assemblies Of Thin Metal Sheets

Moez JOMAA, René BILLARDON

*LMT-Cachan (ENS de Cachan / CNRS (UMR 8535) / Université Paris 6)*
*61, avenue du Président Wilson  F-94235 Cachan Cedex FRANCE*
*E-mail: jomaa@lmt.ens-cachan.fr, billardon@lmt.ens-cachan.fr*

**Abstract.** This paper presents the finite element analysis of the resistance of a clinch point to pullout test -that follows the numerical analysis of the forming process of the point-. The simulations have been validated by comparison with experimental evidences. The influence on the numerical predictions of various computation and process parameters have been evaluated.

**Keywords:** clinch process, pullout test, finite element analysis, non standard initial state.
**PACS:** 02.70.Dh

## INTRODUCTION

Clinching is a mechanical process to assemble two metal sheets by stamping them between a punch and a die. The connection point is formed by local plastic deformation of both sheets. Because of its simplicity of setting and low cost, this assembling process has a fulgurate development in the automotive sector. Besides, it is worth noticing that materials that cannot be spot-welded together, e.g. steel and aluminium alloys, may be assembled by clinching.

However, the use of clinch points remains limited by a lack of standard procedures to design such assemblies as well as a lack of reliable numerical tools to predict their mechanical resistance. Few works have been published on the simulation of the forming process [1-6] and even less on the simulation of the mechanical resistance [5].

Clinching tools are composed of a die, a punch and a blank-holder. The main successive steps (blank-holder approach, point forming and tools removal) of the forming process of a clinch point are schematised on Figure 1.

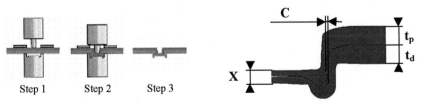

**FIGURE 1.** Clinch process.

**FIGURE 2.** Clinch point geometry

CP907, *10th ESAFORM Conference on Material Forming,* edited by E. Cueto and F. Chinesta
© 2007 American Institute of Physics 978-0-7354-0414-4/07/$23.00

The geometry of a clinch point can be characterised by the remaining thickness of both sheets on the axis of the point, denoted by $X$, and the clinch lock, denoted by $C$ (see Figure 2). Clinch lock determines the resistance of the point but is difficult to measure, whereas remaining thickness can be easily controlled during the process.

For given sheets to assemble, forming process parameters -to be optimised- are the (dimensions of) tools -punch and die- as well as the remaining thickness.

To assert their resistance, clinch -as well as spot-welded or riveted- points are generally submitted to two standard tests, viz. shear test and pullout test (also called pure tension test). During the shear test, the two metal sheets that constitute the specimen (see Figure 3) are subjected to opposite in-plane displacements that are perpendicular to the connection point. During the pullout test, the two metal sheets that constitute the cross shape specimen (see Figure 4) are subjected to opposite displacements that are in the same direction as the axis of the connection point.

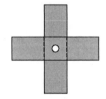

**FIGURE 3.** Shear test specimen.     **FIGURE 4.** Pullout test specimen.

This paper focuses on the finite element analysis of the resistance of clinched points to the pullout test. However, in order to take into account the exact geometry of the connection point as well as the mechanical state of the material after the forming process, viz. the residual stresses and strain hardening, the analysis of the resistance test must be subsequent to the analysis of the forming process.

## EXPERIMENTAL CAMPAIGN

To validate the numerical tool that is built during this study, Renault car manufacturer has conducted an experimental campaign on different materials using TOX™ tools to form circular clinch points.

Results presented herein concern different assemblies refered as $M_p$-$M_d$ ($t_p/t_d$, $X/t_d$), where $M_p$ (resp. $M_d$) denotes the material of the sheet on punch (resp. die) side, whereas the thickness of the sheet on punch side, $t_p$, and the remaining thickness, $X$, are normalised by the thickness of the sheet on die side, $t_d$. The two different steels and the aluminium alloy that are considered are respectively denoted by $A$, $B$ and $C$.

## NUMERICAL SIMULATION OF THE FORMING PROCESS

It has been verified that the pressure applied by the blank-holder is such that the deformation of the sheets is localised in the vicinity of the clinch point. Hence, to save computer cost, the numerical simulation of the forming process of a circular point can be axisymmetric, ignoring the exact geometry of the assembled sheets far from the connection point.

Such simulations of the forming process have been carried out with ABAQUS™/Explicit code using the standard "Mass Scaling" technique [6-7].

For these simulations, tools are modelled as rigid surfaces (discrete rigid R3D elements in ABAQUS™) whereas sheets are modelled as deformable bodies with linear solid elements (4-node CAX4R in ABAQUS™).

Contact between sheets –that plays as a prominent role- and contact between tools and sheets are modelled with standard Mohr-Coulomb model.

Sheet metal behaviour is modelled with a standard elastoplastic model with isotropic hardening.

The die is supposed to be fixed. To avoid shock and wave propagation problems that are common for this kind of simulation, the so-called "SMOOTH" ABAQUS™ procedure is adopted to apply punch and blank-holder displacements.

To avoid finite element distortion during the simulation, remeshing technique is used with a sufficient initial mesh refinement, typically defined by 10 to 12 linear quadrilateral elements through the thickness of the thinnest metal sheet.

By comparing the numerical predictions and the experimental results, this approach has been validated for different mono- or bi-material assemblies in terms of clinch point geometry and punch effort vs. punch displacement.

However, the last stage of the forming process implies the removal of the tools and a large elastic spring-back of the sheets. Since these phenomena cannot be modelled with a dynamic explicit algorithm, the prediction of the mechanical state of the material in the vicinity of the clinch point after the forming process, viz. the residual stresses, strain hardening of the material(s) and inter-sheet pressure, a "static relaxation step" must be performed with an implicit algorithm, typically after a data transfer to ABAQUS™/Implicit code.

## NUMERICAL SIMULATION OF THE PULLOUT TEST

To perform the numerical simulations, subsequently, of the pullout test and of the forming process, two strategies can be adopted as schematised in Figure 5.

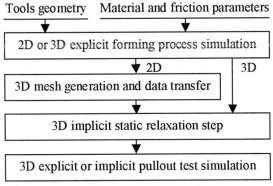

**FIGURE 5.** Simulation strategies.

In the first approach –that has been adopted for the examples presented herein- the same 3D mesh is used to simulate the forming process and the pullout test.

171

Another approach consists in the 2D simulation of the forming process -performed as described in previous section- followed by a data transfer from the 2D mesh to a 3D mesh including the arms of the specimen "far" from the vicinity of the clinch point.

In both cases, the so-called quasi-static relaxation step must be performed before the simulation of the pullout test. Taking advantage of the two symmetry planes of the cross specimen, it is possible to model only one quarter of the structure. Besides, whereas solid elements must be used in the vicinity of the clinch point, shell elements are rich enough to model the "arms" of the specimens.

The complete model that has been used during the forming process for the simulations presented herein is given in Figure 6. Apart from the shell elements, the rest of the model is the 3D version of the 2D model that is briefly described above and discussed in [6]. The tools are modelled by discrete rigid elements (R3D elements in ABAQUS™) whereas the sheets are modelled in the vicinity of the tools by linear solid elements (8-node C3D8R elements in ABAQUS™). The numerical procedures that have been validated in 2D situations are still valid for this 3D model.

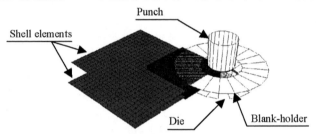

**FIGURE 6.** Geometry and mesh of the model.

To simulate the pullout test, the boundary conditions that are imposed –after removing the tools and the "static relaxation step"- to the ends of the specimen arms, correspond to rigid grips i.e. to uniform displacements -that are parallel to the clinch point axis- without any rotation.

During the pullout test, clinched assemblies generally fail by "unbuttoning" -that follows some sliding and, in some cases, the failure of one of the sheets in the vicinity of the clinch point-. All simulations that are presented below ignore any local failure of the -elastoplastic with isotropic hardening- material, and the failure of the specimen corresponds to the start of a relative finite sliding between the sheets. Hence, although all the results that are presented herein have been obtained by using ABAQUS™/Explicit code using the standard "Mass Scaling" technique, these simulations of the pullout test can also be performed with ABAQUS™/Implicit code that fails to converge only when sliding starts.

## Influence of the implicit static relaxation step

It is recalled that the elastic spring-back of the sheets after removing the tools is modelled by performing a so-called quasi-static relaxation step with ABAQUS™/Implicit code. Figure 7 illustrates in a particular case the influence of this relaxation step on the load vs. displacement plot that is predicted during the subsequent simulation of the pullout test.

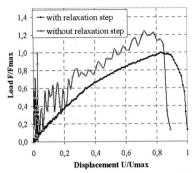

FIGURE 7. Load vs. displacement during pullout test $C - C$ (*1.25, 0.4*).

## Influence of the inter-sheet friction coefficient

Contrary to contact between sheets and tools, inter-sheet contact plays a prominent role during the forming process [6]. Figure 8 illustrates in two different cases the influence of the friction coefficient -as used in standard Mohr-Coulomb model- on the load vs. displacement plot that is predicted during the simulation of the pullout test.

The value of the friction coefficient between aluminium sheets appears smaller than the value of the friction coefficient between steel sheets.

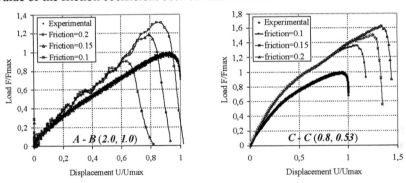

FIGURE 8. Inter-sheet friction influence.

## Influence of the non-standard initial state

To evaluate the influence of the non-standard state of the materials induced by the forming process, simulations of the pullout test have been carried out, with or without ignoring the hardening of the materials and the residual stresses that were predicted by the forming process simulation, the geometry of the clinch point remaining the same for both cases -respectively denoted as "non-standard" and "virgin" initial states-.

Figure 9 illustrates in two different cases the influence of the non standard initial state on the load vs. displacement plot that is predicted during the simulation of the pullout test.

173

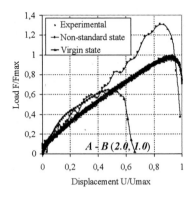 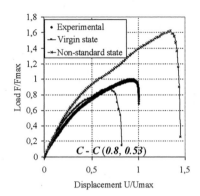

**FIGURE 9.** Influence of the non-standard initial state.

# CONCLUSIONS

Finite element simulations of pullout tests applied to clinch assemblies have been linked to the finite element simulations of the forming process of these assemblies. The predictive capability of these simulations has been validated by comparing the predictions of the simulations with experimental results.

Numerical predictions of the resistance of clinch points to pull-out tests are very sensitive to the point geometry but also to the value of the friction coefficient between sheets and to the non-standard initial state of the sheets after the forming process.

The simulations presented herein do not take into account a possible failure of clinch points by local failure of the material. Implementation of a procedure to predict material damage and failure is under development.

# ACKNOWLEDGEMENTS

Support by Renault Company is gratefully acknowledged.

# REFERENCES

1. V. Hamel, J.M. Roelandt, J.N. Gacel, F. Schmit, "Finite Element Modeling of Clinch Forming with Automatic Remeshing", Computers and Structures 77 (2000) 185-200.
2. J.P. Varis, "The suitability of Clinching as a Joining Method for High-strength Structural Steel", J. of Materials Processing Technology 132 (2003) 242-249.
3. J.P. Varis, J. Lepisto, "A Simple Testing-based Procedure and Simulation of the Clinching Process Using Finite Element Analysis for Establishing clinching Parameters", Thin-Walled Structures 41 (2003) 691-701.
4. J.P. Varis, "The suitability of Round Clinching Tools for High Strength Structural Steel", Thin-Walled Structures 40 (2002) 225-238.
5. C. Pietrapertosa, A.M. Habraken, J.-P. Jaspart, "Clinchage des Tôles Minces: Modélisation du comportement", Techniques de l'Ingénieur, BM 7 820.
6. M. Jomâa, R. Billardon, G. Roux, "Finite Element Analysis of Clinch Process to Assemble Thin Metals Sheets", ESAFORM 2006, N. Juster & A. Rosochowski eds, 863-866.
7. Hibbitt, Karlsson and Sorensen Inc., ABAQUS/Explicit User's manual, v. 6.5.

# Numerical Method to Analyze Local Stiffness of the Workpiece to avoid Rebound During Electromagnetic Sheet Metal Forming

Désirée Risch, Alexander Brosius, Verena Psyk, Matthias Kleiner

*Institute of Forming Technology and Lightweight Construction, University of Dortmund, Baroper Str. 301, 44227 Dortmund, Germany*

**Abstract.** Electromagnetic sheet metal forming is a high speed forming process using pulsed magnetic fields to form metals with high electrical conductivity such as aluminum. Thereby, workpiece velocities of more than 300 m/s are achievable, which can cause difficulties when forming into a die: the kinetic energy, which is related to the workpiece velocity, must dissipate in a short time slot when the workpiece hits the die; otherwise undesired effects, for example rebound, can occur. One possibility to handle this shortcoming is to locally increase the stiffness of the workpiece. In order to be able to estimate the local stiffness a method is presented which is based on a modal analysis by means of the Finite-Element-Method. For this reason, it is necessary to fractionize the considered geometries into a part-dependent number of segments. These are subsequently analyzed separately to determine regions of low geometrical stiffness. Combined with the process knowledge concerning the velocity distribution within the workpiece over the time, a prediction of the feasibility of the forming process and a target-oriented design of the workpiece geometry will be possible. Numerical results are compared with experimental investigations.

**Keywords:** Electromagnetic Sheet Metal Forming, Modal Analysis
**PACS:** 81.20 Hy; 89.20 Kk; 02.60 Cb

## MOTIVATION & FORMING MECHANISM

Electromagnetic sheet metal forming is a high speed forming process using pulsed magnetic fields to form metals with high electrical conductivity such as aluminum. A typical arrangement of tool coil and workpiece for this technology is illustrated in Figure 1a. The regarded setup consists of a spirally wound tool coil above which the workpiece is positioned. The drawing ring in combination with the cover plate represents the die, which, in this case, is a cylindrical cup. The forming machine is simplified, represented by an equivalent circuit consisting of a high current switch, a capacitor $C$, an inner inductance $L_i$, and an inner resistance $R_i$.

Closing the high-current switch affects a sudden discharge of the capacitor battery, thereby creating a highly damped sinusoidal current $I(t)$. The current causes a magnetic field $H(t)$ inducing a second current in the workpiece which is directed in the opposite direction of the coils current. As long as the workpiece is close to the coil, this induced current prevents the magnetic field from penetrating through the

CP907, *10th ESAFORM Conference on Material Forming*, edited by E. Cueto and F. Chinesta

workpiece in drawing-direction. The energy density of the magnetic field within the gap between the sheet and the coil refers to a magnetic pressure $p$ acting nearly orthogonal onto the workpiece's surface. If the yield point of the workpiece material is exceeded, plastic deformation occurs and the distance between workpiece and tool coil increases rapidly. A more detailed description of the process principle is given in [1].

a) Experimental setup

b) Geometrical variants

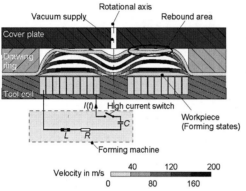
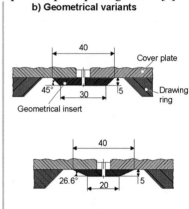

FIGURE 1: a) Experimental setup          b) Geometrical variants

Due to the characteristic forming mechanism local workpiece velocities of up to 300 m/s are achievable, which can cause difficulties when forming into a die. This correlation and the associated importance of the geometrical stiffness of the workpiece in this highly dynamic process can be explained on the basis of the energy transfer during the process, which is described in [2] and briefly summarized in the following. When the charging energy stored in the capacitor of the forming machine is transferred to a magnetic pressure the deformation of the workpiece starts, namely in the area where the windings of the tool coil are located. This means that the magnetic pressure pulse is transferred into kinetic energy on the one hand which is reflected in the velocity distribution of the workpiece, and into forming energy on the other hand. At the moment when the pressure is decreased to zero, this means that no external forces are acting any longer, the complete energy available for the remaining forming process is stored in the workpiece. In a forming operation without a die the process continues until the kinetic energy is completely transferred into deformation energy. When forming into a die the deformation is stopped as soon as the workpieces hits the die. This contact is accompanied by a sudden decrease of the workpiece velocity and the remaining kinetic energy in the workpiece has to be dissipated by the die. If this transfer cannot be entirely guaranteed, the energy remaining in the workpiece can cause further undesired effects, as for example rebound (compare Figure 1a). One possibility to handle this shortcoming is to locally increase the stiffness of the workpiece in order to transfer the kinetic energy into elastic deformation energy without any elastic-plastic deformation. The effect of such a variation of the geometrical stiffness is analyzed on the basis of different geometrical variants, shown in Figure 1b.

# STRATEGY TO ESTIMATE LOCAL STIFFNESS OF SPECIFIC WORKPIECE AREAS

On the basis of an example, illustrated in Figure 2, the general idea to estimate the local stiffness of workpiece areas is explained. At first, the structure is divided into significant sectors of the workpiece and a modal analysis of each segment is done. Thereby, it is important to constrain the nodes on the regarded cutting edges. The simplest possibility is to constrain these nodes to a zero oscillation condition so that any "movement" of the nodes is avoided. With this strategy the calculation of the eigenfrequencies $f$ of each sector can be done in order to estimate the local stiffness $K$ roughly, considering the mass $m$ via the following equation [3]

$$K = (2\pi \cdot f)^2 \cdot m \qquad (1)$$

Comparing the calculated stiffness of adjacent segments allows the detection of any local discontinuity of the stiffness within the workpiece, which could cause difficulties with regard to the desired geometry.

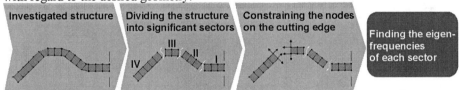

FIGURE 2: Strategy for local stiffness estimation

## DISCUSSION OF THE RESULTS

According to [5], a higher stiffness of the final workpiece geometry is less sensitive to the rebound. Based on this result, a detailed stiffness analysis by means of the described strategy was carried out, considering the local stiffness of different segments of the structure as well as the geometrical stiffness course over the process time. Thereby, the geometry of the intermediate forming states was determined by a coupled electromagnetic-mechanical simulation [4]. Subsequently, the modal analysis was done using the general purpose program ANSYS, whereby the used method to determine the eigenfrequencies was Block Lanczos. In doing so, each segment was defined by those nodes which are located at the cutting position of the final geometry, as indicated in Figure 2. Thus, the mass of each segment was kept constant while the thickness as well as the arc length vary in the different forming states. The results of these investigations are summarized in Figure 3.

Regarding the geometrical stiffness of the segments of the final forming state at the time 200 µs, a discontinuity in the stiffness distribution occurs at the crossing between sector II and sector III. The absolute value of this step is larger in example B, which can be the reason for the rebound in example B.

Furthermore, the geometrical stiffness of the segments in this critical area (indicated as segment III and as squares in the diagram) was higher in case of example A for which the better forming results could be achieved, as shown in Figure 3. In both examples the velocity distribution during the first contact between the workpiece and die is similar (compare the enlarged section in Figure 3), but in case of a lower geometrical stiffness course over time (example B) the required forming energy is lower so that the kinetic energy could not be dissipated completely and rebound occurs. Contrarily, the higher geometrical stiffness (example A) requires more forming energy so that the kinetic energy is totally dissipated.

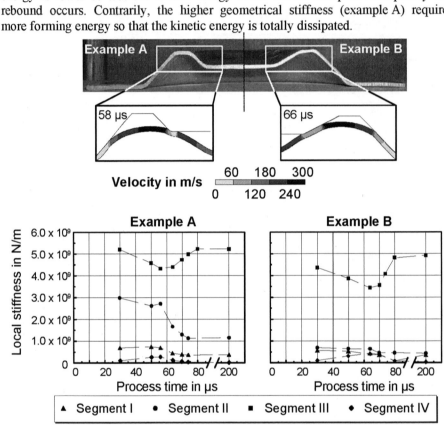

FIGURE 3: Experimental as well as numerical results

Additionally, the course of the geometrical stiffness of each segment over the process time is illustrated in the diagram in Figure 3. Since the desired geometry is well adapted to the intermediate forming states, the consideration of the final geometry is sufficient for the identification of critical areas. But in other part geometries (e.g. Figure 4) the intermediate forming stages have to be taken into account because high local stiffness introduced during the deformation process might be irreversible in the final forming result. This can be avoided by adapting the velocity distribution as shown in Figure 4. Despite the fact, that the stiffness of segment II for a charging energy of 300 J is temporarily higher than the charging energy of 1.200 J the forming result is much better. The reason for that can be seen in the lower stiffness over time course for segment I in case of the charging energy of 300 J.

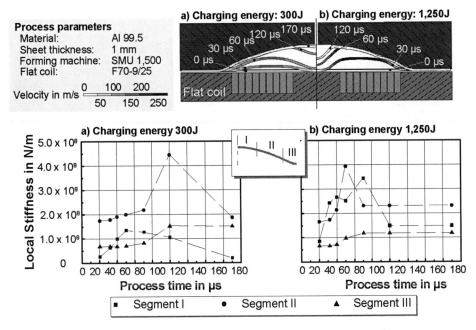

FIGURE 4: Numerical results by forming into a spherical die

## SUMMARY & OUTLOOK

In this contribution a numerical method to determine the local geometrical stiffness has been introduced, on the one hand, and its influence on the forming process has been shown, on the other hand. Moreover, a target-oriented modification of local workpiece stiffness was realized. Thereby, the importance of the process knowledge to interpret the results concerning the local workpiece stiffness has been pointed out.

The constraining within this approach was done by restraining the nodes to zero, which is the simplest method. In order to guarantee compatible modes of oscillation, special efforts have to be given in constraining the boundary of the analyzed segments in the future. Moreover, it is difficult to define the location of the cutting edge. Therefore, a method to position the cutting edges depending on the part geometry is required for further analysing and to standardizing the determination of the significant sectors.

## ACKNOWLEDGMENTS

The authors wish to thank the German Research Foundation (DFG) for the financial support of the research work reported here appurtenant to the research unit FOR 443.

179

# REFERENCES

1. D. Risch, C. Beerwald, A. Brosius, and M. Kleiner: On the Significance of the Die Design for the Electromagnetic Sheet Metal Forming. 1st ICHSF 2004, Dortmund, Proceedings pp. 191-200, ISBN 3-00-012970-7
2. M. Kleiner, D. Risch, C. Beerwald, and A. Brosius: Influence of the Velocity Distribution and Die Geometry on the Deformation Process of Electromagnetic Sheet Metal Forming, Annals of the WGP, Production Engineering XII(2005)2, pp. 95-98, WGP e.V., Hannover, ISBN 3-9807670-7-8.
3. K.-H. Grote; J. Feldhusen.: Dubbel Taschenbuch für den Maschinenbau, Springer-Verlag, 19. Auflage, ISBN 3540221425
4. A. Brosius: A Method for the Determination of Strain-Rate-Dependent Flow Curves by Means of Electromagnetic Tube Forming and Iterative Finite-Element-Analyses, PhD Thesis Dortmund, 2005
5. D. Risch, E. Vogli. I. Baumann, A. Brosius, C. Beerwald, W. Tillmann, and M. Kleiner.: Aspects of Die Design for the Electromagnetic Sheet Metal Forming Process, 2nd ICHSF 2006, Dortmund, Proceedings, pp.189 -200, ISBN 3-00-018432-5

# Measurement of material properties of 6000 Al-sheet for car body application using thermal imaging

Ralf Schleich[1], Alexander Dillenz [2], Manfred Sindel[3] and Mathias Liewald[1]

1)Institute for Metal Forming Technology, Universität Stuttgart,
Holzgartenstraße 17, 70174 Stuttgart, Germany
2) e/de/vis GmbH, Nobelstrasse 15, 70569 Stuttgart, Germany
3) AUDI AG, Postfach 1144, 74148 Neckarsulm, Germany

**Abstract:** The paper presents experimental results of thermo-graphical measurement of material properties. The analytical expression for the temperature variation of the specimen deformed in the elastic state is determined starting from the first law of thermodynamics. The experimental method for determining material properties based on the Joule-Thompson effect is presented in detail. The thermo-graphical method has been used to determine formability in different state of stresses of the AA 6016-T4 aluminium alloys.

**Keywords:** material properties, ductility, material experiments, sheet metal, Joule-Thompson effect

## INTRODUCTION

Modern lightweight construction using high strength materials in automotive industry requires improved technologies for measurement of mechanical material properties. In order to predict formability it is of great importance to measure very precisely the mechanical properties of Al-sheet metal. Not only in today's car development but also in every safety relevant construction, the material forming behaviour because of working load is highly considered. Depending on the respective application on the one side a dimensional accuracy or on the other side a well-tempered compensation of stress peaks is preferred [1]. For the structural designer of aluminium-weight tension structures, ductility is of great importance because of its role in a relief of stress concentrations. To characterize material properties concerning the ductility for forming simulation based on the law of plasticity and crash test basically the yield strength, tensile strength and the ultimate strain are used. Ductility is strictly defined as the ability of a material to be drawn into a wire. More generally, it is used to refer to the amount of plastic deformation a material can endure before failure. Ductility is not a uniquely defined material property, quantifiable by a single test [2]. The uniaxial tensile test is the most common method to determine mechanical properties, which is in conventional usage related with several inaccuracies.

The Joule-Thompson effect gives the opportunity to adequately detect the yield strength, tensile strength and the point of cracking. It is based on the fact that at elastic straining metals extend their volume up to one percent thus the enthalpy decreases and the sample undergoes a reduction in temperature as described in the first law of thermodynamics [3]. In the point of yield strength first plastic deformation produces heat due to inner friction, which leads to a change in derivative of sample temperature with respect to time which can be easily detected by current temperature measuring devices. This phenomenon has first been observed by Tresca in 1870 [4]. Gabryszewski [5] first used this method for the determination of the yield stress in the case of the uniaxial tensile test. The method has been extended for the biaxial tensile test for the determination of the yield locus by Sallat [6] and it was improved by Müller and Pöhlandt [8] and Banabic et al [9, 10].

CP907, 10th ESAFORM Conference on Material Forming, edited by E. Cueto and F. Chinesta
© 2007 American Institute of Physics 978-0-7354-0414-4/07/$23.00

## Theoretical Background

The theoretical background of determining material properties by thermo-graphical measuring devices during the deformation process has been elaborated precisely by Banabic et al [10]. This background is given in the following work and is restricted to small deformations and isotropic material properties. The first law of thermodynamics for reversible processes can be written as:

$$\rho \cdot \dot{e} = \sigma : d + \rho \cdot r \qquad (1)$$

where $\rho$ means mass density, $\dot{e}$ stands for change in the internal energy, $\sigma$ means the Cauchy stress, d stands for the strain rate, and $\rho$ means a heat source. The supplied heat $\rho r$ stands for reversible processes: $\rho r = \rho T \dot{s}$, where $T$ means the absolute temperature and $s$ the entropy. Utilizing such conditions the first law becomes:

$$\dot{e} = T \dot{s} + \frac{1}{\rho} \sigma : d \qquad (2)$$

As e can be regarded as a function of s and $\varepsilon$, the next equation contains:

$$\dot{e} = \left( \frac{\delta e}{\delta s} \right)_{\varepsilon} + \left( \frac{\delta e}{\delta \varepsilon} \right)_{s} : \dot{\varepsilon} \qquad (3)$$

hence,

$$T = \left( \frac{\delta e}{\delta s} \right)_{\varepsilon} \; and \; \sigma = \left( \frac{\delta e}{\delta \varepsilon} \right)_{s} \qquad (4)$$

and (Maxwell relation)

$$\left( \frac{\delta T}{\delta \varepsilon} \right)_{s} = \left( \frac{\delta \sigma}{\delta s} \right)_{\varepsilon} = T \left( \frac{\delta \sigma}{T \delta s} \right)_{\varepsilon} = \frac{T}{c_v} \cdot \left( \frac{\delta \sigma}{\delta T} \right)_{\varepsilon} \qquad (5)$$

The last term is obtained by considering a constant volume in: $(T \cdot \delta s)_{\varepsilon} = (c_v \delta T)_{\varepsilon}$. In solids compressive stresses are observed for heating processes, while keeping the volume constant (suppressing thermal expansion). Consequently, during a tensile test, the last term in equation 5 is negative, and hence

$$\left( \frac{\delta T}{\delta \varepsilon} \right) \varepsilon < 0 \qquad (6)$$

Without consideration of the temperature dependency of the bulk and shear modulus the temperature change under adiabatic conditions can be described as follows:

$$\Delta T = \left( \frac{\delta T}{\delta \varepsilon} \right)_{s} : \Delta \varepsilon = - \frac{T}{\rho \cdot c_v} \cdot \alpha_v C_b tr(\Delta \varepsilon) \qquad (7)$$

These thermo-elastic effects can be used as ductility criterion because of the significant differences in forming mechanism and correlating to this different amount of dissipated heat. The simplest approach in ductility measurement uses the ultimate tensile strain from the uniaxial tensile test as the main criterion of ductility measurement. A larger ultimate tensile strain often corresponds with a more ductile material property [2]. In spite of the easiness of this approach there are two disadvantages to distinguish. On the one hand neither the strain hardening nor the kind of necking is considered and on the other hand there can exist clearly different material properties. These would be seen subjective with different ductilities but are misinterpreted with the same properties.

Main aim in developing new ductility criterions focus on improvement of prediction of ductile material behaviour and estimating a cost and time saving measuring methodology. Referring to former approaches results the consideration of a ductility criterion with regard to the uniform elongation, strain hardening effects, form of necking shape and the ability to compensate local stress peaks by using objective indicators. With consideration of strain hardening effects and closely linked increase of tensile force the logarithmic ratio of yield strains until ultimate tensile strain is used.

$$
D_V = \frac{r \cdot \left[ \left( \left[ \ln \frac{[\ln(\varepsilon_{Gl}+1)]^r}{[\ln(\varepsilon_{Rp0.2}+1)]^r} \right]^2 + \left[ \ln \frac{\varepsilon_{Gl}+1}{\varepsilon_{Rp0.2}+1} \right]^2 \right)^{0.5} + \left( \left[ \ln \frac{R_m}{\sigma_{Bruch}} \right]^2 + \left[ \ln \frac{\varepsilon_{Br}-\varepsilon_{Gl}+1}{\varepsilon_{Gl}+1} \right]^2 \right)^{0.25} \right]}{1+r}
\tag{8}
$$

This approach provides the opportunity of inserting the most suitable material flow criterion. The mentioned examples in this work contain the flow criterion according to Ludwik. Because of using the necking width and anisotropy of the material, this proceeding is similar to the description of formability [12].

## Experiments

According to the recent state of the art, infrared cameras uses cooled InSb or MCT focal plane array detectors with a thermal resolution (NETD) of approximately 10-15 mK to measure minimal thermo-elastic changes. The most simple case of application of improving potentials of thermal imaging for measuring material properties is the uniaxial tensile test. Figure 1 shows the thermal image sequence of the uniaxial tensile specimen starting at the beginning of necking until failure. Correlating to this inhomogeneous forming the temperature increase can be adequately resoluted.

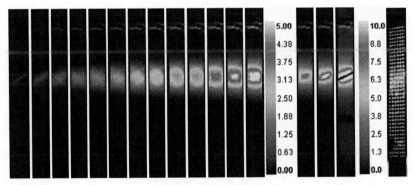

**FIGURE 1:** Time sequence of relative temperature increase in tensile specimen. Time scale ranges from necking to crack in 50 ms steps. Notice initial 45° dislocation (image 1 from left) and overlaid 135° dislocation (image 3 ff). Range of temperature scale was changed from image 13 (5°C) to 14 (10°C). After fracture, a global temperature rise due to thermo-elasticity can be observed. The last image shows the thermal imaging coupled with forming analysis.

The state of stress is well defined and the inaccuracies because of tactile measurement are well known. The detection of yield strength is based on an approximation which reveals yield point to appear at 0.2 percent of plastic strain, independent from used alloy, forming mechanism and state of stress. Additionally, in tactile measurement, all strains beyond the tensile stress are averaged to the measurement range. By necking, the conventional measured strain increasingly diverges from real local strain (see Fig. 2a). Furthermore at the point of failure, the tactile specimens are often displaced by the application flip so error in measurement increases similar to increasing tensile strengths. At the point of specimen discharge caused by failure, the reverse thermo-elastic effect appears and the sample temperature rises uniformly.

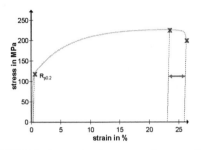 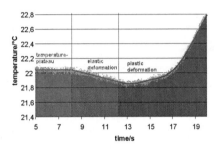

**FIGURE 2:** a) example of a stress-strain-curve with point of insufficient accuracy due to tactile measurement (left); b) temperature sequence at the yield point by using uniaxial tensile test (right)

Different thermo-graphical methods for the experimental determination of the yield loci are successfully used at the Institute for Metal Forming Technology (IFU) from University of Stuttgart. On the one hand a thermal point sensor can detect several yield points at different states of stress. On the other hand the here presented thermal imaging furthermore offers the opportunity of evaluating a high resoluted strain distribution until failure. By varying the longitudinal and transverse forces acting on a cross tensile specimen (see Fig. 3) any point of the yield locus in the range of biaxial tensile stress can be obtained. A description of most effective design of cross tensile specimen, which has been optimised by means of optical stress measurement can be found in [13].

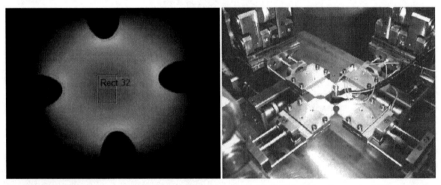

**FIGURE 3:** a) Cruciform specimen for the biaxial tensile test (left); b) Biaxial tensile test device for cross specimens (right)

Using this geometry, the thermo-graphical measurement visualised a zone of homogeneous stress, until first plastic deformation occurs in the notches. Since the geometry depends to some extent on the material properties, it was verified that for the used dimensions a large zone of homogeneous biaxial strain is detectable [10].

In the case of ductility investigation the thermal imagine is an expedient tool for visualising formability. For this purpose crash profiles of several aluminium alloys have been tested by compression until failure. The obviously more brittle specimen shows a more distinctive thermo-elastic effect because of the stronger strain fluctuation at geometric instability. Therefor the summated quantity of produced heat is less than at the more ductile specimen. This can be explained with the fact that most of the forming ends at the early point of cracking while most of the dissipated heat at the ductile specimen is produced during a longer time of deformation. These relationships are shown in Fig. 4.

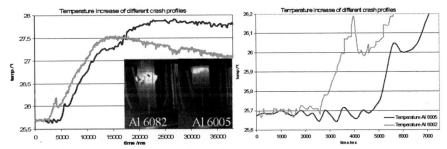

**FIGURE 4:** a) Cruciform specimen for the biaxial tensile test (left); b) Biaxial tensile test device for cross specimens (right)

## Conclusions

The thermo-graphical method used for the experimental determination of material properties, yield loci or indicators of ductility for sheet metals has only been used empirically or has been neglected at all. Experiments made on the AA 6016-T4 aluminium alloy properties have proven the validity of the thermo-graphical method to determine material properties as for example the yield strength or parameters of the new made ductility approach. With this technique it's also easily possible to detect the local resoluted point of yielding. Thus beginning of local necking and tensile strength can be determined precisely. By using a combined temperature and forming analysis as shown in Fig. 1, local strain can be measured and visualised online until cracking, in one step. At the point of sample discharge caused by failure, the reverse thermo-elastic effect appears and the sample temperature rises uniformly. Furthermore local strain distributions and particularly the development of strain distribution by time can measured and visualised as shown in Fig. 5 for the cruciform specimen. The yield point is firstly reached at the smallest area of flange and increase it's surface by relocating to the middle of the specimen. Similarly plastic deformation also starts at the smallest area of flange and take form of a ring about the yielding area. Hence the relocation of plastic deformed zone appears temporal offset to the thermo-elastic yielding-effect.

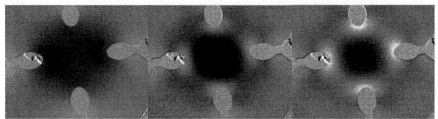

**FIGURE 5:** a) Cruciform specimen at yielding (left); b) Cruciform specimen at beginning of plastic deformation (middle); c) Cruciform specimen at further plastic deformation (right)

This conclusions lead to more precisely determined material properties which have effect on the prediction of the yield curve and the Forming Limit Curve. PLC-effects, necking, and – for biaxial loading – spatial resolved yield can be determined as well.
Regarding the whole product life cycle, product planning, tooling, and production can benefit from this improved determination of mechanical material properties.

# References

1. G. E. Dieter, *Introduction to Ductility*, Paper Seminar der American Society for Metals, Ohio, 1967
2. C. Leppin, H. Shercliff, J. Robson, J. Humphreys, Material Property Measurement: Ductility, Alumatter Aluminium Wissensdatenbank; http://www.alumatter.info; 2006
3. G. Backhaus, *Deformationsgesetze*, Akademie Verlag, Berlin, 1983
4. J. Bell, *The Experimental Foundation of Solid Mechanics*, Springer Verlag, New-York, 1973
5. Z. Gabryszewski, W. Srodka, *Zastosowarie temperatury sprzezonej z odksztalceniamido oceny dysypacji energii i wyznaczania granic plastycznosci*, Mechanica Teoretyczna i Stosowana, 1981
6. G. Sallat, *Theoretische und experimentelle Untersuchungen zum Fließverhalten von Blechen im zweiachsigen Hauptspannungszustand* Diss. TU Karl Marx Stadt 1988
8. W. Müller, K. Pöhlandt, *New experiments for determining yield loci of sheet metal*, 1996
9. D. Banabic, D.S. Comsa, S. Keller, S. Wagner, K. Siegert, *An yield criterion for orthotropic sheet metals*, In: Innovations in processing and manufacturing of sheet materials, 2001
10. D. Banabic, J. Huetink, *Determination of the yield locus by means of temperature measurement*
11. C. Leppin, *Duktilität und Umformbarkeit von Werkstoffen – Werkstoffcharakterisierung, aber wie?*, Forschungsbericht Alcan Technology, 2003
12. R. Schleich, M. Sindel, M. Liewald, *Potentials of new ductility criterions in car development with lightweight materials*, In: International Aluminium Journal, March 2007
13. R. Kreißig, *Theoretische und experimentelle Untersuchungen zur plastischen Anisotropie* Diss. TU Karl-Marx-Stadt 1981

# Ductile Damage Evolution and Strain Path Dependency

C.C. Tasan[1,2], J.M.P. Hoefnagels[2], R.H.J. Peerlings[2], M.G.D.Geers[2]
C.H.L.J. ten Horn[3], H. Vegter[3]

[1]*The Netherlands Institute for Metals Research (NIMR), PO Box 5008, 2600GA, Delft, The Netherlands*
[2]*Eindhoven University of Technology, Department of Mechanical Engineering, PO Box 513, 5600MB, Eindhoven, The Netherlands*
[3]*Corus Research Development & Technology, PO Box 10000, 1970 CA IJmuiden, The Netherlands*
*e-mail: c.tasan@tue.nl, tel:+31 40 247 5169, Fax:+31 40 2447355*

**Abstract.** Forming limit diagrams are commonly used in sheet metal industry to define the safe forming regions. These diagrams are built to define the necking strains of sheet metals. However, with the rise in the popularity of advance high strength steels, ductile fracture through damage evolution has also emerged as an important parameter in the determination of limit strains. In this work, damage evolution in two different steels used in the automotive industry is examined to observe the relationship between damage evolution and the strain path that is followed during the forming operation.

**Keywords:** Damage, strain path, sheet metal, forming limit diagrams.
**PACS:** 91.60.Ba

## INTRODUCTION

Forming limit curves (FLC) are generally used in the industry to define the safe forming strains for any given forming operation. In these diagrams, localized necking of the sheet is the criterion for determining the safe forming regions. However, it is also known that sheet metals may fail due to one or the combination of the following three different mechanisms. These are instability with localized necking (followed by ductile or shear fracture inside the neck area), shear fracture (based on shear band localization) and ductile fracture (based on nucleation, growth and coalesence of voids) [1].

With the rise of the popularity of advanced high strength steels in the last decade in automotive industry, ductile fracture which occurs before the forming of a possible neck has evolved as an important factor in determining the limits of formability. It is known that damage evolution affects the mode of failure very significantly, especially in the biaxial strain paths where void growth is more relevant [2].

The goal of this work is to examine and compare the FLC's, fracture limit curves (FrLC) and fracture surfaces of an interstitial-free (IF) steel and a dual-phase (DP) steel, in order to analyze the relationship between ductile damage and the strain paths followed.

*CP907, 10th ESAFORM Conference on Material Forming, edited by E. Cueto and F. Chinesta*
© 2007 American Institute of Physics 978-0-7354-0414-4/07/$23.00

# EXPERIMENTAL METHODOLOGY

Determination of FLC's are carried out according to the principle of Nakajima test using a hemisphere punch. A measurement grid of 0.5mm spacing is applied to the surface of the undeformed blanks and the blanks with different geometries are deformed until the point of fracture. The strains between the grid points on the samples are measured in the major and minor directions by the use of the photogrammetric software PHAST. To measure the local strains at which necking starts, the strain profile along the major strain direction perpendicular to the crack, but excluding the strain points inside the neck, is fitted to obtain a value for the necking strain. The necking strains of each geometry are combined to form the FLC. To obtain the FrLC curve, the principle strains at fracture are obtained by subtracting the crack opening as measured using a microscope from the distance between grid points across the crack. In addition, to visualize the path followed by each geometry, specimens of the same geometry are deformed to different amounts of final deformation, and local strain values are measured and marked on the strain diagram.

To characterize the failure modes of the different geometries of IF and DP steels, uniaxial and biaxial specimens from each are cut and metallographically prepared for SEM examination in three viewing directions, as schematically shown in Figure 1.

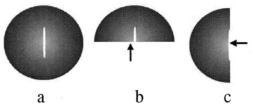

a        b        c

**FIGURE 1.** Viewing directions for fractography analysis (a) full FLC specimen from the top (b) Thickness cross section perpendicular to the crack (c) Fracture surface

# PRELIMINARY RESULTS AND DISCUSSION

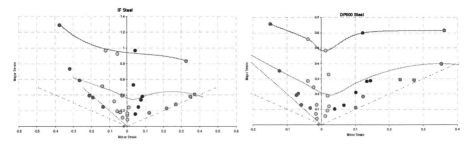

**FIGURE 2.** FLC and FrLC for IF steel (left) and DP600 steel (right). Each geometry is represented with different colored dots. Important strain paths, i.e. shear, uniaxial and biaxial, are shown respectively with the dotted lines from left to right. Note that the FLC-curves have been determined previously with high accuracy from a separate set of Nakazima strips (data points not shown in graphs).

Obtained FLC and FrLC curves of IF and DP600 steels are given in Figure 2. It is seen for all geometries of both steels that the deformation starts with a linear path up to the point of necking. Past the FLC, for the right hand side of the graphs (positive minor strain), a vertical path is followed up to the fracture limit curve. This can be explained by the fact that the neck is constrained in the minor strain direction (along its axis), which yields a plain-stress state in the neck and thus a vertical strain path. For the left hand side of the graphs (negative minor strain), the strain paths between FLC and FrLC are not completely bend towards a vertical path, which may be explained by the decrease in constrain of the neck in the minor direction with decreasing width of the Nakazima strips. Another interesting point is the difference of fracture behaviour of the two alloys with different formability. It is observed that the FrLC of the less formable DP600 steel has the same shape as the FLC, whereas the FrLC of the very formable IF steel is a more linear curve.

The uniaxially-strained IF specimen fails through shear band localization as shown in Figure 3-a. A cross section of the morphology perpendicular to the crack direction is shown in Figure 3-b. The SEM micrographs of the crack reveal three different microstructural zones. In the center of the crack (Figure 3-c: right inset) a severely deformed morphology is observed which shows extensive energy dissipation and also varying sizes of voids. Next to this zone is a region with a relatively less deformed surface and smaller sized void formation (Figure 3-c left inset). The neighbouring region is a region of extensive shear, which shows oriented dimples and a highly deformed morphology (Figure 3-d).

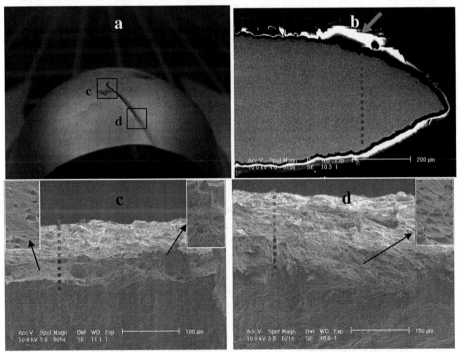

**FIGURE 3.** Uniaxially-strained IF specimen. (a) Macroscopic view of the crack (b) Thickness cross section of the shear region. The red arrow shows the whitening due to charging of the resin used. Necking starts at the blue dotted line. (c,d) Fracture surface at (c) the center of the specimen and (d) the

189

shear zone. The fracture surface is shown by the blue dotted line whereas the red dotted lines show the localized necks. Insert pictures show higher magnification images of voids in different regions.

The biaxially-strained IF specimen fails through severe thinning which is followed by necking and cup-cone type of ductile fracture, see thickness cross section of the fracture surface in Figure 4-b. Void formation and growth in the center of the crack triggers the fracture, while the material fails by shearing towards the surfaces. It is possible to see some void formation on the shear-edges (Figure 4-d)

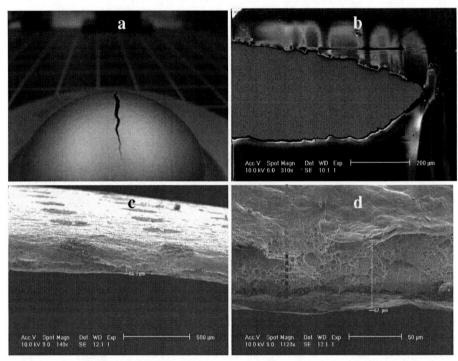

**FIGURE 4.** IF specimen which follows the biaxial path. (a) Macroscopic view of the crack. (b) Thickness cross section of the fracture surface. (c,d) Fracture surface of the specimen. Blue dotted line shows the fracture surface whereas the red dotted line denotes the localized neck.

Uniaxially-deformed DP600 sheet fails by necking and through-thickness shear fracture (Figure 5). The investigation of the microstructure of the crack reveals excessive plastic deformation (Figure 5-c) and damage evolution (Figure 5-b). It is also possible to see examples of a void coalesence mechanism (Figure 5-c, right inset). Another interesting morphology to point out is the local microcracks formed by the coalesence of voids. (Figure 5-c, left inset).

Biaxially-deformed DP600 sheet also fails similarly, although excessive thinning is observed in this case (Figure 6-a). The crack morphology shows a through-thickness shear fracture, with some local changes in shear direction (Figure 6-b). As in the uniaxial DP600 specimen, void formation and local microcracks formed through void coalesence is also observed here; however, the number of voids is observed to be higher (Figure 6-c,d).

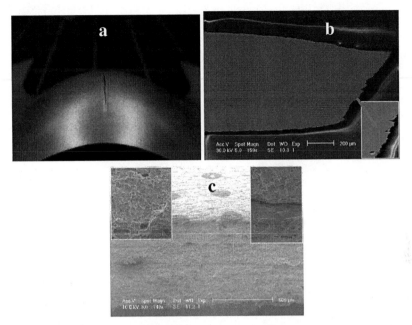

**FIGURE 5.** Uniaxial DP600 specimen. (a) Macroscopic view of the crack. (b) Thickness cross section. Inset showing magnified view of the subsurface voids (c) Fracture surface of the specimen. Insets showing magnified views of the voids (left) and microcracks (right).

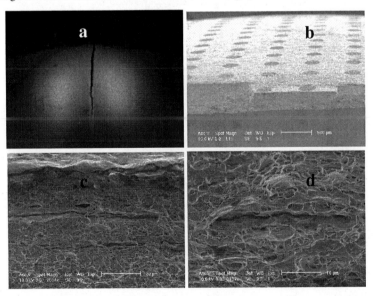

**FIGURE 6.** Biaxial DP600 specimen. (a) Macroscopic view of the crack. (b,c,d) Fracture surface.

Comparing the uniaxial and biaxial for the case of DP600 it is noted that both cases show a through-thickness shear fracture, which seems to be initiated at the specimen surface. This is in contrast with the IF specimens, which both for the uniaxial and biaxial case show that the initial fracture is triggered by void formation and growth in the center of the specimen, i.e. ductile fracture. This difference can be attributed by the higher formability of the IF steel compared to the DP600 steel and also seems to be corroborated by difference in absolute major strain going from FLC to FrLC, which is approximately a factor of 2 higher for IF steel than for DP600. Examining the uniaxial and biaxial case of the IF steel, in both cases the initial ductile fracture is followed by failure due to shearing towards the sample surface, however, in the uniaxial case shearing occurs mainly in the in-plane direction, while in the biaxial case shearing occurs in the thickness direction along the full length of the crack.

## CONCLUSIONS

Nakazima strips of IF and DP600 steel sheets are tested to the point of fracture to obtain both FLC's and FrLC's. The fracture surfaces are also examined to investigate the differences in damage evolution in different strain paths. The preliminary results show distinct differences in size and distribution of voids and microcracks between uniaxial and biaxial strain paths, demonstrating a clear impact of strain path and ductile damage on the failure mechanisms observed.

## ACKNOWLEDGMENTS

This work is funded by The Netherlands Institute of Metals Research (NIMR) (project no: MC2.05205a), which is gratefully acknowledged.

## REFERENCES

1. H. Hooputra, H. Gese, H. Dell, H. Werner, *Int. Journal. Of Crashworthiness* **9**, 449-463 (2004).
2. Z. Marciniak and J.L. Duncan, *Mechanics of Sheet Metal Forming*, London: Edward Arnold, 1992

# Some considerations on force trends in Incremental Forming of different materials

G.Ambrogio[1], J. Duflou[2], L.Filice[1], R. Aerens[2]

[1] Dept. of Mechanical Engineering, University of Calabria – 87036 Rende (CS), Italy
[2] Dept. of Mechanical Engineering, KULeuven - Celestijnenlaan 300 A - 3001 Heverlee, Belgium

**Abstract.** Today, incremental Forming challenges are mainly related to formability limits and precision. High achievable strain levels, together with the possibility to form complex shapes without need for dedicated dies, probably represent the main process advantages. However the attention on material formability is always very relevant. Taking into account both the formability and the process accuracy, the knowledge of the forces generated between the punch and the clamped sheet supplies strategic information to the analyst. In fact, the force level is not only relevant for the equipment deflection but also influences the precision. In fact, in previous publications the authors demonstrated that there is a strict correlation between the force trend and the material failure approaching. In this paper, a broader analysis on AA1050-O, AA3003-O and DC04 drawing steel is carried out, highlighting the force trends depending on the process parameters and the relationship with formability limits.

**Keywords:** Incremental Forming, Material formability, Force trend.
**PACS:** 81.20.Hy

## INTRODUCTION

Analysis of Incremental Forming processes is carried out, today, mainly with the aim to define the formability limits and to determine the achievable precision by comparing the obtained parts with the desired ones. In fact, the high allowable thinning, combined with the possibility to form complex shapes without the need for dedicated dies, probably represent the main process advantages [1,2]. On the other hand, springback, exalted by the presence of large, unsupported surfaces, together with the low stiffness of some forming equipment, tend to decrease the obtained precision when Incremental Forming processes are used [3]. However, some interesting industrial applications of the process showed its suitability for manufacturing of small batches or single prototypes. As far as both the formability and the process precision are considered, the knowledge of the forces generated between the punch and the clamped sheet may supply consistent information to the analyst. In fact, in previous publications some of the authors demonstrated that there is a strict correlation between the force trend and the material failure approaching [4,5]. Starting from the considerations formulated above, a wide study on force trends in Incremental Forming of different materials was carried out, as reported in this paper. In particular, AA1050-O, AA3003-O and DC04 deep drawing steel were chosen to measure the forming

CP907, 10th ESAFORM Conference on Material Forming, edited by E. Cueto and F. Chinesta

force trend when manufacturing a simple truncated cone. As process variables, the wall slope, the material thickness, the punch diameter as well as the tool pitch were varied in a large range. Some relevant guidelines and suggestions have been derived from the analysis of the obtained curves with respect to the different analysed materials, as it is accurately discussed in detail in the paper.

## FORCE TREND FOR INVESTIGATED MATERIALS

A wide experimental campaign was performed both at the Laboratory of Mechanical Engineering of University of Calabria and at the Laboratory of the Department of Mechanical Engineering at the Katholieke Universiteit Leuven. All experiments were conducted utilizing CNC milling machines, equipped with a Kistler piezoelectric dynamometer and a charge amplifier connected with a data acquisition system. Although the utilized measurement instruments are able to measure the three components of the force and three moments, only the component along the axis of the tool (Fz) is considered of strategic importance in the present analysis. The projection of this component (Fz) on the wall, which we shall conventionally refer to as the "tangential force", and which is equal to Fz multiplied by the sine of the wall inclination angle, is the factor responsible for the sheet stretching and, in consequence, for the product failure [4]. Clamping equipment which ensures a high rigidity during the experiments was mounted on both the machine tables (Figure 1).

**FIGURE 1.** The equipment utilized during the two experimental campaigns.

A hardened steel tool completed the equipment and an emulsion of mineral oil was used as lubricant to reduce the friction related phenomena. In order to reduce the number of experiments, some factors were kept constant from both a process parameter and a geometrical point of view, i.e. tool speed rotation and the feed rate. A frustum of a cone was chosen as specimen profile. Even though the initial major base dimension was not equal for the two experimental campaigns, the same locking conditions were ensured, positioning a backing plate at the bottom of the sheet at the process beginning. On the other hand, the experimental campaign was performed changing some process parameters in order to evaluate their influence on the measured force. Three materials were considered, namely two aluminium alloys (AA1050-O and Al3003-O) and a deep drawing steel (DC04), characterized by different mechanical properties as summarized in Table 1.

**TABLE 1.** Mechanical properties of the investigated materials.

| Properties | AA1050-O | Al3003-O | DC04 |
|---|---|---|---|
| Young's Modulus | 70000 N/mm$^2$ | 70000 N/mm$^2$ | 210000 N/mm$^2$ |
| Yield Strength | < 55 N/mm$^2$ | 40 N/mm$^2$ | 210 N/mm$^2$ |
| Tensile Strength | 95 N/mm$^2$ | 103 N/mm$^2$ | 300 N/mm$^2$ |
| Strain Hardening Exponent | 0.140 | 0.224 | 0.240 |
| Failure Strain | 25% | 40% | 37% |

At the same time, taking into account the knowledge base on the investigated process [4,5,6], the experiments were repeated varying the tool pitch (p) and the tool diameter ($D_p$) in a wide range, while the wall inclination angle ($\alpha$) was changed according to the specific material formability limit. Finally, also the influence of the sheet thickness (s) was included into the analysis, executing the experiments in different conditions. The highest and the lowest value on the investigated experimental plane are shown in the following table (Table 2).

**TABLE 2.** Extreme Value of the Experimental Plane.

| Parameters | Lowest Value | Highest Value |
|---|---|---|
| Tool Diameter  ($D_p$) | 10 mm | 25 mm |
| Tool Depth Step  (p) | 0.3 mm | 1 mm |
| Wall Inclination Angle ($\alpha$) | | |
| AA1050-O | 60° | 80° |
| Al3003-O | 20° | 70° |
| DC04 | 60° | 80° |
| Sheet Thickness (s) | | |
| AA1050-O | 1 mm | 2 mm |
| Al3003-O | 0.85 mm | 2.0 mm |
| DC04 | 0.5 mm | 1 mm |

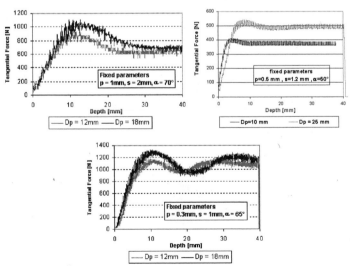

**FIGURE 2.** Force curves for varying tool diameters (1050, 3003, and DC04 respectively).

According to previous results, coherent considerations can be derived for the investigated materials changing the process parameters. First of all, a strong influence on the force distribution is exerted by the tool diameter; more in detail, increasing the tool dimension a double effect was observed: firstly the load peak increases, since the contact surface increases too and secondly, there is a sort of delay in the achievement of stationary conditions (Figure 2). This phenomen can be explained by the longer bending phase which is imposed on the blank when a larger tool is mounted.

Analogous conclusions can be derived with regard to the influence of the tool pitch and the sheet thickness on the forming load. In particular, increasing the tool pitch and the sheet thickness shifts the load curves towards higher values, since higher deformations are locally imposed (Figure 3.a) and, contemporary, a significantly higher force is required to deform a thicker blank (Figure 3.b).

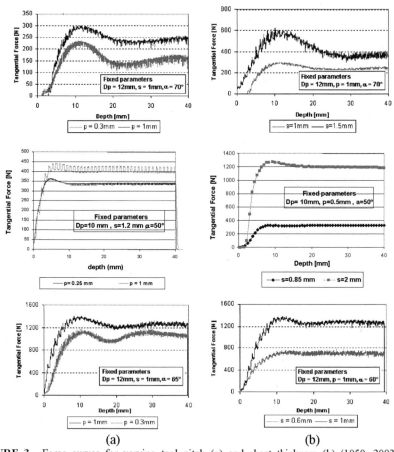

(a)                                     (b)

**FIGURE 3.** Force curves for varying tool pitch (a) and sheet thickness (b) (1050, 3003, DC04 respectively).

Finally, as in the formability analysis, the most relevant parameter, which strongly characterizes the force distribution trend, is the wall inclination angle. In fact, putting aside the material properties, increasing the slope angle the force trend changes from a steady state behaviour to a monotonically decreasing one, which predicts the material failure approaching (Figure 4).

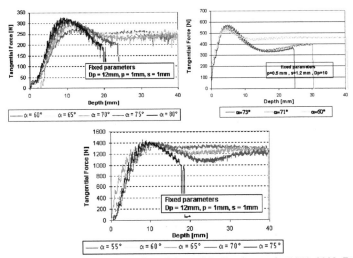

**FIGURE 4.** Force distributions at the varying of the wall inclination angle (1050, 3003, DC04).

## DISCUSSION OF THE RESULTS

According to previous analysis [4,5] and, according to the results presented here, it is possible to asses that the force gradient (K) after the peak levelstrongly depends on the experimental settings and can be recognized as an index of the process stability. In other words, a threshold value can be identified, or, in a more pragmatic approach, an uncertainty region, which discerns a safe condition from an unsafe one. Continuous monitoring of the K-factor could thus ensure a suitable on-line control of the process feasibility.

At the same time, it is easy to understand that the force gradient K is material dependent, as can be concluded when qualitatively observing Figures 2, 3 and 4. The K factor can be calculated using the following Equation:

$$K = \frac{F_j - F_{j-\varepsilon}}{H_j - H_{j-\varepsilon}} \tag{1}$$

where $F_j$ and $H_j$ are the actual values for the measured force and the component height, while $\varepsilon$ is a user-defined buffer for the factor value renewal. Concluding, the analytical analysis of the measured trend allowed to define the uncertainty regions for the investigated materials traced in the following maps (Figure 5).

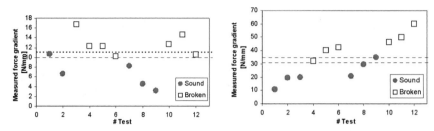

**FIGURE 5.** K-factor distribution for the AA1050-O, and DC04 respectively.

Since it has already been shown that the failure conditions depend on the sheet thickness [4], the diagrams shown above only relate to the experiments carried out on 1mm thick sheets for AA 1050-O and DC04

## CONCLUSIONS

Systematic tests have been carried out in order to show the main influences of the tool diameter, the tool pitch, the sheet thickness and the wall angle on the tangential force component. Increasing each of those parameters will increase the tangential force: the highest sensitivity can be observed for the sheet thickness, but only the increase of the wall angle will cause failure of the material. However, the ultimate wall angle is depending on the sheet thickness. For high values of the wall angle, the intensity of the force gradient, after the maximum level has been detected, can be used as an indication for material failure prediction.

## REFERENCES

1. J. Jeswiet, F. Micari, G. Hirt, A. Bramley, J. Duflou, J. Allwood, "Asymmetric Single Point Incremental Forming of Sheet Metal", *Annals of the CIRP*, **54/2**, 2005, 623-650.
2. G. Ambrogio, L. De Napoli, L. Filice, F. Gagliardi, M. Muzzupappa, "Application of incremental forming process for high customised medical product manufacturing", *Journal of Materials Proc. Tech.*, **162-163**, 2005, 156-162.
3. G. Hirt, J. Ames, M. Bambach, R. Kopp, "Forming strategies and Process Modelling for CNC Incremental Sheet Forming", *Annals of the CIRP*, **52/1**, 2004, 203-206.
4. L. Filice, G. Ambrogio, F. Micari, "On-Line Control of Single Point Incremental Forming Operations Through Punch Force Monitoring", *Annals of the CIRP*, **55/1**, 2006, 245-248.
5. J. Duflou, Y. Tunckol, A. Szekere, P. Vanherck, "Experimental Study on Force Measurements for Single Point Incremental Forming", accepted for the *Internat. Journal of Machine Tools and Manufacture*
6. J.R. Duflou, A. Szekeres, P. Vanherck, "Force Measurements for Single Point Incremental Forming: An Experimental Study",Proc. of the 11th Internat. Conference on Sheet Metal, Erlangen, 2005, 441-448.

# Innovative User Defined Density Profile Approach To FSW Of Aluminium Foam

Dorotea Contorno*, Livan Fratini*, Luigino Filice†, Francesco
Gagliardi†, Domenico Umbrello† and Rajiv Shivpuri°¶

*  *Dept. of Manuf. and Management Eng. – University of Palermo – 90100 Palermo (Italy)*
† *Department of Mechanical Engineering, University of Calabria 87036 Rende (Italy)*
° *Industrial, Welding and Systems Eng. Department – Ohio State University 43210 Columbus (USA)*

**Abstract.** Metallic foams are one of the most exciting materials in the world of mechanical industry due to their reduced mass and the good mechanical, thermal and acoustic characteristics. Consequently, their application, is increasing day by day even with the important drawbacks that reduce their suitability and diffusion such as high manufacturing cost and difficulty in processing. An innovative approach is outlined in this paper that enables the production of complex shapes taking advantage of deformation processing and friction stir welding (FSW). The aim is to create customized tailored manufactured parts. The cellular construction of foams makes this approach rather challenging as the cell walls are extremely thin and deform unpredictably especially in the presence of rotating and moving hard tool. In this paper, an integrated approach to overcome some of the above challenges is proposed. The initial density is modified by using simple deformation processes, in order to obtained the desired "crushed density", customized for the intended application. Then, the panels are joined to specially designed solid blocks by using FSW process with a proper set-up. Finally, the obtained specimens are evaluated for mechanical performance and the quality of the joint..

**Keywords:** Welding, FSW, Aluminium Foam.
**PACS:** 81.20.Vj

## INTRODUCTION

Metal foams made by first compressing mixtures of aluminium alloy powders and Titanium hydride (TiH$_2$) and then melting and foaming the resulting densified compact are now being exploited commercially under trade names such as "Alulight" or "Foaminal" [1]. Preliminary tests [2], for example, have shown that for a given density the "Alulight" closed cell powder-route aluminium foams are amongst the stiffest and strongest of the commercial ones.

However, generally, closed-cell aluminium foam offers a unique combination of properties such as, precisely, high stiffness, low density, and energy absorption [3,4]; there are several technologies to produce this material type and in each of them a series of variables affecting the foam topology has to be considered like size and volume fraction of the solid particles [5,6], foaming temperature, melt viscosity and surface tension, air pressure and flow rate, etc. In the recent years, aluminum foam

CP907, *10$^{th}$ ESAFORM Conference on Material Forming,* edited by E. Cueto and F. Chinesta
© 2007 American Institute of Physics 978-0-7354-0414-4/07/$23.00

properties and possible applications have been deeply investigated in order to allow an always more and better use of this material in several applications [7-9]; only few information can, instead, be found concerning joining techniques for integrating metallic foams in already existing structures, such as in the automotive applications.

Recent studies show that the laser beam welding, with its keyhole effect, is a feasible process for joining aluminum alloys and aluminum foams [10]. However, laser beam welding is not the unique process for joining sheet materials. Recently, a new solid state welding technique, as the friction stir welding (FSW), can be applied for joining sheet materials and aluminum foams; this technique works thanks to a properly shaped rotating pin, characterized by a nuting angle, that is inserted into the blanks with a tool sinking and it is moved along the welding path determining the joining of the two edges. In this joining process, the necessary heat for the welding sequence is generated both by friction action and by deformation work.

In a previous research [11], the authors applied FSW for joining Alulight ® foams, thick plates and sandwiches made up from an internal foam and two external layer of aluminum thick sheet. Several process conditions were investigated and the obtained specimens were tested using a properly developed 3 point bending equipment, showing that there is a strongly influence of the process parameters on the joint quality.

In this work, instead, the joint quality, carried out on simple aluminium foam panels by FSW, was always investigated using a 3 point bending analysis. The density of the joined parts was modified by simple compression tests in order to obtain the desired "crushed density" profiles customized for the intended application. The influence of this variable for different process conditions was highlighted, showing the joints ability to withstand the bending loads all over the investigated cases.

## THE AIM OF THE WORK

The aim of this work is to design an efficient joint made of foam and solid aluminum plate. Actually, in several applications, in fact, the metallic foams have to be welded to solid structures in order to ensure a synergic behavior (Figure 1).

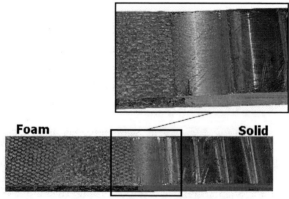

**FIGURE 1.** The Mixed Foam-Solid Structure.

Due to the relevant difference in density between the foam and the solid panel, the joining process usually represents a critical point. To overcome this problem, in this study the FSW process is utilized, with the aim to allow the material flow from the solid to the foam, in order to fill the neighbor cells ensuring a suitable strength after welding.

However, the foam panels have a density distribution along the thickness that strongly depends on the manufacturing process. Actually, density increases slightly close to the faces but, of course, a better mechanical behavior is obtained when the density becomes very high at the maximum distance from the symmetry plane. For this reason, in fact, some foam-sheet panels are manufactured.

In this study an innovative approach was developed. A simple upsetting process caused a cell implosion inducing a customized density increasing. Consequentially, a better mechanical behavior was expected, in the normal applications, even if a certain disadvantage in terms of mass is obtained. The performance increasing was measured executing some bending tests on the joined structure. The experimental evidences is reported in the next paragraphs.

## THE EXPERIMENTAL PROCEDURES

Foam to thick plate welds have been developed by FSW process. The effect of density variation obtained pre-compressing the panels has been investigated. In particular starting from the original thickness of 9 mm, the panels have been compressed up to 7 and 5 mm by simple upsetting on a hydraulic testing machine (Figure 2).

**FIGURE 2.** The Upsetting Phase.

In this way, a different height reduction of 0%, 22% and 44% was obtained respectively.

The original foam specimen was 100x40x9mm, with a mass of about 34 g, showing an apparent density of 929 kg/m3. The mass was equivalent to a solid sheet with a thickness of about 3 mm.

Three different butt joint configurations have been investigated, joining 9 mm, 7 mm, 5 mm thick foam panels with Aluminium of the same thickness, respectively, constituted by the UNI AlMg1Si0.6 alloy. A properly designed clamping fixture was utilized in order to fix the specimens to be welded on a traditional milling machine (Figure 3a). The steel plates, composing the fixture were finished at the grinding machine in order to assure an uniform pressure distribution on the fixed specimens.

The tools, made in UNI-C40 steel, present three geometries depending on the different thickness to be welded, the relevant dimensions are shown in Table 1.

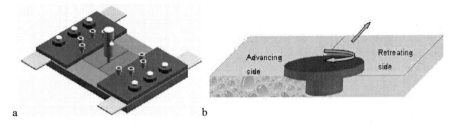

a                                         b

**FIGURE 3.** a The Clamping Fixture; b Sketch of Configuration.

**TABLE 1.** Dimensions of Tools

| Thickness of blanks (mm) | Shoulder diameter (mm) | Pin diameter (mm) |
|---|---|---|
| 5 | 5 | 14 |
| 7 | 7 | 20 |
| 9 | 9 | 24 |

The thick aluminum plates was positioned in the retreating side on the basis of accurate studies on the material flow occurring in FSW processes [12]. Actually material moves from the retreating side towards the advancing one, such flux fulfills the gap between the two blanks to be welded and also the neighbor cells. Previous researches led to the assumption of the utilized process parameters: tool rotation speed (R) equal to 1040 r.p.m., tool feed rate (Vf) of 108 mm/min, tilt angle of 2° and, moreover, the tool axis was set such as the pin was totally inside of the aluminum plate and tangent to the foam-plate contact surface (Figure 4b).

Three repetitions of each test were developed. In order to not consider the transition area, the outer zones have been cut away and just the middle part of the joint was formed. The welded specimens to be tested had dimensions 40x200x9 mm, 40x200x7 mm and 40x200x5 mm.

## DISCUSSION OF THE RESULTS

The average load curves of the bending tests, obtained for the three different typologies of products are reported in the Figure 4.

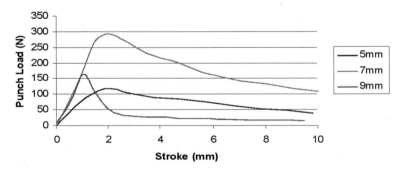

**FIGURE 4.** Load vs. Punch Stroke.

202

However the above data are not directly comparable, since they concern different specimen thickness. Because of the 3 point bending equipment, it is possible to use the well known equation that describes the relationship between the punch stroke and the geometrical and material parameters till the yielding (Equation 1);

$$S = \frac{Fl^3}{48EI}$$ (1)

where S is the punch stroke, F the applied load, l the specimen length, E the Young's modulus and I the moment of inertia.

The ratio F/I describes the joint efficiency (JEC) since it normalizes the behavior for any sheet thickness. The values are shown in table 2; it is possible to verify that the higher joint efficiency coefficient correspond to the 5 mm thick joint.

TABLE 2. The course of Joint Efficient Coefficient

| Thickness (mm) | 5 | 7 | 9 |
|---|---|---|---|
| JEC= $F_{MAX}/I$ (N/mm$^4$) | 0.281 | 0.256 | 0.067 |

On the other hand, the mass role must also be considered since the density strongly affects the Young's modulus (E) and, so, the material behavior. Generally it is possible to state [13] that:

$$E_{foam} = f(E_{solid}, \rho_{foam}, \rho_{solid})$$ (2)

Thus, a more consistent coefficient can be defined to describe the joint performance, named Specific Joint Efficiency Coefficient, defined as in the follow equation 3:

$$SJE = \frac{F}{\rho I}$$ (3)

In fact, density should be increased in order to get two distinct positive effects. The former is the strength increase due to the Young's modulus increase, the latter is related to the better welding performance when FSW is applied to materials characterized by higher density. On the contrary, the designer tends to reduce the quantity of utilized material, to reduce the material costs and the structure weight. These two needs generate a sort of optimum combination as it is shown in the figure 5.

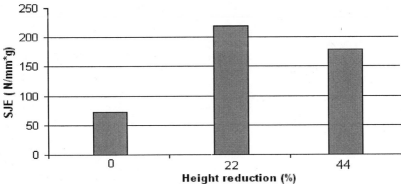

FIGURE 5. Specific Joint Efficiency Coefficient Trend.

# CONCLUSIONS

Aluminum foams were investigated in the paper focusing the attention on the possibility to weld foam-solid structure by FSW. Actually, the welding efficiency strongly depends on the material density and, for this reason, a customized density increasing was developed using a simple upsetting process.

This technique, however, induces a mass increase that can be considered as a drawback. Taking into account both the advantages and disadvantages, a sort of optimum behavior can be found as demonstrated choosing a proper coefficient to measure the joint efficiency. The above considerations have been developed assuming that the modification in the foam density were applied to the whole panel; on the other hand further research activities can be developed in order to join foam panels and aluminum plates superimposing a local pre-deformation to the panel just close to the joint area. In this way the geometry and the characteristics of the foam panel would be maintained and furthermore an improvement in the FSW joint would be obtained.

Overall, further investigations are required in order to obtain a more robust assessment of the above approaches.

# ACKNOWLEDGMENTS

This work was made using MIUR (Italian Ministry for University and Scientific Research) funds.

# REFERENCES

1. B. Matijasevic and J. Banhart, *Scripta Materialia* **54**, 503-508 (2006).
2. M. F. Ashby, *Metal Foams and Honeycombs Data-base*. Granta Design, Cambridge, U.K., 1996.
3. A. F . Bastawros, H. Bart-Smith and A. G. Evans, *J. Mech. Phys. Solids.* **48/2**, 301 (2000)
4. T. Miyoshi, M. Itoh, S . Akiyama and A. Kitahara, *Adv. Engng. Mater.* **2/4**, 179 (2000).
5. W. Deqing, S. Ziyuan, *Mater. Sci. Eng.* A **361/1–2**, 45-49 (2003).
6. S.W. IP, Y. Wang, J.M. Togeri, *Can. Inst. Mining Metall.* **38/1**, 81-92 (1999).
7. J. Banhart, M.F. Ashby and N.A. Fleck, "Cellular metals and metal foaming technology", Proc. 2nd Int. Conf. MIT Press Bremen, Germany, 2001.
8. D.S. Schwartz, D.S. Shih, A.G. Evans and H.N.G. Wadley, "Porous and cellular materials for structural applications", MRS Symp. Proc. (1998).
9. VI. Shapovalov, *MRS Bull.* (19-24) 1994.
10. T. Bernard, J. Burzer and H.W. Bergmann: *J. Mater. Proc. Tech.* **115**, 20-24 (2001).
11. L. Filice, L. Fratini and D. Umbrello, "On the joining of Aluminum Foams and Aluminum Foam Sandwiches", in *Porous Metals and Metal Foaming Technology*, edited by The Japan Institute of Metals, 639-642, (2006),
12. L. Fratini, G. Buffa, D. Palmeri, J. Hua, R. Shivpuri, *ASME J. Eng. Mat. Tech.* **128/3**, 428-435 (2006).
13. G. Lu, G.Q. Lu and Z.M. Xiao, *J. Porous Mat.* **6**, 359-368 (1999).

# Influence of Anisotropy Properties in Finite Element Optimization of Blank Shape Using NURBS Surfaces

R. Padmanabhan[1*], M.C. Oliveira[1], A.J. Baptista[1], J.L. Alves[2]
and L.F. Menezes[1]

*1 CEMUC, Department of Mechanical Engineering, University of Coimbra, Polo II, Pinhal de Marrocos, 3030-201Coimbra, Portugal, * Corresponding author: padmanabhan@dem.uc.pt*
*2 Department of Mechanical Engineering, University of Minho, Campus de Azurém, 4800-058 Guimarães, Portugal*

**Abstract.** Sheet metal forming is a complex process controlled by process parameters and material properties of the blank sheet. The initial anisotropy has influence on the determination of optimal blank shape because it governs the material flow. In this paper, the influence of the initial anisotropy, in achieving an optimal blank shape, is analyzed using mild steel (DC06) blank sheet and two different tool geometries: circular and rectangular cup. The numerical method is based on the initial NURBS surface used to produce the mesh that models the blank and the resulting flange geometry of the deformed part. Different rolling direction orientations were considered in the blanks for deep drawing to investigate their effect on the blank shape optimization procedure. From the numerical study it is evident that the described method is sensitive to the initial anisotropy in the material and can produce optimal initial blank shape within few iterations.

**Keywords:** Blank shape, Anisotropy, NURBS, Optimization, Trimming, FEM.
**PACS:** 46.15.-x, 46.35.+z

## INTRODUCTION

In sheet metal forming, the initial blank shape is one of the important process parameter that has a direct impact on the quality and final cost of the formed part. The initial blank shape for a part is determined either by trial and error method or through any of the numerical approaches proposed by researchers [1-4]. Numerical solutions provide an insight on the deformation behavior of sheet metal blanks subjected to deep drawing. Numerical models are also simple to create and modify depending on the requirements allowing elimination of material wastage. One such solution to determine an optimal blank shape for a sheet metal part is described in this paper. The iterative virtual try out method proposed combines three numerical tools, DD3TRIM [5], DD3IMP [6], and NURBS. The effect of initial anisotropy and rolling direction orientation of the blank sheet on optimal blank shape for a part is investigated in this study. The anisotropy, prevalent in the pre-processed sheet segments due to processing operations such as rolling, influences subsequent deformation such as deep-drawing. It

CP907, *10th ESAFORM Conference on Material Forming*, edited by E. Cueto and F. Chinesta
© 2007 American Institute of Physics 978-0-7354-0414-4/07/$23.00

dictates the shape of the yield surface and strongly affects the strain distributions obtained during sheet metal forming.

## BLANK SHAPE OPTIMISATION PROCEDURE

In deep drawing processes, the average element size influences results like draw-in prediction, depending upon the complexity of the final shape of the part. Generally, to accommodate the continuous variation of the nodal coordinates in optimization procedure, a time-consuming remeshing procedure is employed. In blank shape optimization process, it is important to fix this numerical parameter and avoid remeshing procedures. Hence, a regular and uniform mesh with dimensions large enough to accommodate the probable blank shapes is defined as a base mesh. DD3TRIM is a numerical tool developed to trim solid finite element meshes [5] which is used in the procedure to eliminate remeshing and generate the initial mesh. The trimming operation can be performed based on a NURBS surface. A NURBS surface is a geometric representation extensively used in the design and manufacture of components in aircraft, automobile and shipping industries. The blank shape optimization method is illustrated in figure 1.

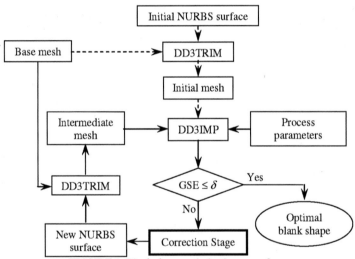

**FIGURE 1.** Blank shape optimization procedure.

An initial blank shape is selected or determined based on empirical formulae [7], and a corresponding NURBS surface is produced. A 3D solid FE base mesh is produced with an average element size according to the deep drawing process. The base mesh is then trimmed with initial NURBS surface, using DD3TRIM, to produce the initial finite element mesh. This mesh is subjected to deep drawing simulation. The flange contour of the formed part is compared with the required target contour. If the flange contour is different of target contour, depending on its deviation, the initial NURBS surface is corrected using a simple push/pull technique applied to a predefined set of points of the surface, and a new NURBS surface is produced [8].

206

This new NURBS surface is used to trim the base mesh to produce intermediate blank shape which is subjected to deep drawing process. The deviation between the flange and the target contours is quantified with a geometrical shape error (GSE). This error is defined as the root mean square of the shape difference between the target shape and the deformed shape in the following equation [9]:

$$GSE = \sqrt{\frac{1}{n}\sum_{i=1}^{n}\left|X^{inter} - X^{final}\right|^2}$$  (1)

where $X^{inter}$ and $X^{final}$ are the coordinates of the predefined set of points at the intersection with the target contour and at the end of the forming process, respectively. $n$ is the number of control points used in the initial NURBS surface. The procedure is repeated until the GSE falls bellow a user defined value of accuracy, $\delta$.

Deep drawing simulations were carried out using the in-house finite element code DD3IMP (contraction of *Deep Drawing 3D Implicit code*) [6]. The deep drawing process parameters like, the tools geometry, the mechanical properties of the blank sheet, the friction conditions and the blank holder force are fixed during the optimization procedure.

## RECTANGULAR AND CIRCULAR CUP EXAMPLES

The blank shape optimization procedure described in the previous section takes the material flow characteristics into account since it is based on the FE simulation results. The material flow characteristic is in turn governed by the initial anisotropy apart from other material properties. The geometry of the rectangular and circular cup is presented in figure 2 (a) and (b) respectively.

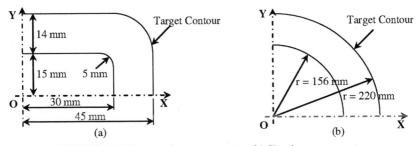

**FIGURE 2.** (a) Rectangular cup geometry, (b) Circular cup geometry.

Mild steel (DC06) blank sheets, 0.8 mm thick for rectangular cup and 3.5 mm thick for circular cup, were used in this study. The material properties are: Young's modulus 210 GPa, yield stress 123.6 MPa, strength coefficient 529.5 MPa, strain hardening coefficient 0.268 and Poisson's ratio 0.3. Swift law describes isotropic work-hardening. Orthotropic behavior is described by the classical Hill´48 quadratic yield criterion [10], with r-values of: $r_0 = 2.53$, $r_{45} = 1.84$ and $r_{90} = 2.72$. Three different blank sheet orientations (0°, 45° and 90°, with respect to OX) were utilized to determine the influence of rolling direction orientation on the optimal blank shape for each part. Although, circular cup is axisymmetric these tests are performed for validation of the optimization procedure. Initial process parameters were chosen based on empirical relations and optimal values. A blank holder force of 800 N and 50000 N was used for rectangular cup and circular cup, respectively, and a friction coefficient of 0.08 was used in both the cases.

## DISCUSSION ON RESULTS

One of the ways that the rolling direction orientation in the blank sheet is best visible in the deep drawn part is through earing, which is correlated with planar anisotropy and may involve extensive trimming to produce the required flange contour. The described procedure determines the optimal blank shape for a rectangular cup and a circular cup according to the rolling direction orientation of the blank sheet. Figure 3 shows the optimal blank shapes obtained for different rolling direction orientations considered. Figure 3 (a) shows the results for rectangular cup and 3 (b) for circular cup. When the rolling direction of the blank sheet is parallel (0°) and perpendicular (90°) to X-axis (OX), the optimal blank shape resembles the same. The optimal blank obtained when the rolling direction orientation is 45° to X-axis (OX) is larger along the axes and shorter along the diagonal. These results are in good agreement with the fact that according to the material properties more flow occurs along 45° to the rolling direction.

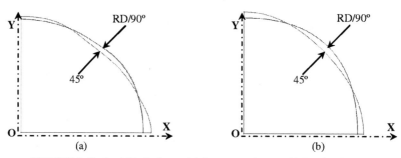

**FIGURE 3.** Optimal blank shapes (a) for rectangular cup (b) for circular cup.

To guarantee a good solution, the thickness variation in the formed part should be minimized. Blank sheet optimization procedure can also contribute for a more uniform flow enhancing the part's quality. Figure 4 illustrates the thickness distribution in the square cup for 0°, 45°, and 90° blank sheet orientations. Figure 4 (a) presents the distribution along OX direction and (b) along OY direction. The thickness remained the same across the bottom of the cup and reduced along the punch radius and the cup wall in all the three cases. The thickness increased at the flange section due to radial compression caused by the draw-in. The thickness increase at the flange is similar for 0° and 90° orientations and lower for 45° orientation of the blank sheet. This is due to increased thickness strain at 45° to the rolling direction of the blank sheet. When the rolling direction of the blank sheet is oriented at 45° to X-axis, material flows more along the X and Y-axes, hence the thickness at the flange is less in this case.

Figure 5 (a) illustrates the thickness distribution in the circular cup along OX direction for 0°, 45°, and 90° orientations of the blank sheet. Similar to square cup example, the thickness remained the same across the bottom of the cup and reduced along the punch radius and the cup wall. The thickness increased at the flange section and is similar for the 0°, and 90° orientations of blank sheet. The thickness at the flange is much less when the rolling direction orientation is 45° to X-axis. Contrary to this, the thickness at the flange along the diagonal of the circular cup is more when the orientation is 45°, figure 5 (c), compared to the other cases. This clearly describes the improved thickness strain distribution along rolling and transverse directions. The thickness remained almost the same for all rolling direction orientations and in both geometries, guaranteeing the part's quality.

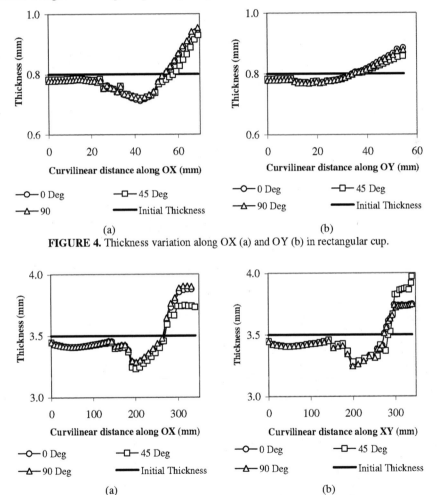

FIGURE 4. Thickness variation along OX (a) and OY (b) in rectangular cup.

FIGURE 5. Thickness variation along OX (a) and XY (b) in circular cup.

In spite of the influence of the initial anisotropy of the blank sheet, the described method is capable of determining optimal blank shape for a formed part within few iterations. Figure 6 shows the geometric shape error evolution during the optimization procedure for each rolling direction orientation of the blank sheet, for the rectangular and circular cup examples. In both the cases, the geometric shape error reduced to less than 0.5 mm within four iterations.

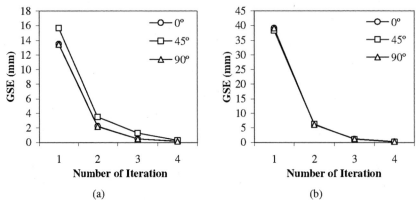

(a)                                           (b)

**FIGURE 6.** Geometric shape error (a) rectangular cup (b) circular cup.

## CONCLUSIONS

The blank shape optimization method described in this paper is capable of determining optimal blank shape for any sheet metal part. Initial anisotropy and thus the orientation of the blank sheet rolling direction has strong influence on the flow characteristics of the blank sheet and in the optimal initial blank for a part. Material flow less along rolling direction and transverse direction compared to diagonal direction. It is evident from the simulation results that the described method is capable of achieving at the optimal blank shape for a part within 4 iterations and the blank shape depends on the orientation of the rolling direction with respect to an axis.

## ACKNOWLEDGMENTS

The authors are grateful to the Portuguese Foundation for Science and Technology (FCT) for the financial support for this work, through the Program POCI 2010.

## REFERENCES

1. Toshihiku Kuwabara, Wen-hua Si, *J. Mater. Process. Tech.*, 1997, pp. 89-94.
2. Park S.H, Yoon J.W, Yang D.Y, Kim Y.H, *Intl. Journal of Mech. Sciences*, 1999, pp. 1217-1232.
3. Lee C.H, Huh H, *J. Mater. Process. Tech.*, 1998, pp. 145-155.
4. Guo Y.Q, Batoz J.L, Naceur H, Bouabdallah S, Mercier F, Barlat F, *Comp. and Structures*, 2000, pp. 133-148.
5. Baptista A.J, Alves J.L, Rodrigues D.M, Menezes L.F, *Finite Elem. Anal. Des*, 2006, pp. 1053-1060.
6. Menezes L.F and Teodosiu C, *J. Mater. Process. Tech.*, 2000, pp. 100-106.
7. Barata da Rocha, Ferreira Duarte J, Tecnologia da Embutidura, Edited by Associação Portuguesa das Tecnologias de Conformação Plástica (APTCP), 1993.
8. Les Piegl, Wayne Tiller, The NURBS Book, 2nd Edition, Springer-Verlag Berlin, 1997.
9. Park S.H, Yoon J.W, Yang D.Y, Kim Y.H, *Int. J. Mech. Sci.*, 1999, pp. 1217-1232.
10. Hill R, Proceeding of the Royal Society, London, 1948, pp. 281-297.

# Numerical Prediction of Elastic Springback in An Automotive Complex Structural Part

**Livan Fratini\*, Giuseppe Ingarao, Fabrizio Micari**

*Dipartimento di tecnologia meccanica, produzione ed ingegneria gestionale University of Palermo,*
*Viale delle Scienze 90128, Palermo, Italia*

**Abstract.** The occurrence of elastic springback phenomena in sheet metal processing operations determines a relevant issue in the automotive industry. The routing and production of 3D complex parts for automotive applications is characterized by springback phenomena affecting the final geometry of the components both after the stamping operations and the trimming ones. In the present paper the full routing of a automotive structural part is considered and the springback phenomena occurring after forming and trimming are investigated through FE analyses utilizing an explicit implicit approach. In particular a sensitivity analysis on process parameter influencing springback occurrence is developed: blank holder force, draw bead penetration and blank shape.

**Keywords:** Sheet metal stamping, Springback phenomena, F.E.M.,

**PACS:** 82.20.Wt

## INTRODUCTION

Taking into account a simple routing of an industrial part, the initial blank is first stamped and then is trimmed, eliminating the in excess flange. It should be observed that after the stamping operation springback occurs since, as the forming tools are removed, the residual internal stress state of the component is not equilibrated by the external actions of the tools anymore and an elastic deformation occurs up to reach a new geometry for which the residual internal stress state is self equilibrated. The inhomogeneous distributions of strains along the stamped sheet thickness together with the elasto-plastic behavior of the workpiece material determine the occurrence of the springback phenomena. Such first deformation causes an increase in the lead time of the component due to difficulties in the positioning of the stamped part in the trimming dies because of its changed geometry. Typically springback phenomena already occur after stamping phase; what is more, as trimming is developed new springback phenomena are observed in the produced part since, again, the internal residual stress state is not equilibrated and a new equilibrium configuration is reached through an elastic distortion of the component. In the last years several investigations on the effectiveness of numerical models to predict springback have been developed, taking into account both the effects of process parameters and numerical ones [1]. In particular the latter aspects have been focused considering the influence of numerical

CP907, *10th ESAFORM Conference on Material Forming*, edited by E. Cueto and F. Chinesta
© 2007 American Institute of Physics 978-0-7354-0414-4/07/$23.00

parameters such as the type of the utilized element, the number of integration points, the hardening rule and so on, with the aim to improve the effectiveness and reliability of the numerical results [1-4].

Several research groups analyzed the effects of the geometry and of the process parameters (such as friction conditions, blank holder forces, punch and die radius and so on) on springback predictions with FEA [5-6]. It should be observed that the final aim of the correct prediction of springback phenomena occurring in complex 3D industrial forming processes is the complete engineering of such processes with the total compensation of at least reduction of the springback distortions. In the latest years some papers have been focused on the die compensation through integration of numerical simulations with optimization techniques in order to minimize reworking cost and time due to springback occurrence [7-8].

In this paper an industrial case study is considered: the manufacturing of an automotive structural part is investigated through numerical simulations. The springback phenomena occurring after forming and trimming are studied through FE analyses utilizing a commercial codes based on explicit loading - implicit unloading analysis. The investigation has been developed at the varying of same relevant process parameter affecting the springback occurrence, namely drawbeads penetration, blank shape and blankholder force. The aim is analyze the process parameter influence in springback amount through numerical simulation, in order to outline same effective and fast strategies in springback reduction at least from an industrial point of view.

## THE INDUSTRIAL CASE STUDY AND THE FE MODEL

The considered component was a structural part quite common in the automotive industry. The component is obtained trough two operation: a stamping step and a trimming one; after the trimming operation two separated components are obtained In the following figure 1, the stamped and the trimmed parts are shown.

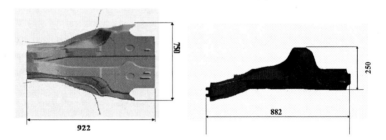

**FIGURE 1.** Sketch of the industrial part (dimensions in mm)

The sheet metal initial thickness was 1.6mm and the used material was ZSTE 300 BH steel characterized by a yield strength of 380MPa, and by the following flow rule:

$$\sigma = 700(\overline{\varepsilon})^{0.16} \qquad (1)$$

obtained through a campaign of tensile tests developed on 0° direction specimens. Furthermore the following Lankford's anisotropy parameters were determined:

$r_{0°}=1,080$; $r_{45°}=1$; $r_{90°}=1,40$.

As far as the industrial routing developed to obtain the proposed component, is regarded, the next steps are commonly followed.

❑ Blanking of the initial sheet.
❑ Stamping of the parts; such process is developed following the sketch reported in figure 2: the blankhoder is first moved to clamp the blank, then punch let the blank reach the movible die surface and a coining phase occur(figure 2b), finally the punch is moved deforming the sheet into the die cavity; in this phase the movable dies is moved steadily to the punch and acts a force of 250 KN. During all the stamping process a blankholder force of 850 KN was superimposed. As far as drawbeads are regarded (shown in the stamped part reported in figure 1), an equivalent unit restraining force of 120N/mm was applied.
❑ Trimming of the flange.

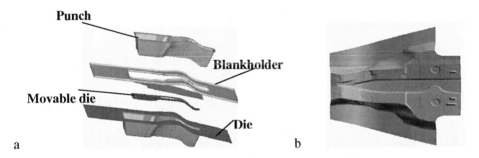

**FIGURE 2.** a) Sketch of the industrial process, b) Formed part after coining

LS-Dyna explicit commercial code was utilized in order to simulate the proposed industrial case study. The code takes into account the spring phenomena following an hybrid approach based on the explicit loading - implicit unloading analysis of the process. The utilized material has been modeled through the Barlat-Lian yield criterion [9], considering an isotropic hardening law. The blank was meshed through about 800 quadrilateral shell elements; a four level geometric remeshing strategy was applied. In particular the Belyschko-Lin-Tsay (B-T) shell element [4], with seven integration points along the thickness, was used. As far as the punch velocity is regarded an artificially increased value equal to 1m/s was utilized, checking that the kinetic energy was below the 10% of the deformation work, in order to avoid any inertia effect. Frictional actions were considered through a Coulomb model with a coefficient equal to 0.12. The utilized numerical code permitted to model the drawbeads through the introduction of equivalent restraining forces per length unit applied on the drawing blank. No geometrical effects of the utilized drawbeads were considered. Finally, trimming operations were developed simply cutting the model along the trimming lines without any further requirement.

## THE FE ANALYSIES

Process parameter values mentioned before, were found taking in to account the feasibility of the part in terms of maximum thinning to avoid ductile fracture insurgence. A first study on springback occurrence was carried out on this reference set up, both after stamping and trimming operation. As shown in figure 3 a maximum thinning of 24.6% is reached and what is more a large amount of springback is obtained in the stamped part (figure 4a). In particular an "opening" of the section is observed; in the most distorted section the difference between the desired geometry and the obtained one reaches 12 mm. This phenomenon has to be taken into account in the design phase, in fact such geometry variations could determine problems in the positioning of the stamped part in the trimming dies, causing some additional costs due to reworking operations and repositioning or even making impossible the trimming operation itself. The amount of springback after trimming appears low enough, in fact a maximum displacement of about 2,7mm is observed, and in particular this displacement is due to a torque effect as shown in figure 4b.

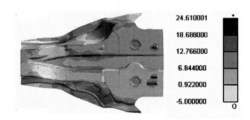

**FIGURE 3.**Reference set up: percentage thinning

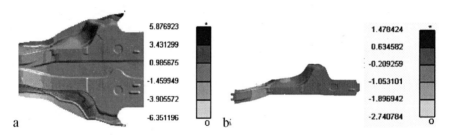

**FIGURE 4.**Springback displacement along the opening direction (mm): a) after stamping b) after trimming

To minimize springback occurrence is necessary to increase the stretching mechanics in the blank during the stamping processes in order to induce greater plastic deformations. In other words the restraining forces have to be increased, in this paper three different options are taken in to account based on the increase of draw beads penetration, the increase of the blankholder force and the change of the blank shape, respectively. The three strategies were followed independently; in other words, no cross effects were considered. Actually, with the increasing of the stretching phenomena the thinning tendency increases too, so a proper calibration of the mentioned process parameters has to be developed in order to get the process engineering.

At first the effect of drawbead penetration was considered; in particular the value of the equivalent unit restraining forces has been varied. Different values were investigated and the best result was reached utilizing a value of 160 N/mm. In fact, in the stamping process, only a slight increasing of the maximum thinning was obtained, on the contrary a significant reduction in springback amount was observed at least after the stamping step (see figure 5 a and b).

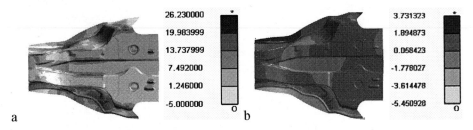

**FIGURE 5.**Icreased draw beads penetration set up: a) thinning, b) springback after stamping (mm)

The thinning map after stamping shows also larger work hardened areas, this leads to obtain a more stiff final component; what is more the maximum lateral walls opening is reduced up to 8.9mm. Furthermore the blank surface was increased, enlarging the flange in order to increase the effect of the blankholding action (see figure 6 a and b): in particular a localized enlargement of the flange was developed in the maximum opening zone (figure 6a).

**FIGURE 6.**a) Enlarged blank, b) reference blank

Again, the obtained maximum thinning was comparable with the reference one, while a springback reduction was obtained both after the stamping and the trimming operations. In fact the maximum lateral walls opening was reduced up to 10mm, and what is more, in the springback after trimming analysis, the maximum displacement along the selected direction was reduced of about 47% of the reference value (figure 7 a and b).

Finally, an increase of blankholder force was considered. Two strategies were followed: an increase during all the stamping processes and an increase just in the last step of the stamping sequence; actually no satisfying solutions were obtained with both the strategies and an excessive thinning was induced in the sheet.

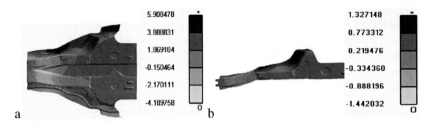

**FIGURE 7.**Enlarged blank set up springback (mm): a) after stamping b) after trimming.

## CONCLUSIONS

In the present paper the full routing of an automotive structural part is considered and the springback phenomena occurring after forming and trimming are investigated. The process parameter influence on springback amount was taken in to account through numerical simulation; same significant improvement were obtained in springback reduction. Further developments regard the investigation of the influence of geometrical parameters, with the aim to minimize springback amount already in die design phase.

## ACKNOWLEDGMENTS

This work was made using MIUR (Italian Ministry for University and Scientific Research) funds.

## REFERENCES

1.  K.Li, W.P.Carden, R.H. Wagoner, International journal of Mechanical Sciences **44** (2002) 103-122.
2.  Z. Cedric Xia, A parametric study of springback behavior, NUMIFORM Conference Proceedings, 2001, pp. 711-716..
3.  S.W. Lee, D.Y. Yang, Journal of Material Processing technology **80-81** (1998) 60-67.
4.  A. Andersson, S. Holmberg, Simulation and verification of different parameters effect on springback results, Numisheet Conference Proceedings, Jejv Island, Korea, 21-25 Oct., 2002 , pp. 201-206..
5.  R. Bahloul, S. Ben-Elechi, A. Potiron, Journal of Material Processing technology **173** (2006) 101-110.
6.  Gang Liu, Zhongqin Lin, Youxia Bao, Finite Element in Analysis and Design **39** (2002) 107-118.
7.  L. Pepelux, J.P. Ponthot, Journal of Material Processing technology **125** (2002) 785 -791.
8.  L. Geng, R. H. Wagoner, International Journal of Mechanical Sciences **66** (2004) 1097-1113.
9.  F. Barlat, F. J. Lian, International journal of plasticity **5** (1989) 51-66.

# Anisotropic and Mechanical Behavior of 22MnB5 in Hot Stamping Operations

A. Turetta[1,a], S. Bruschi[2,b], A. Ghiotti[1,c]

[1]DIMEG, University of Padova, Via Venezia 1, 35131, Padova, Italy
[2]DIMS, University of Trento, Via Mesiano 77, 38050, Trento, Italy
[a]alberto.turetta@unipd.it ,[b]stefania.bruschi@ing.unitn.it, [c]andrea.ghiotti@unipd.it

**Abstract.** The hot stamping of quenchable High Strength Steels offers the possibility of weight reduction in structural components maintaining the safety requirements together with enhanced accuracy and formability of sheets. The proper design of this technology requires a deep understanding of material behavior during the entire process chain, in terms of microstructural evolution and mechanical properties at elevated temperatures, in order to perform reliable FE simulations and obtain the desired characteristic on final parts. In particular, the analysis of technical-scientific literature shows that accurate data on material rheological behavior are difficult to find; while the lack of knowledge about anisotropic behavior at elevated temperatures is even more evident. To overcome these difficulties, a new experimental set-up was developed to reproduce the thermo-mechanical conditions of the industrial process and evaluate the influence of temperature and strain rate on 22MnB5 flow curves through uniaxial tensile tests; an optical strain measurement system was utilized to evaluate the effective strain after necking. From the same data, plastic anisotropy evolution was determined by means of a specially developed procedure. The influence of different cooling rates was taken into account and the rheological properties were correlated with microstructural changes occurring during deformation, previously evaluated through a dilatometric analysis performed in the same range of temperatures.

**Keywords:** Sheet metal forming, Forming Limit Diagram
**PACS:** 81.05.Bx

## INTRODUCTION

In the last years the increasing demand in automotive components characterized by higher and higher strength/mass ratio has forced industries to utilize the new generation of high and ultra-high strength steels. However, as the forming of such steels at room temperature is practically impossible, the utilization of sheet working operations at elevated temperatures is increasing more and more. Due to hot forming, lower loads on tools as well as higher accuracy of the formed component thanks to the great reduction in springback can be easily obtained. In the hot stamping process, the steel blank is heated up above austenitization, then transferred to the press where the whole deforming phase has to take place in fully austenitic conditions; the use of not-pre-heated dies permits a rapid cooling of the sheet during deformation that assures the

CP907, 10th ESAFORM Conference on Material Forming, edited by E. Cueto and F. Chinesta
© 2007 American Institute of Physics 978-0-7354-0414-4/07/$23.00

achievement of a fully martensitic microstructure in the formed component at room temperature.

Numerical simulations are even more important in the optimization of this innovative process in order to obtain the desired characteristic on final product in terms of mechanical strength and microstructure. Recently, many models and yield criteria for anisotropic material have been proposed and implemented in FE codes as the quality of computational results is strongly influenced by the accuracy of the variables implemented to describe the material behaviour (e.g. yield stresses, anisotropic coefficients, yield locus evolution...). In order to obtain material data as function of temperature and strain rate, a new experimental setup to carry out uniaxial tensile tests was developed at DIMEG-University of Padova laboratories; the newly developed apparatus permits to reproduce the thermal cycle of the industrial process by means of an inductive heating system and to acquire the strain field during deformation through the optical measurement system ARAMIS™.

The influence of temperature and strain rate on 22MnB5 flow curves was studied according to different cooling rates after austenitization and plastic anisotropic coefficients as function of process parameters were calculated. The influence of an applied stress on the shifting of CCT curves was evaluated by means of a dilatometric analysis performed on a Gleeble™ machine, in order to justify the trend of flow curves obtained in the temperature range of 22MnB5 phase transformation.

## EXPERIMENTAL APPARATUS

A new experimental apparatus was developed at the Chair of Manufacturing Technology at University of Padua to perform uniaxial tensile tests under imposed thermo-mechanical conditions. The apparatus consists of a 50 kN MTS™ hydraulic testing machine, equipped with an inductive heating system connected to a 30 kW high frequency power supply and with the ARAMIS™ optical measurement system to detect the strain distribution during deformation (Figure 1).

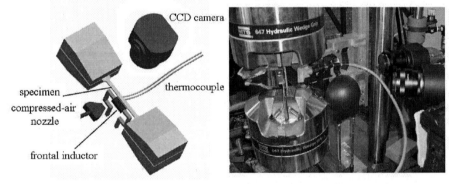

**FIGURE 1.** The new experimental apparatus developed at University of Padova.

The frontal inductor heats the specimen up above the austenitization temperature for a specified holding time in order to obtain a homogenous austenitization, then a rapid cooling up to 100 K/s is applied through a compressed-air jet; once the desired temperature is reached, deformation is performed in isothermal conditions. The inductor shape was previously optimized through an infrared thermocamera analysis in order to obtain a uniform thermal distribution and detect the correct gauge length in the specimen at different testing temperatures. The temperature is controlled through a K-type thermocouple spot-welded in the middle of the gauge length; the specimen geometry was chosen according to the recommendations of ISO 10130.

A LabVIEW™ program guarantees the imposed thermal profile by adjusting the inductive power through a PID controller and synchronizes the tensile test together with the acquisition of the optical measurement system. The system can acquire at 12 Hz and its CCD camera is placed in front of the specimen on which an appropriate stochastic pattern is created in order to resist during deformation at high temperature and assure accurate length and width strain measurements in the zone of interest.

## Analysis Procedure

As an official guideline to determine anisotropic coefficients at elevated temperatures still not exists, a new procedure was developed to increase the accuracy in the analysis of the acquired data: several stage points are taken in correspondence of the previously identified uniform temperature zone of the specimen, their major and minor strain paths are exported and values corresponding to the different points are averaged. Normal anisotropy is calculated for each stage according to Eq. (1) where $\varepsilon_1$ is the longitudinal true strain, $\varepsilon_2$ the true strain in width direction and $\varepsilon_3$ the true strain in thickness direction.

$$r = \frac{\varepsilon_2}{\varepsilon_3} = -\frac{\varepsilon_2}{\varepsilon_1 + \varepsilon_2} \qquad (1)$$

A common trend was noticed in the evolution of normal anisotropy during the entire tensile test: the values fluctuate considerably in correspondence of the initial part of the tensile curve, then they approach nearly a constant value. Plastic anisotropy is therefore calculated averaging the data in the zone characterized by uniform deformation before the necking onset corresponding to the reaching of the maximum force value.

For investigating the material anisotropic behaviour, the r-values are determined for tensile specimens cut at 0°, 45° and 90° with respect to the rolling direction of the sheet. The coefficients of the average normal anisotropy $\bar{r}_n$ and the planar anisotropy $\Delta r$ are then calculated according to Eqs. (2) and (3):

$$\bar{r}_n = \frac{1}{4}(r_0 + r_{90} + 2r_{45}) \qquad (2)$$

$$\Delta r = \frac{1}{2}(r_0 + r_{90} - 2r_{45}) \qquad (3)$$

Afterwards the major strain values acquired through the optical system are correlated with the corresponding force obtained from the MTS™ load cell in order to determine the true stress – true strain curves of the material, thus considering the effective strain after necking.

# RESULTS AND DISCUSSION

Uniaxial tensile tests were performed by means of the new experimental apparatus in the same thermo-mechanical conditions of the industrial hot stamping process in order to determine the influence of cooling rate, temperature and strain rate on the flow curves and on the plastic anisotropy coefficients of the quenchable high strength steel 22MnB5. Specimens, obtained from a 1.75 mm thickness blank, were austenitized at 950°C for 3 minutes, then two different cooling rates equal to 30 and 50 K/s were applied until the desired temperature was reached; isothermal tensile tests were then performed at 500°, 650° and 800°C with strain rates of 0.01, 0.1 and 1s⁻¹. For each test condition at least two test runs were carried out to assure repeatability of results.

## Anisotropic Behaviour

Anisotropy is one important mechanical property influencing sheet metal forming operation and it is a result of the crystallographic structure acquired during the thermo-mechanical processing of the blank. The evolution of average normal and planar anisotropy was determined at different tests conditions through the above described procedure and the obtained results are shown in Figure 2 with their standard deviation.

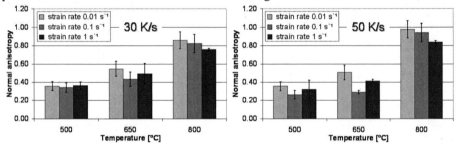

**FIGURE 2.** Average normal anisotropy sensitivity to temperature and strain rate (specimens cooled at 30 K/s and 50 K/s after austenitization).

The average normal anisotropy is strongly affected by deformation temperature; at 800°C the material shows an almost isotropic behaviour, while the anisotropic tendency increases with decreasing temperature. It can be seen that the strain rate has influence only at lower temperatures, where the bainitic transformation may take place; the material exhibits a similar trend with both cooling rates of 30 and 50 K/s.

The planar anisotropy is approximately equal to zero in all testing conditions (Table 1): this implies the crystallographic grain orientation due to the sheet rolling practically disappears after austenitization.

| Planar anisotropy | | Temperature [°C] | | | Planar anisotropy | | Temperature [°C] | | |
|---|---|---|---|---|---|---|---|---|---|
| | 30 K/s | 500 | 650 | 800 | | 50 K/s | 500 | 650 | 800 |
| Strain rate [s⁻¹] | 0.01 | 0.13 | -0.07 | 0.05 | Strain rate [s⁻¹] | 0.01 | 0.13 | -0.11 | -0.06 |
| | 0.1 | -0.02 | 0.06 | 0.06 | | 0.1 | 0.10 | 0.03 | -0.07 |
| | 1 | -0.06 | 0.14 | 0.02 | | 1 | 0.01 | 0.01 | -0.12 |

**TABLE 1.** Influence of temperature and strain rate on planar anisotropy.

# Flow Curves

The experimental data acquired for the anisotropy analysis were used to determine the 22MnB5 flow curves in the same testing conditions; longitudinal strain previously calculated was correlated to load values for the true stress calculation in order to extend the flow curves after necking. The material shows a similar behaviour at both cooling rates of 30 and 50 K/s and exhibits a strong temperature dependency as shown in Figure 3 (a). Strain rate also influences 22MnB5 flow curves: in particular, at lower velocities where microstructural changes may occur during deformation, material strengthening is evident during the tensile test at $0.01 \ s^{-1}$ where the bainitic transformation drastically changes the slope of the curve (Figure 3 (b)).

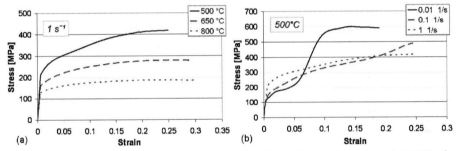

**FIGURE 3.** 22MnB5 sensitivity to temperature (a) and strain rate (b) (specimens cooled at 50K/s after austenitization).

# Dilatometric Analysis

A preliminary dilatometric analysis was carried out in a Gleeble™ machine in order to correlate the rheological behaviour of 22MnB5 at lower strain rates with the onset of phase transformation; the same thermal cycles of the tensile test were applied, maintaining the specimen always in zero-force condition. While the specimen was kept at the testing temperature, its dilatation in the width direction was measured through a diametral dilatometer in order to determine the onset of bainitic transformation. The change of slope in the dilatometric curve corresponds to the transformation beginning, as shown in Figure 4 (a) for two testing conditions. If compared with the change of slope of the true stress-true strain curves, it can be noticed that phase transformation in the tensile tests begins before the corresponding point predicted through the dilatometric measurement indicated (Figure 4 (b)); this behaviour is due to the shift of CCT curves when an external stress is applied. Further investigations are being carried out to quantify this shifting.

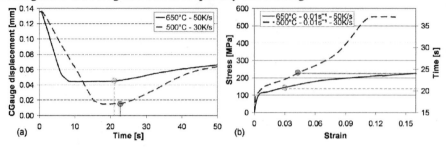

**FIGURE 4.** Dilatometric identification of the bainitic phase transformation onset (a), and correspondent identification in the flow curves for the same testing conditions (b).

## CONCLUSIONS

A new experimental apparatus equipped with an inductive heating system was developed in order to study 22MnB5 rheological behavior in the same thermomechanical conditions of the industrial hot stamping process. Uniaxial tensile tests were performed at different temperatures and strain rates, and material flow curves determined together with the anisotropic coefficients evolution by locally measuring the strain paths with the optical measurement system ARAMIS™. Experimental results show that temperature and strain rate have a strong influence on considered mechanical properties. It's important to underline that the material is almost isotropic at 800°C, while the normal anisotropy decreases with decreasing temperature and the planar anisotropy practically disappears after austenitization. To justify the change in the slope of the flow curves performed at lower strain rates, a preliminary investigation showed that an applied external stress causes a shift in the material CCT curves, anticipating the onset of bainitic transformation. Further studies are being carried out to quantify this effect.

## REFERENCES

1. D. Lorenz and K. Roll, "Simulation of Hot Stamping and Quenching of Boron alloyed Steel" in *7th Int. ESAFORM Conf. on Mat. Forming*, 2004, Trondheim, Norway.
2. P. Hein, "A Global Approach of the Finite Element Simulation of Hot Stamping" in *Sheet Metal 2005 Conference*, Proc. (2005), 2005: p. 763-770.
3. M. Geiger, G. van der Heyd, M. Merklein and W. Hußnatter, "Novel concept of Experimental Setup for Characterisation of Plastic Yielding of Sheet Metal at Elevated Temperatures" in *Advanced Materials Research*, 2005, **6-8**: p. 657-664.
4. A. Turetta, A. Ghiotti, and S. Bruschi, "Investigation of 22MnB5 formability in Hot Stamping operations" in *Journal of Materials Processing Technology*, 2006. **177**: p. 396-400.
5. D. Banabic, H.-J. Bunge, K. Pohlandt, A.E. Tekkaya, *Formability of Metallic Materials,* Berlin: Springer, 2000.
6. M. Geiger, M. Merklein, and C. Hoff, "Basic Investigation on the Hot Stamping Steel 22MnB5" in *Sheet Metal 2005 Conference*, Proc. (2005), p. 795-802.

# Optimization Of Nakazima Test At Elevated Temperatures

## A. Turetta[1,a], A. Ghiotti[1,b], S. Bruschi[2,c]

[1]DIMEG, University of Padova, Via Venezia 1, 35131, Padova, Italy
[2]DIMS, University of Trento, Via Mesiano 77, 38050, Trento, Italy Here
[a]alberto.turetta@unipd.it, [b]andrea.ghiotti@unipd.it, [c]stefania.bruschi@ing.unitn.it

**Abstract.** Nowadays hot forming of High Strength Steel is gaining the strict requirements of automotive producer: in fact deformation performed simultaneously with quenching assures a fully martensitic microstructure at room temperature and thus high strength properties that allow the thickness reduction of the body-in-white components. Basic aspects of hot stamping are still under investigation and supplementary achievements are expected for a successful application of sheet metal forming technologies at elevated temperatures. Among data needed to settle a numerical model of the process, information about material formability may help in better designing and optimizing hot stamping operations. In the first part of the work, a new experimental apparatus based on Nakazima concept is presented; process parameters are optimized in order to accurately replicate the thermo-mechanical conditions typical of the industrial process, paying particular attention to the thermal and microstructural evolution. On the other hand, as commercial FE codes require the implementation of Forming Limit Diagrams at constant temperature, numerical investigations have been performed in order to determine the proper testing conditions to obtain FLD at nearly constant temperature.

**Keywords:** Sheet metal forming, Forming Limit Diagram, Hot Stamping.
**PACS:** 81.05.Bx

## INTRODUCTION

In the last years conventional sheet metal forming operations at room temperature have been deeply investigated considering all the aspects related to process design and optimization as well as material behavior issues; on the other hand, basic aspects of hot stamping are still under investigation. This innovative forming technology has been developed to increase formability of high strength steel sheets which are first heated in a furnace and then formed in austenitic conditions; the use of cold dies assures a rapid cooling simultaneously with deformation in order to obtain a martensitic microstructure at the end of the process, thus reducing sheet thickness while maintaining safety requirements [1].

The work presented in this paper is part of a research project, which aims at developing a general approach that will be able to offer accurate evaluations of the influence of process parameters on the properties of final sheet components in terms of microstructure and strength characteristics [2-4]. However, as the feasibility of producing sound sheet components is strictly related to the material formability, its

CP907, 10th ESAFORM Conference on Material Forming, edited by E. Cueto and F. Chinesta
© 2007 American Institute of Physics 978-0-7354-0414-4/07/$23.00

correct evaluation in temperature still represents a challenging task [5-7]. To investigate material formability in hot conditions, a novel test, based on the Nakazima concept, was designed and build up at DIMEG-University of Padova laboratories. The steel sheets are heated up above the austenitization temperature through induction heating and then deformed by a hemispherical punch that can be cold or pre-heated. Through a careful control of the relevant parameters of hot stamping, the developed test acts like a simulative test, where processing conditions can be carefully reproduced and varied. Results from physical simulation experiments can increase the fundamental knowledge of the process. On the other hand, the careful choice of the test mechanical and thermal parameters (i.e. temperatures of the punch and tools, punch speed, additional cooling during forming, punch coating...) can assure an as much as possible uniform thermal field on the sheet during the deforming phase in order to obtain a formability limit diagram for each of the temperatures the sheet can experience during the actual industrial process. The first part of the paper presents the novel experimental apparatus; then the optimization of the process conditions for simulating the industrial process is reported; while the last part of the paper shows the results of preliminary numerical investigations for determination of FLD at nearly constant temperature.

## THE EXPERIMENTAL APPARATUS

The experimental set-up developed to investigate the sheet formability at elevated temperatures is shown in Figure 1. The test is based on the Nakazima concept, which enables the determination of the whole formability curve by obtaining different strain paths with specimens of different widths, from 200 mm to 25 mm. The rectangular blanks are heated up above the austenitization temperature and then cooled down while deformation takes place.

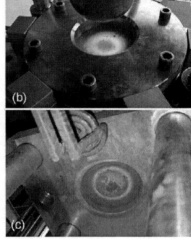

**FIGURE 1.** The experimental apparatus developed for the formability test.

The experimental apparatus is made of a hemispherical punch (whose diameter is fixed in the experimentation), a die, a blank-holder, and a draw-bead which prevents a possible uncontrolled drawing-in of the sheet material. The dedicated machine is an Instron hydraulic press that permits a punch velocity in the range between 10 mm/min and 1500 mm/min. Both the punch and the blank holder are equipped with cartridge heaters to control and vary the thermal field in the sheet and the cooling rate during the deforming phase.

The specimens are heated up to the austenitization temperature through inductor heads, carefully designed in shape and dimensions: pancake for blanks larger than 100 mm and rectangular frontal inductor for smaller specimen in order to keep the temperature homogeneous. A pneumatic system keeps the inductor coil and the specimen at the set distance during the heating phase and removes the inductor to perform the test. During the heating and the deforming phases, the temperature fields in the specimen and in the dies equipment are monitored using both an infrared thermo-camera and spot-welded thermocouples on the sheet surface in the area interested by the deformation.

The strain field in the sheet is measured by an optical deformation measurement system equipped with a proper lighting equipment, providing the possibility to display the 3D-coordinates of the surface [8]. A grid pattern is etched on the specimens by means of an electrochemical system which guarantees the necessary pattern resistance during deformation at elevated temperature.

The material used in the investigation is the boron steel 22MnB5 (well-known with the commercial name of Usibor 1500). In the as-delivered condition, it presents a mixed ferritic-pearlitic microstructure, with standard values of yield and ultimate strengths. However, when deformed and simultaneously quenched in hot stamping operations, its fully martensitic microstructure assures a final yield strength up to 1250 MPa and ultimate strength up to 1500 MPa. This makes the 22MnB5 steel processed under these conditions a proper candidate for those structural automotive parts that have to be characterized by excellent anti-intrusion behavior, like door beams, b-pillar reinforcements and others. The used 22MnB5 sheets are 1.75 mm thick, and they have an aluminum/silicon-based pre-coating to prevent oxidation at elevated temperatures.

The efficiency of the designed inductor heads was tested through heating trials: the objective was to guarantee a complete austenitization in every location of the sheet interested by deformation, as well as a uniform thermal field.

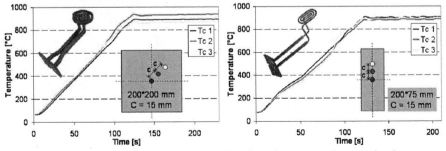

**FIGURE 2.** Temperature evolution during the heating phase for square and rectangular sheets.

Coils of different shape were tested until an uniform temperature field was obtained for every specimen geometry: in particular, the distance between the inductor and the blank as well as the PID coefficients of the control system were properly chosen to assure that. The austenitization temperature was set equal to 900°C, with a soaking time in temperature of 5 min. The choice of these values was supported by heating experiments previously carried out on a Gleeble machine in order to identify the most appropriate heating cycle leading to fully austenitic and homogeneous microstructure [9]. Figure 2 shows the temperature evolution measured by means of three thermocouples in the case of square and thin rectangular sheet using two optimized inductor coils. At the end of the heating phase, the whole sheet is austenitized, and is characterized by an acceptably uniform thermal field.

## RESULTS

The modified Nakazima test designed and set up with the above described features enables the conduction of physical simulation experiments whose aim is to reproduce in a controlled environment those variations of the process parameters that are likely to affect both the material formability and resulting microstructure of the component at room temperature. In particular, the effect of punch speed and temperature is investigated with regard of the microstructure the sheet presents at room temperature after forming. After soaking in temperature, the sheet is cooled in air for 5 s, in order to reproduce the blank moving from the furnace to the press during the industrial practice. Then the punch moves down and deforms the blank, while quenching it.

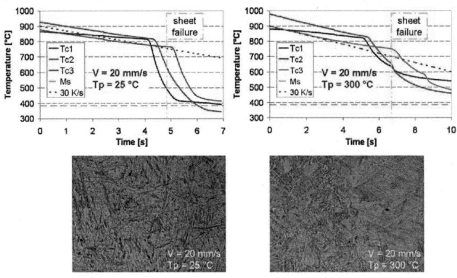

**FIGURE 3.** Temperature evolution during deformation of a square specimen with different punch temperatures (Tp) and ram speeds (V), and microstructure in Tc2 position.

226

The analysis was carried out with 3 values of the punch speed, namely 5-10-20 mm/s, and 2 values of the punch temperature, 25 and 300°C. All the tests were performed until the sheet failure. Figure 3 reports the temperature evolution at the three thermocouples location during the forming phase for punch speed equal to 20 mm/s and punch temperature equal to 25°C and 300°C; the martensite start temperature Ms is reported (dotted line) together with the critical cooling rate responsible of the start of bainite formation (values obtained in phase transformation experiments carried on a Gleeble machine equipped with a diametral dilatometer; more details about the applied procedure and results are reported in [10]). From scientific literature, it's well known that the presence of an applied stress (in this case the deforming action) affects the martensite start temperature and the martensite fraction formed in the supercooled region, and can shift the material CCT curves, increasing the critical cooling rate to bainite formation. The micrographs reported in the same graphs (corresponding to Tc2 location) show that only deforming at a speed of 20 mm/s and keeping the punch at room temperature can assure a fully martensitic microstructure at the end of the test. In all other cases, a mixed microstructure is obtained. This demonstrates that such test can act as physical simulation of the industrial process, giving an insight of the effects of variations of process parameters on the material formability and microstructure.

In a previous work, a numerical model of the new Nakazima apparatus was set to determine the optimum test conditions, by investigating the influence of ram speed and punch temperature on the thermal and microstructural evolution during deformation [10]. The same model is utilised to identify those process conditions that can lead to FLD in quasi-isothermal conditions. The heat transfer coefficient HTC between the punch and the blank is varied, in order to evaluate the effect that different punch materials have on the temperature gradients of the sheet. In particular, the contact steel punch-steel blank and ceramic punch-steel blank are analyzed.

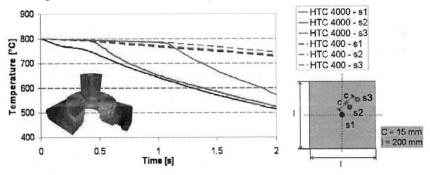

**FIGURE 4.** Influence of different HTCs on the sheet thermal profile during deformation at 20 mm/s with punch at room temperature.

In the numerical analysis, only the deformation phase was taken into account: temperature evolution was monitored by means of 3 lagrangian sensors placed on the specimen surface, 15 mm far from each other. Different test conditions were analysed in terms of punch speed and initial testing temperature; Figure 4 shows that, in case of

227

initial sheet temperature of 800°C, deformation can occur without considerable temperature differences when the HTC at the punch-sheet interface is reduced from 4000 W/m²K, typical of steels, to 400 W/m²K, representative of refractory materials.

## CONCLUSIONS AND OUTLOOK

The formability of high strength steels, simultaneously formed and quenched during hot stamping processes, has to be investigated under the same processing conditions the sheet experiences during the manufacturing stage, especially in terms of cooling rate from austenitization temperature. On the other hand, the evaluation of the blank thinning and eventually fracture through numerical simulations based on FE models requires the implementation of formability limit curves as function of the various process parameters, mainly temperature and strain rate. To this aim a new experimental apparatus was developed and its inductive heating system was optimized to obtain a uniform temperature distribution on the specimens during the heating phase. On one hand, the developed formability test can act as a physical simulation of the industrial hot stamping process, being capable to reproduce the same thermal and mechanical events and offering the possibility to evaluate the influence of testing parameters on thermal and microstructural evolution during deformation. On the other hand, results from numerical simulations seem to offer promising perspectives in setting the proper test conditions to perform FLD in quasi-isothermal conditions. Experimental work is now in progress to investigate the performances of ceramic materials that could represent a suitable substitute to conventional die steels, thanks to their wear resistance and thermal properties.

## REFERENCES

1. E.Lamm, *Advanced Steel Solutions for Automotive Lightweighting*. Windsor Workshop June 5, 2005.
2. D.Lorenz and K. Roll, "Simulation of Hot Stamping and Quenching of Boron alloyed Steel" in *7th Int. ESAFORM Conf. on Mat. Forming*, 2004, Trondheim, Norway.
3. M.Geiger, G. van der Heyd, M. Merklein, W. Hußnatter, "Novel concept of Experimental Setup for Characterisation of Plastic Yielding of Sheet Metal at Elevated Temperatures" in *Advanced Materials Research*, 2005. **6-8**: p. 657-664.
4. M.Geiger, M. Merklein, and C. Hoff, "Basic Investigation on the Hot Stamping Steel 22MnB5" in *Sheet Metal 2005 Conference*, Proc. (2005), p. 795-802.
5. K.Mori, S. Maki, and Y. Tanaka, "Warm and Hot Stamping of Ultra High Tensile Strength Steel Sheets Using Resistance Heating" in *Annals of the CIRP*, **54-1** (2005), 209-212.
6. L.G.Aranda, P. Ravier, and Y. Chastel, "Hot Stamping of Quenchable Steels: Material Data and process Simulations" in *IDDRG 2003 Conference*, Proc. 2003, p. 166-164.
7. P. Hein, "A global approach of the finite element simulation of hot stamping" in *Proceedings of the Sheet Metal 2005 Conf.*, 2005, pp. 763-770.
8. M.Geiger and M. Merklein, "Determination of forming limit diagrams – a new analysis method for characterization of materials´ formability" in *Annals of the CIRP 52/1*, 213, 2003.
9. A.Turetta, A. Ghiotti, and S. Bruschi, "Investigation of 22MnB5 formability in Hot Stamping operations" in *Journal of Materials Processing Technology*, 2006, **177**: p. 396-400.
10. A.Turetta, A. Ghiotti, and S. Bruschi "Testing Material Formability in Hot Stamping Operations" in *IDDRG 2006 Conference*, Proc. 2006, p. 126-132

# Finite Element Calculation of Local Variation in the Driving Force for Austenite to Martensite Transformation

K. Datta *, J. Post **, A. Dinsdale[+], H. J. M. Geijselaers*, J. Huétink*

*The Netherlands Institute for Metals Research, University of Twente, The Netherlands.
**Philips DAP, Advanced Technology Centre B. V., The Netherlands.
[+] National Physical Laboratory, Teddington, U.K.

**Abstract.** The mechanics and thermodynamics of strain induced martensitic transformation are coupled for a metastable alloy steel and implemented in FE models of forming processes. The basic formulations are based on a fifty year old treaty by Patel and Cohen[1]. The variation in Gibbs energy due to local variation in strain, strain rate, temperature and state of stress of a forming part is calculated by FE codes. The local variation in Gibbs energy gives a probabilistic image of the potential sites for strain induced martensitic transformations.

**Keywords:** Martensite, stainless steel, Gibbs energy.
**PACS:** PACS Category 80: Interdisciplinary Physics and Related Areas of Science and Technology

## 1. Introduction

In this work, the formulation proposed by Patel and Cohen in 1953 [1] is adopted. The driving force for the strain induced phase transformation in metastable alloys is an algebraic sum of the chemical Gibbs energy and the mechanical energy (shear and normal components of the applied stress incident on the austenite > martensite habit plane). This formulation is applied to the metastable stainless steel Sandvik Nanoflex. Sandvik Nanoflex is used to make precision parts manufactured by Philips B.V. and its general composition is given in Table 1. The precision parts are formed through multistage stamping and are heat treated to the desired specification. A prototype, three step stamping operation called TIMPLE (Fig. 1) has been modelled using Finite Element Methods to define the local variations in the Gibbs Energy associated with this phase change. The formulation is further extended to calculate the speed of transformation.

Table 1: Chemical composition of Sandvik Nanoflex (weight percent)

| C, N | Cr | Ni | Mo | Ti | Al | Si | Cu |
|------|------|-----|-----|-----|------|------|-----|
| <0.05 | 12.0 | 9.0 | 4.0 | 0.9 | 0.40 | ≤0.5 | 2.0 |

**FIGURE. 1**    The parts stamped after each of the steps in TIMPLE  process.

The chemical thermodynamic properties for the different phases of the steel are calculated using the database MTDATA®, a software from the National Physical Laboratory (NPL), for the calculation of phase equilibria. The Finite Element calculations were carried out both with the Philips Internal Code called Crystal® [2] and with the general purpose commercial code ABAQUS v6.5®[3]. The model was built in a generic manner such that such calculations can be performed on any metastable alloy system.

## 2. Thermodynamics of strain induced martensite formation

Superimposed on the chemical contribution to the thermodynamic properties is the contribution arising from the mechanical energies for defining strain induced martensite transformation. The basic equation by Patel and Cohen [1] has been used to couple mechanical energy with the chemical free energy.

$$F^M - F^A \big\|_{M_s} = F^M - F^A \big\|_{M_s} + U \qquad (1)$$

where U is the mechanical energy for transformation and results from the normal and shear component of stresses resolved along the habit plane of austenite γ> epsilon martensite ε > martensite α′ transformation. Here F respresents chemical free energy, U represents mechanical energy, Ms and Ms' are martensite start temperatures in absence and presence of mechanical energy respectiuvely. The contribution from non-chemical energies i.e. grain boundary, interfacial  energy etc. are neglected since the initial grain size of the sheet metal is at least 10 μ .

When tensile samples of Sandvik Nanoflex are x-rayed [4], the XRD plots show the occurrence of austenite and martensite phases in preferential habit planes. Austenite {111} typically changes to martensite {110} and austenite {311} to martensite {211}. Thus the habit planes for Sandvik Nanoflex are roughly defined from these x-ray results. The stresses are resolved along these habit planes and therefore the mechanical energy U must be expressed as a function of the orientation of the transforming martensite plate.

U = shear stress resolved along a potential habit plane times the transformation shear strain + the normal stress resolved perpendicular to the habit plane times the normal component of the transformation strain.                                (2)

In the case of uniaxial tension or compression, the resolved shear and normal stresses are

$$U = \frac{1}{2}\gamma_o \sigma_1 \sin 2\theta \,{+\!\!\!/\!\!\!-}\, \frac{1}{2}\varepsilon_o \sigma_1 (1 + \cos 2\theta) \qquad\qquad (3)$$

where $\sigma_1$ is the absolute value of the applied stress (tension or compression) and $\theta$ is the angle between the specimen axis and the normal to any potential habit plane. $\gamma_o$ and $\varepsilon_o$ are the shear and principal components of the strain respectively.

## 3. Thermodynamics of Sandvik Nanoflex

Gibbs energies of the different phases of Sandvik Nanoflex relative to austenite is shown in Figure 2. Ferrite data is an extrapolation of higher temperature data to lower temperatures.

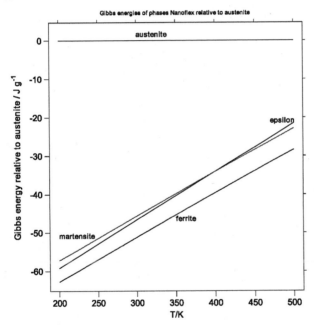

**FIGURE. 2**      Free energies of various phases in Sandvik Nanoflex relative to austenite.

Figure 2 also shows that the h.c.p. phase (epsilon) may become more stable than austenite at low temperatures. The temperature at which epsilon martensite becomes stable will depend on the relative effects of distortion on austenite and the h.c.p. phases.

231

From practical knowledge, it is known that the $M_s$ temperature for Sandvik Nanoflex is about 83 K and the $A_s$ is about 1300 K. Based on this information and with the reference to the paper by Ghosh and Olson [5], it is assumed that the equilibrium temperature between austenite and martensite is given by :

$$T_0 = \frac{A_s + M_s}{2} \approx 700K \tag{4}$$

where $T_0$ is the equilibrium temperature, $A_s$ is the austenite start temperature and $M_s$ is the martensite start temperature. In practice, above 400K, Sandvik Nanoflex has a stable austenitic structure with respect to stress assisted transformations. The strain induced transformation is inhibited above the equilibrium temperature $T_0$. The martensite phase in Sandvik Nanoflex can be aged to precipitation harden the steel. These rhombohedral precipitates are usually very hard to shear and give rise to high wear resistance.

## 4. Implementation in a Finite Element Model

### 4.1 Calculation of the evolution of free energy

The three step prototype TIMPLE stamping process is modelled using Crystal as well as Abaqus 6.5 . Axisymmetric elements are used and the material model described by Post et al. [6] is applied. A subroutine is written which can interface with MTDATA. The chemical free energy is input to this subroutine from MTDATA as a function of temperature, pressure and chemical composition of the steel. Based on Patel and Cohen [1] formulation, the shear and normal components of stresses are extracted from the main FE calculations and at each integration point, a function is calculated which adds U, the mechanical energy (eq. 3) to the chemical free energy. These calculations are repeated throughout the process steps such that the local variations in the energy can be mapped. The crystal orientation is assumed to be random for the sake of simplicity. However, a reference to the work of Videau et al. [7] suggests that such an assumption is realistic since, at a macro level, the response of the applied force to the phase change has a similar trend. With the applied force, it generates maximum martensite in tension, minimum in compression and somewhere intermediate, in shear. Since these calculations are meant to give an impressionist or probabilistic image of the evolution of free energy, and not an exact picture, such an assumption holds good. It is worth, though to couple the analysis with good texture models, if the effect of crystal orientation on the finished properties of the parts are to be studied. Fig. 3 gives the contour plot of the evolution Gibbs energy in the stamped part.

Fig. 4 shows the predicted and actual profile of strain induced martensite content, as calculated by Crystal®. The upper contour plot depicts the predicted values whereas, the lower one is the actually measured values of martensite. The dotted line represents the error in the predicted dimensions. In the inter stamping steps, stress assisted transformations can occur due to presence of residual stresses. This has not been taken into consideration in these calculations.

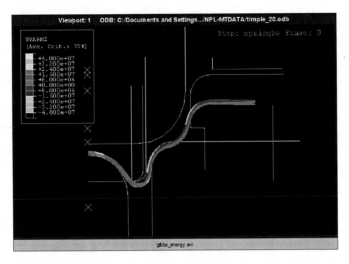

**FIGURE. 3** Evolution of free energy of transformations after Patel-Cohen [1] formulation during deep drawing operation TIMPLE

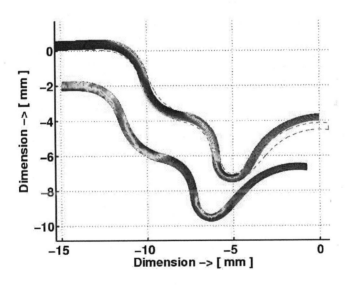

**FIGURE. 4** The actual (lower plot) and the calculated (upper plot) local variation in strain induced martensite formation in parts stamped by TIMPLE process.

# 5. Conclusion

Coupling of thermodynamics of phase change with the mechanics proves to be a flexible yet powerful way of choosing the right alloy chemistry for a specific end use in precision forming of parts. The model is built in a generic fashion such that such calculations can be performed on any metastable alloy system. However, it must be bourne in mind, that the predicted values are not exact but indicative of the potential sites for transformation. In order to study the sensitivity of the material texture, the thermodynamic subroutines must to be coupled with a good texture model.

## ACKNOWLEDGEMENTS

The authors would like to thank R. Delhez and N. van der Pers of TU Delft for all the analyses with XRD and important discussions. They are also thankful to A. Groen of Abaqus Benelux BV, now with SKF Netherlands BV for creating the TIMPLE model. This research was carried out under project number MC5.01101 in the framework of strategic research programme of the Netherlands Institute for Metals Research (www.nimr.nl) in the Netherlands. The authors are grateful to both NIMR as well as Philips BV for their permission to publish this work.

## REFERENCES

1. J. R. Patel and M. Cohen, Acta Metallurgica, Vol. 1, Sept. 1953, 531–538.
2. J. Post, R. J. M. Vonken, NUMIFORM, 2004, Columbus, Ohio, US.
3. ABAQUS 6.5 Users' Manuals, 2006.
4. R. Delhez and N. van der Pers, Measurement of dislocation density in Sandvik Nanoflex using X-Rays, NIMR Scientific Project Report for MC5.01101, January, 2004.
5. G. Ghosh and G. B. Olson, Journal of Phase Equilibria, vol. 22, no. 3, 2001, 199 – 207.
6. J. Post, K. Datta and J. Huetink, NUMIFORM2004, Columbus, Ohio, US.
7. J.-C. Videau, G. Cailletaud and A. Pineau, Experimental Study of the Transformation Induced Plasticity in Cr-Ni-Mo-Al-Ti Steel, *Journal de Physique IV*, 6 (1996) 465-474.

# Control the springback of metal sheets by using an artificial neural network

Axinte Crina

*University of Bacau, Calea Marasesti 157, 600115 Bacau, Romania*

**Abstract.** One of the greatest challenges of manufacturing sheet metal parts is to obtain consistent parts dimensions. Springback is the major cause of variations and inconsistencies in the final part geometry. Obtaining a consistent and desirable amount of springback is extremely difficult due to the non-linear effects and interactions of process and material parameters. In this work, the ability of an artificial neural network model to predict optimum process parameters and tools geometry which allow to obtain minimum amount of springback is tested, in the case of a cylindrical deep-drawing process.

**Keywords:** springback, metal sheets, artificial neural network

## INTRODUCTION

In metal-sheet forming operations, the shape of a product made from metal-sheet is produced by means of its tools representation. But in practice it is often the case that once the loading force is removed, this representation is not ideal – the final unloading configuration of the part includes variations in shape and dimensions, caused mainly by the springback of the drawn parts material.

Nowadays two methods are commonly used to deal with springback phenomenon:

- by controlling the stresses/strains state during the forming process in order to diminish the amount of springback
- by compensating the springback by using an optimum tool geometry and/or optimum process parameters.

The application of these methods in sheet metal industry is typically carried out on a trial-and-error basis, relying entirely on the experience of designers. This traditional procedure is slow, very laborious and costly.

In this paper an alternative approach is proposed which assumes the application of an optimization procedure, based on the Neural Network method coupled with the finite element method, in order to find an optimal relation between the process parameters and geometry of tools, so that the effects of springback on the shape and dimensional accuracy of the virtual obtained part to be minimum.

The analysis is performed in the case of a cylindrical deep-drawing process. The geometrical parameters of parts profile whose variation is investigated in order to quantify the amount of springback and their nominal values are presented in figure 1.

CP907, *10th ESAFORM Conference on Material Forming*, edited by E. Cueto and F. Chinesta

The selected process parameters used in simulation in order to investigate their influence on the springback intensity are indicated in the figure 2.

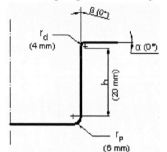

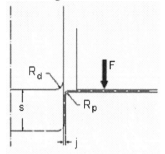

**FIGURE 1** Geometrical parameters of part and their nominal values

**FIGURE 2** Process parameters used in simulation

# IMPLEMENTATION OF THE ANN MODEL INTO THE SPRINGBACK CONTROL

The implementation of the artificial neural network model in order to find the optimum relation between the process parameters, tools geometry and springback parameters assumed the following four steps: data collection, choice of the ANN model, training of the ANN model and generalization.

## Data collection

The data necessary for the training process of ANN model were obtained by quantifying the springback of the virtual parts. The 27 combinations of the process parameters established according to a fractional factorial experiment design were used as input data for the network and the values of the springback parameters resulted from the simulations were used as their associated targets.

Finite element analysis was used to simulate both, the cylindrical deep-drawing process (by using the ABAQUS/Explicit software) and the unloading phase (by using the ABAQUS/Standard software). A three dimensional axis-symmetric model was used in simulation but only a quart of the model was solved due to the symmetry conditions.

## Choice of the ANN model

A two-layer neural network with a sigmoid activation function between the input and hidden layers and a linear activation function between the hidden and the output layers was used. Within the input layer, five neurons - the analyzed process parameters $(R_p, R_d, F, j, s)$ (figure 2) were used; within the output layer, five neurons - the analyzed geometrical parameters of part $(r_p, r_d, \alpha, \beta, h)$ (figure 1) were also used. The number of neurons within the hidden layer must be chosen so that the square mean error to the end of the training process to be minimum.

## Training of the ANN model

The training process was based on the backpropagation algorithm. In order to monitor the correctness of the training process, a cross validation data set of 15% from the total inputs of network was used. The validation process was repeated for different numbers of hidden neurons in order to determine which network topology provides the lowest validation mean square error. For the chosen ANN model, the optimum number of the hidden neurons was set to 5 and, in consequence, this topology was used for the next functional phase of network: generalization.

## Generalization

The generalization phase assumed that the network prescribe correct outputs for inputs that it has not seen yet. In the case of the present analyses, a data set of 25% from the total inputs was given to the network. In figure 3 a comparative analysis of the desired outputs and the outputs prescribed by the ANN model are presented.

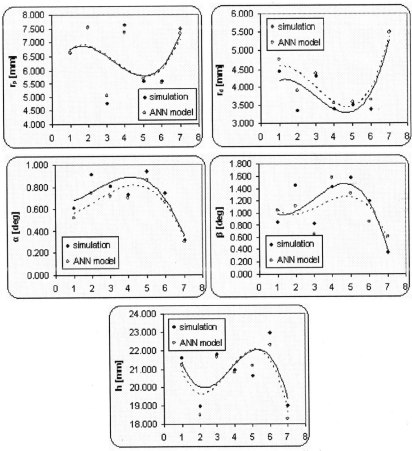

**FIGURE 3** Comparative analysis of the desired/prescribed outputs

237

By analyzing the above diagrams a good concordance between the desired outputs and the prescribed ones could be observed, which allows to validate the ANN model.

## IDENTIFICATION OF THE OPTIMUM PROCESS PARAMETERS/TOOLS GEOMETRY

In order to find the best set of process parameters/tools geometry that lead to a minimum amount of springback, the above validated network was tested for different combinations of process parameters and tools geometry. The neural network had to prescribe the outputs for different inputs without have defined the target values of these inputs. Good results reported to the nominal geometry of part were obtained for the following set of parameters: $R_p = 5.5$ mm; $R_d = 3.4$ mm; $F = 49$ kN; $j = 1$ mm; $s = 30.3$ mm.

In order to validate the optimization procedure, a simulation has been performed using as input data the above set of parameters and the obtained results were compared with the nominal geometry of part. A good agreement between the simulation results and part nominal geometry was observed (table 1). As consequence, the preview mentioned set of parameters could be considered as optimum.

TABLE 1. Comparative analysis of the results

| | $R_p$ | $R_d$ | F | j | s | $r_p$ | $r_d$ | $\alpha$ | $\beta$ | h |
|---|---|---|---|---|---|---|---|---|---|---|
| Values obtained by using initial parameters | 6 [mm] | 4 [mm] | 45 [kN] | 1 [mm] | 30 [mm] | 6.522 | 4.602 | 0.528 | 0.784 | 17.69 |
| Values prescribed by the ANN model | 5.5 [mm] | 3.4 [mm] | 49 [kN] | 1 [mm] | 30.3 [mm] | 6.078 | 4.034 | 0.425 | 0.491 | 20.076 |
| Values resulted from simulation | | | | | | 6.022 | 3.996 | 0.391 | 0.366 | 20.192 |
| Nominal values | | | | | | 6.000 | 4.000 | 0.000 | 0.000 | 20.000 |

## CONCLUSIONS

By applying the optimization procedure based on the neural network method, good improvement was obtained concerning the shape and dimensional accuracy of the formed parts (table 1). As consequence, the results of the present analysis are hopefully in using the artificial neural network to control the springback phenomenon in the case of deep-drawing processes.

## REFERENCES

1. C. Axinte, *Theoretical and experimental researches concerning the springback phenomenon in the case of cylindrical deep-drawn parts*, Ph. D. Thesis, Bucharest, 2006.
2. M. A. Arbib, *The handbook of brain theory and neural networks*, 2nd Edition, 2002
3. Fausett L., *Fundamentals of neural networks*, Prentice-Hall, 1994

# Approaches To Modelling Of Elastic Modulus Degradation In Sheet Metal Forming

Marko Vrh[1], Miroslav Halilovič[2] and Boris Štok[2]

[1] *Kovinoplastika Lož; Stari trg pri Ložu - Slovenia*
[2] *Laboratory for Numerical Modelling & Simulation; Faculty of Mechanical Engineering, University of Ljubljana; Ljubljana - Slovenia*

**Abstract.** Strain recovery after removal of forming loads, commonly defined as springback, is of great concern in sheet metal forming, in particular with regard to proper prediction of the final shape of the part. To control the problem a lot of work has been done, either by minimizing the springback on the material side or by increasing the estimation precision in corresponding process simulations. Unfortunately, by currently available software springback still cannot be adequately predicted, because most analyses of springback are using linear, isotropic and constant Young's modulus and Poisson's ratio. But, as it was measured and reported, none of it is true. The aim of this work is to propose an upgraded mechanical model which takes evolution of damage and related orthotropic stiffness degradation into account. Damage is considered by inclusion of ellipsoidal cavities, and their influence on the stiffness degradation is taken in accordance with the Mori-Tanaka theory, adopting the GTN model for plastic flow. With regard to the case in which damage in material is neglected it is shown in the article how the springback of a formed part differs, when we take orthotropic damage evolution into consideration.

**Keywords:** Springback, damage, orthotropic properties, stiffness degradation.
**PACS:** 62.20

## INTRODUCTION

Dislocations, initial microvoids and microcracks in a material result from the material processing. Exposing materials to further loading enlarges initial microvoids and microcracks into cavities, and from dislocations nucleate new microvoids and microcracks, the last stage in the damage process being void coalescence and eventual rupture. Experimental and theoretical results show that cavities and cracks in materials cause degradation of stiffness and strength of material [2],[4],[7],[9]. To study plastic flow of porous materials Gurson [1] established a yield criterion that depends not only on the von Mises equivalent stress but also on the hydrostatic stress and void volume fraction (or porosity). Chu and Needleman (1980) proposed a model for nucleation of voids, according to which nucleation depends on equivalent plastic deformation of matrix material. The Gurson model was modified by Tvergaard (1990) in order to find better agreement in the finite element analyses of void growth. In this work we adopt a combined ductile damage model for porous materials attributed to Gurson-Tvergaard-Needleman model. Parameters of the GTN model are usually obtained from tensile tests by best fit method, but problems occur because one measurement of force does

CP907, *10th ESAFORM Conference on Material Forming*, edited by E. Cueto and F. Chinesta
© 2007 American Institute of Physics 978-0-7354-0414-4/07/$23.00

not provide enough information for inverse identification of all the GTN model parameters, as well as for identification of hardening characteristic of matrix material. To overcome that problem further investigations regarding measurement of relevant damage quantities, such as porosity, density and elastic modulus have been done, using different methods [5],[6],[7],[10]. Investigations of elastic modulus assumed that porosity in material is constituted by spherical voids, thus causing damaged material to behave isotropically. But, during the deformation process the cavities tend to elongate in the direction of the major principal strain. Also, regarding phenomenological origin (cavities, microcracks, particles with low strength, unbound particles in matrix material etc.) damage in materials can have different forms. All this causes that damaged material becomes to a certain degree anisotropic, even when initially assumed isotropic. Objective modelling of all kinds and shapes of damage, which can occur in materials, is hopeless. In this work damage is defined as a region within a representative volume of material without strength, and all kinds, shapes and sizes of damages are substituted by corresponding ellipsoidal cavities. The behaviour of this average damage model is determined upon equivalence to the elastic response of real damaged material, and it is related, due to homogenization principle, directly to the physical state at a material point. For the prediction of the elastic strain recovery after removal of forming loads two things have important role: first, the stress state which is achieved at the end of the forming process, and second, proper characterization of the elastic response in a formed part considering actual damage. In order to determine precisely the degree of damage in a formed material, and to determine precisely also the associated elastic response we propose an upgraded damage model which takes the orthotropic damage into account.

## CONSTITUTIVE LAW FOR DAMAGED MATERIAL

A region having no strength in the representative cell of volume $V_{cell}$ is represented by an ellipsoidal cavity of volume $V_{void}$. Apart from the porosity $f = V_{void} / V_{cell}$ the ellipsoid is characterized also by its shape. For this purpose we introduce two void aspect ratios $R_1 = b/a$, $R_2 = c/a$, where $a,b,c$ are respective semi-axes. Damage at a material point can be thus characterized by three parameters $\{R_1, R_2, f\}$.

**Plastic potential:** Plastic potential that is associated with the GTN model

$$\Phi = \frac{(\sigma_{eq})^2}{(\sigma_y)^2} + 2 f q_1 \cosh\left(\frac{3 q_2 \sigma_H}{2\sigma_y}\right) - (1 + q_3 f^2) \qquad (1)$$

yields, when associated plasticity is used, for the plastic strain rate

$$d\varepsilon_{ij}^p = \frac{\partial \Phi}{\partial \sigma_{ij}} d\lambda \quad \Rightarrow \quad d\varepsilon_{ij}^p = \left(\frac{1}{3}\frac{\partial \Phi}{\partial \sigma_H}\delta_{ij} + \frac{3}{2\sigma_{eq}}\frac{\partial \Phi}{\partial \sigma_{eq}} s_{ij}\right) d\lambda \qquad (2)$$

Above, $\sigma_{eq}$ is von Mises equivalent stress, $\sigma_y$ is yield stress of the matrix material, $\sigma_H$ is hydrostatic stress, and $q_1, q_2$ and $q_3$ are parameters of the damage model.

**Evolution of porosity:** The law governing the porosity evolution considers two mechanisms, void growth and void nucleation, respectively.

$$\dot{f} = \dot{f}_{growth} + \dot{f}_{nucleation} \tag{3}$$

The first term on the RHS can be formulated by considering mass conservation

$$\dot{f}_{growth} = (1 - f) \cdot \dot{\varepsilon}_{kk}^p \tag{4}$$

whereas the nucleation of voids due to microcracking and decohesion of particle-matrix interface is related to plastic deformation of matrix material. A corresponding equation for void nucleation was proposed by Chu and Needleman

$$\dot{f}_{nucleation} = A_n \, \dot{\bar{\varepsilon}}_m^p \quad ; \quad A_n = \frac{f_n}{s_n \sqrt{2\pi}} \exp\left[ -\frac{1}{2} \left( \frac{\bar{\varepsilon}_m^p - \varepsilon_n}{s_n} \right) \right] \tag{5}$$

where $A_n$ follows a normal distribution about the mean nucleation strain $\varepsilon_n$ with a standard deviation $s_n$. Parameter $f_n$ is maximum nucleated void volume fraction. In this study decrease of strength of material due to extensive void nucleation is omitted.

**Evolution of cavities' shape:** Damage is by assumption represented by ellipsoidal cavities and their shape is determined by void aspect ratios $\{R_1, R_2\}$. Though the cavities shape depends on many mechanisms in material, i.e. void nucleation, void growth and void coalescence, we assume in our approach, that for each material there exist unique relationships $R_1 = R_1(\varepsilon_1, \varepsilon_2, \varepsilon_3)$ and $R_2 = R_2(\varepsilon_1, \varepsilon_2, \varepsilon_3)$, where $\varepsilon_1, \varepsilon_2$ and $\varepsilon_3$ are principal strains.

**Hooke's law:** When the cavities are modelled by ellipsoids, Eshelby's equivalence principle and his solution of an ellipsoidal inclusion can be used [3],[11]. The principle is best combined with Mori-Tanaka's concept of average stress in a matrix [8]. Based on the Mori-Tanaka theory and using effective elastic tensor of material, a constitutive law for elastic deformation response [8] can be expressed as

$$\sigma_{ij} = \overline{C}_{ijkl} \varepsilon_{kl}^e \quad ; \quad \overline{C}_{ijkl} = \overline{C} \quad ; \quad C_{ijkl}^o = C^o \quad ; \quad C_{ijkl}^v = C^v$$

$$\overline{C} = C^o \left( I - f \left[ (C^v - C^o) \cdot (f I + (1 - f) \cdot S + C^o]^{-1} \cdot (C^v - C^o) \right) \right)^{-1} \tag{6}$$

where being elastic tensor of voids $C^v$ is a zero tensor, $C^o$ is the tensor of elastic constants of matrix material and $S$ is Eshelby's transformation tensor.

If undamaged material is assumed to be elastically isotropic, the damaged material becomes orthotropic and the corresponding material principal axes coincide with ellipsoidal axes. Nine engineering constants $\{E_1, E_2, E_3, G_{12}, G_{13}, G_{23}, \nu_{12}, \nu_{13}, \nu_{23}\}$ define the respective orthotropic material behaviour, which can be approximated in case of small porosity $f \leq 0.1$ by

$$
\begin{aligned}
E_i &= E_0(1 - \alpha_i(R_1, R_2)f^n) & i &= 1,3 \\
G_{ij} &= G_0(1 - \beta_{ij}(R_1, R_2)f^m) & i &= 1,3 \; ; \; j = 1,3 \\
\nu_{ij} &= \nu_0(1 - \chi_{ij}(R_1, R_2)f^k) & i &= 1,3 \; ; \; j = 1,3
\end{aligned}
\tag{7}
$$

In the above equations the respective influences of the porosity $f$ and shape of ellipsoidal cavity on degradation of stiffness are separated, which is important for identification of damage. Functions $\{\alpha_i(R_1,R_2),\beta_{ij}(R_1,R_2),\chi_{ij}(R_1,R_2)\}$ are called shape influence factors and they depend on the shape of ellipsoidal cavities.

## IDENTIFICATION OF PARAMETERS

For the presented model the following constants and relations must be identified:

$\sigma_y = \sigma_y(\overline{\varepsilon}_m^p)$        hardening characteristic of matrix material

$q_1, q_2, q_3, f_n, s_n, \varepsilon_n$        parameters of the GTN model

$R_1 = R_1(\varepsilon_1, \varepsilon_2, \varepsilon_3)$ , $R_2 = R_2(\varepsilon_1, \varepsilon_2, \varepsilon_3)$      ellipsoid aspect ratios

Parameters under discussion can be identified using the classical tensile test ($F - \Delta L$ diagram) for a sheet specimen, where Young's modulus must be measured in two directions: in the direction of the principal strain $\varepsilon_1$, and in the direction which is perpendicular to the direction of the principal strain $\varepsilon_2$. On the basis of measured degradation of Young's modulus during loading/unloading we can determine the respective damage parameters $\{R_1, R_2, f\}$ in every moment, extracting them from equations (7). Namely, the degradation of Young's modulus can be defined as

$$\Delta E_i = E_0 - E_i = \alpha_i(R_1, R_2)f^n \qquad i = 1,3 \qquad (8)$$

and from equation (8) it follows

$$\frac{\Delta E_1}{\Delta E_2} = \frac{\alpha_1(R_1, R_2)}{\alpha_2(R_1, R_2)} \qquad (9)$$

Since in the tensile test of isotropic matrix material the principal strain $\varepsilon_2$ equals the principal strain $\varepsilon_3$, it follows $R_1(\varepsilon_1, \varepsilon_2, \varepsilon_2) = R_2(\varepsilon_1, \varepsilon_2, \varepsilon_2) = R(\varepsilon_1, \varepsilon_2, \varepsilon_2)$. Also, the quotient $\alpha_1(R, R)/\alpha_2(R, R)$ being a monotonically decreasing function of variable $R$, and assuming $\varepsilon_2 \cong -1/2\varepsilon_1$, we can determine in accordance with equation (9) the variable $R(\varepsilon_1, -\varepsilon_1/2, -\varepsilon_1/2)$ in every moment, only from the known degradation of Young's modulus in two directions. Based on the identified aspect ratio $R$ we can evaluate further the influence factor $\alpha_1$, and from equation (8) finally extract the actual porosity

$$f = \sqrt[n]{\Delta E_1 /(E_0 \cdot \alpha_1(R, R))} \qquad (10)$$

Considering the just exposed approach we see, that from a tensile test in which Young's modulus is measured in two directions during loading, the average damage, represented with ellipsoidal cavities, can be extracted. The extracted porosity is the additional information which can be used for the identification of parameters of the GTN model, while the identified relation $R(\varepsilon_1, -\varepsilon_1/2, -\varepsilon_1/2)$ can be used for the determination of orthotropic elastic tensor $\overline{C}_{ijkl}$ in the computation of springback.

# NUMERICAL EXAMPLE AND CONCLUDING REMARKS

Stress-strain states in deep drawing are characterized by combined action of tensile and bending loadings, both causing damage in material in its own way. Since usually greatest influence on a part's final shape is experienced in bent regions, it is appropriate, for the purpose of springback study only, to observe an elementary bend. Accordingly, we will consider a simple drawing of a 0.5 mm thick flat sheet, realized by a relative displacement of rigid punch against rigid die for 1.45 mm, as schematically shown in Fig. 1a. A gap between them is 1 mm and respective drawing radii are 0.5 mm. Finite element commercial code ABAQUS is used to build a corresponding numerical model and subsequent computer simulation of drawing. Plain strain continuum elements with incompatible modes CPE4I are used.

Because our laboratory experiment is still in progress, the material parameters used in simulation are adopted for research purpose as follows. Hardening of matrix material is given by the relation $\sigma_y(\bar{\varepsilon}_m^p) = 200 + 1500(\bar{\varepsilon}_m^p)^{0.6}$ and the GTN parameters are taken from literature: $q_1 = 1.5$, $q_2 = 1$, $q_3 = 2.25$, $\varepsilon_n = 0.3$, $s_n = 0.01$, $f_n = 0.04$. Evolution of cavities shape $R_1 = R_1(\varepsilon_1, \varepsilon_2, \varepsilon_3)$ is modelled by using (4) and relations

$$\left(\frac{db}{b}\right) / \left(\frac{da}{a}\right) = \frac{d\varepsilon_2}{d\varepsilon_1} \quad ; \quad \left(\frac{dc}{c}\right) / \left(\frac{da}{a}\right) = \frac{d\varepsilon_3}{d\varepsilon_1} \tag{11}$$

Results of springback obtained by the proposed elastic degradation model, referred as model d) in Table 1, are compared with results of some other models, their specification being given directly in the table. Undamaged von Mises material with no elastic degradation is taken as the reference model. It should be emphasized, that all above models result in the same $F - \Delta L$ diagram when exposed to tensile test loading. But in spite of that, they predict different springback behaviour, as shown in Table 1. Using model b) instead of model a) results in lower stress state at the end of loading, and therefore springback is predicted to be less intensive. But by considering isotropic degradation of elastic properties springback is predicted to be more intensive and comparison of models b) and c) shows even 20.5% difference between predicted final displacements. The springback prediction of the proposed model d) is between results of b) and c) in this case. However, when the strain-stress state is more complex, the elastic response of model d) can be even more intense than the response of model c). By comparing Fig. 1b and Fig.1c it can be concluded, that degradation of elastic properties is strongly correlated with porosity. But decrease of elastic moduli in directions 2 and 3 is greater than in direction 1. The reason for rapid degradation in those directions is in voids shapes, which are almost spheres in intrados and are prolate in extrados, as shown in Fig. 1c. Consequently, springback in those directions would be even more intensive.

**TABLE 1.** Relative deviation of edge displacement with regard to model a) result

| Used model | Deviation [%] |
|---|---|
| a) undamaged material (von Mises plastic potential and initial elastic properties) | / |
| b) damaged material, undamaged stiffness (GTN, initial elastic properties) | −9.13 |
| c) isotropic damaged material (GTN, elastic properties considering $R_1 = R_2 = 1$) | +9.52 |
| d) damaged material (GTN with orthotropic elastic properties degradation) | +4.42 |

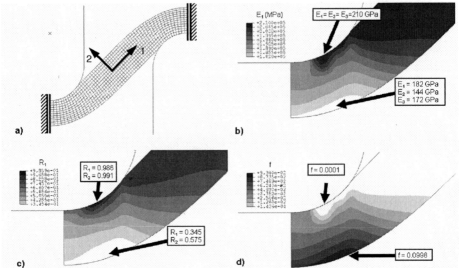

**FIGURE 1.** a) geometry of the forming tool and deformed sheet, b) degraded elastic moduli in observed bend, c) void aspect ratios d) porosity

The discussed model of orthotropic elastic stiffness degradation is strongly correlated with porosity and void aspect ratios evolution. From the comparison of the numerical results of springback behaviour it can be concluded, that not only porosity but also shapes of voids have significant influence.

# REFERENCES

1. A.L. Gurson, "Continuum theory of ductile rupture by void nucleation and growth. I. Yield criteria and flow rules for porous ductile media", *J. Eng. Mater. Technol.* **99** (1977) (1), pp. 2–15

2. Dongye Fei and Peter Hodgson," Experimental and numerical studies of springback in air v-bending process for cold rolled TRIP steels", *Nuclear Engineering and Design,Volume* 236, Issue 18,(2006),Pages 1847-1851

3. Eshelby, J.D.: The determination of the elastic field of an ellipsoidal inclusion, and related problems. *Proc. Roy. Soc., London*, A241, p.376-396 (1957)

4. Fabrice Morestin and Maurice Boivin, "On the necessity of taking into account the variation in the Young modulus with plastic strain in elastic-plastic software", *Nuclear Engineering and Design*, Volume 162, Issue 1,March 1996,Pages 107-116

5. F. Augereau, V. Roque, L. Robert and G. Despaux, "Non-destructive testing by acoustic signature of damage level in 304L steel samples submitted to rolling, tensile test and thermal annealing treatments", *Materials Science and Engineering A*, Volume 266, Issues 1-2,30 June 1999,Pages 285-294

6. Agarwal, A. M. Gokhale, S. Graham and M. F. Horstemeyer, "Void growth in 6061-aluminum alloy under triaxial stress state", *Mater. Sci. Eng. A* 341 (2003), p. 35.

7. Hung-Yang Yeh and Jung-Ho Cheng, "NDE of metal damage: ultrasonics with a damage mechanics model", *International Journal of Solids and Structures*, Volume 40, Issue 26, December 2003, Pages 7285-7298

8. Mori, T., Tanaka, K., "Average stress in matrix and average elastic energy of materials with misfitting inclusions", *Acta Metall*. 21, p. 571-574 (1973)

9. M. Yang, Y. Akiyama and T. Sasaki, "Evaluation of change in material properties due to plastic deformation" *Journal of Materials Processing Technology*, Volume 151, Issues 1-3, 1 September 2004, Pages 232-236

10. W. B. Lievers, A. K. Pilkey and D. J. Lloyd; "Using incremental forming to calibrate a void nucleation model for automotive aluminum sheet alloys", *Acta Materialia*, Volume 52, Issue 10 , 7 June 2004, Pages 3001-3007

11. Y. H. Zhao, G.P.Tandon and G.J.Weng, "Elastic Moduli for a Class of Porous Materials", *Acta Mehanica,* Vol. 76(1989), p. 105-130

# Numerical Simulations and Experimental Researches for Determining the Forces of Incremental Sheet Forming Process

Valentin Oleksik, Octavian Bologa, Gabriel Racz and Radu Breaz

*University "Lucian Blaga" of Sibiu, Engineering Faculty, Department of Machines and Equipment, Emil Cioran, 4, 550025, Sibiu, Romania, Phone: +40-269-216062/450*
*E-mail: valentin.oleksik@ulbsibiu.ro, octavian.bologa@ulbsibiu.ro, gabriel.racz@ulbsibiu.ro, radu.breaz@ulbsibiu.ro*

**Abstract.** The current paper refers to one of the new non-conventional forming procedures for sheets metal, namely incremental forming. Problems occurring during calculation of stress, thinning and the forces in the process of incremental sheet metal forming have been analyzed in this paper. The paper presents a comparison study based on the simulation by the finite element method of incremental sheet metal forming and experimental researches referred on the same process.

**Keywords:** numerical simulation, incremental sheet metal forming, experimental research.
**PACS:** Replace this text with PACS numbers;

## 1. INTRODUCTION

The usage of unconventional metal forming procedures has steadily increased in the automotive, aviation, medical equipment manufacturing and consumer goods manufacturing industries.

The present paper proposes a new forming process through which the parts can be obtained with complex configuration, log-sided without using expensive tools.

The incremental sheet forming process represents a modern method of cold forming of relatively recent date applicability and in an incipient stage because of the lack of some results obtained from systematic researches.

At the process of incremental sheet forming (Fig. 1), the deformation is achieved by a punch that comes in partial contact with the surface of the blank. For the achievement of the form of the part, one of the active elements (generally the punch) will have an axial, continuously or gradual (incremental) movement on vertical direction, while the other (the die) will perform a horizontal plane movement. The flexibility of the process is great as the same punch and the same die, according to the movements imposed to the active elements, using the same machine tool there can be obtained many forms [1].

Through this process, there can easily be achieved parts of complex form and of big dimensions for which the conventional moulds are difficult to achieve. Advantages

CP907, *10th ESAFORM Conference on Material Forming*, edited by E. Cueto and F. Chinesta
© 2007 American Institute of Physics 978-0-7354-0414-4/07/$23.00

devolve also from the possibility to implement the processing on numerical controlled milling machines [2].

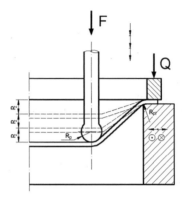

**FIGURE 1.** The principle of incremental sheet forming process

# 2. THE NUMERICAL SIMULATION BY FINITE ELEMENT METHOD

The finite element network associated with the part's geometry is built so that it allows an unfolding of the analysis in good conditions, without necessitating a re-discretisation because of its exaggerated distortions. The part, discretised as a shell, deformable body, is composed of 2441 Thin-Shell-163-type elements. Because of the way in which the elements are connected, the network contains 2490 nodes. The blank and the die, at the final stage, are presented in Fig. 2.

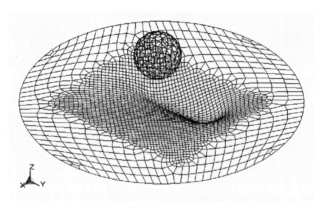

**FIGURE 2.** The die and the blank at final stage

A shear factor of 5/6, and a total of 5 integration points through the thickness were used in order to catch the variation of the stresses and strains through the thickness.

The hourglass control based on Belytschko and Tsay viscous formulation was selected in order to avoid problems with single point Gaussian integration.

The material associated with the part's elements corresponds to a deep-drawing sheet DC03. The LS-Dyna material model 36 (Barlat's 3-parameter plasticity) was chosen. This model combines isotropic elastic behaviour with anisotropic plastic potential developed by Barlat and Lian [3]. This model also includes an isotropic linear strain-hardening rule, which is satisfactory since there are not significant plastic-strain reversals in the model. The considered elasticity modulus is $E = 0.7e+5$ MPa, the transversal contraction coefficient is $\nu = 0.27$, the initial thickness $g = 1$ mm while the yield stress is $\sigma_Y = 195$ MPa, effective elastic strain $\varepsilon_0 = 0,003$, the strength coefficient $K = 465$ MPa and hardening coefficient $n = 0,21$. The anisotropic characteristic's width to thickness strain ratio values $R_{00} = 1,04$; $R_{45} = 0,86$; $R_{90} = 0,72$. The Y coordinate is the rolling direction.

The forming system consisting from hemispherical punch, die and blankholder are considered.

A thin sheet circular blank, placed on an active die with rectangular working zone is considered. The punch is placed unsymmetrical and, in the first stage, has a perpendicularly movement on the sheet level. In the second stage the punch follow a rectangular trajectory around the active die borders. There are not imposed boundary conditions, on the nodes placed on the circumference because the blankholder eliminate this necessity. Two different values of the hemispherical punch diameter and two different deeps of the parts are taken into consideration. The most important geometric parameters of the forming process are presented in Table 1.

TABLE 1. Geometric parameters of the forming process

| Case | C1 | C2 | C3 | C4 |
|---|---|---|---|---|
| Blank sheet diameter - D [mm] | 120 | 120 | 120 | 120 |
| Initial sheet thickness - g [mm] | 1 | 1 | 1 | 1 |
| Active die radius – Rpl [mm] | 6 | 6 | 6 | 6 |
| Clearance between punch and active die - j [mm] | 7 | 7 | 7 | 7 |
| Punch diameter - d [mm] | 12 | 12 | 20 | 20 |
| Deep of the part - h [mm] | 6 | 10 | 6 | 10 |

Four separate different analyses were performed, corresponding to the dimensional combinations from Table 1, in order to depict the changes at the level of material stress, strains and the forces due of these parameters [4].

The numerical results of the simulation were centred on the determination of the equivalent the effective strain, the percentage of sheet thickness reduction and the variation of the forces on two different directions.

A process of plastic deformation of this kind is difficult to control, due to some characteristic aspects such as: a relatively small contact surface of the tools with the blank; the strengths appearing in the material must be superior to those which bring it in the plastic state, but they must be less than the critical ones; the state of biaxial stretching is the least favourable to the plastic deformation processes. Because of this fact, the results of the numeric analyses in the non-linear field are oriented towards the description in a qualitative and quantitative way of the strain process results.

A series of quantitative aspects of the four studied situations are presented in Table 2.

247

**TABLE 2.** Quantitative aspects of the four studied cases

| Case | C1 | C2 | C3 | C4 |
|---|---|---|---|---|
| Maximum equivalent strain $\varepsilon_{VM}$ - [mm/mm] | 0.32 | 0.59 | 0.25 | 0.50 |
| Maximum thickness reduction $s_{max}$ - [%] | 18.64 | 36.20 | 13.81 | 26.92 |
| Maximum forces on vertical direction $F_z$ - [kN] | 5.46 | 8.73 | 5.68 | 11.08 |
| Maximum forces on horizontal direction $F_x$ - [kN] | 1.34 | 3.56 | 1.36 | 3.51 |

Figure 3 present the state of the effective strain (a.) and the thickness reduction (b.) for the case C4. This state is presented at the final stage of the punch trajectory. The effective strain has a non-uniform distribution, as it can be observed a significant localization of the strain in punch trajectory.

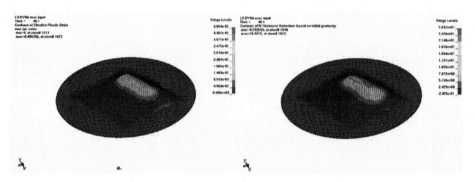

**FIGURE 3.** The distribution of effective strain (a.) and the percentage of sheet thickness reduction (b.)

Figure 4 presents graphically, based on the simulation, the variation of the two force types during the forming process, for the case C4. We presented here the vertical force $F_z$ and the horizontal forces $F_x$.

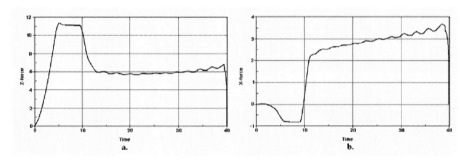

**FIGURE 4.** The variation of the $F_z$ and $F_x$ force during the forming process obtained by simulation

## 3. EXPERIMENTAL RESEARCHES

The experimental test stand is composed of the milling machine, the dynamometric table and the die (Fig. 5). The punch is fastened in the milling machine's chuck. The die has a single part, the inferior one, and is placed on the dynamometric table. It consists of a base plate, two supports, the port-active die and the blankholder. The blank is stiffly fastened between the die and the blankholder by means of four screws.

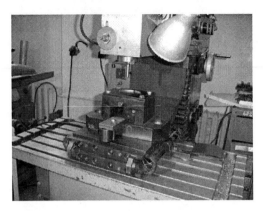

**FIGURE 5.** The employed experimental installations

The data acquisition system is composed of four modules: the transducers placed on the two rings of the dynamometric table, the signal conditioning modules, the analog – digital conversion device (KPCI 3108, Keithley Instruments Inc.) and a software package that controls the acquisition system and processes the collected data (TestPoint).

The same parameters with the numerical simulation were taken into account for determining the maximal forming force on the z direction – perpendicular on the blank's plane - $F_z$ and the maximal forming force on the x directions – in the blank's plane - $F_x$ (Table1).

The data acquisition system allowed the simultaneous recording of the values for both forming forces ($F_z$ and $F_x$), the acquisition frequency employed being 25Hz. The acquisition time for each sample was 40 seconds. The forming forces' real values were calculated with the help of the interdependence relationship between the received electric signal's value and the forces' sizes, determined with the help of the calibration equations.

The experimental program employed, with the natural variables, along with the mean values of the measured responses for each experiment are shown in Table 3.

**TABLE 3.** The mean values of the measured responses

| Case | Independent variables | | Mean values of the response functions | |
|------|------|------|------|------|
| | h [mm] | $D_p$ [mm] | $F_z$ [kN] | $F_x$ [kN] |
| C1 | 6 | 12 | 5.504 | 1.335 |
| C2 | 10 | 12 | 8.932 | 3.596 |
| C3 | 6 | 20 | 5.480 | 1.323 |
| C4 | 10 | 20 | 10.902 | 3.340 |

Figure 6 presents graphically, based on the collected data, the variation of the two force types during the forming process, for the fourth analyzed cases.

Analyzing the graphs above (Fig. 4 and Fig. 6), it can be observed that the force on horizontal direction $F_x$ presents a slight decrease during the punch's vertical stroke, due to the horizontal component of the total forming force, followed by an accented increase during the punch's horizontal stroke at the end of which the maximal value is reached.

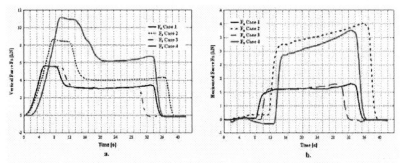

**FIGURE 6.** The variation of the two forces during the forming process, (a.) – for vertical force and (b.) for the horizontal force

The vertical force $F_z$ increases during the punch's vertical stroke, reaches a maximum at the end of this stroke, shows a level line due to the time needed to change the feed to the horizontal direction, and decreases at the beginning of the horizontal stroke and then remains on an almost constant level until the stroke's end.

## 4. CONCLUSIONS

The comparison referring to the numerical simulation of the incremental sheets metal forming and experimental researches allows the emphasizing of following conclusions:

- The forces on vertical direction $F_z$ and the force on horizontal direction $F_x$ have the same variations on the numerical simulation and on the experimental process;
- Both the force on z direction and the force on x direction don't have significant differences for the maximal values between simulation an experiment;
- The deep of the part also influences both the force on vertical direction $F_z$ and the force on horizontal direction $F_x$, as these increase with its increase;
- The punch diameter influences the $F_z$ force linearly, the force increasing with the punch diameter increase.

## REFERENCES

1. H. Iseki, "An Approximate Deformation Analysis and FEM Analysis for the Incremental Bulging of Sheet Metal using a Spherical Roller" in *J. Mat. Proc. Tech.* 111, 2001, pp. 150-154.
2. K. Kitazawa, and A. Nakajima, "Cylindrical Incremental Drawing of Sheet Metals by CNC Incremental Forming Process" in *Advanced Technology of Plasticity. Incremental Forming, In: Proc - ICTP, Vol. II,* edited, M. Geiger, Springer-Verlag, Nürnberg, pp. 1495-1500.
3. F. Barlat, and J. Lian, "Plastic Behavior and Stretchability of Sheet Metals. Part I:A Yield Function for Orthotropic Sheets Under Plane Stress Conditions" in *International Journal of Plasticity.* Vol. 5, 1989, pp.51-66.

4. O. Bologa, V. Oleksik and G. Racz, "Experimental research for determining the forces on incremental sheet forming process" in *Proceedings of the 8$^{th}$ ESAFORM Conference on Material Forming*, 27-29 April 2005, Cluj-Napoca, Romania, ISBN 973-27-1174-4, Volume I, pp. 317-320.

# Analysis Of Dynamic Dent Resistance Of Auto Body Panel

S.S.Deolgaonkar *, Dr.V.M.Nandedkar [Ψ]

* Department of Mechanical Engineering, Gramin Polytechnic, Vishnupuri, Nanded-431606, Maharashtra, India
[Ψ] Department of Production Engineering, S.G.G.S .Institute of Engg. & Tech. , Vishnupuri, Nanded-431606, Maharashtra, India

**Abstract.** In automotive industry there is increasing demand for higher quality exterior panels, better functional properties and lower weight. The demand for weight reduction has led to thinner sheets, greater use of high strength steels and a change from steel to aluminum grades. This thickness reduction, which causes decrease in the dent resistance, promoted examination of the dent resistance against static and dynamic concentrated loads. This paper describes an investigation of the suitability of explicit dynamic FE analysis as a mean to determine the dynamic dent properties of the panel. This investigation is carried out on the body panel of utility vehicle and covers two parts, in first experimental analysis is carried out on developed test rig, which is interfaced with the computer. This test rig measures deflection with accuracy of. 001mm. The experimental results are then compared with the simulation results, which is the second part. Simulation is carried with non-linear transient dynamic explicit analysis using Ansys –Ls Dyna. The experimental results show great accuracy with simulation results. The effect of change in thickness and geometry of the existing fender is then studied with help of simulation technique. By considering the best possible option overall weight of fender is reduced by 7.07 % by keeping the dent resistance of the panel constant.

**Keywords:** Dent resistance, Explicit Dynamic, Test rig. Automotive.

## INTRODUCTION

A localized plastic deformation caused by an impact on a sheet metal is described as dent. Quantitatively a dent is in the terms of physical features such as depth, diameter, and width. Dent resistance is defined as minimum force or load required to initiate a dent and hence is measure of suitability of a particular sheet metal in a given application like automotive panels. Denting in the real world is random. Dent in automotive panels can be produced by in plant handling or in service. Impacting a panel on another or dropping a panel onto a holder or conveyor can occur in fabrication. In service, dents caused by flying stones, door or shopping cart impact in parking lots, palm printing, hail force etc. are also normal. The cost for repairing the damage is relatively high. Thus minimizing this kind of damage has become a goal of the automotive manufacture particularly when thin sheet steel is used. Improvement can also enhance perceived product value. In automotive industry there is increasing demand for higher quality exterior panels, better functional properties and lower weight. The demand for weight reduction has led to thinner sheets, greater use of high strength steels and a change from steel to aluminum grades. These demand have meant

CP907, *10th ESAFORM Conference on Material Forming,* edited by E. Cueto and F. Chinesta
© 2007 American Institute of Physics 978-0-7354-0414-4/07/$23.00

that the stiffness and dent resistance of panels has become more focused, and the need for accurate methods, both experimental and numerical, for pre detecting the stiffness and dent resistance has been emphasized. In present work dynamic dent resistance of auto body panel is estimated both by experimental and numerical methods and the results are validated.

Dent resistance can be measured by static as well as dynamic methods, as per requirements. For static dent testing, data are typically compared by analyzing dent depths caused by a fixed load or by comparing the load necessary to cause a fixed dent depth. Dynamic Dent test of laboratory specimen are run on test systems referred to as Drop-weight-tester. In the drop weight test typically a specimen and an indenter is used, the indenter is dropped from the height and typical load deflection curve is plotted

Many studies regarding the dent and stiffness of automotive panels have been carried out, some with contradictory results. Dicello and George [1] came to the conclusion that the lower stiffness of the panel, the better the dent resistance. Yutori et al [2] found that the higher the stiffness, higher is the dent resistance, there seems to be no simple relation between the stiffness and dent resistance. Werner [3] concluded that for a panel with low stiffness it is beneficial to reduce the stiffness, whereas for a panel with high stiffness it is beneficial to increase the stiffness in order to improve the dent resistance. In [4] it was found that the static dent resistance is directly proportional to the final yield stress of the material (i.e. work-hardening during stamping and bake-hardening during painting). In [5] it was found that bake hardened steels show batter dent resistance than non bake–hardenable steels, even though the materials had a similar thickness and yield stress after forming. A comparison between the dent properties for steel and aluminum was made in [6], it was concluded that there is a principal difference between the appearance of these curves for steel and aluminum. The curves for aluminum shows a local maximum in dent resistance at the panel curvature, where as the curves for steel shows a local minimum.

Dent resistance is complex in nature and governed by various factors such as the panel geometry and curvature, the support conditions, the sheet thickness, the material properties of the sheet material, the load level and load type. In addition, the effect of the stamping process resulting in thickness reduction, work hardening of the sheet material, residual stress and springback also directly or indirectly affecting the dent resistance. In general conclusion that can be drawn from different investigation carried out is that the dent behavior is complex phenomenon depending on several different parameters and that is hard to intuitively assess the dent properties. Another conclusion that can be drawn is that there, in general, is a large scatter in the experimental results reported in the literature, which increases the complexity of the dent behavior of the auto body panel. Today, stiffness and dent resistance of outer panels, such as doors, hoods and lids are normally determined by physical testing. However this testing procedure has several drawbacks. It is both costly and time consuming but most importantly it cannot be carried out until rather late in the car project, in general towards the end of the design process. By then most of the factors governing the stiffness and dent properties are already specified. Thus prediction of panel's strength properties at an early stage in the design process is required, which is possible by simulation technique.

# AUTO BODY PANEL AND ITS MATERIAL

Front fender of public utility vehicle is taken for the analysis. The fender is as shown in figure 1 The material characteristics of the fender are given in table 1. In the table t denotes the initial sheet thickness $\sigma_y$ is the yield strength, $S_{ut}$ is the ultimate tensile strength $r_0, r_{45}, r_{90}$ are the anisotropy coefficient in terms of ratio of width strain to thickness strain, n is the strain-hardening exponent, E is Young's modulus and v is Poisson's ratio

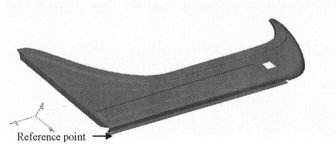

Reference point →

**FIGURE 1.** Front Fender Of Automobile.

**TABLE 1. Material Properties Of Fender.**

| Material | t | σy | Sut | r0 | r45 | r90 | n | E | v |
|----------|-----|-----|------|------|-----|------|------|-----|-----|
|          | mm  | Mpa | Mpa  |      |     |      |      | GPa |     |
| MS       | 1   | 143 | 290  | 1.96 | 2.6 | 2.11 | 0.22 | 210 | 0.3 |

# EXPERIMENTAL AND SIMULATION PROCEDURE

For experimental determination of dynamic dent resistance of auto body panel test rig is developed this test rig consist of two main parts viz. body of test rig and interfacing unit. The body of the test rig is designed so as to test the door, fender, hoods etc. of automobile. Interfacing unit is used for converting analog signals coming from the transducer to digital one for getting data on computer for load deflection curve.

In experimental procedure the dent resistance of fender is calculated at three different points, denter is allowed to fall from the height of 101.2 mm this drop height produces initial drop velocity of 1.38 m/s [7] For denting different loads varying from 19.62 N to 98.10 N were used. The fender is kept on the two supports, the distance of two supports from the dent point is 100 mm from both side. Distance of three different dent points on the fender from the reference point (as shown in figure 1) in x and z direction is given in the table 2. Initially the denter is vertically aligned with the first point, the denter plate is so adjusted that the distance between the tip of the denter and the upper surface of the fender is 101.2 mm, at this point denter plate is fixed on the vertical studs by tightening the screws. The tip of the transducer is kept exactly below the first point touching the bottom side of the fender. The denter is allowed to fall on the fender from the said height, this impact causes plastic deformation of the fender, which gives the value of the dent by calculating the difference between the initial reading and the final reading of the transducer. This procedure is repeated for the different loads and the different points.

Simulation is carried out with Ansys Ls-Dyna, which combines the Ls-Dyna explicit finite element program with the powerful pre and post processing capabilities of the Ansys program. Shell 163, a 4-noded element with both bending and membrane capabilities is used. The element has 12 degrees of freedom at each node: translations, accelerations, and velocities in the nodal x, y, and z directions and rotations about the nodal x, y, and z-axes. For getting accurate results fine meshing is done with the Hypermesh and then model is imported to the Ansys for analysis.

**TABLE 2. Distance of denting points from reference point.**

|  | Point 1 | | Point 2 | | Point 3 | |
|---|---|---|---|---|---|---|
| Axis | Z | Y | Z | Y | Z | Y |
| Distance (mm) | -160 | -120 | -70 | -400 | -120 | -550 |

# EXPERIMENTAL AND SIMULATION RESULTS FOR THE FENDER

The experimental as well as numerical investigation of dynamic dent resistance is carried out at three different points on the front fender.

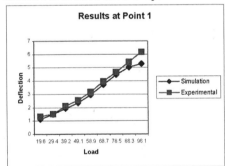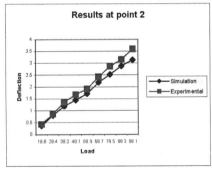

**FIGURE 2.** Comparison of Simulation and Experimental results at point 1 and 2.

Figure 2 and figure 3 shows the comparison of load deflection curve at point 1, 2 and 3 calculated by the experimental and the simulation procedure. As can be seen from figure 2 and 3 experimental result shows close accuracy with simulation results. At indentation points 1, 2 and 3 it can be seen that the deformation is slightly more the experimental work., however the overall behavior is captured with variation of 5%. The three indentation points were selected from different parts of the geometry where there are different values of the strain levels

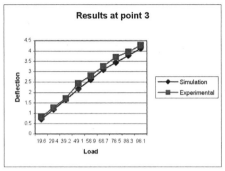

**FIGURE 3.** Comparison of Simulation and Experimental results at point 3.

## SIMULATION RESULTS OF MODIFIED FENDER

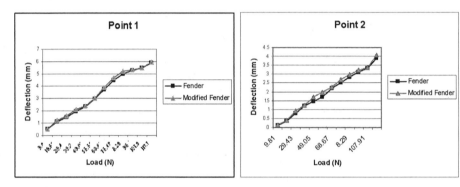

**FIGURE 4.** Comparison of simulation results for fender and modified fender at point 1 and 2.

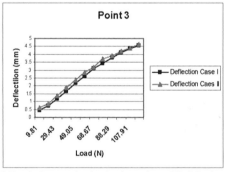

**FIGURE 5.** Comparison of Simulation results for fender and modified fender at point 3.

The existing geometry of the fender is modified by sweeping the panel with curvature of 7 mm as shown in the following figure 6. Numerical analysis is carried out on modified fender for which same boundary conditions, same material properties and same denter with its material module is used. Simulation results of modified

256

fender at different points and different loading conditions from 101.2 mm height is calculated

**FIGURE 6.** Modified fender with sweep of 7 mm.

In numerical analysis of fender while defining the real constant the thickness of the fender is assigned as 1 mm. In case modified fender denting values are calculated by changing the thickness values, at thickness of 0.91mm dent values of modified fender are closely matching with the dent values of the fender as shown in figure 4 and 5, this thickness reduction gives 7.07 % of weight reduction in modified fender.

## CONCLUSION

An Experimental test method for determining dynamic dent resistance was developed. The results from the test carried out shows small scatter within ±5% of average value.. Experimental results show greater accuracy with the simulation results. So Ansys Ls-Dyna can be used to predict the dent behavior of auto body panel in early stage of design process. Dent resistance of auto body panel can be improved by changing the geometry of the panel. In present case dent resistance is improved by sweeping the panel geometry of the fender by 7mm. Weight reduction of 7.07% of fender is also possible by changing the geometry of the panel with keeping dent resistance in the limit.

## REFERENCES

1. Dicello A.,George RA, " Design criteria for dent resistance of auto body panels" in SAE paper 740081,1974
2. Yutori Y.,Normura S., Kokubo I.,Ishigaki H., " Studies on static dent resistance" , IDDRG,1980,pp.561-569
3. Warner M.F.," Finite element simulation of steel body panel performance for quasi –static dent resistance", Automotive body material IBEC,1993
4. Chen C.H.,Rastogi P.Horvarth C.," Effect of steel thickness and mechanical properties on vehicle outer performance: stiffness,oil canning and dent resistance", Automotive body material IBEC,1993
5. Shi M.F.Michel P.E., Brindza J.A.,Bucklin P., Belanger P.J. ,Principe J.M., " Static and dynamic dent resistance performance of automotive steel body panels" SAE paper 970158, 1997
6. Van Veldhuizen, Kranendonk W., Ruifrok R. " The relation between the curvature of horizontal automotive panels, stiffness and the static dent resistance" Material and Body testing IBEC 1995
.7. M.F.Shi, " Dynamic dent resistance performance of steel and aluminum" SAE paper 930786,1993

# Prediction of Limit Strains in Limiting Dome Height Formability Test

Amir A. Zadpoor[*†], Jos Sinke[†], and Rinze Benedictus[†]

[*]Netherlands Institute for Metals Research ,Mekelweg 2,Delft 2628CD, The Netherlands
[†]Faculty of Aerospace Engineering, Technical University of Delft (TU Delft), Kluyverweg 1, Delft 2629HS, The Netherlands

**Abstract**. In this paper, the Marciniak-Kunczynski (MK) method is combined with the Storen-Rice analysis in order to improve accuracy of the predicted limit strains in Limiting Dome Height (LDH) test. FEM simulation is carried out by means of a commercial FEM code (ABAQUS) and FEM results are postprocessed by using an improved MK code. It has been shown that while original MK method considerably misspredicts the limit strains, a combination of MK method and Storen-Rice analysis can predict the dome height with a very good accuracy.

Keywords: Limiting dome height, formability limit, Marciniak-Kunczynski, FEM
**PACS:** 81.20.Hy, 81.40.Np

## INTRODUCTION

Forming Limit Diagrams (FLDs) consist of curves (forming limit curves) plotted in the plane of principal strains. The FLD is presented as a "material-property", although it is influenced by process parameters as well. Among the methods used for determination of FLDs, Limiting Dome Height (LDH) is one of the most commonly used ones specifically in the automotive industry.

Experimental determination of FLDs is associated with many difficulties. First, the required tests can be quite time-consuming, expensive, and elaborative specifically when the number of involving parameters is relatively high. Second, required equipment is not available in all workshops. Therefore, much effort has been put in theoretical prediction of FLDs. A variety of methods have been proposed by numerous researchers since the 60's. Classic works of Swift [1] and Hill [2] for theoretical prediction of diffuse and localized necking in plane stress conditions give the first theoretical methods.

Banabic has given a classification of the most important classical methods used in theoretical prediction of FLDs [3]. According to that classification, methods used for calculation of FLDs can be categorized as either theoretical or semi-empirical. Theoretical models can be divided into methods for the homogeneous sheet metals and methods for the non-homogeneous sheet metals. There are a large number of sub-classes of the methods based on the homogenous sheet metal assumption. The

CP907, 10th ESAFORM Conference on Material Forming, edited by E. Cueto and F. Chinesta
© 2007 American Institute of Physics 978-0-7354-0414-4/07/$23.00

methods based on non-homogenous sheet metals are, indeed, variants of the method first proposed by Marciniak and Kuczynski [4].

This paper is concerned with precise theoretical prediction of the dome height in simulated LDH test. Yang and Hsu [5] have previously used the Storen-Rice criterion for determination of the dome height. However, some studies have found that this criterion under-predicts the limit strain for the left hand side of the forming limit diagram [6]. In this paper, we use an improved version of the Marciniak-Kuczynski (MK) method for accurate prediction of the limit strains in simulated LDH test.

FEM modeling of the LDH test is carried out by using a commercially available FEM package (ABAQUS). It is shown that practical application of MK method for determination of the limit strains in forming processes with complicated geometry and very large deformations, such as in the case of LDH test, is associated with some difficulties which can result in highly miss-predicted limit strains. In this paper, we have made a combination of the MK method and Storen-Rice analysis [7] in order to improve accuracy of the predicted limit strains.

## MARCINIAK-KUNCZYNSKI THEORY

In the Marciniak-Kunczynski (MK) method, it is assumed that an initial imperfection is present in the sheet metal. The imperfection is modeled by a band of smaller thickness (see Figure 1 for a schematic presentation of the assumption). The initial imperfection can originate from an actually smaller thickness, a local variation of the strength, $K$, or a combination of both. Physical meaning of this assumption is well described in reference [8].

Originally, it was assumed that the imperfection zone is perpendicular to the major principal stress, $\sigma_1$, axis. This assumption can give right-hand side of FLDs. In order to be able to produce both sides of FLDs, Hutchinson and Neale [9], [10] modified the assumption to allow the imperfection lie at an angle, $\theta$, to the minor principal stress, $\sigma_2$, axis.

The minimum formability limit in the negative minor strain region is obtained by varying the imperfection angle from 0 to 90. However, it is revealed that in the positive minor strain region, the minimum formability limit is obtained when imperfection zone is perpendicular to the major principal strain axis (i.e. the original geometry).

During a biaxial straining process, the imperfection zone deforms more than the uniform zone. Therefore, the strain path of the imperfection zone (plotted in principal strains plane) is continuously ahead of the strain path of the uniform zone. At a certain point, when the strain localization takes place, the difference between the strain path of the imperfection and the uniform zone begins to increase drastically. In the MK analyses, strain paths of the both zones are traced and a criterion is used for detection of that high degree of discrepancy which is presumed to be an indicator of the strain localization. Once the strain localization is detected, the sheet metal is assumed to have failed.

The initial imperfection parameter, $f_0$, is defined as

$$f_0 = \frac{K_b t_{b,0}}{K_a t_{a,0}} \tag{1}$$

First, let's postulate following assumptions:
1. Plane stress condition applies.
2. Damage phase is excluded and, thus, the constant volume assumption can be adopted. The constant volume assumption is stated as follows

$$\varepsilon_{33,b} = -\left(\varepsilon_{11,b} + \varepsilon_{22,b}\right), \dot{\varepsilon}_{33,b} = -\left(\dot{\varepsilon}_{11,b} + \dot{\varepsilon}_{22,b}\right)$$
$$\varepsilon_{33,a} = -\left(\varepsilon_{11,a} + \varepsilon_{22,a}\right), \dot{\varepsilon}_{33,a} = -\left(\dot{\varepsilon}_{11,a} + \dot{\varepsilon}_{22,a}\right) \tag{2}$$

3. The anisotropy coordinate system is identical to the principal stress and strain coordinate systems
4. There is an initial imperfection in the sheet metal with an imperfection parameter equal to $f_0$.
5. Initial imperfection axis is initially oriented at a $\theta$ angle to the minor principal stress, $\sigma_2$, axis.

The compatibility condition can be stated as

$$\varepsilon_{tt,b} = \varepsilon_{tt,a}, \dot{\varepsilon}_{tt,b} = \dot{\varepsilon}_{tt,a} \tag{3}$$

The force equilibrium condition can be formulated as

$$\sigma_{nn,b} f\left(f_0, \varepsilon_a, \varepsilon_b\right) = \sigma_{nn,a}$$
$$\sigma_{nt,b} f\left(f_0, \varepsilon_a, \varepsilon_b\right) = \sigma_{nt,a} \tag{4}$$

In each instance, parameters of the imperfection zone in that particular instance should be used in the calculations. Transformation from anisotropy (1-2-z) coordinate systems to groove coordinate system (t-n-z) is performed by using the coordinate rotation matrix as follows

$$\boldsymbol{\sigma}_{ntz} = \mathbf{T} \boldsymbol{\sigma}_{12z} \mathbf{T}^T$$
$$\boldsymbol{\varepsilon}_{ntz} = \mathbf{T} \boldsymbol{\varepsilon}_{12z} \mathbf{T}^T \tag{5}$$

Where

$$\mathbf{T} = \begin{bmatrix} \cos(\theta(\theta_0, \varepsilon_a)) & \sin(\theta(\theta_0, \varepsilon_a)) & 0 \\ -\sin(\theta(\theta_0, \varepsilon_a)) & \cos(\theta(\theta_0, \varepsilon_a)) & 0 \\ 0 & 0 & 1 \end{bmatrix},$$

$$\boldsymbol{\varepsilon}_{12z} = \begin{bmatrix} \varepsilon_{11} & 0 & 0 \\ 0 & \varepsilon_{22} & 0 \\ 0 & 0 & -\left(\varepsilon_{11} + \varepsilon_{22}\right) \end{bmatrix}, \boldsymbol{\sigma}_{12z} = \begin{bmatrix} \sigma_{11} & 0 & 0 \\ 0 & \sigma_{22} & 0 \\ 0 & 0 & 0 \end{bmatrix}$$

Evolution of the angle $\theta$ is described by the following equation

$$\tan(\theta + \delta\theta) = \tan(\theta)\frac{1 + \delta\varepsilon_{11,a}}{1 + \delta\varepsilon_{22,a}} \tag{6}$$

Evolution rule of the imperfection parameter, $f$, can be expressed as

$$f\left(f_0, \varepsilon_a, \varepsilon_b\right) = f_0 \exp\left(\varepsilon_{33,b} - \varepsilon_{33,a}\right) \tag{7}$$

The code used for MK analysis was first validated by using it for a biaxial straining problem where limit strains were well known from the literature. It was concluded that

260

results of the MK code are in a good agreement with the reference limit strains. Initial imperfection factor was initially assumed 0.98 for all imperfection zone angles.

The sheet was assumed to have been failed when strain increment severity factor (ratio of the strain increment in the region $b$ to that of the region $a$) reached 10. The Hollomon's hardening rule and von Mises yield criterion were used in this study.

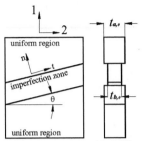

FIGURE 1. A schematic presentation of the geometry used in MK analysis.

## LIMITING DOME HEIGHT TEST

The LDH test description given in the NUMISHEET 96 benchmark documents was used in our analysis. Geometry, materials characteristics, and boundary-loading conditions are given in NUMISHEET 96 documents [11]. A FEM model was built by using an ABAQUS modeler. The model was first simulated for a punch travel of 30 mm to validate the model. Strain distribution of the model was compared with the experimental and simulation results provided by the NUMISHEET 96 documents. Draw quality mild steel (IF) was used in all of the simulations carried out in this study. Figure 2a depicts the problem geometry. Figure 2b compares calculated strains of the current simulation (along x-axis) with those of the NUMISHEET 96 documents. Several entries of the NUMISHEET 96 are used here. All the entries are identified by the name used by the NUMISHEET 96 documents. Names starting with 'E' stand for experimental studies and names starting with 'S' stand for simulation studies. One may notice that results of the current simulation are lying within experimental and computational benchmark strains provided by the NUMISHEET 96 documents.

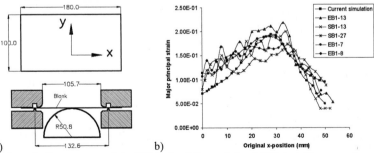

FIGURE 2. a) Geometry used in FEM simulation of the LDH test b) comparison between experimental and computational strain distribution from the NUMISHEET 96 documents and current simulation

# RESULTS AND DISCUSSION

Once the model was validated, the MK method was applied to detect the dome height. The strain severity factor was computed for all the involving elements of the blank. The strain severity factors for a particular imperfection zone angle is defined as follows

$$f_{nn} = \frac{\delta\varepsilon_{nn,b}}{\delta\varepsilon_{nn,b}}, f_{nt} = \frac{\delta\varepsilon_{nt,b}}{\delta\varepsilon_{nt,b}}, f_{eq} = \frac{\delta\varepsilon_{eq,b}}{\delta\varepsilon_{eq,b}} \qquad (8)$$

The effective strain severity factor can be, then, defined as

$$f_{eff} = \max(f_{nn}, f_{nt}, f_{eq}) \qquad (9)$$

The maximum effective strain severity factor is calculated by varying the imperfection zone angle. The punch travel was set to 0 at the beginning of the contact with the blank and maximum punch travel was set to 40 mm. Shell elements (4500 elements) were used for discretization of the blank. The LDH test was first simulated by ABAQUS explicit solver and, then, the MK code was used as a built-in post-processor of the FEM results. In the first trial, it was seen that the MK model highly under-predicts the dome height. The predicted dome height was 6 mm which highly deviates from 30.5 mm average benchmark dome height (see NUMISHEET 96 documents [11]). Besides that, the limit strains at 6 mm were too far from the theoretical and experimental FLDs and failure location was not matching the benchmark results. Figure 3a depicts the maximum effective strain severity factor vs. punch travel for the element first failed.

There was a hypothesis that parameters of the MK method such as initial imperfection factor and strain severity thresholds are not appropriate. The hypothesis was examined by changing the parameters. It was found that even when all parameters of the model are varied in a virtually wide range, no significant change takes place in the limit strains. Therefore, it was concluded that this miss-prediction can not be attributed to the improper parameter selection. In order to improve accuracy of the limit strain predicted by the MK method, we slightly modified the original MK method by combining it with the Storen-Rice method. In the modified version, the failure criterion is updated so as the strain severity factor is no longer the sole factor in determination of the strain localization onset. It is assumed that the Storen-Rice model can give a good approximate prediction of the limit strains. When strain severity factor reaches the presumed threshold, the strains are first compared with the Storen-Rice limit strains, i.e. $\varepsilon_1^*$ and $\varepsilon_2^*$.

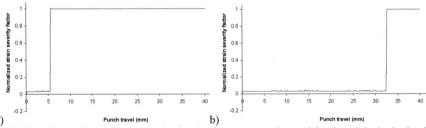

**FIGURE 3.** a) Normalized effective strain severity factor vs. punch travel for the original criterion b) Normalized effective strain severity factor vs. punch travel for the modified criterion

An element is considered to have failed only when the principal strains are close to the Storen-Rice limit strains. Therefore, an additional condition, which can be stated as follows, need to be satisfied

$$\left|\frac{\varepsilon_1 - \varepsilon_1^*}{\varepsilon_1^*}\right| \le \Omega_1, \left|\frac{\varepsilon_2 - \varepsilon_2^*}{\varepsilon_2^*}\right| \le \Omega_2 \quad (10)$$

Where $\Omega_1$ and $\Omega_2$ are the maximum allowed discrepancies between principal limit strains predicted by the MK method and those of the Storen-Rice model. In this study, $\Omega_1$ and $\Omega_2$ were assumed 0.25. The simulations were repeated again by using the modified criterion. The effective strain severity factor of the element, which first met the modified failure conditions, is depicted in Figure 3b. Table 1 compares the dome height and failure location predicted by the modified MK model (current simulation) and the benchmark experimental results. It can be seen that the modified strain localization criterion results in a much better prediction compared to the original one. Therefore, it is suggested that the modified criterion can improve accuracy of the MK method when complex straining takes place.

TABLE 1. Comparison of the results with experimental results of the NUMISHEET 96 benchmark

| Method | Dome height (mm) | Failure x-position (mm) | Failure y-position (mm) |
|---|---|---|---|
| Current simulation | 32.6 | 38.0 | 0 |
| Benchmark | 30.5 (average) | 36.0 (EB1-10) | 0 (EB1-10) |
| | | 36.4 (EB1-04) | 0 (EB1-04) |

## ACKNOWLEDGMENTS

This research was carried out under projectnumber MC1.05224 in the framework of the Strategic Research program of the Netherlands Institute for Metals Research in the Netherlands (www.nimr.nl).

## REFERENCES

1. Swift, H. W., *J. Mech. Phys. Solids* **1**, 1-18 (1952).
2. Hill, R., *J. Mech. Phys. Solids* **1**, 19-30 (1952).
3. Banabic, D., "Forming Limits of Sheet Metals," in *Formability of Metallic MAterials*, edited by D. Banabic, Berlin: Springer Verlag, 2000, pp. 173-214
4. Marciniak, Z., and Kuczynski, K., *Int. J. Mech. Sci.* **9**, 609-620 (1967).
5. Yang, T., and Hsu, T., *J. Mat. Proc. Tech.*, **117**, 32-36 (2001).
6. Zhu, X., Weinmann, K., and Chandra, A., *J. Engg. Mater. Tech. - ASME Trans.* **123**, 329-333 (2001).
7. Storen, S., and Rice, J. R., *J. Mech. Phys. Solids* **23**, 421-441 (1975).
8. Marciniak, Z., Kuczynski, K., and Pokora, T., *Int. J. Mech. Sci.* **15**, 789-800 (1973).
9. Hutchinson, J.W., Neale, K.W., and Needleman, "Sheet Necking-II. Time-Independent Behavior," in *Mechanics of Sheet Metal Forming*, edited by D.P. Koistinen and N.M. Wang, New York: Plenum, 1978, pp. 111-126.
10. W Hutchinson, J.W., and Neale, K.W., "Sheet Necking-III. Strain-Rate Effects," in Mechanics of Sheet Metal Forming, edited by D.P. Koistinen and N.M. Wang, New York: Plenum, 1978, pp. 269-283.
11. NUMISHEET 1996 benchmark documents available at http://rclsgi.eng.ohio-state.edu/~lee-j-k/Numisheet96/

# Development of aluminium-clad steel sheet by roll-bonding for the automotive industry

M. Buchner*, B. Buchmayr*, Ch. Bichler[†] and F. Riemelmoser[†]

*Chair of Metal Forming, University of Leoben, Franz-Josef-Strasse 18, A-8700 Leoben, Austria
[†]ARC Leichtmetallkompetenzzentrum Ranshofen GmbH, Postfach 26, A-5282 Ranshofen, Austria

**Abstract.** The objective of the present work is a basic study of production, modelling and validation of sheet composites of AA6xxx-automotive alloy and IF-steel. In this context the influence of surface preparation, pre-heating temperature of aluminium and steel plate, and thickness reduction on the bond strength of the composites as well as on the formation of intermetallic interface layers is analysed by shear tests and metallographic evaluations of the interface.

**Keywords:** aluminium, clad steel, roll bonding
**PACS:** 06.60.Mr, 06.60.Vz

## 1. INTRODUCTION

The development of new composites is a consequence of new processes or improvement of existing processes in the automotive industry. The trend in automotive manufacturing is to produce components of different sheet materials. The advantage of these composites is the combination of the favourable properties of both materials. The number of join patches is increased steadily because of the use of multi-materials like aluminium and steel. However, conventional fusion welding causes very hard and brittle intermetallic phases when joining aluminium and steel. As an alternative technology cold welding like roll cladding is considered where materials are joined by pressure and moderate heat treatment as well as sufficient plastic deformation.

## 2. EXPERIMENTAL PROCEDURE

To quantify the main parameters on bonding quality of a 2 mm thick AA6xxx-automotive alloy and a 2.5 mm thick IF-steel rolling experiments are performed according to the matrix in TABLE 1. Due to Al-Fe-intermetallic phases being very brittle, blank and zinc plated steels are chosen to verify the influence on the bond strength. The surfaces of aluminium and steel sheets are sanded and degreased with ethanol. To analyse influence of the surface preparation some sheets are only degreased or no preparation is done. After the surface preparation all aluminium sheets are pre-heated for 20 min and some of the steel sheets are pre-heated for 30 min at temperatures between 350 °C and 450 °C in air atmosphere without prevention of oxidation. Then the composites are rolled immediately in a 250 mm diameter two-high mill with a roll temperature of about 130 °C at 3.9 m/min. After the rolling process some composites are heated again at temperatures between 400 °C and 500 °C for 1 to 10 h to activate

CP907, *10th ESAFORM Conference on Material Forming*, edited by E. Cueto and F. Chinesta
© 2007 American Institute of Physics 978-0-7354-0414-4/07/$23.00

**TABLE 1.** Matrix of main parameters on bonding quality

| Parameters | Variants | | |
|---|---|---|---|
| steel surface | blank | zinc plated | |
| surface preparation | degreasing / sanding / degreasing | only degreasing | no preparation |
| thickness reduction (surface expansion) | | 18-45 % | |
| pre-heating temperature | 350 °C | 400 °C | 450 °C |
| bonding time | | no variation | |
| post-heat treatment (temperature / time) | 400 °C / 1-10 h | 450 °C / 1-10 h | 500 °C / 1-10 h |

diffusion. The bond strength is measured using shear tests where the cross section of the steel is weakened to prevent early aluminium fracture.

# 3. RESULTS

## 3.1. Effects of main parameters on bond strength

**Steel surface** The bond strength of the zinc plated composites is lower than the bond strength of the blank composites due to brittle $\zeta$-Fe-Zn-intermetallic phases (see FIGURE 1 (a)).

**Surface preparation** The surface preparation is a significant parameter. The best result is achieved with sanding and degreasing. Degreasing without sanding is not much better than no preparation (see FIGURE 1 (b)).

**Thickness reduction** The threshold thickness reduction is identified to be about 20 % at 400 °C. Beyond this threshold thickness reduction the bond strength increase with increasing thickness reduction (see FIGURE 1 (c)).

**Pre-heat treatment** Pre-heat treatment is also a significant parameter because the bond strength is doubled from about 40 MPa to about 80 MPa by pre-heating of steel. However, the bond strength increases only slightly by varying the pre-heat temperature between 350 °C and 450 °C (see FIGURE 1 (d)).

**Post-heat treatment** Post-heat treatment is an important factor to improve the "green" bond strength if the composite is processed with cold steel because of diffusion the bond strength rises to the value of composites with pre-heated steel (see FIGURE 1 (e)).

## 3.2. Metallographic observations of the interface

Aluminium oxide layers are usually too thin to observe via light optical microscopy, so the constitution of the interface layers of the polished and ionic etched aluminium-steel-composites is analysed by SEM and EDX (line scan, mapping). At the interface of composites with blank steel there are ferrous and very thin aluminium oxides which are

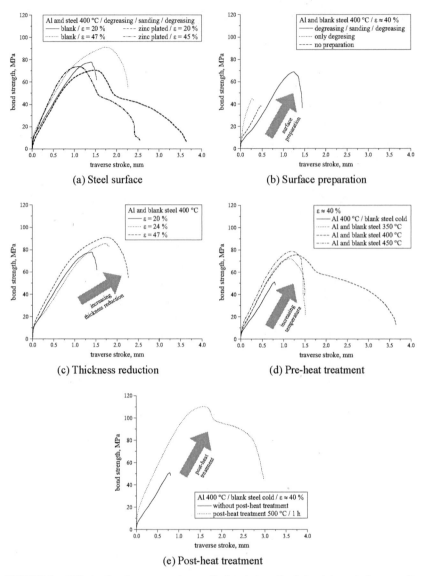

**FIGURE 1.** Effects of various parameters on the bond strength of aluminium-steel-composites

broken and separated by metal joints. The percentage of the oxides at the interface and their average length decreases with increasing thickness reduction (see FIGURE 2 and 3). At the interface of composites with zinc plated steel there are no ferrous oxides but zinc agglomerations and complex intermetallic phases at higher temperatures due to the low melting point of zinc (see FIGURE 4).

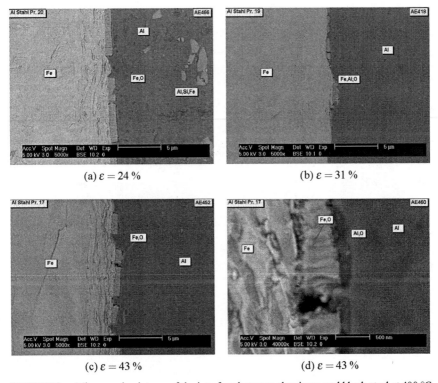

(a) $\varepsilon = 24\,\%$          (b) $\varepsilon = 31\,\%$

(c) $\varepsilon = 43\,\%$          (d) $\varepsilon = 43\,\%$

**FIGURE 2.**   Microscopic pictures of the interface between aluminum and blank steel at 400 °C

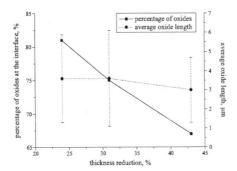

**FIGURE 3.**   Percentage of oxides at the interface and average oxide length against thickness reduction of aluminum and blank steel at 400 °C

## 3.3. Bond strength model

The aim of this work is to develop a bond strength model which offers an enhancement of the bond strength using FE-calculations on the basis of a better understanding of the

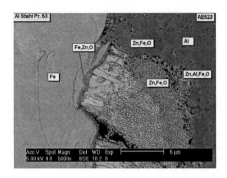

**FIGURE 4.** Microscopic pictures of the interface between aluminum and zinc plated steel at 450 °C

bonding process and the significant parameters. A couple of exitising models are:

1. Bond strength model by Vaidyanath et al. [1] (modified by Wright et al. [2])
2. Bond strength model by Bay [3] (modified by Zhang [4])

An analysis of existing bond strength models showed that there is a lack of models for higher processing temperatures. So an existing bond strength model will be adopted and modified where the influence of the main parameters in TABLE 1 is taken into consideration.

## 4. CONCLUSIONS

To achieve a good bond strength surface preparation like sanding and degreasing is necessary. Furthermore both sheets should be pre-heated. If the steel sheet is not pre-heated a post-heat treatment is essential. A minimum thickness reduction of about 20 % should be applied, but further thickness reduction is of minor importance for bond strength improvement.

## ACKNOWLEDGMENTS

The authors would like to thank the Austrian National Foundation for funding this research work in the frame of the project Austrian Light Weight Structures.

## REFERENCES

1. L. R. Vaidyanath, M. G. Nicholas, and D. R. Milner, *British Welding Journal* 6, 13–28 (1959).
2. P. K. Wright, D. A. Snow, and C. K. Tay, *Metals Technology* 5, 24–31 (1978).
3. N. Bay, *FRICTION AND ADHESION IN METAL FORMING AND COLD WELDING*, Dissertation, Technical University of Denmark (1985).
4. W. Zhang, *Bond Formation in Cold Welding of Metals*, Dissertation, Technical University of Denmark (1994).

# Behavior Laws And Their Influences On Numerical Prediction

Xavier Lemoine[1]

1 Centre Automobile Produit, Arcelor Research S.A., voie Romaine B.P. 30320,
57283 Maizières-lès-Metz, France

**Abstract.** Many studies show that the improvement of the forming numerical prediction for rolled sheets is done through laws of increasingly complex behavior, in particular by combination of the isotropic and kinematic hardening (mixed hardening) to take account of the Baushinger effect. This present work classifies the steel grades compared to the Baushinger effect. For some forming cases, it shows also the influence of a mixed hardening law on this numerical prediction, in term of deformation, thinning, residual stresses, and punch force...

**Keywords:** Effect Baushinger, Steel, Springback, Isotropic and kinematics hardening.
**PACS:** 62.20.Fe, 81.05.Bx, 83.10.Ff, 83.10.Gr, 83.50.-v, 83.60.-a

## INTRODUCTION

About the forming numerical prediction for rolled steel sheets, it is necessary to use the constitutive law. This behavior law is generally constituted by a yield criterion to define the yield stress in all directions and plastic anisotropy (Lankford), and by hardening laws (isotropic, kinematics...) to take into account the evolution of this yield locus during the plasticity. In the objective to improve the numerical predictions of codes EF from a point of view rheology or forming, it was developed increasingly complex models and requiring a number significant of parameters to be identified. This present paper points out some behavior models and when it is necessary to use them.

## BEHAVIOR MODELS

Concerning yield surfaces, the model more used for steels is the Hill48 model for its good taking into account of plastic anisotropy. Another yield criterion, we find in FE code for stamping is Hill 90:

$$\left|\sigma_1 + \sigma_2\right|^a + \frac{\sigma_b^a}{\tau^a}\left|\sigma_1 - \sigma_2\right|^a +$$
$$\left|\sigma_1^2 + \sigma_2^2\right|^{\frac{a}{2}-1}\left\{-2A\left(\sigma_1^2 - \sigma_2^2\right) + B\left(\sigma_1 - \sigma_2\right)^2 \cos 2\alpha\right\}\cos 2\alpha = \left(2\sigma_h\right)^a \tag{1}$$

CP907, 10th ESAFORM Conference on Material Forming, edited by E. Cueto and F. Chinesta
© 2007 American Institute of Physics 978-0-7354-0414-4/07/$23.00

where a, A, B are material parameters, $\sigma_b$ the equi-biaxial yield stress and $\tau$ the shear yield stress [1].There is other criteria (Barlat91 [2]…) interesting but will be to study later on if necessary. For the moment, we will limit ourselves to the 2 criteria above and the isotropic criterion of Von Mises.

In the panoply of the isotropic hardening laws, we will limit to the law of Swift, Hockett-Sherby and the linearization between these 2 laws (combined S-H) [3].

$$R = (1-\alpha)\underbrace{K(\varepsilon + \varepsilon_0)^n}_{Swift\ law} + \alpha\underbrace{(\sigma_{sat} - (\sigma_{sat} - \sigma_0)\exp(-m\varepsilon))}_{Hockett-Sherby\ law} \qquad (2)$$

In general, we are satisfied with isotropic work hardening for the forming numerical prediction, but the figure 1 shows the Baushinger effect (asymptote value of exponential curve (in dotted line)) deduced of reversible simple shear tests of steel according to Ultimate Tensile Strength UTS. The same mechanisms (precipitate, multiphase…) which increase the Ultimate Tensile Strength increase the Baushinger effect.

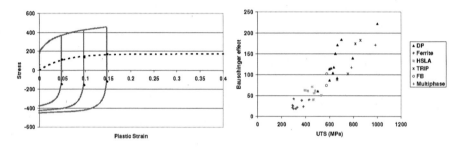

**FIGURE 1.** Steels classification of effect Baushinger .

To take into account the Baushinger effect of steels, it is necessary to introduce the kinematics hardening law. We use in this paper 2 models. The first model is the model most usually to establish in codes EF.

$$dX_{ij} = C_x X_{sat} d\varepsilon_{ij}^p - C_x X_{ij} d\bar{\varepsilon}^p \qquad (3)$$

The second model is a light modification of the first model to bring by Teodosiu [4,5].

$$dX_{ij} = C_X (X_{sat} \frac{\sigma'_{ij} - X_{ij}}{\bar{\sigma}} - X_{ij}) d\bar{\varepsilon}^p \qquad (4)$$

Others aspects of behavior could be to take into account as strain rate, work hardening stagnation (Yoshida-Uemori model [6], Teodosiu-Hu model [4,5]…), orthogonal paths (TeodosiuHu model[4,5]), damage..., but the study will become too large.

## RHEOLOGICAL PREDICTION

In general, the parameters of an isotropic hardening law are identified starting from a uniaxial tensile test between 1% to uniform elongation. But for High Strength Steels, where the uniform elongation is weak (5% for DP1000), the identification becomes difficult. However the numerical prediction of forming often requires deformations more important, from where questions concerning extrapolations of stress-strain curve.

2 ways are possible, either we continue to identify the parameters on the uniaxial tensile test and we choose the isotropic hardening model from as rheological tests being able to go to greater deformation like the bulge or shear test, or the parameters are identified directly on these last tests. For the reason which it is easier to have the tensile test, we use the first way. The figure 2 shows clearly the model "combined S-H" is very interesting about extrapolation of curve for here a DP. That is true for the majority of steels.

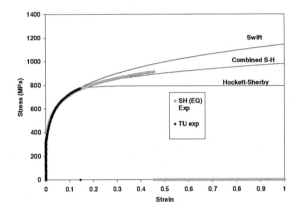

**FIGURE 2.** Extrapolation of strain-stress curve.

Figure 3 recommends that the surface of load (Hill48) more used in codes EF for steels for reasons of plastic anisotropy is not correct for the response in stress in particular for the paths of biaxial expansion obtained here in experiments by the machine of Pr. T. Kuwabara [7] or plane traction with regard to steels to high Lankford. On the other hand the criterion of Von Mises is better in constraint but not in plastic anisotropy. A criterion as Hill90 is appropriate better in this case. For HHS steels where the Lankford coefficient is moer isotrope, the preconisation is similar. the improvement of behavior don't carry in the same rheological path (uniaxial tensile test in different direction).

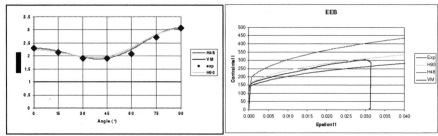

**FIGURE 3.** Rheological predicitons for mild steel.

The figure 4.a presents that the mixed model (isotrope and kinematic) (first kinematic model eq 4) is not always an improvement of behavior prediction, with a bad compromise between the uniaxial tensile and shear tests. But other models (figure 4.b) take better into account the elbow of the Baushinger tests like second model (eq 5), Yoshida-Uemori model [6]...

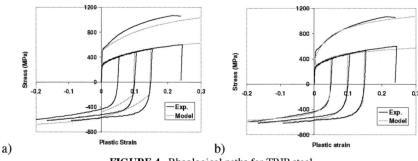

a)             b)

**FIGURE 4.** Rheological paths for TRIP steel.

According to steels, the first order differs moreover it is necessary to take account of the processes of working. Indeed, if the equivalent deformation maximum is lower than the deformation of uniform elongation, it is not necessary to seek an extrapolation of the stress-strain curve. For the mild steels it is especially the yield loci is important. On the other hand for Hight Strength Steel, the extrapolation of the curve becomes important then the yield criterion.

## FORMING AND SPRINGBACK PREDICTION

Like different authors on forming prediction similar or using others yield criteria [9], we find that the yield loci which give simultaneous a good prediction in plastic anisotropy and stress-strain curves for different rheological paths (cf. figure3) as Hill90 improve the prediction of thickness and the prediction of cup height profiles, about cup drawing test. Indeed, the figure 5 shows for same condition of process and same prediction of thickness distribution, the yield criterion of Hill90 gives a better prediction than the criterion of Hill48.

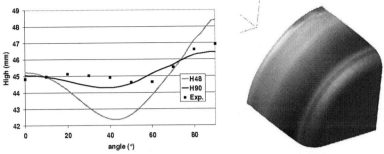

**FIGURE 5.** prediction of cup height profiles for mild steel.

About the mixed hardening law, an analytical model developed by I. Inkine [10] shows clearly an improvement of springback prediction for different cases of U-form process (various friction conditions, different Blank-Holder forces, 2 die rays (2mm and 6mm)...) for DP steel whom the thickness is 1 mm. In the figures (6.b to 6.e), the prediction values of angle b (cf. figure 6.a) are plotted in function of experimental value. The figure 6.b plots all configurations with 2 type of hardening laws. The first type is isotropic hardening law (Swift law), the second type is mixed hardening law (Swift law with $2^{nd}$ kinematics hardening). The figure 6.d where there are the results only of mixed hardening law, gives a better prediction than the figure 6.c (results of isotropic hardening law). The figures 6.e and 6.f show clearly that the model based on shell element is not sufficient for combination of 2 mm die ray and 1 mm thickness sheet.

## CONCLUSIONS

This present paper shows the interest to take into account at best the material behavior, in term of yield criteria or hardening laws (isotropic or mixed) in point of view rheological. It is present also the impact of choice of these models in different forming applications.

## REFERENCES

1. R. Hill, *Journal Mech. Phys. Solids* **38**, pp. 405-417 (1990)
2. F.Barlat, D.J. Lege and J.C. Brem, *Int. J. Plasticity* **7**, pp. 693-712 (1991)
3. Autoform 4.0, Release notes
4. S. Bouvier, V. Tabacaru, M. Banu, C. Maier, C. Girjob, B. Gardey, H. Haddadi and C. Teodosiu, *Digital Die Design Systems (3DS)Contract IMS 1999 000051, 18 Months Progress Report, November 2001.-"Selection and Identification of Elastoplastic Models for the Materials used in the Benchmarks"*
5. S. Bouvier, C. Teodosiu, H. Haddadi and V. Tabacru, *Journal de Physique IV* **105**, pp. 215-222 (2003)
6. F. Yoshida and T. Uemori, *International Journal of Mechanical Sciences* **45**, pp. 1687-1702 (2003)
7. T. Kuwabara, S. Ikeda and K. Kuroda, *International Journal of Materials Processing Technology* **80-81**, pp. 517-523 (1998)

8. Material parameters for sheet metal forming simulations by means of optimisation algorithms. RFCS Project 7210-PR-244, 2003
9. J.W. Yoon, F. Barlat, K. Chung, F. Pourboghrat, D.Y. Yang, *Int. J. Plasticity* **16**, pp. 1075-1104 (2000)
10. I. Inkin, "Etude expérimentale et modélisation de la flexion cyclique des tôles en aciers stables et métastables", Ph.D. Thesis, Institut National Polytechnique de Lorraine, 2004

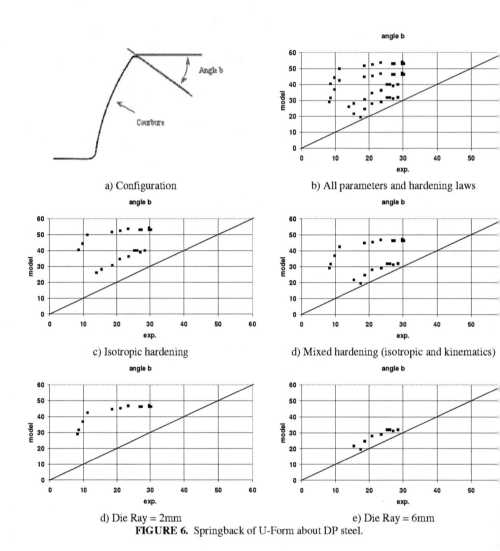

a) Configuration

b) All parameters and hardening laws

c) Isotropic hardening

d) Mixed hardening (isotropic and kinematics)

d) Die Ray = 2mm

e) Die Ray = 6mm

**FIGURE 6.** Springback of U-Form about DP steel.

# Experimental Device to Study Surface Contacts in Forming Processes

A. Attanasio [1], A. Fiorentino[1], E. Ceretti [1], C. Giardini [2]

[1] University of Brescia – Dept. of Mech. & Industrial Engineering, via Branze 38, 25123 Brescia, Italy
[2] University of Bergamo – Dept. Of Design & Technology., viale Marconi 5, 24044 Dalmine (BG), Italy

**Abstract.** Friction plays a fundamental role in forming processes since it influences many aspects of the process such as: material flow, forces, die wear and part quality. There are many different models for representing friction and many different tests to evaluate it. Friction is very different when considered in cold, warm and hot forming operations. To have an equipment giving the possibility of performing simple tests is fundamental to evaluate friction as a function of contact pressure, sliding velocity, piece roughness, tool and piece materials, tool coating, temperature. The present paper illustrates the equipment realized at the Brescia University Lab based on a large scale pin-on-disk and some preliminary tests to evaluate friction forces under different process conditions. The equipment covers a wide range of process parameters and can be fully controlled in normal force and sliding speed.

**Keywords:** Friction, Lubrication, Wear, Deformation.
**PACS:** 81.40.Pq, 81.40.Lm

## INTRODUCTION

To study contact phenomena taking place during plastic deformation processes is quite difficult since it requires the knowledge of a lot of variables that cannot be easily evaluated and that change locally in actual processes. Friction is one of the most important aspects concerned with contact between two parts, namely workpiece and die. Many models and many tests have been developed to describe and to evaluate friction [1] which does not depend only on the roughness and on the material of the parts into contact, but it is strictly related to other process parameters too, such as contact pressure, sliding velocity, temperature, lubrication conditions, coating of the tool.

The main problem arising when trying to study how friction changes according to these parameters is to realize a test where all these parameters are under control and kept constant. For this reason a pin-on-disc tribometer has been designed and built at Brescia University Lab on the base of an existing equipment present at the WZL of Aachen University [2]. This tribometer is able to simulate contact phenomena under controlled conditions so realizing a simpler tribological case.

The expected results are: 1. to investigate the influence of the several process and material parameters on friction; 2. to compare two or more lubricants under different working situations (pressure, velocity, temperature); 3. to study the influence of tool

CP907, 10th ESAFORM Conference on Material Forming, edited by E. Cueto and F. Chinesta
© 2007 American Institute of Physics 978-0-7354-0414-4/07/$23.00

coatings; 4. to analyze the die wear as function of time for different working conditions. Moreover, the realized equipment can be used to study the effects of tool coatings and lubrication in cutting operations. It is also expected that the data scatter could be kept lower in these tests with respect to actual processes, since it is possible to keep the process conditions stable and under control. The equipment has a wide range of variation for the investigated figures so to carry out tests at very similar conditions to the actual ones.

## THE EQUIPMENT DESCRIPTION

The realized pin-on-disk tribometer is able to physically simulate the actual process in under control conditions. The main idea is to put into contact two different elements, called pin and plate (the first representing the die and the second the workpiece), pushed together at constant and controlled normal force and sliding one respect to the other at constant and controlled velocity. A system of load cells gives the possibility of sampling the tangential force, that is the friction force generated during the relative motion between the two elements. The idea is that, according to Coulomb friction law, the ratio between the tangential and the normal loads is the friction coefficient: $\mu = T/F$.

The realized equipment is shown in Figure 1. The main components of the device are: the brushless motor and the epicyclical gear unit (1) that allow to move the disk (2) at variable angular velocity; the plate (3) is mounted on this disk by means of screws and its radial position defines the sliding velocity which ranges from 0 to 420 mm/s; the pin (4) is pushed by means of a hydraulic cylinder (5) against the plate. The maximum reachable normal load is 12.5 kN, that corresponds to a contact pressure equal to 250 MPa being the pin a cylinder with a diameter equal to 12 mm and a corner radius equal to 2 mm.

The brushless motor is a servomotor studied for high performance machine tool and controls the sliding velocity between pin and plate continuously. The electronic system control allows the definition of a zero starting point and to stop the tests after a given length. The force steady state is reached very quickly and therefore a short sliding length can be chosen (about 35 mm). The hydraulic actuator is controlled by means of a closed loop (a pressure transducer is provided) in order to control the normal force. The sampled pressure is then used to calculate the effective normal force. The pin movement against the plate requires the two parts to be perfectly perpendicular. This is guaranteed by the disk and bearings dimensioning together with the realization of a prismatic coupling between the pin holder and the fixed part of the equipment (6). The pin holder and its base are placed on the equipment frame by means of three axial load cells (7) (two are placed in the front and one in the back). The output signals of the load cells are added. The resulting signal is therefore proportional to the force applied along the axis of the load cells, that is the friction force ($T$) generated between pin and plate.

It is also possible to make tests at different temperatures using a cartridge heater placed in a hole under the plate. A feedback system, based on thermocouples, allows to set and maintain the temperature at a constant value. As mentioned before this equipment can be used to study lubricant or tool coating effects in metal cutting

276

operations by simply rotating the pin holder of 90° and placing it radially with respect to the disk, so reproducing orthogonal cutting operations.

The signals of pressure and force transducers are sampled at 1 kHz by a PCMCIA device and processed by a software developed in LabView®. Figure 2 shows the Data Acquisition System scheme and the transducer output.

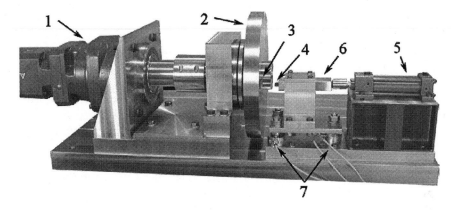

**FIGURE 1.** The main components of the equipment.

## FIRST RESULTS

As a first application, the developed equipment was tested to compare the effects of tool coating on friction forces and friction coefficient variation. In particular, two sets of tests, using as plate material a sheet of aluminum alloy (3004 series) with an external lining on both the surfaces and as pin material an HSS steel (uncoated and TiN coated), were conducted. The sheet thickness was 0.2 mm, the normal force was 3 kN (corresponding to a normal pressure of 60 MPa) and the sliding velocity was 42 mm/s. Figure 2 shows the graph of acquired normal and tangential forces together with the friction coefficient value for HSS uncoated tool. It is possible to see the transient initial period followed by the steady-state phase.

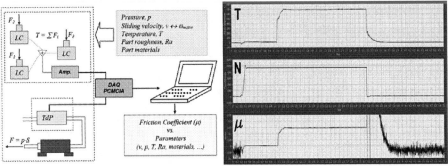

**FIGURE 2.** The data acquisition system and the transducer output.

277

The average value of friction coefficient in the steady-state is taken as reference value. The experimental friction coefficient values were 0.16 for the tests with uncoated tool and 0.12 for the tests with coated tools, that is a friction coefficient reduction equal to 25 %.

## CONCLUSION AND FUTURE WORKS

In the present paper the pin-on-disk tribometer developed at Brescia University Lab is described together with the first results obtained when studying the influence of TiN coating on HSS steel when working 3004 series Al sheet. The equipment furnishes the value of the friction coefficient for given pin and plate materials, contact pressure, temperature and sliding velocity. In order to improve the knowledge on the contact phenomena a more extensive test campaign will be conducted varying all the above said process parameters. The knowledge of friction dependence on materials, pressure, temperature and velocity will allow to apply a numerical model able to relate these parameters with the local friction coefficient value [3]. In particular, the following equation shows the proposed relationship whose coefficient can be evaluated with a limited number of experimental tests:

$$\mu = \alpha(p,v) \cdot \frac{\sqrt{Ra_{PIN}^2 + Ra_{PLATE}^2}}{\sqrt{3} \cdot \left( Ra_{PIN} \cdot \frac{Ep_{PLATE}}{Ep_{PIN}} \cdot \frac{HV_{PIN}}{HV_{PLATE}} + Ra_{PLATE} \cdot \frac{Ep_{PIN}}{Ep_{PLATE}} \cdot \frac{HV_{PLATE}}{HV_{PIN}} \right)} \tag{1}$$

where $Ra$ is the part roughness, $Ep$ is the plastic modulus and $HV$ is the Vickers hardness of the material, $\alpha(p,v)$ is a coefficient function of local pressure and sliding velocity and can be calculated with the experimental tests. The developed equipment will be used in the next future to conduct tests on Titanium and Magnesium alloys under different working temperature (warm processes), contact pressure and sliding velocity. The final goal will be to identify an equation or a model representing how friction locally varies according to formula (1) for the new tested alloys. The identified model will be implemented in an FE code so to modify friction coefficient value as the process parameters locally change and to improve the FEM results reliability.

## ACKNOWLEDGMENTS

This work has been made possible thanks to PRIN 2005 MIUR funds.

## REFERENCES

1. A. Doege, C. Kaminsky, A. Bagaviev, "A new concept for the description of surface friction phenomena" in *Jou. of Materials Processing Technology*, Elsevier **94**, pp. 189-192 (1999).
2. F. Klocke, G. Messner, C. Giardini, E. Ceretti,. "FE-simulation of micro-tribological contacts in cold forming: experimental validation" in Proc. of 2° ICTMP, Nyborg (D), 2004, pp. 395-402.
3. E. Ceretti, C. Contri, C. Giardini,. "Study on micro-tribological contacts in cold forming: simulations and experimental validation" in Proc. of VII AITEM, Lecce (I), 2005 pp. 117-118.

# Experimental and Numerical Investigations to Extend the Process Limits in Self-Pierce Riveting

*J. Eckstein\*, E. Roos\*\*, K. Roll\*, M. Ruther\*, M. Seidenfuß\*\**

*\*DaimlerChrysler AG, Sindelfingen, \*\* Materialprüfungsanstalt Universität Stuttgart*

**Abstract**. The work characterizes possible types of failure in the Self-Pierce Riveting Process. It illustrates mechanical material investigations and the simulation of fracture according to a micro-mechanical model of Rousselier.

**Keywords:** Self-Pierce Riveting Process, characterization fracture types, Rousselier,

**PACS:** 81.20.Vj

## INTRODUCTION

The systematic implementation of innovative lightweight designs in automobile manufacture is placing increasingly complex demands on joining technology. This is evident from the current Mercedes-Benz S-Class for example, where high-strength steels are used with various sheet thickness combinations, not only in homogenous combinations but also in composite designs. Whereas thermal joining technologies are prioritized for homogenous steel connections, connections between steel and aluminum employ mechanical joining technologies such as clinching and self-pierce riveting. The self-pierce riveting process represents a complex method of massive forming. The auxiliary joining part, the rivet, pierces one or more parts being joined during the process and creates a strength-relevant undercut in the lowest part being joined. Quality-determining features such as the rivet head end position, vertical and horizontal undercut or residual bottom thickness of the die-side sheet determine the demands made of a joint. Mechanical joining processes are subject to limitations due to the strength of the materials being joined. In self-pierce riveting, taking into account some component design guidelines [1], the joining of sheets with strengths of less than 600 MPa is regarded as process-consistent. From that strength radial cracks occur sporadically in the base of the pierce-rivet element, and more so when joining even higher-strength sheets (see Figure 1). When joining high-strength materials the appropriate selection of parameters such as the die, rivet geometry, materials being joined, etc. is crucial for a good riveted joint. In order to satisfy the necessary quality conditions the process window when joining high-strength steels is often very small. Minimal variations or deviations in the joining parameters, for example, can cause material breaking out of the die-side sheet. The rivet reacts very sensitively to lateral offsets in joining tools and can fail due to forced spread or upset. This problem calls

CP907, *10th ESAFORM Conference on Material Forming*, edited by E. Cueto and F. Chinesta
© 2007 American Institute of Physics 978-0-7354-0414-4/07/$23.00

for new methods of systematic, detailed analysis of the joining process and for a specification of joining parameters or process windows. The Finite Element Method (FEM) should be used increasingly as the method of analysis and development.

## PROCEDURE

The aim is to enhance FEM to make it a practicable tool. For this purpose the following procedure has been accomplished: First, characterization of the Failure Modes in Self-Pierce Riveting: For simulation of the joining process there is still insufficient detailed understanding of the failure modes, especially those of the rivet. Only when one has detailed knowledge of the fracturing mechanisms can an appropriate damage model or material model be integrated into the FEM simulation in order to define failure. Second, integration of a Suitable Material Model: Micromechanical material models, which describe the principle of void initiation, are suitable for ductile fracture behavior. A well-known example is the Rousselier model [3]. Applicability to SPR (self-pierce riveting) process simulation was investigated in the present paper.

## MATERIALS AND MODELS

ALMAC-coated rivets manufactured by HENROB were used for the present paper. Joining high-strength steels restricts the selection of a suitable self-piercing rivet. The C5x5H6 rivet used proves ideal with regard to hardness, adequate forming capability, and uncritical length (C5x5H6 = C-rivet geometry; length 5 mm; shaft diameter 5.3 mm; hardness 6, which is equivalent to approx. 570 HV10). Since in this paper a high crack probability was desirable, the selection of materials being joined went in favor of a high-strength TRIP steel (TRIP = Transformation Induced Plasticity) with a tensile strength of 800 MPa (HT800T) and a DP steel (DP = Dual Phase Steel) with 1000 MPa (HT1000X). For the composite design a 5XXX- aluminum alloy was used. The rivet joints were made with a U80 type tool from the Eckold/RiBe Company with automatic force control. Preparation for crack detection was performed by abrasive cutting followed by embedding in granules at 180°C for 13 min. Tensile tests and compression tests to determine the material properties of the rivet and sheets were performed at Stuttgart University Materials Testing Institute using a universal testing machine, type MTS Sintech 65 G. Cross tension tests and tensile shear tests were performed at DaimlerChrysler AG using the Zwick 1484 tensile testing machine. All the Finite Element Analyses (FEA) were calculated using the program MSC.Superform 2005 Release 3 with axisymmetric simplification for 2D analysis. The meshers and tools for mesh separation are from femutec Ingenieurgesellschaft mbH Hamburg.

## CHARACTERIZATION OF FAILURE IN MATERIALS BEING JOINED

This paper differentiates four main fracture patterns in the rivet (see Figure 1):

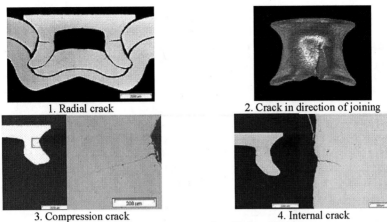

| 1. Radial crack | 2. Crack in direction of joining |

| 3. Compression crack | 4. Internal crack |

**Figure 1.** Fracture patterns in self-pierce riveting

1. The radial crack mentioned briefly at the outset: In many investigations conducted so far this type of fracture is regarded as the cause of reaching the process limit. Of all the critical joints approx. 80% of rivets fail with this fracture. The cause of the radial cracks is preparation [2]. This fact was discovered by conducting cross tension tests and tensile shear tests with HT1000X/HT800T riveted joints. The aim of these tests was to reveal the crack surface and determine the residual load-carrying capacity. Surprisingly, the rivets did not fail at the supposed radial cracks (although with this joint 80% of rivets manifest cracks on the micrograph). The rivets cracked below the usual shaft cracks with ductile/transcrystalline combination fractures, detected under a scanning electron microscope (SEM). There are much higher proportions of ductile dimples. Consequently the rivet still possesses residual forming capacity. This fact suggested that the radial cracks do not arise during the joining process but at a later point in time. Micrographs of the rivet fractured under cross tension load do not manifest any cracks either and confirm the theory. In order to avoid the high temperatures prevailing in abrasive cutting and carefully resolve the internal stress state a different method of preparation was sought. After the following steps preparation was conducted by hand: The first step is to saw into the sheets with a handsaw, up to the rivet from two opposite sides, the second step is to release the rivet from the remaining sheet bond by hand carefully and the final step is to grind the rivet half way on an abrasive disk with water cooling. The rivet half was then embedded by the usual method and polished. As the cross tension test and the tensile test had suggested already, there were no longer any cracks in the rivets prepared in this way. Preparations with cut-off grinding and by hand are compared in Figure 2.

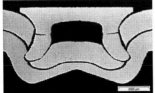 

| HT1000X-HT800T cut-off grinding | HT1000X-HT800T prepared by hand |

**Figure 2.** The same joint compared after different methods of preparation

2. Crack in the direction of joining: If the void of the self-piercing rivet is filled when joining soft materials and if the rivet is simultaneously forced to spread substantially, the rivet may crack open in the direction of joining. Under the SEM the resulting crack displays a ductile dimple fracture. The failure matches the expected fracture behavior that the base material also demonstrates in a tensile test under load to fracture. The fracture can easily be explained by the high circumferential stresses when maximum forming capacity has been reached.

3. Compression cracks: If tools are applied with an angular or lateral offset, if the sheets are too hard relative to the rivet, or if the friction between the materials being joined is too high, the rivet base is compressed and fails along the compression folds. The shear bands can also be seen here on etched micrographs. This fracture behavior was also evident after compression tests with the base material, after exceeding the plastic forming capacity. Here the specimen also fails along shear bands that had formed.

4. Internal cracks in the notch area of the rivet shaft: This is the failure mode that is expected for the self-pierce riveting process if there is excessive load in the rivet shaft. Evidence of this was provided by FEA. The joining process was replicated in 2D axisymmetrically. The highest levels of stress were found in the last analytical step of tool relief inside of the shaft in the cold-formed notch. In the shaft area the direction of the first principal stress is approximately that of the direction of joining. The gradient of stresses across the shaft width is high. After initial compressive stress in the riveting process the tensile stress on the inside, just below the notch, increases continuously after separation of the upper sheet. On tool relief (removal of punch, holder, and die) a brief increase in stresses occurs in the shaft area followed by relief. Despite this fact this failure mode was only discovered in approx. 3% of the critical joints. Revealed crack surfaces only manifest fractured surfaces in the direct vicinity of the parting plane. The fracture surface is an intercristalline/transcristalline brittle fracture. The fracture morphology is not the same as that of the ductile failure of the base material or the crack in the direction of joining. Preparation may be exerting an influence here as well. Ductile fractures, as in the direction of joining, can be reproduced by a micromechanical material model in simulation. In simulation failure due to compression can also be forecast using the calculated contour with deformation. The cause of the radial crack is to be found in preparation. For this reason it can not occur in the joining process. Response of a ductile material model to an internal crack is not to be expected due to fracture morphology. An analysis and forecast of internal cracks may be possible by evaluating the stresses arising.

The piercing of the upper sheet, which gave the process its name, takes place in almost every joint. The only exception is very soft aluminum. Here it can happen that the material remains intact like a film round the rivet base. Piercing the die-side sheet is not permissible for meeting quality criteria. All tested material showed a ductile shear dimple fracture in the piercing area. Use of the Rousselier the material model is validated for the piercing of sheets due to verification of ductile fracture.

# MATERIAL MODEL INTEGRATION

In accordance with the fracture patterns, suitable damage models had to be found for reproduction in simulation. In addition to macromechanical fracture criteria that can be easily and readily integrated the Rousselier micromechanical material model was investigated for usability in SPR process simulation. The Rousselier damage model [3] describes the ductile fracture by assuming the physical process of void initiation [4]. This is divided into three phases, initiation of the voids, their growth, and finally their coalescence. Initiation of the voids takes place on particles of a second phase, on grain boundaries, or, in high-carbon steels, also in the pearlite due to detachment of particles from the surrounding matrix or particle fracture. The resulting void becomes larger as the load increases and ultimately causes the entire component to fracture due to coalescence of the voids. This described phenomenon can be seen clearly in the investigated rivet material. Rousselier implements growth of the voids by extending the von Mises flow function:

$$\Phi\left(\sigma_v, \sigma_m, f, k_f\right) = \frac{\sigma_v}{(1-f)} - k_f + 2 \cdot \sigma_k \cdot f \cdot e^{\left(\frac{\sigma_m}{(1-f)\cdot\sigma_k}\right)} = 0 \tag{1}$$

In the equation $\sigma_v$ is von Mises equivalent stress, $\sigma_m$ is hydrostatic stress, $k_f$ is actual yield stress, f is the void volume fraction, and $\sigma_k$ is the material parameter (Rousselier constant). In order to extend the model the critical void parameter $f_c$ was introduced [4]. If the void volume f is given that value, the element stiffness is assumed to equal zero. In a tensile test on a notched circular specimen, checked by calculation, the parameters $k_f$, $f_0$, and $f_c$ are coordinated by outputting the force-progression characteristic and comparison with the experiment.

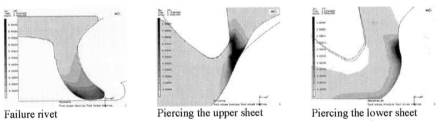

Failure rivet        Piercing the upper sheet        Piercing the lower sheet

**Figure 3.** Reproduction of void growth according to Rousselier

In a simulation of the self-pierce riveting process the rivet material shows growth of the void volume in the self-piercing rivet base (Figure 3, left). If the failure case of the rivet when joining very thick aluminum, and hence wide spreading of the rivet base, is simulated, the simulation shows an increase in void volume. However, the required critical value to initiate fracture is not reached. The explanation of this can be found in the model. The simplification for 2D axisymmetric analysis generates a uniform distribution of damage in a circumferential direction. The damage case is in the plane of the analysis, simplified for the two-dimensional analytical case. In reality there are radial scores in the self-piercing rivet base due to the cold-forming process. These defects cause notching, at which the base is split open due to an increase in stress. This remains a case for 3D analysis. Using a type-same joint with TRIP sheets the Rousselier model was evaluated with regard to piercing of the upper sheet. Compared to the experiment, separation is reproduced very well with regard to failure location, time, and direction of cracking. Figure 3, center, illustrates this process. In the black areas the element stiffness is already assumed to equal zero. Simulation of the rupture of the die-side sheet was performed using a combination joint comprised of aluminum and TRIP steel. For the die-side steel sheet the critical area is also illustrated very clearly (Figure 3, right). As in the experiment a failure occurs on the outside of the sheet. Again it is possible to make reliable statements about failure location and time. If a different scaling is selected for the void volume, a direction of failure can be recognized that is similar to that of the experiment [5].

## CONCLUSION

The cause of radial cracks was demonstrated in the preparation procedure. The cracks do not arise in the joining process, as was assumed to date, so they no longer represent the current process limit. By preparing manually this fact was demonstrated on over 50 specimens without exception. The cross tension tests and tensile shear tests confirm this proof. As a result, this research produced joints of HT800T-HT800T (1.5 mm each) and HT1000X-HT800T (1.4 mm and 1.5 mm) with sufficient undercut and without cracks! Failure occurred only due to sporadic unilateral upset of individual rivets with slight lateral offset in the joining machine. Attention should be drawn to the fact that no importance was attached to achieving ideal parameters in the joining process – the focus was on provoking radial cracks. The ultimate process limit for self-pierce riveting now has to be determined, conducting the correct preparation. The cause of cracking is assumed to be interaction between the stress state redistributing itself, temperature during the separating process, and the chemical influences of the ALMAC coating. The Rousselier material model produces very good results when the sheets are separated. In a simplified axially symmetrical two-dimensional case the damage to the rivet can not be calculated with Rousselier. Rousselier parameter determination in the tensile test proved to be relatively quick and unproblematic.

## REFERENCES

1. Merkblatt, DVS/EFB 3410: Stanznieten – Überblick. DVS- Verlag, Düsseldorf 2005
2. J. Eckstein, M. Ruther, K. Roll, E. Roos, M. Seidenfuß: Analyse der Versagensformen beim Halbhohlstanznieten, Schweißen und Schneiden 58 Heft 11, DVS-Verlag, Düsseldorf 2006
3. G. Rousselier: Ductile fracture models and their potential in local approach of fracture, Nuc. Eng. Design 105, pp. 97-111 (1987)
4. M. Seidenfuß: Untersuchungen zur Beschreibung des Versagensverhaltens mit Hilfe von Schädigungsmodellen am Beispiel des Werkstoffes 20 MnMoNi 5 5, Staatliche Materialprüfanstalt (MPA) Universität Stuttgart, ISSN 0721-4529
5. J. Eckstein, K. Roll, E. Roos: APPLICABILITY OF DAMAGE CRITERIONS FOR SIMULATION OF THE SELF- PIERCE RIVETING PROCESS, Proceedings of the FLC Zurich 2006, IVP, ETH Zurich, Switzerland

# 4 – ANISOTROPY AND FORMABILITY

## *(D. Banabic, F. Barlat, O. Cazacu, and T. Kuwabara)*

# A Comparison Between Geometrical and Material Imperfections in Localized Necking Prediction

Holger Aretz

*Hydro Aluminium Deutschland GmbH, Research & Development,*
*Georg-von-Boeselager-Str. 21, D-53117 Bonn, Germany*
*E-mail: holger.aretz@hydro.com*

**Abstract.** The Marciniak-Kuczyński model [8] is classically used in conjunction with the assumption of a pre-existing thickness imperfection in form of a narrow groove. Alternatively, one may replace the thickness imperfection by a material imperfection. In the present work the material imperfection is realized by assuming an initially soft groove. Both imperfection concepts are compared to each other under linear and non-linear strain-paths. It turns out that the concepts are only equivalent under linear strain-paths while the forming limit stresses are insensitive to the type of imperfection and the strain-path.

**Keywords:** Formability, M-K model, Localized necking, Sheet metal forming
**PACS:** 81.05.Bx, 81.20.Hy, 81.40.Ef, 81.40.Jj, 81.40.Lm, 81.40.Np, 83.10.Gr, 83.60.-a, 83.60.Wc

## INTRODUCTION

One of the most prominent localized necking models is certainly the one proposed by Marciniak and Kuczyński [8] ('M-K model'). In this model, an imperfection in form of a narrow groove is assumed. Classically, the groove represents a region of reduced sheet thickness so that strain localization takes place first in the groove when a critical strain level is reached. Alternatively, one may replace the thickness imperfection by a material imperfection. In the present work both concepts are compared to each other under linear and non-linear strain-paths.

## THEORETICAL BASIS

### Constitutive Model

An elastic-plastic constitutive model based on the associated flow-rule and isotropic hardening is used. For the sake of brevity the constitutive framework is not fully repeated here and it is referred to [3, 4]. In this work, the additive split of the total strain-rate into an elastic and a plastic part is assumed, along with the rate form of Hooke's law of linear elasticity:

$$\dot{\boldsymbol{\varepsilon}} = \dot{\boldsymbol{\varepsilon}}^e + \dot{\boldsymbol{\varepsilon}}^p, \quad \dot{\boldsymbol{\sigma}} = \boldsymbol{C} : \left( \dot{\boldsymbol{\varepsilon}} - \dot{\boldsymbol{\varepsilon}}^p \right) \tag{1}$$

CP907, *10th ESAFORM Conference on Material Forming*, edited by E. Cueto and F. Chinesta

with $C$ being the isotropic elastic tensor. The boundary of the elastic regime in stress space, i.e. the yield surface, is described by means of a phenomenological yield function $F$ of the form

$$F(\boldsymbol{\sigma}, a_i, \bar{\varepsilon}) = \bar{\sigma}(\boldsymbol{\sigma}, a_i) - Y_{\text{ref}}(\bar{\varepsilon}) \leq 0, \ \bar{\varepsilon} = \int \dot{\bar{\varepsilon}} \, dt \qquad (2)$$

$F < 0$ refers to elastic loading while $F = 0$ corresponds to plastic loading. $\bar{\sigma}$ is the equivalent stress while $a_i = const$ are material dependent shape parameters that enable the yield function to represent features of initial plastic anisotropy. $Y_{\text{ref}}$ is an arbitrary reference flow-stress of the material. According to the isotropic hardening theory it is assumed that $Y_{\text{ref}}$ depends on the accumulated equivalent plastic strain $\bar{\varepsilon}$. $\dot{\bar{\varepsilon}}$ is plastic stress-power conjugate to $\bar{\sigma}$, i.e.

$$\boldsymbol{\sigma} : \dot{\boldsymbol{\varepsilon}}^P = \bar{\sigma}(\boldsymbol{\sigma}, a_i) \cdot \dot{\bar{\varepsilon}}(\dot{\boldsymbol{\varepsilon}}^P) \qquad (3)$$

The evolution of plastic strains is governed by the associated flow-rule which is expressible as (see [5])

$$\dot{\boldsymbol{\varepsilon}}^P = \dot{\bar{\varepsilon}} \, \partial \bar{\sigma} / \partial \boldsymbol{\sigma} \qquad (4)$$

The Cauchy stress tensor can be written as

$$\boldsymbol{\sigma} = \sigma_{ij} \boldsymbol{e}_i \otimes \boldsymbol{e}_j, \quad i, j = 1, 2, 3 \qquad (5)$$

whereby 1, 2, 3 denote the original rolling, sheet transverse and sheet normal direction, respectively.

In order to account for the initial plastic orthotropy of the sheet metal the non-quadratic plane-stress yield function 'Yld2003' [1, 2] is applied. This yield function contains eight shape parameters $a_i$, $i = 1, \ldots, 8$, that can be fitted to experimental input data, see reference [1]. In this work, Yld2003 is fitted to the three directional uniaxial yield stresses and associated $r$-values ($0°$, $45°$, $90°$ to the original rolling direction) as well as to the equibiaxial yield stress $Y_b$ and the associated $r$-value denoted as $r_b$.

For the elastic-plastic stress computation the semi-implicit 'Tangent Cutting Plane' algorithm [9, 10] with automatic sub-incrementation of the imposed strain increment is used. For details on the implemented procedure it is referred to [3].


## Localized Necking Models

The forming limit diagram (FLD) corresponding to localized necking is obtained by recording and plotting the major and minor principal plastic in-plane strains at the onset of necking. In the present work, a $\varepsilon_{\text{RD}}$-$\varepsilon_{\text{TD}}$-plot is considered, whereby 'RD' and 'TD' denote the sheet rolling and transverse direction, respectively. Alternatively, one may plot the associated stresses, which forms the forming limit stress diagram. For the range where one of the in-plane strains is negative Hill's localized necking model [7] is adopted. In the regime where both strains are positive the M-K model [8] is applied. For details on this procedure it is referred to the references [3, 4].

288

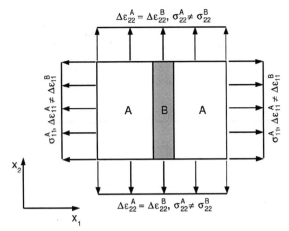

**FIGURE 1.** Geometry and boundary conditions of the M-K model.

The geometry and the boundary conditions of the M-K model are depicted in Fig. 1. In the present work, the imperfection 'B' represents either a region of reduced thickness (concept of geometrical imperfection) or a region that is initially softer than the surrounding matrix 'A' (concept of material imperfection). The material imperfection is realized by initializing the equivalent plastic strain in 'A' with a small initial equivalent plastic strain $\bar{\varepsilon}_0^A$. Hence, region 'A' takes an initial yield strength being slightly higher than that of region 'B'. In contrast, the concept of geometrical imperfection is realized by assuming an initially existing thickness imperfection $\omega_0 = t_0^B/t_0^A < 1$ with $t_0^{A,B}$ being the initial thicknesses in regions 'A' and 'B', respectively. The groove fails in a strain mode close to plane strain. In the M-K model used here the sheet patch depicted in Fig. 1 is incrementally stretched under a linear path of total (elastic + plastic) strain until the criterion for necking defined as

$$\text{If } (\Delta\varepsilon_{33}^p)^B/\Delta\varepsilon_{33}^A \geq 100.0 \text{ then necking} \tag{6}$$

is fulfilled. Note that $\Delta\varepsilon_{33}^A$ denotes a *total* thickness strain increment while $(\Delta\varepsilon_{33}^p)^B$ is a *plastic* strain increment. The accumulated plastic strains $(\varepsilon^p)^A$ are reported in the FLD. For details on the solution procedure it is referred to [3, 4].

## APPLICATION EXAMPLE

The material considered here is the aluminum alloy AA6022-T43 of initial thickness 1.0mm whose material data was taken from [6]. The hardening curve obtained by a bulge test is given by

$$Y_{\text{ref}}(\bar{\varepsilon}) = 363.44 - 234.67 \exp(-7.278\bar{\varepsilon}) \text{ MPa} \tag{7}$$

**TABLE 1.** Experimental anisotropy data for the aluminum alloy AA6022-T43 [6]. The equibiaxial yield stress is not explicitly given in reference [6].

|  | Uniaxial 0° | Uniaxial 45° | Uniaxial 90° | Equibiaxial |
|---|---|---|---|---|
| $r$-Value [−] | 1.029 | 0.532 | 0.728 | 1.149 |
| Yield Stress [MPa] | 136.0 | 131.2 | 127.6 | 128.77 |

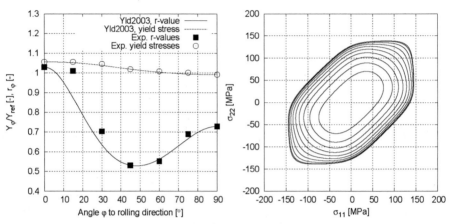

**FIGURE 2.** Left: Predicted and experimental directional yield stresses and $r$-values for AA6022-T43. The experimental data was taken from [6]. Right: Predicted yield surface in $\sigma_{11}, \sigma_{22}, \sigma_{12}$-space (contours of constant $\sigma_{12}$) for AA6022-T43.

The experimental data used for calibration of the utilized yield function Yld2003 is listed in Table 1. Since in reference [6] the equibiaxial yield stress $Y_b$ is not explicitly given $Y_b = Y_{ref}(\bar{\varepsilon} = 0)$ was pragmatically used.

The predicted directional data is displayed in Fig. 2, along with the experimental data found in [6]. Yld2003 was fitted to the data given in Table 1 by means of the Newton solver outlined in [1]. Using $Y_{ref} = 128.77$ MPa and the yield function exponent $m = 8$ the so-calculated Yld2003 parameters are as follows: $a_1 = 1.08872$, $a_2 = 0.97359$, $a_3 = 0.96788$, $a_4 = 1.03363$, $a_5 = 0.95113$, $a_6 = 0.98710$, $a_7 = 0.93391$, $a_8 = 0.91122$. The associated yield surface is displayed in Fig. 2 as well.

Localized necking simulations adopting the above mentioned material data were conducted using (i) a geometrical imperfection of $\omega_0 = 0.996$ and (ii) a material imperfection of $\bar{\varepsilon}_0^A = 0.0035$. The material imperfection corresponds to an increase of the reference yield stress of $\Delta Y_{ref} = 5.9$ MPa which is about 4.6% of the initial yield stress. The elastic constants that were used are $E = 70$ GPa and $\nu = 0.33$. The size of the material imperfection was iteratively adapted so that the localized necking predictions agree (at least to some extend) with the corresponding ones using the geometrical imperfection under linear strain-paths. The simulated forming limit strains and stresses are shown in Fig. 3. From the good agreement of both forming limit curves one might conclude that the material imperfection is equivalent to the geometrical one. This is, however, not generally the case as will be demonstrated next. Prior to the localized necking simulation

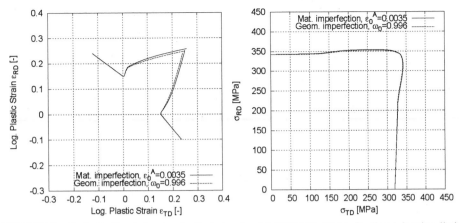

**FIGURE 3.** Left: Predicted forming limit strains for AA6022-T43. Right: Predicted forming limit stresses for AA6022-T43.

the material is now pre-strained under equibiaxial tension using a major plastic strain of 0.05. The resulting predictions are shown in Fig. 4. One recognizes that the predictions based on the material and the geometrical imperfection concept agree in plane strain tension, but show remarkable differences in the region around equibiaxial tension. Thus, both imperfection concepts are, in general, *not* equivalent. It is speculated that this is attributable to the different evolution equations in which the initial imperfections are involved.

Not unexpected, one recognizes by comparing Fig. 4 with Fig. 3 that the forming limit stresses are, at least visually, insensitive to the strain-path. This confirms that the forming limit stresses are path-insensitive when the strain-path changes without intermediately elastic unloading. Fig. 4 suggests that the forming limit stresses are, at least visually, even insensitive to the type of imperfection. This is remarkable as the limiting strain curves in Fig. 4 agree in plane strain tension while spreading towards equibiaxial tension. Thus, one might expect that the corresponding forming limit stress curves in Fig. 4 show a spread towards equibiaxial tension as well, but this is not observed. In other words, irrespective of the applied imperfection concept and the imposed strain-path the forming limit stresses show, at least visually, a uniqueness in the present case.

## CONCLUSIONS

The two concepts of material and geometrical imperfections were compared to each other under linear and non-linear strain-paths. It was shown that both concepts are only equivalent under linear strain-paths. Moreover, it was shown that irrespective of the applied imperfection concept and the imposed strain-path the forming limit stresses show, at least visually, a uniqueness in the present case. The concept of material imperfection is believed to be justifyable since variations of yield strength or, equivalently from the perspective of the isotropic hardening theory, of equivalent plastic strain of the or-

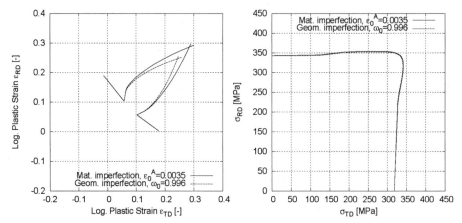

**FIGURE 4.** Left: Predicted forming limit strains for AA6022-T43. Right: Predicted forming limit stresses for AA6022-T43. The material was pre-strained under equibiaxial tension.

der $1 \cdot 10^{-3}$ appear realistic in a rolled sheet metal. Such variations might result, for instance, from natural variations regarding grain orientations. However, this statement is purely speculative as the author is at the time this paper is being written not aware of any research work dedicated to this issue. Moreover, one should bear in mind that pure material imperfections are as unrealistic as pure geometrical imperfections while a *combination* of both imperfection types is believed to be more realistic. However, if a geometrical and a material imperfection are superimposed both imperfections might take values that are even smaller than assumed in the present paper. This needs to be respected when conducting experimental investigations regarding this issue.

# REFERENCES

1. H. Aretz: Applications of a new plane stress yield function to orthotropic steel and aluminium sheet alloys, Modelling and Simulation in Materials Science and Engineering 12 (2004) 491-509
2. H. Aretz: A non-quadratic plane stress yield function for orthotropic sheet metals, Journal of Materials Processing Technology 168 (2005) 1-9
3. H. Aretz: Numerical analysis of diffuse and localized necking in orthotropic sheet metals, International Journal of Plasticity (in press)
4. H. Aretz: Impact of the equibiaxial plastic strain ratio on FLD prediction, Proc. ESAFORM 2006 Conference, 2006
5. H. Aretz: A less hypothetical perspective on rate-independent continuum theory of metal plasticity, Mechanics Research Communications 33 (2006) 734-738
6. J.C. Brem, F. Barlat, R.E. Dick, J.W. Yoon: Characterization of aluminum alloy sheet materials, Numisheet'2005
7. R. Hill: On discontinous plastic states, with special reference to localized necking in thin sheets, Journal of the Mechanics and Physics of Solids 1 (1952) 19-30
8. Z. Marciniak, K. Kuczyński: Limit strains in the process of stretch-forming sheet metal, International Journal of Mechanical Sciences 9 (1967) 609-620
9. M. Ortiz, J.C. Simo: An analysis of a new class of integration algorithms for elastoplastic constitutive relations, International Journal for Numerical Methods in Engineering 23 (1986) 353-366
10. J.C. Simo, T.J.R. Hughes: *Computational Inelasticity*, Springer-Verlag, 1998

# Stochastic simulations of forming limit diagrams

Ø. Fyllingen, O.S. Hopperstad, M. Langseth

Structural Impact Laboratory, Department of Structural Engineering, Norwegian University of Science and Technology, Rich. Birkelands vei 1 A, NO-7491 Trondheim, Norway

**Abstract.** Finite element simulations of square patches of material with spatial thickness variations subjected to a set of proportional strain paths have been performed. Localization is detected when the thickness increment in one element is much larger then the average thickness increment of the whole patch. Gaussian random fields with a Matérn covariance function have been used to model the spatial thickness variations and by use of Monte Carlo analysis stochastic forming limit diagrams have been estimated. The effect of changing the effective range and sill of the covariance function has been investigated.

**Keywords:** FLD, FE, Monte Carlo
**PACS:** 62.20.-x, 62.20.Fe, 61.82.Bg

## INTRODUCTION

The forming limit diagram (FLD) is used to determine the formability of metallic materials. In experimental tests, a sheet is subjected to proportional strain paths with different strain ratios, and the strain at incipient localized necking is taken as the formability limit. Experiments performed by van Minh et al. [1] show that "the variation in forming limits is much greater than that due to the experimental error" and it is proposed that this scatter reflects an intrinsic property of the material which is important in determining material formability. In the current study, a method will be proposed for calculating the scatter band in the FLD by use of a finite element based Marciniak-Kuczynski approach. The thickness variations are modelled by use of random fields, and the effect of changing the field characteristics will be investigated.

## MODEL

The model which will presented here is a further development of a model proposed by Lademo et al. [2,3], which adopts the ideas introduced by Marciniak and Kuczynski [4] to the finite element method environment. They consider a square patch of material, as depicted in **FIGURE 1** discretized into a grid of $n_{el}$ x $n_{el}$ square shaped membrane elements. The initial width of the patch is $w_0$ and the initial thickness is described by the field $t_0(x,y)$. The patch is fixed along side 1-2 and 1-4 and displacements are imposed along side 2-3 and 3-4 such that proportional straining is obtained. The ratio between the strain increment in the thickness direction for element i, $\Delta\varepsilon_{3,i}$, and the average strain increment in the thickness direction,

CP907, *10th ESAFORM Conference on Material Forming*, edited by E. Cueto and F. Chinesta
© 2007 American Institute of Physics 978-0-7354-0414-4/07/$23.00

$\Delta\varepsilon_{3,\text{avg}}$, is $\beta_i = \Delta\varepsilon_{3,i} / \Delta\varepsilon_{3,\text{avg}}$. Localized necking is assumed to occur if any $\beta_i$ exceeds the critical value $\beta_{cr} = 2.0$. Hence, the average true strains at the time of necking may be calculated.

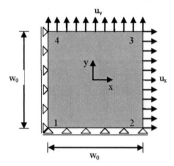

**FIGURE 1.** Finite element model

If the procedure is repeated for several different strain paths a FLD can be constructed. The necking criterion has been implemented into a user-defined material model in the FEM-code LS-DYNA by Berstad et al [5], where the yield criterion is defined as:

$$f(\boldsymbol{\sigma},R) = \left(\frac{1}{2}\left(\left|\sigma_1\right|^m + \left|\sigma_2\right|^m + \left|\sigma_1 - \sigma_2\right|^m\right)\right)^{\frac{1}{m}} - \left(\sigma_0 + R\right) \leq 0 \qquad (1)$$

where $(\sigma_1, \sigma_2)$ are the principal stress components in the plane, m is a material parameter and $\sigma_0$ is the yield stress. The isotropic strain hardening variable R is defined by:

$$R\left(\bar{\varepsilon}\right) = \sum_{i=1}^{2} Q_i \left(1 - \exp\left(-C_i \bar{\varepsilon}\right)\right) \qquad (2)$$

where $\bar{\varepsilon}$ is the equivalent plastic strain and $(Q_1, C_1, Q_2, C_2)$ are work hardening parameters. The chosen material parameters are given in **TABLE 1**. Further, the density, Young's modulus and Poisson's ratio were set to 2700 kg/m$^3$, 70 000 MPa and 0.33, respectively.

**TABLE 1.** Material parameters for isotropic hardening rule

| $\sigma_0$ [MPa] | $Q_1$ [MPa] | $C_1$ | $Q_2$ [MPa] | $C_2$ | m |
|---|---|---|---|---|---|
| 87.00 | 11.40 | 724.4 | 133.6 | 15.50 | 8 |

## NUMERICAL STUDY

It will be investigated how different randomly varying inhomogeneities will effect the variation and shape of FLDs. It is assumed that the inhomogeneities can be represented by random fields. Here, the thickness field $t_0(x,y)$ will be described by a stationary Gaussian random field with a Matérn covariance function:

$$C(h) = \begin{cases} \dfrac{s^2}{2^{\nu-1}\Gamma(\nu)}(h/R)^\nu K_\nu(h/R), & h > 0 \\ s^2, & \text{otherwise} \end{cases} \quad (3)$$

where h is the distance between two locations. $K_\nu$ and $\Gamma$ are the modified Bessel function of order one and the Gamma function, respectively. $s^2$ is called the sill or variance parameter, $\nu$ is the smoothness parameter and R is referred to as the range parameter. Further on the effective range is defined as the distance $h_0$ at which the correlation, $\rho(h) = C(h) / s^2$, has dropped to 0.05. Isotropic fields may be converted into anisotropic fields by use of coordinate transformations. By defining $\phi$ as the ratio between the maximum and minimum ranges this may be done. The effect of varying $h_0$, s and $\phi$ will be studied.

In the study a plate of 100 mm x 100 mm discretized into 40 x 40 square shaped membrane elements will be considered. The thickness field will have a mean of 2.0 mm, and $\nu$ of the covariance function will be equal to 5.0. For each case, 50 samples of the corresponding thickness field will be generated. Simulations along 4 linear strain paths, $\theta = (0°, 15°, 30°, 45°)$, will be done for each of the samples for the isotropic field, while 7 linear strain paths, $\theta = (0°, 15°, ..., 90°)$, will be considered for the anisotropic fields. Further on, the direction of maximum range will be parallel to the y-direction for the anisotropic fields. At each direction, the mean and standard deviation of $\varepsilon = \sqrt{\varepsilon_x^2 + \varepsilon_y^2}$ at localization will be estimated.

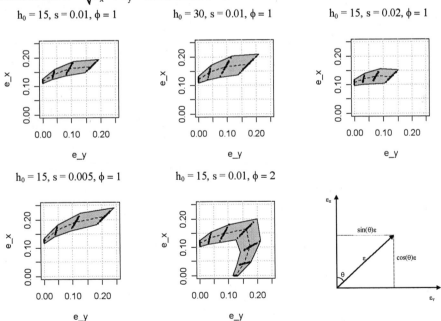

**FIGURE 2.** Stochastic FLDs for different combinations of $h_0$, s and $\phi$

In **FIGURE 2**, FLDs for different combinations of $h_0$, s and $\phi$ are presented. The FLDs have been constructed by drawing a stapled line through the mean values of $\epsilon$ and by shading the areas of the mean $\pm$ 3 standard deviations. The shaded area contains all the simulation data, and hence it is expected that a large proportion of all the possible outcomes are contained within this area. The range of the estimated mean of the FLDs presented here corresponds approximately to Marciniak-Kuczynski based FLD calculations with an inhomogenity factor in the range 0.98-0.995. These are typical values used in the literature. For example Barlat and Jalinier [6] relate the inhomogenity factor to measurements. From **FIGURE 2** it may be observed that an increase in $h_0$ from 15 mm to 30 mm leads to an increased variance. Further on, by increasing s the forming limit is increased. This increase is largest at $\theta = 45°$. An anisotropic field leads to an unsymmetrical FLD about a line aligned at $\theta = 45°$. The variance seems to be larger for $\theta > 45°$, which corresponds to the direction of the maximum ranges or "wavelengths". In **FIGURE 3** the bands of localization are shown for one sample at different strain paths.

| $\theta = 0°$ | $\theta = 15°$ | $\theta = 30°$ | $\theta = 45°$ |

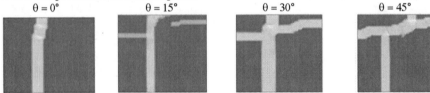

**FIGURE 3.** Pictures of localisation for one sample with $h_0 = 15$, s = 0.01 and $\phi = 1$

## CONCLUSIONS

The assumed thickness variations resulted in a quite wide scatter band. Further on, a change of the characteristics of the thickness field resulted in a change of the shape and variance of the FLD. By imposing an anisotropic field the FLD were not symmetric about $\theta = 45°$. In order to compare the model to experimental data it is desirable to relate the field characteristics to measured inhomogeneities. This could eventually lead to non-Gaussian field and a different covariance function.

## REFERENCES

1.  H. Van Minh, R. Sowerby, J. L. Duncan, Variability of forming limit curves, Int. J. mech. Sci. 16 (1974) 31-44
2.  O. G. Lademo, T. Berstad, O.S. Hopperstad, K.O. Pedersen, A numerical tool for formability analysis of aluminium alloys. Part I: Theory, Steel Grips 2 (2004) 427-431
3.  O.-G. Lademo, T. Berstad, O.S. Hopperstad, K.O. Pedersen, A numerical tool for formability analysis of aluminium alloys. Part II: Experimental validation, Steel Grips 2 (2004) 433-437
4.  Z. Marciniak and K. Kuczynski, Limit strains in the processes of stretch-forming sheet metals, Int. J. Mech. Sci. 9 (1967) 609-620
5.  T. Berstad, O.-G. Lademo, O.S. Hopperstad, Weak and strong texture models in LS-DYNA, SINTEF REPORT STF80MK F05180 (2005)
6.  F. Barlat and J.M. Jalinier, Formability of sheet metal with heterogeneous damage, Journal of Materials Science 20 (1985) 3385-3399

# Two Prediction Methods For Ductile Sheet Metal Failure

BRUNET Michel [1], CLERC Patrice[2]

*1-INSA de Lyon, LaMCoS, 20 avenue Einstein, 69621 Villeurbanne, France.*
*2-MECANIUM, 66 boulevard Niels Bohr, 69603 Villeurbanne, France*

**Abstract.** Two analytical approaches are detailed for the determination of Forming Limit Diagrams (F.L.D.) and compared with experimental results. The first one is the "Enhanced Modified Maximum Force Criterion EMMFC" and the second one is the "Through-Thickness Shear Instability Criterion TTSIC". The criteria are both written in an intrinsic analytical form and are applicable to linear and non-linear given strain paths as it occurs in any FEM codes for sheet-metal forming simulation. Finally, the two methods are complementary depending on the nature of failure and the predicted curves are in reasonable agreement with the trend of experimental results for a wide range of materials.

**Keywords:** Necking, Sheet-Metal, Plasticity, Forming Limit, Failure Criteria.
**PACS:** 62.20.Fe

## INTRODUCTION

When dealing with the necking in sheet metals one must distinguish between *diffuse* and *localized* necking but also with the sudden transition between *ductile* and *brittle mode of fracture*. If the loading conditions and material properties allow a neck to develop gradually after the onset of diffuse necking, plastic instability theories may be applied. From this last point, Marciniak and Kuczynski [1] have developed a FLD model which contains an initial thickness imperfection in the form of a band. This model is known as the "M-K model" and is widely used. A problem with the M-K model is that the imperfection has to be chosen unrealistically large which is not in agreement with experimental measurements. Moreover, it has been observed experimentally for several alloys that failure can take place by *shear localization* through the *thickness* at an angle close to the direction of *maximum shear stress*. The instability may take place before any local necking in the plane of the sheet is visible or inside a developing local neck. In order to account for the shear fracture mode, a through-thickness shear instability criterion based on the work of Bressan and Williams [2] is presented here in an intrinsic general form. The modified maximum force criterion first introduced by Hora *et al* [3] and developed in an intrinsic form by Brunet *et al* [4] is also used by seeking a direction of zero extension within the *plane* of the sheet rather than within the *thickness*. An enhanced formulation is presented here where the *strain rate sensitivity* and the *thickness effect* are taken account.

CP907, *10th ESAFORM Conference on Material Forming*, edited by E. Cueto and F. Chinesta
© 2007 American Institute of Physics 978-0-7354-0414-4/07/$23.00

# ENHANCED MODIFIED MAXIMUM FORCE CRITERION

In order to improve Swift's criterion, Hora *et al* [3] took into account the experimentally confirmed fact that the onset of necking depends on the strain ratio: $\beta = \Delta\varepsilon_2/\Delta\varepsilon_1$. In the following, Hora's necking model is *enhanced* in a form containing the thickness effect by the fact that the diffuse necking is written with the major *uniform* principal stress outside the starting neck with:

$$d\sigma_{1u}/d\varepsilon_1 \leq \sigma_{1u} \tag{1}$$

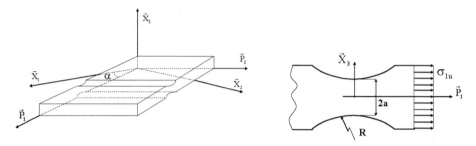

**FIGURE 1.** Orthotropic axes $\vec{X}_i$, principal stresses axes $\vec{P}_i$ and assumed neck profile.

It is assumed that the stress state becomes tri-axial in the neck. By analyzing the tri-axial stress state in the neck of a sheet undergoing plane strain plastic deformation, an approximate correction factor is obtained as:

$$\sigma_{1u} = \sigma_1 - B\bar{\sigma}\frac{a}{3R} \qquad \text{where} \qquad B = \sqrt{\frac{(1+r_0)(1+r_{90})}{(1+r_0+r_{90})}} \tag{2}$$

After diffuse necking, the state of strain evolves towards the plane strain state in the band, due to the related stress state change, there is an additional hardening effect where the stress dependency of $\sigma_{1u} = \sigma_{1u}(\varepsilon_1, \beta, a)$ is such that:

$$d\sigma_{1u} = \frac{\partial\sigma_{1u}}{\partial\varepsilon_1}d\varepsilon_1 + \frac{\partial\sigma_{1u}}{\partial\beta}d\beta + \frac{\partial\sigma_{1u}}{\partial a}da \tag{3}$$

Using Eq. (1), (2), (3), the "EMMFC" criterion takes the expanded form:

$$\frac{\partial\sigma_1}{\partial\varepsilon_1} + \frac{\partial\sigma_1}{\partial\beta}\frac{d\beta}{d\varepsilon_1} \leq \sigma_1 - \frac{B\bar{\sigma}}{3R}[1 + \frac{\beta}{2}]t_0\exp(\varepsilon_3) \tag{4}$$

The *strain rate sensitivity factor m* can be introduced in the first term of Eq. (4) as:

$$\frac{\partial\sigma_1}{\partial\varepsilon_1} = H'\left(\frac{\dot{\bar{\varepsilon}}}{\dot{\bar{\varepsilon}}_0}\right)^m \frac{\partial\sigma_1}{\partial q}\frac{\partial\bar{\varepsilon}}{\partial\varepsilon_1} \tag{5}$$

where H' is the *tangent modulus* of the *hardening curve* $\bar{\sigma} = H(\bar{\varepsilon})$ obtained at the reference strain rate $\dot{\bar{\varepsilon}} = \dot{\bar{\varepsilon}}_0$ and q the effective stress of the chosen yield surface.

298

# THROUGH THICKNESS SHEAR INSTABILITY CRITERION

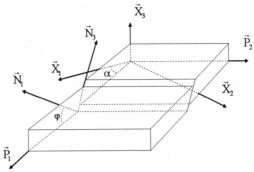

**FIGURE 2.** Orthotropic axes $\vec{X}_i$, principal stresses axes $\vec{P}_i$ and shear fracture axes $\vec{N}_i$.

Sheet    metal experimental observations at the failure site show that the *fracture plane* lies in a direction near to that of *maximum shear stress*. These observations have suggested to Bressan and Williams [2] that the application of a shear criterion may be useful in determining the onset of failure in sheet metal forming. The shear criterion is presented here in a more general form where the principal directions of the stresses $\vec{P}_i$ do not generally coincide with the orthotropic anisotropy axes $\vec{X}_i$ as it can be seen on figure (2). For a given angle $\alpha$ and values of the plastic strain increments in the orthotropic axes, the inclination $\varphi$ of the shear plane is obtained with the relation:

$$(\sin^2 \varphi \cos^2 \alpha - \cos^2 \varphi)d\varepsilon_{11} + (\sin^2 \varphi \sin^2 \alpha - \cos^2 \varphi)d\varepsilon_{22} + 2\sin^2 \varphi \cos \alpha \sin \alpha \, d\varepsilon_{12} = 0 \quad (6)$$

The formula given by Bressan and William [2] is recovered with $\alpha$=0 such that:

$$\cos 2\varphi = -\beta/(\beta + 2) \quad (7)$$

Introducing the critical shear stress on the inclined plane, the through-thickness shear instability criterion (TTSIC) is:

$$\sigma_{1CR} \geq \frac{2\tau_{CR}}{\sin 2\varphi} \quad (8)$$

The critical shear stress can be obtained from either a uni-axial test or an equi-biaxial test. In the case of the uni-axial test, it may be difficult to measure precisely the uni-axial limit strain which exists inside the diffuse necking region. The critical shear stress may be determined more accurately from the biaxial test or by a necking experimental point taken on the positive side of the FLD, $0 \leq \beta \leq 1$. An other possibility for the critical shear stress is to use the previous "EMMFC" by assuming that the onset of shear instability predicted by the "TTSIC" criterion follows the prediction of the "EMMFC" criterion for a given $\beta$ strain ratio value.

## CRITERIA COMPARISONS

Typically, figure (3) shows a theoretical comparison between the two criteria for direct strain path and a virtual isotropic material of hardening curve $\overline{\sigma} = B(c + \overline{\varepsilon})^n$

where B=564.9 MPa, c=0.00723 and n= 0.242. The critical shear stresses 254 MPa and 240 MPa are calculated with the critical necking strain point obtained with "EMMFC" criterion for $\beta=1$ and $\beta=0$ (plane strain) respectively and without strain rate or thickness effects. It can be seen that the "TTSIC" criterion strongly depends upon the critical shear stress and the "EMMFC" curve lies between the two "TTSIC" curves from the negative to the positive sides of the FLD's. It may be noticing that if the minimum of the "EMMFC" curve is at plane strain ($\beta=0$), the "TTSIC" curves exhibit a minimum for a positive strain ratio: $0.15 \leq \beta \leq 0.2$ .

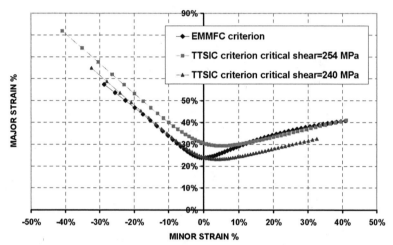

**FIGURE 3.** Theoretical FLD's comparison

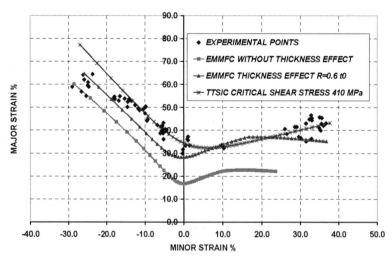

**FIGURE 4.** FLD's Titanium Alloy, $t_0 = 1.05$ mm; $R_0 = 1.46$, $R_{45} = 3.18$, $R_{90} = 3.09$

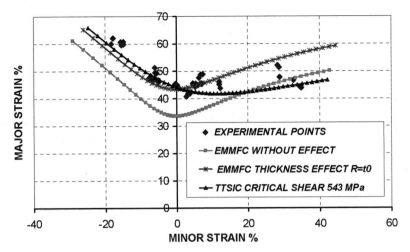

**FIGURE 5.** FLD's Stainless Steel, $t_0 = 1.5$ mm; $R_0 = 0.94$, $R_{45} = 0.87$, $R_{90} = 0.83$

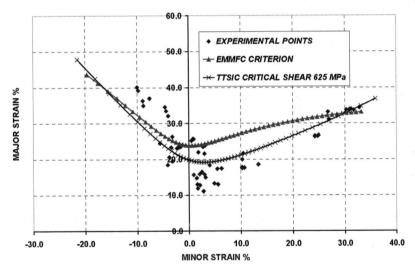

**FIGURE 6.** FLD's TRIP Steel, $t_0 = 1.3$ mm; $R_0 = 1.06$, $R_{45} = 0.87$, $R_{90} = 0.88$

The examples shown in Fig. (4) and (5) are for a Titanium alloy and a Stainless steel respectively for a thickness of 1.05 mm and 1.5 mm. The hardening curves are assumed to have the form $\bar{\sigma} = B(c + \bar{\varepsilon})^n$ with B=693.7 MPa, c=1.106e-02, n=0.168 for the Titanium alloy and B=1211 MPa, c=2.336e-02 and n=0.352 for the Stainless steel. The "TTSIC" are calibrated with $\beta$=1 on experimental points and it can be seen that without thickness effect the "EMMFC" criterion underestimates the limit strains. Taking account the radius of curvature R of the circle osculating the neck profile in the order of magnitude of the thickness, typically $0.5t_0 \leq R \leq t_0$, the "EMMFC" is in a better agreement with the experimental points.

It is not always the case as it can be seen in figure (6) for a TRIP steel with a hardening curve of the form $\overline{\sigma} = B(c+\overline{\varepsilon})^n \exp(-d\overline{\varepsilon})$ where B=1728.5 MPa, c=0.00963, n=0.2784 and d=0.2253. As it can be seen on the pictures of figure (7), the failure mode of this TRIP steel is clearly a shear fracture mode where the necking is not visible or it was not possible to identify any signs of a start of necking. In all experiments, the strain distribution measurement has been performed with our Digital Image Correlation (DIC) method, see Ref. [4]. The use of data from the fractured area gives a large variability in the experimental points and must be handled with caution. Despite of this, the "TTSIC" criterion calibrated on the biaxial points of the FLD shows a better trend than the "EMMFC" without effects, this result is consistent with the experimental observations of the shear fracture mode of the TRIP steel.

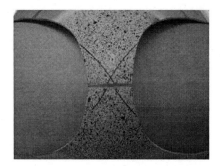 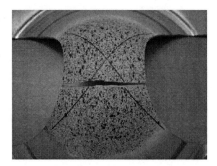

**FIGURE 7.** TRIP steel, examples of shear fracture planes on test samples

## CONCLUSION

Bearing in mind the scatter of the experimental data, the two criteria present a reasonable correlation with experimental points and have the advantage of simplicity in the mathematical calculation required. When necking occurs and is clearly identified, the "EMMFC" criterion should be used with the thickness effect and also with the strain rate effect not presented in this paper.

## REFERENCES

1. Z. Marciniak and K. Kuczynski. *Int. J. Mech. Sci.* **.9,** 609-620, (1967).
2. J. D. Bressan and J. A. Williams. *Int. J. Mech. Sci.* **25,** 155-168, (1983).
3. P. Hora, L. Tong and J. Reissner, "A prediction method for ductile sheet metal failure in FE-simulation" in *Proceedings NUMISHEET'96 Conference*, Dearborn, MI, USA, 1996, pp. 252-256.
4. M. Brunet and F. Morestin. *J. Mater. Process. Technol.* **112,** 214-226, (2001).

# Modelling of Damage Reduction and Microstructure Evolution by Annealing

H. Li[*], J. Lin[*], T. A. Dean[*], S. W. Wen[†], A. C. Bannister[†]

[*]Department of Mechanical and Manufacturing Engineering, School of Engineering, University of Birmingham, Edgbaston, Birmingham B15 2TT, UK

[†] Corus R, D & T, Swinden Technology Centre, Moorgate, Rotherham, South Yorkshire, S60 3AR, UK

**Abstract.** The main aim of the research described in this paper is to develop a set of mechanism-based unified viscoplastic constitutive equations to model the damage and dislocation development in cold deformation and their recovery by annealing. In addition, recrystallisation of the deformed material during the subsequent annealing process is modelled. The annealing temperature was 700°C. The effects of annealing time on damage and dislocation reduction during annealing are also investigated. Tensile tests were performed on a low carbon ferritic steel before and after annealing. The experimental results are used to characterise the unified constitutive equations using an Evolutionary Programming (EP)-based optimization method. Using these equations, the stress-strain relationships for interrupted constant strain rate tests, were predicted with good accuracy.

**Keywords:** Cold forming, damage reduction, dislocation recovery, constitutive modelling
**PACS:** 62.20.Fe, 61.72.Cc, 61.72.Hh, 61.72.Qq, 62.20.-x

## INTRODUCTION

Dislocation density and plasticity-induced damage are two competing mechanisms tending to strengthen and weaken material respectively, during cold forming. Cold forming processes normally cause a concentrated build-up of micro-damage within the working material in regions where hydrostatic stress is positive [1]. The accumulation of this damage can cause reduction in quality, reliability and longevity of the formed products and lead to material failure during forming and/or in service [2]. Steel contains inclusions, which are generally harder compared to the matrix material at room temperature [3]. This creates a possibility for microvoids to form around inclusions due to deformation and dislocation accumulation. Ageing and heat treatment, after forming, may reduce the level of defects and recover the mechanical properties. Recrystallisation and recovery take place under ageing and heat treatment. For cold working in industrial operations, annealing is sometimes used during or after deformation to reduce residual stress and work hardening.

Damage reduction is a relatively new area of study. Small amount of damage could be reduced by matrix evolution such as dislocation recovery, recrystallisation and grain size change during annealing. Much research has been carried out on static recrystallisation and the relationship between recovery, recrystallisation and grain size

CP907, 10th ESAFORM Conference on Material Forming, edited by E. Cueto and F. Chinesta
© 2007 American Institute of Physics 978-0-7354-0414-4/07/$23.00

change [4]. The work described in this paper was undertaken to produce a mathematical method which can be used to describe metallurgical change and residual damage in steel that has been cold formed and subsequently annealed.

## EXPERIMENTAL INVESTIGATIONS

### Material and Experimental Program

The raw material used in the experiments was a plate steel code STC-C used for large diameter pipes and supplied by Corus R, D & T Swinden Technology Centre. The plate steel was produced through casting and hot rolling. It is a low carbon ferritic pipe steel with very fine grain sizes at an average of 10.5 $\mu m$ after pre-heat treatment, hence the strength and ductility can be higher than normal low carbon steel.

Interrupted uniaxial tensile tests were carried out at room temperature (20°C) and the overall test conditions are shown schematically in Figure 1. Step 1: the testpieces were deformed at a strain rate of $0.008s^{-1}$ at room temperature. On reaching a pre-specified strain, the test was stopped. The testpieces were annealed for different periods between 8-60 minutes at 700°C and cooled in air. Step 2: once cooled to room temperature, the testpieces were deformed again to failure; C-gauge was used to measure the diameter changes at cross section area.

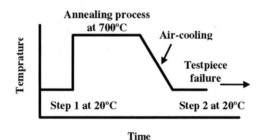

**FIGURE 1.** A diagram showing the experiment program.

### Experimental Results

The symbols in Figure 2 show the experimental stress-strain relationships for the interrupted tensile tests. Figure 2(a) shows the results for the simple one step loading case. Figure 2(b), (c) and (d) show the results with first step of which strain are 0.14, 0.11 and 0.12 respectively and annealing time are 8, 30 and 60 min respectively. The results shown in Figure 3, are total (sum of first and second step strain) and second step strain against annealing period. Failure strain is strain measured at failure point.

These experimental results have been used to determine material constants for the set of constitutive equations described below. Softening due to annealing is considerable in the flow curves, and the highest flow stress occurs at 8 minutes annealing time. The annealing process could reduce the rate of plasticity induced damage nucleation and growth due to recovery and recrystallisation by annealing, as seen in these flow stress test results. All testpieces subjected to annealing have higher ductility than the one-step test piece.

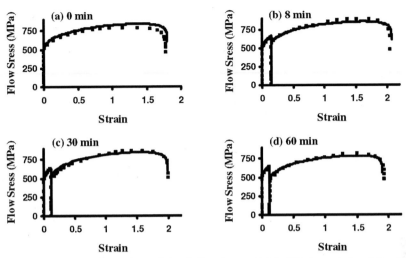

**FIGURE 2.** Comparison of experimental (symbol) and computed (solid) stress-strain relationships for the interrupted tests. The annealing periods are (a) 0, (b) 8, (c) 30 and (d) 60 min.

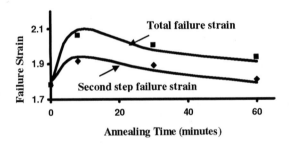

**FIGURE 3.** Comparison of experimental (symbols) and computed (solid) strain at failure against annealing time at 700°C.

# MODELLING OF DAMAGE REDUCTION IN ANNEALING

## Unified Constitutive Equations

During room temperature deformation stage, dislocation density and damage accumulated. As the deformed material is annealed at 700°C, dislocation recovery and damage reduction occur and also recrystallisation may take place, which cause grain size change. For the convenience of modelling these phenomena and the elastic-plastic behaviour of the steel deformed at room temperature, a unified approach [5, 6] is employed and the mechanical properties represented by internal variables, is modelled using rate equations, which are summarised below:

$$\dot{\varepsilon}_p = \left[\left(\frac{\sigma}{(1-D)} - R - k\right)\middle/ K\right]^n \cdot d^{\gamma} \tag{1}$$

$$\dot{S} = H \cdot [x \cdot \bar{\rho} - \bar{\rho}_c \cdot (1-S)] \cdot (1-S)^{\lambda_1} \tag{2}$$

$$\dot{x} = X_1 \cdot (1-x) \cdot \bar{\rho} \tag{3}$$

$$\dot{d} = (G_1/d)^{\psi_1} - G_2 \cdot \dot{S} \cdot (d/d_0)^{\psi_2} \tag{4}$$

$$\dot{\bar{\rho}} = k_1 \cdot (1-\bar{\rho}) \cdot |\dot{\varepsilon}_p|^{\delta_1} - C_r \cdot \bar{\rho}^{\delta_2} - \frac{C_S \cdot \bar{\rho}}{1-S} \dot{S} \tag{5}$$

$$\dot{R} = 0.5 \cdot B \cdot \bar{\rho}^{-1/2} \cdot \dot{\bar{\rho}} \tag{6}$$

$$\dot{\omega}_1 = a_1 \cdot (1-\omega_1) \cdot \dot{\bar{\rho}} \tag{7}$$

$$\dot{\omega}_2 = \left[a_2 \cdot \frac{D \cdot d^{n_3}}{(1-D)^{n_1}} \cdot |\dot{\varepsilon}_p|^{n_2}\right] \tag{8}$$

$$\dot{D} = \dot{\omega}_1 + \dot{\omega}_2 \tag{9}$$

$$\sigma = E \cdot (\varepsilon_T - \varepsilon_p) \tag{10}$$

To minimise the viscoplastic effect of the material deforming at 20°C, a high $n$ value is chosen ($n = 30$) in equation (1). $E$ ($E = 2.00 GPa$) is the Young's modulus in equation (10) and $k, K, n, \gamma, B, k_1, \delta_1$ are material constants related to viscoplastic deformation and $a_1, a_2, n_1, n_2, n_3$ are constants related to damage evolution during cold forming process. On the other hand, constants related to microstructure evolution during annealing process are $C_r, \delta_2, C_S, H, \bar{\rho}_c, \lambda_1, X_1, G_1, \psi_1, G_2, d_0$ and $\psi_2$.

Viscoplastic flow of the material, represented by strain rate $\dot{\varepsilon}_p$, is modelled using equation (1), which is a function of flow stress $\sigma$, plasticity induced damage $D$, isotropic hardening $R$, initial yield stress $k$. The effect of grain size $d$ on the material flow is characterised by constant $\gamma$. The hardening of the material due to plastic deformation is directly related to the dislocation density and its evolution is given by equation (6). The normalized dislocation density varies from 0 to 1, is defined by $\bar{\rho} = (\rho - \rho_i)/\rho$, where $\rho$ is the current dislocation density and $\rho_i$ is the dislocation density for the virgin material. The first term in equation (5) represents the evolution of dislocation density due to plastic deformation and dynamic recovery [7]. The second and the third terms in equation (5) express the static recovery of $\bar{\rho}$ and the effect of recrystallisation on the reduction of $\bar{\rho}$, respectively, during annealing process. Equations (2-4) describe the microstructure evolution due to annealing.

306

Recrystallisation is directly related to dislocation density. During annealing, when the dislocation density reaches a critical value $\bar{\rho}_c$, accumulated during the first step deformation, giving sufficient time, recrystallisation takes place. The evolution of recrystallisation is represented in equation (2) and equation (3) describes the onset of recrystallisation. Equation (4) models static grain growth and grain refinement due to recrystallisation. Further details related to microstructure evolution equations are described in [6]. Damage nucleation and growth rates are represented by equations (7) and (8), respectively. Plasticity induced damage is created by the accumulation of dislocations around hard inclusions. Damage nucleation rate thus is strongly related to the dislocation evolution rate, and the equation (7) also indicates that damage will be reduced due to reduction of dislocation density during annealing. Damage growth is associated with plastic deformation and grain size, along with the amount of damage that has been created. The total damage evolution is given by equation (9).

## Computational Procedures

To determine the material constants, the unified viscoplastic constitutive equations were numerically integrated using an implicit integration method. Thus, according to experimental procedures (Figure 1), the computational procedures include 3 key steps. During the first step deformation at 20°C, recrystallisation and grain size evolution would not take place. And also, static recovery of $\bar{\rho}$ would not occur either. To model these features, set $C_r = 0, \dot{S} = 0, \dot{x} = 0$ and $\dot{d} = 0$. The second step of the computational procedure was to model the phenomenon during annealing process. At this stage, due to the fact that there is no further deformation, $\dot{\varepsilon}_T = 0$ and $\sigma = 0$. The last step is related to the cold deformation to failure after annealing of the material. The computational setting was carried out under the same condition as described for the first step.

## Determination of Equations and Computed Results

Evolutionary programming (EP) optimisation techniques for determining the material constants arising in the constitutive equations are based on minimising the residuals between the experimental and computed data. The interrupted deformation experimental data (Figure 2) were used here to optimise the constitutive equations. The determined material constants are listed in Table 1 and the results are shown in Figure 2.

The solid curves in Figure 2, which are plotted using the constitutive equations with the optimised material constants listed in Table 1, are able to approximate the experimental data, symbols in Figure 2, for different annealing times 0, 8, 30 and 60 min, respectively. The general trend exhibited by the strain at failure during reloading is correctly predicted (Figure 3), with the predicted annealing time leading to maximum ductility falling only marginally short of the experimentally determined value. The complex relationship between the annealing period and the resulting reloading curve is interrupted within the model as the result of damage and dislocation recovery resulting from static recovery, recrystallisation and grain size change. This indicates that the constitutive equations are able to model the material deformation behaviour.

**TABLE1.** Material constants determined for the set of unified viscoplastic constitutive equations

| $k$ $(MPa)$ | $K$ $(MPa)$ | $\gamma$ $(-)$ | $n$ $(-)$ | $B$ $(MPa)$ | $k_1$ $(-)$ |
|---|---|---|---|---|---|
| 382 | 160 | 0.413 | 30 | 758 | 1.77 |
| $\delta_1$ $(-)$ | $a_1$ $(-)$ | $a_2$ $(\mu m^{-1})$ | $n_1$ $(-)$ | $n_2$ $(-)$ | $d_0$ $(\mu m)$ |
| 1.31 | 0.186 | 1.83 | 14.2 | 1.92 | 8.01 |
| $H$ $(s^{-1})$ | $\bar{\rho}_c$ $(-)$ | $\lambda_1$ $(-)$ | $X_1$ $(s^{-1})$ | $G_1$ $(\mu m)$ | $\psi_1$ $(-)$ |
| 3.34 | 0.015 | 10.8 | 96.7 | 1.58 | 4.08 |
| $G_2$ $(s^{-1})$ | $\psi_2$ $(-)$ | $C_r$ $(s^{-1})$ | $\delta_2$ $(-)$ | $C_S$ $(-)$ | $n_3$ $(-)$ |
| 9.91 | 3.50 | 1.61 | 2.92 | 1.29 | 0.80 |

# CONCLUSIONS

A set of constitutive equations has been formulated to predict the damage and dislocation density recovery features observed in interrupted experiments for a low carbon steel during cold deformation. In addition, the constitutive equations developed enable the modelling of the evolution of recrystallisation and grain size during annealing. The materials constants of the equations are determined from the experimental data using an EP based optimization method.

# ACKNOWLEDGMENTS

The support for this project from Corus R, D & T Swinden Technology Centre is gratefully acknowledged.

# REFERENCES

1. L. Gurson, Continuum theory of ductile rupture by void nucleation and growth: Part I - Yield criteria and flow rules for porous ductile media, *Journal of Engineering Materials and Technology*, 99 (1977) 2-15.
2. J. L. Chaboche, Continuum damage mechanics: present state and future trends, *Nuclear Engineering and Design*, 105 (1987) 19-33.
3. R. W. K. Honeycombe and H. K. D. H. Bhadeshia, Steels microstructure and properties, 2nd edition. *Edward Arnold publishing house*, London (1995).
4. F. J. Humphreys, A unified theory of recovery, recrystallisation and grain growth, based on the stability and growth of cellular microstructures II - The effect of second-phase particles, *Acta Materialia*, 45-12 (1997) 5031-5039.
5. A. D. Foster, J. lin, Y. Liu, D. Farrugia and T. A. Dean, Constitutive modelling of damage accumulation during the hot deformation of free-cutting steels, *Proceedings of the 8th Esaform conference*, (2005), I, 201-204.
6. J. Lin, Y. Liu, A set of unified constitutive equations for modelling microstructure evolution in hot deformation, *Journal of Materials Processing Technology*, 143-144 (2003) 281-285.
7. Y. Estrin, Dislocation theory based constitutive modelling: foundations and applications, *Journal of Materials Processing Technology*, 80-81 (1998) 33-39.

# Forming Limit Predictions for Single-Point Incremental Sheet Metal Forming

A. Van Bael[1,2], P. Eyckens[1], S.He[1,3], C. Bouffioux[4], C. Henrard[5], A.M. Habraken[5], J. Duflou[6] and P. Van Houtte[1]

[1] Dept. MTM, Katholieke Universiteit Leuven, Kasteelpark Arenberg 44, B-3001 Heverlee, Belgium
[2] Dept. IWT, Katholieke Hogeschool Limburg, Campus Diepenbeek, Agoralaan Gebouw B, bus 3, B-3590 Diepenbeek, Belgium
[3] now at Technology Center, Baoshan Iron and Steel Co., Ltd., Fujin 655, 201900 Shanghai, China
[4] Dept. MEMC, Vrije Universiteit Brussel, Pleinlaan 2, B-1050 Brussels, Belgium
[5] Dept. ArGEnCo, Université de Liège, Chemin des Chevreuils 1, B-4000 Liège, Belgium
[6] Dept. PMA, Katholieke Universiteit Leuven, Celestijnenlaan 300B, B-3001 Heverlee, Belgium

**Abstract.** A characteristic of incremental sheet metal forming is that much higher deformations can be achieved than conventional forming limits. In this paper it is investigated to which extent the highly non-monotonic strain paths during such a process may be responsible for this high formability. A Marciniak-Kuczynski (MK) model is used to predict the onset of necking of a sheet subjected to the strain paths obtained by finite-element simulations. The predicted forming limits are considerably higher than for monotonic loading, but still lower than the experimental ones. This discrepancy is attributed to the strain gradient over the sheet thickness, which is not taken into account in the currently used MK model.

**Keywords:** formability, Marciniak-Kuczynski model, finite-element simulations.
**PACS:** 81.05.Bx, 81.40.Ef, 81.40.Lm, 81.40.Np, 83.10.Ff, 83.10.Gr, 83.80.Ab.

## INTRODUCTION

Single-point incremental forming (SPIF) has emerged in the past few years as a potential alternative to conventional sheet metal deep-drawing to meet the increasing need for rapid prototyping and small batch productions at low cost (see, e.g., review paper [1]). A smooth ended tool on a CNC machine is used to create a local indentation in a clamped sheet, and by moving the point of contact around the sheet according to a programmed tool path, 3-dimensional shapes can be formed. The process is characterized by highly localized deformations. Also, the achievable deformations are much larger than conventional forming limits [2]. Previous studies have revealed that during incremental forming the sheet material is subjected to highly non-monotonic, serrated strain paths [3]. In the present paper, it is investigated to which extent these strain paths may contribute to the increased formability. Forming limits are predicted with a modified version of a model that has been used for forming limit curves of aluminium alloys [4]. Various descriptions of the yield surface and the hardening behaviour are used, including a texture-based yield locus and a microstructure-based work hardening/softening model.

CP907, *10th ESAFORM Conference on Material Forming*, edited by E. Cueto and F. Chinesta
© 2007 American Institute of Physics 978-0-7354-0414-4/07/$23.00

# EXPERIMENTAL DATA

A cold rolled and annealed aluminium AA3003-O sheet of 1.2mm thickness is considered in the present study. A three-axis CNC vertical milling machine is used as the platform for the SPIF process. A sheet of 225 x 225 x 1.2mm is clamped on a four-sided steel fixture using a backing plate with circular orifice, in which cones with different wall angles are formed. During the process, a steel tool with a diameter of 10.0mm follows a series of circular contours. After each contour, the tool moves 0.5mm deeper in a stepwise fashion. This process is repeated until a partial cone with a depth of 40mm is formed [5].

The limiting wall angle is found to be 72°. Cones up to 71° were made without failure. One out of two experiments with 72° cones resulted in a failed piece, as did the cone with a 73° wall angle.

In order to characterize the material behaviour, tensile tests have been performed in a standard tensile test machine, while the bi-axial machine developed at the University of Liège has been used for plane strain and simple shear tests [6]. The experimental procedures to obtain the material parameters for various hardening models are documented in [3]. Also, the crystallographic texture of the initial sheet has been measured using X-ray diffraction on a plane specimen at 50% depth of the sheet. It shows a strong cube component {001}<100>, which is typical for recrystallisation textures, and weaker other components: S {123}<412>, Goss {011}<100> and Brass {011}<211> . The homogenisation procedure described in [7] was used to derive the shape of the texture-based yield locus, starting from the crystallographic orientation distribution function and the Taylor theory while assuming {111}<110> slip systems.

# NUMERICAL MODELLING

## Finite-Element Simulations

The three-dimensional elastic-plastic finite-element model described in [8] is used to simulate the single-point incremental forming of truncated cones with wall angles of 50° and 73°. The simulations are performed with the commercial FEM package Abaqus/Standard. A 40° pie of the blank is considered since a full model is computationally too demanding. The sheet is modelled with three layers of brick elements. The elastic-plastic material behaviour is assumed to be isotropic, using a von Mises yield criterion with the Swift-type hardening law $\sigma=184(\varepsilon+0.00196)^{0.224}$ (MPa). The strain-path history throughout the forming process was extracted for an element in the outer layer of the cone wall.

## Forming Limit Predictions

A previously developed model for the prediction of forming limits [4] has been modified and extended in order to deal with the serrated strain paths that occur during incremental forming. It is based on the Marciniak-Kuczynski theory [9]. The method simulates the evolution of a pre-existing groove in a metal sheet which is being deformed according to a given strain path. For each imposed deformation increment, strain increments in the groove are calculated on the basis of geometrical compatibility and force equilibrium between the groove and the surrounding sheet material. In practice, this requires the solution of a set of two non-linear equilibrium equations. This is solved as a minimisation problem: the residual $R=f_1^2+f_2^2$ is minimised (with $f_1=0$ and $f_2=0$ the force equilibrium equations). In the previous software, the onset of necking was assumed to take place when the strain rate in the groove exceeds a critical limit (in particular when the thinning strain rate in the groove is more than 10 times that in the surrounding sheet). It was necessary to implement a second forming limit criterion to account for the possibility that the force equilibrium equations cannot be satisfied any more at the end of the (finite) increment. Further details will be presented at a forthcoming conference [10].

Three material models were considered for the forming limit calculations: isotropic plasticity with a von Mises yield locus and Swift-type hardening (i.e. the assumptions used for the finite-element simulations), an anisotropic texture-based yield locus in combination with isotropic hardening, and the texture-based yield locus combined with a microstructure-based anisotropic hardening/softening model. These models will be referred to as VON-ISO, TEX-ISO, and TEX-MIC, respectively, in this paper. Forming limits have been predicted for both serrated and monotonic strain paths.

## RESULTS AND DISCUSSION

The strain-path histories obtained with the finite-element simulations of the truncated cones with wall angles of 50° and 73° are given in Fig. 1 for an element at the outer side of the cone wall. These show the evolutions of the major and minor principal true plastic strains in the plane of the sheet. Highly non-monotonic, serrated strain-path changes occur. The direction of the major principal strain slightly oscillates around the radial direction along the cone wall. Correspondingly, the minor principal strain in the plane of the sheet is approximately aligned with the circumferential direction of the cone, i.e. the direction of the tool movement. During the process the sheet is severely stretched along the radial direction, whereas near plane strain conditions apply along the circumferential direction. At the same time, the sheet thickness is severely reduced. At the end of the process, the simulated principal true strains along radial and circumferential directions are 0.434 and 0.016, respectively, for the 50° cone, and 1.15 and 0.013, respectively, for the 73° cone. The indicated dots at the extreme points of the serrated paths are used as input for the Marciniak-Kuczynski code, which further assumes linear paths between these corner points.

Figure 2 shows predicted traditional forming limit curves for the three material models, i.e. assuming monotonic loading paths. A value of 0.998 has been used for the initial ratio between the thickness of the groove and the sheet, which in view of the hardening coefficient $n = 0.224$, produces realistic forming limits of about 22% in monotonic plane strain. The right-hand sides of the forming limit curves are strongly influenced by the adopted yield locus. Lower limit strains are obtained there with the texture-based yield surface, which results from the curvatures of the yield loci in the region between the points of equibiaxial strain and plane strain.

The predicted forming limits for the strains encountered in incremental forming are given in Fig. 3. The lowest forming limits are obtained when monotonic loading in the direction of the final total strains is assumed throughout the process (i.e. the linear strain paths indicated by the dashed lines).

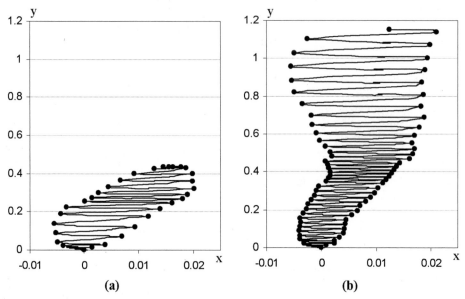

(a)                                                (b)

FIGURE 1. Simulated strain path for cones with wall angles of (a) 50° and (b) 73°, showing the major and minor true strains in the sheet plane along the y- and the x-axis, respectively, in an element of the outer layer across the sheet thickness.

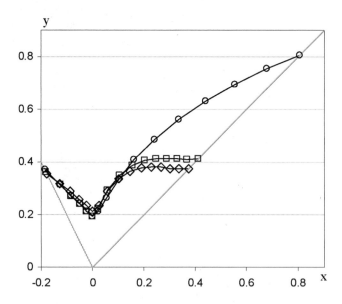

FIGURE 2. Predicted forming limit curves in case of monotonic loading paths for the three material models (circles for VON-ISO, squares for TEX-ISO, diamonds for TEX-MIC).

312

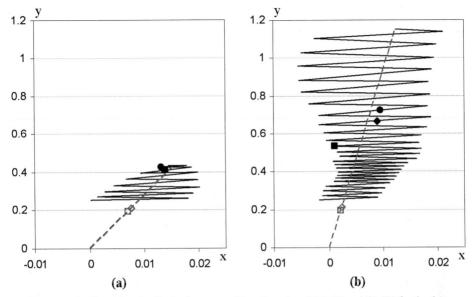

**FIGURE 3.** Predicted forming limits for cones with wall angles of (a) 50° and (b) 73° for the three material models (circles for VON-ISO, squares for TEX-ISO, diamonds for TEX-MIC). The closed symbols are obtained for the serrated strain paths derived from Fig. 1, the open symbols for the monotonic strain paths indicated by the dashed lines.

For the three material models, the predicted forming limits are considerably higher if the serrated strain paths derived from Fig. 1 are considered instead of monotonic loading. In case of the 50° cone, necking is predicted at 41% major true plastic strain for the TEX-ISO model and 43% for the VON-ISO model, whereas no necking is found in case of the TEX-MIC model. For the 73° cone, necking is predicted for all three material models, with forming limits of 53%, 67% and 72% in case of TEX-ISO, TEX-MIC and VON-ISO, respectively.

It is clear that a reliable model should predict the absence of necking for the cone with wall angle of 50°, since this is well below the experimentally obtained limiting wall angle of 72°. Such necking is only just avoided with the TEX-MIC model. For the 73° wall angle, which is only slightly higher than the limiting wall angle, all models predict necking long before the end of the process. This indicates that even taking strain path changes, the texture yield locus and anisotropic hardening into account is insufficient to predict the forming limits in incremental forming.

One of the assumptions in the current Marciniak-Kuczynski model is that the whole sheet is subjected to a uniform deformation. In particular, the strain path for an element in the outer of 3 layers across the sheet thickness is considered. This is the strain history at a depth of about 1/6 of the sheet thickness from the outer surface. In reality, the deformations are not uniform at all across the sheet thickness. For example, the simulations reveal that as the outer layer is elongated along the circumferential direction, the inner layer is compressed, and vice versa. Thereby, a sort of bending and reverse bending occurs twice each time the forming tool passes nearby. It is surmised

313

that the corresponding strain gradients across the sheet thickness play an important role for the occurrence of necking.

## CONCLUSIONS

Partial cones with wall angles of 50° and 73° have been incrementally formed using aluminium 3003-O of 1.2mm thickness, and these processes have been simulated with an implicit elastic-plastic finite-element program. The simulated strain paths at a depth of about 1/6 of the outer sheet surface have been used as input for a Marciniak-Kuczynski type prediction of forming limits, accounting also for the initial texture-based anisotropy and anisotropic hardening. The predicted forming limits are considerably higher than for monotonic loading, but nevertheless underestimate the experimentally observed formability during incremental forming.

It is expected that this discrepancy is caused by bending and reverse bending effects and corresponding strain gradients across the sheet thickness, which are not taken into account in the present model. It is planned to apply the current approach also for strain paths in the other layers across the sheet thickness, in order to study the necking tendencies at different depths. In reality, the different layers must interact with each other, and it is expected that more accurate necking predictions will require new models that take these interactions into account.

## ACKNOWLEDGMENTS

The authors gratefully acknowledge the financial support by the Federal Government of Belgium (Belgian Science Policy Contract P5/08) and by the Institute for the Promotion of Innovation by Science and Technology in Flanders (IWT). As Research Director of the Fund for Scientific Research (FNRS, Belgium), A.M. Habraken also thanks this research fund for its support.

## REFERENCES

1. J. Jeswiet, F. Micari, G. Hirt, A. Bramley, J. Duflou and J. Allwood, *CIRP Annals* **54/2**, 623-650 (2005).
2. M.S. Shim and J.J. Park, *Journal of Materials Processing Technology* **113**, 654-658 (2001).
3. P. Flores, L. Duchêne, C. Bouffioux, T. Lelotte, C. Henrard, N. Pernin, A. Van Bael, S. He, J. Duflou, A.M. Habraken, *International Journal of Plasticity* **23**, 420-449 (2007).
4. S. He, A. Van Bael and P. Van Houtte, Materials Science Forum **495-497**, 1573-1578 (2005).
5. J. Duflou, A. Szekeres and P. Vanherck, *Advanced Materials Research* **6-8**, 441-448 (2005).
6. P. Flores, P. Moureaux, A.M. Habraken, *Advances in Experimental Mechanics IV* **3-4**, 91-97 (2005).
7. P. Van Houtte and A. Van Bael, *Int. J. Plasticity* **20**, 1505-1524 (2004).
8. S. He, A. Van Bael, P. Van Houtte, A. Szekeres, J. Duflou, C. Henrard and A.M. Habraken, *Advanced Materials Research* **6-8**, 525-532 (2005).
9. Z. Marciniak and K. Kuczynski, *Int. J. Mech. Sci.* **9**, p.609 (1967).
10. P. Eyckens, S. He, A. Van Bael, P. Van Houtte and J. Duflou, "Forming limit predictions for the serrated strain paths in single point incremental forming", to be presented at NUMIFORM07.

# Non normal and non quadratic anisotropic plasticity coupled with ductile damage in sheet metal forming: Application to the hydro bulging test

Houssem Badreddine*, Khémaïs Saanouni*, Abdelwaheb Dogui**

*ICD/LASMIS, FRE : CNRS N° 2848 – Université de Technologie de Troyes
12, rue marie curie BP 2060, 10000 Troyes – France.
**LGM/GM-MA05–Ecole Nationale d'Ingénieurs de Monastir
Avenue Ibn El Jazzar, 5019 Monastir – Tunisie.

**Abstract.** In this work an improved material model is proposed that shows good agreement with experimental data for both hardening curves and plastic strain ratios in uniaxial and equibiaxial proportional loading paths for steel metal until the final fracture. This model is based on non associative and non normal flow rule using two different orthotropic equivalent stresses in both yield criterion and plastic potential functions. For the plastic potential the classical Hill 1948 quadratic equivalent stress is considered while for the yield criterion the Karafillis and Boyce 1993 non quadratic equivalent stress is used taking into account the non linear mixed (kinematic and isotropic) hardening. Applications are made to hydro bulging tests using both circular and elliptical dies. The results obtained with different particular cases of the model such as the normal quadratic and the non normal non quadratic cases are compared and discussed with respect to the experimental results.

**Keywords:** Finite deformations, elastoplasticity, anisotropy, non normality, non quadratic yield function, ductile damage, finite elements, numerical simulation, sheet metal forming, hydro bulging.
**PACS:** 83.60.–a

## INTRODUCTION

The main objective of this work is to propose "an advanced" anisotropic finite elastoplastic model fully coupled with an isotropic ductile damage for the numerical simulation of defects in various sheet forming processes. In order to improve the material anisotropy description, non associative and non normal formulation is used [1], [7], [10]. Consequently, two different equivalent stresses are used in the yield criterion and the plastic potential. For the yield function the quadratic Hill 1948 equivalent stress is chosen and for the plastic potential the Karafillis and Boyce non quadratic equivalent stress is taken in considering [5]. For both, the non linear isotropic and kinematic hardenings are taken into account. The full coupling with the ductile damage is made in the framework of continuum damage mechanic with effective state variables based on the total energy equivalence [8]. The objectivity requirement is ensured by using an appropriated rotating frame formulation [3], [9].

The model has been implemented into ABAQUS ® FE code for metal forming simulations discussed in [1], [6], [8].

Applications are made to uniaxial tension and deep drawing hydro bulging tests with spherical and elliptical punches. In order to show the ability of the non normal plasticity formulation in the description of the plastic anisotropy, both the normal and non normal versions of the proposed model are used and their respective results are compared.

## CONSTITUTIVE EQUATIONS

The fully coupled model is formulated assuming the large plastic deformation together with small elastic strain. This assumption leads to the additive decomposition of the total strain rate into elastic (Jaumann rate) and plastic parts. Dealing with finite plastic deformation, the objectivity requirement is ensured by using a rotating frame formulation [3], [9] in which each second order tensor $\underline{T}$ is rotated according to:

$$\overline{\underline{T}} = \underline{Q}^T \, \underline{T} \, \underline{Q} \tag{1}$$

where the orthogonal tensor $\underline{Q}$ is deduced from an appropriate constitutive equation [1].

The state relations or force-like variables derive from the state potential as:

$$\overline{\underline{\sigma}}(\overline{\underline{\varepsilon}}_e, d) = (1-d) \, \overline{\underline{\Lambda}} : \overline{\underline{\varepsilon}}_e \tag{2}$$

$$\overline{\underline{X}}(\overline{\underline{\alpha}}, d) = \frac{2}{3}(1-d) C \, \overline{\underline{\alpha}} \tag{3}$$

$$R(r, d) = (1-d) \, Q \, r \tag{4}$$

$$Y(\overline{\underline{\varepsilon}}_e, \overline{\underline{\alpha}}, r, d) = \frac{1}{2} \overline{\underline{\varepsilon}}_e : \overline{\underline{\Lambda}} : \overline{\underline{\varepsilon}}_e + \frac{1}{3} C \, \overline{\underline{\alpha}} : \overline{\underline{\alpha}}_\alpha + \frac{1}{2} Q \, r^2 \tag{5}$$

where $\overline{\underline{\Lambda}}$ is the fourth order elastic operator of the nondamaged Representative Volume Element (REV), which in the isotropic case is equal to $2\mu_e \, \underline{I} + \lambda_e \, \underline{1} \otimes \underline{1}$. $C$ is the kinematic hardening modulus and $Q$ is the linear isotropic hardening modulus.

The plastic yield function $\overline{f}$ and plastic potential function $\overline{F}$ are:

$$\overline{f}(\overline{\underline{\sigma}} - \overline{\underline{X}}, R ; d) = \frac{\overline{\sigma}_c(\overline{\underline{\sigma}} - \overline{\underline{X}})}{\sqrt{1-d}} - \frac{R}{\sqrt{1-d}} - \sigma_y \leq 0 \tag{6}$$

$$\overline{F}(\overline{\underline{\sigma}}, \overline{\underline{X}}, R ; d) = \frac{\overline{\sigma}_p(\overline{\underline{\sigma}} - \overline{\underline{X}})}{\sqrt{1-d}} - \frac{R}{\sqrt{1-d}} + \frac{3a}{4(1-d)C} \overline{\underline{X}} : \overline{\underline{X}} + \frac{b}{2(1-d)Q} R^2$$
$$+ \frac{S}{s+1} \left\langle \frac{Y - Y_0}{S} \right\rangle^{s+1} \frac{1}{(1-d)^\beta} \tag{7}$$

where $\sigma_y$ is the initial yield stress in simple tension; $a$ and $b$ characterize the kinematic and isotropic hardening non linearity respectively; $\beta$, $S$, $s$ and $Y_0$ characterize the ductile damage evolution. The stress norms $\overline{\sigma}_p$ and $\overline{\sigma}_C$ are the equivalent stresses entering the yield and the plastic potential functions. The first one $\overline{\sigma}_p$ is taken as quadratic as proposed by Hill 1948:

$$\bar{\sigma}_p{}^2 = (\bar{\sigma} - \underline{X}) : \bar{H}' : (\bar{\sigma} - \underline{X}) \tag{8}$$

where $\bar{H}'$ is an anisotropic fourth order operator characterized by the material constants $F'$, $G'$, $H'$, $L'$, $M'$, $N'$. The second one, $\bar{\sigma}_C$ is taken as non quadratic as proposed by Karafillis and Boyce 1993:

$$2\left[\bar{\sigma}_C(\bar{\sigma} - \underline{X})\right]^m = (1-p)\sum_{i \neq j}^{3} |\bar{q}_i - \bar{q}_j|^m + p\frac{3^m}{2^{m-1}+1}\sum_{i}^{3} |\bar{q}_i|^m \tag{9}$$

$$\bar{q} = \bar{H} : (\bar{\sigma} - \bar{X}) \tag{10}$$

where $\bar{H}$ is the Karafillis and Boyce fourth order anisotropic operator of the plastic yield function, it's characterized by material constants $F, G, H, L, M$ and $N$. The material constant $m$ is the yield function shape parameter ($m \geq 1$ to ensure the convexity of the yield function) and $p$ ($0 \leq p \leq 1$) represents a balance parameter who allows to mix the Barlat's equivalent stress [4] and the lower bound. We find the case of the normal plasticity by taking $\bar{\sigma}_C = \bar{\sigma}_p$ then $\bar{H} = \bar{H}'$ for Hill equivalent stress and $\frac{3}{2}\bar{H}^2 = \bar{H}'$ and $m = 2$ or $4$ for the Karafillis and Boyce equivalent stress.

The complementary or evolution relations derive from the plastic potential thanks to the normality rule with respect to the plastic potential, as following:

$$\bar{D}_p = \dot{\lambda}\frac{\partial \bar{F}}{\partial \bar{\sigma}} = \frac{\dot{\lambda}}{\sqrt{1-d}}\bar{n}_p \tag{11}$$

$$\dot{\bar{\alpha}} = -\dot{\lambda}\frac{\partial \bar{F}}{\partial \underline{X}} = \dot{\lambda}\left[\frac{1}{\sqrt{1-d}}\bar{n}_p - a\bar{\alpha}\right] = -\dot{\lambda}\bar{l}_p \tag{12}$$

$$\dot{r} = -\dot{\lambda}\frac{\partial \bar{F}}{\partial R} = \dot{\lambda}\left[\frac{1}{\sqrt{1-d}} - br\right] = -\dot{\lambda}\bar{l}_p \tag{13}$$

$$\dot{d} = \dot{\lambda}\frac{\partial \bar{F}}{\partial Y} = \dot{\lambda}\frac{1}{(1-d)^\beta}\left\langle\frac{Y-Y_0}{S}\right\rangle^s = \dot{\lambda}\,\bar{y} \tag{14}$$

where $\dot{\lambda}$ is the plastic multiplier which is determined by the consistency condition applied to the yield function:

$$\dot{\lambda} = \begin{cases} 0 & \bar{f} < 0 \\ \dfrac{1}{\bar{H}_\lambda}\left\langle\bar{a}_\lambda : \bar{D}\right\rangle & \bar{f} = 0 \end{cases} \tag{15}$$

$$\bar{a}_\lambda = \sqrt{1-d}\,\bar{A} : \bar{n}_c \tag{16}$$

$$\bar{H}_\lambda = \bar{n}_c : \left(\bar{A} + \frac{2}{3}C\bar{I}\right) : \bar{n}_p - \frac{a}{\sqrt{1-d}}\bar{n}_c : \bar{X} + Q - \frac{bR}{\sqrt{1-d}} + \frac{\sigma_y}{2}\frac{1}{(1-d)^{\beta+1}}\left\langle\frac{Y-Y_0}{S}\right\rangle^s \tag{17}$$

Note that the deviator second order tensors $\bar{n}_p$ and $\bar{n}_c$ entering the relations (11), (12) and (17) are the normal tensors to the plastic potential and to the yield criterion surfaces respectively.

For the sake of shortness, the associated numerical aspects can be found in [1]. It is worth noting that the model is implemented in the finite element code

ABAQUS/Standard via the UMAT routine and ABAQUS/Explicit via the VUMAT routine.

## APPLICATION TO THE HYDRO – BULGING TESTS

The hydro bulging tests is able to characterize the metal sheet formability in expansion with out important influence of the friction contact. During this test, a circular blank with radius 66.5mm and 1mm thickness is clamped at its external boundary between a die and blank holder. A linearly increasing (between 0 Bar and 400 Bar) hydraulic pressure is applied in the bottom surface of the blank. Circular and elliptic dies are used both with 164 mm outer diameter, 6 mm corner radius and a 25mm height; while 91 mm inner diameter for the circular die; and 110 mm major axis and 74 mm minor axis for the elliptic die. The experimental tests are performed at the LGM-ENIM using the X6CrNi18-09 steel. The results are expressed in term of the applied pressure versus the pole displacement and the sheet profile (along meridian for circular die and minor and major axis for the elliptic die). In the simulations, different versions of our model are considered namely: normal quadratic plasticity (model1), quadratic non normal plasticity (model2) and non quadratic non normal plasticity (model3). These models are identified using uniaxial tension tests conducted up to the final fracture of the X6CrNi18-09 steel sheet specimens. These specimens are cut with an angle $\psi_0$ according to the rolling direction (chosen as direction 1 in the following). The angle $\psi_0$ varies from 0° to 90° with a constant step of 15°. In the Table 1 the identified material parameters are summarized.

In the Figures 1.a, 1.b, 1.c and 1.d the results obtained with the various versions of our model for hydro bulging test with the circular die are given. In the Figure 1.a we compare the experimental curve, in term of the applied pressure versus the pole displacement, to the numerical results. This figure shows that the results obtained with the models 2 and 3 (with non normal plasticity) are the same and more close to the experimental curve compared to the model 1 which uses the normal formulation. The rupture profiles predicted by the 3 models are shown in the Figures 1.c and 1.d. For the model 1 (with classical normal plasticity) the propagation of the crack is done with 4 crossed branches intersecting at the dome pole. For the models 2 and 3 (with non normal plasticity) this is done with only 3 crossed branches intersecting at the dome pole. Results of the model 2 and 3 are similar indicating that the yield criterion form (quadratic or not) don't have any significant influence for this loading path. However, value of the pressure at fracture is better predicted by the non normal models (model 2 and 3) while the model 1 predicts $P=324$ Bar with 33.8% deviation from the experimental value (see Figure 1.d).

In the Figures 2.a, 2.b, 2.c and 2.d the results obtained with the various models for hydro bulging test with the elliptical die are presented. The Figure 2.a shows that the model 1 induces the greatest pressure at fracture about 345 Bar with 40% deviation from the experimental value. The models 2 and 3 predict about 254 Bar for the fracture pressure with an error of 10 %. The Figure 2.a shows that the non normal models give approximately the same pressure- displacement curves close to the experimental one.

**TABLE 1.** Material parameters identified for the three models.

| | $E$ | $\nu$ | $\sigma_j$ | $F$ | $G$ | $H$ | $L$ | $M$ | $N$ |
|---|---|---|---|---|---|---|---|---|---|
| unite | MPa | - | MPa | - | - | - | - | - | - |
| Model1 | 195000 | 0.3 | 290 | 0.496 | 0.405 | 0.595 | 1.500 | 1.500 | 1.885 |
| Model2 | 195000 | 0.3 | 290 | 0.861 | 0.800 | 0.200 | 1.500 | 1.500 | 1.502 |
| Model3 | 195000 | 0.3 | 290 | 0.439 | 0.421 | 0.239 | 1.000 | 1.000 | 1.001 |

| | $m$ | $p$ | $F'$ | $G'$ | $H'$ | $L'$ | $M'$ | $N'$ | $Q$ |
|---|---|---|---|---|---|---|---|---|---|
| unite | - | - | - | - | - | - | - | - | MPa |
| Model1 | - | - | 0.496 | 0.405 | 0.595 | 1.500 | 1.500 | 1.885 | 2682 |
| Model2 | - | - | 0.417 | 0.403 | 0.500 | 1.500 | 1.500 | 1.222 | 2500 |
| Model3 | $\forall$ | $\forall$ | 0.417 | 0.403 | 0.500 | 1.500 | 1.500 | 1.222 | 2500 |

| | $b$ | $C$ | $a$ | $s$ | $S$ | $\beta$ | $Y_0$ |
|---|---|---|---|---|---|---|---|
| unite | - | MPa | - | - | MPa | - | MPa |
| Model1 | 1.65 | 3650 | 75 | 1 | 600 | 4 | 0 |
| Model2 | 1.65 | 3650 | 75 | 1 | 1000 | 8 | 0 |
| Model3 | 1.65 | 3650 | 75 | 1 | 1000 | 8 | 0 |

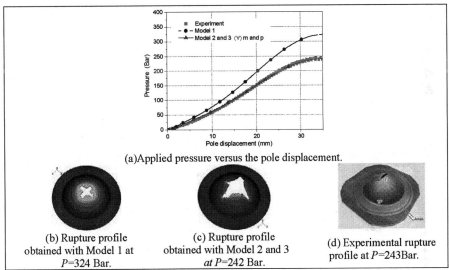

(a)Applied pressure versus the pole displacement.

| (b) Rupture profile obtained with Model 1 at $P$=324 Bar. | (c) Rupture profile obtained with Model 2 and 3 at $P$=242 Bar. | (d) Experimental rupture profile at $P$=243Bar. |
|---|---|---|

**FIGURE 1.** Results of the hydro bulging test with circular die.

# CONCLUSION

In this paper an anisotropic elastoplastic model strongly coupled with isotropic ductile damage is formulated in finite plastic deformations. A non associative and non normal plasticity theory is used with quadratic and non quadratic equivalent stresses. After the implementation in the ABAQUS/Explicit F-E code, the results obtained for uniaxial tension as well as for biaxial extension (hydro bulging) with X6CrNi18-09 steel sheet show a good agreement with experience for the non normal plasticity models compared to the normal formulation. These results have also demonstrated that for this sheet material the non quadraticity of the yield surface doesn't has an important influence for the studied loading paths. We conclude that the non normality is the major factor to obtain a realistic description of the steel sheet anisotropy.

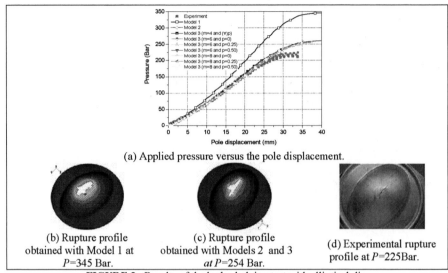

(a) Applied pressure versus the pole displacement.

| (b) Rupture profile obtained with Model 1 at $P$=345 Bar. | (c) Rupture profile obtained with Models 2 and 3 at $P$=254 Bar. | (d) Experimental rupture profile at $P$=225Bar. |

**FIGURE 2.** Results of the hydro bulging test with elliptical die.

## REFERENCES

1. H. Badreddine,"formulation générale selon une approche macroscopique d'un méta modèle élastoplastique couplé à l'endommagement ductile en grandes déformations: intégration dans un code de calculs par élément finis et application à la mise en forme", Ph.D. Thesis, Université de Technologie de Troyes (France) – Ecole Nationale d'Ingénieurs de Monastir (Tunisie), 2006.
2. J.L. Chaboche, "Continuum damage mechanics", Parts I and II, J. of Ap. Mechanics **55**, 59-72 (1988).
3. A. Dogui, " Plasticité anisotrope en grandes déformations", Ph.D. Thesis, Université Claude Bernard, Lyon (France), 1989.
4. F. Barlat and J. Lian, "A yield function for orthotropic sheet under plane stress conditions ", Int. J. Plasticity **5**, 51-56 (1989).
5. A. P. Karafillis and M. C. Boyce, "A general anisotropic yield criterion using bounds and a transformation weighting tensor ", J. Mech. Phys. Solids **41**, 1859-1886 (1993).
6. M. Khelifa, H. Badreddine, M.-A. Gahbiche, K. Saanouni, A. Cherouat, and A. Dogui, "Effect of anisotropic plastic flow on the ductile damage evolution in sheet metal forming. Application to the circular bulging test", International Journal of Forming Processes **8**, 271-289 (2005).
7. K. Runesson and Z. Mroz, "A note on nonassocieted plastic flow rules ", Int. J. Plasticity **5**, 639-658 (1989).
8. K. Saanouni and J.L. Chaboche, "Computational damage mechanics. Application to metal forming ", Chapter 7 of Vol. 3 in Numerical and Computational methods, Elsevier Oxford, I. Miline, R.O. Ritchie and B. Karihaloo, ISBN 0-08-043749-4, 2003, pp. 321-376.
9. F. Sidoroff and A. Dogui, "Some issues about anisotropic elastic-plastic models at finite strain ", Int. J. Sol. Str. **38**, 9569-9578 (2001).
10. T. B. Stoughton, " A non associated flow rule for sheet metal forming ", Int. J. of plasticity **18**, 687-714 (2001).

# Formulation of anisotropic Hill criteria for the description of an aluminium alloy behaviour during the channel die compression test

## A. GAVRUS[1], H. FRANCILLETTE[2]

[1]*LGCGM (EA 3913), INSA de RENNES, 20, Av. des Buttes de Coësmes, 35043, Rennes, France*
[2]*SCR/CM-INSA (UMR CNRS 6226), INSA DE RENNES*
*tel: (00)(33)(0)223238666, fax: (00)(33)(0)223238726*
e-mail: adinel.gavrus@insa-rennes.fr

**Abstract.** During the last years the study of the plastic deformation modes and the anisotropic mechanical behaviour of aluminium alloys have been the subject of many investigations. This paper deals with a phenomenological identification of an anisotropic Hill constitutive equation of aluminium AU4G samples using a channel die compression device at room temperature. By considering the different possible orientations of the samples in the channel die device, three initial textures, named ND (normal direction Z), LD (longitudinal direction X) and TD (transverse direction Y), were defined with the corresponding stresses $\sigma_{ND}$, $\sigma_{LD}$ and $\sigma_{TD}$. To describe the anisotropy of the material, a quadratic Hill criteria is used. An Avrami type equation based on the mixture of the hardening and softening phenomena is used to describe variation of each stress component with the equivalent plastic strain. The identification of the parameters of the law is made using an identification software (OPTPAR) and a good correlation between the experimental stresses and computed ones is obtained. The variation of the Hill parameters with a proposed equivalent strain, describing the deformation history of the material, is analysed. Finally, using the expressions of $F$, $G$, $H$ and $N$, the constitutive equation of the normal anisotropy in the plane XY is obtained.

**Keywords:** Anisotropic Hill criteria, Channel Die compression test, Aluminum alloy.

## INTRODUCTION

The analysis of the anisotropic behavior of materials has been the subject of various studies, generally in the field of aluminum alloys forming processes [1-2]. The majority of works search to define a mathematical description of the anisotropy starting from a modified Hill criteria or Banabic [3], Cazacu-Barlat [4] models and using classical uniaxial tensile tests. More recently, researches were interested on the description of the zirconium behavior [5-8]. The main particularity of this material is the easiness of prismatic glide in polycrystalline samples linked to the hexagonal structure. To describe more accurately the stress-strain curves for the three principal directions, named LD (longitudinal direction X), TD (transverse direction Y) and ND (normal direction Z), channel die compression tests were used in [9]. The main

CP907, *10th ESAFORM Conference on Material Forming*, edited by E. Cueto and F. Chinesta
© 2007 American Institute of Physics 978-0-7354-0414-4/07/$23.00

advantage of this experimental test is the possibility to define a rigorous model able to describe the plastic deformation of the material specimen. The aim of the present study is to analyze the macroscopic behavior of an anisotropic rolling plate using the channel die compression test at room temperature. After the description of the experimental device and the general analytical methodology to identify the Hill coefficients, an application of the formulation on an aluminum alloy rolled plate is presented for a normal anisotropy.

## EXPERIMENTAL DEVICE

The experimental test is defined by a compression along a Z0 direction with an imposed velocity V, a free flow along a X0 direction and a blocked displacement along a Y0 direction (Figure 1).

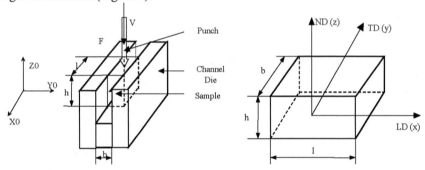

**Figure 1.** Experimental channel die compression test and experimental sample.

By considering a parallelepiped sample (lxbxh=11x8x10 mm) cut out from an initial rolled plate  (Fig. 1), the anisotropy is defined by three initial textures named LD (longitudinal direction), TD (transversal direction) and ND (normal direction).

## NORMAL ANISOTROPY OF AN ALUMINUM PLATE

According to a 3D anisotropy, the quadratic Hill criterion is written as a function of the stress tensor $\sigma$ in the form:

$$f(\sigma) = F\left(\sigma_{yy} - \sigma_{zz}\right)^2 + G\left(\sigma_{zz} - \sigma_{xx}\right)^2 + H\left(\sigma_{xx} - \sigma_{yy}\right)^2 + 2L\sigma_{yz}^2 + 2M\sigma_{xz}^2 + 2N\sigma_{xy}^2 = 1$$

where F, G, H, L, M and N represent the Hill parameters which must be determined. Starting from the theory developed in [10] we obtain:

$$\sigma_{LD\|Z0}^2\left(\frac{FG + FH + GH}{F + H}\right) = \sigma_{ND\|Z0}^2\left(\frac{FG + FH + GH}{G + H}\right) = \sigma_{TD\|Z0}^2\left(\frac{FG + FH + GH}{F + G}\right) = 1$$

where: $F = \dfrac{A}{AB + AC + BC}, G = \dfrac{B}{AB + AC + BC}, H = \dfrac{C}{AB + AC + BC}$

$$A = \frac{1}{2}\left[\sigma_{LD\|Z0}^2 + \sigma_{TD\|Z0}^2 - \sigma_{ND\|Z0}^2\right], B = \frac{1}{2}\left[\sigma_{ND\|Z0}^2 + \sigma_{TD\|Z0}^2 - \sigma_{LD\|Z0}^2\right] C = \frac{1}{2}\left[\sigma_{LD\|Z0}^2 + \sigma_{ND\|Z0}^2 - \sigma_{TD\|Z0}^2\right]$$

In reference [10] $\sigma_{LD\|Z0}$, $\sigma_{ND\|Z0}$, $\sigma_{TD\|Z0}$ correspond to compression along the initial LD, ND and TD directions of the sample respectively. The stresses in the Z0 direction can be computed directly from the measured forces in this direction. The main advantage of the channel die compression test here, in comparison with the tensile one, is the fact that we eliminate the striction phenomenon linked to a localized plastic flow, very difficult to be analyzed by an analytical model. Furthermore the friction is eliminated using a Teflon plaque between the specimen and the tool. Then for a normal anisotropy in the plane of a rolled plate we have F=G i.e. $\sigma_{LD\|Z0} \approx \sigma_{ND\|Z0}$ and N = F+2H. Using the relationships of F, G and H in function of A, B and C we obtain:

$$F = G = \frac{2}{4\sigma_{ND\|Z0}^2 - \sigma_{TD\|Z0}^2}, H = \frac{2\left[2\sigma_{ND\|Z0}^2 - \sigma_{TD\|Z0}^2\right]}{\sigma_{TD\|Z0}^2\left[4\sigma_{ND\|Z0}^2 - \sigma_{TD\|Z0}^2\right]}$$

The behavior equation of the normal anisotropy in the xy plane is obtained in the finally form:

$$\sigma_{xx}^2 + \sigma_{yy}^2 - \frac{2r}{r+1}\sigma_{xx}\sigma_{yy} + \frac{2(2r+1)}{r+1}\sigma_{xy}^2 = \sigma_0^2 = \frac{2r+1}{2(r+1)}\sigma_{TD\|Z0}^2 \quad with \quad r = 2\frac{\sigma_{ND\|Z0}^2}{\sigma_{TD\|Z0}^2} - 1$$

## EXPERIMENTAL RESULTS

The previous experimental tests were performed for an aluminum alloy polycrystalline plate (AU4G) up to approximately 25%-30% plastic strain. Starting from the variation of the loads and using the analytical description presented in the second part, the corresponding macroscopic stress-strain curves are pictured in Fig. 2.

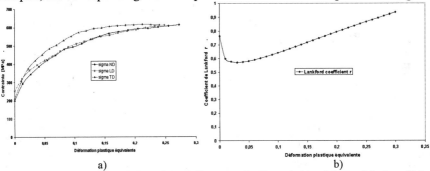

a)                                    b)

**Figure 2.** a) Experimental variation of Von-Mises stress b) The variation of the Lankford coefficient with the equivalent plastic strain.

These macroscopic curves show important stress level differences due to the anisotropy of the specimens induced by the three different initial textures (named LD, ND and TD). In order to describe the stress-strain variation, a non-linear regression model was used from the following phenomenological constitutive equation [11]:

$$\sigma = \sigma_{00} + \sigma_{pl}\left[1 - \exp(-n\bar{\varepsilon})\right]^{n_2} W + \sigma_{sat}\left[1 - W\right] \quad with \quad W = \exp\left(-r\bar{\varepsilon}^s\right)$$

323

The regression is based on the minimization of an error function defined in a least square sense (OPTPAR package). This model take into account the hardening and the dynamic recovery phenomena (described by the $\sigma_{pl}$ and $n$ parameters) and the fraction of the material $W$ which accommodates the plastic deformation. The comparison between the experimental and computed stresses show a very good agreement for the three different stress-strain curves (the computed curves are practically superposed with the experimental ones). Results of the identified coefficients are presented in Table 1.

**Table 1.** Computed parameters of the phenomenological law for the LD, ND and TD textures.

| | LD | ND | TD |
|---|---|---|---|
| $\sigma_{00}$ | 254. | 200. | 213. |
| $\sigma_{pl}$ | 714.260 | 538.519 | 459.817 |
| n | 1.239 | 3.607 | 7.563 |
| $n_a$ | 0.5 | 0.5 | 0.5 |
| W | 1. | 1. | 1. |
| Error ($E_r$) | 1.60% | 2 % | 2.60% |

According to the stress-strain variation described the variation of the Lankford coefficient with the proposed equivalent plastic strain is pictured in Figure 2b.

## NUMERICAL VALIDATION

In order to validate the proposed model a simulation of the tensile test is presented. The Voce law of the LD texture is introduced via the ABAQUS tensile model (Fig 3).

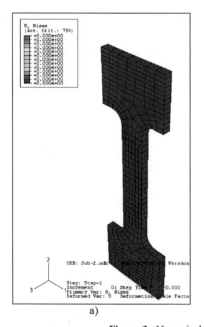 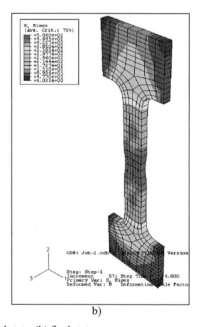

a)  b)

**Figure 3** : Numerical results: (a) initial state, (b) final state.

A comparison between the forces-displacement curves is presented in Fig 4.

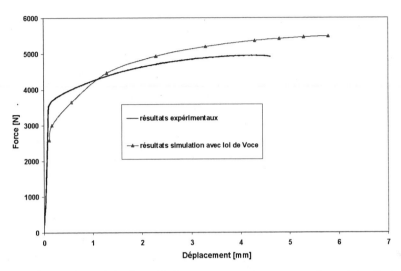

**Figure 4**: Comparison of the force-displacement curves between experimental tensile test and numerical simulation using the Voce law identified from a channel die compression test.

A small difference is observed due to different microstructure phenomena linked to the tensile test, because the dislocation glide is different in tensile test compared to the dislocation glide in channel die compression test.

## CONCLUSIONS

Starting from the channel die compression test it is then possible to describe the 3D anisotropy of materials. A new criteria based on the Hill formulation is introduced via an evolution of the Hill coefficients as a function of the proposed equivalent plastic strain. This approach leads to a more realistic description of the anisotropic behavior. The proposed constitutive equation for analyzing the stress-strain variation can correctly describe various types of stress-strain curves shape, taking into account hardening, dynamic recovery and softening phenomena.

## REFERENCES

1. Malo K.A., Hopperstad O.S., Lademo O.G., *J. of Mat. Processing Technology,* **80-81,** 538-544, (1998).
2. Lademo O.-G., Hopperstad O.S., Langseth M., *International Journal of Plasticity,* **15,** 191-208, (1999).
3. Banabic D. et al., "Description of anisotropic behaviour of AA3103-0 aliminium alloy using two recent yeld criteria", *6th EUROMECH-MECAMAT,* Liège, Belgium, 9-12 September 2002, edited by S. Cescotto, C. Teodosiu, A.-M. Habraken, R. Billardon and I. Doghri, pp. 297-304.
4. Cazacu O., Barlat F., *Mathematics and Mechanics of Solids,* **6** 613-630 (2001).
5. Christodoulou N., Turner P.A., Ho E.T.C., Chow C.K., Resta Levi M., *Metallurgical and Materials Transactions* A **31,** 409-420, (2000).

6. Francillette H., Bacroix B., Lebensohn R.A., Béchade J.L., J. *Phys. IV* **11** 83-90 (2001).
7. Fundenberger J.J., Philippe M.J., Wagner F., Esling C., "Microstructures et aptitude au formage de différents alliages de zirconium", *Journées d'étude: le zirconium, propriétés-mictrostructure*, Saclay, France, 25-26 avril 1995, edited by G. Cailletaud and P. Lemoine, pp. 87-96.
8. Allais L., Vaubert V., Tournié I., "Anisotropie de comportement d'une tôle de zirconium α 702", *Journées d'étude: le zirconium, propriétés-mictrostructure*, Saclay, France, 25-26 avril 1995, edited by G. Cailletaud and P. Lemoine, pp. 257-266.
9. Francillette H., Bacroix B., Gaspérini M., Béchade J.L., *Materials Science and Engineering* **234-236** 974-977 (1997).
10. Gavrus A., Francillette H., " Identification of anisotropic hill criteria from the channel die compression test. Application to a normal anisotropy of zirconium 702α", *Journal de Physique IV*, **105**, 11-18, (2003).
11. Francillette H, Gavrus A., Lebensohn R. A. – "A constitutive law for the mechanical behaviour of zr 702α", *Journal of Materials Processing Technology*, **142/1**, 43-51, (2003).

# Impact of the Parameter Identification of Plastic Potentials on the Finite Element Simulation of Sheet Metal Forming

M. Rabahallah[1,2], S. Bouvier[1], T. Balan[2], B. Bacroix[1], C. Teodosiu[1]

*1 LPMTM-CNRS, UPR 9001, University Paris13, 99 Av. J-B. Clément, 93430 Villetaneuse, France*
*2 LPMM, UMR 7554, ENSAM Metz, 4 rue A. Fresnel, 57078 Metz Cedex 3, France*

**Abstract.** In this work, an implicit, backward Euler time integration scheme is developed for an anisotropic, elastic-plastic model based on strain-rate potentials. The constitutive algorithm includes a sub-stepping procedure to deal with the strong nonlinearity of the plastic potentials when applied to FCC materials. The algorithm is implemented in the static implicit version of the Abaqus finite element code. Several recent plastic potentials have been implemented in this framework. The most accurate potentials require the identification of about twenty material parameters. Both mechanical tests and micromechanical simulations have been used for their identification, for a number of BCC and FCC materials. The impact of the identification procedure on the prediction of ears in cup drawing is investigated.

**Keywords:** Strain rate potentials, Plastic anisotropy, sheet metal forming simulation, parameter identification.
**PACS:** 62.20.Fe, 83.50.-v, 47.11.Fg, 02.60.Cb, 81.20.Hy

## INTRODUCTION

The finite element simulation of sheet metal forming requires an accurate description of the plastic anisotropy. The quadratic yield surface of Hill (1948), commonly available in commercial finite element codes, does not always provide a sufficient accuracy to describe the initial anisotropy of sheet metals (especially aluminum alloys). Consequently, more flexible mathematical functions have been proposed since then to describe plastic yield anisotropy. The main drawbacks of such functions are a) the increased number of material parameters to be identified and b) the numerical implementation in general purpose finite element codes. The benefit from their use is a better description of the plastic anisotropy and, implicitly, a better prediction of plastic flow during the forming process simulation.

In this work, plastic anisotropy is described using strain rate potentials. This description has some advantages in terms of parameter identification using micromechanical models based on texture measurements. On the other hand, it requires a specific stress update algorithm, due to the lack of an explicit yield function to be used as loading/unloading condition. Several such potentials are considered and their parameters identified using micromechanical simulations. Alternatively, when mechanical tests are used for the parameter identification, different sets of material

CP907, *10th ESAFORM Conference on Material Forming*, edited by E. Cueto and F. Chinesta
© 2007 American Institute of Physics 978-0-7354-0414-4/07/$23.00

parameters are obtained. In order to explore the effect of these parameters on the numerical simulations, the different plastic potentials have been implemented in a finite element code. Then, they are applied to the numerical simulation of cylindrical cup drawing.

## STRAIN-RATE POTENTIALS AND THEIR PARAMETER IDENTIFICATION

As it has been shown *e.g.* by Hill [1], for many models of material behavior (including plasticity) two convex, dual potentials exist from which the stress tensor can be derived as a function of the strain-rate tensor and vice-versa. In plasticity, the most classical formulation is the one that uses the yield criterion:

$$\phi(\boldsymbol{\sigma}) = \tau, \quad \mathbf{D}^P = \dot{\lambda} \frac{\partial \phi}{\partial \boldsymbol{\sigma}}, \tag{1}$$

where $\boldsymbol{\sigma}$ is the stress tensor and $\tau$ is a positive scalar with the dimension of stress. The plastic strain-rate tensor $\mathbf{D}^P$ is defined by the associated flow rule, in terms of the gradient of the potential $\phi$ and the plastic multiplier $\dot{\lambda}$.

The dual potential $\psi$ of this yield criterion is then simply written as

$$\psi(\mathbf{D}^P) = \dot{\lambda}, \quad \boldsymbol{\sigma}' = \tau \frac{\partial \psi}{\partial \mathbf{D}^P}, \tag{2}$$

where $\boldsymbol{\sigma}'$ is the deviatoric part of $\boldsymbol{\sigma}$.

For the quadratic von Mises and Hill'48 yield functions, the dual (plastic strain-rate) potential can be derived analytically. More complex plastic potentials have been proposed, e.g. the "Quartus" model by Arminjon and Bacroix [2] to better describe the plastic anisotropy of steel sheets. On the other hand, Barlat and co-workers developed the "Srp93" model [3] and later the "Srp2004-18p" model [4], mainly dedicated to aluminum alloys. Recently, it was shown [5] that Srp2004-18p has the best ability to accurately describe the plastic anisotropy of both steel and aluminum alloy sheets.

Quartus and Srp3004-18p require the parameter identification of about twenty material parameters. These parameters can be identified to fit the micromechanically predicted yield surface of the material, as revealed after texture measurement [5]. Alternatively, material parameters can be determined using mechanical test results [6].

## FINITE ELEMENT IMPLEMENTATION

The finite element implementation of strain rate potentials follows that of classical yield functions [7]. Nevertheless, some specific developments are required. In particular, the yield criterion is no longer explicit and the elastic / plastic status of a trial stress state has to be determined numerically. In order to solve this problem, Bacroix and Gilormini [8] have demonstrated that:

$$\min_{\mathbf{N}} \{ \tau \Psi(\mathbf{N}) - \mathbf{T} : \mathbf{N} \} \begin{cases} < 0 & \text{if} \quad \mathbf{T} \quad \text{inside the yield surface,} \\ = 0 & \text{if} \quad \mathbf{T} \quad \text{on the yield surface,} \\ > 0 & \text{if} \quad \mathbf{T} \quad \text{outside the yield surface.} \end{cases} \tag{3}$$

In eq. (3), $\mathbf{T}=\boldsymbol{\sigma}'-\mathbf{X}$ designates the so-called effective stress, the backstress tensor $\mathbf{X}$ describes the kinematic hardening and $\mathbf{N}=\mathbf{D}^P/|\mathbf{D}^P|$ is the plastic strain rate direction. Finally, this property is used to determine the elastic/plastic status of a stress point.

The incremental constitutive algorithm based on strain-rate potentials is given below in a compact form. The internal variable $R$ describes the isotropic hardening; $K$ and $G$ are the elastic bulk and shear moduli, respectively. The subscripts "$n$" and "$n+1$" denote the beginning and the end of an increment. The fourth-order elastic and elastic-plastic tangent moduli are denoted by $\mathbf{C}^e$ and $\mathbf{C}^{ep}$, respectively. Finally, $\Delta\varepsilon$ designates the imposed strain increment:

0) Input data: $\Delta\boldsymbol{\varepsilon}$, $\boldsymbol{\sigma}_n$, $R_n$, $\mathbf{X}_n$

1) Elastic trial: $\boldsymbol{\sigma}'^{ee} = \boldsymbol{\sigma}'_n + \mathbf{C}^e : \Delta\boldsymbol{\varepsilon}'$

2) Detect elastic / plastic state of trial stress – eq.(3)

3) If trial stress is elastic: $\boldsymbol{\sigma}_{n+1} = \boldsymbol{\sigma}'^{ee} + \boldsymbol{\sigma}_n^{sph} + K tr(\Delta\varepsilon)\mathbf{I}$,

$$\mathbf{X}_{n+1} = \mathbf{X}_n, \quad R_{n+1} = R_n \text{ and } \mathbf{C}^{ep} = \mathbf{C}^e.$$

4) Else: solve for $\Delta\boldsymbol{\varepsilon}^P$: $\mathbf{X}_{n+1} + \tau_{n+1}\dfrac{\partial\Psi}{\partial\Delta\boldsymbol{\varepsilon}^P} - 2G\left(\Delta\boldsymbol{\varepsilon}' - \Delta\boldsymbol{\varepsilon}^P\right) - \boldsymbol{\sigma}'_n = 0$

      update stress and internal variables, compute algorithmic modulus

5) Return

This algorithm has been implemented in the finite element code Abaqus as a UMAT routine. The investigated plastic potentials have been incorporated in this routine, together with their first and second order derivatives.

## FINITE ELEMENT SIMULATION OF CUP DRAWING

The numerical simulation of cylindrical cup drawing is commonly used to validate the accuracy of plastic potentials. Indeed, the earing profile depends mainly on plastic anisotropy, while it is less influenced by other material properties. A unique geometry has been simulated here, corresponding to a punch diameter of 40 mm and a drawing ratio of two. The considered material is an AA6022-T43 aluminum alloy. The material parameters have been identified for both Quartus and Srp2004-18p using the texture-based identification approach. Additionally, the material parameters identified with respect to mechanical tests are also available [6]. The predicted earing profiles corresponding to these different situations are depicted in Figure 1. The earing profiles (number and position of the ears) are similar in all situations. Nevertheless, the ears are higher when simulated with Srp2004-18p. On the other hand, for each potential, one obtains higher ears when the parameters are identified with respect to texture, as compared to the more classical experimental identification. Consequently, the parameter identification procedure can strongly influence the conclusions about the accuracy of a given potential.

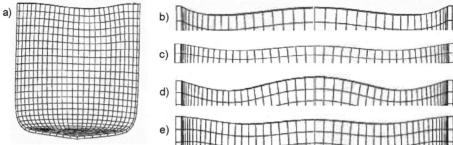

**FIGURE 1.** Results of the finite element simulations of cup drawing. Left: a) mesh and final geometry. Right: ears profile (180°): b) using Quartus with texture-based identification and c) with mechanical-tests-based identification; c) using Srp2004-18p with texture-based identification and d) with mechanical-test-based identification.

## CONCLUSIONS

A complete set of numerical tools has been developed to allow the use of advanced strain-rate potentials for the finite element simulation of forming processes (parameter identification techniques, constitutive update algorithm). As expected, the results of the finite element simulations are sensitive to the plastic potential used. Also, the parameter identification technique is shown to have a non-negligible impact on the simulation results. Cup drawing experiments are required in order to further evaluate the accuracy of the different potentials and identification techniques.

## ACKNOWLEDGMENTS

The authors are grateful to Dr. Frédéric Barlat from the Alcoa Technical Center for fruitful discussions and for providing the materials and the mechanical test results. The Région Lorraine provided financial support to the first author.

## REFERENCES

1. R. Hill, *J. Mech. Phys. Solids* **35**, 22-33 (1987).
2. M. Arminjon and B. Bacroix, *Acta Mechanica* **88**, 219-243 (1991).
3. F. Barlat, K. Chung and O. Richmond, Int. J. Plasticity **9**, 51-63 (1993).
4. F. Barlat and K. Chung, "Anisotropic strain rate potential for aluminum alloy plasticity", in Proc. of the 8th Esaform Conference on Material Forming, edited by D. Banabic, Cluj-Napoca, Romania, 27-29 April, 2005, pp.415-418.
5. M. Rabahallah, B. Bacroix, S. Bouvier and T. Balan, "Crystal plasticity based identification of anisotropic strain rate potentials for sheet metal forming simulation", IIIrd European Conference on Computational Mechanics, edited by C.A. Mota Soares et al., Lisbon, Portugal, 2006.
6. D. Kim, F. Barlat, S. Bouvier, M. Rabahallah, T. Balan and K. Chung, "Non-quadratic anisotropic potential based on linear transformation of plastic strain rate", Int. J. Plasticity, in press
7. S.Y. Li, E. Hoferlin, A. Van Bael, P. Van Houtte and C. Teodosiu, Int. J. Plasticity **19**, 647-674 (2003).
8. B. Bacroix and P. Gilormini, Modelling Simul. Mater. Sci. Eng. **3**, 1-21 (1995).

# A depth dependent analytical approach to determine material breaking in SPIF

G.Ambrogio[1], L.Filice[1], L.Manco[1], F.Micari[2]

[1] Dept. of Mechanical Engineering, University of Calabria – 87036 Rende (CS), Italy
2 Dept. of Manufacturing, Production and Management Engineering, University of Palermo, 90100 Palermo, Italy

**Abstract.** Formability is a relevant issue in Single Point Incremental Forming (SPIF) process since it is one of the main point of strength together to the possibility to avoid any dedicated die. Several researches agree that, depending on working material and process parameters, in SPIF operations there is a threshold slope of the wall that cannot be overcame without material breaking. If deep Incremental Forming is taken into account, despite the previous statement it is possible to demonstrate that, when the threshold angle is imposed, there is a relation between the actual workpiece depth and the material breaking approaching. In this paper, the latter relationship was investigated and formally derived by using a proper statistical regression.

**Keywords:** Incremental Forming, material formability, ANOVA.
**PACS:** 81.20.Hy

## INTRODUCTION

Incremental Forming is one of the most innovative solutions in the field of sheet metal forming processes due to its well-known advantages in terms of flexibility and set-up costs reduction [1,2]. Higher is application growing in industry, higher is the required improvement of the process understanding both from a process control and a basic research points of view. Today, in fact, mainly a "trial and error" approach is pursued in new processes set-up [3]. However, it is a diffused opinion that material formability is one of the most relevant issues in SPIF processes [4,5].

Usually $FLD_0$ point is used in Incremental Forming to define the formability; sometimes it is defined more simply from a geometrical point of view, by means of the maximum wall inclination angle $\alpha_{max}$. However, some applications have shown that slope angles, higher than the critical one, can be safely obtained specially when complex geometry, characterized by a variable slope wall, has to be manufactured [6]. The latter, for instance, are usually characterized by different slopes, sometimes higher than the critical one, with a depth associated to each slope usually low. In this case, also vertical walls can be manufactured imposing a depth of few millimeters since the low sheet stiffness and the punch radius induce a transient behavior that avoids the severe thinning which theoretically corresponds to the imposed slope [7].

CP907, 10th ESAFORM Conference on Material Forming, edited by E. Cueto and F. Chinesta
© 2007 American Institute of Physics 978-0-7354-0414-4/07/$23.00

In other words, a different behavior can be observed with respect to the common pieces, so that the simple knowledge of the $\alpha_{max}$ value results not adequate for the design of the manufacturing step in the above specified conditions.

For this reasons, it is easy to image that a correlation between the critical slope angle and the workpiece depth could be established, for given material and process parameters. In other words, when the critical angle is adopted, it is reasonable to think that material damage occurs only after a certain depth.

An experimental campaign was carry out in order to acquire consistent data related to the discussed issue. Subsequently, a statistical analysis was carried out, obtaining a formal law able to predict which is the allowable depth when the critical slope angle is reached or overcome, varying the other process parameters.

## EXPERIMENTAL CAMPAIGN

Starting from the available base of knowledge on the considered process, a set of experiments was properly designed and executed in order to deeply analyze the role played by the process parameters correlation with respect to the component soundness.

According to the formability limit for the utilized material, namely an Aluminum Alloy 1050–O, some tests were done varying those process parameters which directly influence the formability, such as the tool diameter ($D_p$), the tool depth step (p), the wall inclination angle ($\alpha$) and, finally, the sheet thickness (s). Naturally, others variables, like the tool speed rotation or the tool feed rate, were kept constant to reduce the problem complexity and because they have not a significant influence on the material breaking [8]. A frustum of cone, having a major base diameter ($D_0$) equal to 80 mm, was chosen as specimen shape for the whole experimental campaign and, as consequence, each experiment was carried out up to reach the material breaking or, vice versa, a final height equal to the major base dimension. In this way, for the experiments which led to breaking, the final height reached by the product at the necking conditions ($H_{max}$) was measured.

All the experiments were performed on a 3-axis CNC milling machine, equipped with a clamping device and a baking plate to ensure the sheet locking. The tool speed rotation and the feed rate were fixed equal to 120 r.p.m and 1000 mm/min respectively. An emulsion of mineral oils was utilized to lubricate the interface between the punch and the sheet.

## STATISTICAL ANALYSIS

The experiments were analyzed following the Response Surface Methodology, which corresponds to a set of mathematical and statistical methods able to highlight as few independent variables can describes an output factors. In this way, as above introduced, four independent factors were considered, while the dependent one was derived by the measure of the component final height. In Table 1 the lowest and the highest values for the investigated range are displayed. Since the analysis was executed at the formability limit, the lowest wall inclination angle was fixed equal to the critical slope for the investigated material, i.e. 70°, with the aim to better specify

332

the influence of the other parameters with respect to the maximum achievable height [8].

Each experiment was firstly designed belonging to a CCD face center plane in order to reduce the number of tests; subsequently, during the analysis, the plane was reduced neglecting the experiments which allowed a sound component. Furthermore, an external bound to generalize the analytical model applicability (Figure 1) also when maximum depth is allowed was introduced.

**TABLE 1.** Extreme Value of the Experimental Plane.

| Parameters | Lowest Value | Highest Value |
|---|---|---|
| Wall Inclination Angle ($\alpha$) | 70° | 80° |
| Tool Diameter ($D_p$) | 12mm | 18mm |
| Tool Depth Step (p) | 0.3mm | 1mm |
| Sheet Thickness (s) | 1mm | 2mm |

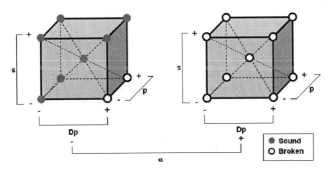

**FIGURE 1.** The investigated experimental plane.

The experimental results were analyzed by the ANOVA in order to highlight the effects of the single independent factors and their interactions on the dependent one (see Table 2).

**TABLE 2.** Influence of input variables and their interactions on $H_{max}$.

| | Factors | | | | | | |
|---|---|---|---|---|---|---|---|
| | $\alpha$ | $D_p$ | P | s | $\alpha*D_p$ | $\alpha*p$ | $\alpha*s$ |
| **Influence** | High | High | High | High | Medium | Medium | Medium |
| **on $H_{max}$** | $\alpha^2$ | $D_p^2$ | $p^2$ | $s^2$ | $D_p*p$ | $D_p*s$ | $P*s$ |
| | No | Low | No | No | No | No | No |

As it can be easily observed, the first order factors are the most significant when determining the product final height. This confirms the results recognizable from the state of the art [1]. On the other hand, only the interaction between the wall inclination angle and the other process parameters is partially significant to influence the output value, confirming again the relevant role played by the former parameter on formability in Incremental Forming process. ANOVA methodology allowed also to identify a quadratic model as the most suitable solution to describe the investigated phenomenon and to predict the maximum achievable height. As above introduced, a bound was defined to identify the not critical conditions which allow to obtain deep

component, with a final height equal to its major base. In fact, as it is already known, the slope angle equal to the critical one, namely $\alpha_{max}=70°$ for the AA 1050–O, represents a threshold value which divides the super safe conditions ($\alpha<\alpha_{max}$) and the unsafe ones ($\alpha\geq\alpha_{max}$). At the same time, observing only those experiments which led to sound product, it is easy to understand that a strong influence is played by the tool diameter and the sheet thickness; more in detail, the material formability, in terms of achieved depth, decreases increasing the tool diameter $D_p$ and reducing the sheet thickness s. This result can be easily explained considering that the same portion of material repeatedly undergoes to the punch action, determining a sort of material removal by wear. In this way, it can be assessed that formability is proportional to the tool diameter and inversely proportional to the sheet thickness, so that the ratio $D_p$/s can be proposed to classify the safe and unsafe cases for $\alpha=70°$. More in detail, if $D_p$/s is higher than a critical value K* or $\alpha>\alpha_{max}$, the achievable specimen height can be predicted according to the following equation (1):

$$H_{max} = 130.93 - 12.61D_p + 54.56p - 0.89\alpha + 45.64s + 0.068D_p \cdot \alpha - 0.62p \cdot \alpha$$
$$- 0.488\alpha \cdot s + 0.254D_p^2 \tag{1}$$

On the contrary, safe configurations, which corresponds to slope angle lower than $\alpha_{max}$ or equal to $\alpha_{max}$ but characterized by a ratio $D_p$/s<K*, can be manufactured up to maximum investigated depth ($H_{max} = 100\% \cdot D_0$). Due to the number of experiments carried out, it was not possible to fix an exact value for the K*-factor. However, an uncertainty region, ($13 \leq K* \leq 18$, see Figure 2), divides safe and un-safe conditions.

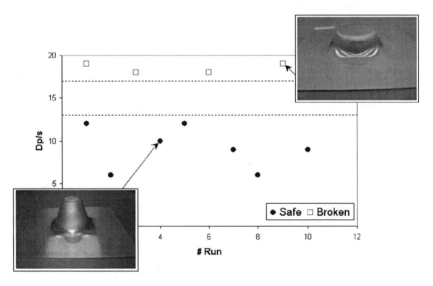

**FIGURE 2.** $D_p$/s values for experiments carried out for $\alpha=\alpha_{max}$.

Naturally, the model suitability was firstly statistically evaluated taking into account some performance indexes, such as the $R^2=98\%$, the $R^2_{corr}= 96\%$, the $R^2_{Pred}=$ 89%, rather than the Normal Probability Plot and the Prediction Capability Diagram (see Figure 3).

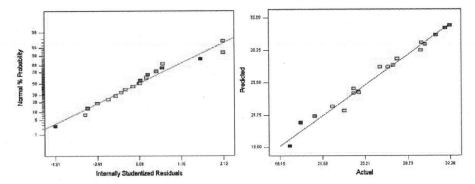

**FIGURE 3.** Normal Probability Plot and Prediction Capability for the investigated analysis .

## MODEL VALIDATION

The analytical model was finally assessed executing tests characterized by three new parameters configurations. For sake of simplicity the geometry was always the same. Only critical conditions were considered during this phase ($\alpha \geq \alpha_{max}$) and, according to the model prediction, the experiments were properly designed to reach the material breaking condition or the fixed upper-bound ($H_{max} = 100\% \cdot D_0$). The analyzed cases and the comparison between the experimentally obtained depth and the predicted one are shown in the next Figure 4 and Table 3.

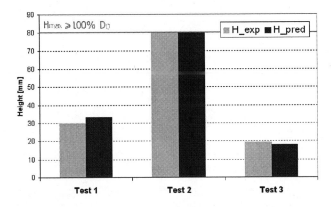

**FIGURE 4.** Comparison between the predicted height and the obtained one during the validation tests.

A suitable model behaviour was observed both in the prediction of safe conditions than in the unsafe ones, thus confirming the model capability to be used as design tool

335

when slope angle, higher then the critical one for a specific material, has to be manufactured.

TABLE 3. Validation tests.

|        | $D_p$ [mm] | p [mm] | $\alpha$ [°] | s [mm] | H_exp [mm] | H_Pred [mm] |
|--------|------------|--------|--------------|--------|------------|-------------|
| Test 1 | 12         | 1      | 77           | 2      | 30         | 33.2        |
| Test 2 | 15         | 0.5    | 70           | 1.5    | 80         | 80          |
| Test 3 | 18         | 0.5    | 90           | 1.5    | 19.5       | 18.2        |

# CONCLUSIONS

In this paper, a simple model to relate the maximum component depth and the process parameters when the critical slope angle in incremental forming process is overcome was proposed and tested. In particular, it supplies good results when single slope geometries are manufactured by Incremental Forming. However, it was stated that it is not correct to define a critical slope angle for given material and process parameters; on the contrary, there are some conditions in which a certain depth is obtainable even if the critical slope is overcome.

The applicability of this logic is very important in the practice because real parts are usually characterised by several slopes, but for a limited depth. Thus, it is possible to think that a robust design has to take into account the investigated phenomenon in order to allow the possibility to extend the process applicability also to geometries having punctual severe wall slopes.

Naturally, further investigations are required to definitively asses and complete the highlighted model. First of all, the variability due to a more complex geometry and the material influence have to be investigated and introduced into the analysis.

# ACKNOWLEDGMENTS

The authors would like to thank Mr. F. Pulice, for his technical support.

# REFERENCES

1. J. Jeswiet, F. Micari, G. Hirt, A. Bramley, J. Duflou, J. Allwood, *Annals of the CIRP*, **54/2**, 623 (2005).
2. G. Hirt, J. Ames, M. Bambach, R. Kopp, *Annals of the CIRP*, **52/1**, 203, (2004).
3. A. Attanasio, E. Ceretti, C. Giardini, *Journal of Materials Proc. Tech.*, **177/1-3**, 409-412, (2006)
4. M.S. Shim, J.J. Park, *Journal of Materials Proc. Tech.*, **113/1-3**, 654-658, (2001).
5. J. Jeswiet, E. Hagan, A. Szekeres, *Journal for Manufacturing*, **216**, 1367-1371, (2002).
6. G. Ambrogio, L. De Napoli, L. Filice, F. Gagliardi, M. Muzzupappa, *Journal of Materials Proc. Tech.*, **162-163**, 2005, 156-162.
7. G. Ambrogio, L. Filice, F. Silvestri, F. Micari, "Rapid Prototyping through the application of AISF technique", 9th Esaform Conference Procedings, University of Strathclyde, Glasgow, UK, 2006, 875-878.
8. G. Ambrogio, L. Filice, L. Fratini, F. Micari, "Some relevant considerations between the process parameters and process performance in incremental forming of metal sheet", 6th Esaform Conference Procedings, University of Salerno, Salerno, Italy, 2003, 175-178.

# Experimental and Numerical Investigation of Kinematic Hardening Behavior in Sheet Metals

Hang Shawn Cheng*, Wonoh Lee*, Jian Cao*, Mark Seniw[+]
Hui-ping Wang [¶] and Kwansoo Chung[#]

*Department of Mechanical Engineering, Northwestern University,
2145 Sheridan Road, Evanston, IL 60208, U.S.A.
[+]. Department of Material Science and Engineering, Northwestern University,
2220 Campus Drive, Evanston, IL 60208, U.S.A
[¶]General Motors Corporation, Warren, MI 48090, U.S.A.
[#]Department of Mechanical Engineering, Intelligent Textile System Research Center,
Seoul National University, 56-1 Shinlim-dong, Kwanak-gu, Seoul 151-742,South Korea

**Abstract.** Characterization of material hardening behavior has been investigated by many researchers in the past decades. Experimental investigation of thin sheet metals under cyclic loading has become a challenging issue. A new test fixture has been developed to use with a regular tensile-compression machine (for example, MTS machine). Experimental results of tension-compression tests are presented followed by a review of existing testing methods. Numerical modeling of the tested data is presented using a new kinematic hardening model.

**Keywords:** Tension-compression Test, Kinematic Hardening Law, Permanent Softening Behavior, Non-symmetric Transient Behavior
**PACS:** 62.20.Fe

## INTRODUCTION

It has been known for over a century that many materials' properties depend on their loading paths. The phenomenon was first discovered by Bauschinger [1]. For these materials, simple material models such as isotropic hardening are not able to model their reverse loading properties correctly. This is a vital issue for accurate springback simulation for formed parts.

Various experimental techniques have been developed to test materials along reverse loading paths. Reverse torsion and shear tests can achieve high strains. However, they face the problem of non-uniform strain distribution as well as difficulties to interpret the testing results. Bulk compression tests and in-plane compression tests provide more uniform strain distribution with appropriate length-to-diameter/thickness ratio. But large strain is not easy to obtain due to the specimen's tendency to buckle under compression. Boger et al. [2] used solid flat plates as buckling constraints and applied normal pressure through a hydraulic clamping system. Kuwabara et al. [3] developed a device with two pairs of combs to reduce the unsupported area. The comb device is expensive to machine. None of the methods

CP907, 10th ESAFORM Conference on Material Forming, edited by E. Cueto and F. Chinesta
© 2007 American Institute of Physics 978-0-7354-0414-4/07/$23.00

mentioned above could completely eliminate the unsupported area, which is one of the main reasons that the specimen buckles. In this paper, we introduce a novel device to eliminate unsupported area of the in-plane compression specimen. For the purpose of accurate prediction, the nonlinear kinematic hardening law based on the modified Chaboche model [4, 5] was utilized with the softening parameter to represent the permanent softening during the reverse loading [6]. Furthermore, in order to describe different transient behavior during the re-loading period, the transient parameter has been introduced in this paper.

## DOUBLE-WEDGE IN-PLANE COMPRESSION TEST

The conceptual design of the new in-plane compression test device is shown at Fig. 1. The specimen is clamped by the upper and lower clamps installed on the tensile machine. Two pairs of wedge plates are simply placed on top of the bottom clamp, one pair on each side of the specimen. The wedge plates can slide against each other as well as against the top and bottom clamps. A spring under tension is installed on each side of the device to ensure the wedge plates on the same side of the specimen being in firm contact with each other. This mechanism allows the wedge plates to cover the specimen surfaces completely, no matter whether the specimen is under tension or compression.

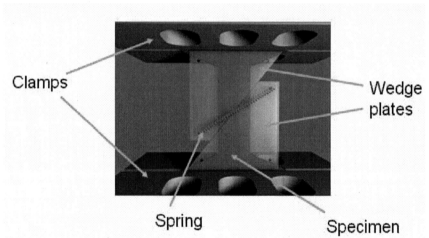

**FIGURE 1.** Conceptual design of the double-wedge in-plane compression test

The actual device is made from aluminum blocks. As shown in Fig. 2, wedge plates are very easy to machine. Each plate is 3-inch tall, 2.5-inch wide and 0.75-inch thick. Three long bolts are used to fix the relative position of the plates in each pair. Teflon sheets are attached on the inner surface of the plates to reduce friction between the plates and the specimen. User can adjust the blots to accommodate specimens with different thickness as well as to apply a different amount of normal pressure on the specimen. There is a steel spring on each side of the device to ensure firm contact between the plates. During trial tests, it was found that when a thick specimen tried to buckle, the alignment of the top and bottom plates is difficult to hold. Therefore, we modified the device by adding sliding keyways as shown in Fig. 2.

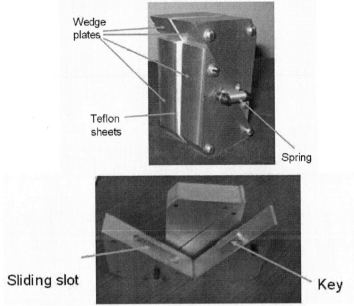

**FIGURE 2.** Double-wedge in-plane compression test device

Standard ASTM E8M-00 subsize specimen geometry [7] is adopted to ensure the uniformity of the strain distribution. An Instron tensile machine is used to measure the displacement and load. Fig. 3 shows the load-vs-displacement curve for a void run. The loading are in the negative values due to the spring force in the setup, which is the reason that the wedge plates are firmly against the clamps to ensure a good contact. The load due to the springs and friction for a void run will be deducted from the testing results. Strain will be calibrated as a function of crosshead displacement using a regular tensile test with an extensometer. Lubricants are applied on the Teflon sheets attached on the wedge plates as well as on the specimen surface to reduce friction. Before the test, the gap between wedge plates is adjusted carefully to ensure the specimen can slide freely along the plates and maintain firm contact. Test results for 1.6mm thick DP600 steel materials will be presented in the next section along with calculated curves.

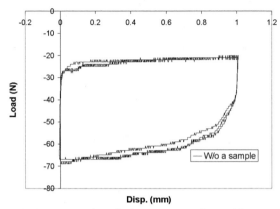

**FIGURE 3.** Load-vs-displacement curve for a void run

## MATERIAL CHARACTERIZATION

The nonlinear kinematic hardening constitutive law based on the modified Chaboche model [4,5] is given by

$$f(\boldsymbol{\sigma} - \boldsymbol{\alpha}) - \sigma_{iso} = 0 \qquad (1)$$

where $\boldsymbol{\alpha}$ is the back stress for the kinematic hardening and the effective stress, $\sigma_{iso}$, is the size of the yield surface. In the Chaboche model, the back-stress increment is composed of two terms, $d\boldsymbol{\alpha} = d\boldsymbol{\alpha}_1 - d\boldsymbol{\alpha}_2$ to differentiate transient hardening behaviors during loading and reverse loading. To complete the constitutive law, hardening behavior describing back-stress movements should be provided for $d\alpha_1$ and $d\alpha_2$. From Fig. 4, it can be seen that the yield surface of this particular material does not expand after unloading/reloading. Therefore, the change in the yield surface size, $d\sigma_{iso}$ is zero and only kinematic hardening behavior is considered in this work.

In order to describe the permanent softening in reverse loading, hardening parameter, $h_1(=d\alpha_1/d\varepsilon)$, was modified by introducing the softening parameter, $\xi$; i.e.,

$$h_1^s = h_1 \cdot \xi(\varepsilon^*) \qquad (2)$$

where $\varepsilon^*$ is the accumulative effective strain and the superscript 's' stands for 'softening'. Details on the consideration of softening behavior are referred to the previous works [6].

Since there is no softening behavior in the previous during the re-loading, the softening parameter $\xi$ can be applied only for the reverse loading. To represent non-symmetric transient behavior during the re-loading from the reverse loading, the hardening parameters $h_1$ and $h_2(=d\alpha_2/d\varepsilon)$ are modified here by using the transient parameter $\eta$,

$$h_i^t = h_i \cdot \eta(\varepsilon^*) \tag{3}$$

where $i$ is 1 and 2, and the superscript 't' stands for 'transient'. The transient parameter was considered as

$$\eta = b_6 \exp(-c_6 \varepsilon^*) \tag{4}$$

where $b_6$ and $c_6$ are values dependent on the total accumulative effective strain during the previous reverse loading, $\varepsilon^*_{pre}$. The values of $b_6$ and $c_6$ were parameterized as

$$b_6 = b_6^1 + b_6^2 \exp(-b_6^3 \varepsilon^*_{pre}), \qquad c_6 = c_6^1 + c_6^2 \exp(-c_6^3 \varepsilon^*_{pre}) \tag{5}$$

In order to measure the nonlinear kinematic hardening behavior, uni-axial tension-compression tests were performed. The DP600 material showed softening behavior during the reverse loading and non-symmetric transient behavior during the re-loading. The hardening behavior calculated with the softening and non-symmetric transient parameters for DP600 base sheets are plotted in Fig. 4, which confirms that the modified Chaboche model with softening and non-symmetric transient parameters well represents the permanent softening as well as the Bauschinger and non-symmetric transient behaviors.

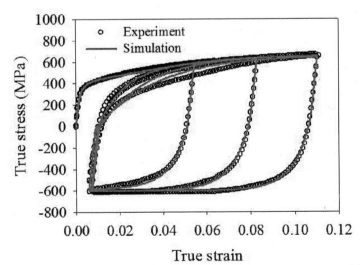

**FIGURE 4.** Calculated and measured hardening behavior in tension-compression tests with softening and non-symmetric transient parameters for DP600

341

# SUMMARY

In this paper, we introduced a novel device for in-plane compression tests for sheet materials. This double-wedge device is easy to be fabricated and able to cover the specimen surface completely. Therefore, potential buckling of sheet specimen can be prevented. Test results for DP600 steel are presented. To characterize mechanical properties, uni-axial tension-compression tests were performed. The nonlinear kinematic hardening law based on the modified Chaboche model successfully described the hardening behavior including the Bauschinger behavior with the permanent softening during reverse loading and the non-symmetric transient behavior during re-loading stage.

# ACKNOWLEDGMENTS

The authors would like to thank General Motors for its support.

# REFERENCES

1. J. Bauschinger, "On the Change of the Elastic Limit and the Strength of Iron and Steel, by Drawing out, by Heating and Cooling, and by Repetition of Loading (Summary)" *Minutes of Proceedings of the Institution of Civil Engineers with Other Selected and Abstracted Papers* LXXXVII: 463. 1986
2. R.K. Boger, R.H. Wagoner, F. Barlat, M.-G. Lee, and K. Chung, *Int. J. Plasticity* **21**. 2319-2343 (2005).
3. T. Kuwabara, "Advances of Plasticity Experiments on Metal Sheets and Tubes and Their Applications to Constitutive Modeling", in Proceedings of NUMISHEET 2005, edited by L.M. Smith et al., 2005, pp. 20-39.
4. J.L. Chaboche, *Int. J. Plasticity* **2**. 149-188 (1986).
5. K. Chung, M.-G. Lee, D. Kim, C. Kim, M.L. Wenner and F. Barlat, *Int. J. Plasticity* **21**. 861-882 (2005).
6. W. Lee, J. Kim, D. Kim, C. Kim, M.L. Wenner and K. Chung, *Int. J. Plasticity*. Submitted (2007)
7. Annual Book of ASTM 2000, Vol 03.01, P 77

# A Modified Split Sleeve For The Cold Expansion Process And Application To 7085 Plate

F. Barlat[1], M.E. Karabin[2], and R.W. Schultz[3]

[1]Alloy Technology and Materials Research Division, [2]Process Mechanics Division, [3]Product Manufacturing Division, Alcoa Technical Center, 100 Technical Drive, Alcoa Center, PA 15069-0001, USA

**Abstract.** In this work, the split sleeve cold expansion process is studied numerically and experimentally. The finite element (FE) simulations show that a strain concentration occurs in the material near the sleeve split. The results are not very sensitive to macroscopic plastic anisotropy. The maximum strain produced during this process is influenced more by the sleeve geometry than by the material properties. It is also shown numerically that a different sleeve design can reduce the strain concentration drastically. Therefore, this sleeve design is investigated experimentally on a 7085-T7651 aerospace aluminum alloy plate. Although the new design is not optimized, experimental evidence shows that it can significantly reduce cracking near a sleeve split.

**Keywords:** Cold expansion; Split sleeve; Overlap sleeve; Numerical simulation; Plastic anisotropy; Cracking.
**PACS:** 62.20.-x, 62.20.Mk

## INTRODUCTION

The cold expansion process is used by the aircraft industry to improve fatigue life in structural components. Holes are drilled into plates in order to accommodate fasteners. These holes serve as stress concentrations that can lead to fatigue cracks. This problem can be mitigated by the generation of circumferential compressive stresses near the holes. Several methods can be used to introduce a compressive stress field around a fastener hole [1-5]. In the split sleeve cold expansion process [1, 6], a solid mandrel with a bulge in conjunction with a split sleeve that has parallel walls [6] is pulled through the hole. Expansion is produced sequentially along the axial length. Because of the split, plastic deformation is not axisymmetrically distributed around the hole and cracks can occur, generally in the vicinity of the split in the sleeve.

The objective of this work is to suggest solutions to reduce the likelihood of cracks near the split in the sleeve for this process. Two possible solutions for resolving crack formation are investigated numerically using a finite element (FE) code. The first deals with an optimum orientation of the split with respect to the material symmetry axes. For this purpose, an anisotropic response is provided as a constitutive material description in the FE code. The second solution deals with a modification of the split sleeve geometry. These two approaches may be used concurrently but they are studied

CP907, 10th ESAFORM Conference on Material Forming, edited by E. Cueto and F. Chinesta

separately in this work. Since the second solution (change in split sleeve geometry) is more promising, it is investigated experimentally.

## COLD EXPANSION MODELING

The alloy used in this study is 7085-T7651, which was recently developed for aerospace structure applications. For characterization, material is taken from the $T/2$ and $T/4$ locations of a 101.6 mm thick plate where $T$ corresponds to the thickness. It is well known that thick plates exhibit a microstucture gradient through the thickness, or short transverse (ST) direction, due to temperature and deformation gradients during processing. Many microstructural features that produce heterogeneous plastic flow, such as constituent particles or grain boundaries, can initiate a crack locally but the purpose of this work is to assess the influence of macroscopic properties on cracking tendency. The anisotropy of the material and its gradient are the factors investigated. The strong crystallographic texture gradient leads to different properties at $T/4$ and $T/2$. An anisotropic yield function [7] that describes the behavior of aluminum alloys particularly well is implemented in ABAQUS, though the user-defined material subroutine. This yield function is applicable to general stress states and contains 18 anisotropy coefficients [7]. The influence of all the other microstructural features is assumed to be reflected by the stress-strain behavior of the material, which is characterized by a Swift law. In this work, the anisotropy coefficients are not identified using mechanical test results as input but using properties computed with the Taylor-Bishop and Hill (TBH) crystal plasticity model.

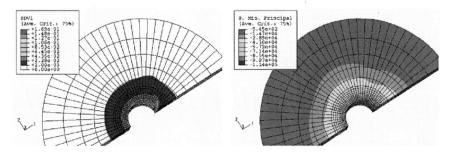

**FIGURE 1.** 3D finite element model for cold expansion: Effective strain distribution (left) and compressive residual stress (right) around the hole after expansion.

The influence of the material properties on cold expansion process is investigated using a three-dimensional (3D) formulation. The tool is not modeled in these simulations. The expansion is achieved through the application of a radial displacement and a free surface at the split. There is no distinction between the plate entry and exit surfaces in the model. Figure 1 shows the equivalent plastic strain contours in the 3D model after the cold expansion process. Strains are elevated around the split in the sleeve and slightly lower on the free surface than at the mid-plane. Figure 1 shows also a contour plot of the minimum principal stress after the process. High compressive stresses exist around the entire periphery of the hole to a distance of at least $2R_o$, the diameter of the hole. There is some evidence of material anisotropy effect in the stress contours around the hole. Since the objective in this study is to reduce the potential for cracking from cold expansion due to material anisotropy, there may be an optimum positioning of the split sleeve with respect to the material axes. In this case, optimum corresponds to the smallest maximum equivalent plastic strain. Plastic strain can be linked to damage and it might be the driving force for crack initiation and propagation. The FE results indicate that with a split oriented at 45° from the longitudinal direction (L), the amount of plastic strain is the least but differences are small. Nevertheless, there may be some advantageous orientations for the location of the split in the sleeve with respect to the material axes. However, since values for resistance to crack initiation as a function of material orientation depends on microstuctural features and are not given in this paper, no valid conclusions can be made about the optimum sleeve orientation.

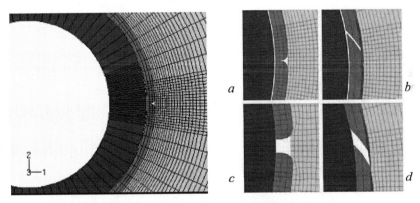

**FIGURE 2.** 2D finite element models for cold expansion with the plate (green), mandrel (blue) and sleeve (red). Model of plate disc with hole (left) and close up of two sleeve geometries before (*a* and *b*) and during (*c* and *d*) cold expansion (right).

The influence of the geometry of the split sleeve is investigated using a 2D model (Figure 2). Two different sleeve versions are shown in this figure; the conventional sleeve (*a*) where the cut is radial and the modified one (*b*) where the cut is more helical and the edges of the sleeve overlap. The elements in the sleeve are somewhat skewed in the overlap version. It is believed that the element shape should not have any significant effect on the results. The sleeve is just an elastic body that is used for contact purposes. The deformed geometries during cold expansion of the two sleeves are shown in Figure 2 (*c* and *d*). With the traditional sleeve geometry, the sleeve edges separate with increasing expansion allowing the material to extrude into the gap. The material free surface does not expand as much as the surrounding metal that is in full contact with the tool. The edges of the sleeve act as a punch, causing high strains at the corner of the sleeve gap. This produces large strain concentrations in the vicinity of the split that can lead to cracking. While the nominal strain on the hole surface is 2%, the peak strains in the aluminum at the split sleeve location are ten to 20 times the nominal depending upon the sleeve geometry.

Instead of having a conventional sleeve where the two edges of the split are parallel, the modified sleeve is oversized in the circumferential direction such that the edges overlap. The thickness in the overlapping regions is tapered to provide a relatively uniform thickness overall. With an overlapping sleeve, up to a limited amount of expansion, no gap develops at the sleeve-aluminum interface and only a small indentation occurs in the region where the two edges touch. There is, perhaps, a slight discontinuity in the contacting surface of the sleeve. Equivalent plastic strain for the two split sleeve geometries is plotted in Figure 3. High levels of plastic strain are seen near the split for both configurations. However, the magnitude in the conventional sleeve is about twice that in the overlap version. This would indicate that less damage occurs to the material for the latter. This reduction of plastic strain by 50% with the overlap sleeve is typical for all orientations that were modeled. Also, the high strain pattern in the conventional sleeve is symmetric with respect to the split plane defined by the longitudinal and the short transverse directions (L-ST) of the plate, whereas no symmetry exists in the overlapping sleeve. Therefore, the orientation of the sleeve with respect to the material axes may play a larger role in the overlapping sleeve. The level of compressive stress surrounding the hole is not reduced through the use of the overlap sleeve. In fact, the compressive is uniformly greater in the vicinity of the split with the overlap sleeve.

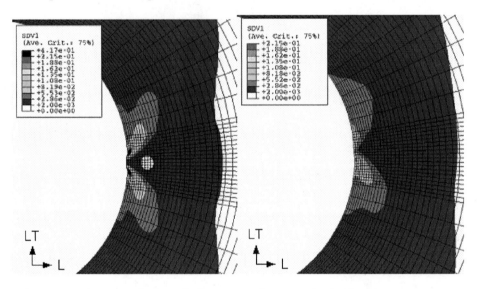

**FIGURE 3.** Strain contours for two split sleeve geometries, conventional and overlap.

The numerical study showed that changes to the sleeve geometry have a greater impact on the accumulation of plastic strain than macroscopic plastic anisotropy. The numerical simulations show that the modified sleeve can reduce the strains in the vicinity of the split by 50%, while providing an equivalent and more uniform compressive stress field. Therefore, based on the simulations of the cold expansion process, the overlapping sleeve approach bears further study through experiments.

# COLD EXPANSION EXPERIMENTS

Conventional sleeves are purchased from Fatigue Technology Inc. (FTI). In order to produce the overlapping sleeve, an oversized sleeve with the same shape is used. In order to trim the excess material from the blank, a cut is made on both sides of the split. The initial machining layout is shown in Figure 4. The cuts in this figure are circular while the previous numerical model has cuts that are more stepped. With the stepped cuts, the change in slope at the split on the outer surface in the expanded configuration is less than that with the circular cuts. The pertinent dimensions of the modified sleeves are given elsewhere [8]. The fixture used to hold the blank while cutting is shown in Figure 4. Clamps help to maintain the proper position. The method of machining is wire-cut electrical discharge machining (EDM). Cutting is conducted using standard burning settings with a 0.25 mm diameter wire.

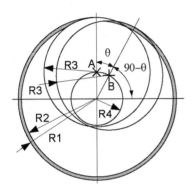 

**FIGURE 4.** Layout of the machining cuts for the modified split sleeve (left) and fixture used to hold sleeve blanks (right): Left-clamp, center-fixture, right sleeve after burning

Three test samples, each measuring $101.6 \times 25.4 \times 304.8$ mm, are machined from five different lots of 7085-T7651 for a total of 15 samples. Three holes are drilled and reamed in each plate, two at $T/4$ and one at $T/2$. The nominal hole diameter is 21.13 mm and this is maintained through the hole depth and from hole to hole to within a tolerance of 0.013 mm. With a maximum mandrel size of 20.84 mm diameter, and a sleeve thickness of 0.46 mm, the nominal expansion of the holes is 3%. For each test sample, one plate is tested with three conventional sleeves, one plate is tested with three modified sleeves, and the third plate has a mix (one type at $T/2$ and another at $T/4$). Prior to testing, all modified sleeves have powdered graphite applied to the newly machined surfaces.

In summary, 23 holes use the modified sleeve and 22 use the conventional sleeve. A summary of the results is shown in Table 1, which indicates that the modified split sleeve significantly reduces the occurrence of cracking during cold expansion. It also results in a shorter crack length, when cracking does occur, suggesting the existence of a lower crack driving force relative to the control sleeve. A statistical contingency analysis of cracking is done for 3 cases: 1) $T/2$; 2) $T/4$; and 3) $T/2$ and $T/4$ combined.

In all three cases, the difference in cracking by sleeve type is significant. This experiment also confirmed previous observations. Namely, holes at $T/4$ are more prone to cracking than those at $T/2$. Additionally, more cracks develop on the exit side of the hole than on the entrance side and the exit side (top) has slightly greater permanent expansion.

TABLE 1. Cold expansion results.

| Position | Modified Sleeve | | Control Sleeve | |
|---|---|---|---|---|
| | % Cracked | Average Crack Length | % Cracked | Average Crack Length |
| T/2 | 0 | NA | 57 | 2.9 mm |
| T/4 | 46 | 2.8 mm | 85 | 8.6 mm |

## CONCLUSIONS

Numerical simulations of the split sleeve cold expansion process showed that the influence of plastic anisotropy on the strain distribution around the hole is small. Strains are slightly less at $T/2$ than at $T/4$, which could indicate lower damage and a lower driving force for crack initiation and propagation at $T/2$. In contrast, the simulations results showed that changes to the sleeve geometry have a greater impact on the accumulation of plastic strain. FE simulations were carried out with the traditional sleeve geometry and with an overlapping sleeve, which resulted to a 50% differences in maximum strain around the hole. The results of the experiment confirm the numerical simulations. The modified split sleeve design leads to a dramatic reduction of cracking during the cold forming process and a substantial decrease in length of the cracks when these are observed.

## ACKNOWLEDGMENTS

The authors are grateful to Dhruba Chakrabarti and Harry Zonker (Alcoa Technical Center) for reviewing this manuscript.

## REFERENCES

1. J.L. Phillips, Fatigue improvement by sleeve coldworking, SAE paper No. 730905, 1973.
2. L. Reid, Split sleeve cold expansion as a rework process for previously cold expanded holes, in: International Committee on Aeronautical Fatigue ICAF'93, June 7–11, Stockholm, Sweden, 1993.
3. M.O. Lai, Y.H. Siew, Fatigue properties of cold worked holes, J. Mater. Proc. Techn. 48 (1995) 533-540.
4. J.T. Maximov, Spherical mandrelling method implementation on conventional machine tools, Int. J. Mach. Tools Manuf. 42 (2002) 1315–1325.
5. T.N. Charkherlou, J. Vogwell, A novel method of cold expansion which creates near-uniform compressive tangential residual stress around fastener holes, Fatigue Fract. Eng. Mater. Struct. 27 (2004) 343-351.
6. Fatigue Technology Inc., Cold expansion of holes using the standard split sleeve system and countersink cold expansion (CsCx), FTI Process Specification 8101D, June 7 2002.
7. Barlat, F., Aretz, H., Yoon, J.W., Karabin, M.E., Brem, J.C., Dick, R.E., 2005. Linear transformation-based anisotropic yield functions. Int. J. Plasticity 21, 1009–1039.
8. Karabin, M.E., Barlat, F., Schultz, R.W., 2006. Numerical and experimental study of the cold expansion process in 7085 plate using a modified split sleeve. Submitted for publication in J. Mater. Proc. Technology.

# Mechanical Anisotropy of Stainless Steel Sheets Depending on Texture and Grain Misorientation

Yuriy Perlovich, Vladimir Fesenko, Margarita Isaenkova
and Vladimir Goltcev

*Moscow Engineering Physics Institute, Kashirskoe shosse 31, 115409 Moscow, Russia*

**Abstract.** The anisotropy of mechanical properties of sheets from stainless steel is considered as a function of their texture and structure features, revealed by use of X-ray study. The work was conducted as applied to thin sheet stainless steel of austenitic type.

**Keywords:** tensile testing, texture, deformation incompatibility, anisotropy
**PACS:** 62.20.-x; 68.55.jk

## INTRODUCTION

When predicting mechanical properties of metal materials, one ought to have in mind, that testing of samples consists in their specific plastic deformation, accompanied by reorientation of grains and texture changes. Depending on the initial texture and the used deformation scheme, texture changes by measurement of mechanical properties can be more or less significant, but in all cases they submit to the clear regularities. These regularities underlie the general theory of texture formation [1], which allows to connect grain reorientation with strain hardening and fracture probability. The given paper illustrates the texture approach to interpretation of mechanical properties.

## TEXTURE APPROACH TO TENSILE TESTING

Since any plastic deformation is accompanied by grain reorientation, tensile testing of sheet samples results in essential texture changes, which are different in cases of tension along rolling direction (RD) and transverse direction (TD). The crystalline lattice of grains tends to rotate to the definite final orientations, stable relative to deformation by tension. Therefore, measurement of mechanical properties is associated with processes, controlling attainment of stable orientations by grains of used samples. These properties show, how the transition develops from the initial texture to the final one, how high there are the yield point and the following strain hardening, how critical there is the stress concentration due to mutual misorientation of neighboring grains. The behavior of samples under tension can be analyzed on the basis of known models of texture formation, supplemented by concepts, concerning initial structure features of tested samples, which are sequent from obtained texture data.

The texture of samples is characterized by inverse texture pole figures (IPF), constructed for sections, perpendicular to the tension axis. IPF is a crystallographic certificate of the studied surface, showing inputs of all possible crystallographic planes in its formation in units of pole density p within the elementary stereographic triangle. Value p(hkl) shows, by how many times an input of plane {hkl} in formation of the studied surface is higher than in the case of the textureless sample. When the input of plane {hkl} is equal to that in the textureless sample, p(hkl)=1; when this plane dominates in the surface, p(hkl)>1, and when this plane is met rarer than in the textureless sample, p(hkl)<1 [1]. When comparing IPF for RD- and TD-sections of the sheet, one can see differences in the initial crystallographic situation by RD- and TD-tension, responsible to the significant degree for the anisotropy of measured mechanical properties.

A capacity of sample to deform under tensile testing without fracture is determined by two factors, depending on the texture character. The first of them is presence of grains, in which easy slip in one or two systems takes place with low strain hardening. Such grains have orientations, intermediate relative to final texture components, stable due to operation of several mutually symmetric slip systems [1]. These orientations are situated at the periphery of stable texture maxima and correspond to zones of their scattering. As the texture become more perfect, a volume fraction of grains with intermediate orientations decreases and strain hardening grows. The higher is an initial fraction of grains with unstable orientations, capable to deform with low strain hardening, the higher deformation degree can be attained by tensile testing of sample without fracture.

The second factor characterizes a deformation incompatibility of grains under testing of samples and reflects different behavior of grains, belonging to stable components of the final texture of tension. When grains of two different components are neighbors, they are separated by a high-angle boundary, being a locus of stress concentration and a probable origin of the crack under tensile testing. The higher is a probability to meet in the sample the high-angle boundary, the sooner will occur its fracture. When between stable texture maxima in IPF there is a continuous series of intermediate components, the number of high-angle boundaries in the sample and the average misorientation of neighboring grains are lower, than in the case of perfect texture with mutually isolated maxima.

By tensile testing of notched samples the plastic deformation and accompanying texture changes are localized within layers, adjacent to the fracture surface, but the dependence of measured mechanical properties on texture and structure features remains in this case the same, as by testing of plain samples.

The mechanical anisotropy is characterized by ratios $\Psi$ of properties, measured by testing along RD and TD. For example, $\Psi_\delta = \delta_{RD}/\delta_{TD}$, where $\delta$ – relative elongation. The texture approach to prediction of mechanical anisotropy includes the following stages, considered as applied to deformation by tensile testing:

(1) Final stable orientations of grains relative to the axis of tensile testing are specified for the crystalline lattice of studied material on the basis of known models of texture formation (in practice the trajectory of this axis in IPF is considered) [2].

(2) The initial texture of tested samples is measured and IPF are constructed for the

sections, perpendicular to the axis of tension.

(3) The parameter S1 is chosen for estimation of the fraction of grains, which can reorient to the stable orientations due to operation of one or two slip systems, that is by relatively low strain hardening.

(4) The parameter S2 is chosen for taking into account a degree of deformation incompatibility, connected with initial misorientation of grains in samples.

(5) The texture anisotropy of sheet is characterized by ratios K1=S1(RD)/S1(TD) and K2=S2(RD)/S2(TD), which are determined by the corresponding IPFs.

(6) A correlation is sought between coefficients of mechanical anisotropy $\Psi$ and coefficients of texture anisotropy K1 and K2.

## EXPERIMENTAL PROCEDURE AND OBTAINED RESULTS

The anisotropy of sheets from stainless steel was studied. This steel consists of the single phase with FCC crystalline lattice. Plain samples from sheets in the annealed condition and after cold rolling up to deformation degrees $\varepsilon$ of 20, 30 and 40% were tested by tension along both RD and TD. Measured mechanical properties of samples (yield stress $\sigma_{0.2}$, tension strength $\sigma_B$, maximal elongation $\delta$) and coefficients $\Psi$ of their anisotropy are presented in upper lines of Table 1. Annealed and cold-rolled sheets differ in structure condition, and this circumstance exerts decisive influence on the level of their mechanical properties, but as for anisotropy of these properties, it is connected with texture features of material.

TABLE 1. Mechanical and texture anisotropy of sheets from stainless steel

| | annealed cond. | | $\varepsilon = 20\%$ | | $\varepsilon = 30\%$ | | $\varepsilon = 40\%$ | |
| --- | --- | --- | --- | --- | --- | --- | --- | --- |
| | RD | TD | RD | TD | RD | TD | RD | TD |
| $\sigma_{0.2}$,MPa | 210 | 208 | 630 | 750 | 800 | 843 | 880 | 960 |
| $\Psi_{\sigma_{0.2}}$ | 1.0 | | 0.8 | | 0.95 | | 0.95 | |
| $\sigma_B$, MPa | 595 | 565 | 725 | 800 | 840 | 885 | 907 | 1020 |
| $\Psi_{\sigma_B}$ | 1.05 | | 0.9 | | 0.95 | | 0.9 | |
| $\delta$, % | 55 | 45 | 12.5 | 21 | 9 | 14 | 4 | 7 |
| $\Psi_\delta$ | 1.2 | | 0.6 | | 0.65 | | 0.6 | |
| S1 | 0.83 | 0.59 | 0.48 | 0.56 | 0.63 | 1.34 | 0.63 | 0.83 |
| K1 | 1.41 | | 0.86 | | 0.47 | | 0.76 | |
| S2 | 0.3 | 10 | 0.5 | 30* | 0.15 | 50* | 0.2 | 50* |
| K2 | 0.03 | | 0.015* | | 0.003* | | 0.004* | |

* Values were determined with a poor precision.

IPF were constructed by the X-ray method [3], which includes diffractometric measurement of tilting curves for reflections from crystallographic planes {001}, (011} and {111} by quick rotation of the sample about the axis, perpendicular to the studied surface. Typical IPF for RD- and TD-sections of sheet samples before (a,c) and after (b,d) tensile testing are shown in Fig. 1.

IPF for RD-section of the initial annealed sheet (Fig. 1-a) contains two texture maxima, corresponding to grains, in which RD coincides with axes <001> or <111>, and the whole spectrum of intermediate components is present as well. Tensile testing causes redistribution of pole density to the component <111> (Fig. 1-b), testifying that under tension this component is more stable, than <001>. The initial texture of TD-section (Fig. 1-c) contains only one component with axis <011> in parallel with TD, and under tensile testing along TD grain reorientation results in growing of the component <111> (Fig. 1-d), as under testing along RD. These changes in IPF for RD-section agree with the known regularities of texture development in FCC metals [1].

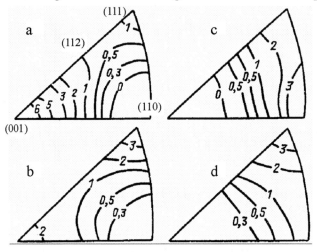

**FIGURE 1.** Typical IPF for RD- (a,b) and TD-sections (c,d) of sheet samples before (a,c) and after (b,d) tensile testing up to fracture.

The capacity of sample for easy deformation is estimated by parameter S1, which characterizes scattering of the most stable texture component <111>: S1 = p(C)/p(111) for RD-section or p(C')/p(111) for TD-section, where points C and C' are distanced from pole (111) by 25° and situated at different sides of the stereographic triangle, that is (111-001) and (111-011), respectively. The deformation incompatibility is estimated by parameter S2, characterizing a relationship of components, stable under tensile deformation: S2 = p(111)/p(001). Under deformation grains of different components impede one another most of all, when their ratio S2 is close to 1, and by any deviation of S2 from 1 the deformation incompatibility weakens. The used texture parameters and coefficients of texture anisotropy for all studied samples are presented in lower lines of Table 1. When due to smallness of p(001) the calculation accuracy of values S2(TD) and K2 was too low, these values were indicated with sign *.

Figure 2-a shows changes of $\Psi_\delta$, K1 and S2(RD) by passing from one sample to another. Mutual nearness of values $\Psi_\delta$ and K1 testifies, that the plasticity resource of samples is controlled by texture parameter S1 and the anisotropy of relative elongation as a first approximation is determined by the texture anisotropy. When taking into account, that under TD-tension the deformation incompatibility is negligibly small and in this connection the calculation accuracy of K2 is very low, we use parameter S2(RD) by analysis of variations in the mechanical anisotropy. Judging from values S2(RD), the deformation incompatibility is maximal in the sheet rolled by 20% with S2=0.5 and minimal - in the sheet rolled by 30% with S2=0.15. Therefore, after 20%-

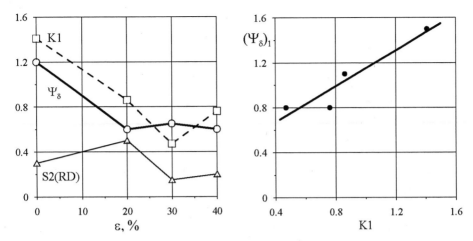

**FIGURE 2.** The effect of texture on the mechanical anisotropy of sheets from stainless steel:
(a) changes of $\Psi_\delta$, K1 and S2(RD) by passing from one sample to another;
(b) diagram of correlation between coefficients of mechanical and texture anisotropy.

rolling both elongation $\delta_{RD}$ and anisotropy $\Psi_\delta$ prove to be understated and after 30%-rolling – overstated relative to those for the annealed sheet and after 40%-rolling. We can take into account the effect of deformation incompatibility by shifting upwards all values $\Psi_\delta$ in proportion to S2(RD). Then value $(\Psi_\delta)_1 = \Psi_\delta + \Delta$, where $\Delta = k \cdot S2(RD)$, characterizes the estimated anisotropy of relative elongation $\delta$ in the absence of deformation incompatibility. In Fig. 2-b the diagram of correlation between the corrected coefficient of mechanical anisotropy $(\Psi_\delta)_1$ and the coefficient of texture anisotropy K1 is constructed by k=1. The correlation has the simplest character and can be used by prediction of properties without complicated calculations.

It is evident, that behavior of $\Psi_\sigma$ submits to the same texture dependence. This result shows efficiency of the texture approach to prediction of plasticity anisotropy as applied to stainless steel.

## REFERENCES

1. G. Wassermann und J. Grewen, "Texturen metallischer Werkstoffe", Springer-Verlag, Berlin/Heidelberg, 1962.
2. Ya.D.Vishnyakov, A.A. Babareko, S.A. Vladimirov and I.V. Egiz, "The Theory of Texture Formation in Metals and Alloys", Publishing House "Nauka", Moscow, 1979 (in Russian).
3. Yu. Perlovich, S. Struev and S. Kapliy, "A new precise X-ray method for building of inverse pole figures" in *Tenth International Conference on Textures of Materials. Clausthal 20-24 Sept. 1993. Abstracts.* TU Clausthal, Deutsche Gesellschaft fur Materialkunde, 1993, p. 137.

# Assessment Of Mold-Design Dependent Textures In CIM-Components By Polarized Light Optical Texture Analysis (PLOTA)

Frank Kern[1], Johannes Rauch[2] and Rainer Gadow[3]

[1, 2, 3] *Institute for Manufacturing Technologies of Ceramic Components and Composites (IMTCCC)- Universität Stuttgart, Allmandring 7b, D-70569, Germany*

**Abstract.** By thermoplastic ceramic injection moulding (CIM) ceramic components of high complexity can be produced in a large number of items at low dimensional tolerances. The cost advantage by the high degree of automation leads to an economical mass-production. The structure of injection-moulded components is determined by the form filling behaviour and viscosity of the feedstock, the machine parameters, the design of the mold and the gate design. With an adapted mold- and gate-design CIM-components without textures are possible. The "Polarized Light Optical Texture analysis" (PLOTA) makes it possible to inspect the components and detect and quantify the textures produced by a new mold. Based on the work of R. Fischer (2004) the PLOTA procedure was improved by including the possibility to measure the inclination angle and thus describe the orientation of the grains in three dimensions. Sampled thin sections of ceramic components are analysed under the polarization microscope and are brought in diagonal position. Pictures are taken with a digital camera. The pictures are converted in the L*a*b*- colour space and the crystals color values a* and b* in the picture are measured. The color values are compared with the values of a quartz wedge, which serves as universal standard. From the received values the inclination angle can be calculated relative to the microscope axis. It is possible to use the received data quantitatively e.g. for the FEM supported simulation of texture-conditioned divergences of mechanical values. Thus the injection molding parameters can be optimized to obtain improved mechanical properties.

**Keywords:** Texture Analysis, Polarized Light, CIM-Components
**PACS:** 78.20.Fm, 81.40.Ef, 83.50.Uv

## INTRODUCTION

Thermoplastic ceramic injection molding (CIM) is characterized by the production of components with net shape dimensions, high surface quality, high number of units produced and low cycle times. These features and a high automation level of the machine equipment employed lead to economic efficiency due to low cost per part (Gadow and Süssmuth [1]).

The component quality is influenced by the viscosity and form filling behavior of the CIM feedstock, the injection parameters, the mold and the gate design, all these parameters need to be coordinated in order to obtain components of high quality. Most commercially available ceramic powders used for CIM process are irregularly and splintery shaped and they differ more or less from the ideal spherical shape. The tendency of the particles to reorient in a position with lowest possible flow resistance, but with a very limited time available for this process, causes flow textures during the form filling at high velocities and injection pressures (Eisbacher [4] and Fischer [5]). The preorientation of particles by texture may be amplified or annihilated during the sintering process.

Most ceramics are anisotropic in their physical and mechanical properties. In non textured microstructures the effects of anisotropy of each individual particle is leveled out by the random arrangement of the particles in the volume observed. This leads to average values of physical and mechanical properties (Randele and Engler [6] and Wessermann and Grewen [7]).

PLOTA (polarized light optical texture analysis) is a new method for qualitative and quantitative flow texture analysis, which combines traditional transmitted light polarization microscopy with modern electronic color analysis methods based on electronic image processing and the L*a*b* - color standard (DIN 6174 [8] and Kipphan [9]). This is a modern succession method for the traditional Fedorow universal stage method (Sarantschina [10]).

The procedure briefly summarized below has been described in detail by Fischer [11]. The color analysis method uses the fact that the majority of ceramic materials are optically birefringent. For the analysis a ceramic thin section with a thickness of approx. 25μm is examined with transmitted light polarization microscopy.

Due to the birefringence characteristical interference colors occur by transillumination. The interference colors depend on the angle between the crystal (grain) orientation and the reference direction. The sample is commonly prepared and placed in such a way that the injection direction is orthogonal to the microscope axis which is the reference direction. The angle dependent interference colors are correlated to the position of the crystal lattices. This fact enables to obtain information about the location and the degree of the texture orientation. The interference colors are detected with a CCD - camera which works in combination with a transmitted light polarization microscope. The red-green-blue (RGB) - image information is converted into standardized L*a*b* color values in an electronic picture processing unit. The standardized L*a*b*- color information are assigned to accurately defined angle positions. The colors of interference observed, are correlated to the thickness of a quartz wedge, used as a reference standard (Gadow and Rauch [12]). The measured colors vectors a* and b* in the samples were compared with this standard. Distinct angle differences in local and global grain orientation signify a broad degree of texture. To describe the orientation of a grain in a thin section two angles are needed. First, the angle of declination measured with the injection direction as a reference. It ranges between 0° and 180°. To obtain the angle of declination the polarizers are crossed and the full-wave (first order) retardation plate is inserted to add path difference. The angle of inclination is measured referring to the optical axis of the microscope. It has a range from 0° up to 90°. To get the angle of inclination the interference color in the special position is compared with a standard table which contains the a* and b* values of a quartz wedge measured with the same microscope and the same conditions of magnification and aperture. This permits to calculate the path difference in the crystal which correlates with the angle of inclination. The textures in the samples are visualized in polar diagrams. This makes it easy to compare the textures in different samples.

## REALIZATION

Two different feedstocks for CIM with maximized solid content were prepared in order to provoke the formation of textures. One contained spheroidal particles of alumina with a primary grain size of 160nm and a solid content of 82 mass-% (sample 1) the other contained spheroidal alumina with a primary grain size of 350nm and a solid content of 85 mass-% (sample 2). The parameters of the CIM-process, maximum filling speed, maximum filling pressure, plastification temperature and mold temperature, were the same for both samples. The components produced were simple test bars (dimension 4.5 x 5.5 x 60 mm³). A mold with a flinching piston was used to avoid free jet injection and to minimize texture effects in the component. The feedstock pushed back the piston while filling the mold. After sintering of the test bars thin section with a thickness of 25μm were prepared. The sections were parallel to the direction of injection. Two perpendicular sections of each bar were prepared. Both samples showed strong textures after sintering. In the next step PLOTA was used to quantify the samples textures (fig 1).

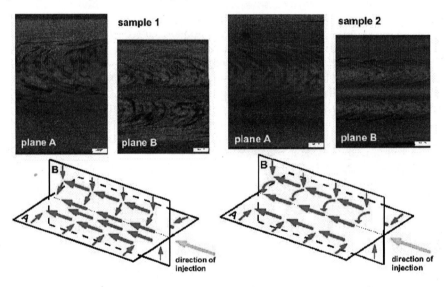

**FIGURE 1.** Microscope images of the samples 1 and 2 under polarized light and red I plate. The drawings under the images gives the orientation of the c-axis (red arrows) obtained by PLOTA.

## RESULTS

Sample 1 showed a strong turbulent texture in the middle of plane A and two turbulent areas in plane B. At the rim of the sample the c-axis of the grains is oriented perpendicular to the direction of injection. The orientation in the turbulent areas was parallel to the direction of injection. The grain orientation to the planes A and B was coplanar in all cases. The transition between both areas with different orientations was well defined.

Sample 2 showed a strong turbulent texture in the middle plane A and two turbulent areas in plane B. At the rim of the sample the c-axis of the crystals were oriented perpendicular to the direction of injection. The orientation of the turbulent areas was parallel to the direction of injection. The grain orientation to the planes A and B was coplanar in all cases. The transition between the areas is not well defined, the orientation of the crystals changed smoothly from a parallel orientation to a perpendicular. This is shown with the circular arrows in fig 1.

## DISCUSSION

Both samples showed heavy textures. The grains are orientated during the CIM-process in a position with lowest possible flow resistance. It was shown that the the two feedstocks with similar rheological properties but different grain size have very different form filling behavior. The formation of textures led to voids and disruptions in the microstructure of the components. It was surprising that spheroidal grains are able to orientate by fluiddynamic processes during the filling of the mold. In order to obtain a deeper understanding of the processes more detailed rheological studies are required as well as FEM-simulations of the mold filling process of feedstocks with submicron size or nanosize fillers (Wenzelburger et al. [13]).

## SUMMARY

The suitability of the optical texture analysis method (PLOTA) as a test method to check the component quality and to analyze component damages is shown. It is furthermore a promising tool of optimization in R & D of CIM-products in mechanical engineering. It was shown that not only milled powders with sharp edges but also almost spheroidal particles are able to orient during the CIM process.

### References

1. Gadow, R., and Süssmuth, G., *Spritzgießen in der technischen Keramik,* Vulkan Verlag, Essen, Germany, 1988.
4. Eisbacher, G.H., *Einführung in die Tektonik,* Ferdinand Enke Verlag, Stuttgart, Germany, 1991.
5. Fischer, R., *Charakterisierung herstellungsbedingter Texturen in spritzgegossenen $Al_2O_3$- Keramikbauteilen,* Diplomarbeit, Institut für Mineralogie und Kristallchemie, Universität Stuttgart, Germany, 1991.
6. Randele, V., and Engler, O., *Introduction to Texture Analysis,* Gordon and Breach Science Publischers, The Netherlands, 2000.
7. Wessermann, G., and Grewen, J., *Texturen metallischer Werkstoffe,* Springer Verlag, Berlin, Germany 1939.
8. DIN 6174: *Farbmetrische Bestimmung von Farbabständen nach der CIELABFormel,* Germany, 1979.
9. Kipphan, H., *Handbuch der Printmedien,* Springer Verlag, Berlin, Heidelberg, New York, Germany, USA, 2000.
10. Sarantschina, G.M., *Die Fedorow- Methode,* VEB Deutscher Verlag der Wissenschaften, Berlin, Germany, 1963.
11. Fischer, R., *Optimierung des Fertigungsprozesses von keramischen Spritzgießbauteilen durch lichtoptische Texturanalyseverfahren,* Ph.D. Thesis, Faculty of mechanical Engineering, University of Stuttgart, Germany, 2004.
12. Gadow, R. and Rauch, J. „*Progress in Polarization Light Optical Texture Analysis (POLTA) of Injection Molded Ceramic Components*" Proceedings of the 10th European Inter-regional Conference on Ceramics, CIEC 10, eds. B. Wilshire, M. R. Bache, Swansea Materials Research Centre, ISBN 978-0-9546104-1-8, 2006, p. 65 - 75
13. Wenzelburger, M., Fischer, R., and Gadow R., *Texture Analysis and Finite Element Modeling of Operationel Stresses in Ceramic Injection Molding Components for High-Pressure Pumps,* Int. J. Appl. Ceram. Technol., 2 [4] 278-284, 2005.

# Accurate hardening modeling as basis for the realistic simulation of sheet forming processes with complex strain-path changes

Vladislav Levkovitch, Bob Svendsen

*Deprtment of Mechanical Engineering, University of Dortmund,*
*D-44227 Dortmund, Germany*
*e-mail: Vladislav.Levkovitch@udo.edu*
*e-mail: Bob.Svendsen@udo.edu*

**Abstract.** Sheet metal forming involves large strains and severe strain-path changes. Large plastic strains lead in many metals to the development of persistent dislocation structures resulting in strong flow anisotropy. This induced anisotropic behavior manifests itself in the case of a strain path change through very different stress-strain responses depending on the type of the strain-path change. While many metals exhibit a drop of the yield stress (Bauschinger effect) after a load reversal, some metals show an increase of the yield stress after an orthogonal strain-path change (so-called cross hardening). To model the Bauschinger effect, kinematic hardening has been successfully used for years. However, the usage of the kinematic hardening leads automatically to a drop of the yield stress after an orthogonal strain-path change contradicting tests exhibiting the cross hardening effect. Another effect, not accounted for in the classical elasto-plasticity, is the difference between the tensile and compressive strength, exhibited e.g. by some steel materials. In this work we present a phenomenological material model whose structure is motivated by polycrystalline modeling that takes into account the evolution of polarized dislocation structures on the grain level – the main cause of the induced flow anisotropy on the macroscopic level. The model considers besides the movement of the yield surface and its proportional expansion, as it is the case in conventional plasticity, also the changes of the yield surface shape (distortional hardening) and accounts for the pressure dependence of the flow stress. All these additional attributes turn out to be essential to model the stress-strain response of dual phase high strength steels subjected to non-proportional loading.

**Keywords:** induced flow anisotropy, distortional hardening, cross hardening, strain-path changes, pressure dependent plasticity, sheet forming.

## INTRODUCTION

Metal forming processes involve large plastic strains and severe strain path changes. Large plastic strains lead in many metals to strong flow anisotropy. This induced anisotropic behavior manifests itself in the case of a strain path change by very different stress-strain responses depending on the type of the strain path change [1]. To describe two-stage strain path changes, Schmitt et al. introduced in [2] the scalar parameter

$$\theta = N_1 \cdot N_2 \qquad (1)$$

CP907, *10th ESAFORM Conference on Material Forming*, edited by E. Cueto and F. Chinesta
© 2007 American Institute of Physics 978-0-7354-0414-4/07/$23.00

Here, $N_1$ and $N_2$ are the strain-rate direction before and after the strain-path change, respectively. For example, if $\theta$ equals -1, we have a load reversal, and if $\theta$ is 0, the strain-path change is referred to as orthogonal.

While many metals exhibit a drop of the yield stress (Bauschinger effect) after a load reversal, some metals show an increase of the yield stress after an orthogonal strain path change (so-called cross hardening effect). The reason for this induced flow anisotropy is the development of persistent dislocation structures during large deformations [3]. These consist of walls of high dislocation density separating low dislocation density areas. The one side of each wall contains excess dislocations of the same sign, and the other side such dislocations of the opposite sign. After a load reversal, plastic deformation takes place due to the slip on the same slip systems but in opposite direction. Excess dislocations, since they repel each other, facilitate this slip, resulting in the Bauschinger effect. After an orthogonal strain path change, new slip systems are activated and the existing dislocation walls act as obstacles, resulting in the cross hardening effect.

To model the Bauschinger effect, the concept of kinematic hardening has been successfully used for years. However, the presence of kinematic hardening results in a drop of the yield stress after an orthogonal strain path change, which contradicts tests results on materials exhibiting the cross hardening effect. Accordingly, the combined isotropic-kinematic hardening ansatz is insufficient to describe the constitutive behaviour of materials, exhibiting both the Bauschinger and the cross hardening effects, and has to be extended. Another effect, not accounted for in the classical elasto-plasticity, is the difference between the tensile and compressive strength (strength differential effect), exhibited e.g. by some steel materials [4].

In this work we analyze the mechanical response of a dual phase high strength sheet steel in one- and two-stage loading processes. It turns out that the material exhibits the strength differential effect and that the deformation induced flow anisotropy is to complex to be modeled by the combined hardening ansatz. To describe the constitutive response of the investigated material, we extend the isotropic-kinematic-distortional hardening model, presented at the ESAFORM 2006 [5], by incorporating the first invariant of the stress tensor into the yield condition.

## MODEL DESCRIPTION

In [5] we presented a macroscopic material model whose structure is motivated by polycrystalline modelling [6], that takes into account the evolution of polarized dislocation structures on the grain level, representing the main cause of the induced flow anisotropy on the macroscopic level. Besides a shift of the yield surface and its proportional expansion as in the case of conventional plasticity, the presented model also accounts for the changes of the yield surface shape (distortional hardening). To be able to describe the pressure dependent yielding behavior, exhibited e.g. by some steel materials, we incorporate the trace of the stress tensor into the yield function [4]

$$\phi = \sigma_e + a\,tr(\boldsymbol{T}) - Y \qquad (2)$$

359

where $Y$ is the yield stress, whose evolution is determined by Voce isotropic hardening, $a$ is a material parameter, governing the plastic flow pressure sensitivity, and

$$\sigma_e = \sqrt{\frac{3}{2}\Sigma \cdot (\boldsymbol{I}_{dev} + \boldsymbol{H})\Sigma} \tag{3}$$

represents the equivalent stress measure. Here,

$$\Sigma := \boldsymbol{T}' - \boldsymbol{X} \tag{4}$$

is the difference between the deviatoric part of the Cauchy stress $\boldsymbol{T}$ and the back stress $\boldsymbol{X}$, $\boldsymbol{I}_{dev}$ is the deviatoric part of the fourth-order identity tensor and $\boldsymbol{H}$ the fourth-order tensor, introduced to represent distortion of the yield surface. The plastic strain rate is given by the associative flow rule

$$\boldsymbol{D}_p = \dot{\lambda}\frac{\partial \phi}{\partial \boldsymbol{T}'}. \tag{5}$$

Note, that taking the derivative of the yield function with respect to the deviatoric stress part insures the plastic incompressibility of the material, which is a good approximation also for steel materials exhibiting the strength differential effect [4].

Ignoring any texture effects for simplicity, the Jaumann rate of the Cauchy stress is given by the isotropic hypo-elastic relation

$$\overset{\circ}{\boldsymbol{T}} = 2\mu\big(\boldsymbol{D} - \boldsymbol{D}_p\big) + \lambda tr\big(\boldsymbol{D} - \boldsymbol{D}_p\big)\boldsymbol{I} \tag{6}$$

Further, the evolution of the kinematic hardening is given by the Armstrong-Frederick relation

$$\overset{\circ}{\boldsymbol{X}} = C_X\left(\sqrt{\frac{2}{3}}X_{Sat}\boldsymbol{N} - \boldsymbol{X}\right)\dot{\lambda} \tag{7}$$

with $\boldsymbol{N}$ being the direction of the plastic strain-rate. The fourth-order tensor $\boldsymbol{H}$, representing distortional hardening, is described by the following evolution equation

$$\overset{\circ}{\boldsymbol{H}} = C_D(D_{Sat} - H_D)(\boldsymbol{N} \otimes \boldsymbol{N})\dot{\lambda} + C_L[L_{Sat}(\boldsymbol{I}_{dev} - \boldsymbol{N} \otimes \boldsymbol{N}) - \boldsymbol{H}_L]\dot{\lambda} \tag{8}$$

Here, $H_D$ is the projection of $\boldsymbol{H}$ onto the direction of the plastic strain-rate

$$H_D = \boldsymbol{H} \cdot (\boldsymbol{N} \otimes \boldsymbol{N}) \tag{9}$$

Further, $H_L$ is the part of $H$ orthogonal to $N$

$$H_L = H - H_D(N \otimes N) \tag{10}$$

In addition, $D_{Sat}$ and $C_D$ represent the saturation value and saturation rate parameters, respectively, for $H_D$, while $L_{Sat}$ and $C_L$, respectively, represent these parameters for the latent part $H_L$.

To understand the behavior of the current model, we consider an initially annealed material state with vanishing $H$ and assume $L_{Sat} < 0$ as well as for simplicity $D_{Sat} = 0$. If the material is subjected to proportional loading in the direction $N_1$, the directional part $H_D$ does not evolve since $D_{Sat} = 0$. Consequently, only $H_L$ evolves, saturating to the value $L_{Sat}(I_{dev} - N_1 \otimes N_1)$, which is orthogonal to $N_1$. Such an evolution of $H$ does not influence the yield strength of the material in the loading direction $N_1$, while the strength in all directions, being orthogonal to that of the loading direction, increases. After an orthogonal strain-path change with re-loading direction $N_2$, the directional part as the projection of $H$ onto $N_2$ takes on a negative value and saturates toward zero during re-loading due to $D_{Sat} = 0$. This results in shrinkage of the yield surface in the direction of re-loading. On the other hand, the yield surface expands in direction $N_1$ since $N_1$ is now orthogonal to the current loading direction. This behavior corresponds qualitatively to that obtained by Peeters et al. in [6], using a microstructural model that accounts on the grain level for the evolution of polarized dislocation sheets.

With the presented model it is possible to describe both the Bauschinger and the cross hardening effects, simultaneously. The model was implemented into commercial FE code ABAQUS via UMAT interface.

## MODEL APPLICATION

To demonstrate the capability of the presented model to describe the complex pressure dependent hardening behavior under non-proportional loading, we use it to simulate the constitutive behavior of a dual phase high strength sheet steel. (Due to commercial dependency, the experimental results can be given only in scaled format.)

Figure 1 left shows clearly that the investigated material exhibits the difference between the tensile and compressive yield strength. Similar behavior was reported in [4] for bulk high strength steels. Tension and compression experiments in rolling (RD) and transverse (TD) directions after 10% pre-tension in RD (Fig. 1 right) shows that the material exhibits both the Bauschinger and the cross hardening effect. Accordingly, such a behavior cannot be described by classical pressure independent combined hardening models.

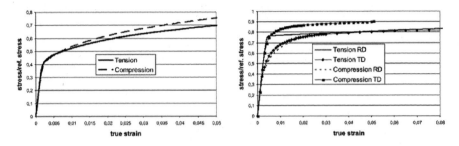

**FIGURE 1.** Experimental results. Left: tensile and compressive yield curves in RD on the as-received material. Right: tensile and compressive yield curves after 10% tensile pre-strain in RD.

In contrast, applying the model, described in the previous section, it was possible to simulate with a good accuracy the whole set of experimental data with only one set of material parameters (Fig. 2). Accordingly, the mixture of isotropic, kinematic and distortional hardening in combination with the yield function, depending on the trace of the stress tensor, seems to be an adequate constitutive ansatz to model the pressure dependent hardening behavior of dual phase high strength steels. (Due to commercial dependency, it is not possible to give the material parameters.)

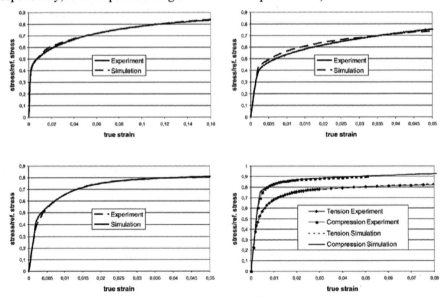

**FIGURE 2.** Comparison of experimental and numerical results. Upper left: uniaxial tension in RD. Upper right: uniaxial compression in RD. Bottom left: uniaxial compression in RD after 10% tensile pre-strain in RD. Bottom right: uniaxial tension and compression in TD, respectively, after 10% tensile pre-strain in RD.

The accurate modeling of the deformation induced flow anisotropy is crucial for the spring back simulation of processes with complex strain path changes. The application of the model to the simulation of the spring back after strip drawing on material, pre-deformed in different directions, represents work in progress and will be reported on during the conference.

## ACKNOWLEDGMENTS

The author would like to thank ThyssenKrupp Steel AG for providing the experimental data for the dual phase high strength steel investigated in this work.

## REFERENCES

1. E.F. Rauch and S. Thuillier, "Rheological behaviour of mild steel under monotonic loading conditions and cross-loading", *Mater. Sci. Engng.* A164 (1993), pp. 255-259.
2. J.H. Schmitt, E. Aernoudt and B. Baudelet, "Yield loci for polycrystalline metals without texture", *Mater. Sci. Eng.* 75 (1985), pp. 13-20.
3. S. Bouvier, J.L. Alves, M.C. Oliveira and L.F. Menezes, "Modelling of anisotropic work-hardening behaviour of metallic materials subjected to strain path changes" *Comp. Mater. Sci.* 32 (2005), pp. 301-315.
4. W.A. Sptitzig, R.J. Sober and O. Richmond, "The effect of hydrostatic pressure on the deformation behavior of maraging and HY-80 steels and its implication for plasticity theory" *Metal. Trans. A* 7A, (1976), pp. 1703-1710.
5. Levkovitch, V., Svendsen, B., Wang, "Micromechanically motivated phenomenological modeling of induced flow anisotropy and its application to metal forming processes with complex strain path changes", *The 9th International ESAFORM Conference on Material Forming*, Glasgow, UK, April 26-28, 2006.
6. B. Peeters, S.R. Kalidindi, C. Teodosiu, P. Van Houtte and E. Aernoudt, "A theoretical investigation of the influence of dislocation sheets on evolution of yield surfaces in single-phase b.c.c. Polycrystals", *Journal of the Mechanics and Physics of Solids* 50 (2002), pp. 783-807.

# Orthotropic Model For Metallic Sheets In Finite Elasto-Plasticity

## Sanda Cleja-Tigoiu

*Faculty of Mathematics and Computer Science, University of Bucharest*
*str.Academiei 14, 010014 Bucharest, Romania*

**Abstract.** In this paper we develop a model, with internal variables, small elastic strains, large elastic rotations and plastic deformations, within the constitutive framework of finite (multiplicative) elasto-plasticity, in Eulerian description. The orthotropic initial anisotropy for sheets and evolving anisotropy are involved in the model. We pay attention to the full description of the model, not only on the anisotropic yield criteria. The boundary value problems, which describe at every time t the metal forming processes for sheets, lead to solve an appropriate variational inequality, for the velocity field and the plastic multiplier considered to be unknowns of the quasi-static problem.

**Keywords:** orthotropy, large deformation, yield condition, plasticity
**PACS:** 62.20.Fe; 46.35.+z

## INTRODUCTION

In this paper we develop a model to describe the behavior of sheet metals, within the constitutive framework of finite elasto-plasticity, proposed by Cleja-Tigoiu [3], and Cleja-Tigoiu and Soos [4], for crystalline materials. The mathematical description of the elasto-plastic behaviour, as well as the material symmetry concept proposed in [3] and [4], were experimentally motivated by crystal plasticity. We consider a model with internal variable and relaxed configurations, based on the multiplicative decomposition $\mathbf{F} = \mathbf{EP}$, of the deformation gradient $\mathbf{F}$ into its elastic $\mathbf{E}$ and plastic parts $\mathbf{P}$, with small elastic strains $\varepsilon^e$ but large elastic rotations $\mathbf{R}^e$ and plastic deformations, a feature generally accepted in metal forming plasticity. Thus

$$\mathbf{E} = \mathbf{R}^e \mathbf{U}^e, \quad \mathbf{U}^e = \mathbf{I} + \varepsilon^e, \quad |\varepsilon^e| << 1 \tag{1}$$

The complete set of the constitutive representation, first described with respect to a local set of relaxed configuration, see [3], [4], consists of: (1) the *elastic type constitutive equation*, (2) *current yield condition*, (3) *evolution equations* for $\mathbf{P}$ and for (4) internal variables. The evolution equations are associated to the current yield condition.

CP907, *10th ESAFORM Conference on Material Forming*, edited by E. Cueto and F. Chinesta
© 2007 American Institute of Physics 978-0-7354-0414-4/07/$23.00

For a general orthotropic model, with non-symmetric Mandel's type stress measure, we developed a constitutive description in [5]. We pay attention to the full description of the model, not only on the anisotropic yield criteria, which describes accurately the material behavior.

## CONSTITUTIVE FRAMEWORK

Here we start with the Eulerian description of the model, subsequently derived from the model with relaxed configurations by the push forward procedure. The initial *othotropic anisotropy* is characterized by the group $g_6 := \{\mathbf{Q} \in Ort \mid \mathbf{Q}\mathbf{n}_i = \mathbf{n}_i, \text{ or } \mathbf{Q}\mathbf{n}_i = -\mathbf{n}_i, \ i = 1,2,3.\}$, where $Ort$ is the set of orthogonal tensors and $\{\mathbf{n}_1, \mathbf{n}_2, \mathbf{n}_3\}$ is the orthonormal basis of the symmetry directions. The anisotropy directions in the actual configuration are given by $\mathbf{m}_i = \mathbf{R}^e \mathbf{n}_i$.

• The *orthotropic linear elastic constitutive equation* can be represented (see [5] and [6]) in a tensorial representation, invariant with respect to $g_6$

$$\mathbf{T} = \mathcal{E}\left(\mathbf{m}_1 \otimes \mathbf{m}_1, \mathbf{m}_2 \otimes \mathbf{m}_2\right)[\overline{\varepsilon^e}], \quad \overline{\varepsilon^e} = \mathbf{R}^e \varepsilon^e (\mathbf{R}^e)^T, \tag{2}$$

$\mathcal{E}$ — fourth order elastic tensor, $\overline{\varepsilon^e}$ — small elastic strain in the actual configuration.

• The yield criteria applicable to metals is *pressure insensitive* and thus only $\mathbf{S} := \mathbf{T} - \frac{1}{3}\mathbf{I}$— the deviator of Cauchy stress, enters the yield function and it is sensitive to the **sign of the stress**, see [1] and [2]. The yield criteria used in (3) was proposed in [2], and three numerical sequences of the yield surface determined from the reported experimental data, reproduce very well the observed assymetry in appropriate yielding stresses. No internal variables effectively appear in [2].

• The *evolving anisotropy* is characterized (see [3] and [4]) by a set of scalar internal (hardening) variables $\kappa$ and by a tensorial field $\mathbf{a}$, respectively, in order to describe the change in shape and the motion in the appropriate stress space of the yield surface.

• The *yield criteria* of the model is proposed directly in the actual configuration:

$$\mathcal{F}(\mathbf{T}, \mathbf{a}, \kappa) := (f_2)^{3/2} - f_3 - \tau_Y^3 = 0, \tag{3}$$

with $\tau_Y = \tau_Y(\kappa)$— *the yield stress in shear*, and with $f_2$ and $f_3$ homogeneous functions of degree 2 and 3, respectively, invariant relative to $g_6$. The representation for $f_2$ is given by

$$f_2 := f_2(\mathbf{T}, \mathbf{a}, \kappa) = C_1 \overline{\mathbf{S}} \cdot \overline{\mathbf{S}} + C_2 \overline{\mathbf{S}}^2 \cdot (\mathbf{m}_1 \otimes \mathbf{m}_1) + C_3 \overline{\mathbf{S}}^2 \cdot (\mathbf{m}_2 \otimes \mathbf{m}_2) +$$
$$+ C_4 (\overline{\mathbf{S}}\mathbf{m}_1 \cdot \mathbf{m}_1)^2 + C_5 (\overline{\mathbf{S}}\mathbf{m}_2 \cdot \mathbf{m}_2)^2 + C_6 (\overline{\mathbf{S}}\mathbf{m}_1 \cdot \mathbf{m}_1)(\overline{\mathbf{S}}\mathbf{m}_2 \cdot \mathbf{m}_2), \quad \overline{\mathbf{S}} = \frac{\mathbf{S}}{\rho} - \mathbf{a}. \tag{4}$$

The constitutive functions $f_3$ is not dependent on tensorial (kinematic) variable

$$f_3 := B_1(\mathbf{S} \cdot \mathbf{S})(\mathbf{S}\mathbf{n}_1 \cdot \mathbf{n}_1) + B_2(\mathbf{S} \cdot \mathbf{S})(\mathbf{S}\mathbf{n}_2 \cdot \mathbf{n}_2) + [\mathbf{S}^2 \cdot (\mathbf{n}_1 \otimes \mathbf{n}_1)](B_3(\mathbf{S}\mathbf{n}_1 \cdot \mathbf{n}_1) +$$
$$+ B_4(\mathbf{S}\mathbf{n}_2 \cdot \mathbf{n}_2)) + [\mathbf{S}^2 \cdot (\mathbf{n}_2 \otimes \mathbf{n}_2)](B_5(\mathbf{S}\mathbf{n}_1 \cdot \mathbf{n}_1) + B_6(\mathbf{S}\mathbf{n}_2 \cdot \mathbf{n}_2)) + B_9(\mathbf{S}\mathbf{n}_1 \cdot \mathbf{n}_1)^3 + \tag{5}$$
$$+ (\mathbf{S}\mathbf{n}_1 \cdot \mathbf{n}_1)(\mathbf{S} \cdot (\mathbf{n}_2 \otimes \mathbf{n}_2))[B_7(\mathbf{S}\mathbf{n}_1 \cdot \mathbf{n}_1) + B_8(\mathbf{S}\mathbf{n}_2 \cdot \mathbf{n}_2)] + B_{10}(\mathbf{S}\mathbf{n}_2 \cdot \mathbf{n}_2)^3.$$

The material coefficient functions $C_i, B_j$ are dependent on $\kappa$.

**Remark 1.** *The initial conditions* have to be introduced $\mathbf{P} = \mathbf{I}, \mathbf{a}(t_0) = 0, \kappa(t_0) = 0$. The *initial yield condition* is characterized by $(f_2((\mathbf{T}, 0, 0)))^{3/2} - f_3((\mathbf{T}, 0, 0)) - \tau_Y(0)^3 = 0$, with $f_2(\mathbf{T}, 0, 0)$ supposed to be positive for all $\mathbf{T}$. Thus $f_2$ has locally positive values along a deformation process starting from the initial values. The failure of the material is produced when $f_2(\mathbf{T}, \mathbf{a}, \kappa) = 0$.

• The *rate of plastic deformation* is given by the *flow rule associated* to the yield condition:

$$
\begin{aligned}
\mathbf{R}^e \{\dot{\mathbf{P}}\mathbf{P}^{-1}\}^s (\mathbf{R}^e)^T &= \mu \mathbf{N}^p(\mathbf{T}, \mathbf{a}, \kappa), \\
\mathbf{N}^p = \partial_{\mathbf{T}} \mathcal{F}(\mathbf{T}, \mathbf{a}, \kappa) &\equiv \frac{3}{2} \sqrt{f_2} \partial_{\mathbf{T}} f_2(\mathbf{T}, \mathbf{a}, \kappa) - \partial_{\mathbf{T}} f_3(\mathbf{T}, \mathbf{a}, \kappa)
\end{aligned}
\tag{6}
$$

The **plastic spin** is accepted to be zero, $\mathbf{W}^p \equiv \{\dot{\mathbf{P}}\mathbf{P}^{-1}\}^a = \mu \Omega^p = 0$, in order to have $\mathbf{R}^e \equiv \mathbf{I}$ for the *without spin motion* of the body, $\mathbf{W} = \{\dot{\mathbf{F}}(\mathbf{F})^{-1}\}^a = 0$, as it follows from $\dot{\mathbf{R}}^e(\mathbf{R}^e)^T = \mathbf{W} - \mu \Omega^p(\mathbf{T}, \mathbf{a}, \kappa)$.

**Remark 2.** The *plastic multiplier* is defined through *Khun-Tucker conditions* $\mu \geq 0$, $\mathcal{F} \leq 0$, $\hat{\mu} \mathcal{F} = 0$, and the *consistency condition* $\mu \dot{\mathcal{F}} = 0$.

**Remark 3.** We emphasize that the associated flow rule (6) describes a compressible effect, since *mass density* $\tilde{\rho}$ evolves through $\dot{\tilde{\rho}} + \tilde{\rho} \mu \, \text{tr} \, \mathbf{N}^p(\mathbf{T}, \mathbf{a}, \kappa) = 0$. Hence the *plastic incompressibility* is assured if $\text{tr} \, \mathbf{N}^p = 0$.

• In order to have a good agreement with experimental data, the material parameters which enter the yield functions have been determined, using yield stresses measured in uniaxial tensile tests and the coefficients of plastic anisotropy $r_\theta$ in different directions $\mathbf{e}_\theta$, for *pre-deformed* metallic sheets, see [1], [2]. We describe $r_\theta$ by

$$
r_\theta := \frac{\mathbf{N}^p \mathbf{e}_\perp \cdot \mathbf{e}_\perp}{(\mathbf{N}^p)\mathbf{e}_h \cdot \mathbf{e}_h}, \quad (\mathbf{e}_\theta, \mathbf{e}_\perp, \mathbf{e}_h), \quad \text{orhogonal axes}
\tag{7}
$$

corresponding to *direction* $\mathbf{e}_\theta$, *plane transverse* $(\mathbf{e}_\perp \perp \mathbf{e}_\theta)$ and *short transverse directions*.

• The *scalar hardening variable* $\kappa$ can be characterized for instance, like in [3],

$$
\begin{aligned}
\dot{\kappa}_1 = \mu \eta_1 \, |\mathbf{N}^p \mathbf{m}_1 \cdot \mathbf{m}_1|, \quad \dot{\kappa}_2 = \mu \eta_2 \, |\mathbf{N}^p \mathbf{m}_2 \cdot \mathbf{m}_2|, \\
\dot{\kappa}_3 = \eta_3 \{\mathbf{N}^p \mathbf{m}_1 \cdot \mathbf{N}^p \mathbf{m}_1 - (\mathbf{N}^p \mathbf{m}_1 \cdot \mathbf{m}_1)^2\}^{1/2}, \quad \eta_i \quad \text{material constants.}
\end{aligned}
\tag{8}
$$

• The evolution equation for the *tensorial variable* $\mathbf{a}$ is defined by an appropriate *kinematic hardening rule*, like in [4].

## RATE BOUNDARY VALUE PROBLEM

Our goal is to formulate an appropriate *variational inequality*, related to the rate quasi-static boundary value problem. At every time t, *the velocity field*, $\mathbf{v} \in \mathcal{V} - admissible vector fields*, as well as the *plastic multiplier* $\beta$ are the unknowns of the quasi-static problem, for the stress and deformation state, as well as the extend of the plastically deformed zone, supposed to be known. Let us consider that the body occupies at the moment t the domain $\Omega_t$ and $\Omega_t^p = \{\mathbf{x} \in \Omega_t \mid \mathcal{F}(\mathbf{T}(\mathbf{x},t), \mathbf{a}(\mathbf{x},t), \kappa(\mathbf{x},t)) = 0\}$ is a current plastic domain.

**Theorem.** *Find* $\mathbf{U} = (\mathbf{v}, \beta) \in \widetilde{\mathcal{K}}$, *solution of the variational inequality*

$$a[\mathbf{U}, \mathbf{V} - \mathbf{U}] \geq f[\mathbf{V} - \mathbf{U}] \quad \forall \mathbf{V} \in \widetilde{\mathcal{K}} \tag{9}$$

$a[\cdot, \cdot]$ *is the bilinear and symmetric form*

$$a[\mathbf{V}, \mathbf{W}] := K[\mathbf{v}, \mathbf{w}] - B[\mu, \mathbf{w}] - B[\gamma, \mathbf{v}] + A[\mu, \gamma] \tag{10}$$

*defined* $\forall \mathbf{V} = (\mathbf{v}, \mu), \mathbf{W} = (\mathbf{w}, \gamma)$ *and* $f[\mathbf{V}] := \int_{S_T} \dot{\mathbf{S}}\mathbf{n} \cdot \mathbf{v} da, \quad S_T \subset \partial\Omega_t.$

Here $\widetilde{\mathcal{K}}$ *is the convex* $\widetilde{\mathcal{K}} := \{(\mathbf{w}, \delta) \mid \mathbf{w} \in \mathcal{V}_{ad}, \quad \delta : \Omega_t \longrightarrow \mathbf{R}_{\geq 0}\} \subset \mathcal{H}_{ab}.$

*The bilinear forms are defined* $\forall (\mathbf{v}, \mu)$ *and* $(\mathbf{w}, \gamma) \in \mathcal{H}_{ab}$ — *an appropriate vector space,* through

$$K[\mathbf{v}, \mathbf{w}] = \int_{\Omega_t} \left(\nabla\mathbf{v}\frac{\mathbf{T}}{\rho} \cdot \nabla\mathbf{w} + \mathcal{E}[\{\nabla\mathbf{v}\}^s] \cdot \{\nabla\mathbf{w}\}^s\right) dx$$
$$A[\mu, \gamma] = \int_{\Omega_t^p} \mu\gamma dx, \quad B[\mu, \mathbf{v}] = \int_{\Omega_t^p} \mu\mathcal{E}[\partial_{\mathbf{T}}\mathcal{F}(\mathbf{T}, \mathbf{a}, \kappa)] \cdot \{\nabla\mathbf{v}\}^s dx \tag{11}$$

We could exemplify the possibilities of the model to describe the behavior of the sheets in different applied loading conditions, in connection with certain industrial applications.

## ACKNOWLEDGEMENTS

The author acknowledges support from the Romanian Ministry of Education and Research through CEEX program (Contract RELANSIM No. 163/20/07/2006).

## REFERENCES

1. C. Banabic, H. Aretz, D.S. Comsa, L. Paraianu, An improved analytical description of orthotropy in metallic sheets. *Int. J. Plasticity*, **21**, 493-512 (2005).

2. O. Cazacu, F. Barlat, A criterion for description of anisotropy and differential effects in pressure-insensitive metals. *Int. J. of Plasticity*, **20** , 2027-2045 (2004).

3. S. Cleja-Țigoiu, Large elasto-plastic deformations of materials with relaxed configurations-I.Constitutive assumptions, II. *Role of the complementary plastic factor, Int. J. Engng. Sci.*, **28**, 171-180, 273-284 (1990).

4. S. Cleja- Țigoiu and E. Soós, Elastoplastic models with relaxed configurations and internal state variables. *Appl. Mech. Rev.*, **43**, 131–151 (1990).

5. S. Cleja-Tigoiu , Orthotropic Σ– Models in finite elasto-plasticity, *Rev. Roum. Math. Pures Appl.*, **45**, 2, 219- 227 (2000).

6. S. Cleja-Tigoiu, Anisotropic Elasto-plastic Model for Large Metal Forming Deformation, *Int.J. of Forming Processes*, (2007).

7. S. Cleja-Tigoiu, Anisotropic and dissipative finite elasto- plastic composite, *Rend. del Seminario Matematico dell'Universita et Politecnico di Torino*, **58**, 1, 69- 82 (2000).

# Theoretical Prediction of the Forming Limit Band

D. Banabic[1a], M. Vos[2], L. Paraianu[1], P. Jurco[3]

[1] Centre of Research for Sheet Metal Forming Technology "CERTETA"
Technical University of Cluj Napoca, 15 C.Daicoviciu, Cluj Napoca, Romania
[2] ELKEM, Hoffsveien 65 B, Majorstuen, NO-0303 Oslo, Norway
[3] FORTECH, 6 C. Brancoveanu, 40467 Cluj Napoca, Romania
[a] Corresponding author: banabic@tcm.utcluj.ro

**Abstract.** Forming Limit Band (FLB) is a very useful tool to improve the sheet metal forming simulation robustness. Until now, the study of the FLB was only experimental. This paper presents the first attempt to model the FLB. The authors have established an original method for predicting the two margins of the limit band. The method was illustrated on the AA6111-T43 aluminum alloy. A good agreement with the experiments has been obtained.

**Keywords:** Formability, Forming Limit Band, Theoretical Model, Aluminum Alloy.
**PACS:** 81.20.Hy; 62.20.Fe.

## INTRODUCTION

Forming Limit Curve (FLC) concept, introduced by Keller [1], represent the boundary between two domains: the region corresponding to strains which do not cause defects (points located below the curve), and the region to dangerous strains (points located above the curve). In industrial applications, the variability of the parameters of the raw material (mechanical characteristics, thickness, surface quality, etc.) [2] is a source of uncertainty in relation with the position of the FLC. At present, the engineers take into account the variability in a purely empirical manner. We notice that the aim of obtaining a full robustness imposes a better evaluation of the incertitude strip affecting the position of the FLC. Taking into account this fact, several researchers [3, 4] have proposed a more general concept, namely the Forming Limit Band (FLB) as a region covering the entire dispersion of the FLC. Until now, the study of the dispersion was only experimental. The present paper intends to perform the transition from the experiment to the theoretical approach.

## MODELLING OF THE FORMING LIMIT BAND

The authors proposed an original method for FLB modeling. The stages of this method are as follows:
1. Select the theoretical model for the FLC prediction

CP907, 10th ESAFORM Conference on Material Forming, edited by E. Cueto and F. Chinesta
© 2007 American Institute of Physics 978-0-7354-0414-4/07/$23.00

2. Define the input parameters in the FLC modeling program (the mechanical parameters, the process parameters, the calibration parameters of the modeling program etc).
3. Determine the mechanical and process parameters (experimentally) and set the calibration parameters of the program
4. Analyze the scattering of the input parameters
5. Analyze the sensitivity of the FLC to the variation of the input parameters
6. Select the input parameters which significantly influence the FLC
7. Define the variation domain of the significant input parameters
8. Define the way in which the significant parameters influence the FLC
9. Define the set of input parameters for the upper and lower FLC predictions
10. Predict the upper and lower FLCs and, implicitly, the FLB (the area between the two)
11. Determine the FLB experimentally
12. Compare the predicted FLB with the experimental one
13. Correct the model by selected more input parameters as significant ones
14. Validate the FLB theoretical model.

In the following we will exemplify the method on the AA 6111-T43 aluminum alloy. To simplify the presentation, we will group these stages in the following manner: Theoretical model (steps1 and 2); Material characterization (steps 3 and 4); Sensitivity analysis (steps 5-8); FLB prediction (steps 9 and10); Validation of the model (steps 11-14).

## Theoretical model

As it is well known [5], there is a large number of theoretical models developed for predicting the FLC. Among them, the most frequently used is the model proposed by Marciniak and Kuczynski (M-K theory) [6]. The strain-localisation model proposed by Marciniak is based on the assumption that the sheet metal has a slightly different thickness in some regions. In the presented model the BBC2003 plasticity criterion, developed by the authors [7], and Hollomon and Voce strain hardening laws were used. Details concerning the solving method of the model can be found in paper [8]. Based on this model, the authors have developed the FORM-CERT [9] program, which was used in the research presented here for the calculus of limit strains. The input data in this program are [9]:

•  mechanical parameters: the uniaxial yield stresses in the rolling, diagonal and transversal directions, respectively, $\sigma_0$, $\sigma_{45}$, $\sigma_{90}$; the biaxial yield stress, $\sigma_b$; anisotropy coefficients in the rolling, diagonal and transversal directions, respectively, $r_0$, $r_{45}$, $r_{90}$; the biaxial anisotropy coefficient, $r_b$; the strain hardening coefficient, n; the coefficient K (in the Hollomon law).
•  process parameter: the non-homogeneity coefficient, $f_0$.
•  calibration parameter: the increment in the numerical procedure.

# Material characterization

An aluminum 6111-T43 alloy of 0,95 mm thickness was used. The testing and analysis was done at the Alcoa Technical Center in Pittsburgh, after approximately 5 months natural ageing.

The uniaxial tensile testing has been used to determine the uniaxial yield stresses, anisotropy coefficients and strain-hardening law coefficients. A hydraulic bulge test [10, 11] was used to determine the biaxial yield stress. The hydraulic bulge test allows measurements of the hardening behavior up to strains of about twice those achieved in uniaxial tension. The compression disks test has been used to determine the biaxial anisotropy coefficient [10,11]. The material data of the tested material determined are summarized in the Table 1.

TABLE 1. Uniaxial tension test data

| Dir. | YS | UTS | % Elong | | r | Hollomon | | | Voce | | |
|------|-----|------|------|------|-------|-------|------------|-----------|-----------|-------|
| | MPa | MPa | Unif. | Tot. | | n | K MPa | A MPa | B MPa | C |
| 0° | 140.1 | 279.2 | 22.0 | 24.9 | 0.884 | 0.272 | 533.31 | 372.25 | 230.36 | 9.901 |
| 45° | 133.9 | 272.1 | 22.8 | 29.0 | 0.708 | 0.272 | 516.11 | 368.11 | 228.22 | 9.158 |
| 90° | 131.4 | 269.8 | 26.4 | 26.4 | 0.720 | 0.276 | 512.81 | 367.63 | 231.11 | 8.950 |

Based on these results, both the average values and the standard deviation of the mechanical parameters have been determined. Table 2 presents these variations for the mechanical parameters which have a significant influence on the FLC. By using the average values and the standard deviations the +/-3 Sigma values have been calculated for each parameter. These values are presented in the last two columns of Table 2. These values will serve as input data for the determination of the Lower (LFLC) and Upper (UFLC) Forming Limit Curves.

TABLE 2. Normal distribution of $\sigma_0$, $\sigma_{90}$, $\sigma_b$ and n.

| Parameter | Av. value | Standard dev. | Base case(- 3-Sigma) | Base case(+ 3-Sigma) |
|-----------|-----------|---------------|----------------------|----------------------|
| $\sigma_0$ | 140,1 MPa | +/- 2% | 137,3 MPa | 142,9 MPa |
| $\sigma_{90}$ | 131,4 MPa | +/- 6% | 123,99 MPa | 139,81 MPa |
| $\sigma_b$ | 140,0 MPa | +/- 5% | 133,0 MPa | 147,0 MPa |
| n | 0,27 | +/- 2% | 0,26 | 0,28 |

# Sensitivity analysis

As it is known from the literature [5], a series of factors influence the FLC. The weight of their influence and the way their growth displaces the FLC can be determined by sensitivity analysis. In this analysis, all input parameters from the FLC calculation program have been taken into consideration: four values of the yield stresses, four values of the anisotropy coefficients, two values of the strain hardening law and a value of the process. Due to the simplicity of modifying input data and the fact that it is user-friendly, FORM-CERT program [9] has been used for this analysis.

# Forming Limit Band prediction

Through the sensitivity analysis the parameters which significantly influence the FLC and the way they influence it were selected. Therefore, the input data for the LFLC and UFLC could be selected (see Table 3).

Due to the fact that the FLC moves upwards $\sigma_0$, $\sigma_{90}$ and n grow and downward as $\sigma_b$ grows, as it can be seen in Table 3, for determining the LFLC the minimum values of the first three parameters and the maximum of the fourth were used and vice versa for the UFLC. The LFLC and UFLC were determined using the values in Table 3 and the basic case of the FLC (so-called average) (see Figure 1) using the average values presented in Table 2. The area between the two extreme limit curves, LFLC and UFLC, determined based on the values of +/-3-Sigma, represents the so-called Forming Limit Band (FLB).

TABLE 3. 3-Sigma limits of $\sigma_0$, $\sigma_{90}$, $\sigma_b$ and n used for FLD prediction.

|  | LFLC | UFLC |
|---|---|---|
| $\sigma_0$ | 137,3 MPa | 142,9 MPa |
| $\sigma_{90}$ | 123,99 MPa | 139,81 MPa |
| $\sigma_b$ | 147 MPa | 133,0 MPa |
| n | 0,26 | 0,28 |

# Validation of the model

The theoretical model of the FLB has been validated through the comparison with experimental results obtained by the authors. Many methods have been developed for the experimental determination of the FLC (see [5, 12, 13, 14]). The Experimental Forming Limit Curve has been determined at the ALCOA Technical Centre, Pittsburgh, US, using the method proposed by Hecker [14]. A detailed presentation of the experimental method used can be found in paper [11]. According to this method, three types of points can be defined in the rupture area of the sample: safe, necking and fracture affected. The points corresponding to these three situations are shown in Figure 1. A good prediction of the experimental values of the limit strain has been obtained in the right area of the diagram. In the left area the predicted limit curves under-estimate the experimental values. This fact is largely due to the measurement errors of the limit strain close to the plain-strain (minor strain values are very small and consequently affected by errors). It can also be observed that in the plane strain area the UFLC is under the basic case curve. This is due to the computation errors, which are larger in the plane strain area. An improvement of the prediction precision of the model can be achieved by including more parameters in the list of significant parameters (such as the anisotropy coefficients, the thickness etc.). Knowing the fact that the variation of the anisotropy coefficients mainly affects the right area of the FLD one can expect that their inclusion will improve the precision of the prediction in this area.

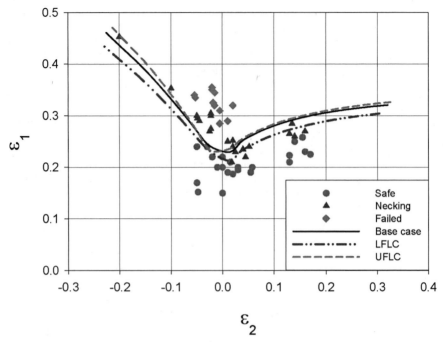

**FIGURE 1.** Forming Limit Band prediction as the area between LFLC and UFLC.

## CONCLUSIONS

To improve the prediction of production robustness, variability and noise have to be incorporated into the simulation. In this purpose, the new concept of Forming Limit Band is a very useful tool. This paper presents the first attempt to model the FLB. The authors have established an original method for predicting the two margins of the limit band. The method was illustrated on the AA6111-T43 aluminum alloy. A good agreement with the experiments has been obtained. However, some deviations could be noticed. In order to improve the accuracy of the theoretical model further research is necessary.

## ACKNOWLEDGMENTS

The authors would like to thank Dr. F. Barlat for his very valuable advices and Dr. J. Brem for his assistance during the experiments at ALCOA Technical Centre.

# REFERENCES

1. S.P. Keeler and W.A. Backofen, *Trans. of the A.S.M.* **56**, 25-48 (1963).
2. B. Carleer and M. Sigvant, Process Scatter with Respect to Material Scatter. In: *New Developments in Sheet Metal Forming,* edited by M. Liewald, Institute for Metal Forming Technology, University of Stuttgart, 2006, pp. 225-239.
3. K. Janssens, F. Lambert, S. Vanrostenberghe and M. Vermeulen, *J. of Materials Proc. Techn.* **112** 174-184 (2001).
4. M. Strano and B.M. Colosimo, *International Journal of Machine Tools & Manufacture* **46**, 673-682 (2006).
5. D. Banabic, H.J. Bunge, K. Pöhlandt and A.E. Tekkaya, *Formability of Metallic Materials,* edited by D. Banabic, Berlin: Springer Verlag, 2000.
6. Z. Marciniak and K. Kuczynski, *Int. J. Mech. Sci.* **9** 609-620 (1968).
7. D. Banabic, H. Aretz, D.S. Comsa and L. Paraianu, *Int. J.Plasticity* **21** 493–512 (2005).
8. D. Banabic, D.S. Comsa, P. Jurco, G. Cosovici, L. Paraianu and D. Julean, *J. of Materials Proc. Techn.* **157-158** 23-27 (2004).
9. P. Jurco and D. Banabic, "A user-friendly program for calculating Forming Limit Diagrams" in *Proc. of the 8th ESAFORM Conference on Material Forming* edited by D. Banabic, Cluj Napoca, 2005, pp.423-427.
10. M. Vos, D. Banabic, P. Jurco, J. Brem and F. Barlat, "Forming limit prediction using BBC2003 yield criterion for aluminium automotive alloy" in *Proc. of the IDDRG Conference,* edited by A. Barata da Rocha, Porto, 2006, pp.51-58.
11. M. Vos, "Modern material laws used in the simulation of forming of aluminium alloys", PhD Thesis, Technical University of Cluj Napoca, 2006.
12. W. Hotz, "European efforts in standardization of FLC", in: *Numerical and experimental methods in prediction of forming limits in sheet forming and tube hydroforming processes* edited by P. Hora, ETH Zürich, Zürich 2006, pp. 24-25.
13. D. Banabic, F. Barlat, O. Cazacu and T. Kuwabara, "Formability and Anisotropy", in *ESAFORM-10 Years,* edited by F. Chinesta, Berlin Heidelberg: Springer Verlag, (in press).
14. S.S. Hecker, *Sheet Metal Industrie* **52** 671-676 (1975).

# Consistent Parameters for Plastic Anisotropy of Sheet Metal (Part 1-Uniaxial and Biaxial Tests)

K. Pöhlandt[1], K. Lange[2], D. Banabic[3], J. Schöck[1]

[1]Institut für Statik und Dynamik der Luft -und Raumfahrtkonstruktionen, Universität Stuttgart,
Pfaffenwaldring 27, D-70569, Germany, URL: www.isd.uni-stuttgart.de,
email: poehlandt@isd.uni-stuttgart.de; email: schoeck@isd.uni-stuttgart.de
[2]Institut für Umformtechnik, Universität Stuttgart, Holzgartenstr. 17, D-70174 Stuttgart,Germany
email: lange@ifu.uni-stuttgart.de
[3]Technical University of Cluj-Napoca, C. Daicoviciu nr. 15, 400020 Cluj-Napoca, Romania
URL: www.utcluj.ro, email: banabic@tcm.utcluj.ro

**Abstract.** The anisotropy parameters for sheet metal used hitherto are mainly determined by uniaxial tensile tests. Such tests, however, do not give sufficient information about the yield locus and the forming behaviour in that range where the two principal tensile stresses are of similar magnitude like in stretch forming. The same applies for combined tensile and compressive stress like in deep-drawing. To fill these gaps, new parameters are defined. Their experimental determination is briefly discussed.

The "equibiaxial yield stress" and "equibiaxial anisotropy" which refer to equibiaxial tensile stress can be determined by cross tensile tests. However, these require a special apparatus. Alternatively experiments for obtaining plane strain can be applied for determining the equibiaxial parameters indirectly. This is possible using conventional tensile testing machines. In this case also anisotropy parameters for plane-strain deformation, the "semibiaxial anisotropy" in rolling and transverse direction, can be determined.

**Keywords:** continuum mechanics, plasticity, anisotropy, metal forming.
**PACS:** 62.20.Fe.

## UNIAXIAL STRESS

The most commonly applied measure of plastic anisotropy of sheet metal is the well-known r-value or normal anisotropy

$$r = \frac{\varphi_w}{\varphi_s} \qquad (1)$$

where the suffixes w and s refer to the width and the thickness direction, respectively (the suffix "t" is not used for the thickness because "T" will be used for the transverse direction).

CP907, 10th ESAFORM Conference on Material Forming, edited by E. Cueto and F. Chinesta
© 2007 American Institute of Physics 978-0-7354-0414-4/07/$23.00

By convention, these quantities are determined for an elongation of 20% in tensile tests (for a rigorous treatment the strains in Eq. (1) should be replaced by the strain rates; because of experimental difficulties this is not applied in practice).

Since the r-values vary with the direction in the plane of sheet it is usual to determine r for various directions and to calculate the mean normal anisotropy:

$$r_m = \frac{r_0 + 2r_{45} + r_{90}}{4} \tag{2}$$

As a measure of the variation of r in the plane of sheet, the planar anisotropy is used:

$$\Delta r = \frac{r_0 + r_{90}}{2} - r_{45} \tag{3}$$

For a given value of $\Delta r$, the parameters $r_0$ and $r_{90}$ can be quite different if they only satisfy Eq. (3). Therefore it has been proposed in [1*] to calculate another independent parameter from $r_0$, $r_{90}$ and $r_{45}$, the so-called "orthoplanar anisotropy" as:

$$\delta r = (r_{90} - r_0). \tag{4}$$

For common metals it is often of equal order of magnitude as $\Delta r$ [1].

Whereas $\Delta r$ is a measure of the variation of the yield locus through a rotation of the principal axes by $45^0$, $\delta r$ is a measure of the asymmetry of the yield locus for a given orientation of the principal stresses in rolling and transverse direction.

Since $r_m$ is a measure of the resistance of the sheet metal against a variation of its thickness it has been related to deep-drawability. The quantity $\Delta r$ is related to earing [2]. However, deep-drawing takes places under combined tensile and compressive stress. Deep-drawability cannot be evaluated from results of uniaxial tensile tests only.

The same argument applies to the behaviour of a sheet metal during stretch-forming by biaxial tensile stress.

Using the parameters $r_m$, $\Delta r$ and $\delta r$ all cases of material behaviour can be described in which anisotropic behaviour results from the normal anisotropy. However, the following case is also possible: normal isotropy ($r_m = 1$; $\Delta r = \delta r = 0$) and despite this anisotropy of a kind which cannot be described by the above parameters. This phenomenon can only be understood by considering biaxial states of stress.

In the following text at first only the orientation of the principal stresses parallel to the rolling and transverse directions is considered.

---

* - The references are listed at the end of the second part of this paper.

# PARAMETERS FOR EQUIBIAXIAL STRESS

## Yield Criteria

The plastic behavior of a metal under multiaxial stress is described by a statement about the possible stress tensors during plastic flow, the so-called yield criterion. Any stress tensor is represented by a point in the space defined by the three principal stresses. In this space stress states which enable plastic yield lie on a surface, the so-called "yield locus". For incompressible materials the yield locus is a cylinder which is well-defined by its intersection with the plane $\sigma_3 = 0$.

In case of isotropy the v. Mises criterion applies:

$$(\sigma_1 - \sigma_2)^2 + (\sigma_2 - \sigma_3)^2 + (\sigma_3 - \sigma_1)^2 = 2\sigma_f^2 \tag{5}$$

where $\sigma_f$ is the yield stress under uniaxial load.

For anisotropic materials various "phenomenological" yield criteria have been proposed which can be determined by mechanical parameters. Most commonly used is the quadratic criterion by Hill [3] which (for $\sigma_3 = 0$) can be written as:

$$\sigma_L^2 - \frac{2r_0}{1+r_0}\sigma_L\sigma_T + \frac{r_0(1+r_{90})}{r_{90}(1+r_0)}\sigma_T^2 = \sigma_{f0}^2 \tag{6}$$

The suffixes 0 and 90 refer to the angle to the rolling direction; $\sigma_{f0}$ is the yield stress in rolling direction.

By the associated flow rule of plasticity, the r-values are related to the tangents to the yield locus:

$$\frac{d\sigma_T}{d\sigma_L} = \frac{1+r_0}{r_0}; \sigma_T = 0; \quad \frac{d\sigma_L}{d\sigma_T} = \frac{r_{90}}{1+r_{90}}; \sigma_L = 0 \tag{7}$$

since only parameters for uniaxial stress are used, Eq. (7) is a poor approximation in cases where both principal stress are of similar magnitude. Therefore, in more recent yield criteria the know-ledge of the equibiaxial yield stress is required. This is the stress which has to act equally in both principal directions for enabling plastic flow.

## Equibiaxial Yield Stress

The equibiaxial yield stress $\sigma_{be}$ can differ strongly from the uniaxial one. *The effect of the ratio $\sigma_{be}/\sigma_{f0}$ on plastic flow is illustrated by Figure 1.*

# Equibiaxial Anisotropy

The equibiaxial anisotropy $r_{be}$ can be determined by various methods [4, 5]. When determining the equibiaxial yield stress by *cross tensile tests* it is possible to determine the equibiaxial anisotropy as well by measuring the distortion of a reference circle in the centre of the specimen (Figure 2).

From the strains in longitudinal and transverse direction the equibiaxial anisotropy is obtained:

$$r_{be} = \frac{\varphi_T}{\varphi_L}; \qquad \varphi_L = \ln\frac{d'_L}{d}; \qquad \varphi_T = \ln\frac{d'_T}{d} \tag{8}$$

By the associated flow rule of plasticity $r_{be}$ is related to the tangent on the yield locus curve in the point of equibiaxial stress:

$$\frac{d\sigma_T}{d\sigma_L} = -\frac{\varphi_L}{\varphi_T} = -\frac{1}{r_{be}} \tag{9}$$

The equibiaxial anisotropy is a measure of the asymmetry of the yield locus. If Eq. (7) could be pre-assumed one would obtain

$$r_{be} = \frac{r_0}{r_{90}} \tag{10}$$

The effect of the equibiaxial parameters on metal forming processes can be illustrated by the example of hydraulic bulging whereby the equibiaxial yield stress influences the sheet thickness distribution in radial direction and the equibiaxial anisotropy influences the thickness distribution in circumferential direction.

The higher the equibiaxial stress, the higher is the resistance of the sheet metal against a variation of thickness (similar to a high r-value).

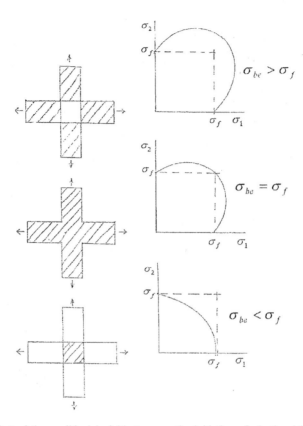

**FIGURE 1.** Effect of the equibiaxial yield stress on the initiation of plastic yield (hatched) under equibiaxial stress ($\sigma_1 = \sigma_2$) in a cross tensile specimen (assuming $\sigma_{f1} = \sigma_{f2} = \sigma_f$ and neglecting notch effects).

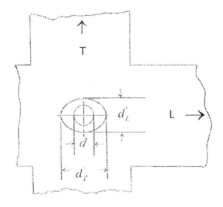

**FIGURE 2.** Effect of equibiaxial anisotropy on the distortion of a circle in the center of a cross tensile specimen under equibiaxial stress

# Other Methods for Determining Equibiaxial Parameters

Since a special apparatus is needed fore cross tensile tests other experiments have been applied.

The *hydraulic bulge test* only enables to determine the equibiaxial parameters averaged over all directions in the plane of sheet. Furthermore, similarly to the cross tensile test the experiment re-quires a special facility.

*Compression of circular specimens* as proposed by Barlat [5] may give good results for the equibiaxial anisotropy, however, friction may cause a large error of the equibiaxial yield stress.

There is still a demand of experiments to determine anisotropy parameters without needing more than a conventional tensile testing machine. In the experiments described in the second part of this paper the stress components of the plane-strain points are determined directly or indirectly.

# Consistent Parameters for Plastic Anisotropy of Sheet Metal (Part 2- Plane-strain and Compression Tests)

K. Pöhlandt[1], K. Lange[2], D. Banabic[3], J. Schöck[1]

[1]*Institut für Statik und Dynamik der Luft -und Raumfahrtkonstruktionen, Universität Stuttgart,*
*Pfaffenwaldring 27, D-70569, Germany*
*URL: www.isd.uni-stuttgart.de, email: poehlandt@isd.uni-stuttgart.de;*
[2]*Institut für Umformtechnik, Universität Stuttgart, Holzgartenstr. 17, D-70174 Stuttgart, Germany*
[3]*Technical University of Cluj-Napoca, C. Daicoviciu 15, 400020 Cluj-Napoca, Romania*

**Abstract.** To include the case of deep-drawing (without blank-holder), states of combined tensile and compressive stress have to be considered whereby it is necessary to define two more anisotropy parameters. They are called "tensile-compressive anisotropy" in rolling and transverse direction. Finally, a new consistent system of "true" anisotropy parameters is presented. They are defined as the difference between the experimentally determined anisotropy parameters and the values which would be obtained in case of isotropy. They all are zero for isotropic materials.

**Keywords:** continuum mechanics, plasticity, anisotropy, metal forming.
**PACS:** 62.20.Fe.

## EXPERIMENTS AND PARAMETERS FOR PLANE-STRAIN DEFORMATION

### Experiments

In the *plane strain compression test* [6, 7] two punches of identical geometry are pressed into the specimen from both sides. If the length of the punches is large compared to the width, because of the hindrance of material flow by friction, plane strain is obtained. Unfortunately the experiment can only be applied on sheet metal of a minimum thickness of about 5 mm. Plane strain can also be achieved by *bending* using specimens of a width large compared to their thickness [8]. *Tensile tests with suppressed lateral contraction* can be obtained by using notched specimens [9]. The methods are not accurate enough to determine the shape of the yield locus. Consequently, their purpose is more academic rather than practical.

### Anisotropy Parameters for Plane Strain

Using the parameters obtained by uniaxial tests and the stress components of points $P_0$ and $P_2$ (see Figure 3), the yield locus can be calculated assuming any yield criterion. The determination of the yield locus by means of Bezier functions has been

CP907, *10th ESAFORM Conference on Material Forming,* edited by E. Cueto and F. Chinesta
© 2007 American Institute of Physics 978-0-7354-0414-4/07/$23.00

described in [10]. This also enables to determine the equibiaxial parameters indirectly. However, in the four stress components of $P_0$ and $P_2$ more information is contained than in the two equibiaxial parameters. It is possible to calculate two more anisotropy parameters which describe the material behaviour for stress states in the vicinity of $P_0$ and $P_2$. Such parameters enable to estimate the formability of sheet metal in processes like bending. After Marciniak and Kuczinski [11] they are also related to the forming limits. Since for an isotropic material the ratio of stress components in $P_0$ is ½, the parameters to be defined are called "semibiaxial anisotropy" for the rolling direction ($P_0$) or transverse direction ($P_2$). They are defined as a measure of the deviation of the tangents to the yield locus from horizontal or vertical orientation for a given "semibiaxial" ratio of the principal stresses (1:2 or 2:1).

In case of isotropy the relation applies at point $P_0$

$$\frac{d\sigma_L}{d\sigma_T} = 0 \tag{11}$$

By analogy to Eq. (7) it is now written

$$\frac{d\sigma_L}{d\sigma_T} = \frac{r_{sL}}{1+r_{sL}} \; : \qquad \frac{\sigma_T}{\sigma_L} = \frac{1}{2} \tag{12}$$

Here the suffix s refers to "semi"; $r_{sL}$ is the semibiaxial anisotropy in longitudinal (= rolling) direction. It is related to the strains by

$$\frac{d\sigma_L}{d\sigma_T} = -\frac{\varphi_T}{\varphi_L} \tag{13}$$

From the law of volume conservation it follows

$$r_{sL} = \frac{\varphi_T}{\varphi_L}; \qquad \frac{\sigma_T}{\sigma_L} = \frac{1}{2} \tag{14}$$

Though this equation is similar to (1), for isotropy the relation applies

$$r_{sL} = 0 \tag{15}$$

The semibiaxial anisotropy in transverse direction is defined by

$$r_{sT} = \frac{\varphi_L}{\varphi_T}; \qquad \frac{\sigma_L}{\sigma_T} = \frac{1}{2} \tag{16}$$

This definition is favorable to the one given by Pöhlandt in [1] for defining consistent parameters.

# TENSILE-COMPRESSIVE STRESS

Two more points of the yield locus have to be considered for the case $\sigma_T = -\sigma_L$ (Figure 3a).

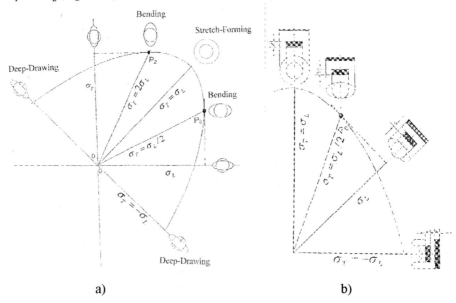

a)                                                      b)

**FIGURE 3. a)** Distortion of a circle of a given equal radius by deformation at distinguished points of the v. Mises yield locus. The figure includes all states of stress which are possible in sheet metal forming without blank-holder ($\sigma_3 = 0$). **b)** Variation of the initial sheet thickness $s_0$ by deformation (schematic). Fig. 3a) has been rotated by $45^0$; because of symmetry of the figure to the line $\sigma_T = \sigma_L$ only the right section is shown. For deep-drawing the ideal case is considered whereby the sheet thickness remains exactly constant. In practice, however, the thickness will vary over the height of deep-drawn cups.

These points are much closer related to deep-drawing than the r-values $r_0$ and $r_{90}$ for uniaxial stress which are usually considered as responsible for the deep-drawing behaviour. However, the word "deep-drawing" should not be used in their denominations explicitly since this might cause associations with $r_0$ and $r_{90}$. They are called "t-c anisotropy" (tensile-compressive) in rolling and transverse direction and defined by

$$r_{tcL} = \varphi_s / \varphi_T; \qquad \sigma_T = -\sigma_L; \qquad \sigma_L > 0 \qquad (17)$$

$$r_{tcT} = \varphi_s / \varphi_L; \qquad \sigma_T = -\sigma_L; \qquad \sigma_L < 0 \qquad (18)$$

These parameters could be determined by shear deformation tests (e. g. by inclined tensile tests with suppressed lateral contraction as proposed in [12]). Their determination would clearly enable a better understanding of deep-drawing.

# CONSISTENT PARAMETERS

The above-given definitions are summarized by

$$r = \begin{cases} \varphi_s / \varphi_T; & 0 > \sigma_T / \sigma_L \geq -1 \\ \varphi_T / \varphi_s; & 0 \leq \sigma_T / \sigma_L < 1 \\ \varphi_T / \varphi_L; & \sigma_T = \sigma_L \\ \varphi_L / \varphi_s; & 1 > \sigma_L / \sigma_T \geq 0 \\ \varphi_s / \varphi_L; & 0 > \sigma_L / \sigma_T \geq -1 \end{cases} \tag{19}$$

These parameters, however, are not yet consistent.

Any parameter called "anisotropy" should be zero for an isotropic material. Therefore a general new definition of parameters is proposed:

$$R = r_{\exp} - r_{iso} = \quad true \quad anisotropy \tag{20}$$

Here $r_{\exp}$ is calculated from (19) using test results, and $r_{iso}$ is calculated from (19) after v. Mises:

$$\sigma_L^2 - \sigma_L \sigma_T + \sigma_T^2 = \sigma_f^2 \tag{21}$$

It follows

$$\frac{d\sigma_T}{d\sigma_L} = \frac{1}{2} + \begin{cases} +|F(\sigma_L)|; & -1 \leq \sigma_T / \sigma_L \leq 1/2 \\ -|F(\sigma_L)|; & 1/2 \leq \sigma_T / \sigma_L \leq 1 \end{cases} \tag{22}$$

$$F(\sigma) = \frac{3\sigma}{4\sqrt{\sigma_f^2 - \frac{3}{4}\sigma^2}} \tag{23}$$

(see Figure 4). Because of the symmetry of Figure 3a) to the line $\sigma_T = \sigma_L$ the relations for the other states of stress are not given explicitly in (22). From Eqs. (13), (19) to (23) the values in Table 1 are obtained. Here the values of the ratio $\sigma_T / \sigma_L$ are computed for the isotropic case. These values have been used to define the distinguished points on the yield locus. Of course, in the anisotropic case these values may differ from those calculated for the isotropic case. This fact is clearly presented by Kuwabara in paper [13]. The anisotropic yield criterion does not predict the plane strain state for $\delta\sigma_T/\delta\sigma_L=0.5$ and $\delta\sigma_T/\delta\sigma_L=2$ and the experimental data obtained by Kuwabara [13] [14] closely agrees with the prediction.

The definition of the true anisotropy can also be applied to states of biaxial stress between the distinguished points of the yield locus, see Figure 5.

# DIAGONAL PRINCIPAL STRESSES

The parameters in Table 1 must be completed by parameters for orientations of principal stresses different from the rolling and transverse direction, in particular under +/- $45^0$. For this case the tensile-compressive, normal and semibiaxial anisotropy are equal for $+45^0$ and $-45^0$; only four parameters are needed and defined by analogy to Table 1.

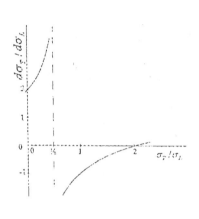 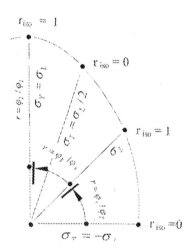

**FIGURE 4.** Illustration to Eqs. (22), (23)   **FIGURE 5.** Section of Fig. 3a) rotated by $45^0$ showing ranges of validity of the various definitions of anisotropy

**Table 1.** Anisotropy Parameters at Distinguished Points of the Yield Locus (L = longitudinal, T = transverse direction)

| $\frac{\sigma_T}{\sigma_L}$ | Name | Symbol | Definition | $r_{iso}$ | R | Forming process | Know-ledge |
|---|---|---|---|---|---|---|---|
| -1 | t-c anisotropy in L | $r_{tcL}$ | $r = \varphi_z / \varphi_T$ | 0 | $\varphi_z / \varphi_T$ | deep-drawing | - |
| 0 | normal anisotropy in L | $r_0$ | $r = \varphi_T / \varphi_z$ | 1 | $\varphi_z / \varphi_T$ -1 | (tensile test) | good |
| 1/2 | semibiaxial anis. in L | $r_{sL}$ | $r = \varphi_T / \varphi_z$ | 0 | $\varphi_T / \varphi_z$ | Bending | poor |
| 1 | equibiaxial anisotropy | $r_{be}$ | $r = \varphi_T / \varphi_L$ | 1 | $\varphi_T / \varphi_z$ -1 | Stretching | poor |
| 2 | semibiaxial anis. in T | $r_{sT}$ | $r = \varphi_L / \varphi_z$ | 0 | $\varphi_L / \varphi_z$ | Bending | poor |
| ∞ | normal anisotropy in T | $r_{90}$ | $r = \varphi_L / \varphi_z$ | 1 | $\varphi_L / \varphi_z$ -1 | (tensile test) | good |
| -1 | t-c anisotropy in T | $r_{tcT}$ | $r = \varphi_z / \varphi_L$ | 0 | $\varphi_z / \varphi_L$ | deep-drawing | - |

# CONCLUSIONS

The paper represents a consistent review of the sheet metal anisotropy coefficients define mode. Until now, these coefficients were determined by uniaxial and biaxial tensile tests. However, these parameters are not sufficient for anisotropy characterization. This is why the authors have proposed the definition of new parameters which would characterize the anisotropic behavior of sheet metals in tension-compression and plane strain states. Experimental tests for determining these parameters are also proposed. A new consistent system of "true" anisotropy parameters is presented. They are defined as the difference between the experimentally determined anisotropy parameters and the values which would be obtained in case of isotropy. They all are zero for isotropic materials. The proposals presented in the paper can represent the basis for a standardisation of the determination of sheet metal anisotropic coefficients.

# REFERENCES

1. K. Pöhlandt, A.E. Tekkaya and J. Schöck, *"Concepts for characterizing Plastic Anisotropy of Sheet Metal"* in: Proc. 24th IDDRG Congr., edited by N. Boudeau, Besancon, France, 2005, pp. 22-24.
2. J. Reissner, H. Mülders and R. Plänker, *Bänder Bleche Rohre* **21** 484-488, (1980).
3. R. Hill, *Proc. Roy. Soc.* **A 193**, 281-297 (1948).
4. K. Pöhlandt, D. Banabic and K. Lange, "Equibiaxial anisotropy coefficient used to describe the plastic behavior of sheet metal" in Proc. 5th Esaform Conf., edited by M. Pietrzyk, Krakow, Poland, 2002, pp. 723-726.
5. F. Barlat, et al. *Int. J. Plasticity* **19**, 1297-1319 (2003).
6. A.B. Watts and H. Ford, *Proc. Inst. Mech. Eng.* **169**, 1141-1150 (1965).
7. N. Becker, Development of Methods for Determining Flow Curves in the Range of High Strains (in German), Berlin Heidelberg:Springer Verlag, 1994.
8. K.A. Malo, O.S. Hoppenstedt and O.G. Lademo, *J. Mater. Process. Tech. Technol.* **80-81,** 538-544 (1998).
9. R.H. Wagoner, *Metall. Trans.* **11A**, 165-175 (1980).
10. K. Pöhlandt, D. Banabic and K. Lange, "Determining Yield Loci of Sheet Metal from Uniaxial and Plane-Strain Deformation Data" in *Proc. 6th Esaform Conference on Material Forming*, edited by V. Brucato, Salerno/Italy, 2003, pp. 223-226.
11. Z. Marciniak and K. Kuczynski, *Int. J. Mech. Sci.*, **9**, 609-620 (1967).
12. W. Müller and K. Pöhlandt, *J. Mater. Process Technol.*, **60**, 643-648 (1996).
13. T. Kuwabara, A Van Bael and E. Iizuka, *Acta Materialia*, **50**, 3717-3729 (2002).
14. T. Kuwabara, *Private communication*, (2007).

# The Forming of AISI 409 sheets for fan blade manufacturing

F. D. Foroni[1], M. A. Menezes[2] & L. A. Moreira Filho[3]

[1, 2, 3] *ITA – Aeronautic Technological Institute, IEM*

*Praça Mal. Eduardo Gomes, 50 – Vila das Acácias – S. J. Campos, Brasil – CEP 1228-900*
*e-mail: fernandoforoni@rocketmail.com[1], miguelm@ita.br[2] & lindolfo@ita.br[3]*

**Abstract.** The necessity of adapting the standardized fan models to conditions of higher temperature has emerged due to the growth of concern referring to the consequences of the gas expelling after the Mont Blanc tunnel accident in Italy and France, where even though, with 100 fans in operation, 41 people died. The objective of this work is to present an alternative to the market standard fans considering a new technology in constructing blades. This new technology introduces the use of the stainless steel AISI 409 due to its good to temperatures of gas exhaust from tunnels in fire situation. The innovation is centered in the process of a deep drawing of metallic sheets in order to keep the ideal aerodynamic superficies for the fan ideal performance. Through the impression of circles on the sheet plane it is shown, experimentally, that, during the pressing process, the more deformed regions on the sheet plane of the blade can not reach the deformation limits of the utilized sheet material.

**Keywords:** Aeronautic profiles, Fan blades, Deep-drawing, Plastic instability.
**PACS:** 81.20.Hy

## 1. INTRODUTION

In a fan project, blades are considered the most important parts, because determinates the fan efficiency, noise level, durability and appearance. The blades were projected considering aeronautic profiles and the duty loads were calculated by an aerodynamic analysis which considers torsion momentum and axial and tangential load on each profile sections [1]. The calculation of stresses in blade during the job, made with the finite element method, will be presented in another paper. The blade was projected by using two deep-drawn sheets, for the pressure and suction side, and a cast iron structure in order to provide the interface for the panels to the fan hub as shown in Fig 1. Due to that, an alternative manufacturing fan process, focused on blades, can be proposed by panel deep-drawing formings and posterior edge welding as considered on this work. This way, the manufacturing process defines limitations on costs and technical aspects.

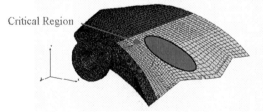

Critical Region

Fig. 1. Critical Regions of the blade in operation.

CP907, *10th ESAFORM Conference on Material Forming*, edited by E. Cueto and F. Chinesta

In sheet metal forming, the understanding of the mechanics of deformations, sheet failures (by necking or fracture) and the strain distribution are important to improve all processes. The final strain distribution on metallic parts is shown to vary by material properties such as: strain-hardening index, $n$, strain-rate sensitivity index, $m$, normal plastic anisotropy coefficient, $R$, initial strain-hardening of material, $\varepsilon_o$ and fracture strain, $\varepsilon_f$, even though the strains are more influenced by friction, work pressures or the sheet thickness. Therefore, in order to control the forming process, it is necessary to know the plastic behavior of the sheet under the particular process conditions. In this work, in order to verify if the maximum strains, specially at the most deformed areas, reach the limit; classic instability theories and also the necking angle property theory are considered. So, this works aims to analyze, using pre-existent theories and using forming limiting curves, if the maximum strains reach its limits on the deep-drawing process.

## 3. DEEP-DRAWING PROCESS

The common challenges on a deep-drawing process are: avoid ripple marks, sheet break during the forming process, localized necking, the use of more steps than necessary, the guarantee of the complete forming of the sheet on the desired final shape and determining the necessary forming load in order to select the correct equipment. In order to avoid these problems, it was used a deep-drawing process with a blank holder to avoid ripple marks, a material with good behavior on this process and two millimeter thickness plate, as the use of a load system with 50% extra load.

The set up process, as the results of each deep-drawing test are presented in order to establish a trustful sense of the viability of the deep-drawing process.

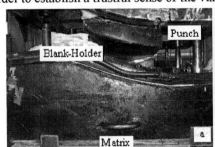 

Fig. 2. Forming of Sample 1 by punch load - Details (a) side; (b) top.

Firstly, using a blank holder, an initial conforming was made in order to guarantee a pre-stress existence. After that, the conformation continued by a punch movement made by a hydraulic cylinder. When finalized the forming, the punch was lifted up and a hole was made on the sheet. This hole was in order to check if lubricating oil or even air was stuck between the sheet and the matrix in order to lead to a complete forming on the deep-drawing process.

By the other hand, Fig. 3 shows the suction side of the blade after forming that even at its most critical region, with the bigger deformation, there was no appearance of ripple marks, which indicates the lubricating and friction conditions between sheet and matrix and sheet and blank-holder were satisfactory to avoid them. In addition, there were no big localized deformations, in particular, at the most deformed regions where there is a reduction of thickness as indicated on the aeronautic profile blades by using finite element method [6]. However, after form the sheet and looking at the grid used in order to measure the strains, it was concluded there was no reliability on the strains measured due to the grid size. This means, the grid was not enough sensible to measure the strain gradients. Due to those observations, it was necessary to make a new kind of grid, and also a new test.

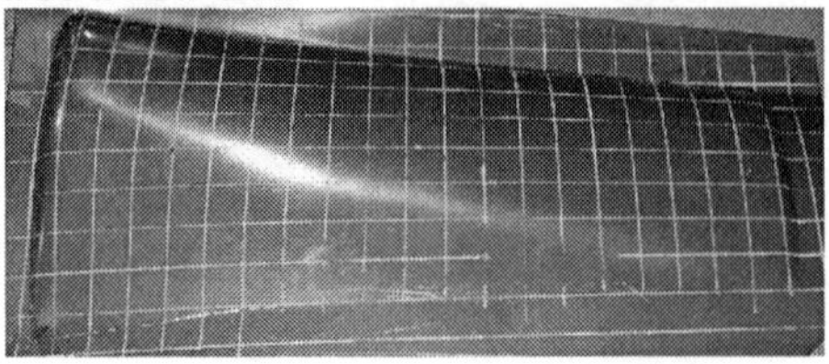

Fig. 3. Formed sheet.

## 4- FORMING LIMITING CURVES AND DISCUSSIONS

The process done up to this point measuring the limit deformation on the sheet plane was not reliable. Due to this, a new kind of grid, more refined, and with circles much lower than the previous grid squares, was used in order to show the strain gradients on the sheet plane. The refined limit strain measurement made us able to verify the reliability of the deep-drawing process of the pressure side of the blade because makes possible to verify the reliability of the process comparing the strains measured with the limit strains calculated by using the theories of Dorn, Hill, Swift-Hill and TPAE. These theories indicate, with dependability, if there is or not failure during the deep-drawing process by observing if the measured strains are lower that the calculated theoretically by them. These theories also allow the definition of the limit strains on AISI 409 stainless steel by plastic properties such as strain-hardening index, $n$, normal plastic anisotropy coefficient, $R$ and initial strain-hardening, $\varepsilon_o$. This way, the plate was marked using an electrolytic process with circle printings forming a circular grid to measure strains on the entire blade area and show the strain gradients, mainly at the most deformed areas. The circles were printed with initial diameter 8mm followed by posterior deep-drawing according to previous tests procedure.

Strains are calculated by dividing the final ellipse and the initial circle diameters, in log scale. This way, the localized deformations were measured in positions where the deformations were bigger (mainly at blade root). These were compared with the calculated limit strains theories on a forming limit curve, as shown on Fig. 5, considered the material anisotropic, with ($R=1,2$); and anisotropic with behavior similar to isotropic with ($R=1,0$), initial strain $\varepsilon_o = 0$ and the strain-hardening index of sheet materials, $n=0,22$.

Therefore, it is verified by Fig.4 that the deformations on the most drawn regions, at the blade root did not reached the limit strains for the AISI 409 stainless steel, which lead to an inexistence of failure of these sheets during the deep-drawing process. Thus, the deep-drawing process was considered reliable, and the experimental tests could stainless be concluded.

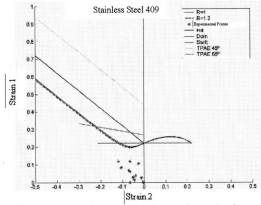

Fig. 4. Limit strain theories applied with experimental points measured.

# 5- CONCLUSIONS

It was verified by the experimental tests that the deep-drawing process was feasible and controlled as the most deformed areas, at the blade root where the plate is thinner, did not reach the limit strains for the stainless steel AISI409. This confirms the inexistence of failure during the deep-drawing process. Besides that, at these most deformed regions, the sheet will be cut for posterior edge welding, what will naturally reinforce these areas according to the blade panel project.

It can also be observed, at Figure 4, the effect of plastic anisotropy on diffuse and localized necking at the sheet plane. It was verified that, for a material with initial strain $(\varepsilon_o=0)$, the influence of normal plastic anisotropy is only significant for the limit strains calculated by the theory of Swift (diffuse necking), and that is more expressive for the deep-drawing region of the forming limit curve.

*Acknowledgements*: The authors would like especially to thank to the Tecsis Tecnologia e Sistemas Ltda and Petrobras by the experimental support.

REFERENCES

1- J. R. Anderson, Fundamentals of Aerodynamics, McGraw-Hill, 2nd Edition, New York (1991).
2- J.D.Lubahn, Failure of Ductile Metals in Tension, In: Transactions ASME, Vol. 68, May, (1946) 271-276.
3- H. W. Swift, Plastic Instability Under Plane Stress, In: Journal of the Mechanics and Physics of Solids, vol. 1, (1952) 1-18.
4- J. E. Dorn , E. G. Thomsen, J. Janilek, Plastic Flow in Metals, Research Report, War Production Board, (1945), In: Transactions ASME, (1969) 659-663.
5- R. Hill, On Discontinuous Plastic States, With Special Reference to Localised Necking in Thin Sheets, In: Journal of the Mechanics and Physics of Solids, vol. 1, (1952) 19-30.
6- F.D.Foroni, Desenvolvimento de Processo de Conformação de Pás Metálicas de Alto Desempenho para Aplicação em Sistemas de Metrô e Túneis Rodoviários, MSc. Thesis, Instituto Tecnológico de Aeronáutica-ITA, São José dos Campos (2005).
7- M.A.Menezes, A New Theory to Assess Strain Limits of Anisotropic Sheet Metals, In: 2nd Esaforming, Proceedings of the 2nd Esaforming, Guimarães, Portugal (1999) 141-146.

# The Influence of Strain Rate Variations on the Appearance of Serrated Yielding in 2024-T3 Al-Clad Aluminium Alloy

Alan G. Leacock\*, Robert J. McMurray\*, D. Brown\* and Ken Poston†

\*Advanced Metal Forming Research Group, Nanotechnology and Advanced Materials Research
Institute, University of Ulster, Shore Road, Newtownabbey, Co. Antrim, N. Ireland, BT37OQB
†Bombardier Aerospace, Airport Road, Co. Antrim, N. Ireland, BT3 9DZ

**Abstract.** To avoid failure during the stretch forming process using manual control, machine operators tend to achieve the final form using a stop-start approach. It was observed that when approaching full form, stretcher-strain marks appeared on the surface of the part if the operator stopped and restarted the forming operation. In order to investigate this phenomenon, a series of tensile tests was conducted using two batches of 2024-T3 aluminium alloy. The specimens were tested using several different strain rates, representative of those used on the shop floor. Additional tests were conducted involving a series of pauses under displacement control at differing levels of strain and strain rate. In the uninterrupted tests for the two batches of 2024-T3 material tested, serrated yielding was observed just prior to failure. However for the tests in which there was a pause in displacement, the material consistently exhibited serrated yielding when the crosshead began to move again. These results indicate that the pause provides an opportunity for strain ageing and pinning of the dislocations resulting in serrated yielding of this alloy. In order to avoid serrated yielding, stretch forming operations using 2024-T3 aluminium should be conducted at a constant strain rate without interruption. This also has far reaching implications for those involved in the production and testing of these alloys. The test programme described represents an initial attempt to investigate a phenomenon noted during an industrial forming process and should be extended to analyse the affect of strain path changes on the occurrence of serrated yielding.

**Keywords:** Stretch Forming, Portevin–Le Chatelier effect, Al-Cu, 2024-T3, Strain Ageing
**PACS:** 81.40.Lm

## INTRODUCTION

The appearance of Portevin–Le Chatelier (PLC) bands during the stretch forming of solution heat treated Al-Cu alloys was previously noted as a problem [1]. More recent work has shown that increasing the natural ageing time to more than 420 minutes for these alloys eliminates the appearance of PLC bands [2]. However previous work[3] with Al-Cu in a similar temper condition illustrated than an additional cold working process after ageing produced an alloy that exhibited the PLC effect. The additional cold work was an attempt to produce a T3 temper condition. It should be noted that the current processing steps for the T3 temper are solution heat treat, cold work and naturally age to a substantially stable state [4], not age then cold work. In the case presented by Apacoda [3] the problem was solved by slowing the forming process from 14 inches/min to 0.4 inches/min. Although detracting from the visual appearance of the final part, these bands do not adversely affect the final strength [3]. Nevertheless, the high quality surface finish required in aerospace skin components often results in the rejection of such components.

CP907, *10th ESAFORM Conference on Material Forming,* edited by E. Cueto and F. Chinesta
© 2007 American Institute of Physics 978-0-7354-0414-4/07/$23.00

While conducting a stretch forming production trial at Bombardier Aerospace, the formation of PLC bands, near the end of the forming process, was noted. In an effort to avoid failures, rising–table stretch forming machine operators often use a *stop–start* control to approach final form. These parts were free of PLC bands until the point at which this control method was employed, whereupon the bands consistently appeared

## EXPERIMENTAL PROCEDURE

In order to test for this effect in a controlled fashion, the uniaxial tensile test was employed. Previous work has shown the strain path in stretch–draw formed components to be predominately uniaxial [5].

Two batches of commercial 2024-T3 sheet material, 0.508 mm thick, was used for the test programme. Specimens were removed in the rolling direction of the sheet material. Each specimen was manufactured as per ASTM E8M, having a gauge-length of 75 mm and a reduced width of 12.5 mm.

The tests were conducted at constant crosshead speeds of 5, 10 and 15 mm/min on a screw driven 5500R Instron. These crosshead speeds resulted in initial strain rates of approximately $1.110 \times 10^{-3}$ s$^{-1}$, $2.220 \times 10^{-3}$ s$^{-1}$ and $3.328 \times 10^{-3}$ s$^{-1}$ respectively. For each of the three strain rates, two test control methods were used; strain until failure, or strain until 4, 8 or 12% total engineering strain, then pause the crosshead motion for 6 seconds before restarting and continuing at the previous strain rate until failure. There was only one pause per test specimen. Three specimens from each batch were used to repeat each test, giving a total of 72 specimens, 36 from each batch. The strain was measured throughout the test using a dynamic strain gauge extensometer (2620-600 series) with a 25 mm gauge-length.

## RESULTS

For clarity, only the results from a representative sample of the specimens are shown in Fig. 1 and 2. The six repeats of each test were found to exhibit minimal variation.

Figure 1 (a) shows the uninterrupted test results. For each strain rate considered the specimens did not exhibit any serrated yielding until beyond the maximum tensile strength. Between the maximum tensile strength and failure some serrated yielding was observed along with PLC band formation on the specimen surface. The initiation of these bands was accompanied with an audible '*ping*'.

In each of the interrupted tests, at all strain rates, serrated yielding appeared immediately upon restarting the crosshead motion after the pause. These serrations changed from type A[1] to type B[2] for strain levels approximately equal to 12% of the total engineering strain. The formation of the latter (type B) bands was accompanied by an audible '*ping*'.

---

[1] Continuously propagating bands moving from one end of the specimen to the other
[2] Jumping bands appearing in a stepped fashion

The severity of the stress–strain steps in each of these plots is seen to increase with increasing strain rate. Furthermore, upon reaching the transition strain from type A to B bands, the strain steps also increase with increasing strain rate.

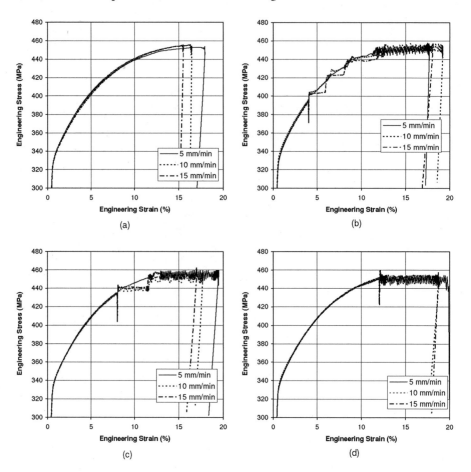

**FIGURE 1.**  Engineering stress strain curves, (a) without a pause and pause for 6 seconds at (b) 4%, (c) 8% and (d) 12% total engineering strain

In order to determine the cause of the resulting serrated yielding after a pause, the stress–time relationship was analysed. Figure 2 (a) shows the relationship between engineering stress and total time. It is evident that some form of strain ageing occurred during the pause. A closer view of the stress drop–pause time is provided in Fig. 2 (b), (c) and (d). These plots were created by subtracting the stress at the start of the pause from the stress during the pause. The time scale was also referenced to zero at this point. The commencement of the pause was determined using a conditional statement to ensure uniformity of selection. The stress drop is a logarithmic function of time that increases with strain magnitude and strain rate.

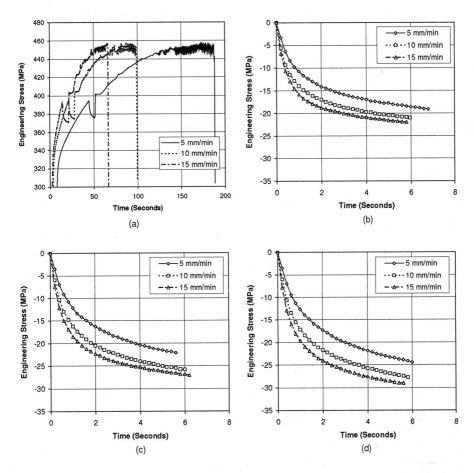

**FIGURE 2.** (a) Stress–time plot for a 6 second pause at 4% total engineering strain, and relative stress drop–time plots for (b) 4%, (c) 8% and (d) 12% total engineering strain

# DISCUSSION

The characteristic stress drop noted in Fig. 2(b)–(c) has already been attributed to copper in solid solution causing strain ageing [6]. The increasing stress drop as a function of both strain magnitude and rate can be attributed to increased diffusion rates caused by temperature and dislocation density changes, though not exclusively. For the lower strain rates the process of deformation tends towards an isothermal condition, since there is time to dissipate some of the energy within the specimen via heat transfer. However at the higher strain rates there is less time for heat transfer and the process tends towards adiabatic conditions. The added complication of higher dislocation densities as the material work hardens also increases the available paths for diffusion [7]. Therefore the increasing temperature and dislocation density both serve to increase the overall diffusion rate. This interaction is demonstrated by the increasing stress drops as a function of strain rate and magnitude during the pause and by the change from type A to type B serrations. Higher diffusion rates are required to create the type B serrations [7].

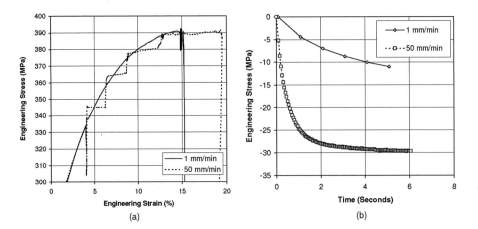

**FIGURE 3.** Additional strain rate tests (a) Engineering stress–strain plot and (b) Engineering stress drop–time plot

As mentioned previously, Apacoda [3] encountered a similar problem, though current evidence suggests that this may have been due to the pre-straining after ageing used to obtain the 'T3' temper. He identified three key factors that contributed to the appearance of PLC bands:

1. The degree of previous strain hardening while processing the sheet (Neither the pre-strain rate nor magnitude were specified)
2. Ageing time after solution heat treatment (again no exact figures were presented), and,
3. Strain rate during stretch forming.

The first two factors have been addressed in this paper and previous research[2]. The third factor deserves further attention. The stress-strain steps shown in Fig. 1(b)–(d), are the least obvious for a crosshead speed of 5 mm/min. Indeed the band propagation during tests at this rate was not visually apparent until the change to type B bands. Furthermore Tzur et al. [6] did not report any serrated yielding during their test of Al-Cu during multiple pauses and strain ageing. These tests were conducted at a strain rate of $1.6 \times 10^{-4}$ s$^{-1}$. This corresponds to an approximate crosshead speed of 0.75 mm/min for a 75 mm gauge-length.

Therefore, in order to extend the current work, a new series of tests was conducted to extend the strain rate range. Crosshead speeds of 1 and 50 mm/min were chosen to test the effects of both decreasing and increasing the strain rate. Due to material availability limitations, a new batch of material was required. The results for a pause at 4% total engineering strain are shown in Fig. 3(a).

For the lower crosshead speed of 1 mm/min this material did not exhibit any PLC bands upon restarting the crosshead. The 50 mm/min test speed did however result in the previous behaviour noted for the other test speeds. At this high speed the test machine was incapable of stopping the crosshead without a slight overshoot. As a result there is a significant change in the shape of the stress drop–time plot (Fig. 3(b)). The lower magnitude of the stress drop for the 1 mm/min test indicates less strain ageing during the pause. The degree of practical application of such a low strain rate will depend upon the cost implications of PLC band formation. Forming in a single non–stop process would be a more practical solution for industry. Using a geometric approach to the machine setup, a non–stop forming operation can be achieved with confidence [5].

## CONCLUSIONS

From the limited test programme presented in this paper we can conclude that for commercial 2024-T3 aluminium alloy:

1. Interrupting the straining of the material by a pause will result in PLC band formation for strain rates ranging from $1.110 \times 10^{-3}$ s$^{-1}$ to $1.105 \times 10^{-2}$ s$^{-1}$
2. For a strain rate of $2.222 \times 10^{-4}$ s$^{-1}$, a test interruption at 4% engineering strain does not result in PLC band formation
3. Strain ageing plays a significant role in the likelihood of PLC band formation.
4. The most practical industrial application would require a non–stop forming operation.

This test programme should be extended to include additional strain rate combinations and strain path changes.

## ACKNOWLEDGMENTS

This work was supported by funding from the EPSRC (GR/R41125/01), and all materials were supplied by Bombardier Aerospace.

## REFERENCES

1. R. D. Edwards, *Journal of the Institute of Metals*, **84**, (1955-56), pp.199–209
2. H. Jiang, Q. Zhang, X. Wu, J. Fan, *Scripta Materialia*, **54**, (2006), pp.2041–2045
3. D. R. Apacoda, *SME Technical Paper No. MR77-249*, 1977
4. J. R. Davis, "Alloy and Temper Designation Systems," in *Aluminium and Aliminium Alloys*, ASM International, 1993, pp. 29
5. A. G. Leacock, "Numerical simulation of anisotropic plasticity in stretch formed aluminium alloys", *PhD Thesis*, University of Ulster, 1999.
6. M. Tzur, S. Dirnfeld, A. Rosen, *Materials Science and Engineering*, **11**, (1973), pp.219–222
7. F. B. Klose, A. Ziegenbein, J. Weidenmüller, H. Neuhüser, P. Hähner, *Computational Materials Science*, **26**, (2003), pp.80–86

# Orthotropic Yield Criteria for modeling the combined effects of anisotropy and strength differential effects in sheet metals

## B. Plunkett[a], O. Cazacu[b], F. Barlat[c]

[a] Air Force Research Laboratory, Munitions Directorate, Eglin Air Force Base, FL 32542, USA
[b] Department of Mechanical and Aerospace Engineering, University of Florida/REEF, Shalimar, FL 32579-1163, USA
[c] Alloy Technology and Materials Research Division, Alcoa Inc., Alcoa Technical Center, 100 Technical Drive, Alcoa Center, PA 15069-0001, USA

**Abstract.** In this paper, yield functions describing the anisotropic behavior of textured metals are proposed. These yield functions are extensions to orthotropy of the isotropic yield function proposed by Cazacu et al. [1]. Anisotropy is introduced using linear transformations of the stress deviator. It is shown that if two linear transformations are considered, the proposed anisotropic yield function represents with great accuracy both the tensile and compressive anisotropy in yield stresses and r-values of materials with hcp crystal structure and of metal sheets with bcc crystal structure that exhibit asymmetry between tensile and compressive behavior. Furthermore, it is demonstrated that the proposed formulations can describe very accurately the anisotropic behavior of metal sheets whose tensile and compressive stresses are equal.

**Keywords:** Anisotropy, Yield Criterion, Sheet metal.

## INTRODUCTION

This paper presents full stress 3-D yield criteria for describing the anisotropic plastic response of textured metals. The aim is to develop models that can be applicable to materials that exhibit strength differential effects as well as to materials for which not noticeable difference exist between the behavior in tension and compression under monotonic loading. Key in this development is the use of the isotropic yield function proposed in [1]. Anisotropy is introduced using several linear transformations. In the next section, the isotropic yield criterion is succinctly presented. After reviewing general aspects of linear transformations operating on the Cauchy stress tensor, in Section 3, a new anisotropic yield function involving two linear transformations is introduced. The input data needed for the calculation of anisotropic yield function coefficients are discussed. Illustrative examples of application of this yield function to the description of anisotropy and tension/compression asymmetry of materials with cubic and hexagonal close packed crystal structure are presented. Moreover, it is shown that the 3-D yield criterion involving two linear transformations captures with high accuracy the anisotropy in yielding of metals with no tension/compression asymmetry.

CP907, 10th ESAFORM Conference on Material Forming, edited by E. Cueto and F. Chinesta
© 2007 American Institute of Physics 978-0-7354-0414-4/07/$23.00

# Proposed Model

If a material only deforms by a reversible shear mechanism such as slip, yielding depends only on the magnitude of the resolved shear stress. Thus, the yield locus in the deviatoric $\pi$ plane (plane which passes through the origin and is perpendicular to the hydrostatic axis) must have six-fold symmetry. If the material deforms by twinning, yielding depends on the sign of the applied shear stress, i.e., yield in tension and compression should be different. Hosford and Allen [2] used a modified Taylor polycrystal model to show that for randomly oriented fcc and bcc crystals deforming solely by twinning, the ratio between the yield stress in tension and compression should be 0.78 and 1.28, respectively. To account for this strength differential effect, the following isotropic yield function was proposed in [1]:

$$\phi = \left\| S_1 \right| - kS_1 \big|^a + \big\| S_2 \big| - kS_2 \big|^a + \big\| S_3 \big| - kS_3 \big|^a \tag{1}$$

where $S_1, S_2, S_3$ are the principal values of the deviatoric stress tensor $\mathbf{S}$, $a$ is the degree of homogeneity and $k$ is a parameter that allows for differences in the yield stress in tension and compression.

To extend this isotropic criterion to orthotropy (see [1]), a linear transformation was applied on the deviatoric stress tensor $\mathbf{S}$, i.e. in Equation (1) $S_1, S_2, S_3$ were substituted by the principal values of a transformed tensor $\mathbf{\Sigma}$ defined as:

$$\mathbf{\Sigma} = \mathbf{C} : \mathbf{S} \tag{2}$$

Thus, the orthotropic yield criterion (denoted in the following as CPB06) is of the form:

$$F = \phi\left(\Sigma_1, \Sigma_2, \Sigma_3\right) = \left(\left|\Sigma_1\right| - k\Sigma_1\right)^a + \left(\left|\Sigma_2\right| - k\Sigma_2\right)^a + \left(\left|\Sigma_3\right| - k\Sigma_3\right)^a \tag{3}$$

where $\Sigma_1, \Sigma_2, \Sigma_3$ are the principal values of $\mathbf{\Sigma}$. The only restrictions imposed on the fourth-order tensor $\mathbf{C}$ (see Equation (2)) are: (i) to satisfy the major and minor symmetries and (ii) to be invariant with respect to the orthotropy group. Thus, for 3-D stress conditions CPB06 involves nine independent anisotropy coefficients and it reduces to the isotropic criterion (1) when $\mathbf{C}$ is equal to the fourth-order identity tensor. It is worth noting that although the transformed tensor is not deviatoric, the orthotropic criterion is insensitive to hydrostatic pressure and thus the condition of plastic incompressibility is satisfied (see [1] for details). For $k \in [-1,1]$ and any integer $a \geq 1$, the anisotropic yield function is convex in the variables $\Sigma_1, \Sigma_2, \Sigma_3$.

To increase the number of anisotropy coefficients in the formulation, instead of one linear transformation, $n$ linear transformations ($n \geq 2$) can be considered [3]. Such a methodology was used in [4], [5], and [6] to describe the pronounced anisotropy displayed by certain aluminum alloys. For example, Yld2004-18p [5] involves two linear transformations. When the two transformations are equal, Yld2004-18p reduces to Yld91 orthotropic yield criterion [7]. Here, we demonstrate that additional linear transformations can be incorporated into the CPB06 criterion for an improved representation of the anisotropy of the yield surface of certain alloys.

The following analytic yield function, denoted CPB06ex2, is proposed:

$$F(\boldsymbol{\Sigma}, \boldsymbol{\Sigma}') = \left(\left|\Sigma_1\right| - k\Sigma_1\right)^a + \left(\left|\Sigma_2\right| - k\Sigma_2\right)^a + \left(\left|\Sigma_3\right| - k\Sigma_3\right)^a$$
$$+ \left(\left|\Sigma_1'\right| - k'\Sigma_1'\right)^a + \left(\left|\Sigma_2'\right| - k'\Sigma_2'\right)^a + \left(\left|\Sigma_3'\right| - k'\Sigma_3'\right)^a \tag{4}$$

where $k$ and $k'$ are material parameters that allow for the description of strength differential effects, $a$ is the degree of homogeneity, while

$$\boldsymbol{\Sigma} = \mathbf{C}\colon \mathbf{S} \text{ and } \boldsymbol{\Sigma}' = \mathbf{C}'\colon \mathbf{S}. \tag{5}$$

Thus, for 3-D stress conditions the orthotropic criterion (4) involves 18 anisotropy coefficients. When $C_{ii} = 1$ and all other $C_{ij} = 0$ ($i \neq j$) and $k = k'$, this yield function reduces to the isotropic yield function (1). Note that when $\mathbf{C} = \mathbf{C}'$ and $k = k'$ the proposed criterion reduces to the CPB06 yield criterion (see Eqs. 3).

The anisotropy coefficients involved in the yield criteria can be found through the minimization of an error function. The experimental data in the error function may consist of flow stresses and r-values in tension and compression corresponding to different orientations in the plane of the sheet, biaxial flow stress in tension and compression, as well as out of plane yield stresses and r-values.

$$\text{Error}(\mathbf{C}, \mathbf{C}') = \sum_i w_i \left(\frac{\sigma_i^{th}}{\sigma_i^{ex}} - 1\right)^2 + \sum_j w_j \left(\frac{r_j^{th}}{r_j^{ex}} - 1\right)^2 \tag{6}$$

In the above equation $i$ represents the number of experimental yield stresses, $j$ represents the number of experimental r- values available while the superscript indicates whether the corresponding value is experimental or predicted.

## Applications

In Fig. 1 are shown the comparison between model predictions and data for sheets of textured AZ31B (data after [8]) while in Fig. 2 and Tables 1 and 2 a comparison for a medium carbon low alloy steel sheet (data after [9]) is presented. The anisotropy coefficients involved in the expression of the theoretical yield loci as well as the constants $k$ and $k'$ were determined using the initial yielding data reported for each material. Note that the proposed theory reproduces very well the observed asymmetry in yielding.

Although the CPB06 yield criterion (Cazacu et al., 2006) and the extensions proposed in this paper were formulated to capture the strength differential effects most often associated with materials that display tension/compression asymmetry in yielding, the proposed formulations are not limited to such materials. If the yield in tension is equal to the yield in compression, the parameters k and k' associated with strength differential effects are automatically zero. In Figs. 3 and 4, the CPB06ex2 yield criterion (4) is applied to aluminum alloy sheets of 2090-T3 and 6111-T4 (data after [5]). Tensile yield stresses and r-values along seven directions in the plane of the sheet (i.e. from rolling to transverse directions in 15° increments) were measured from uniaxial tensile tests. Additionally, the experimental values of the balanced biaxial yield stress as well as the corresponding r-value obtained using a disk compression test were reported. The model is capable of capturing the highly anisotropic data for each material with a high degree of accuracy.

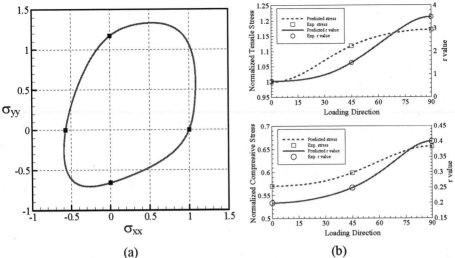

(a)                                      (b)

**FIGURE 1.** (a) Plane stress yield loci and (b) anisotropy of the yield stress (normalized by the tensile stress in the rolling direction) and the r-values for AZ31B Mg, measured and predicted with the CPB06ex2. yield criterion. Experimental data after [8].

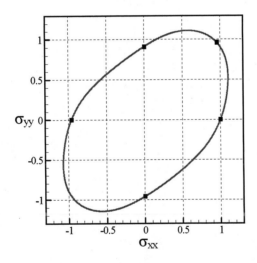

**FIGURE 2.** Projection of the CPB06ex2 yield surface in the xy plane for a medium carbon low alloy steel. Symbols represent normalized experimental data of [9].

**TABLE 1.** Measured and theoretical normalized yield stress values for a medium carbon low alloy steel. Experimental data after [8].

|  | Tension | | Compression | | |
|---|---|---|---|---|---|
|  | $X_T$ | $Y_T$ | $X_C$ | $Y_C$ | $Z_C$ |
| Experiment | 1 | 0.914 | 0.963 | 0.960 | 0.958 |
| CPB06ex2 | 1 | 0.914 | 0.963 | 0.960 | 0.958 |

**Table 2.** Measured and predicted r -values for a medium carbon low alloy steel. Experimental data after [8].

| Loading direction | Tension | | Compression | | | | | |
|---|---|---|---|---|---|---|---|---|
| | x | y | z | y | z | 45°in (xy) plane | 45°in (xz) plane | 45°in (yz) plane |
| Experiment | 0.67 | 1.4 | 0.71 | 1.3 | 1.05 | 0.91 | 1.45 | 1.3 |
| CPB06ex2 | 0.67 | 1.4 | 0.71 | 1.3 | 1.05 | 0.91 | 1.45 | 1.3 |

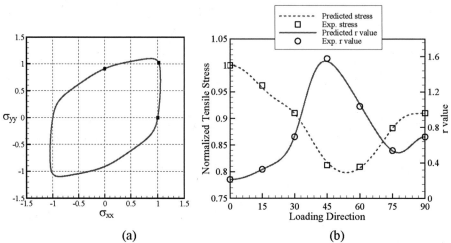

(a)                                                        (b)

**FIGURE 3.** (a) Plane stress yield loci and (b) anisotropy of the yield stress (normalized by the tensile stress in the rolling direction) and the r-values for 2090-T3 aluminum sheet, measured and predicted with the CPB06ex2. yield criterion. Experimental data after [5].

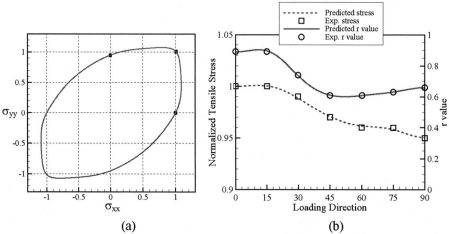

(a)                                                        (b)

**FIGURE 4.** (a) Plane stress yield loci and (b) anisotropy of the yield stress (normalized by the tensile stress in the rolling direction) and the r-values for 6111-T4 aluminum sheet, measured and predicted with the CPB06ex2. yield criterion. Experimental data after [5].

400

# REFERENCES

1. O. Cazacu, B. Plunkett, F. Barlat, *Int. J. Plasticity* **22**, 1171-1194 (2006).
2. W. F. Hosford and T. J. Allen, Met. Trans. **4**, 1424-1425 (1973)
3. F. Barlat,, J.W. Yoon., O. Cazacu, *Int. J. Plasticity*, In Press (2006).
4  F. Barlat, J.C. Brem, J.W. Yoon, K. Chung, R.E. Dick, D.J. Lege, F. Pourboghrat, S.H. Choi, , E. Chu, *Int. J. Plasticity* **19**, 1279-1319 (2003).
5. F. Barlat, H. Aretz, J.W. Yoon, M. E. Karabin, J. C. Brem, R.E. Dick, *Int. J. Plasticity* **21**, 1009-1039 (2005).
6. F. Bron, J. Besson, *Int. J. Plasticity* **20**, 937-963 (2004).
7. F. Barlat, D.J. Lege, J.C. Brem, *Int. J. Plasticity* **7**, 693-712 (1991).
8. X. Y. Lou, M. Li, R.K. Boger, S. R. Agnew, R. H. Wagoner, *Int. J. Plasticity* **23**, 44-86 (2007).
9. A. A. Benzerga, J. Besson, A. Pineau, *Acta Mater* **52**, 46232638 (2007).

# 5 – HYDROFORMING

## *(J. C. Gelin)*

# Tube Bulge Process : Theoretical Analysis And Finite Element Simulations

Raphaël Velasco and Nathalie Boudeau

*Femto-ST Institute, Department of Applied Mechanics, ENSMM Besançon, 24 chemin de l'Epitaphe, 25000 Besançon, France*

**Abstract:** This paper is focused on the determination of mechanics characteristics for tubular materials, using tube bulge process. A comparative study is made between two different models: theoretical model and finite element analysis. The theoretical model is completely developed, based first on a geometrical analysis of the tube profile during bulging, which is assumed to strain in arc of circles. Strain and stress analysis complete the theoretical model, which allows to evaluate tube thickness and state of stress, at any point of the free bulge region. Free bulging of a 304L stainless steel is simulated using Ls-Dyna 970. To validate FE simulations approach, a comparison between theoretical and finite elements models is led on several parameters such as: thickness variation at the free bulge region pole with bulge height, tube thickness variation with z axial coordinate, and von Mises stress variation with plastic strain.

**Keywords:** Tube bulge test, Flow stress curve, Finite elements simulation, Error calculation
**PACS:** 81.20.Hy

## INTRODUCTION

Tube bulge process is a useful test to establish precisely tubular material behavior [1]. It is surprising not usually used, although tube hydroforming processes become more popular every year [2-4]. Fuchizawa and Narazaki [5] have published a reference paper on the subject, developing an analytical model and leading experimental tests with aims of comparing uniaxial tensile tests and bulge test. Stress-strain characteristics for tubular material should be determined using bulge test according to them, because roll forming process and welding change the structure of the tubes during their forming. This study was completed with finite element simulations [6], to have a chain of analysis, from experimental tests to FE simulation. It allows to simulate tube behavior in an hydroforming application, using a flow stress curve issued of experimental bulge tests, which is accurate to describe this behavior in a bi-axial pressure stress state. Strano and Altan [7] put forward the advantages of bulge tests too, compared to tensile tests. They implemented a completely different analytical approach, based on a very simple inverse energy approach for the determination of flow stress curve. Hwang and Lin [8] suggested another mathematical analysis to study the plastic strain of a tube during bulge hydroforming process. Coupled with a finite element analysis, the influence of some parameters such as initial tube length and thickness, friction coefficient, or die entry radii on the

CP907, *10th ESAFORM Conference on Material Forming*, edited by E. Cueto and F. Chinesta
© 2007 American Institute of Physics 978-0-7354-0414-4/07/$23.00

pressure variation with bulge height is studied. This analysis is completed with experimental tests [9], and compared to Fuchizawa's model. It is so noticed that finite element simulations using stress-strain characteristics obtained issued of this model give a pressure variation during bulging which is closer to experimental results than using tensile tests or Fuchizawa's model.

## ANALYTICAL MODEL

The geometric model that will be used all along this paper is first described. Fig. 1 and Table 1 shows a schematic view of tube bulging and the configuration adopted. Few assumptions have been made:
  - ➤ In the free bulge region, the tube is supposed to strain forming two arcs of circle whose centers moved along the Y axis during the bulging.
  - ➤ Bulging is supposed to be perfectly symmetrical regarding Y axis
  - ➤ Results found with this model will be acceptable "far" from borders of the free bulge region
  - ➤ Calculations are made on thin tubes (1mm)

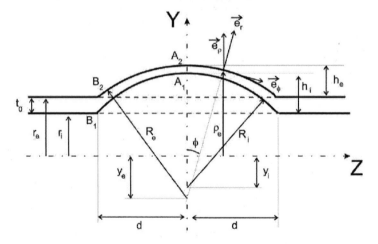

**FIGURE 1.** Configuration of the bulged

$r_i$, $r_e$ and $t_0$ are geometrical parameters defined by the chosen tube. For the whole investigations, the length of the free bulge region (2.d) is equal to 50mm. $h_e$ is the known parameters that defined the bulging of the tube. $R_e$ and $y_e$ are functions of the bulge height $h_e$. $h_i$ is the unknown parameter during tube bulging. Knowing $h_i$ allows to define the thickness $t(z)$ of the tube, first at the centre of the bulging ($t(0)$), and by extension at any point of the bulging. This model isn't focused on the centre of the bulging, to permit to take into account errors made on the measurements of $h_e$.

**TABLE 1.** Geometrical configuration for tube bulging process

| Parameter | Description |
|---|---|
| $r_i$ | Initial internal radius of the tube |
| $r_e$ | Initial external radius of the tube |
| $t_0$ | Initial thickness of the tube |
| $2.d$ | Length of the free bulge region of the tube |
| $R_i$ | Radius of the internal curvature of the strained tube in the $(Y,Z)$ plan |
| $R_e$ | Radius of the external curvature of the strained tube in the $(Y,Z)$ plan |
| $\rho$ | Radius of curvature of the strained tube in the $(X,Y)$ plan |
| $y_i$ | Distance between the centre of the internal curvature and the tube axis |
| $y_e$ | Distance between the centre of the external curvature and the tube axis |
| $h_i$ | Bulge height of the internal part of the tube |
| $h_e$ | Bulge height of the external part of the tube |
| $t(z)$ | Thickness of the tube at z coordinate |

# Determination Of The Tube Thickness

Thickness of the tube is determined using volume calculations and incompressibility of the material in the free bulge region. A Newton-Raphson algorithm is used to solve the non-linear equation (1) to determine $h_i$, which is necessary to calculate thickness of the tube.

$$V_e - V_i(h_i) = V_0 \tag{1}$$

$V_e$ and $V_i$ stand respectively for the external and internal volume of the free bulge region. $V_0$ is the initial material volume of that region. Solving this equation permits to determine $t(0)$, which is given by (2), and then $t(z)$, which is calculated geometrically (3) (Fig 2).

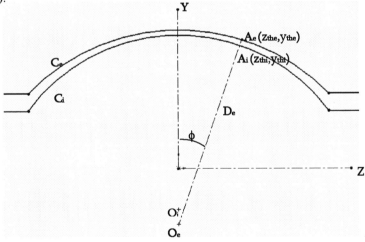

**FIGURE 2.** Geometrical configuration for the calculation of $t(z)$

$$t(0) = t_0 + h_e - h_i \tag{2}$$

407

$$t(z) = t(z_{the}) = \overline{A_e A_i} = \sqrt{(z_{the} - z_{thi})^2 + (y_{the} - y_{thi})^2}$$ (3)

## Determination Of Mechanical Strain And Stress

The strain state of a tube under internal pressure is represented by the tensor (4):

$$\underline{\varepsilon} = \begin{pmatrix} \varepsilon_r & 0 & 0 \\ 0 & \varepsilon_\theta & 0 \\ 0 & 0 & \varepsilon_\phi \end{pmatrix}$$ (4)

The three principal strains are explained on (5).

$$\varepsilon_r = \ln\left(\frac{t(z)}{t_0}\right), \varepsilon_\theta = \ln\left(\frac{\rho(z)}{r_e}\right), \varepsilon_\phi = -\varepsilon_r - \varepsilon_\theta$$ (5)

The stress state of a thin tube ($t_0 \ll r$) under internal pressure is represented by the tensor (6):

$$\underline{\sigma} = \begin{pmatrix} 0 & 0 & 0 \\ 0 & \sigma_\theta & 0 \\ 0 & 0 & \sigma_\phi \end{pmatrix}$$ (6)

The two principal stresses at any point of the tube are explained on (7).

$$\frac{\sigma_\phi}{R} + \cos(\phi).\frac{\sigma_\theta}{\rho} = \frac{p}{t}, \sigma_\phi = \frac{\rho.p}{2.t.\cos(\phi)}$$ (7)

## COMPARISON BETWEEN FE SIMULATIONS AND ANALYTICAL RESULTS

### Finite Element Model

A finite element model was implemented in Ls-Dyna 970. The calculation has been led on a 304L stainless steel, using a Swift law (8) to describe the material behavior.

$$\overline{\sigma} = k.(\varepsilon_{yp} + \overline{\varepsilon})^n$$ (8)

Pressure variation with time is linear, from 0 to 38MPa at the end of the bulging. Fig. 3 shows plastic strain of a bulged tube at the end of the simulation.

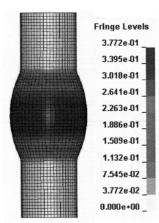

**FIGURE 3.** Plastic Strain on a bulged tube (Max. Strain : 38%)

## Results Comparison

The analytical model, implemented in Scilab, needs a law linking pressure and bulge height to run. This law is curve fitted from FE simulation results (Fig.4.a). Flow stress curves are plotted on Fig. 4.b. The analytical flow stress curve fit very well the Swift law used in the FE simulations. This fact indicates that the Scilab model allows to retrieve exactly a material behavior from a pressure/bulge height law only. To validate properly the FE model, a comparison with experimental results on the pressure variation during bulging has to be led. Fig 5.a. shows a comparison on the pole thickness variation during bulge process. It can be noticed that the analytical results fit well the FE model. At the highest stage of straining (38%), a slight difference between the two models seems to appear, due to the border effects that become probably too influencing. Fig.5.b illustrates the comparison made on thickness variation with z coordinate at the bulging end. A zone of relevance can be defined between z=-15mm and z=15mm, where the error noticed between FE simulation and analytical model stays under 1%.

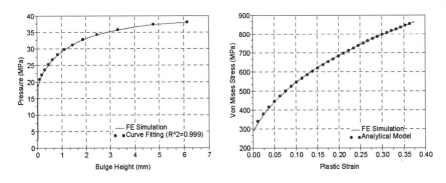

**FIGURE 4.** a)Pressure variation with bulge height ; b)Von Mises stress variation with Plastic Strain

409

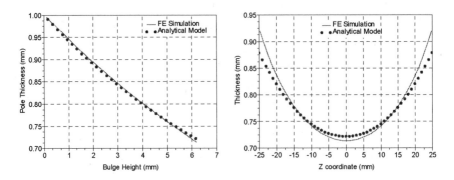

**FIGURE 5.** a)Pole thickness variation with Bulge Height ; b)Thickness variation with Z Coordinate

# CONCLUSION

This paper explains the theoretical model that has been developed to well describe tube bulge process. Based on a geometrical approach necessary to evaluate pole thickness during bulging, it allows to determine flow stress curves for a chosen tube. To study the relevance of the model, finite element simulations of bulge process has been led on a 304L tube. Until 40% strain, a very good correlation between the two models has been noticed, considering several parameters such as pole thickness, thickness in the free bulge region and flow stress curves. Experimental tests will be led in the future. They will permit to collect data on the pressure variation with bulge height.

# REFERENCES

1. M. Koç, Y. Aue-u-lan, T. Altan, On the characterictics of tubular materials for hydroforming – experimentation and analysis, Int. Journal of Machine Tools & Manufacture 41 (2001) 761-772.
2. M. Ahmetoglu, K. Sutter, X.J. Li, T. Altan, Tube Hydroforming: current research, applications and need for training, Journal of materials Processing Technology 98 (2000) 224-231
3. M. Ahmetoglu, T. Altan, Tube hydroforming: state-of-the-art and future trends, Journal of Materials Processing Technology 98 (2000) 25-33
4. K. Siegert, M. Häussermann, B. Lösch, R. Rieger, Tube hydroforming: current research, applications, and need for training, Journal of Materials Processing Technology 98 (2000) 251-258
5. S. Fuchizawa, M. Narazaki, Bulge test for determining Stress-Strain characteristics of thin tubes, Advanced Technology of plasticity – Proceeding of the Fourth International Conference on Technology of Plasticity (1993) 488-493
6. T. Sokolowski, K. Gerke, M. Ahmetoglu, T. Altan, Evaluation of tube formability and materials characteristics: hydraulic bulge testing of tubes, Journal of Materials Processing Technology 98 (2000) 34-40
7. M. Strano, T. Altan, An inverse energy approach to determine the flow stress of tubular materials for hydroforming applications, Journal of Materials Processing Technology 146 (2004) 92-96
8. Y.M. Hwang, Y. K. Lin, Analysis and finite element simulation of the tube bulge hydroforming process, Journal of Materials Processing Technology 125-126 (2002) 821-825
9. Y.M. Hwang, Y.K. Lin, T. Altan, Evaluation of tubular materials by hydraulic bulge test, International Journal of Machine Tools & Manufacture (2006)

# Process parameters calibration in 3D tube hydroforming processes

Di Lorenzo R., Ingarao G. and Micari F.

*Dipartimento di Tecnologia Meccanica, Produzione e Ingegneria Gestionale*
*Università di Palermo, viale delle Scienze 90128, Palermo/Italy*

**Abstract.** In tube hydroforming the concurrent actions of pressurized fluid and mechanical feeding allow to obtain tube shapes characterized by complex geometries such as different diameters sections and/or bulged zones. What is crucial in such processes is the proper design of operative parameters aimed to avoid defects (for instance shape defects or ductile fractures). The main process parameters are material feeding history (i.e. the punches velocity history) and internal pressure path during the process. In more complex three dimensional processes, also the action of a counterpunch is generally useful to reduce thinning in particular in expansion zones of the tube (i.e. T or Y shaped tubes). The good calibration of these parameters allows the optimal design of the process; in fact many researches have proposed different approaches to the optimization of these parameters. Generally, the main goals in the optimization approaches concern the control of thinning and the reaching of the desired final shape. In this paper, a fully three dimensional tube hydroforming operation is studied, aimed to produce a T-shaped tube. A numerical simulations campaign was developed in order to analyze the influence of process parameters on the final product quality; in particular, the influence of the pressure path was considered in order to evaluate the effects of wrinkling phenomenon on thickness distribution. The basic idea is that the possibility to determine useful wrinkles at an early stage of the process may lead to better results in terms of maximum thinning on the final part. The numerical investigations led to a knowledge base about the process mechanics and the influence of wrinkling behavior which is very effective in order to implement an optimization procedure on the process parameters.

**Keywords:** Tube hydroforming, Finite Element Method, Wrinkling, Gradient method
**PACS:** 81.20.Hy

## INTRODUCTION

In the last years, tube hydroforming processes have undergone a significant evolution both with regard to the knowledge of process mechanics and parameters and as their industrial applicability issues are concerned. In fact, tube hydroforming technologies avoid the sequence of stamping and welding operations which characterizes the traditional tube production, guaranteeing at the same time, complex shape components production with high mechanical properties.

The tube hydroforming operations were widely investigated and it is well known that they are based on the concurrent actions of pressurized fluid and mechanical feeding which allow to obtain tube shapes characterized by complex geometries [1,2].

CP907, *10th ESAFORM Conference on Material Forming*, edited by E. Cueto and F. Chinesta
© 2007 American Institute of Physics 978-0-7354-0414-4/07/$23.00

The calibration of both material feeding history and internal pressure path during the process is the crucial task in the design of tube hydroforming operations since the prevention of bursting or buckling strongly depends on such calibration.

Moreover, in three dimensional processes, also the action of a counterpunch is generally useful to reduce thinning, in particular in expansion zones of the tube (i.e. T or Y shaped tubes). Many researches have proposed different approaches to the optimization of the typical tube hydroforming parameters. Some researchers proposed gradient based optimization methods [3,4] integrated with numerical simulations while other approaches have been presented based on statistical tools such as Taguchi method [5]. Also some applications of two dimensional finite element analysis to calibrate pressure histories were presented whose results were validated through experimental evidences [6]. Some authors also dealt with a typical phenomenon occurring in hydroforming processes namely a sort of intermediate beneficial wrinkling effect which, if occurring at an early stage of the process, is very helpful in obtaining safe components [7]. As complex three dimensional processes are concerned, some numerical investigations on T-shape tube hydroforming aimed to analyze the effects of operating parameters on the final part quality were presented [8,9]. As well, optimization of loading paths was investigated [10,11] basing on finite element method or utilizing fuzzy logic [12].

Recently, the authors developed some procedures to optimize pressure paths and punch velocity histories with the application of an integrated method FEM - Gradient based optimization tools [13]; such studies regarded axy-symmetrical components production.

In this paper, a fully three dimensional tube hydroforming operation is studied, aimed to produce a T-shaped tube. A numerical simulations campaign was developed in order to analyze the influence of process parameters on the final product quality.

In particular, the influence of the pressure path was considered in order to evaluate the effects of beneficial wrinkling phenomenon on thickness distribution. The basic idea is that the possibility to determine useful wrinkles at an early stage of the process may lead to better results in terms of maximum thinning on the final part. The numerical investigations led to a knowledge base about the process mechanics and the influence of wrinkling behavior which is very effective in order to implement an optimization procedure on the process parameters. Moreover a preliminary application of gradient based optimization techniques was carried out in order to determine optimal pressure paths to reduce fracture danger in a T-shape tube hydroforming operation also with the aim to increase the height of the tube bulged zone as it will be discussed in the following sections. In particular, the optimization technique and the investigated case studies will be presented in the following.

## THE GRADIENT BASED OPTIMISATION METHOD

The gradient based approaches proved their usefulness in the optimization problems in which the analytical links between problem objective and problem design variables are unknown. The formulation of such problems generally consists of the following steps: the definition of the set of design variables, the identification of an objective function, the solution of a certain number of direct problems (i.e. numerical simulations aimed to evaluate the values assumed by the objective function at the varying of the design variables) and finally the reaching of optimum values of the design variables. The general formulation of a minimization problem can be summarized as follows: 1) identify the design variables by a vector $x$; 2) choose the initial values: $x_k \in R_n$ with $k=0$ ($k$ denotes the method iteration number); 3)calculate the gradient of the objective function $\nabla f(x_0)$: if the convergence is reached the algorithm can be stopped; 4) else calculate an updated value of the design variables $x_{k+1} = x_k + \alpha_k d_k$ (where the scalar $\alpha_k \geq 0$ is called "step size" or "step length" at iteration $k$ and indicates the entity of design variables adjustment at iteration $k$; $d_k$ is the direction of movement i.e. the direction along which the objective function goes towards a minimum; 5) verify that $f(x_{k+1}) < f(x_k)$; 6) repeat step 4 and 5 until convergence is reached. Such general approach can be refined according to different techniques with respect both to the gradient calculation and to the definition of step size and step direction. Among the different possibilities available in the literature, the procedure proposed in the research project was the steepest descent method. Such method is based on the hypothesis that if a minimum of the objective function is required then the search direction is given by the opposite of the function gradient. In this way, a finite difference method was utilized in order to calculate the gradient; fixing a perturbation of the design variables $\varepsilon$ it was possible to calculate the gradient as follows:

$$\nabla f(x_k) = \frac{f(x_k + \varepsilon) - f(x_k)}{\varepsilon}. \tag{1}$$

The calculation of the gradient required the evaluation of the objective function values for each value and for each perturbation of the design variables.

In the application here presented the evaluation of the objective function was obtained through an integration with the numerical simulations: the FEM code provided the desired values of the objective function at each iteration of the applied method. As the step size evaluation is concerned a line search procedure was utilized in order to determine the most performing value of $\alpha_k$ with respect to the minimization of the function. The method is stopped when the convergence is reached, i.e. when the function gradient is equal to zero and the objective function is minimized. It has to be highlighted that this kind of method guarantees the reaching of a minimum (even a local one) for any initial values of the design variables. From a technological point of view this means that if the reached optimum is satisfying in terms of product quality, it can be considered a good solution for the given design problem.

## THE INVESTIGATED CASE STUDIES

The aim of this paper was the optimization of a T-shape tube hydroforming operation. The geometrical details of the utilized die are shown in Figure 1. The main goal of the optimization was the minimization of maximum thinning and the increasing of the bulge height obtained at the end of the process.

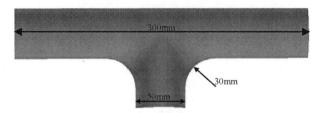

**FIGURE 1.** Die geometry of he analyzed case studies.

Thus, the objective function utilized in this application consisted of two terms: the former is the maximum thinning on the tube walls ($t\%$) and the latter measuring the distance ($d$) between the desired bulge height ($h_{tot}$)and the obtained one.

Moreover, since the basic idea of the optimization strategy was the exploitation of the beneficial wrinkling effect, the operation consisted of two phases: in the former phase both material feeding and internal pressure were kept constant, in the latter phase the material feeding was constant while the internal pressure had a linear trend reaching a maximum value. In this way, in the early phase of the operation a gathering of a proper amount of material in the bulge zone was obtained (beneficial wrinkling) while in the latter phase the desired bulge height was reached by a calibration action given by a pressure peak. According to the above considerations, two design variables were chosen to implement the optimization procedure, namely the pressure value in the early phase of the operation ($p_i$) and the punch stroke value corresponding to the beginning of the latter phase ($s_f$). In other words the pressure vs. stroke curve was optimized. A total punch stroke ($s_{tot}$) value was also fixed for each of the investigated cases. Moreover also a preliminary numerical campaign was performed to determine the most proper value of the final pressure (maximum pressure reached at the end of the process) which was equal to 35MPa. The chosen tube material is AA6060-T6 with a tube initial thickness equal to 2mm and an external diameter equal to 50mm.

The following three cases were investigated utilizing the LS-DYNA explicit commercial code:

case1) $h_{tot}$=34mm; $s_{tot}$=48mm; $p_{i0}$= 8MPa; $s_{f0}$= 28%$s_{tot}$;
case2) $h_{tot}$=40mm; $s_{tot}$=52mm; $p_{i0}$= 13MPa; $s_{f0}$= 60%$s_{tot}$;
case3) $h_{tot}$=40mm; $s_{tot}$=55mm; $pv_0$= 13MPa; $s_{f0}$= 53%$s_{tot}$

*Case1)* The first step of the investigation was focused on a case study characterized by a lower value of the desired bulge height with respect to the other cases. The initial values of the design variables ($p_{i0}$ and $s_{f0}$)led to a final maximum thinning equal to 9% and a final bulge height lower than the desired one. Moreover a wrinkled tube was obtained at the end of the process. The gradient based optimization procedure above described was applied and 2 iterations of the method were necessary to reach the optimal solution which guarantees a final thinning lower than 8% and the reaching of the desired final bulge height. Moreover, the obtained results were compared with two linear pressure paths: the first comparison (pressure path indicated by linerar1) led to higher thinning (11%) with respect to the optimal solution obtained with the optimization procedure; the second comparison (pressure path indicated by linerar2) led to a bulge height of 32,6mm worse than the one obtained with the optimal solution. Figure 2 shows: (a) the pressure paths (initial, iteration1 of the method and optimal) evolutions during the application of the optimization procedure and also the linear trends utilized for the comparisons; (b) an intermediate step of the FEM simulation utilizing the optimal pressure path and showing the beneficial wrinkling.

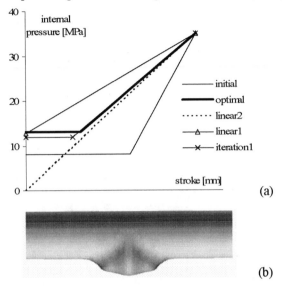

(a)

(b)

**FIGURE 2.** Pressure paths (a) and beneficial wrinkling effect (b) in case1.

*Case2)* The second case study was aimed to obtain an higher value of bulge height with respect to case1. In this way, an higher $s_{tot}$ value and a different pressure path were chosen in the initial step of the optimization, in order to reduce thinning and to obtain a 40mm bulge height. The optimization procedure led to a final thinning of about 8%, which is satisfactory, but a final bulge height of 38,6mm which was not the desired one. For these reasons a further case was analyzed (case3) in which the $s_{tot}$ value was increased and the pressure path was slightly modified.

*Case3)* In this case study only one iteration of the optimization procedure was necessary to reach an optimal solution. In fact a final thinning of 8% was reached together with a final bulge height of 40mm which was the desired one. Also in this case the beneficial wrinkling effect was observed proving that the chosen strategy to calibrate pressure path during the process was successful. Figure 3 shows: (a) the pressure paths (initial and optimal) evolutions in the optimization procedure; (b) the final component obtained shape with the thinning distribution.

## CONCLUSIONS

The early results of a research project on 3D tube hydroforming processes were presented. The investigated case studies aimed to build up a knowledge base on T-shape tube hydroforming in order to reach a proper optimization strategy for the operative parameters in such operations. The following developments will concern a deeper study of the utilised optimisation technique and the experimental verifications of the obtained results.

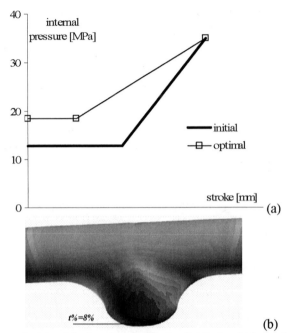

**FIGURE 3.** Pressure paths (a) and final component thinning distribution (b) in case3.

# REFERENCES

1. M. Ahmetoglu and T. Altan,, *Journal of Material Processing Technology* **98**, 25-33 (1998).
2. D. Schmoeckel, M. Geiger, C. Hielscher and R. Huber, *Annals of CIRP* **48**, 497-513 (1999).
3. J. Yang, B. Jeon, and S. Oh, *Journal of Material Processing Technology* **113**, 666-672 (2001).
4. C. Labergere and J.C. Gelin, "New strategies for optimal control of command laws for tube hydroforming processes", 8th ICTP Conference Proceedings, 2005.
5. B. Li and T. Y. Nye, *International Journal of Advanced Manufacturing Technology* **28**, 23-30 (2006).
6. M. Koc and T. Altan, *International Journal of Machine Tools & Manufacturing* **42**, 1285-1295 (2002).
7. L. H. Lang, Z.R. Wang, D.C. Kang, S.J. Yuan, S.H Zhang, J. Danckert and K.B. Nielsen, *Journal of Material Processing Technology* **151**, 165-177 (2004).
8. C.T. Kwan and F.C. Lin, *International Journal of Advanced Manufacturing Technology* **21**, 420-425 (2003).
9. H.K. Zadeh and M.M. Mashhadi, *Journal of Material Processing Technology* **177**, 684-687 (2006).
10. S. Jirathearanat and T. Altan, "Optimization of loading paths for tube hydroforming", 8th ICTP Conference Proceedings, 2005.
11. M. Imaninejad, G. Subhash and A. Lokus, *International Journal of Machine Tools & Manufacturing* **45**, 1504-1514 (2005).
12. P. Ray and B.J. Mac Donald, "Intelligent control of tube hydroforming processes using finite element analysis", 8th ICTP Conference Proceedings, 2005.
13. R. Di Lorenzo, G. Ingarao and F. Micari, "Optimization Internal pressure and material feeding optimisation in tube hydroforming", 9th Esaform Conference Proceedings, 2006, pp. 383-386.

# Magnesium Tube Hydroforming

M. Liewald, R. Pop, S. Wagner

Institute for Metal Forming Technology, Holzgartenstraße 17, 70174 Stuttgart, Germany
URL: http://www.ifu-stuttgart.de          e-mail:  mathias.liewald@ifu.uni-stuttgart.de
robert.pop@ifu.uni-stuttgart.de
stefan.wagner@ifu.uni-stuttgart.de

**Abstract.** Magnesium alloys can be considered as alternative materials towards achieving light weight structures with high material stiffness. The formability of two magnesium alloys, viz. AZ31 and ZM21 has been experimentally tested using the IHP forming process. A new die set up for hot IHP forming has been designed and the process experimentally investigated for temperatures up to 400 °C. Both alloys exhibit an increase in formability with increasing forming temperature. The effect of annealing time on materials forming properties shows a fine grained structure for sufficient annealing times as well as deterioration with a large increase at the same time. The IHP process has also been used to demonstrate practicability and feasibility for real parts from manufacture a technology demonstrator part using the magnesium alloy ZM21.

**Keywords:** Magnesium, Hydroforming, Alloy, AZ31, ZM21.
**PACS:** 01.10.Fv; 01.50.Pa

## 1. INTRODUCTION

Metals used in the automotive body parts are being increasingly recycled in order to reduce the strain on non-renewable energy resources. One of the various approaches aimed at decreasing the use of non-renewable energy resources is the use of light materials such as magnesium. Such materials meet demands of part designers and engineers with advantages such as lower part weight and higher part strengths [1]. Magnesium sheets and profiles have been proposed as candidates for future automotive steel and aluminium car body parts.

The value of weight reduction can be best judged by the amount of money a certain industry sector is willing to spend to achieve it [2]. As reported in [2], a weight gain of one kilogram cost about $ 6 in the automotive sector. The density of magnesium is the lowest among conventional structural metallic materials, which is about two third of the density of aluminium and one fourth that of steel [3], [4]. Furthermore, magnesium alloys exhibit higher specific strengths and stiffness, better damping capability, high dimensional stability and good machinability.

At room temperature as also at moderately elevated temperatures, hcp structured magnesium has a limited number of slip systems for deformation. This results in poor formability characteristics. However, under hot deformation conditions, prismatic and pyramidal slip systems are activated and it becomes deformable within certain limits.

CP907, 10th ESAFORM Conference on Material Forming, edited by E. Cueto and F. Chinesta
© 2007 American Institute of Physics 978-0-7354-0414-4/07/$23.00

This paper concerns Internal High Pressure (IHP) forming of magnesium alloy tubes at elevated temperatures. The objective is to investigate the formability of different magnesium alloy tubes by using the IHP forming process.

## 2. TEST MATERIALS

Investigations have been carried out with magnesium alloy AZ31 tubes extruded over a spider die and for tubes extruded over a moving mandrel. Further investigations have been made on ZM21 tubes extruded over a moving mandrel.

### 2.1.  Chemical analysis

In order to get information about material composition in the tube wall and the welding line (in the case of AZ31 tubes [3]), additional alloying additions are determined by spectroscopic analysis.

**TABLE 1.** Chemical Analysis of AZ31 and ZM21 tubes

| Element | Tubes extruded over a spider die | | Tubes extruded over a moving mandrel | |
| | AZ31 | | AZ31 | ZM21 |
| | Concentration in % in the bulk material | Concentration in % in the welding line | Concentration in % in the bulk material | Concentration in % in the bulk material |
| --- | --- | --- | --- | --- |
| Al | 2.61 | 2.61 | 2.88 | <0.05 |
| Zn | 0.82 | 0.87 | 0.98 | 1.7 |
| Mn | 0.59 | 0.53 | 0.26 | 0.7 |
| Fe | 0.005 | <0.002 | 0.002 | 0.004 |

Table 1 shows that in case of AZ31 extruded over a spider die, the Zn concentration is higher in the welding line than in the non-welded areas whereby the Mn concentration is lower in the welding line compared to that in the non-welded areas. The manganese concentration of the AZ31 tube extruded over a spider die is higher that in the seamless AZ31 tube. According to the ASTM B275 standard, the manganese concentration should be in a range between 0.15% and 1%, manganese improves weldability and Zn supports a fine-grained structure [4].

### 2.2.  Extrusion process

The objective of this work is to extrude magnesium alloy tubes which are suitable for hydroforming applications. To determine a suitable extrusion technique, magnesium alloy tubes extruded over a spider die as well as over a mandrel are used for the investigations.

Sophisticated profiles with very high dimensional- and shape accuracy are possible by extrusion over a spider die. The spider die tooling is usually fragmented in two parts. The shaping process of the inner contour of the extruded profile occurs in the upper tool, the external one in the lower tool [5]. The material that separates in more streams in the upper tool is bonded back together in the "welding chambers" of the lower tool. Hence, welding lines occur which can be observed even by a macroscopic examination. Extrusion over a moving mandrel, on the other hand, results in seamless hollow profiles of high strength alloys.

## 2.3. Material characterisation

To determine the mechanical properties after extrusion, tensile tests have been conducted for both alloys at 20°C, 235°C, 300°C, 350°C und 400°C forming temperature and average strain rate of 0.02 s$^{-1}$.

**TABLE 2.** Mechanical Properties of AZ31 and ZM21 seamless tubes at room temperature

|  | AZ31 | ZM21 |
|---|---|---|
| Yield Stess: | 208 MPa | 175 MPa |
| UTS: | 256 MPa | 250 MPa |
| Elongation to fracture | 8% | 19% |

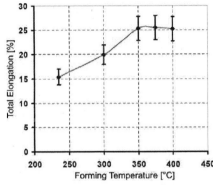

Figure 1 shows the achieved elongations at elevated forming temperatures. It can be seen that the achievable total elongation is 25% at 350°C, as against ca. 8 % at 20°C. For higher temperatures, no significant improvement of formability can be observed. Depending on specimen, the maximum elongation scatters in the range of ca. 5 %. Also, the ultimate tensile strength at elevated temperatures decreases from 105 MPa at 200°C to 35 MPa at 400°C.

**FIGURE 1.** Total elongation versus forming temperature for AZ31 seamless tubes

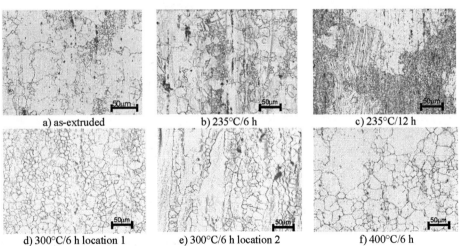

a) as-extruded    b) 235°C/6 h    c) 235°C/12 h

d) 300°C/6 h location 1    e) 300°C/6 h location 2    f) 400°C/6 h

**FIGURE 2.** Structure of AZ31 tube (longitudinal direction) in as-extruded state and after annealin

The influence of heat treatment and thereby material formability has been investigated for AZ31 magnesium tubes. The tubes were annealed at 235°C, 300°C, 350°C and 400°C for 6 h and 12 h respectively. The investigations show that no significant improvement in the structure occurs from the as-extruded state (Figure 2a) when the specimen is annealed for 6 hours at 235°C (Figure 2b). On the other hand, annealing for 12 hours at 235°C results in areas of fine grains (Figure 2c), whereby some big grains can be seen as well. At 300°C and annealing time of 6h, a most uniform structure can been observed (Figure 2d). However, the structure is not spread out uniformly all over across the longitudinal tube section. In the proximity of the outer tube surface, long stretched grains can be noticed (Figure 2f). Grain growth can also be observed for temperatures higher than 300°C.

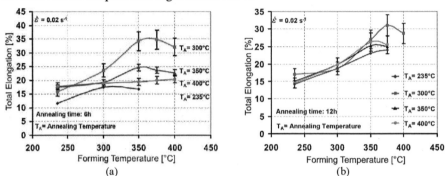

(a)                                               (b)

**FIGURE 3.** Elongation to rupture at different forming temperatures for different annealed specimens

Tensile tests have been conducted for the annealed specimens in order to determine the influence of structure on formability in axial direction. The alloy AZ31 has been considered here. Figure 3a shows the elongation curves over forming temperature for specimens annealed at different temperatures for 6 h. As against the elongation curves with annealing for 12 h at different temperatures, the highest total elongation has been observed for tubes annealed at 300°C for 6 h.

Annealing at different temperatures, viz. 235°C, 350°C and 400°C show no positive influence on the formability compared to the as-extruded state. In the case of specimens annealed for 12 h, a small improvement can be observed for annealing at 300°C. Likewise, as in case of 6 h annealing period, an increase in annealing time to 12 h does not improve the formability as compared to the as-extruded state.

The fact that the total elongation increases from 22 % at strain rate of 0.2 s$^{-1}$ to 46% at strain rate of 0.002 s$^{-1}$ (Figure 4) may support the assumption of grain boundary sliding as well. Furthermore, the tensile tests have shown that only the strain rate influences the UTS (Figure 5). For example, the UTS increases from 42 MPa at a strain rate of 0.002 s$^{-1}$ to 72 MPa at a strain rate of 0.2 s$^{-1}$ for a forming temperature of 350°C.

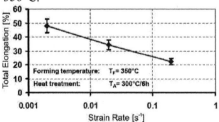
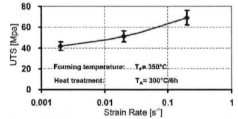

**Figure 4.** Achievable total elongations at different strain rates

**Figure 5.** UTS at 350°C forming temperature and different strain rates

## 3. EXPERIMENTAL SETUP FOR IHP FORMING AT HIGH TEMPERATURES

Figure 6 exhibits the lower die setup for IHP warm forming tooling designed and developed at the Institute of Metal Forming Technology, Stuttgart, Germany [6]. The tooling consists of two units; the clamping- and the forming unit. The pressure medium (nitrogen gas) is supplied through the sealing punch.

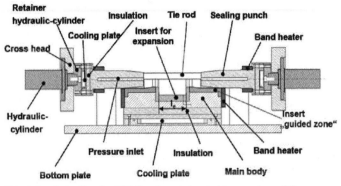

**Figure 6.** Principle of a tooling for IHP-forming at elevated temperatures [6]

The investigations have shown that the formability of magnesium in plane strain is limited. It is known that the formability can be improved if a compressive stress is superimposed in the material being formed. Furthermore, it is known that the range of feasible parts can be enlarged if the material can be fed forward into die cavity. The geometry of the sealing punches for IHP-forming with axial feeding needs to be different from the conical ones (used only for sealing). A new punch design was employed in order to avoid sticking of the tube material in the tool feeding area.

## 4. EXPERIMENTAL RESULTS

### 4.1. AZ31 tubes extruded over a spider die

The maximum achievable circumferential strain in ruptured parts can be found to be at 8 % at 300 °C under forming pressures of 0.73 MPa, whereby a pressure slope of 800 Pa s$^{-1}$ is selected. Figure 7a shows the formed tube with a rupture along the welding line. Further investigations have shown that independent from the forming temperature and strain rate, bursting occurs along the welding lines. One reason for part failure along the welding line may be attributed to the cavities in this area (Figure 7b). Since these cavities can be observed even after the extrusion process, the influence of forming temperature and strain rate on the maximum achievable circumferential strain is limited.

To avoid the cavities which appear at the welding lines and permit higher circumferential strains, further investigations have been conducted with tubes extruded over a moving mandrel.

421

a) tube ruptured on the welding line        b) microstructure in the welded portion

**Figure 7.** Magnesium tube formed at 300°C with an increase of pressure of 800 Pa s⁻¹

## 4.2. ZM21 deformed with axial feeding

IHP-bursting tests using magnesium alloy ZM21 with axial feed have been conducted at temperatures up to 350 °C. The maximum circumferential strain was measured to be 116 %. It is however to be noted that the formed part contacts the tooling in certain regions, leaving room for further extension of the mentioned limit. Figure 8a shows a demonstrator part formed at 350°C in which the maximum circumferential strain was measured to be 94%. Figure 8b shows the corresponding part thickness in the transverse direction for all four sides. The initial tube wall thickness amounts to 2 mm. A non-linear pressure curve has been selected with maximum value of 4.2 MPa. A non-linear pressure curve has been selected in order to increase forming pressure and calibrate the small contours after the initial forming of the part geometry.

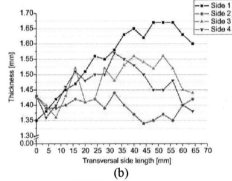

(a)                          (b)

**Figure 8.** Magnesium alloy ZM21 demonstrator part formed at 350°C using a non-linear pressure curve (a) and the corresponding part wall thickness (b)

## 5. SUMMARY

Magnesium alloy has been recognised as an alternative towards achieving light-weight structures with high material stiffness. The formability of two magnesium alloys, viz. AZ31 and ZM21 has been experimentally tested using the IHP-forming process. A new die set for hot IHP-forming has been designed and the process experimentally investigated for temperatures up to 400°C. The alloys exhibit an increase in formability with increasing forming temperature. The effect of annealing time on the forming properties shows a fine grained structure for sufficient annealing times and deterioration with a large increase in the same. The forming limit is thus found to be a maximum for annealing times of ca. 6 h as against annealing times of ca. 12 h. The forming limit increases from 19 % at 20°C to 116 % at 350°C in the case of ZM21 even under conditions of tool contact during the IHP forming process. The IHP process has also been used to manufacture a technology-demonstrator part using the magnesium alloy ZM21.

# REFERENCES

[1] H. Palkowski et al., "A Study of the Microstructure of Strip Cast Magnesium" in *Magnesium*, edited by K.U. Kainer, ISBN 978-3-527-31764-6, Proceedings of the 7th International Conference on Magnesium Alloys and Their Applications, Dresden, Germany, 2006, pp. 357-363
[2] M. Kettner et al., "Wrought Mg Alloy for Civil Aircraft Application – A Process Chain Approach" in *Magnesium*, edited by K.U. Kainer, ISBN 978-3-527-31764-6, Proceedings of the 7th International Conference on Magnesium Alloys and Their Applications, Dresden, Germany, 2006, pp. 305-317
[3] Jäger, S. and Wizemann, C., Internal High Pressure Forming of Magnesium AZ31 Tubes, Internal Report, Institute for Metal Forming Technology, Stuttgart, Germany, 2004
[4] C. Kammer, "Normung der Magnesiumlegierungen," in *Magnesium Taschenbuch*, edited by Aluminium-Zentrale Düsseldorf, 2000, pp. 141-153, ISBN 3-87017-264-9
[5] G. Sauer and A. Ames , "Werkzeuge zum Strangpressen" in *Strangpressen*, edited by M. Bauser et al., Düsseldorf, 2001, pp. 675-684, ISBN 3-87017-249-5
[6] Jäger, S. and Wizemann, C., "Internal High Pressure Forming of Magnesium AZ31 Tubes", Internal Report, Institute for Metal Forming Technology, Stuttgart, Germany, 2004

# Crash Performance Evaluation of Hydro-formed DP-steel Tubes Considering Welding Heat Effects, Formability and Spring-back

Kyung-Hwan Chung*, Junehyung Kim*, Wonoh Lee*, Ji-Ho Lim[¶],
Chongmin Kim[#], Michael. L. Wenner[§] and Kwansoo Chung*

*Department of Material Science and Engineering, Intelligent Textile System Research Center,
Seoul National University, 56-1,Shinlim-dong, Kwanak-ku, Seoul 151-742, Korea
[¶]Automotive Steel Applications Research Group, Technical Research Labs., POSCO,
699 Gumho-dong, Gwangyang-si, Jeonnam, 545-090, Korea
[#]Materials & Processes Lab., GM R&D and Planning, General Motors Corporation,
Warren, MI 48090-9055, USA
[§]Manufacturing Systems Research Lab., GM R&D and Planning, General Motors Corporation,
Warren, MI 48090-9055, USA

**Abstract.** In order to numerically evaluate hydro-formed DP-steel tubes on crash performance considering welding heat effects, finite element simulations of crash behavior were performed for hydro-formed tubes with and without heat treatment effects. Also, finite element simulations were performed for the sequential procedures of bending and hydro-forming of tubes in order to design process parameters, particularly for the boost condition and axial feeding, considering formability and spring-back. Effects of the material property including strain-rate sensitivity on formability as well as spring-back were also considered. The mechanical properties of the metal active gas (MAG) weld zone and the heat affected zone (HAZ) were obtained utilizing the continuous indentation method in this work.

**Keywords:** DP-steel, tube bending, hydro-forming, crash test, strain-rate sensitivity, formability, spring-back, welding heat effect, continuous indentation method.
**PACS:** 62.20.Fe

## INTRODUCTION

In an effort to reduce the weight of automobiles, significant efforts are being put forth in the automotive industry to replace the conventional steel with the advanced high strength steel such as the dual-phase (DP) steel. The DP-steel is particularly suitable for structural components such as the engine cradle because of its superior combination of ductility and strength. In optimizing the forming process of the DP-steel engine cradle however, the proper evaluation of the crash performance of the formed engine cradle is important in addition to the formability and spring-back evaluation. After being formed, the engine cradle is typically metal active gas (MAG) welded to the main body using brackets. Since the DP-steel undergoes significant mechanical property changes when welded due to welding heat effects, it might be

CP907, *10th ESAFORM Conference on Material Forming*, edited by E. Cueto and F. Chinesta
© 2007 American Institute of Physics 978-0-7354-0414-4/07/$23.00

necessary to properly account for the property change in order to accurately evaluate the crash performance.

In this work, the welding heat effects on crash performance have been evaluated for bent and hydro-formed DP590 tubes with 2.0mm thickness in efforts to improve the process design procedure of the engine cradle in the process design stage. Also, FEM simulations have been performed for sequential procedures of bending and hydro-forming of tubes in order to design process parameters, particularly for the boost condition and axial feeding, considering formability and spring-back. As for the constitutive law, the isotropic hardening law was used along with the non-quadratic anisotropic yield function Yld2000-2d [1]. Forming limit diagrams were calculated based on Hill's bifurcation theory [2] and the M-K theory [3]. The mechanical properties of the MAG weld zone and the heat affected zone (HAZ) were obtained by conjointly utilizing the continuous indentation method [4] and its FEM analysis.

## MATERIAL CHARACTERIZATION

In order to describe the initial anisotropic yield stress surface, the non-quadratic yield stress function, Yld2000-2d, was considered. This yield function for the plane stress condition is defined as

$$f^{\frac{1}{M}} = \left\{\frac{\Phi}{2}\right\}^{\frac{1}{M}} = \bar{\sigma} \tag{1}$$

where

$$\Phi = \left|\tilde{S}'_I - \tilde{S}'_{II}\right|^M + \left|2\tilde{S}''_{II} + \tilde{S}''_I\right|^M + \left|2\tilde{S}''_I + \tilde{S}''_{II}\right|^M. \tag{2}$$

In Equation (2), $\tilde{S}'_k$ and $\tilde{S}''_k$ ($k = I \sim II$) are the principal vales of tensor $\tilde{s}'$ and $\tilde{s}''$, which are deviatoric stresses modified by linear transformations, respectively, having eight anisotropic coefficients in total. Based on crystal plasticity, the exponent $M$ is recommended to be 6.0 for BCC materials.

The anisotropy of the DP590 base sheet was characterized using uni-axial yield stresses and $R$-values obtained from tensile tests along the directions of $0°$, $45°$ and $90°$ off the rolling direction. Also, the normalized balanced biaxial yield stress was assumed to be 1.0 and the condition, $L'_{12} = L'_{21}$ has been used for the balanced biaxial $R$-value [1]. From these results, the anisotropic coefficients were calculated and the resulting anisotropic yield characteristics are shown in Fig. 1. The anisotropic properties of the DP590 base tube, which was made from the DP590 base sheet, were assumed to be the same as those of the DP590 base sheet. On the other hand, the ERW zone (the weld zone made to fabricate the tube), MAG weld zone and HAZ were assumed to have the isotropic property.

The isotropic hardening constitutive law and the hardening equation of the Hollomon type, $\bar{\sigma} = K(\varepsilon_0 + \bar{\varepsilon})^n$, were applied for all the materials in this work. For the DP590 base sheet, the uni-axial tension data along the rolling direction, shown in

425

Table 1 was considered as the reference state. The hardening behaviors of the DP590 base tube and ERW zone were measured from uni-axial tension tests whose sub-sized specimens were cut out from the tube at various locations. For the convenience of numerical calculations, hardening curves except for the ERW zone were averaged for the base tube, and the hardening parameters determined are also listed in Table 1. As for the material properties of the MAG weld zone and HAZ, the continuous indentation method and its FEM analysis were conjointly utilized. Indentation tests were conducted at various locations near the MAG arc-weld zone, utilizing a sphere indenter, from which the experimental load-depth curves were obtained. Then, various hardening parameters were tried out for finite element simulations of the indentation test until the calculated load-depth curves show good agreement with experimental ones at every location, to determine the weld zone hardening parameters.

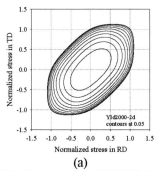
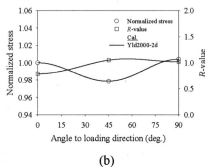

(a)                                (b)

**FIGURE 1.** Characteristics of Yld2000-2d for the DP590 base sheet: (a) yield surface contour and (b) anisotropies of normalized stress and $R$-value

**TABLE 1.** Hardening parameters of the DP590 sheet and tube

| Materials | $K$(MPa) | $n$ | $\varepsilon_0$ |
|---|---|---|---|
| DP590 base sheet | 1042.3 | 0.231 | 0.0224 |
| DP590 base tube (averaged) | 1003.8 | 0.222 | 0.0378 |
| ERW zone | 927.8 | 0.067 | 0.0034 |

Strain-rate sensitivity tests were carried out for the DP590 base sheet, using the specialized specimen whose gauge length is 32mm, at four constant engineering strain-rates: 1, 10, 100 and 200/s. The strain-rate sensitivity was described by the power law relationship, $\bar{\sigma} = C\dot{\bar{\varepsilon}}^m$, with the strain-rate sensitivity index, $m$=0.0145. The same strain-rate sensitivity property was applied for all materials.

Forming limit diagrams for all materials were calculated utilizing Hill's bifurcation theory and the M-K theory. Note that, since Hill's bifurcation theory is not so effective for strain-rate sensitive materials, the initial defect parameters were obtained so that the plain strain formability of Hill's bifurcation theory matches with that of the M-K theory without strain-rate sensitivity.

## NUMERICAL SIMULATION

FEM simulations were performed for the sequential procedures of tube bending and hydro-forming in order to design the process parameters, considering formability and spring-back: boost condition and axial feeding. Crash simulations were then carried out for hydro-formed tubes with and without considering heat treatment effects for comparison purposes. Three cases were considered in handling the initial tube material properties: (i) DP590 base sheet, (ii) DP590 base tube, (iii) DP590 base tube + ERW zone. For all cases, simulations were performed with strain-rate sensitivity.

## Tube Bending with a Flexible Mandrel

Figure 2(a) shows a schematic view of tools for the tube bending with a flexible mandrel. Here, two $R_c/d$ ratios were considered: $R_c/d$=2.0 and $R_c/d$=2.5. The straight tube is bended up to 90°, before unloaded for spring-back, by rotationally moving the bend and clamp dies together with respect to the bend die center as shown in Fig. 2(b). The bending velocity of the clamp and bend dies is 39.2 rad/s. An effective way to reduce wrinkling is to apply a tensile boost condition besides providing a flexible mandrel. The amount of boost is defined as the ratio of the pressure die velocity ($V_{pd}$) to the tangential velocity of the clamp and bend dies ($R_c \cdot \dot{\phi}$) in rotational motion (see Fig. 2(b)). Spring-back was quantified by the angle difference of $\theta$ for tube bending before and after spring-back ($\theta$ is defined in Fig. 2(b)).

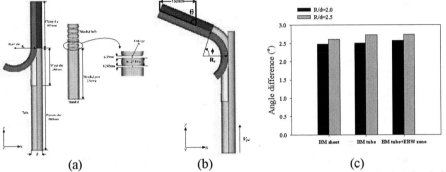

(a)  (b)  (c)

**FIGURE 2.** Tube bending with a flexible mandrel: (a) tube blank and bending tools (b) tube during bending (c) spring-back angle differences for the tube bending

The ABAQUS/Explicit code with the user-defined subroutine was first used to simulate the bending process and then the ABAQUS/Standard code was used to simulate spring-back. Simulation results without either wrinkling on the compressive side of the tube or failure on the tensile side of the tube could be obtained at the 85% boost level for both $R_c/d$ ratios. The simulated spring-back angle difference for the tube bending is shown in Fig. 2(c).

## Tube Bulging in Hydro-forming

After the bending process, tube bulging is performed sequentially by the hydro-forming tools shown in Fig. 3(a). In the hydro-forming operation, pressurization and axial feeding must be combined properly to prevent failure and wrinkling on the tube. For simplicity, in this work, the influence of only axial feeding on formability was numerically investigated.

Hydro-forming simulations were performed using the ABAQUS/Explicit code, after importing the optimized bending results with spring-back analysis. As for the simulation conditions, the internal pressure and three different axial feeding displacements were applied to the tube as their histories are illustrated in Fig. 3(b) and Fig. 3(c). The results with 25mm axial feeding condition showed that failures occurred in all cases, however those with 37.5mm axial feeding condition showed neither failure nor wrinkling.

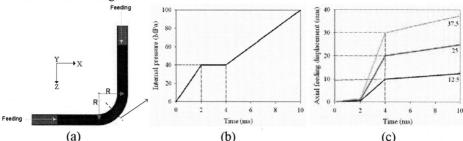

(a)　　　　　　　　　(b)　　　　　　　　　(c)

**FIGURE 3.** Hydro-forming process: (a) schematic view of the tools (b) internal pressure (c) axial feeding displacement curves

## Crash Simulations

In order to numerically evaluate the bent and hydro-formed tube, which has neither failure nor wrinkling, on crash performance considering welding heat effects, FEM simulations were performed with and without welding heat effects. After the material properties of the MAG weld zone and HAZ were imposed on the formed tube, crash simulations for two different cases were performed using the ABAQUS/Explicit code as illustrated in Fig. 4. Both cases have the boundary condition of one fixed bottom, while for case I the other end is free, and for case II the other end is constrained in the X-direction.

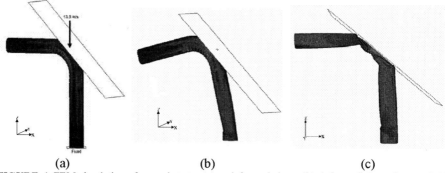

(a)　　　　　　　　　(b)　　　　　　　　　(c)

**FIGURE 4.** FEM simulations for crash tests: (a) undeformed shape (b) deformed shape for case I (c) deformed shape for case II

As for the simulation results, the maximum deceleration values for all cases are compared in Fig. 5. The maximum deceleration values with considering welding heat effects are lager than those without considering welding heat effects due to the increased strength of the MAG weld zone. However, the differences of the maximum deceleration values for the cases with and without welding heat effects were small even though the difference for case II was slightly larger than that for case I.

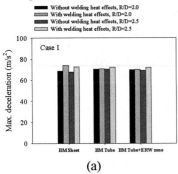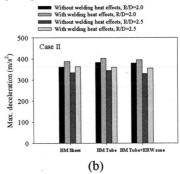

(a)                                                        (b)

FIGURE 5..Maximum deceleration values for (a) case I and (b) case II

# SUMMARY

In this work, the material characterizations for the DP590 sheet and tube, MAG weld zone, HAZ were carried out for the proper evaluation of crash tests. The continuous indentation method and its FEM simulations were used to obtain the material properties of the MAG weld zone and HAZ in particular. Based on the determined material characteristics, FEM simulations were performed for the sequential procedures of bending and hydro-forming in order to design the boost condition and axial feeding considering formability and spring-back. Crash performance evaluations were then numerically carried out for the tube, formed with optimized process parameters, with and without welding heat effects. With this work, numerical methods to properly optimize tube hydro-forming and also to properly implement the welding heat effects on the tube crash performance were established, which will be useful to evaluate the welding heat effects on the tube crash performance under realistic crash conditions.

# ACKNOWLEDGMENTS

This work was supported by General Motors & POSCO, which are greatly appreciated. Gratitude also goes to Dr. Dong-Seong Ro at Frontics for performing the indentation tests.

# REFERENCES

1.  F. Barlat, J. C. Brem, J. W. Yoon, K. Chung, R. E. Dick, S-H. Choi, F. Pourbograt, E. Chu and D. J. Lege, *Int. J. Plast.* **19**, 1297-1319 (2003).
2.  R. Hill, *J. Mech. Phy. Solids* **1**, 19-30 (1952).
3.  Z. Marciniak and K. Kuczynski, *Int. J. Mech. Sci.* **9**, 609-620 (1967).
4.  J.-H. Ahn and D. Kwon, *J. Mater. Res.* **16**, 3170-3178 (2001).

# Resulting material and structural properties after hydroforming processes

## J.C. Gelin*, S. Thibaud and C. Labergere

*FEMTO-ST Institute, Department of Applied Mechanics*
*ENSMM Besançon, 26 Rue de l'Epitaphe, 25030 Besançon, France*
*\*Corresponding author: jean-claude.gelin@univ-fcomte.fr*

**Abstract:** The paper is concerned both with new developments associated to optimization and control of hydroforming processes for the manufacturing of metallic liners used for hydrogen storage, and with the analysis of resulting material properties after the sequence of forming and hydroforming processes. In the first approach, an optimization procedure is proposed, consisting to determine the hydroforming pressure through an incremental step by step optimization scheme. In the second approach, one proposes a control loop running overall the process based on the adjustment of the volume of fluid in the inner tube cavity. Then an application corresponding to the hydroforming of metallic liners for hydrogen storage is related, and finally the resulting material properties of a sequence involving forming, trimming and hydroforming are related.

**Keywords:** Hydroforming, Fluid flows, Pressure Control, Structural Properties.
**PACS:** 81.05 Bx, 89.20 Bb

## INTRODUCTION

The hydroforming processes for flanges or tubes are now increasingly used in different fields as automotive and aeronautic industries, as well as for manufacturing tanks for storage of gas under high pressure [1, 2, 3]. These new application fields imply the use of incoming materials [4, 5] as high strength steels and other high strength metallic alloys.

Among hydroforming processes, one classically distinguishes two different technologies. In tube hydroforming technology, a tube generally with a circular cross section is first shaped by bending and clamped in a tooling system, and then filled by a fluid and finally pressurized [6]. It results that the tube takes the inner shape of the tools that allows producing complex seamless parts. The flange hydroforming technology consists to take benefit from hydraulic pressure to assist flange deep drawing by replacing the die cavity by a cavity filled by an hydraulic fluid [7], leading to the increase of Limit Drawing Ratio comparatively to classical deep drawing processes. The modeling and simulation of hydroforming processes through the finite elements method is now well established using accurate models accounting properly deformation mechanisms and plastic behaviors of thin structural parts [8, 9]. Pioneer finite simulations of flange hydroforming processes are related in [10] whereas numerous authors investigated the tube hydroforming processes in using analytical approaches [11] or numerical ones. The identification of material properties involved

CP907, *10ᵗʰ ESAFORM Conference on Material Forming*, edited by E. Cueto and F. Chinesta
© 2007 American Institute of Physics 978-0-7354-0414-4/07/$23.00

in hydroforming was investigated in [11] and the role of a proper material modeling has been underlined [12, 13].

One another important aspect is related to the prediction of formability limits, i.e. the necking and fracture in sheet hydroforming and bursting in tube hydroforming [14]. In that field, some authors apply the classical Forming Limit Diagrams (FLD) [14], but it has been underlined that the pressure strongly influences the FLD [15]. This fact can be accurately accounted through the linear stability analysis as mentioned in [16].

The way to control hydroforming processes to get components with required thickness variation and absence of local buckling and fracture was investigated in [17, 18] where the authors proposed to combine FEM simulations and optimization in order to optimize inner pressure vs. time. Finally the application of hydroforming processes for automotive fuel thanks manufacturing was investigated in [19].

## OPTIMIZATION AND CONTROL OF THE TUBE HYDROFORMING PROCESSES

The use of the finite element method coupled to optimization algorithms is a solution to obtain the optimum parameters for the hydroforming processes, as the FEM is used to evaluate the effects of a set of parameters on the process. In a first approach, one defines an objective function (eq.1) that intends to minimize thickness variation during hydroforming, resulting in an expression as:

$$F_{obj}(p) = \frac{1}{n}\sqrt{\sum_{i=1}^{n}\left(\frac{h_i(p) - h_0}{h_0}\right)^2} \tag{1}$$

where $h_0$ is the initial tube thickness, $h_i$ stands for the thickness at node i and n is the total nodes number involved in the problem. The optimization method (OM) proposed by the authors [17] consists to combine the simulation of the process and the evaluation of the cost function and then to minimize this cost function with constraints respect in using a SQP algorithm. One also proposes [18] a strategy based on a control algorithm to determine the command laws for tube hydroforming to get a better quality parts. It consists to include a control loop inside the incremental loop associated to the finite element solution procedure. The employed methodology consists in combining the theory of optimal control with the response surface method (Moving Least Square Approximation). The strategy adopted here consists to build an approximation for the pressure law vs. time. This approximation can be specified in a simple polynomial form or a more complicated one allowing possibilities to get the form of the command law. In a first stage, one choose a linear piecewise pressure approximation expressed in the following form:

$$p(t) = \frac{p_{i+1} - p_i}{t_{i+1} - t_i}t + \frac{p_i t_{i+1} - p_{i+1}t_i}{t_{i+1} - t_i} \tag{2}$$

In summary the hydroforming process control follows the diagram in figure 1.

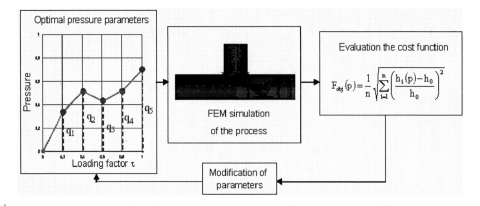

**Figure 1.** Optimization loop used for process control in tube hydroforming

In addition to the optimization and control procedure above described, one propose here to control the hydroforming process starting from the volume of injected fluid in the inner tube cavity. The internal pressure is often defined as a time dependent function denoted as P(t), but the pressure drops cause stability problems in certain cases. This aspect can be avoided by imposing the volume of injected fluid according to V(t).

It is thus necessary to specify the compressibility modulus K for the fluid (K=2200 MPa for the water at room temperature).

The pressure is then related to the fluid volume with the following equation:

$$P(t) = K \ln\left(\frac{V_0(t)}{V(t)}\right) + L(t) \tag{3}$$

where P(t) is the pressure, V(t) is the volume of fluid in the compressed state, $V_0(t)$ is the volume of fluid in the uncompressed one and L(t) is an added pressure as a function of time, where the volume of fluid is calculated from the mass of injected fluid related to the density.

## APPLICATIONS
### Hydroforming of Metallic Liners

As an application of the proposed approach, one search to obtain a metallic liner for hydrogen storage where the geometry of the hydroforming die is described in figure 2. The geometry of the initial tube corresponds to an external diameter equal 40mm and a thickness equal to 5mm. The material corresponds to an austenitic stainless steel SS304. The elastic properties correspond to E =196000 MPa, $v$= 0.3 and the hardening law is expressed as $\sigma_y = K \overline{\varepsilon}^{-n}$ with $K$ =1250 MPa and $n = 0.4$.

The contribution of pressure is simply used to increase the inner volume of the part. When the tube and the die are completely in contact, the pressure increases quickly, as one observes at t = 0.0018s, see figure 3. It has to be noted that the evolution of fluid volume follows a regular and quasi-linear increasing curve from t=0.01s. The study has been completed with the analysis of the risks of failure and bursting that could occur during hydroforming. In figure 4, one relates the straining paths and the FLDs obtained from the linear stability analysis as related in [16] accounting for internal pressure effects. It clearly appears no risks of failure or bursting during the hydroforming process.

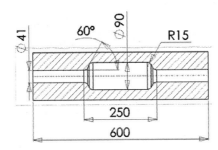

**Figure 2.** Geometry of the die cavity (dimensions in mm)

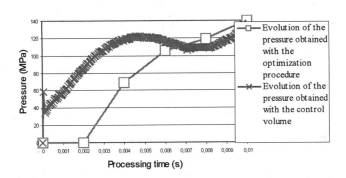

**Figure 3.** Comparison of the pressure curve vs. time obtained with optimization procedure and control volume

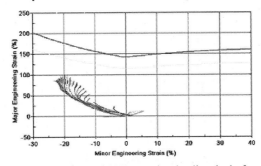

**Figure 4.** Strain paths and CLFs associated to liner hydroforming

# APPLICATION TO THE SIMULATION OF A CRASH TEST

The example that one consider corresponds to the nowadays preoccupation in automotive industry, as it concerns the crash of a passive security part obtained from forming-assembling sequences. Again, the results obtained with and without the TRIP steel model are compared. This example consists in a crash test of a passive security part inspired from [20, 21]. One studies the impact of 850 kg mass with an initial velocity equal 54km/h on a thin walled structural part. In figure 5, the simulation procedure is presented. In a first time, it consists to taking into account forming, springback and assembling processes. In a second time, crash simulations are carried out and then investigated.

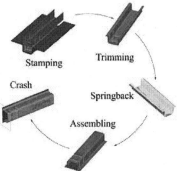

**Figure 5.** Numerical procedure used for complete crash simulation with metal forming effects.

Different simulations have been done. First, the crash test has been carried out on the CAD model. No initial stress exists in the part and this case is called model 1. Secondly, the crash test has been conducted on a security part obtained by the complete simulations chain: forming, springback, assembly. The TRIP model has been used for the simulations; this is model 2. The comparisons between models 1 and 2 show a difference in the deformed shape after crash (figure 6). Taking into account the forming effects (model 2), a crushing can be observed first and then localization appears. A plastic kneecap clearly appears after the embossed part that breaks the wave propagation. In model 1, without forming effects, a crushing followed by a buckling is observed.

*(a)*                                                                                    *(b)*

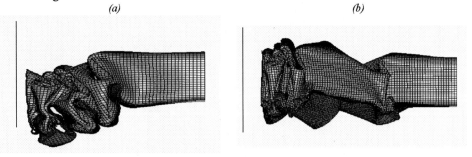

**Figure 6**: Deformed shape obtained after crash test: (a) model 1 and (b) model 2

For energy absorption, the same observations can be done. Between 0 and 8 ms, the deformation is localized in the embossments. After that the behaviour is different. The results obtained with model 2 are in agreement with experimental works [20] indicating that the dynamic buckling doesn't appear. Taking into account the forming effects (model 2), a crushing can be observed first and then localization appears. A plastic kneecap clearly appears after the embossed part that breaks the wave propagation. In model 1 without accounting forming effects, a crushing followed by a buckling is observed. For energy absorption, the same observations can be done. Between 0 and 8 ms, the deformation is localized in the embossments and then the behaviour is different. The results obtained with model 2 are in agreement with experimental works [20] indicating that the dynamic buckling doesn't appear.

## CONCLUSIONS

This paper summarizes different approaches for the optimal control of process parameters in flange and tube hydroforming. One has first underlined the necessity to properly account pressure variations as well as fluid flows arising either in flange hydroforming or in tube hydroforming. One has also evaluated the standard optimization methods that are used in flange or tube hydroforming, and one propose a new promising method based on the control of volume of fluid in the inner tube cavity, that avoid pressure drops and abrupt thickness variations. The proposed methodology has been applied to the hydroforming of metallic liners used for manufacturing hydrogen storage tanks. Finally the effects of the phase transition on elastic-plastic properties are described and adapted to the simulation of the behaviour of stamped parts and in-use properties.

## REFERENCES

1.  M. Kleiner, M. Geiger, A. Klaus, Annals of the CIRP **52/2**, 521-542(2003).
2.  M. Merklein, M. Geiger, J. Mater. Process. Technol. **125-126**, 532-536(2002).
3.  Nouvelles technologies pour l'énergie, CLEFS CEA **44** (2001).
4.  E.J. Vinarcik, Light Metal Age **60**, 38-41(2002).
5.  N. Alberti, A. Forceliese, L. Fratini, F. Gabrielli, Annals of the CIRP **47/1**, 217-220(1998).
6.  F. Dohmann, C. Hartl, J. Mater. Process. Technol. **71**, 174-186(1997).
7.  J. Tirosh, P. Konvalina, Int. J. Mech. Sci. **27**, 595-607(1985).
8.  P. Boisse, J.C. Gelin, J.L. Daniel, Computers and Structures **58/2**, 249-261 (1996).
9.  L. Boubakar, L. Boulmane, J.C. Gelin, Eng. Comput. **13/2-3-4**, 143-171 (1996).
10. J.C. Gelin, P. Delassus, Annals of the CIRP **42/1**, 305-308 (1993).
11. N. Asnafi, Thin-walled Structures **34**, 295-330 (1999).
12. T. Sokolowski, K. Gerke, M. Ahmetoglu, J. Mater. Process. Technol. **92**, 34-40, (2000)
13. K.I. Manabe, M. Amino, J. Mater. Process. Technol.**123**, 285-291 (2002).
14. M. Koç, T. Atlan, Int J. Mach. Tool Des. Res. **42**, 123-138 (2002).
15. L.P. Lei, B.S. Kang, S.J. Kang, J. Mater. Process. Technol. **113**, 673-679 (2001).
16. A. Lejeune, N. Boudeau, J.C. Gelin, J. Mater. Process. Technol.**143-144**, 11-17(2003).
17. J.C. Gelin, C. Labergere, J. Mater. Process. Technol. **125-126**, 565-572(2002).
18. J.C. Gelin, C. Labergere, Int. J. Forming Processes. **7/1-2**, 141-158(2004).
19. B.S. Kang, B.M. San, J. Kim, Int. J. of Machine Tools & Manufacture **44**, 87-94(2004).
20. S. Thibaud, N. Boudeau, JC Gelin, J. Mater. Process. Technol. **177**, 433-438(2006).
21. J.C. Gelin, C. Labergere, S. Thibaud, J. Mater. Process. Technol. **177**, 697-700(2006).

# 6 – FORGING AND ROLLING

## *(J. L. Chenot and J. Kusiak)*

# Localisation of damage zones with respect the final part in forging process for different geometric parameters

Ibrahim Khoury, Pascal Lafon, Laurence Giraud-Moreau, Carl Labergère

*Charles Delaunay Institute (FRE CNRS 2848)*
*Laboratory of Mechanical System and Concurrent Engineering*
*University of Technology of Troyes*
*12 Rue Marie Curie, BP2060, 1000 TROYES France.*
*Phone: 0033325715671 – fax: 0033325715675*
*ibrahim.khoury@utt.fr*

**Abstract.** The important criterion to obtain a good forged piece is the correct filling and the damage. A good filling means that the contour of the rough forged piece contains the contour of the final shape. If the damage occurs in zones that will be machined we consider that a good forged piece has been obtained. The goal is to differ between a correct and a defected rough forged depending on the obtained form of the rough forged and the location of the damage.

**Keywords:** Geometric parameters, objective functions, forging process, numerical simulation, damage zones.
**PACS:** Forming, 81.20.Hy

## INTRODUCTION

In an environment of increasing international competition where countries with lower production costs quickly catch up technologically, new thinking is required in order to meet the competition [1]. Most of time, specialists have some liberties to obtain the desired forged parts, that's why numerical simulations are applied in most of engineering offices and manufacturing industries to evaluate the forming difficulties in metal forming process.

The most important driving force in a manufacturing process is to produce the best product at the lowest cost [3]. The simulation process requires the introduction of some input data such as the material behaviour laws, the tools geometry, the initial part geometry and the process parameters (velocity, material friction laws), etc. To differ between a correct and a defected rough forged depending on the obtained form of the rough forged and the location of the damage, an automatic procedure has been developed in order to compare the geometry of the rough forged and the geometry of the machined forged part [7]. In this paper, an example of numerical simulation presents the optimal geometric parameters obtained by the experimental design and allows obtaining a correct forged part.

CP907, *10th ESAFORM Conference on Material Forming,* edited by E. Cueto and F. Chinesta
© 2007 American Institute of Physics 978-0-7354-0414-4/07/$23.00

## TOOLS IN LASMIS

A finite element package has been developed in the Mechanical System and Concurrent Engineering Laboratory to solve elastoplasticity problems with ductile damage in large deformation [9]. This package allows realizing numerical simulation of metal forming processes. The global resolution of the coupled system is carried out by using Abaqus/Explicit. For the simulation of 2D metal working processes a special procedure (shell script) has been developed in order to execute automatically Abaqus step by step together with an adaptive meshing methodology [4, 8].

## MECHANICAL ANALYSIS OF FORGING PROBLEM

As the strain rate varies along the deformation stroke during the actual forging step, hot compression tests at continuously varying strain rate were also performed, aimed at evaluating the effects of strain rate histories on the material instantaneous flow strength [10].

### Basic Mechanical Equations

In hot forging processes the elastic part of strain is negligible when compared with the large plastic strains present and the temperature variations do not significantly influence the material flow. So, the isothermal flow formulation can be used to model the process. The material considered in the simulation process is an elastoplastic material with ductile damage having the same specification of Z12CNDV12 at 950°C [6] with 100% of elastoplastic deformation. A user material Umat routine integrated with Abaqus/Explicit is activated. The Young's Modulus $E$ and the maximum yield Stress $\sigma_y$ of the elastoplastic material depend on the forging temperature.

$$E = 230883 - 30000 * e^{0.00145.T} = 111933 MPa \qquad (1)$$
$$\sigma_y = 407.5 - 0.376 * T = 51.25 MPa \qquad (2)$$

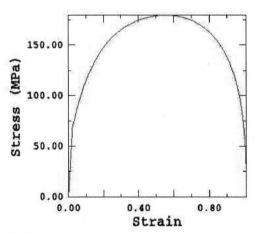

FIGURE 1. Stress versus Strain

# Localisation of Zones of Damage

The goal of an experimental design is to know the influence of some geometric parameters defining the tools on different objective functions, and also to differ between a correct and a defected rough forged depending on the obtained form of the rough forged (correct filling) and the location of the damage. A damaged part is not taking into account in the experimental design.

A good forged piece is a piece with no default, with correct filling and without crack and shrinking. Before localising the zone of damage (if there is) in the rough forged, we should test the filling.

Therefore, an automatic procedure has been developed in order to compare the geometry of the rough forged and the geometry of the machined forged part. A second automatic procedure has been developed in order to localise the damage in the forged part. The methodology consists of two main steps: first (correct filling), and second, with a priori knowledge of location of the damage, analyse of the effect of the damage on the rough forged damage (no crack and shrinking).

If the contour of the rough forged doesn't contain the contour of the forged part desired after machining we consider that the forged part is not useful, if not [FIGURE 2], we pass to the second step, in this step, we test the localisation of the damage, if the damage occurs in zones that will be machined, we consider that we obtained a correct forged part. On the contrary, if the damage occurs in zones that are supposed to be inside the machined forged part, we consider that the forged part is damaged.

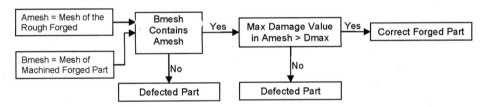

**FIGURE 2.** Actigram of the Procedure of Localisation of Damage

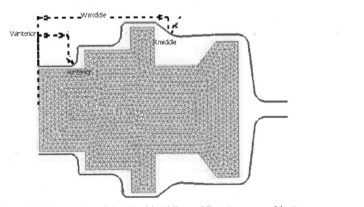

**FIGURE 3.** The Rough Forged (red contour) and the Machined Forged Part (green meshing)

441

In details the procedure tests all elements of the meshed rough forged and the meshed machined forged part. The position of all nodes of the meshed machined forged part should be localised inside the elements or mistaken with the node of meshed machined forged, if yes we test the damage in the meshed rough forged, if the value of damage exceed the given value by the client the script is able to localise it and to inform us if the element damaged is localised in the machined forged part or no, if yes we consider that the rough forged is defected.

Only geometrical forms allowing obtaining a correct forged part are taking into account to find the optimal geometric parameters of the process.

## EXPERIMENTAL DESIGN

Lots of parameters have an influence on the optimisation of the forming process. The type of material can not be taken as an optimisation parameter because it's generally imposed by the designer. During the forging process we notice the presence of an extra material included and a residual material called burr in the rough forged, this extra material is subsequently machined than the burr is deburred to get the final disk shape.

The geometry of the tools is the most interesting parameter to study because blacksmith make lots of errors in choosing the form of the tool. In this paragraph, an operation of forging of a turbine disk is simulated in 2D using Abaqus/Explicit (axisymetric model).

Two experimental designs are presented; in each one we vary two parameters (a distance and a radius) [FIGURE 3] to show the influence of the geometry on the total strain energy $E_T$ [2, 11] and the average of the damage D.

$$E_T = \int\limits_0^t \int\limits_V \left( \sigma^c : \dot{\varepsilon}.dV \right) dt \tag{3}$$

$\sigma^c$ : The Cauchy stress matrix

$\dot{\varepsilon}$ : The total (elastic and plastic) strain rate

The billet has a cylindrical shape with a diameter of 170mm and an elevation of 80mm.

From FIGURE 5 we conclude that the lowest value of the total strain energy $E_T$ corresponds to Winterior=12.5mm and Rinterior=3.5mm with a difference of 23% with the higher one, and the lower value of $E_T$ corresponds to Wmiddle=62mm and Rmiddle=13mm with a difference of 35% with the higher one.

For theses specific experimental designs, for all simulations the highest value of damage occurs in the area in the centre of the machined forged part and it doesn't reach 0.3, Which means the critical zone is located on the centre.

The maximum value of damage in the rough forged occurs more than Dmax=0.3 given by the client but the zone that reached this maximum value will be machined (it's the external surface of the rough forged) [FIGURE 4], which means that all geometric forms taken in the two experimental designs satisfy the criterion.

To resume lots of criterions are taken into account to obtain the optimal parameters. The first important step is to respect the specified constrains (the maximum value of

442

damage). In second step we look for the optimal parameters that give the minimum values of $E_r$.

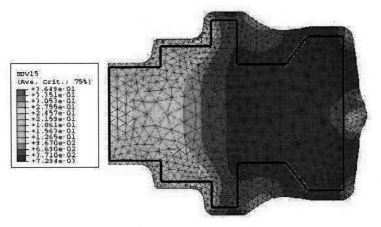

**FIGURE 4.** Distribution of the Damage in the Rough Forged

**TABLE 1.** Total Strain Energy $E_r$ (Joules) for different values of Winterior and Rinterior

| $W_{interior}$ , $R_{interior}$ in mm | $R_{interior} = 3.5$ | $R_{interior} = 4$ |
|---|---|---|
| $W_{interior} = 12$ | 1.187E+07 | 1.181 E+07 |
| $W_{interior} = 12.5$ | 1.006 E+07 | 1.063 E+07 |
| $W_{interior} = 13$ | 1.052 E+07 | 1.244 E+07 |
| $W_{interior} = 13.5$ | 1.098 E+07 | 1.125 E+07 |
| $W_{interior} = 14$ | 1.152 E+07 | 1.180 E+07 |

**TABLE 2.** Total Strain Energy $E_r$ (Joules) for different values of Wmiddle and Rmiddle

| $R_{middle}$ , $W_{middle}$ in mm | $W_{middle} = 60$ | $W_{middle} = 62$ |
|---|---|---|
| $R_{middle} = 9$ | 1.162 E+07 | 1.182 E+07 |
| $R_{middle} = 11$ | 1.199 E+07 | 1.214 E+07 |
| $R_{middle} = 13$ | 1.026 E+07 | 0.895 E+07 |
| $R_{middle} = 15$ | 1.046 E+07 | 0.994 E+07 |
| $R_{middle} = 17$ | 1.056 E+07 | 0.978 E+07 |

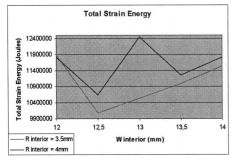 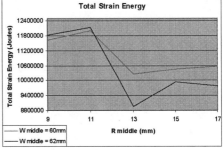

**F IGURE 5.**Total Strain Energy $E_r$ in Joules

# CONCLUSION

An experimental design has been done on an example of a turbine disk.

In order to be able to determine the optimal geometric form of the tools, the total strain energy has been calculated the experimental designs.

An automatic procedure has been developed in order to compare the geometry of the rough forged and the geometry of the machined forged part. A second automatic procedure has been developed in order to localise the damage in the forged part. These two procedures allow us to differ between a correct and a defected rough forged. Therefore the geometric forms allowing obtaining a correct forged part are taking into account to find the optimal geometric parameters of the process.

Our main objective in the future is to develop a step of forming integration coupled with an optimization procedure in order to obtain the optimal parameters of the process with respect to the constrains given by the client, The objective function will be a combination between the $E_T$ and the area that should be machined to obtain the machined forged part because the smaller the machining, the higher the quality of the forging [12].

# REFERENCES

1. I. Khoury, P. Lafon, L. Giraud-Moreau and C. Labergére, *Study of Simulation of Forming Process Toward an Optimization Procedure of the Process*, Publisher City: Publisher Name, 2005, pp. 25-30.
2. I. Khoury, L. Giraud-Moreau, P. Lafon and C. Labergére, Towards an Optimisation of Forging Process using Geometric Parameters, *Journal of Materials Processing and Technology 177*, 2006, pp. 224-227.
3. D. Vieilledent, "Optimisation des Outils en Forgeage à Chaud par Simulation Éléments Finis et Méthode Inverse Applications à des Problèmes Industriels", Ph.D. Thesis, Ecole Nationale Supérieure Des Mines De Paris, France, 1999.
4. Y. Hammi, "Simulation Numérique de l'Endommagement dans les Procédés de Mise en Forme ", Ph.D. Thesis, University of Technology of Troyes, France, 2000.
5. P.J. Frey, P.L. Georges, "Maillage Application aux Eléments Finis," edited by Institut National de Recherche en Informatique et en Automatique, INRIA. Publisher Hermès, 2004.
6. J. F. Mariage, "Simulation Numérique de l'Endommagement Ductile en Formage de Pièces Massives", Ph.D. Thesis, University of Technology of Troyes, France, 2003.
7. J.F. Boujut, "Un exemple d'intégration des fonctions métier dans les systèmes de CAO : la conception des pièces forgées tridimensionnelles", Institut National Polytechnique de Grenoble, France 1993.
8. K. Saanouni, "Numerical Modelling in Damage Mechanics" in *NUMEDAM'00, Revue Européene des Eléments Finis-2001*, N°10 pp. 2-4.
9. A. Cherouat, H. Borouchaki, K. Saanouni, P. Laug. "An Adaptive Remeshing Procedure in Elastoplastic Deformation with Ductile Damage: Application in Metal Forming Processes", *ESAFORM, Proceedings of 6th International Conference on Material Forming, 2003*.
10. M. P. Brown and K. Austin, *The New Physique*, Prod. Eng. 5/1, 1998, pp. 127-130.
11. L. Fourment, J.L. Chenot, Optimal design for non-steady-state metal forming processes. II. Application of shape optimization in forging, *International Journal Numerical Methods in Engineering N°39* ,1996. pp. 33–65.
12. X. Zhao, G. Zhao, G. Wang, T. Wang, Sensitivity Analysis Based Multiple Objective Preform Die Shape Optimal Design in Metal Forming, *International Journal of Material Science technology Vol.22 N°2* ,2006. pp. 273–278.

# The Results Of The Investigation Of Thermomechanical Processing Of PM Steel

Prof. dr hab.inż. Stefan Szczepanik, Mgr inż. Bartosz Wiśniewski, Dr inż. Jerzy Krawiarz

*AGH University of Science and Technology, Faculty of Metals Engineering and Industrial Computer Science, al. Mickiewicza 30, 30-059 Kraków, Poland,*
szczepan@metal.agh.edu.pl

**Abstract.** Hot die forging of PM steel is used to obtain products with high densities. The combination of this process with heat treatment of forgings directly after their forming is researched in order to reduce energy consumption in the manufacture of PM steel products. This work determined the influence of the cooling ratio directly after hot forging of PM steel samples on their structure and mechanical properties. The properties of the PM preforms were examined after sintering and after sintering, quenching into water and tempering for 1 h at 250, 350 and 550°C, respectively, as well as after forging at given temperatures and cooling in water and air, respectively. Forged steel after quenching was tempered at the same temperature as the sintered samples. Good mechanical properties were obtained by hot forging at 1100 °C. Sintered steel with 0.6 % $C_{graphite}$ is characterized by good hardenability and is susceptible to plastic forming at 1100 – 940 °C. During its cooling in air a bainitic-martensitic structure is obtained, whereas after cooling in water the structure is martensitic. The properties of the forged steel are strongly dependent on deformation temperature and cooling conditions. The tensile strength of the forged PM steel with 0.6 Cgraphite after forming at 1100°C is much higher than that of the same heat-treated as-sintered steel. Traditional heat treatment applied to materials after deformation at 1100 °C slightly increases properties in comparison to the material directly quenched into water. The best strength was $1585 \pm 193$ MPa, bending strength $3364 \pm 142$ MPa and hardness $588 \pm 43$ HB. Application of controlled cooling of sintered PM steel directly after close-die forging diminishes the energy consumption during product manufacture.

**Keywords**: PM steel, forging, heat treatment, mechanical properties, structure.

## INTRODUCTION

The primary factors that determine the properties of products manufactured from metal powders are, apart from their chemical composition, their density and microstructure. This is the reason for the research and development of methods that allow to diminish the porosity of the products manufactured by the PM route, such as e.g. double pressing and sintering, sintering with the liquid phase, high temperature sintering, hot densification, surface burnishing and forming, as well as for the optimization of the conditions of heat treatment [1-4]. Heat treatment as an additional process after sintering is used to increase the tensile properties of such products as,

CP907, *10th ESAFORM Conference on Material Forming,* edited by E. Cueto and F. Chinesta
© 2007 American Institute of Physics 978-0-7354-0414-4/07/$23.00

among others, sintered constructive elements of engines and gear boxes [2-3]. The enhancement of these properties is also possible with a shortened technological cycle of heat treatment thanks to the sinter hardening [3].

Hot die forging of the sintered material allows for its densification up to ca. 0,97 ÷ 0,99 of the density of the solid material [2-5]. A proper selection of the chemical composition and the conditions of conventional hot treatment of the products from sintered steel that underwent hot forging results in their high tensile strength (Rm over 1500 MPa) and hardness over 450 HB [6]. A favourable effect of the changes of the properties of sintered products was obtained in the work [7, 8] as a result of direct joining of hot forming with quenching. Examination of the possibilities of using such joint procedure in the manufacture of products from sintered steel obtained on the basis of Distaloy DC-1 iron powder, with special focus on the evolution of the properties of the material, is the subject of this paper.

## EXAMINATION OF THERMOMECHANICAL PROCESSING OF SINTERED STEEL

### Subject, scope and aim of the research

The subject of the research was to obtain experimental data on the influence of the conditions of hot die forging and cooling from the austenite range with and without preheating before forging.

The scope of the research covered:
- dilatometric determination of critical temperatures of phase changes in the sintered steel with heating rate 3°C/min.
- obtaining data about the kinetics of the transformation of superfused austenite of sintered steel at continuous cooling and their analysis in the form of CCT - diagram,
-examination of density, microstructure, fractures topography and selected mechanical properties in several variants of die forging and heat treatment.

The object of the research is sintered steel on the basis of Distaloy DC-1 iron powder (content of 2.0 % Ni, 1.5 Mo) with the addition of $0,6\%C_{grafit}$ and a lubricant. Compacts with $\phi$ 48 mm weighing 0,15 kg were pressed at 450 MPa. Sintering was conducted in an industrial furnace of POLMO Łomianki S.A. company, at the temperature 1120°C in 40 min in endogas atmosphere.

The results of the dilatometric examination for the steel obtained in this way, using the dilatometer DT-1000 by Adamel, are presented in the form of CCT- diagram (fig. 1). The chart also shows the cooling curves registered during their cooling in water and air, directly after the end of their forging or after austenitizing. Sintered steel after sintering and slow cooling in industrial furnace with the speed of 0,3°/s should, according to the data in fig. 1, have a bainite structure. In the microstructure of the steel in this form (fig. 2) there also occur, against the bainite background, bright areas that are not etched easily in nital. X-ray microanalysis showed that they are significantly enriched in nickel.

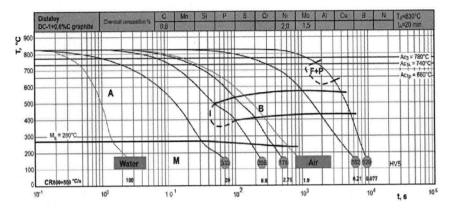

**FIGURE 1.** CCT - diagram of sintered steel on the basis of iron powder Distaloy DC-1, with the addition of $0,6\%C_{grafit}$, and the cooling curves from the austenite range (in water or air) after forging .

**FIGURE 2.** Microstructure of the sintered steel on the basis of the iron powder Distaloy DC-1 with $0,6 \% C_{graphite}$, etched with nital.

## The course and results of the research

Hot forming of sintered steel was performed in one-phase range γ (above $Ac_3$) and two-phase range (α + γ).The former variant involved heating the compacts to 1100°C and forming in closed-die with the deformation ca. 22 %. Immediately after forming the forgings were cooled in air or in sand. As a result of these procedures products with the density of solid material were obtained. After quenching, the material was tempered at 250°C in 1 h. To have a comparison, the products cooled in air were subjected to heat treatment: austenitizing in 870°C in 20 min and cooling in water.

As a result of forging at 1100°C, and the heat treatment conducted, the obtained materials had a structure of low-temperature tempered fine martensite, both in the case of the samples after plastic and heat treatment, and after quenching and tempering (fig.3) The material after forging and cooling in air has a bainite-martensitic structure.

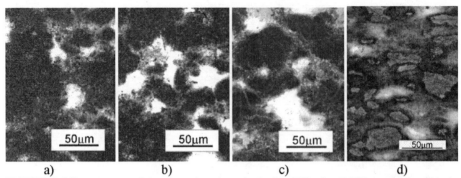

|          a)          |          b)          |          c)          |          d)          |

**FIGURE 3.** Microstructure of the sintered steel on the basis of Distaloy DC-1iron powder with the addition of 0,6 % C graphite : a- after forging at 1100°C, cooling in water and tempering (250oC, 1h), b- after forging at 1100°C, cooling in air, austenitizing 870 oC in 20 min, cooling in water and tempering (250oC, 1h), c- after forging at 1100°C, cooling in air, d- as in c) and additional forming at 760°C, cooling in water and tempering (250 °C/1h). Etched with nital.

In the latter variant the material was obtained by forming the sintered steel at 1100°C and cooling in air. It was then heated to 760°C and formed in a two-phase range $\alpha + \gamma$. Forming was done between two flat dies, after which the material was immediately quenched in water. The total deformation that was achieved during the two forming procedures was 0.41 logarithmic strain. After quenching the material was tempered at 250 °C in 1 h. The result was a product with 99% of solid material density and a martensitic-ferritic structure (fig. 3 d).

The results of the research of the properties of the initial material and those of the material after forging and heat treatment are assembled in tab. 2. After forging at 1100°C and cooling in air a significant improvement of the properties was obtained: bending strength was doubled, tensile strength was thrice as high as that of sintered steel, and its hardness was 364 HB. The materials obtained according to the second variant had a higher density, which influenced the mechanical properties. They are higher except for hardness in comparison to the material after one forging. Forging at 1100°C allows to obtain a material with 98% solid material density.

**TABLE 1.** Results of the examination of physical and tensile properties of the materials obtained from sintered steel on the basis of Distaloy DC-1 iron powder with the addition of 0.6 C $_{graphite}$

| Forming temperature, °C – state of the material | Density, g/cm$^3$ | Bending strength, MPa | Tensile strength , MPa | Brinell hardness, HB |
|---|---|---|---|---|
| Sintered steel after quenching and tempering 250°C/1h | 6.78 ± 0,03 | 998 ± 83 | 357 ± 49 | 201 ± 10 |
| Forging 1100 °C – cooling in air | 7.74 ± 0,07 | 2228 ± 92 | 1057 ± 65 | 364 ± 19 |
| Forging 1100 °C – direct quenching, tempering 250 °C /1h | 7.74 ± 0,07 | 2891 ± 135 | 1613 ± 54 | 541 ± 13 |
| Forging 1100 °C – cooling in air, austenitizing, quenching, tempering 250°C /1h | 7.74 ± 0,07 | 3364 ± 142 | 1585 ± 193 | 588 ± 43 |
| Forging 1100 °C – cooling in air, forging 760 °C – direct quenching, tempering 250°C/1h | 7.82 | 3404 ± 129 | 1603 ± 84 | 525 ± 9 |

Fig. 4 shows the destruction surface of the material in various states. In the sintered material the destruction surface goes through the primary powder particles and is locally ductile (fig. 4a). The plastic areas are more developed in the material after forging at 1100°C, quenching and tempering despite the presence of martensite in the structure of the material (4b). Transcrystalline destruction surface appears in the material after forging at 1100°C, cooling in air, additional forging at 760°C, quenching and low-temperature tempering (fig. 4c).

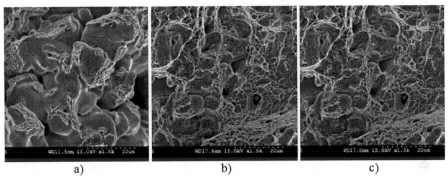

a)                                 b)                                 c)

**FIGURE 4.** Topography of the destruction surface after bending of the sintered steel on the basis of Distaloy DC-1 iron powder with the addition of 0,6 % C $_{graphite}$ : a- in the initial state as a sinter, b- after forging at 1100°C, direct quenching and tempering (250°C/ 1h), c- after forging at 1100°C, cooling in air, additional forming at 760°C, direct quenching and tempering (250 °C/1h). SEM observation.

## CONCLUSION

Sintered steel on the basis of Distaloy DC-1 iron powder, with the addition of 0.6 C $_{graphite}$ demonstrates good hardenability and formability. During its cooling in air from the austenite range a bainite-martensite structure occurs, and a martensite one after quenching in water.

The properties of forged steel depend strongly on forming temperature and cooling conditions. Tensile strength of forged PM steel on the basis of Distaloy DC-1 iron powder with the addition of 0.6 C $_{graphite}$ after hot forming at 1100°C is significantly better than in the case of the steel after sintering subjected to the same heat treatment procedures. Conventional heat treatment applied for the material formed at 1100°C slightly enhances the properties in comparison to the material after direct quenching in water from this temperature. The bending strength was tripled, and the tensile strength increased five times in comparison to the properties of the sintered steel.

## Acknowledgments

The research was conducted as a statutory work at AGH, project no. 10.10.110.563 and Ministry of Science and Higher Education project no. 3 T08D 025 29

# REFERENCES

1. A. Hendrickson, P. Machmeier, D. Smith: Powder Metallurgy, 43, 2000, 4, pp. 327-344.
2. Krebsöge advances powder forging – Metal Powder Report. January 1995, No. 1, pp. 34 -39.
3. R. Ratzi, P. Orth: Metal Powder Report. July/August,2000,7/8, pp. 20 -25.
4. S. Szczepanik: Teoretyczne i technologiczne aspekty kucia spiekanych stali. ZN AGH nr 137, Kraków 1991.
5. S. Szczepanik: Steel Research International 76, 2005,/3, pp. 219-225.
6. T.F.Stephenson, T.Singh, Campbell: Metal Powder Report. 2004, 3, pp. 26-30.
7. B. Wiśniewski, S. Szczepanik: Rudy i Metale Nieżelazne 50, 2005,10-11, pp. 629-634.
8. S. Szczepanik, B. Wiśniewski, T. Wójtowicz: Rudy i Metale Nieżelazne. 49, 2004, 9, pp.462-467.

# High Temperature Straining Behaviour Of High FeSi Electrical Steel By Torsion Tests

P. R. Calvillo*, N. Lasa García** and Y. Houbaert*

* Department of Metallurgy and Materials Science, Ghent University
Technologiepark 903, B-9052 Gent-Zwijnaarde, Belgium.
** Escuela Superior de Ingeniero, Campus Tecnológico de la Universidad de Navarra, Paseo de
Manuel Lardizabal 13, 20.018 San Sebastián, Spain.

**Abstract.** Steel with an increased Si-content has better magnetic properties in electrical applications in terms of high electrical resistivity, reduced energy losses and low magnetostriction. Nevertheless, the oxygen affinity of this element at high working temperatures and the poor ductility observed at room temperature caused by order structures make the thermomechanical processing of these alloys rather difficult. Since these materials do not present a phase transformations from ferrite to austenite, a fundamental study of their workability using torsion tests will help to understand and to optimise their production process. Important critical temperatures in these materials are $T_{ord}$ (the temperature above which the material is disordered), $T_{nr}$ (the temperature below which static recrystallisation is not taking place any more) and other restoration temperatures appearing during processing.

Fe-Si electrical steels, with silicon concentrations of 2, 3 and 4 wt.-%, were tested according to a multi-deformation torsion schedule under continuous of cooling conditions in 18 passes, with temperature ranges from 1150 to 810°C, at a strain rate of 1 $s^{-1}$, the interpass time and the amount of plastic deformation were varied from 20 to 5 sec and from 0.1 to 0.3, respectively. Different critical temperatures, important for the processing of these alloys, were calculated from the dependence of the mean flow stress (MFS) on inverse temperature, based on their changes of slope. The temperatures at which the restorations mechanism, the recrystallization and the recovery stops, $T_{nr}$ were determined and can be described using the relation developed here, based on their dependence on composition, deformation parameters and cooling rate. The metallographic analysis of quenched samples is in good agreement with the critical temperatures obtained through the measurement of the MFS.

**Keywords**: electrical steel, torsion test, straining and restoration

## INTRODUCTION

Fe-Si steels are often used in electrical applications because they present an improvement in the magnetic properties due to the addition of silicon to iron, which reduces the magnetostriction and power losses, hence increasing the electrical efficiency of the steel and reducing the ecological impact when used in electrical motors or transformers. Although it is well known that optimal magnetic properties are obtained for 6.5 wt.-% Si, the steel industry is not likely to produce electrical steel with concentrations above 3.5 wt.-% by a conventional route, mainly because of an ordering phenomenon appearing around 4 wt.-% Si, which makes steel extremely brittle[1, 2, 3, 4].

Depending on the amount of carbon added, steels containing more than 1.95 wt.-% Si do not undergo any allotropic transformation, as the bcc lattice is stable from room temperature up to the melting point, complicating the normalisation of the grain size in these steels. Alloys containing Si in the range of 1.95 to 4.5 wt.-% present normally a

CP907, 10th ESAFORM Conference on Material Forming, edited by E. Cueto and F. Chinesta
© 2007 American Institute of Physics 978-0-7354-0414-4/07/$23.00

disordered structure (called $A_2$). The stacking fault energy (SFE) of these alloys has been reported to be as high as 200 mJ/m$^2$, increasing with temperature, ordering degree and it seems to decrease slightly with the addition of silicon [5]. This high value of SFE indicates that a substructure of dislocation tangles and cells are formed during the hot working process [6]. Due to these phenomena, not all yet well understood, conventional warm and cold working is difficult to perform for high Si steel [3, 4, 5].

Hot torsion tests, consisting of multiple deformations under continuous cooling conditions, were performed in order to find a relation between processing parameters and material behaviour. Knowledge of the deformation behaviour of Si-steels at hot working temperatures is important in manufacturing processes, as well as for the effect it will exert during consequent cold deformation [1, 2].

## EXPERIMENTAL PROCEDURE

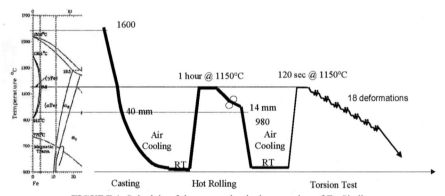

**FIGURE 1.** Schedule of thermomechanical processing of Fe-Si alloys.

The chemical composition of the experimental alloys is shown in Table 1. The alloys were produced through melting ultra low carbon (ULC) steel together with a given amount of Fe-75%Si in an open-air furnace. The alloys were hot charged to avoid cracking; the molten metal was poured at 1,600°C in sand moulds of 40 mm thickness, the solidified slabs were then placed in a furnace at 1,150°C without delay and finally hot rolled to a thickness of 14 mm. The finishing rolling temperature (FRT) always exceeded 950°C. The material was cooled to room temperature in air after hot rolling.

| Element wt.- % | % C | % Si | % Al | % Mn | % P | % S | % N |
|---|---|---|---|---|---|---|---|
| Steel A | 0.002 | 1.88 | 0.075 | 0.048 | 0.016 | 0.009 | 0.003 |
| Steel B | 0.004 | 2.78 | 0.098 | 0.046 | 0.015 | 0.009 | 0.002 |
| Steel C | 0.003 | 4.06 | 0.096 | 0.066 | 0.016 | 0.009 | 0.007 |

**TABLE 1.** Chemical composition (wt.- %) for the Fe-Si alloys.

452

Samples for torsion tests were taken from the slabs, the axis of these samples was parallel to the rolling direction. The tests were performed using a computerized torsion machine, which uses an induction coil to heat the specimens. Samples for multiple deformation torsion tests were heated at 4°C/sec until 1,150°C and held at this temperature for 120 sec. The specimens were subjected to a series of 18 consecutive twists or until fracture at a strain rate of 1 sec$^{-1}$, the time between deformations was 5, 10 and 20 sec and strain passes of 0.1 and 0.3 were given. The specimens were cooled down under an argon protective atmosphere to minimise oxide formation, at a constant rate, calculated by maintaining a constant 20°C decrease in temperature after each pass, independently of the interpass time. Several samples were tested in order to ensure reproducibility of the results. The measured torque (T) and twist (θ) were converted to von Misses effective stress (σ) and strain (ε), using equations reported in literature [6], the mean flow stress (MFS) at each pass was calculated from the σ- ε curve, using the procedure described by Jonas and Borato [7].

Microstructures from multiple deformation tests were analysed in a scanning electron microscope (SEM), implemented with equipment for texture determination. Misorientations below 2° were not considered in data post-processing. Boundaries with misorientation between 2-15° were defined as low angle grain boundaries (LAGB) and those with misorientation higher than 15° were considered as high angle grain boundaries (HAGB).

## RESULTS AND DISCUSSION

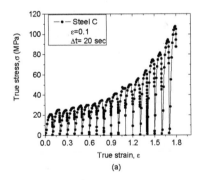 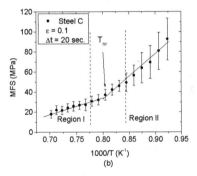

(a)        (b)

**FIGURE 2**. a. Typical stress-strain curve of Steel C (4.06 wt.-% Si) with a pass strain of 0.1, cooling rate of 1 °C/sec and 5 sec of interpass time at strain rate of 1 sec$^{-1}$; b. MFS as a function of the inverse pass temperature used to determine $T_{nr}$ by linear regression.

Fig. 2a shows the flow curves plotted vs. strain for complete multi pass torsion tests of Steel C (4.06 wt.-% Si)) which is an alloy without allotropic transformation. A typical dependence of the MFS on inverse temperature is shown in fig. 2b as a plot of the MFS values vs. 1000/T (K$^{-1}$). A first approximation by linear regressions makes clear that two different regions can be recognised. In region I deformations are carried out in a range of temperature where full restoration mechanisms as recovery and/or recrystallisation are taking place until the change in slope, which corresponds to deformation below $T_{nr}$, the temperature of no-recrystallisation, determined at the intersection of both linear regressions. From this point on, the MFS values increase more steeply when lowering the temperature than in region I, because the strain is being accumulated from pass to pass and a pancake microstructure is formed [7, 8].

Fig. 3a shows data of the MFS vs. 1000/T regarding the effect of the interpass time in the Steel B (2.7 wt.-% Si). Shorter interpass times correspond to higher cooling rates between the deformation steps. $T_{nr}$ increases with increasing interpass time. The increase of the MFS below $T_{nr}$ is steeper for longer interpass times, not only associated with the temperature decrease, but also with the promotion of recovery when shorter cooling rates are applied [8].

Fig. 3b shows the effect of silicon on the $T_{nr}$ at a strain of 0.1. The $T_{nr}$ values are always higher for steels with more Si content and the reason is the substitutional solid solution hardening influence of Si, but also it might be associated to the influence of the higher degree of cell formation on the microstructure with Si addition, retarding the recrystallisation phenomena [6, 9].

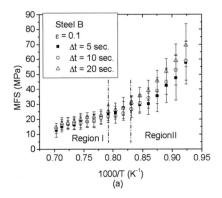 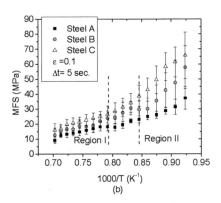

FIGURE 3. a. Effect of interpass time on the MFS/inverse pass temperature dependence of steel B (2.7 wt.-% Si). Strain $\varepsilon = 0.1$/pass, strain rate of 1 s$^{-1}$; b. Effect of Si on the MFS/inverse pass temperature. Strain $\varepsilon = 0.1$/pass, strain rate of 1 s$^{-1}$ and interpass time of 5 s (4°C/s).

Fig. 4a shows how the $T_{nr}$ for the steels studied generally increases for increasing interpass times [10], although for the steel with higher silicon composition and longer interpass time $T_{nr}$ is clearly reduced. It is also shown that increasing the silicon content results in a higher $T_{nr}$, therefore the lowest and the highest curves correspond for the steel with 1.88 and 4.06 wt.-% Si, respectively.

Although the variation of the $T_{nr}$ with the pass strain will be presented in a further publication, it can already be concluded that the results obtained here are in agreement with the few data available for Fe-Si steels [10, 11] and they were used for the calculation of the $T_{nr}$.

The relationship between $T_{nr}$ and deformation parameters in the range studied can be described by:

$$T_{nr} = (A*(wt.-\%Si)+B)\varepsilon^{0.03}\dot{\varepsilon}^{-0.001}t^{0.031} \tag{1}$$

where A is a parameter dependent on chemical composition equal to 22.8 (°C/ wt.-% Si), B = 842 °C is, ε is the strain per cycle, $\dot{\varepsilon}$ is strain rate, in $s^{-1}$ and t the time between cycles (in s). It should be remarked that as the strain rate in the presented experiments has not been varied, the value of its exponent (–0.001), has been obtained by iterating the one found in literature. As strain per cycle and/or time between deformations increase, they seem to have the same effect on restoration process, increasing the $T_{nr}$. The interpass time exponent found in literature was of the same

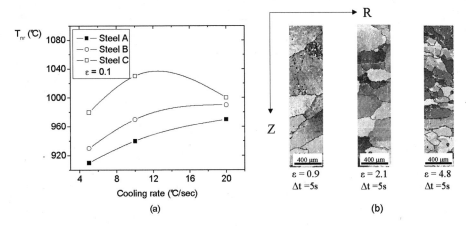

(a)                                    (b)

FIGURE 4. a. Represents the dependence of Tnr on cooling rate as a function of silicon, 0.1 strain per pass and strain rate of $1s^{-1}$. (b) Microstructures obtained by OIM of Steel B at different levels of accumulated deformation in hot torsion at a strain rate of $1sec^{-1}$, interpass time of 5 and strain per pass 0.3. Dark and white lines correspond to high and low angle grain boundaries, respectively.

order of magnitude, but with a negative value, although its graphic representation shows a similar tendency than in the present publication [11]. It is further observed that the contribution of silicon alone is around 50°C in the range of compositions studied.

The initial microstructure of all samples is characterised by equiaxed grains free of LAGB and with an average size larger than 800 μm. Fig. 4b shows images obtained from samples of steel B quenched at different values of accumulated strain from tests conduced at a strain rate of 1 $s^{-1}$, interpass time of 5 s and strain per pass of 0.3, at a 75% of the radius far from the sample's centre. In samples strained 0.9, which belong to region I, no clear evidence of recrystallisation is observed although subgrain formation starts developing, corroborated by the fraction increase of the LAGB. Hence recovery is considered to be the main restoration mechanism, the increase in stress is due to the decrease in temperature and strain hardening. In the interval between region I and II, where $T_{nr}$ is defined, the mean flow stress increases almost progressively due to the generation of new grains by static recrystallisation, which are free of LAGB. In the samples strained 2.1, a slight grain refinement and a high degree of recovery is observed in the non-recrystallised grains, as indicated by the higher density of LAGB.

Accumulation of strain is observed to take place in region II as static recrystallisation is totally suppressed, with the consequent rapid increment in mean flow stress, because of the decrease of temperature, strain hardening and deformation of a finer microstructure than in previous region. Recovery becomes now the only restoration mechanism.

## CONCLUSIONS

The high temperature straining behaviour of three Fe-Si steels has been studied. Special attention was paid to determine the influence of deformation parameters on the $T_{nr}$.

- The flow stress behaviour of Fe-Si steels can be divided in at least two regions, in which recovery is the main restoration mechanism.
- A relationship between $T_{nr}$, the deformation parameters and the chemical composition was worked out.
- The influence of silicon on $T_{nr}$ was estimated as an increase by 22.8 °C per wt.-% Si, while increasing the strain and the inter pass time increases $T_{nr}$, increasing the strain rate appears to decrease $T_{nr}$.

## ACKNOWLEDGEMENTS

The authors will like to thank the Erasmus-Socrates European exchange program and the sponsoring through the BOF-funding at Ghent University.

## REFERENCES

1. G. Lyudkovsdy, P. K. Rastogi and M. Bals, *Journal of Metals*, 18-25 (1986).
2. Y. Houbaert, O. Fisher, J. Schneider and Wuppermann, German Patent No 102 20 282 (2003).
3. O. Kubaschewski, *Iron-Binary phase diagrams*, New York: Springer—Verlag Berlin Heidelberg, 1982, pp. 136-139.
4. P.R. Swann, L. Granäs and B. Lehtinen, *Metal Science* 9, 90-96 (1975).
5. H. J. McQueen, *Metallurgical Transactions A* 8A, 807-824 (1977).
6. G. E. Dieter, *Mechanical Metallurgy*, New York: McGraw-Hill Book Company, 1988.
7. F. Boratto, R. Barbosa, S. Yue, and J. J. Jonas, *"Effect of chemical composition on the critical temperatures of microalloyed steel"* in Proc. Int. Conf. Physical Metallurgy of Thermomechanical Processing of Steels and Other Metals (THERMEC'88), ISIJ, Tokyo, 1988, pp. 383-390.
8. D.Q. Bai, S. Yue, W.P. Sun and J.J. Jonas, *Metallurgical Transactions A* 24A, 1993-2151 (1993).
9. J. Ball and G. Gottstein, *Intermetallics* 1, 171-183 (1993).
10. P. R. Calvillo, T. Ros-Yanez, D. Ruiz and Y. Houbaert, *ISIJ- Internacional* 46, 1685-1692 (2006).
11. P. R. Calvillo, R. Colás and Y. Houbaert, *Advanced Materials Research* 15-17, 708-713 (2007).

# Tool Design and Manufacturing for Bulk Forming of Micro Components

R. S. Eriksen\*, M. Arentoft[†] and N.A. Paldan[†]

*Department of Manufacturing Engineering and Management, Technical University of Denmark, DK-2800 Kgs. Lyngby, Denmark*
[†]*IPU-Production, Produktionstorvet, building 425, DK-2800 Lyngby, Denmark*

**Abstract.** The central part of the metal forming processes are the forming tools. In the macroscale, these tools are manufactured by means of machining- and drilling operations and subsequently hardened. Due to the required precision, this approach is not possible in the microscale. An investigation of alternative tool manufacturing techniques has been carried out, including $\mu$-EDM, electroforming and machining operations. A series of universal forming tools for simple microcompoents has been manufactured using different techniques, allowing side-by-side comparison. A series of ejectors and punches have also been manufactured using a combination of grinding and $\mu$-EDM processing. Critical tooling parameters including geometry, yield strength, surface roughness and production time has been recorded. A silicon replica method has been utilized to measure the internal geometry of the microforming tools

**Keywords:** Tooling, bulk forming, micro components, microEDM
**PACS:** 81.20HY

## INTRODUCTION

The future need for smaller and more integrated devices in the consumer and professional markets is evident. Led by the electronic and digital entertainment industry; pocket-size cameras, MP3-players, Personal Digital Assistants and cell-phones has set production volumes of more than 100 million pieces yearly.

Within the medical and dental industry, the introduction of millimeter size biocompatible metallic components, holds a promise of functioning as spare parts for the human body, enabling groundbreaking treatment of severe bone and dental damages.

These applications all require millimeter size metallic components with excellent precision, low production cost and challenging materials. The idea of utilizing metal bulk forming for production of such components is not new and has been researched by several groups [1, 2, 3]. It is well know, that a critical part of the bulk forming process is the manufacturing of the forming tools. Even for macro-size components, the design phase, material selection and manufacturing process of bulk forming dies is known to be challenging.

In this work we consider two manufacturing process of tooling parts for cold forging of an axel for a hearing aid device.

CP907, *10th ESAFORM Conference on Material Forming*, edited by E. Cueto and F. Chinesta
© 2007 American Institute of Physics 978-0-7354-0414-4/07/$23.00

# COLD FORGING OF AXEL FOR POTENTIOMETER

The micro axel shown in Fig. 1 originates from an industrial application where it makes up a part of a micro potentiometer. The part is manufactured in brass using turning, milling and drilling operations. The largest diameter of the axel is Ø3 mm and the length is 2.95 mm. All tolerances of the component must be within ±10 μm and the corner radii must be kept below 0.03 mm. The part is rotating in the potentiometer application, which leads to strict requirements on the perpendicularity of the guiding surfaces. The two holes, located on the top and on the bottom of the component, are both drilled, thus there are no configured geometry in the bottom of these holes.

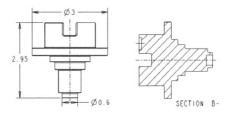

**FIGURE 1.**   Sketch of potentiometer axel

# TOOL DESIGN

The total cost of the central forming tool system for a bulk forming machine is significant, when compared with the price of the complete process system. In order to make the forging process economically attractive for bulk-forming of micro-components, new design methods for tooling parts has to be investigated. It is important that these methods satisfy requirements for low production cost, tool performance, flexibility and quality of the produced specimens.

In the macro scale, the central tooling dies are machined using soft machining and subsequently hardened. This approach is not applicable in the micro scale due to the increased tolerance requirements. This entails that the tooling parts will have to be manufactured directly in a hardened material.

In this work, a standard ISO 8020 punch needles of Vanadis 23 powdered metal, hardened to 64 HRC, is used as bulk material for the tool-set punch, ejector and die. The ISO punch needles are standard mass-produced items with good precision and a form suitable for uncomplicated integration with supporting tooling elements. The punch needle is cut into a number of cylindrical blanks with a height of 6 mm and a diameter of Ø8 mm. These blanks are then surface grinded and make up the raw material for the forging dies. As raw material for punches and ejectors, standard punch needles or simple standard HSS drills will be used.

The use of CAD/CAM, standard tooling components and fast and flexible tool manufacturing processes will make it possible to reduce the lead time for micro forming tool-sets to a matter of days. This in term means that the flexible production line can be used to run smaller batches of components with a short delivery time.

# TOOL DESIGN PROCESS

The tool design is initiated with a CAD-model of the final component. From this model, a number of Finite Element Method (FEM) simulations is conducted, utilizing the Deform-3D software from Scientific Forming Technology Corporation, to investigate the required number of stages and forging strategies. It is known that grain structure and friction have increased influence in micro forming, making the standard size invariant FEM approach inadequate for micro forming simulations [4]. However, since the dimensions on the potentiometer axel treated here is not strictly micro-size, the simulations is expected to show good agreement with the final forming process. The FEM-simulations is processed on a trial-and-error basis, where various die-geometries are evaluated and the flow pattern of the bulk material is observed. Based on the simulations, an optimized die geometry is established, which concludes the tool design phase. It was found that a two-step process is needed for forging of axel. The first step is an extrusion with an neck feature for aligning in the second step die, shown in Figure 3(b). The second step is a heading operation which completes the form of the component.

The forging dies for the micro axel are manufactured using a Sarix Micro Electrical Discharge Machining ($\mu$-EDM). The computer controlled machine is capable of wire eroding hard metal with high precision. The machine features a CAM extension, which makes it possible to reproduce complex geometries from a CAD-model. The machine works much like a traditional milling machine but uses a solid carbide electrode down to 60 $\mu$m in diameter as the active eroding tool. The small scale of this process allows point-flooding with dielectric fluid of the sparking area, avoiding complete submersion of the workpiece and electrode. The central elements of the $\mu$EDM-process is depicted in Figure 2(a).

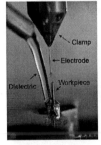

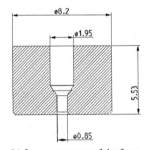

(a) closeup image of the $\mu$-EDM machine working

(b) Inner geometry of the forming die for step 1.

**FIGURE 2.**

The $\mu$-EDM process is divided into three operations. First, a drilling operation is executed using a hollow 300 $\mu$m brass electrode. The purpose of this operation is to ensure proper flushing of debris through the hole, when subsequently machining the internal geometry of the die. The processing time of the drilling operation is about an hour. Secondly, a rough sparking operation is performed. This operation removes the bulk of the material inside the die using a 300 $\mu$m solid carbide electrode. The roughing

459

operation leaves only a $50\,\mu$m layer of stock material along the wall-sides of the die and has a processing time of about eight hours. Finally, a the fine sparking operation is executed with at running time of about eight hours. The purpose of the operation is to ensure good precision of the final geometry and to reduce surface roughness. The final surface roughness after the fine-sparking operation has a value of $R_a = 0.14$ and the precision is within $\pm 1\,\mu$m. Ultimately, the surface roughness should have a $R_a$ value which is about ten times lower (about 0.01), but this is not realizable using the $\mu$-EDM process. Due to nature of the electrical discharge technology it is not possible to achieve corner roundings of less than $\sim 60\,\mu$m, but generally this figure scales with the diameter of the electrode. The reason for this is bound in the fact that a spark-channel is more likely to open at the point of a discontinuity [5].

A alternative option for die manufacturing in hard materials is hard milling. This process is already widely used for die manufacturing in the industry today, but it has not been applied in the micro scale. With these machines, it is possible to high speed machine material with hardness up to 68 HRC directly. The high depth-width aspect ratio in combination with millimeter size dimensions, of a typical micro forming die, make it difficult to directly transfer the existing hard-milling technology to micro scale. The field of micro scale milling is rapidly progressing and the technology should be on the marked within few years.

# MANUFACTURING OF DIES BY ELECTRO FORMING OF HARD NICKEL ALLOYS

Electro forming is a replica process where an external geometry substrate can be converted into an internal geometry, a forming die as an example. The substrate can be machined in a soft material, such as aluminium or steel, using conventional techniques. The substrate is then submerged into a bath with a special conducting solution of water, additives and a block of bulk metal. The substrate is then connected to a current source as the cathode and the metal block is connected as the anode, as depicted in Figure 3(a). Using electro forming it is possible to replicate a external rod-type geometry into an

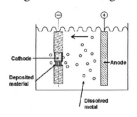
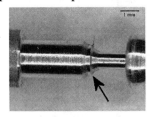
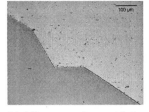

| (a) Schematic showing the electroplating process | (b) Preform turned in aluminium alloy EN AW-6060 | (c) Microscope image of the neck feature on the preform |

**FIGURE 3.**

internal geometry with high precision. Even sharp corners and small features with dimensions around a micron can be replicated because the process works on atom level. Figure 3(b) shows the inverse geometry substrate of the die for the potentiometer axel. The geometry was machined on the $\varnothing$3mm aluminium alloy EN AW-6060 by a partner using a high precision turning lathe. Please note the marked neck, used for alignment purposes in the second step of the forming process.

The substrate was plated during a 48 hours period which covered the rod with a 3 mm Nickel layer. The aluminium substrate was etched away and the form was casted in a epoxy solution. After flat grinding, the internal shape of the electro forming can be observed. Figure 3(c) shows a microscope image of the neck feature marked with an arrow in Figure 3(b). It can be seen that this small feature is very well replicated and that the sharp corner is well preserved. The yield strength of the electro formed die was tested to be about $400 \frac{N}{mm^2}$. This is not sufficient for cold forming purposes as even with soft materials the die stresses will exceed $600 \frac{N}{mm^2}$.

Recent studies have show that electro forming with a Nickel-Cobalt is feasible. The first test-items showed problems with cracks and uneven plating of the substrate, but these problems where solved by using special addictive in the plating-bath. Nickel-Cobalt has a yield strength of about $800 \frac{N}{mm^2}$, thus being better suited as die material.

## ANALYSIS OF MANUFACTURED DIES

Validation of the internal geometry of the forming die is very challenging. Due to the small die dimensions, it is not possible to get a measuring probe of a coordinate measuring machine inside the the die. Instead, a silicon replica set, manufactured by Struers, was used. The silicone pase is capable of replicating features down to $0.1\,\mu$m. Figure 4 show two microscope images of replicas taken of a step 1 die manufactured with $\mu$-EDM and electro forming respectively. It can be seen from the pictures, Figure 4(a) and

(a) Replica of die manufactured using EDM

(b) Replica of electroformed die

**FIGURE 4.**

4(b), that the electro formed die features sharper edges, where the dimensions of the die-geometry shifts. This is especially pronounced around the neck feature marked on the images. The dimensions of both dies are within $\pm 2\,\mu$m, based on optical measurements of the two replica.

# CONCLUSION

Looking at the silicone replicas of the previous section, it can be noted that electro forming is best at reproducing micro features. However, since electro forming is currently only possible with Nickel, there is no practical application for the technology yet. The use of Nickel-Cobalt has show promising result, but it is still too early to draw conclusions from these. Electro forming has shown excellent surface finish quality but further studies are need to determine if dies can be used without subsequent polishing or coating

$\mu$-EDM technology has shown to be very flexible and fast. With this technology it is possible to manufacture tooling parts directly in hard metal on a day to day basis, while maintaining good precision of the manufactured parts. EDM technology is not capable reproducing sharp edges and the surface finish it not practicable for forming purposes.

Surface finish of the manufactured tool parts is an area which needs to be studied further. Is is known that the surface roughness is of importance for friction, tool lifetime and specimen quality in the forming process. The area of polishing and surface finish is not very well researched, especially not in connection with the manufacturing processes. Most work within this area is based on subjective opinions of the person conducting the work and several difference can be found.

An European round-robin benchmark-test has been initiated, involving several partners, with the objective of determining capability of the EDM process within micro scale. The test is performed in ultra-fine-grained carbide, where a simple punch and die is to be manufactured according to best effort requirements. The manufactured parts will then exchanged between partners that will measure and test the tools. The results will be collected in a survey that will be published.

# ACKNOWLEDGMENTS

This work is supported by the European Commission and the Danish Ministry of Science Technology and Innovation under the research projects MASMICRO, no. 500095-2 and the Innovation Consortium MIKROMETAL, no. 66334 respectively.

# REFERENCES

1. C.P. Withen, J.R. Marstrand, M. Arentoft, N.A. Paldan, "Flexible tool system for cold forging of micro component", *First International Conference on Multi-Material Micro Manufacture*, 29June-1 Julty 2005, Forschungszentrum Karlsruhe, Germany, Elsevier, pp. 143-146
2. M. Geiger, M. Kleiner, R. Eckstein, N. Tiesler, U. Engel, "Microforming," keynote paper *Annals of CIRP*, 50 (2) 2001, 445-462
3. K. Osakada, T. Nakamure, "Research and development of precision forging in Japan", *Advanced Technology of Plasticity 2005*, Springer-Verlag, 2005
4. U. Engel, S. Geissdörfer, M. Geiger, "Simulation of microforming processes", *Adv tech. of plasticity*, 2005
5. C. Jameson, "Electrical Discharge Machining", *Society of Manufacturing Engineers*, 2001

# Production Equipment and Processes for Bulk Formed Micro Components

## N.A. Paldan[1], M. Arentoft[1], R.S.Eriksen[2]

[1]IPU, Produktionstorvet, Building 425, DK-2800 Kgs. Lyngby, Denmark, www.ipu.dk
[2]Dept. of Manuf. Eng. and Managem., Techn. Univ. of Denmark, DK-2800 Kgs. Lyngby,Denmark

**Abstract.** Manufacturing techniques for production of small precise metallic parts has gained interest during recent years, an interest led by an industrial demand for components for integrated products like mobile phones, personal digital assistants (PDAs), mp3-players and in the future for spare parts for the human body. Micro components have also found several applications within the medical, audiological and dental industry, applications that impose increased demands for biocompatible and corrosion-resistant materials and cleanness. So far these micro components have mainly been manufactured by traditional machining techniques or chemical etching. However, these traditional machining and etching techniques are generally not well suited for mass production of advanced micro components, due to handling problems, waste of expensive material and long machining times. This calls for development of a novel production system that can meet the demands for high productivity, high reliability, low cost, while being environmental acceptable. Bulk metal forming meets these demands to a great extent, but the technology cannot directly be transferred to the micro scale. A flexible machine system for bulk micro forming has been developed and used to form a number of industrial micro parts in aluminium and silver, with ongoing work on forming of titanium. Manufacture of billets by cropping has been examined using a simple test rig and an automatic cropping device has been designed, manufactured and tested.

**Keywords:** Micro forming, production machines
**PACS:** 81.20.Hy

## INTRODUCTION

The central part of a cold forging line for mass production of micro metal components is the machine system for the forming process. To compete with existing processes like machining and etching, productivity becomes an important factor. The system is therefore planned to run up to 300 strokes a minute. With this speed, a standard production batch at 100.000 pieces can easily be produced in an 8 hour day, with the result that the flexibility of the tool system becomes a key issue. However, in order to establish such production line, a number of supporting techniques and methods need to be developed. This includes a billet preparation module, a lubrication technique, a transfer system with corresponding grippers and manufacturing methods for micro tools.

This paper will describe the design of the test machine for micro bulk forming and the billet preparation module.

CP907, 10th ESAFORM Conference on Material Forming, edited by E. Cueto and F. Chinesta
© 2007 American Institute of Physics 978-0-7354-0414-4/07/$23.00

# Introduction to micro bulk forming

The common definition for a micro component is a part with 2 dimensions below 1mm [1,2]. The component shown on figure 1, is not strictly a micro component according to the definition, but it was one of the components used in the initial forming trials, because it is an industrial component with features below 0.5mm and most tolerances in the range of ±10µm.

**FIGURE 1.** Left: Test component for bulk forming together with the machined component, Right: the forming steps and the formed component.

## TEST MACHINE FOR MICRO BULKFORMING

The aim of the bulk forming machine will be fast and flexible production of micro metal parts. Conventional machining of micro parts is slow, from a few seconds to 30-45 seconds for a more complex part. Production of 100.000 parts using 45 second a part will require 1250 hours of machine time. Compared with this, conventional progressive tools for stamping of small sheet metal parts can run with speeds up to 2000 strokes a minute, but require very complex tooling. Micro bulk forming will be in between these two extremes, with a combination of high productivity and simple tooling. The goal is as mentioned to manufacture parts with a speed up to 300 pieces a minute and with limited costs for development and manufacture of required tools. Beside this a general focus has been on the working environment for the machine operator with regards to safety, noise, vibrations and the use of hazardous chemicals.

More then one step is usually required to manufacture complex part by forming, and this can either be realised by using a press with room for several sets of tooling (the advanced solution) or using a simple setup with one tool in the press and then either having several presses or switching the tool in the press between each forming step [4]. The last solution requires large storage space for the intermediate components, but for micro components this will not be a problem, due to small volume of each part. To ensure as high productivity as possible, the decision was to build a bulk forming test machine with room for two forming processes, and suitable for all 8 basic bulk forming processes [3].

The general design philosophy behind this machine system has been that small components should be manufactured on small machines, so the goal has been to design a benchtop machine [5].

# Design of Test Machine for Micro Bulk Forming

A machine for bulkforming can be split into different main parts which are independent of each other, but the combination of which decides the performance and accuracy of the final machine for bulk forming. These parts are: Frame, drive, sensors, control, transfer system and the active tooling.

There are two general types of press frames, the partly open C-frame and the closed O-frame. The stiffness of the frame and guideways are critical for the production of high quality components in a multiple step setup, and the O-frame is preferred for large presses due to the symmetric construction which reduces the deflections of the press under forming and possible deflections due to thermal gradients. Press frames can be manufactured from sheets or pillars, but for the test machine a prestressed pillar design was chosen since it provides easy access to the tooling and it is relatively simple to manufacture with high accuracy.

Drives for industrial presses can be divided into 3 groups, position, force and energy controlled. High accuracy parts can be manufactured on all three types of presses, but the force and the energy controlled presses require the use of a hardstop, to ensure the final position of the tool. The combination of high productivity and accuracy is best reached by a position controlled drive and the working environment of such a press system is also preferable since there are no hazardous chemicals (hydraulic press) and a very limited amount of noise and vibrations (energy controlled presses). A position controlled drive can be based on many different principles, but the choice for the test machine, is a linear brushless DC motor from California Linear Devices, since this drive is silent, capable of 40 G accelerations, contains only one moving part and can be front mounted. The largest drives available today can provide a maximum force 5300 N, such one was used to power the test machine. A drive based on a rollerscrew will be used for future machines if they requiring higher process forces. The drive is equipped with a $1\mu$ linear encoder, which is used to control the position of the press ram. Piezoelectric force sensors from Kistler type 9313 and 9323, were chosen to measure the dynamic forces acting on the tools, because they have the right combination of small size, high rigidity and precision together with very good dynamic characteristic.

**FIGURE 2.** Left: CAD drawing of the forming machine system, right: The finished machine.

The drive and sensors are connected to an Ultra 5000 motion controller from Allen-Bradley, which is controlled by programs written in the C language. The drive is equipped with digital and analog I/O for reading the position and force sensor data. A master curve control strategy is applied to monitor the loads on punches during the forming process, so the drive can be stopped if the deviation in the load/stroke curve exceeds an acceptable value. The motion controller is connected to a PC running Labview 8, which is used to program the movement of the press, to display load/stroke data, and allows the logging of data for post-process control and datamining .

The active tool elements are mounted in a dieset, which ensures the proper alignment between punch and die, and between ejector and die. The ram is mounted on ultra precision linear ball guideways from Bosch Rexroth, the stiffness of this guideway is acceptable since the dieset is designed to minimise off-center loading on the ram. This is realised by positioning the $2^{nd}$ forming step in line with the centre axis of the drive and the $1^{st}$ step 20 mm from this centre axis. Transfering micro components in multi step forming operations is not a trivial exercise, and a test rig has been designed to test different handling strategies [6]. The translatory movement is realised by a moving coil actuator capable of 25mm of movement with a reproducibility of 1µ. Different handling strategies using passive gripper systems have been tested. Based on this the optimal design will be implemented on the test machine in the near future. ISO punch needles were used as raw material for the active tool elements, because of reduced lead time and cost. Punch and dies has been manufactured from ISO 8020 punches in HSS, Vanadis 23 and Wolfram Carbide. The die cavities were manufactured on a Sarix Micro 3D EDM milling machine, together with any fine geometry details on punches and ejectors. Electroforming of die cavities has been tested but the required quality has not yet been met [7].

## BILLET PREPARATION MODULE

The quality, measured by shape and volume, of the raw material or billets is a critical part of the production of micro components by bulk forming. Bulk forming processes are not based on material removal, so if the volume of the slug is wrong this will influence the geometry of the final component. This is a well known problem in conventional bulk forming and current machine can produce slugs in steel with a deviation of volume down to 0.5% for diameters between 10-50mm. The basic problem is the same for micro bulk forming, but since the dimensions of the billet are scaled down and the acceptable tolerance is a fraction of the absolute dimensions, the tolerances on the billet will be in the range of µm. The manufacture of wires with tolerances in the range of µm is standard, but cutting the wire in length of around a millimetre with tolerances of µm is not common practice.

A high effective, high precision method is therefore required for manufacturing small cylindrical billets. For this purpose, two different versions of the cropping process has been investigated: simple and with axial compression. By applying axial pressure during the cropping process, the fracture zone is limited and the shape of the final component will be better defined. To test this, a test rig is developed, in which a Ø1.9 mm aluminium rod is axial loaded with more than 2 times the flow stress. The test device for cropping with axial compression is shown in Figure 3.

466

**FIGURE 3.** Left: Test device for cropping with or without axial compression, Center: the end of a billet cropped with a closed shear, right: the end of a billet cropped with a closed shear and axial compression.

The best result is achieved by cropping under axial pressures, giving raise to low volume variance, which is the key quality parameter of the cropping process. Also, the roundness of the billet is well preserved when cropping under axial pressure. The cropped volume is in all 3 cases within 1% of the specified volume, which is acceptable for most processes. By cold forging of cropped- and machined billets, no measurable difference is found. Due to the promising results of the cropping test, an automated version is designed and manufactured. This machine is a pure mechanical solution based on cams, and is driven by hand for the initial trials, however the hand wheel can be replaced by a motor.

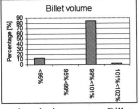

**FIGURE 4.** Left: The automated cropping device, center: Billet distribution, right: a selection of cropped slugs.

The device can crop 1 billet a second with 85% of the billets within 1% of the desired volume. Fine tuning of the springs and tool clearances in the setup should allow the speed to be increased, and remove the amount of billets which are less the 95% of the desired volume.

## FUTURE DESIGN OF EQUIPMENT FOR MICRO BULK FORMING

Valuable knowledge has been gained during the design and testing of the bulk forming machine and automated cropping device and based on this, future designs of the equipment for micro bulk forming will focus on flexibility and modularisation to enable the easy switch between production of different components. Due to the low volume required for storage of micro components, the optimal design could be single step forming machines, with flexible tooling. Machines with different production rates and maximum forming loads would make it possible to choose the optimal machine configuration for a specific step in the manufacture of a component.The forming machine is, as mentioned in the introduction, only a part of the machine system that is required to realise industrial production of micro components by bulk forming, the support equipment: billet preparation, lubrication and tool manufacture are all so required.

## CONCLUSION

The industry requires high volume, high precision micro metal components. For this purpose, cold forging is proven to be a suitable process [8,9]. The requested productivity causes demands of a flexible production line, and in this paper a concept for a flexible cold forging machine is developed and tested and a device for the production of billets is presented. The billets can be prepared by cropping, where the smoothes surfaces are obtained by applying an axial pressure. An industrial micro component for a miniature electronic device is cold forged in two steps, and valuable experience is gained for future design of micro cold forging processes. Having such process, the basis for a production line for mass manufacturing of micro metal parts using the cold forging technology is established.

## ACKNOWLEDGMENTS

The authors wish to thank the European Commission and the Danish Ministry for Science Technology and Innovation for supporting the work by the EU project MASMICRO, no. 500095-2 and the Innovation Consortium MIKROMETAL, no. 66334.

## REFERENCES

1. M. Geiger, M. Kleiner, R. Eckstein, N. Tiesler, U. Engel: "Microforming", keynote paper, Annals of CIRP, 50 (2) 2001, 445-462
2. N. Tiesler., U. Engel, M. Geiger: Forming of microparts-effects of miniaturization on friction, Advanced Technology of Plasticity, Vol. II, Proceedings of the 6th ICTP, sept. 19-24 (1999) pp 889-894.
3. J. Grønbæk: Koldflydepresning. Temarapport: Massivformgivning og pulverteknologi, p. 53. Jernet (in Danish) 1981.
4. C.P. Withen, J.R. Marstrand, M. Arentoft, N.A. Paldan: "Flexible tool system for cold forging of micro components", First International Conference on Multi-Material Micro Manufacture, 29June-1 July 2005, Forschungszentrum Karlsruhe, Germany, Elsevier, pp. 143-146
5. M. Arentoft, N.A. Paldan: "Production equipment for manufacturing of micro metal components", 9th ESAFORM 2006, Glasgow, UK, pp 579-582
6. M. Trevisan: "Transfer system for mass-production of micro components", M.Sc. thesis, Institute for Manufacturing and Management, Techn. Uni. of Denmark, 2006
7. R.S. Eriksen, M. Arentoft and N.A. Paldan, "Tool Design and Manufacturing for Bulk Forming of Micro Components," 10th ESAFORM Conference proceeding, Zaragoza, Spain, 2007
8. M. Arentoft, N. Paldan, R.S. Eriksen: "Cold forging of industrial micro components", Proceedings of the 39th ICFG plenary meeting, Changwon, Korea, 2006
9. H.N. Hansen, T. Eriksson, M. Arentoft, N. Paldan: "Design rules for Microfactory Solutions", proc. of 5th Int. workshop on microfactories, Besancon, France 2006

# Modelling of Viscoplastic Behaviour of IN718 Under Hot Forging Conditions

Y. P. Lin[*], J. Lin[*], T. A. Dean[*] and P. D. Brown[¶]

[*]Department of Mechanical and Manufacturing Engineering, University of Birmingham, Edgbaston, Birmingham B15 2TT, UK
[¶]Rolls-Royce plc, PO Box31, Derby DE24 8BJ, UK

**Abstract.** The mechanical properties of IN718 are directly related to microstructure such as grain size and the hardening mechanisms, which are effective during thermomechanical processing and subsequent heat treatment. In this study a set of unified viscoplastic constitutive equations were determined for IN718 from experimental data of hot forging conditions. Techniques were developed to analyse the reliability and consistency of the experimental data derived from different previous publications. In addition to viscoplastic flow of the material, the determined material model can be used to predict the evolution of dislocations, recrystallisation and grain size occurring during thermomechanical processing.

**Keywords:** IN718, Constitutive Equations, Materials Modelling, Hot Forging.
**PACS:** 81.05.-t.

## INTRODUCTION

The nickel-based superalloy, IN718, is widely used in aerospace industry. Aero-engine parts, such as disks, shafts and blades, are made of IN718 due to its good corrosion resistance and mechanical properties in an extreme working environment [1]. These good mechanical properties at high temperatures are largely related to the microstructure of formed parts. For example, many hot forming operations, such as extrusion, forging, and heat treatment are involved in the manufacturing of gas turbine blades. The microstructure of the material changes at each operation and heat treatment. Experimental and predicted techniques [2-4] should be developed to enable the microstructure and mechanical properties during the forming process can be observed.

The microstructure and mechanical properties of the turbine blade are significantly altered by metallurgical transformations during thermomechanical processing. To obtain specified mechanical properties and microstructure in IN718 turbine blades, it is necessary to be able to predict the microstructure evolution of the superalloy during hot forging conditions. Hence, an accurate material model is an essential tool to obtain trustworthy results. Although significant experimental research has been carried out for IN718 to investigate the viscoplastic flow and microstructure evolution during thermo-mechanical processes, there is no appropriate model available to simulate the evolution. In this study a set of constitutive equations proposed by Lin et al [3] was adapted to model the viscoplastic behaviour of IN718 under hot forging conditions.

CP907, 10th ESAFORM Conference on Material Forming, edited by E. Cueto and F. Chinesta

The material constants within the unified constitutive equations were determined from experimental data derived from various previous publications.

## MODELLING VISCOPLASTIC FLOW AND MICROSTRUCTURE EVOLUTION OF IN718

### Unified Viscoplastic Constitutive Equations

A set of unified viscoplastic constitutive equations has been proposed by Lin *et al* [3] and also has been successfully used to model recrystallisation, grain size and dislocation density evolution in hot rolling of steels [4]. In hot forging IN718, similar phenomena of viscoplastic flow and microstructure evolution have been observed. In this work the constitutive equations developed by Lin *et al* [3] were employed by considering the particular features of IN718 during thermomechanical processing. For ease of discussion, the equations are listed first.

$$\dot{\varepsilon}_p = A_1 \sinh\left[A_2(\sigma - R - \sigma_y)\right](d/d_0)^{-\gamma_1} \qquad 1$$

$$\dot{S} = H_1\left[x\overline{\rho} - \overline{\rho}_c(1-S)\right](1-S)^{\lambda_1} \qquad 2$$

$$\dot{x} = X_1(1-x)\overline{\rho} \qquad 3$$

$$\dot{\overline{\rho}} = (d/d_0)^{\delta_1}(1-\overline{\rho})\left|\dot{\varepsilon}_p\right|^{\delta_2} - C_r\overline{\rho}^{\delta_3} - (C_s\overline{\rho})/(1-S)\dot{S} \qquad 4$$

$$\dot{R} = 0.4B\overline{\rho}^{-0.6}\dot{\overline{\rho}} \qquad 5$$

$$\dot{d} = (G_1/d)^{\psi_1} - G_2\dot{S}(d/d_0)^{\psi_2} \qquad 6$$

$$\sigma = E(\varepsilon_T - \varepsilon_p) \qquad 7$$

Equation 1 models viscoplastic flow of the material, where R represents isotropic hardening which is directly related to dislocation density. $A_1$ is a temperature dependent material constant. The parameter $\gamma_1$ characterises the effect of grain size on viscoplastic flow. $\sigma_y$ is the initial yield stress of the material, which varies with temperature, E (7000MPa) is Young's modulus. In forging processes, recrystallisation is an important feature [5]. Recrystallisation can be dynamic, occurring during deformation, or static, happening after forging or between forging steps. The affected features of the change of recrystallisation, such as grain size and dislocation density, alter the mechanical properties of the material. Thus unified viscoplastic constitutive equations should be able to capture these features and their interactive effects. In the equations, a normalised dislocation density ($\overline{\rho}$) concept is introduced by defining, $\overline{\rho} = 1 - \rho_i/\rho$, where $\rho_i$ is the initial dislocation density and $\rho$ is the dislocation density after deformation. The normalised dislocation density varies from 0 (initial state) to 1 (saturated state). Softening mechanisms, dynamic recovery, static recovery (annealing) and recrystallisation are included in the normalised dislocation density rate equation. Once the normalised dislocation density accumulates to a critical value, $\overline{\rho}_c$, given sufficient time, which is controlled by the onset parameter, x, in Equation 3, recrystallisation takes place. The evolution of recrystallised volume fraction is controlled by Equation 2, related to the evolution of dislocation density and grain size, which are modelled in Equations 4 and 6 respectively. Therefore, the interactive relationship of the physical variables and viscoplastic flow of the material can be fully described. The corresponding temperature parameters in the equations are listed in TABLE 1.

TABLE 1. List of temperature dependent parameters. $\kappa = 8.31 \text{ J} \cdot \text{mol}^{-1} \cdot K^{-1}$ (Universal gas constant); $T$ is absolute temperature in $°K$.

| $\sigma_y = \sigma_{y0} \exp(Q_p /(\kappa T))$ | $C_r = C_{r0} \exp(-Q_r /(\kappa T))$ |
|---|---|
| $\overline{\rho}_c = \overline{\rho}_{c0} \exp(Q_c /(\kappa T))$ | $G_1 = G_{10} \exp(-Q_{gg} /(\kappa T))$ |
| $\psi_1 = \psi_{10} \exp(Q_\psi /(\kappa T))$ | $X_1 = X_{10} \exp(-Q_x /(\kappa T))$ |
| $B = B_0 \exp(Q_b /(\kappa T))$ | $A_1 = A_{10} \exp(-Q_a /(\kappa T))$ |

## Experimental Data Analysis

Thermomechanical tests have been carried out for IN718 in different laboratories worldwide. A total of thirty six Stress-Strain curves were obtained for a temperature range of 900°C to 1066°C and strain rate range from $0.0005 \, s^{-1}$ to $5 \, s^{-1}$ [6-8], which are summarised in TABLE 2. The experimental data come from different laboratories and the tests were carried out using different equipment. The aim of this study is to investigate the consistency and reliability of the experimental data. At low strains, the change of microstructure is limited. Thus the temperature dependent viscoplastic flow of the material can be approximately described with $\dot{\varepsilon}_p \exp(Q/(RT)) \approx A_1 \sinh A_2 \sigma$ or $\ln \dot{\varepsilon}_p + (Q/(RT)) \approx \ln A_1 + A_2 \sigma$. The analysis is carried out to investigate the effect of temperature on flow stresses first. The symbols in FIGURE 1 (a) show the flow stresses against $1/T$, which are obtained from experimental data listed in TABLE 2 for $\dot{\varepsilon} = 0.0005 \, s^{-1}$, $0.01 \, s^{-1}$ and $0.1 \, s^{-1}$ at a strain of 0.2. It can be seen that the data for a particular strain rate can be fitted approximately with straight lines. These lines are approximately parallel, which indicates that the Q value is about consistent. If the selected data from TABLE 2 are plotted on scales of $\log \dot{\varepsilon}$ and $\sigma$ for different temperatures, FIGURE 1 (b) can be obtained. Again, these experimental data can be approximated to parallel straight lines, the slopes of which are directly related to the $A_2$ value. According to FIGURE 1 (a) and (b), the experimental data are quite consistent at low strains.

**TABLE 2.** Experimental Stress-Strain curves obtained from various publications for IN718; 1 and 2 indicate the number of experimental Stress-Strain curves at specified deforming condition. A total of thirty six curves were derived from three sources [6-8].

| $\dot{\varepsilon}$ (1/S) \ T(°C) | 900 | 950 | 975 | 982 | 1000 | 1010 | 1038 | 1050 | 1066 |
|---|---|---|---|---|---|---|---|---|---|
| 5 | | | | 1 | | 1 | 1 | | 1 |
| 1 | | | 1 | | | | | 1 | |
| 0.1 | 1 | 1 | 1 | | 1 | | | 2 | |
| 0.05 | | | | 1 | | 1 | 1 | | 1 |
| 0.01 | 1 | 1 | 1 | | 1 | | | 2 | |
| 0.001 | 1 | 1 | 1 | | 1 | | | 2 | |
| 0.0005 | 1 | 1 | | 1 | 1 | 1 | 1 | 1 | 1 |

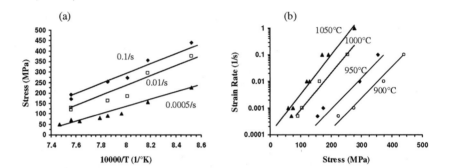

**FIGURE 1.** (a) Stress at $\varepsilon =0.2$ against 1/T for different strain rates; (b) log $\dot{\varepsilon}$ against stress at $\varepsilon =0.2$ for different temperatures. The solid lines are the fittings for the corresponding experimental data (symbols), which are obtained from references [6-8].

## Determination of Constants within the Constitutive Equations

The procedure for determining the values of constants within the constitutive equations is divided into two steps. The first step is to determine the constants relating to static grain growth, which is modelled using the first term of Equation 6. Experimental data of static grain growth derived by Zhang *et al* [9] are shown with the symbols in FIGURE 2 (a) and were used for the work. The data were obtained for aging temperatures of 980°C and 1020°C for IN718 deformed at $\dot{\varepsilon} = 0.01/s$ to a strain of 0.4. The constants, $G_{10}$, $\psi_{10}$, $Q_\psi$ and $Q_{gg}$ within the static grain growth term were determined from the experimental data using an Evolutionary Algorithm~EA-based optimisation method, which has been developed for this purpose by Li *et al* [10, 11]. The determined values of the constant are listed in TABLE 3 and the fitting results are shown with the solid curves in FIGURE 2 (a). Normally, the hot forging temperatures for IN718 are within the range of 950°C to 1050°C. The dashed curves are the predicted results from the equation for the low and high bounds of forging temperatures. The second step is to determine the rest of the constants within the Equations 1-7 and TABLE 1. Selected experimental Stress-Strain curves from TABLE 2 are used for the determination, which are shown with symbols in FIGURE 2 (b) for temperature varying from 950°C to 1050°C and strain rate from 0.001/s to 0.1/s. The temperature and strain rate values selected are those commonly used for hot forging of IN718 components. The same technique [10, 11] was used for this work. The determined values of the constant are listed in TABLE 3. The fitting results are shown with solid curves in FIGURE 2 (b), where fairly good agreements between computed and experimental results are obtained. If the determined constitutive equations were implemented into FE codes, microstructure evolution during hot forging and subsequent change of microstructure during heat treatment could be predicted.

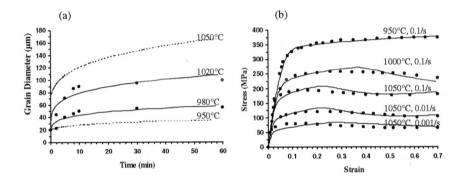

**FIGURE 2.** Comparison of experimental (symbols) and computed (solid curves) results for (a) static grain growth and the dashed curves are predicted results. (b) Stress-Strain relationships for IN718 deforming at different temperatures and strain rates.

**TABLE 3. The determined constants within the unified viscoplastic model for IN718.**

| $Q_p$ (J·mol$^{-1}$) | $Q_c$ (J·mol$^{-1}$) | $\sigma_{y0}$ (MPa) | $C_{r0}$ (s$^{-1}$) | $\rho_{c0}$ (-) |
|---|---|---|---|---|
| 319000.000 | 6998.150 | $9.000 \times 10^{-11}$ | $5.790 \times 10^{13}$ | 0.002 |
| $A_{10}$ (s$^{-1}$) | $A_2$ (MPa$^{-1}$) | $\gamma_1$ (-) | $H_1$ (s$^{-1}$) | $\lambda_1$ (-) |
| $9.600 \times 10^{28}$ | 0.032 | 3.540 | 61.000 | 9.000 |
| $\delta_1$ (-) | $\delta_2$ (-) | $\delta_3$ (-) | $C_s$ (-) | $B_0$ (MPa) |
| 2.950 | 2.460 | 3.400 | 2100.000 | 0.010 |
| $\psi_2$ (-) | $Q_a$ (J·mol$^{-1}$) | $Q_{gg}$ (J·mol$^{-1}$) | $G_{10}$ ($\mu$m) | $\psi_{10}$ (-) |
| $8.500 \times 10^{-3}$ | $6.350 \times 10^5$ | 246057.259 | $7.504 \times 10^{11}$ | 0.331 |
| $Q_r$ (J·mol$^{-1}$) | $Q_\psi$ (J·mol$^{-1}$) | $Q_b$ (J·mol$^{-1}$) | $Q_x$ (J·mol$^{-1}$) | $G_2$ (s$^{-1}$) |
| $2.620 \times 10^5$ | 28320.452 | $1.090 \times 10^5$ | 53500.000 | 262.000 |
| $X_{10}$ (s$^{-1}$) | $d_0$ ($\mu$m) | | | |
| 2.290 | 2.400 | | | |

# CONCLUSIONS

According to the investigations, the experimental data from different publications are fairly consistent at low strain level in terms of flow stresses. The determined unified viscoplastic constitutive equations are able to model the viscoplastic deformation of IN718 under hot forging conditions if the microstructure evolution during the process is considered.

# ACKNOWLEDGMENT

The authors appreciate the support of Rolls-Royce plc, UK.

# REFERENCES

1. Sims, C. T. and Stoloff, Norman S., *Superalloys II: High Temperature Materials for Aerospace and Industrial Power*, John Wiley & Sons Inc, 1987.
2. Feng, J. P. and Luo, Z. J., *Journal of Materials Processing Technology* 108, 40-44 (2000).
3. Lin, J., Liu, Y., Farrugia, D. C. J. and Zhou, M., *Philosophical Magazine A* 85, 1967-1987 (2005).
4. Lin, J., Foster, A. D., Liu, Y., Farrugia, D. C. J. and Dean, T. A., *Journal of Engineering Transactions (In press)*.
5. Na, Y. S., Yeom, J. T., Park, N. K. and Lee, J. Y., *Journal of Materials Processing Technology* 141, 337-342 (2003).
6. Thomas, A., El-Wahabi, M., Cabrera, J. M. and Prado, J. M., *Journal of Materials Processing Technology* 177, 469-472 (2006).
7. Park, N. K., Kim, I. S., Na, Y. S., and Yeom, J. T., *Journal of Materials Processing Technology* 111, 98-102 (2001).
8. Medeiros, S. C., Prasad, Y. V. R. K., Frazier, W. G., and Srinivasan, R., *Materials Science and Engineering A* 293, 198-207 (2000).
9. Zhang, J. M., Gao, Z. Y., Zhuang, J. Y. and Zhong, Z. Y., *Journal of Materials Processing Technology* 101, 25-30 (2000).
10. Li, B., Lin, J. and Yao, X., *International Journal of Mechanical Sciences* 44, 987-1002 (2002).
11. Lin, J. and Yang, J., *International Journal of Plasticity* 15, 1181-1196 (1999).

# Reconstruction of the Deformation Process by Loading Curves for Uniaxial Compression of Zr-based Alloys at Increased Temperatures

Yuriy Perlovich[1], Margarita Isaenkova[1], Olga Krymskaya[1],
Vladimir Filippov[2], Sergey Kropachev[2] and Mikhail Shtutca[2]

[1]*Moscow Engineering Physics Institute, Kashirskoe shosse 31, Moscow, 115409 Russia*
[2] *Chepetckiy Mechanical Plant, Belov str.7, Glazov, 427620 Russia*

**Abstract.** As applied to commercial Zr alloys for nuclear industry, deformed by uniaxial compression at increased temperatures, participation of different processes in plastic deformation is considered on the basis of loading curves and X-ray data on the crystallographic texture of compressed samples.

**Keywords:** Zr-based alloys, loading curve, texture, phase transformation, dynamic recrystastallization.
**PACS:** 62.20.-x; 68.55.jk

## INTRODUCTION

A deformation behavior of commercial Zr-based alloys varies in very wide limits depending on the temperature-rate regime of deformation and the type of alloy. It is convenient to analyze fundamental physical aspects of the deformation behavior by the example of loading scheme as simple as possible. Uniaxial compression is the simplest deformation mode, having close analogues among widespread technological procedures, such as forging or pressing. Its symmetry allows to restrict the consideration to the one-dimensional case and, as a first approximation, to ignore effects of anisotropic structure formation. In the given paper high-temperature deformation processes in Zr-based alloys are reconstructed by the combined analysis of loading curves, measured in the course of uniaxial compression, and data of X-ray texture measurements, conducted at room temperature after deformation. A memory of deformation processes, recorded in the crystallographic texture of studied samples, helps to interpret features of loading curves in a way, connecting mechanical properties of material with its structure.

## EXPERIMENTAL TECHNIQUE AND OBTAINED RESULTS

Model samples $\varnothing 16$ x 24 mm in size from alloys Zr-1%Nb and Zr-2.5%Nb were deformed by uniaxial compression up to $\varepsilon=0.6$ at nominal temperatures within ($\alpha+\beta$)- and $\beta$-regions of Zr-Nb phase diagram [1]. The used deformation rates were equal to

*CP907, 10$^{th}$ ESAFORM Conference on Material Forming*, edited by E. Cueto and F. Chinesta
© 2007 American Institute of Physics 978-0-7354-0414-4/07/$23.00

0.1, 0.4, 0.7 and 1.0 s$^{-1}$. The glass powder was used as a lubricant. After deformation samples were cooled in water.

A prehistory of the used samples goes back to the ingot, consisted of coarse β-grains, undergone β(BCC)→α(HCP) phase transformation (PT) by subsequent cooling in accordance with the orientation relationship $\{011\}_\beta \| (0001)_\alpha$, $<111>_\beta \| \{11.0\}_\alpha$ [2], and within each sample families of α-Zr grains of several definite orientations prevail. By heating for deformation, PT α→β develops by the same orientation relationship, so that in the most probable variant a single β-grain is restored from derivative α-grains.

In the course of compression the loading curves were recorded and recalculated into coordinates "true deformation degree – true stress". Some of them, demonstrating most noteworthy features, are shown in Fig. 1. Among these features there are:

(1) general level of plastic deformation stress;

(2) gradual or sharp transition from elastic to plastic deformation;

(3) presence of the yield drop;

(4) hardening or weakening in the course of plastic deformation;

(5) plastic deformation by the constant stress;

(6) dependence or independence of the measured stress on the deformation rate;

(7) changes in deformation features with growing deformation degree.

When passing from one sample to another, above tendencies in deformation development can become less legible or change to different ones, as it is seen in Fig. 1.

Any plastic deformation is accompanied by regular grain reorientation and in the case of sufficiently high deformation degree results in formation of the crystallographic texture, whose type depends on the loading scheme and operating deformation mechanisms in accordance with the known models of texture formation [3]. When the deformation texture develops in the high-temperature β-phase, the studied texture of the low-temperature α-phase arises as a result of β→α PT by the above orientation relationship. Texture direct pole figures (PF) [3] were measured by the automated X-ray diffractometer DRON-3M of Russian manufacture. The typical distributions of basal axes in α-Zr matrix, that is PF(0001)$_\alpha$, for samples, compressed at different temperatures, are presented in Fig. 2. Inner regions of PF with angular radius 70° were constructed. It is expected that the axial symmetry of the deformation scheme by uniaxial compression would predetermine the same axial symmetry of the arising texture. If this is not the case, it signifies that the number of reflecting grains is non-representative statistically or that the initial ingot is characterized only by several orientations of β-grains. There are the following types of PF(0001)$_\alpha$ depending on the deformation regime [4]:

(1) a very sharp texture, formed by separate intense narrow maxima, which originate from the distribution of axes $<011>_\beta$ in the β-matrix, consisting of coarse grains with orientations, stable under compressive deformation (Fig. 2-a);

(2) a texture with features of axiality, formed by separate small sharp maxima, which originate from the β-matrix, experienced dynamic recrystallization (Fig. 2-b);

(3) a texture, similar to (1) in positions of maxima, but differing from it by their greater scattering (Fig. 2-c);

(4) an extremely scattered texture down to its practical disappearance (Fig. 2-d);

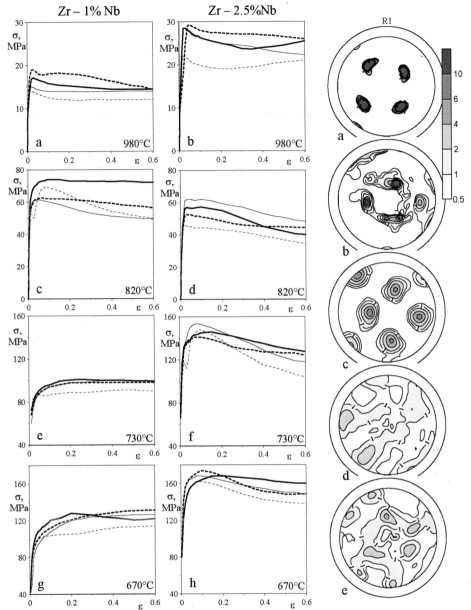

FIGURE 1. Loading curves for alloys Zr – 1%Nb (a,c,e,g) and
Zr – 2.5%Nb (b,d,f,h), deformed by uniaxial compression
at temperatures 980° (a,b), 820° (c,d), 730° (e,f) and 670°C (g,h).
······ $\dot{\varepsilon}$ =0.1s⁻¹, ——— $\dot{\varepsilon}$ =0.4 s⁻¹, ■■■■ $\dot{\varepsilon}$ =0.7 s⁻¹, ——— $\dot{\varepsilon}$ =1.0 s⁻¹ .

FIGURE 2. Typical PF(0001)$_\alpha$ for
samples from Zr-2.5%Nb alloy,
compressed of temperatures
980° (a), 910° (b), 820° (c),
730° (d), 670°C (e).
Angular radius of PF – 80°.

(5) a multicomponent irregular texture, produced by significant compressive deformation of α-phase (Fig. 2-e).

Texture disappearance is a consequence of changes in deformation mechanisms: when grains disintegrate by PT, the intragranular crystallographic slip proves to be difficult and operation of the non-crystallographic mechanism, consisting in mutual displacements of crystallites by interphase boundaries, becomes predominant [4]. Whereas the crystallographic slip is accompanied by regular rotation of the grain lattice and by development of deformation textures, displacements of crystallites along boundaries result into accidental lattice rotations and texture scattering. Combined consideration of loading curves and texture features of samples, deformed at high temperatures, allows to reveal some additional details of the deformation process.

## DISCUSSION

(1) *The general level of plastic deformation stress* rises as the temperature of compression comes down (Fig. 1). At the same time, against a background of this natural tendency the definite features of studied alloys become apparent. The alloy with 2.5% Nb differs from that with 1% Nb by the higher content of β-phase and by finer grains at any temperatures within α– and (α+β)-regions of Zr-Nb phase diagram. Therefore, the equiphase composition, most favorable for operation of the non-crystallographic grain slippage by interphase boundaries, is attained in Zr-2.5%Nb alloy at the lower temperature, than in Zr-1%Nb alloy. Mutual positions of loading curves, measured for two studied alloys at same temperatures, are determined by the combined effect of several structure factors, among which there are grain size, content of β-phase, Nb content in β-phase, critical stress for activation of crystallographic and non-crystallographic deformation mechanisms in the matrix with the concrete structure. Thus, the curves of loading at temperatures of the lower part of (α+β)-region, i.e. 670° and 730°C, for Zr-2.5%Nb alloy are situated at the higher stress level, than for Zr-1%Nb alloy (Fig. 1). But at 820°C, when the former alloy consisted already from β-phase only, whereas the latter alloy – still from two phases, loading curves for the single-phase alloy prove to be lower, than for the two-phase alloy. At last, by compression at temperatures of β-region curves for Zr-2.5%Nb alloy due to the higher Nb content in β-phase are situated again higher, than curves for Zr-1%Nb alloy.

(2) *Gradual or sharp transition from elastic to plastic deformation* is determined by non-simultaneous or simultaneous attainment of the critical shear stress in grains of the tested sample. At the temperature of β-phase deformed samples of both alloys consist of coarse grains with three or four stable orientations. Their behavior under deformation due to the perfect texture is similar to that of a quasi-single crystal, i.e. the critical shear stress proves to be attained simultaneously within macro-regions, comparable in size with the sample. Therefore, in this case the transition from elastic to plastic deformation is sharp. When samples are compressed at temperatures of (α+β)-region, the deformation texture of β-phase is rather scattered, as it is seen by the inherited texture of derivative α-phase, and the deformation texture of α-phase is multi-

component. In this case the critical shear stress is attained in mutually misoriented fine grains non-simultaneously with gradual transition from elastic to plastic deformation.

(3) *Presence of the yield drop* is observed in cases, when the matrix is too perfect and the dislocation density is insufficient initially for maintenance of deformation with the given rate. It is natural, that in Zr-based alloys the yield drop arises only by deformation in β-phase (Fig. 1-a,b).

(4) *Hardening or weakening in the course of plastic deformation* indicates to the predominant tendency in structure formation. Since the temperature of deformation in all cases was sufficiently high for intense dynamic recovery, effects of dislocation hardening could not be essential. Local deformation heating is connected with violation of the phase equilibrium near slip bands, development of the deformation induced PT $\alpha \rightarrow \beta$ and arising of additional portions of β-phase. Thus, deformation heating of material is accompanied by its weakening and at the same time can be connected with hardening due to comminution of grains. As the phase composition of the fine-grained sample approaches to equiphase, its deformation requires the higher stress. At that, activation of the deformation mechanism of grain slippage by interphase boundaries becomes probable in the case of the proper deformation rate. And vice versa, when the content of β-phase exceeds 50% and the phase composition of the sample under deformation moves away from equiphase, its compression becomes easier due to repeated activation of the crystallographic slip. The resulting effect of these mutually discrepant tendencies is apparent by the loading curves. Weakening of samples under compression is observed more often, than hardening. Dynamic recrystallization also causes comminution of grains with all its accompanying effects, including hardening and/or activation of intergranular slippage. One more reason of hardening is cooling of the sample by heat removal at the late deformation stages.

(5) *Plastic deformation by the constant stress* up to rather high deformation degrees shows, that the structure condition of material remains the same at all deformation stages, i.e. plastic deformation has no effect on mechanical behavior of samples. There are two variants of such deformation process: (a) dislocations easily traverse the grain by their way to boundaries, (b) deformation develops by means of intergranular slippage along boundaries. Variant (a) takes place by high-temperature deformation of coarse β-grains, as in the case of Zr-1%Nb alloy, deformed at 980°C with rate 0.4 s$^{-1}$ (Fig. 1-a). Variant (b), having mainly diffusion nature, is realizable under conditions of PT or dynamic recrystallization, creating the fine-grained misoriented structure. Texture data help us to specify the concrete processes, responsible for the stress constancy by compression of samples. In particular, weak scattered textures of samples from Zr-1%Nb alloy, deformed at 730°C (Fig. 2-d) and the corresponding horizontal loading curves (Fig. 1-e) are connected with grain slippage by interphase boundaries under approximately equiphase composition. Similarly, multiple small texture maxima in PF(0001) for samples of both alloys, deformed with some rates at temperatures above 900°C (Fig. 2-b), indicate to dynamic recrystallization of β-phase, which promotes mutual slippage of new-originated fine crystallites by intergranular boundaries and appearance of the horizontal loading

curves (Fig. 1-a,b). However, the horizontal loading curve can be interpreted also as a manifestation of mutual equilibrium of hardening and weakening effects, which are connected with different processes, induced by deformation.

(6) *Dependence or independence of the measured stress on the deformation rate* also indicates both to the material condition and the operating deformation mechanism. According to the dislocation theory of plastic deformation, the yield stress grows with the deformation rate in order to create the necessary density of free dislocations. However, in studied alloys this principle is observed only sometimes, testifying about action of some additional factors, which upset the monotonic character of the above dependence. Among these factors there are features of heat generation and PT under deformation at temperatures of ($\alpha$+$\beta$)-region of Zr-Nb phase diagram. At that, deformation heating of the sample depends on the acting mechanisms and the volume ratio of $\alpha$- and $\beta$-phases. Thus, by compression of Zr-1%Nb alloy at 670°C all loading curves follow one after another in the expected order only up to $\varepsilon$=0.3, when the curve for deformation rate 1.0 s$^{-1}$ goes down, intersecting other curves (Fig. 1-g). This occurs because of increased deformation heating of sample by the maximal deformation rate.

Various inevitable inhomogeneities and, first of all, the texture inhomogeneity of the original coarse-grained ingot also violates the expected dependence of loading curves on the deformation rate. Since the number of initial $\beta$-grains within our rather small samples is non-representative statistically, $\beta$-grains in supposedly identical samples by compression are drawn predominantly to different stable orientations (the axis of compression in BCC-metals coincides finally with <001>, <111>, <011> or <112>) and their further deformation develops in different manners as well as deformation of the derivative grains of $\alpha$-Zr, where, depending on the initial orientation, operate those slip systems or others.

(7) *Changes in deformation features by increase of the deformation degree* are seen by the complicated, non-monotonic character of many loading curves, testifying about successive prevalence of different processes. The change of hardening by weakening in the course of plastic flow at temperatures of ($\alpha$+$\beta$)-region shows, that effects of deformation heating begin to dominate over effects of grain fragmentation, accompanying deformation induced processes of dislocation reorganization and PT (see loading curves for 670° and 730°C in Fig. 1). A wavy character of some loading curves indicates to alternate domination of opposite processes, among which, in particular, there are the following: (a) heating and cooling, associated with $\alpha$→$\beta$ and $\beta$→$\alpha$ PT; (b) operation of the intragrained crystallographic slip and replacement of crystallites by interphase boundaries; (c) intensifying and damping of dynamic recrystallization.

# REFERENCES

1. D.L.Douglass, *The Metallurgy of Zirconium*, International Atomic Energy Agency, Vienna, 1971.
2. R.A. Holt and S.A. Aldridge, *J. Nucl. Mat.* **135,** 246-259 (1985).
3. U.F. Kocks, C.N. Tome and H.-R. Wenk, *Texture and Anisotropy*, Cambridge University Press, 1998.
4. Yu. Perlovich, M. Isaenkova, S. Akhtonov et al. in *The 9$^{th}$ International Conference on Material Forming ESAFORM 2006, Glasgow, UK,* edited by N. Juster and A. Rosochowski, Publishing House "Akapit", Krakow, Poland, 2006, pp. 439-442.

# Effects of Forging Process Parameters on Microstructure Evolution of Aluminum Alloy 7050

Youping YI [†], Yan SHI, Jihui YANG, Yongcheng LIN

*School of Mechanical and Electrical Engineering, Central South University,
410083 Changsha, P.R. China*

**Abstract.** The objective of this work is to investigate the behavior of microstructure evolution of aluminum alloy 7050 under the condition of different forging process parameters by means of combining materials physical model with finite element code. For the purpose of establishing constitutive equation and physical model of microstructure evolution, the isothermal compression test were performed by machine Gleeble 1500 on the condition of temperatures ranging from 250°C to 450°C and constant strain rates of $0.01s^{-1}$, $0.1s^{-1}$, $1s^{-1}$ and $10s^{-1}$. The behaviors of microstructure evolutions of aluminum alloy 7050 under difference process parameters were studied by metallographic observations. The experiment results showed that recrystallization during forming process occurred at the critical strain and the volume fraction of recrystallization changed with the temperature and strain rate. According to the results of isothermal compression test, a constitutive equation and an empirical model of DRX were obtained. A finite element code DEFORM 3D was used to analyze the influence of different forging process parameters on the behavior of microstructure evolution in details. The present model and simulation method can be served as a useful tool to predict and control the properties and shape of aluminum alloy 7050 components during forging.

**Keywords:** Forging; Simulation; Aluminum alloy 7050; Dynamic recrystallization; Grain size
**PACS:** 81.10.Jt; 81.40.Lm; 81.40.Gh

## INTRODUCTION

The modern metal forming process would assure obtaining of high quality products with simultaneous production cost minimization [1]. This problem is significant in a forging process of 7050 aluminum alloy, which is wildly used in the manufacture of airplane components due to its high strength and high toughness. It is well known that the properties of components depend on the micro-structure of materials. Due to the highly alloyed element, the micro-structure of this material is sensitive to the process parameters and undergoes a very complicated evolution during forging such as dynamical recrystallization(DRX), dynamic recovery(DRV) and grain growth[2-4]. Besides, the behavior of micro-structure evolution and the influence of forging process parameters can not been described by accurate theoretical model. Many researches have performed both experimental and theoretical investigations on DRX and DRV in the past[5-10].Nevertheless, little attentions were paid to simulating the relationship

CP907, *10th ESAFORM Conference on Material Forming,* edited by E. Cueto and F. Chinesta
© 2007 American Institute of Physics 978-0-7354-0414-4/07/$23.00

between the process parameters of practical forging and the micro-structure evolution of this alloy.

In the present work, we attempt to use a new approach of combining the semi-empirical model with commercial finite element code Deform 3D to simulate the forging process. The basic material models (flow stress model, recrystallization model) are obtained by the experiment of isothermal compression test on the machine Gleeble 1500. The effects of deformation parameters, e.g. strain rate and deformation temperature, on the features of dynamical recrystalliztion (DRX) and average grain size of 7050 aluminum alloy, were discussed.

## EXPERIMENT PROCEDURES

The simulation models, e.g. flow stress equation and DRX model, are crucial for investigating accurately 7050 alloy deformation behavior at different process parameters. In this work, these models were based on the isothermal compression test on the machine Gleeble 1500. The materials for the research were taken from industrial aluminum alloy 7050 ingot, which was characterized by contents of 0.12Si, 0.15Fe, 2.60Cu, 2.60Mg, 6.70Zn, 0.06Ti, 0.13Zr, and balance Al (wt.%). The cylindrical specimens, 10 mm in diameter and 12 mm in length, were machined with flat bottomed groves on the end faces. In order to reduce friction during compression, the graphite mixed with machine oil was used to fill in the bottomed groves. The isothermal compression tests were performed on the machine Gleeble 1500 in the range of temperature of 250-450°C, at the strain rate of 0.01, 0.1, 1, 10 s$^{-1}$, followed by a water quench to observe the deformed micro-structure. The specimens were sectioned along the longitudinal compression axis, ground and polished. The optical microstructure in the center region of the section plane was examined using a Leica DMIRM image analyzer.

## MATERIAL MODEL FOR HOT DEFORMATION

### Flow Stress Equation

Based on the analysis of the load-displacement data recorded by the thermal simulator during hot compression tests, the flow stress curves of 7050 Aluminum alloy can be obtained. Typical true stress-strain curves are shown in Fig. 1.

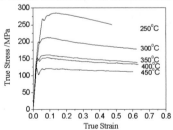

**FIGURE 1.** Flow stress curves at strain rate of 10s$^{-1}$ and different temperatures.

It is observed that the flow stress is composed of different stages(e.g. strain-hardening, strain-softening and steady state-strain) and strongly by affected by the deformation temperature and strain rate. The presence of a distinct peak in the flow stress indicts the beginning of recrystallization softening. The DRX process depends on time and deformation temperature. Higher strain rate and lower temperature lead to a higher peak stress and thus need longer deformation time for nucleation and growth of DRX grains. As expected, the deformation resistance increases with the increasing in strain rate and decreasing in deformation temperature.

As discussed above, the features of flow stress are strongly influenced by the conditions of deformation (temperature, strain rate), which can be described using the Zener-Hollomon parameter as following:

$$Z = \dot{\varepsilon} \exp(Q/RT), \tag{1}$$

where $\dot{\varepsilon}$ is the strain rate [s$^{-1}$], $R$ the mole gas constant [8.31 J·mol$^{-1}$·K$^{-1}$], $T$ the deformation temperature[K], $Q$ the activation energy of hot deformation [kJ mol$^{-1}$].

Various empirical equations have been proposed to describe the thermally activated process of hot deformation. One generally accepted hyperbolic sine function that expresses the dependence of the flow stress on the deformation temperature and strain rate is given as

$$\dot{\varepsilon} = A[\sinh(\alpha\sigma_p)]^n \exp(-Q/RT), \tag{2}$$

where $\sigma_p$ is the peak stress(MPa), $A$, $\alpha$ and $n$ the material constants.

The material constants in Eq.(1) can be determined by regression on the experiment data and the flow stress equation for 7050 aluminum alloy can be expressed as

$$\dot{\varepsilon} = 5.83 \times 10^{18} \left[\sinh\left(1.239 \times 10^{-2}\sigma_p\right)\right]^{7.538} \exp\left(-2.6406 \times 10^5 / RT\right) \tag{3}$$

## Model of DRX

DRX is a common phenomenon in forging process and occurs only when the strain or dislocation density reaches a critical level, which depends on hot process parameters, such as deformation temperature and strain rate [1]. The kinetic of DRX is usually described using critical strain, recrystallized volume fraction and dynamically recrystallized grain size.

For most metal materials, the onset of DRX usually occurs at a critical strain $\varepsilon_c$, which is expressed as

$$\varepsilon_c = 0.8\varepsilon_p, \tag{4}$$

where $\varepsilon_p$ is the peak strain, which can be written from analysis of the experiment data as following

$$\varepsilon_p = 4.107 \times 10^{-3} \dot{\varepsilon}^{0.06} \exp\left(1.318 \times 10^4 / RT\right) \tag{5}$$

For strains exceeding the critical value, the DRX volume fraction $X$ is assumed to follow an Avrami kinetics equation[6]:

$$X = 1 - \exp\left[-0.693 \times \left(\frac{\varepsilon - 0.8\varepsilon_p}{\varepsilon_{0.5}}\right)^2\right], \tag{6}$$

where $\varepsilon_{0.5}$ is the strain for 50% DXF and is given by

$$\varepsilon_{0.5} = 1.214 \times 10^{-5} d_0^{0.13} \dot{\varepsilon}^{0.04} \exp(5.335 \times 10^4 / RT), \tag{7}$$

where $d_0$ is the initial grain size which can be determined as 90 μm by Fig.2(a).

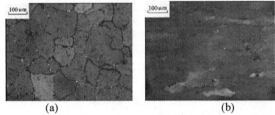

(a)                                        (b)

**FIGURE 2.** Optical mircograph of 7050 Al alloy: (a) initial grain size; and (b) at a temperature of 450°C and strain rate of 0.01s⁻¹.

Fig.2(b) shows the typical optical mircograph of 7050 aluminum alloy obtained from isothermal compression test. DRX grain size can be expressed by Yada model as following:

$$d_{rex} = 78.6022 \dot{\varepsilon}^{-0.03722} \exp(-1902.72 / RT). \tag{8}$$

The average grain size $d_{ave}$ is given as

$$d_{ave} = d_0 (1 - X) + d_{rex} X. \tag{9}$$

## RESULTS AND DISCUSSION

It is known that fine grain is desirable for forging process and the grain size is related closely with DRX volume fraction. The experiment research results for 7050 aluminum alloy indicate the great influence of deformation parameters (temperature and strain rate) on DRX behavior.

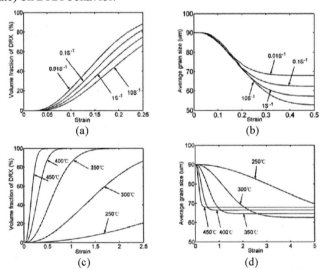

(a)                                        (b)

(c)                                        (d)

**FIGURE 3.** Effect of temperature and strain rate on the DRX behaviors: (a) DRX volume fraction ; (b) average grain size at a given temperature of 450°C ; (c) DRX volume fraction and (d) average grain size at a given grain rate of 0.01s⁻¹. The initial grain size is 90 μm for the all cases.

484

As shown in Fig.3(a), there exists a critical strain value under the condition of a given temperature, below which there is little DRX, but beyond which DRX occurs rabidly. Furthermore, an increase of the strain rate leads to a decrease of the DRX fraction volume. The corresponding average grain size decreases with an increase of strain, and the higher grain rate is favorable for reduction of grain size, see Fig.3(b).

On the condition of higher temperature with a given strain rate of $0.01s^{-1}$, the critical dislocation density can be reached more easily, which results in the initiation of DRX and an increase of the DRX volume fraction. On the contrary, it takes a longer time for dislocation density to reach the critical value at a lower temperature(see Fig.3(c)). The corresponding average grain size decreases much more rapidly at higher temperature than the case at lower temperature(see Fig.3(d)). It is indicated that higher temperature is an important factor to influence the average grain size.

Substituting the above obtained semi-empirical models(from Eq.(1) to Eq.(9)) into the finite element code DEFORM 3D, we can analyze the micro-structure evolution of 7050 aluminum alloy on the conditions of different deformation parameters. In this work, a typical upsetting process of 7050 Al alloy component with diameter of 2200mm and length of 800mm is simulated by using DEFORM 3D, with a constant friction factor of 0.3, a constant heat transfer coefficient of $2kW/m^2/C$ and dies temperature of 300°C.

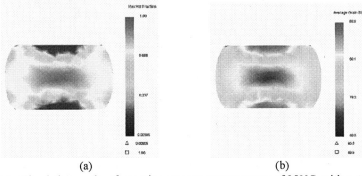

(a)                                        (b)

**FIGURE 4.** Simulation results of upsetting process at temperature of 350°C with a strain rate of $0.01s^{-1}$: (a) volume fraction of DRX; (b) average grain size.

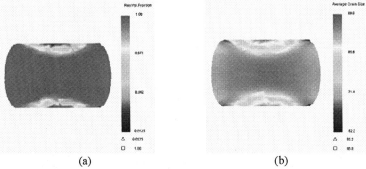

(a)                                        (b)

**FIGURE 5.** Simulation results of upsetting process at temperature of 450°C with a strain rate of $0.01s^{-1}$: (a) volume fraction of DRX; (b) average grain size.

Fig.4 and Fig.5 show the FE simulation results of cross-section of the billet after compression test at a strain rate of $0.01s^{-1}$ with a temperature of 350°C and 450°C, respectively. It is indicated that the DRX volume fraction in the centre area is greater than that on the edge of cross-section, and the average grain size in this area is correspondingly smaller.

As shown in Fig.4, the area which the DRX occurred completely is about the total area of 15% and the area which the grain size reached about 60 μm is only 10%. By contrast, the area which the DRX occurred completely reaches the total area of 90% and the area which the grain size reached about 60 μm is about 80% as shown in Fig.5. This phenomenon demonstrates that the higher temperature is favorable for promoting the occurrence of DRX and refinement of grain size. Also, strain rate is another factor which influences the DRX behavior and grain size. Obviously, the simulation results agree well with the experiment results, and we can obtain desired microstructure of 7050 aluminum alloy by controlling the deformation temperature and strain rate during practical forging.

## CONCLUSIONS

A group of semi-empirical models for analysis of microstructure evolution of 7050 aluminum alloy were developed by means of hot compression test at Gleeble-1500 machine, including flow stress equation and DRX model such as critical strain, volume fraction, DRX grain size and average grain size. The influences of the forging process parameters e.g. deformation temperature and strain rate on DRX behavior of 7050 aluminum alloy were investigated by using experiment research and numerical simulation. Increasing the temperature and decreasing the strain rate can effectively enhance the DRX volume fraction and refine grain size during practical forging. The methods and results in this present work can be utilized in forging process design for 7050 aluminum alloy component.

## ACKNOWLEDGMENTS

This work was financially supported by the State Basic Research Key Projects (973) of China (Grant No.2005CB724105).

## REFERENCES

1. W. Weronski and A. Gontarz, *J. Mater. Process. Technol.* 138, 196-200(2003).
2. D. Kuc, G. Niewielski and J. Cwajna, *Mater. Character.* 56, 318-324(2006).
3. G. Zhenyan and V. G. Ramana, *Int. J. Mach. Tools Manuf.* 40, 691-711(2000).
1. J. Kusiak, R.Kawalla, M.Pietrzyk and H. Pircher, *J. Mater. Process. Technol.* 60, 455-461(1996).
5. R. DING and Z. X. GUO, *Acta Mater.* 49, 3163-3175(2001).
6. M. A. Miodownik, *J. Light Metals* 2, 125-135 (2002).
7. J.D. Robson, *Mater. Sci. Eng. A* 382, 112-121(2004).
8. J. H. Bianchi and L. P. Karjalainen, *J. Mater. Process. Technol.* 160, 267-277(2005).
9. T. Sheppard and X. Duan, *J. Mater. Process. Technol.* 130-131,250-253 (2002).
10. E. A. HOLM, G. N. HASSOLD and M. A. MIODOWNIK, *Acta mater.* 49, 2981-2991(2001).

# A New Definition of Shape Complexity Factor in Forging

R.Hosseini Ara[1], M.Poursina[2], H.Golastanian[2]

*1 IUT, M.Sc. Student, Department of Mechanical Eng, Isfahan University of Tech, 84156, Isfahan, Iran.*
*2 SKU, Assistant Prof, Mechanical Group, Faculty of Engineering, University of Shahre-Kord, 8818634141, Iran.*

**Abstract.** One of the main objectives of forging process design is to ensure adequate metal flow in the dies so that the desired finished part geometry can be obtained without any internal or external defects. This paper presents a preform design method which employs a new criterion based on shape complexity factor to determine the necessity of preform stages for axisymmetric forging parts. The presented criterion was tested on several examples using finite element method to verify the models. Comparison of the new shape complexity factor with the other ones shows that the new criterion is more accurate in estimating the number of preform stages.

Keywords: Shape Complexity Factor, Forging, Finite Element Method.
**PACS:** 62.20.FE

## INTRODUCTION

Closed hot die forging is one of the most applicable methods used in manufacturing parts. In this process, material flows and plastic deformation causes the work piece to fill the dies. However, in some cases the complexity of the final part causes some defects to form in the part. Some examples of these defects are: inadequate die filling, non uniform flow of material, die surface wear, increasing the forces applied on dies and folding defects in forged parts. Thus, it is not possible to perform the process in only one stage and preform stages are necessary [1]. Unfortunately determination of the number of preform stages is a very important question that researchers have not found an answer for.

Several methods and criteria have been established for estimating the required number of preform stages. Shape complexity factor is one of the most applicable one. In fact, shape complexity of the final part affects the flow of material and like the other parameters has its own measure that depends on the definition of shape complexity.

Hot forging of axisymmetric parts have been investigated in this article. New criterion based on "Dead Area Ratio" and "Active Area Ratio" was applied to determine the number of preform stages. Also, the new criterion was examined on several examples using finite element method.

CP907, *10th ESAFORM Conference on Material Forming*, edited by E. Cueto and F. Chinesta
© 2007 American Institute of Physics 978-0-7354-0414-4/07/$23.00

# SHAPE COMPLEXITY FACTOR

Teterin defined the first applicable Shape Complexity Factor (SCF) for axisymmetric parts on the basis of geometry [2]:

$$S_T = \frac{P_F^2 / A_F}{P_C^2 / A_C} \times \frac{2R_{gF}}{R_C} \tag{1}$$

Where $P_F$ and $A_F$ are the perimeter and surface area of the axial cross section, $P_C$ and $A_C$ are the perimeter and area of a circumscribing cylinder, $R_{gF}$ is the distance of the center of gravity from the axis of symmetry and $R_C$ is the radius of the circumscribing cylinder. For a multi-stage operation, the SCF of the final stage was proposed by Zhao [3]:

$$S_F = \frac{S_{FF}}{S_{FP}} \tag{2}$$

Where $S_{FP}$ and $S_{FF}$ are the Shape Complexity Factors for preform shape and final forged part. It is obvious that if the final shape is a cylinder, then the SCF for this part is one and for more complex shapes, the factor will be larger so that the necessity of preform stages will be larger for a defect free forging.

After that, Tomov [4] presented a new SCF for axisymmetric forging parts by calculating the work done for plastic deformation. Tomov believed that the presence or lack of the preform step(s) depend on the condition (3).

$$W^* = (1 - K_1)\varphi_A + K_1 > \varphi_H \tag{3}$$

Where $K_1$ describes the amount of the transformed volume during two arbitrary stages of forging, $\varphi_H$ is the logarithmic height strain and $\varphi_A$ is the logarithmic strain on the cross sectional area of the part. On the basis of Tomov's criterion if condition (3) holds for the part, then preform stage is necessary.

## NEW CRITERION

There are some different definitions and criteria for SCF but these criteria can not estimate the necessity of per-form stages accurately in the field of closed die forging. In fact, this weakness is because of the lack of an exact investigation of all effective parameters in the area of SCF. In the present study, some other important parameters that can influence SCF are investigated. By the means of these effective parameters, new criterion is defined. For this purpose, it is necessary to determine the concept of SCF and its effective parameters.

Basically in forging, shape complexity is one of the most important factors that can influence material flow. On the basis of some research and geometrical rules, cylinder is the least complex shape among the 3-dimensional parts [3]. Because cylinder is an axisymmetric part, its reflection, which is a rectangle, could be used for 2-D analysis. So the key point is to find out that rectangle has the least complexity in 2-D shapes and for axisymmetric parts, the amount of deviation of the section from its circumscribing rectangle is one of the most important parameters to determine the SCF. Maybe that is why Teterin offered his criterion. Therefore, an appropriate scale for measuring this kind of deviation is 'Dead Area Ratio' and 'Active Area Ratio'. As seen in figure 1, the ratio of the hatched area to the circumscribing rectangle area is named 'Dead Area Ratio' or 'DAR' in brief. In this research it is proved that the SCF of a certain section will be larger when 'DAR' is increased. In the similar way, the ratio of the section area to the area of circumscribing rectangle is named 'Active Area Ratio' or 'AAR'. On the other hand the scale of these two ratios is varied for different types of sections, so these ratios are not enough to determine the complexity of a part.

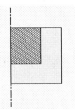

**FIGURE 1.** Determining Dead Area and Active Area for an axisymmetric H-shape.

For example in figure 2, DA and AA ratios of both sections are the same but their complexities are not, so in this case there must be another parameter that can affect SCF of one part, and this could be the distribution of dead area and active area on the section.

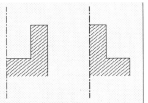

**FIGURE 2.** Example of two different sections with same 'DAR' and 'AAR' ratios.

For this purpose, two factors are defined to measure this kind of distribution. These factors are named 'Axial Distribution Ratio (AR)' and 'Longitudinal Distribution Ratio (LR)'. These two factors are defined as the ratio of the height (Hc) and width (Wc) of section's centroid to the height (H) and width (W) of circumscribing rectangle as illustrated in figure 3.

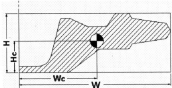

**FIGURE 3.** Determination of 'LR' and 'AR' ratios.

In fact, the increase in these ratios means that material has to flow a longer distance and that means the shape complexity is greater. These ratios are determined using Eqs. 4 and 5 as discussed below. In order to show this result, different cross sections were also examined.

$$LR = \frac{W_C}{W} \tag{4}$$

$$AR = \frac{H_C}{H} \tag{5}$$

After determining the effective parameters in SCF, new criterion based on these factors is shown as Equation 6:

$$\frac{DAR}{AAR} \times LR \times AR \times 2 > DAR \tag{6}$$

Where 'DAR/AAR' is the ratio of the dead area to active area, 'LR' and 'AR' are the distribution ratios and the constant two is used for equalizing the effect of distribution ratios in the equation. In fact, this new criterion is based on these influential factors and also geometrical dimensions of the section and it is not related to the material properties because of the similar behavior of most metals and alloys at high temperatures used in hot forging.

Finally, if eq. 6 holds, then at least one preform step will be necessary.

In this case, it is required to calculate eq. 6 for preform design and then if this equation holds, another preform step will be necessary. Also another application for equation 6 is for optimizing the preform shape. In fact, if one preform design does not match in eq. 6, the number of preform stages will be decreased.

## EXAMPLES

For the first example, an axisymmetric H-shaped part is chosen as shown in figure4 and by varying the geometry of the section in order to have different shape complexity factors. The results are illustrated in table 1. The simulation is axisymmetric and 2-dimensional. Also because of the type of interaction and using Adaptive Mesh Control device, ABAQUS/Explicit was used.

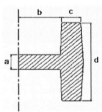

**FIGURE 4.** Shape and parametric dimensions of simulated part.

**TABLE1.** Results of Teterin, Tomov and New criterion.

| Case | Geometry | | | | Teterin | Tomov | | New Criterion | |
|---|---|---|---|---|---|---|---|---|---|
| Number | a | b | c | d | $S_T$ | $W^*$ | $\varphi_H$ | Eq. 6 | DAR |
| 1 | 90 | 100 | 60 | 210 | 1.77 | 0.84 | 0.90 | 0.25 | 0.38 |
| 2 | 45 | 100 | 60 | 165 | 2.24 | 0.89 | 0.92 | 0.41 | 0.45 |
| 3 | 45 | 100 | 60 | 245 | 2.61 | 0.92 | 0.97 | 0.55 | 0.50 |

As seen in table 1, according to Teterin's criterion, results show that preforming steps are necessary for cases 2 and 3. Also Tomov's criterion shows that preforming steps are not necessary for these cases. Finally results of new criterion show that preforming step is necessary for the last case.

In the finite element simulation, the workpiece is steel 1.3505 and the material of the die is H13. The initial temperature of billet and dies are $1200\,^{\circ}C$ and $400\,^{\circ}C$, respectively. The results of finite element simulation are shown in figures 5 and 6.

FIGURE 5. Finite element simulation for the second case.

FIGURE 6. Finite element simulation for the third case.

As seen in figure 5 and 6, case 2 does not require preform step and the initial billet can directly deform in final die, but for case 3 the die was unfilled and the necessity of preform step is unavoidable. So according to the FEM simulations, only the last case requires a preform step.

As the second example the disk turbine analyzed by Zhao [3] and Tomov [4] was selected (figure 7). This example was selected for validating the new criterion since published results were available. Tomov indicated that this piece could be forged with a single preform step. The results of the current investigation also suggest that a single preform is adequate for forging of this IBR disk turbine. The reason is that the inequality given by Eq. 6 holds for this part as shown in Table 2. Preform shape was shown in figure 8. Now the results are verified by the use of FEM simulation. Material flow in the die was modeled in ABAQUS finite element software. The preform shape and the final piece are shown in Figure 9. It can be seen that the preform completely fills the final die in a single stage, right figure.

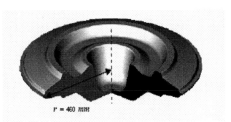

$r = 480\ mm$

FIGURE 7. IBR Disk Turbine.

**TABLE 2.** Result of new criterion for preform and final shape of IBR disk turbine.

| SHAPE | DAR | AAR | LR | AR | EQ. (6) | RESULT |
|-------|------|------|------|------|---------|--------|
| FINAL | 0.5659 | 0.4341 | 0.5105 | 0.5352 | 0.7124 | Preform step is necessary. |
| PREFORM | 0.4676 | 0.5324 | 0.4969 | 0.4949 | 0.4320 | Preform step is not necessary. |

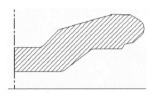

**FIGURE 8.** Preform for IBR disk turbine.

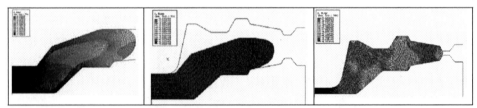

**FIGURE 9.** Finite element simulation for IBR disk turbine.

## CONCLUSIONS

1. The new criterion is more accurate for the estimation of the number of preform stages in comparison with the other criteria in the case of H-shaped parts.
2. New criterion can predict the necessity of the preform stages in axisymmetric hot die forging and there is a good correlation between FEM simulation results and new criterion using equation 6.
3. Flashless design of IBR disk turbine was investigated by the means of new criterion for validation.
4. Tomov's criterion and the current criterion both predict one preform stage for IBR disk turbine forging which validates the new criterion.

## REFERENCES

1. T.Altan, G.Ngaile and G.Shen, "Cold and hot forging Fundamentals and Application", ISBN: 0-87170-805-1, ASM International Press, 2004.
2. G.P.Teterin, I.J.Tarnovsky, A.A.Chechik, "Criterion of complexity of the configuration of forgings", Kuznechno-Shtanmpovochnoe Proizvodstvo vol. 7, pp. 6-8, 1966.
3. G.Zhao, E.Wright, R.V.Grandhi, "Forging preform design with shape complexity control in simulating backward deformation", *International Journal of Machine Tools Manufacturing*, vol. 35, pp. 1225-1239, 1995.
4. B.Tomov, "A new shape complexity factor", *J. Materials Processing Technology*, vol. 92-93, pp. 439-443, 1999.

# Influence Of Forming Machine Stiffness On Product Quality In Hot Forging Operations

M. Croin[1,*], A. Ghiotti[1], S. Bruschi[2]

[1]DIMEG University of Padova, via Venezia 1 – 35131 Padova Italy
[2]DIMS University of Trento, via Mesiano 77 – 38050 Trento Italy
*corresponding author: marco.croin@unipd.it

**Abstract.** The quality of forged components is strictly dependent on the micro-structural phenomena occurring during the manufacturing chain that may cause significant changes in the physical and mechanical features of the final product. Some of these changes take place during the deformation phase and depend on the forming machine characteristics in term of machine kinematics, total elastic deflections of the press system and contact-time. The press stiffness influences the velocity-versus-time curve under load, varying the contact pressure and the heat transfer between the tools and the workpiece and, consequently, the global process conditions. This paper deals with the evaluation of the influence that the forging machine exerts on the process parameters and the quality of the final component in terms of its microstructure distribution after forging. The approach is based on a simulative environment that combines experimental measurements, FE simulation and inverse analysis techniques. A case study, in which operating conditions approximate hot forging of a turbine aerofoil section, is presented. Results of experiments show the influence of machine characteristics on the quality of final product.

**Keywords:** Hot forging, Forging machine, Stiffness, Contact-time, Microstructure
**PACS:** 81.20.Hy, 61.82.Bg.

## INTRODUCTION

In new hot forging process design, the accuracy of the formed component is considered as priority hardware requirements. Furthermore, a trade-off between product accuracy and die service life is usually accepted since it is well known that a stiff press guarantees higher precision and lower scatter in product dimension but, at the same time, higher loads resulting in dies deterioration due to fatigue and wear [1-5]. Despite of their influence, the correlation between the time-dependent characteristics and the forging machine stiffness has not been sufficiently investigated yet. In particular, the heat exchange at the tools-workpiece interface, where the temperature gradient between hot workpiece and cooler dies is significant [6], is strongly influenced by a larger contact-time under pressure with consequent effects on both the tools service life and the mechanical characteristics of forged products [7-9].

In this paper the influence that the heat exchange rate has on the microstructure distribution after a forging operation is presented. The first part focuses on the machine stiffness modelling and the developed experimental approach. Hot forging

CP907, *10th ESAFORM Conference on Material Forming*, edited by E. Cueto and F. Chinesta
© 2007 American Institute of Physics 978-0-7354-0414-4/07/$23.00

experiments, which aim at reproducing the operating conditions of turbine airfoil sections manufacturing [8], were performed on a screw press laboratory plant. In the second part, the influence of process parameters on the final microstructure of the forged component is examined.

## STIFFNESS MODELLING

Screw presses are energy restricted machines as their ability to deform the workpiece depends on the amount of energy stored in their flywheel. The different contributions to press stiffness are described in Figure 1(a). The structural elements of the forging machine can be considered as springs connected in series loaded by the reaction load F generated by the plastic deformation of the workpiece. The system acts as an equivalent spring, whose elasticity is given by:

$$k_{zz,Screw} = \frac{1}{\frac{1}{k_{DS}} + \frac{1}{k_D} + \frac{1}{k_F} + \frac{1}{k_R} + \frac{1}{k_S}} \tag{1}$$

where $k_{DS}$ is the contribution of die set, $k_D$ of dies, $k_F$ of the press frame, $k_R$ of the ram and finally $k_S$ of the spindle. In order to vary the elasticity of the forming press and simulate different forging conditions, interchangeable annular elements with different materials and geometries are embedded in the lower part of the equipment. The equivalent vertical stiffness of the whole forging system changes according to the relation:

$$k_{zz,Screw} = \frac{1}{\frac{1}{k_{DS}} + \frac{1}{k_D} + \frac{1}{k_F} + \frac{1}{k_R} + \frac{1}{k_S} + \frac{1}{k_{AE}}} \tag{2}$$

where $k_{AE}$ is the spring constant of the calibrated annular element. By changing material and geometry, the spring constant of this element changes and so does the vertical stiffness of the system leading to different degrees of press stiffness. Figure 1(b) shows the details of the annular elements that can be used to reduce the frame stiffness.

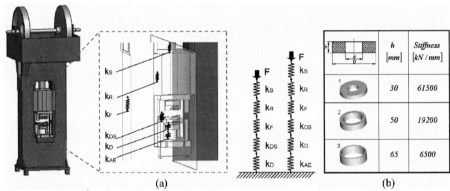

(a)            (b)

**FIGURE 1.** (a) The different contributions to the total stiffness in a screw press and (b) dimensions of spring elements

# EXPERIMENTAL WORK

A set of hot forging experiments where the vertical stiffness of the press frame is varied is the experimental basis to evaluate the influence that the elasticity of a forging system has on the final product quality. The experiments on the Ti-6Al-4V alloy, approximating the deformation in a cross-section of a turbine blade airfoil, were carried out on a 2300kN screw press laboratory plant. The particular case study was chosen as reference because of the thickness variations in the final section, the consequent distributions of contact pressure and the contact-time under pressure typical of the industrial process [8]. The investigated alloy presents a small forgeability window near its $\beta$-transus temperature (which is approximately equal to 995°C); if this temperature is exceeded during the process, the alloy microstructure rapidly changes, affecting the final mechanical properties of the component. The vertical stiffness of the press was varied by using annular elements made of AA6061 in order to simulate different forging machine stiffnesses (see Figure 1(b)). In the following paragraphs, the details about the developed equipment and the experimental plan are outlined.

## Experimental Apparatus

In the experiments, a cylindrical specimen was deformed under plane-strain conditions assured by a closed die cavity, in order to approximate the deformation occurring in the cross-section of a turbine blade far enough from the ends. Figure 2 shows the details of the forging equipment. The dies, which present a characteristic convex profile, are closed by thick plates in order to assure plane strain deformation during the test. The plates are kept together by an external frame able to stand up to the high separating forces due to hydrostatic pressure. The billet used in the experiments is a 30mm diameter and 35mm long cylinder in Ti-6Al-4V alloy, heated through an induction heating system up to 980°C and then forged at 890°C. The dies were preheated by a resistance cartridge heating system up to 180°C.

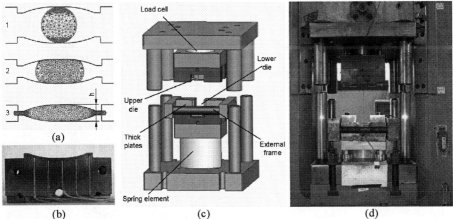

**FIGURE 2.** (a) The plane-strain deformation, (b) the dies with the embedded thermocouples, (c) the model of the dies configuration and the set up during experiments (d)

The temperature during the deformation phase was measured by 10 k-type thermocouples embedded inside the upper and the lower dies. The wires were positioned inside the grooves machined in one of the two halves (see Figure 2(b)) and the hot junctions made by electro-discharge welding the two wires onto the die material at the end of the groove. Both upper and lower dies are made by two identical parts fixed with pins and bolts to make splitting possible and access easier when positioning, welding and inspecting the thermocouples.

## Experimental Plan

The experimental plan, summarized in Table 1, was designed in order to study press stiffness and interface conditions influence on final as-forged microstructure; therefore two sets of tests can be outlined, the first (test conditions #1, #2, #3 ) carried out in dry conditions, the latter (test conditions #4, #5, #6 ) adopting a graphite based lubricant (Aquadag®) in addition to a glass coating applied to the specimen. Within each tests set, system stiffness was varied by means of the special die set and annular elements, resulting in three conditions with an equivalent stiffness of 100%, 56% and 40% respectively. Tests were performed by using 100% of the flywheel energy of the press and a ram stroke of 200 mm assuring an impact velocity of the ram of about 550 mm/s. During the experiments, the main process parameters such as ram stroke, forming load, tools sub skin temperatures and contact time under pressure were monitored at a scan rate of 10 kHz and then processed by a LabVIEW$^{TM}$ software.

Finally, the forged specimens were sawed in half along the longitudinal symmetry plane, ground through silicone carbide papers using water as coolant, etched using the Kroll's reagent and observed.

**TABLE 1.** The experimental plan for the Ti-6Al-4V laboratory tests

| Test condition | # 1 | # 2 | # 3 | # 4 | # 5 | # 6 |
|---|---|---|---|---|---|---|
| System stiffness | 100 % | 56 % | 40 % | 100 % | 56 % | 40 % |
| Lubrication | dry | dry | dry | Aquadag® | Aquadag® | Aquadag® |
| Coating | no | no | no | glazed | glazed | glazed |

## RESULTS

The load versus time curves for the investigated system stiffnesses are shown in Figure 3(a) and 3(c), for the dry and lubricated conditions respectively.

On one hand, the system stiffness reduction leads to a decrease of the forming load with a variation of about 20% between the most and least stiff conditions. This behaviour is observed in both dry and lubricated tests. Moreover, as one might expect, the introduction of lubricant and coating determines a reduction of the forming load of about 17%.

On the other hand, the reduction in system stiffness results in an extension of contact time under pressure. As a consequence in the least stiff test condition, where the workpiece remains in contact with the tools for about the 20% more than the most stiff condition, a larger heat exchange occurs at die interface determining a higher heating of the die, as one can note observing the sub-skin temperature peaks shown in Figure 3(b) and 3(d), for the dry and lubricated conditions respectively.

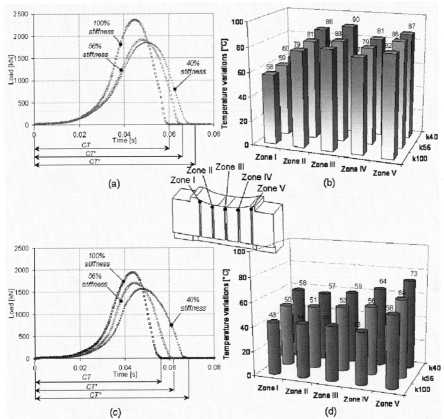

**FIGURE 3.** Forming load curves and dies sub-skin temperature peaks for tests in dry (a),(b) and lubricated (c),(d) conditions.

Furthermore, by comparing the peaks at the same press stiffness level, the influence of interface conditions is inferred. In fact, the remarkable temperature decrease noticed in the lubricated tests is determined by three concurring factors: first of all, the enhanced lubrication condition permits a better material flow and thus a shorter contact time and heat exchange. Secondly the lubrication reduces significantly the heat generated by friction and lastly the glass coating layer works as a thermal barrier slowing down the heat transfer.

Figure 4 shows that in dry conditions, the final as-forged microstructure is affected by system stiffness. In particular, the specimen forged in the stiffest condition exhibits a β microstructure; this means that most of the deformation work was carried out above the beta transus temperature suggesting that the lower heat exchange with tools implies a higher heating of the material. By contrast, the intermediate and less stiff conditions, characterized by an higher heat transfer, show an α+β final as-forged microstructure, with most of the deformation work carried out below the beta transus temperature. Finally in coated-lubricated conditions, the reduction of heat generation and the enhancement of thermal insulation counters the stiffness influence, thus giving an α+β final as-forged microstructure for all tested conditions.

497

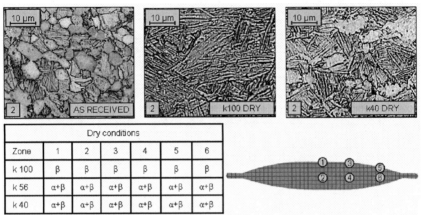

**FIGURE 4.** As received and final as forged microstructure for the tests in dry conditions

## CONCLUDING REMARKS

An approach was developed to study the influence that forming machine stiffness and interface conditions exert on final as-forged microstructure of a multi-phase alloy. Hot forging experiments were carried out with different stiffness and lubrication conditions monitoring the main process parameters. Finally the forged specimens were hacksawed and their microstructure observed.

The tests carried out in dry conditions outlined a strong correlation between forming machine stiffness and final as-forged microstructures. On the contrary the use of lubricant and coating, slowing down the heat transfer between workpiece and tools, countered the effect of machine parameters determining a homogeneous microstructure distribution aside from system stiffness.

## REFERENCES

1. K. Chodnikiewicz, R. Balendra, "The calibration of metal-forming presses" in Journal of Materials Processing Technology 106 (2000) 28-33.
2. E. Doege, G. Silverbach, "Influence of various machine tool components on workpiece quality" in Annals of the CIRP, Vol. 39/1(1990) 209-213.
3. O. Brucelle, G. Bernhart, "Methodology for service life increase of hot forging tools" in Journal of Materials Processing Technology 87 (1999) 237-246
4. Z. Malinowski, J.G. Lenard, M.E. Davies, "A study of the heat transfer coefficient as a function of temperature and pressure" in Journal of Materials Processing Technology 41-2 (1994) 125-142.
5. T. Iwama, Y. Morimoto, "Die life and lubrication in warm forging" in Journal of Materials Processing Technology 71 (1997) 43-48. 7.
6. X. Lu, R. Balendra, "Temperature-related errors on aerofoil section of turbine blade", in Journal of Materials Processing Technology, Volume 115, Issue 2, 4 September 2001, Pages 240-244.
7. K. Lange, "Handbook of Metal Forming", Oxford University Press, New York – Oxford (1989).
8. P.F. Bariani, G. Berti, T. Dal Negro, S. Masiero, "Experimental evaluation and FE simulation of thermal cycles at tool surface during cooling and deformation phases in hot and warm operations", in Annals of the CIRP (2002), Vol. 51, pp. 219-222.
9. A. Ghiotti, M. Croin, P.F. Bariani, "Experimental and numerical modelling of the press stiffness influence on temperatures at die surface in hot forging", in: Proc. 9th Esaform, Glasgow (2006).

# Comparison between 2D and 3D Numerical Modelling of a hot forging simulative test

M. Croin[1], A. Ghiotti[1], S. Bruschi[2]

[1]DIMEG University of Padova, via Venezia 1 – 35131 Padova Italy
[2]DIMS University of Trento, via Mesiano 77 – 38050 Trento Italy

**Abstract.** The paper presents the comparative analysis between 2D and 3D modelling of a simulative experiment, performed in laboratory environment, in which operating conditions approximate hot forging of a turbine aerofoil section. The plane strain deformation was chosen as an ideal case to analyze the process because of the thickness variations in the final section and the consequent distributions of contact pressure and sliding velocity at the interface that are closed to the conditions of the real industrial process. In order to compare the performances of 2D and 3D approaches, two different analyses were performed and compared with the experiments in terms of loads and temperatures peaks at the interface between the dies and the workpiece.

**Keywords:** hot forging, FEM, heat transfer.
**PACS:** 81.05.Bx, 87.18.Bb

## INTRODUCTION

In numerical simulation of hot forming processes, the accurate modelling of thermal events represents one of the most challenging objectives due to the highly heterogeneous temperature fields and the complexity of the heat transfer phenomena that occur. A number of authors has spent large efforts in investigating both experimentally [1-3] and numerically [4-5] several aspects involving temperature in hot forming processes, such as material [6-7], interface conditions between workpiece and tools during deformation [8]. Most of these efforts have been addressed to extend the dies life by introducing new typologies of hard coatings [9] and optimizing the process design [10]. However, the determination of input data for the simulation of thermal events in forging is strictly related to the different models that can be used to investigate the hot forging process.

In this paper the temperatures at the interface between the workpiece and the tools in a hot forging operation is studied by integrating experimental and numerical techniques. In the first part the approach and the application case are outlined. Laboratory tests, capable to reproduce the operating conditions of turbine airfoil forging, were performed on a forging pilot plant based on a screw press. The second part of the paper describes the comparison between two different FE models, respectively 2D plane-strain and 3D, implemented for inverse analysis determination of temperature peaks during the plastic deformation of the workpiece. The presented

CP907, *10th ESAFORM Conference on Material Forming*, edited by E. Cueto and F. Chinesta
© 2007 American Institute of Physics 978-0-7354-0414-4/07/$23.00

results show significant discrepancies in the temperature field and HTC determination when using 2D and 3D numerical models.

## APPROACH

The input parameters for the calibration of numerical simulations may change significantly depending on the simplifications introduced in the used FE model. With regards to these effects and, in particular, to the thermal parameters used in FE simulations, a benchmark test to evaluate the accuracy of numerical simulation was set-up. The test is based on the simulation of a hot forging experiment, which approximates the deformation in the cross-section of a turbine blade airfoil. This particular set-up was chosen as reference case because of the thickness variations in the final section and the consequent distributions of contact pressure and contact-time under pressure typical of a hot forging process. The approach followed in carrying out the study combines laboratory experiments and numerical inverse analysis to calibrate the FE code. The experiments were carried out on a laboratory screw press plant where the process temperatures were measured. Then, two different numerical models, respectively 2D and 3D, were developed and optimized for the evaluation of temperature peaks at the interface between the workpiece and the dies. In the following paragraphs, the results of the experiments, the developed FE models and the application of inverse analysis are described.

## CASE STUDY

Figure 1 shows a schematic representation of the case study reference geometry and the experimental set-up of the hot forging equipment. During the experiments, a cylindrical specimen is deformed under plane-strain conditions assured by the closed dies architecture [11], in order to approximate the deformation occurring in a cross-section of a turbine blade far enough from the ends. The dies used in experiments were manufactured in AISI H11 with a surface hardness after grinding in the range of 48-52 HRC. The used billet is a 30mm diameter and 35mm long cylinder in Ti-6Al-4V titanium alloy, which was heated by an induction heating system up to 980°C and then

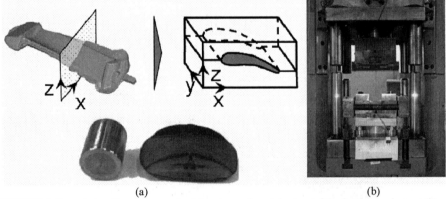

(a)                                                                 (b)

**FIGURE 1.** Case study reference geometry (a) and experimental set-up of the forging equipment (b)

forged at approximately 890°C. The dies were preheated by a resistance cartridge heating system up to 180°C. The temperature distribution during the workpiece deformation was measured by 10 k-type thermocouples embedded inside the upper and the lower dies. Both upper and lower dies are made by two identical parts fixed with pins and bolts to make splitting possible and access easier for the thermocouples application. The wires (0.25mm diameter) were positioned inside the grooves machined in one of the two halves and the hot junctions made by electro-discharge welding the two wires onto the die material at the end of the groove. During the experiments the temperatures in the upper die, the forming load, the ram stroke and the signal from contact time transducer [11] were acquired.

## NUMERICAL MODEL

In order to calibrate the data of the heat exchange at the tools-workpiece interface, an inverse analysis procedure was set up. The change of temperature at a given element depends upon the change in energy in the same time step:

$$\delta T = \frac{1}{cV}\left(\delta Q^d + \delta Q^f + \delta Q^c\right) \qquad (1)$$

where $c$ is the thermal capacity, $V$ the volume, $\delta Q^d$ the increase in energy of the element due to work of deformation, $\delta Q^f$ the increase due to the frictional work and $\delta Q^c$ the change in energy due to conduction. In the evaluation of (1), the most critic contribution is represented by the heat flux due to interface conditions and in particular the determination of the heat transfer coefficient (HTC) between the dies and the workpiece. An inverse analysis procedure was applied to identify the set of heat transfer coefficients for the different die zones that fitted better the temperatures measured during the experiments. The commercial software Deform™ was used in the application case. In the simulations, both the deformation under pressure and the heat conduction inside the dies after deformation were considered. The FE-models implemented for inverse analyses have as input the calculated thermo-mechanical and the experimental conditions and return as output the temperature histories at the location of the hot junctions. By using the Gauss-Newton algorithm the HTC values in the different zones of the tools are identified by minimizing the objective function:

$$Q = \sum_T \sum_{l=1}^{r} \beta_l \left(T_l^{calc}(HTC) - T_l^{exp}\right)^2 \qquad (2)$$

where $\beta_l$ are weighting coefficients, $r$ the total number of experimental measurements, $T_l^{calc}(HTC)$ the temperature values of the profile calculated by the software and $T_l^{exp}$ the temperatures of the experimental profile. Since the plastic deformation of the billet consists of both a loading and an unloading phase [11], the experimental stroke curve of the upper die was implemented in the FE simulation using a point by point curve.

The 2D plane strain model of the forging operation is shown in Figure 2. The geometry in a plane strain model represents a cross section that cuts a very long or infinite depth in such a way that any end effects can be ignored. The plane strain application mode solves for the global displacements (u, v) in the *x*- and *y*-directions. In a state of plane strain, the $\varepsilon_z$, $\varepsilon_{yz}$, and $\varepsilon_{xz}$ components of the strain tensor are assumed to be zero. The temperature evolution inside the physical system is the result of the balance between the internal heat conduction and the internal heat dissipation, under the constraints defined on the area boundary in terms of exchange. Due to the approximation of the 2D assumption, the heat flux is restricted only to the *x*- and *y*-directions, while no thermal transfer is calculated in the other directions. The initial thermal condition of the workpiece was obtained by cooling simulations, calibrated with experimental temperature measurements performed in the central section of the specimen. The tools interfaces were divided into ten equal length zones with a step-wise distribution for the HTC, assumed to be invariable with time. The friction at the interface between the workpiece and the deformable tools was described by the Tresca friction law and it was considered constant in all the zones of the dies.

A more accurate representation of the experimental set up was obtained in the 3D analysis. The thick plates, used to close the die cavity in order to reproduce plane strain conditions in the entire volume of the billet during the experiments, were modelled in the 3D simulation. Figure 2 shows the numerical model for the simulation. As in the 2D model, the tools interfaces were divided into ten equal length zones with a step-wise distribution for the HTC. In the case of 3D simulation, the heat flux is not considered approximated in a plane, but the dissipated power at the interface by irradiation, conduction, convection and friction phenomena is integrated in all the directions of the space. In order to take into account these effects, the machined grooves that embed the thermocouples wires and the hot junctions were modelled in the 3D simulation. The initial thermal conditions were calibrated with experimental temperature measurements performed in the entire surface of the specimen using an infrared thermal-camera. The friction at the interface between the workpiece and the deformable tools was described by the Tresca friction law as in the case of 2D simulation.

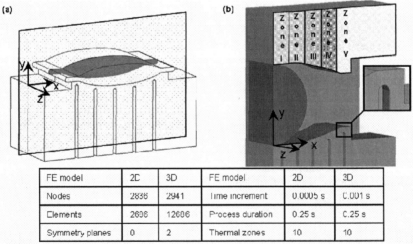

| FE model | 2D | 3D | FE model | 2D | 3D |
|---|---|---|---|---|---|
| Nodes | 2836 | 2941 | Time increment | 0.0005 s | 0.001 s |
| Elements | 2606 | 12606 | Process duration | 0.25 s | 0.25 s |
| Symmetry planes | 0 | 2 | Thermal zones | 10 | 10 |

**FIGURE 2.** 2D (a) and 3D (b) FE models with machine grooves details and HTC zone at the tools-workpiece interface

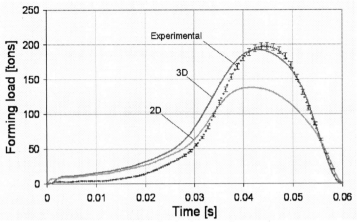

**FIGURE 3.** Experimental and numerical forming load versus time curves

## RESULTS AND DISCUSSION

In Figure 3 the experimental load versus time curve is compared with the ones given by the 2D and 3D numerical simulations. As it can be seen, both models are accurate in the evaluation of the loading and unloading time extension, but differ in the prediction of the forming load peak. In fact, the 2D model leads to a significant underestimation of the maximum force (about 30% whereas the 3D model deviation is around 2%); at the interface this results in normal pressure and sliding velocity profiles below the ones estimated by the 3D model, being the difference increasing with the magnitude of these variables. Furthermore, the inverse analyses, although characterized by a proper consistency between the measured and the computed temperatures for both models, show remarkable discrepancies in the evaluation of heat transfer coefficients at tool-workpiece interface. Figure 4, exploiting the symmetry of the test, shows the distribution of these coefficients on half of the die. The deviations that can be observed ranges from about the 10% in the central zone of the die to about

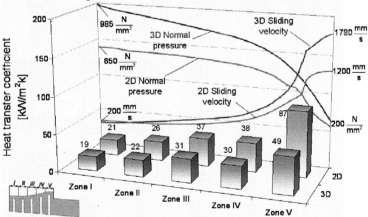

**FIGURE 4.** Calculated heat transfer coefficient distribution at die-workpiece interface

the 80% in the flash land zone. These variations are due on one hand to the difference in the normal pressure, which is recognized as strongly affecting the heat exchange [8] and sliding velocity at die-workpiece interface, and, on the other hand, to the heat flux calculation which, in the 2D model, is restricted to a plane and thus influenced by the grooves, considered as of infinite depth.

## CONCLUSIONS

A combined experimental and numerical analysis was carried out to study the influence of the numerical model on the accuracy of FE simulations on temperature prediction in hot forging operations. Experimental measurement and inverse analysis were used to determine heat transfer coefficient and temperature distribution at die interface during hot forging experiments approximating the deformation in a cross-section of a turbine blade airfoil. The obtained results show significant differences in the results determined in case of 2D and 3D FE models, especially as regards the load and heat transfer calculation.

## ACKNOWLEDGMENTS

The work of this paper is part of the EU research project "VIF-CA". The authors are grateful to Pietrorosa TBM S.p.A for providing the details on industrial process.

## REFERENCES

1. E. Doege, R. Seidel, "Precision forging changes the view: new cooling and lubricating technologies for hot die forging", IV, 1997, 9-12.
2. M.A. Kellow, A.N. Bramley, F.K. Bannister, "The measurement of temperatures in forging dies", International Journal of Machines Tool Design, 9, 1969, 239-260.
3. J.G. Lenard, M.E. Davies, "An experimental study of heat transfer in metal-forming processes" Annals of CIRP, 1992, 307-310.
4. T.A. Dean, Z.M. Hu, J.W. Brooks, "An exercise in the investigation of boundary conditions and evaluation of numerical simulation applied to the net-shape forging of Ti-6Al-4V aerofoil blades", International Conference on Forging And Related Technology, 1998, 73-82.
5. M. Raudensky, J. Horsky, A. Tseng, "Heat transfer evaluation of impingement cooling in hot rolling of shaped steels", Steel Research, 1994, 375-381.
6. P.B. Bariani, S. Bruschi, T. Dal Negro, "Integrating physical and numerical simulation techniques to design the hot forging process of stainless steel turbine blades", in International Journal of Machine Tools and Manufacture, Volume 44, Issue 9, July 2004, Pages 945-951.
7. T.A. Dean, Z.M. Hu, J.W. Brooks, "Aspects of forging of titanium alloys and the production of blade forms", in Journal of Materials Processing Technology, Volume 111, Issues 1-3, 25 April 2001, Pages 10-19.
8. Z. Malinowski, J.G. Lenard, M.E. Davies, "A study of the heat transfer coefficient as a function of temperature and pressure", Journal of Material Processing Technology, 1994, 125-142.
9. C.M.D. Starling, J.R.T. Branco, "Thermal fatigue of hot work steel with hard coatings", in Thin Solid Films, 308-309,1997, 436-442.
10. R. Lapovok, "Improvement of die life by minimization of damage accumulation and optimization of perform design", in Journal of Materials Processing Technology, 80-81, 1998, 608-612.
11. A. Ghiotti, M. Croin, P.F. Bariani, "Experimental and numerical modelling of the press stiffness influence on temperatures at die surface in hot forging", in: Proc. 9th Esaform, Glasgow (2006).

# Estimation of friction under forging conditions by means of the ring-on-disc test

Bernhard Buchner, Andreas Umgeher and Bruno Buchmayr

*Chair of Metal Forming, University of Leoben, Franz-Josef-Strasse 18, A-8700 Leoben, Austria*

**Abstract.** In order to understand the tribological processes and interactions in the tool-workpiece-interface systematically, basic experiments that allow an independent variation of influencing parameters are necessary. The ring-on-disc test is a popular model experiment that is often used in tribological analyses at low normal contact pressures.

The scope of the paper is an analysis of the applicability of the ring-on-disc test for high normal pressures as used in forging processes, using aluminium AA6082 as workpiece material. It turned out, that this test is a convenient method to measure friction under forging conditions.

**Keywords:** friction testing, die forging, aluminium
**PACS:** 06.60.Mr, 06.60.Vz

## 1. INTRODUCTION AND OBJECTIVES

Aluminium forged parts play a significant role as components of light weight constructions in the automotive and aerospace industry. The fundamental knowledge of the friction behaviour at the tool-workpiece interface is necessary due to the fact, that it influences material flow, die filling, wear and workpiece quality. For the simulation of forging processes, more advanced friction models are required in order to increase the prediction accuracy.

The most popular method for friction estimation under forging conditions is the well known ring-compression test [1, 2], where friction is determined by the change of the inner diameter of a compressed ring (see FIGURE 1 (a)). Nowadays, the ring-compression test is mostly used to calibrate finite element analysis, i.e. to make the simulation result look like the experimental result by the calculation of a "friction coefficient" that compensates all uncertainties of modelling. Of course, this crude approach does not allow the detailed investigation of the tribological interactions in the tool-workpiece interface.

Another well known model experiment that allows the determination of interface

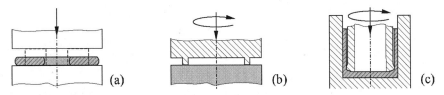

FIGURE 1. Friction testing under forging conditions: (a) ring-compresseion test, (b) ring-on-disc test, (c) twist-compression test proposed by Hansen and Bay [3] (the specimen are marked gray).

friction typically at low contact pressures is the ring-on-disc test: A ring-shaped tool is pressed against a flat specimen in relative motion around the ring's axis and friction is determined by measuring the transferred torque (see FIGURE 1 (b)). The objective of this work was to utilise the ring-on-disc test for the estimation of friction under aluminium forging conditions.

## 2. CONCEPTIONAL ASPECTS OF THE NEW TESTING DEVICE

The tribological conditions in forging operations are characterised by high contact pressures (many times over the workpiece yield stress), high surface enlargement and -modification and high relative velocities as well as high temperatures and temperature gradients. However, the sliding distances are comparatively short.

The main criterion for each model experiment is that its results are transferable to the real process. Thus, besides the friction partners also the loads and contact conditions have to represent the real process. An independent setting of the process-related parameters has to be ensured due to the wide range of contact conditions present in the forging of complex components. Furthermore, the contact area has to be large and homogeneous enough to allow reliable measurements and should not change during the test. Finally, steady lubrication conditions (no lubricant exchange, etc.) in the die-workpiece interface are required.

The ring-on-disc test allows an independent variation of interface pressure, relative velocity and movement as well as temperatures. However, surface enlargement is not considered in the original configuration, and although the specimen deforms plastically under sufficiently high pressures, the operation has the character of hardness testing and the surface roughness remains almost unchanged [4]. A method to overcome this problem was presented by Hansen and Bay [3] by combining backwards can-extrusion with twist-compression in oder to characterise lubricants for the cold forging of steel (see FIGURE 1 (c)).

The new testing device to be developed should allow investigations under cold forging, warm forging and isothermal forging conditions of aluminium alloys, and to model the real process as exact as possible the usually disposed graphite-based lubricants should be also applied in the model experiments. This means that the construction has

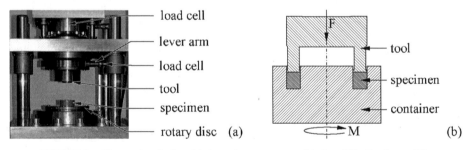

**FIGURE 2.** New testing device: (a) the main components, (b) closed die-forging toolkit.

to withstand high temperatures, high forces, high torques and rough environmental conditions. In order to enable reproducible results the testing sequences (i.e. heating, application of lubricant and testing) should be controlled automatically. Last but not least a cost-effective prototype machine should be designed to clarify the technical feasibility.

## 3. EXPERIMENTAL WORK AND RESULTS

FIGURE 2 (a) shows the Rotational Forging Tribometer (RFT) that was designed and manufactured at the Chair of Metal Forming. The circular motion is supplied from the bottom side by means of an asynchronous motor (7,5 kW) and an angular gear box, whereas the compression force is applied by a 10 kN hydraulic cylinder from the top side. The specimen (workpiece) is mounted on a rotary disc and transmits the frictional torque to the pivot-mounted ring-shaped tool. The tool is supported by a load cell via a lever arm which enables an accurate measurement of the frictional torque. The acquisition of the compression force is realised by another load cell, the rotational speed of the specimen is calculated from the speed of the asynchronous motor. Frictional speed as well as interface pressure is dependent on the tool geometry.

The specimen is brought to forging temperature by means of an inductive heating system that allows temperatures up to 1200 °C, whereas the tool is heated to temperatures of about 250 °C to 450 °C by a heating sleeve. The lubricant is sprayed onto the tool by an automatic application system that allows a reproducible dosage.

The testing device is controlled by a programmable logic control unit (PLC), the data acquisition is realized by a measurement amplifier that is connected to a commercial personal computer.

In a first trial the functionality of the tribometer was tested by performing classical ring-on-disc experiments under cold forging conditions with an outer ring diameter $d_o = 40$ mm and an inner ring diameter $d_i = 36$ mm. As in all other cases reported in this paper AA6082 was used as specimen material and the tools were made from commercial mild steel. The normal pressures were set from 30 MPa up to 430 MPa, the average velocity was varied in the range between 22.5 and 180 mm/s and the friction length was 80 mm. The results revealed, that the intentional function of the testing device could be completely met and the system serves its purpose.

In a next step, classical ring-on-disc experiments under warm forging conditions were

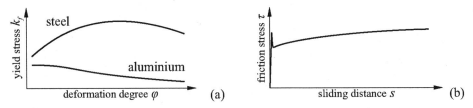

**FIGURE 3.** Qualitative plots: (a) flow curves for hot forging of steel and aluminium, (b) testing result obtained with closed die-forging toolkit for hot aluminium forging.

507

performed, where the specimen were heated to 200, 300 and 400 °C, respectively. The outer ring diameter remained constant with $d_o$ = 40 mm and the inner ring diameter was set to $d_i$ = 32 mm because of the lower yield stress. The normal pressure was varied from 95 MPa up to 230 MPa, the average velocity was set to 5 mm/s and the friction length was 100 mm. It was observed that the tool sank into the specimen at workpiece temperatures greater than 300 °C after short sliding distances (10 – 20 mm) due to frictional heating in the contact zone and the specific flow curve of warm aluminium alloys (see FIGURE 3 (a)). From these experiments it emerged that warm forging experiments with aluminium alloys have to be performed in closed dies. Further experiments proved that also the test proposed by Hansen and Bay [3] does not work with hot aluminium.

FIGURE 2 (b) shows the toolkit for closed die forging which is actually used for friction analysis at the Chair of Metal Forming. The outer ring diameter is still $d_o$ = 40 mm, the inner ring diameter has been reduced to $d_i$ = 30 mm because of manufacturing conditions. Ridges in the bottom surface of the container prevent the ring-shaped specimen from sliding at this interface. The gap between punch and die is kept very small to prevent specimen material from getting extruded, thus there is no surface expansion during the test. This configuration allows experiments with pressures up to 200 MPa and sliding speeds up to 150 mm/s, respectively. A typical friction-stress–displacement curve is plotted in FIGURE 3 (b). The next step will be the development of a similar toolkit which allows surface expansion to a fixed level.

## 4. CONCLUSION

It can be concluded that the ring-on-disc test is an appropriate approach for friction testing under aluminium forging conditions when the toolkit is adapted to the flow behaviour of aluminium alloys. However, to meet the real process conditions, surface enlargement has to be considered in further investigations.

## ACKNOWLEDGMENTS

The authors want to thank the federal province of Styria ("Zukunftsfonds Steiermark", Project 19) and Fuchs Schmiermittel GmbH for financing the project as well as Lenze Antriebstechnik GmbH, Böhler Edelstahl GmbH, Rockmore International Inc. and Acheson Industries for the provision of materials and contributions in kind.

## REFERENCES

1. M. Kunogi, *Reports of the Scientific Research Institute (Kagaku Kenkyujo Hokoku)* **30**, 63–92 (1954).
2. A. T. Male, and M. G. Cockcroft, *Journal of the Institute of Metals* **93**, 38–46 (1965).
3. B. G. Hansen, and N. Bay, *Journal of Mechanical Working Technology* **13**, 189–204 (1986).
4. J. A. Greenwood, and G. W. Rowe, *Journal of Applied Physics* **36**, 667–668 (1965).

# Liquid State Forging: Novel Potentiality to Produce High Performance Components, Process, Plant and Tooling.

M. Rosso*, A. Zago*, P. Claus**, P. Motoiu***

*Politecnico di Torino – Sede di Alessandria – Via Teresa Michel, 5 – I 15100 Alessandria – Italy
**Tecnopress – Cheri - Italy
***IMNR Bucarest - Romania

**Abstract.** The paper deals about a new patented process able to the production of high resistance and high toughness parts, taking into consideration also the tooling need. The molten alloy is introduced into the die cavity at low pressure, then the alloy is forged. The forging action takes place during the alloy solidification process, favouring the reduction of the duration of the process and the production of parts characterised by very high mechanical and ductility properties. The very high mechanical characteristics of the produced parts are obtained thank to their very low porosity content, as well as to their unique microstructure features. Moreover, the process allow the optimisation of the yield of the alloy, in fact the feeding system and the risers are practically absent, this means minimum production of scraps to be recycled.

After a short description of the main features of the equipment and of the process, the work take into consideration the aluminium based produced parts, in their as cast state and after T6 heat treatment, comparing their characteristics with those obtainable by the most traditional low pressure and gravity casting processes.

In particular, samples for the evaluation of the mechanical properties have been machined from the produced parts to obtain their tensile strength, together with their ductility characteristics. The maximum attained hardness values have also been evaluated. Tensile strength higher than 440 MPa, with elongation up to 18% with hardness higher than 125 HB are easily attainable on Al alloys type A356.

Light microscopy observations performed on the transverse section of polished samples and the analysis of the fracture surfaces after mechanical tests allowed to focus the attention on the microstructure details and to highlight the ductile aspects of the fracture to confirm the high quality and high performance of the produced parts.

**Keywords:** Liquid forging, Novel technology, Low pressure die casting, Aluminium alloys, Microstructure, Porosity.
**PACS:** 80; 81.05; 81.40; 83.50

## INTRODUCTION

In recent years many improvements were achieved in the development and processing of light metal alloys following the stringent demand of automotive industries, to save weight and to increase the performance of vehicles. Light metal alloys such as aluminium or magnesium are thus an interesting alternative to ferrous alloys; especially under consideration of the recent developments in casting technologies that allow to obtain nearnet shape components with good mechanical

CP907, 10th ESAFORM Conference on Material Forming, edited by E. Cueto and F. Chinesta
© 2007 American Institute of Physics 978-0-7354-0414-4/07/$23.00

properties. However, the need for more efficient components, high performance and sustainable cost is always increasing and the current limits of standard process need to be overcome. Until now the only process allowing to obtain aluminium safety and / or high performance components is forging. Nevertheless forging is an energy demanding process (most of the time more successive forming operations are needed to obtain the final shape, resulting in a high energy consumption of the production phase) and as it does not allow to obtain complex shaped component a lot of machining is then usually necessary (thus material waste and further energy consumption).

## CASTING TECHNOLOGIES

The traditional alternative to forging is casting, the process is used in the foundries for producing cast products. The casting route represents the cheapest and the most concurrent technology and its evolution is continuous. Different technological approaches are possible to classify casting processes, however the main interesting and adopted are based on the type of used mould or pattern [1-6]. The cheapest way concerns with the use of permanent moulds with the adoption of different range of pressure during casting, from vacuum to gravity or to high pressure die casting. In particular, the high pressure die casting is a well established and reliable process, modification of this process are related to the use of vacuum or low pressure techniques. Moreover, some relatively new and developing casting process, like thixoforming and rheocasting technologies [7] or squeeze casting [8] are suitable for high strength, high ductility, lightweight structural aluminium castings needed for advanced components.

Thixoforming and rheocasting processes work with alloys in the semisolid state and with globular microstructure obtained by a previous specific solidification process and successive heating of the billets at the semisolid state (thixoforming) or by suitable cooling conditions of the molten alloy maintained t a temperature between liquidus and solidus conditions, both minimize the contraction effects caused by the solidification process, as well as the porosity content [9].

By squeeze casting the liquid metal is introduced into an open die, just as in a closed die forging process, the dies are then closed. During the final stages of closure, the liquid is displaced into the further parts of the die. No great fluidity requirements are demanded of the liquid, since the displacements are small. Thus forging alloys, which generally have poor fluidities which normally precludes the casting route, can be cast by this process.

Nowadays manufacturing industries are facing a strong global competition which forces them to bring product and process innovations in the shortest possible period of time. This dynamic industrial context is driven by flexibility and need of excellence and the success of a product is strongly linked to its versatility, which makes it able to reach new markets and to raise new challenges. On another hand, the life time of the same products is now shorter, due to the continuous technological innovations constantly bringing new solutions. In such a perspective, the manufacturer shall be able to use extremely flexible production systems suitable to follow the technological evolution of products, and to vary easily the production rates or to ease the introduction and manufacture of innovative products. This is the main technological driver for research and development in for the industries of today together.

## LIQUID FORGING

On the basis of the previous consideration, a new process was recently developed and patented: Liquid forging process [10]. This innovative process combine the advantages of one of the most efficient casting method: low pressure die casting and of forging, until now the most efficient process for mechanical characteristics. The new Liquid Forging process allows the production of near-net-shape components in one operation, obtaining very high material quality and mechanical properties.

In figure 1 the main sequences of the process are illustrated. During processing, the liquid metal is injected bottom up from the furnace into the closed mould at very low velocity; once the filling is complete, thanks to the use of pistons, a high pressure is applied on the liquid material thus "forging" the component. The high pressure is maintained throughout the component solidification allowing to obtain a final component with very fine microstructure and without defects, making it a very interesting process for the automotive industry, for safety and high performance components such as suspension components and engine components. Moreover this process could allow the use of wrought alloys, usually high performance alloys that until now could only be processed by forging.

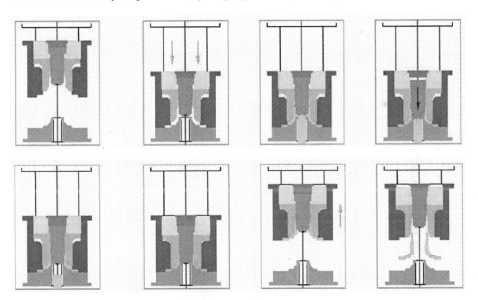

**FIGURE 1.** Main sequences of liquid forging process. From left to right the steps are:
1) Opening of the die; 2) Closing of the die; 3) Introducing of the liquid metal; 4) Closing of the feeding gate; 5) Removal of the feeding pressure; 6) Forging; 7) Opening of the die; 8) Ejection of the component

The acquired experience at Politecnico di Torino since long time in the field of traditional and innovative casting technologies with evaluation of produced parts quality and performances [11 – 14] is profitably used to study and optimize the liquid forging process. The paper aims to introduce the technological backgrounds and to show the encouraging results obtained with these preliminary studies, moreover some design aspects and operating details will also discussed in order to avoid or limit possible defects.

## EXPERIMENTAL PART AND DISCUSSION

The liquid forging process was performed using an industrial-scale 250 metric ton press, equipped with suitable tooling. Tools and die were manufactured with hot working steel grade and some parts were nitrided and coated with plasma spray process to improve there heat checking and wear resistance. To start the study of the process two different parts having cylindrical symmetry were selected, precisely the winch and the wheel illustrated in figure 2, using different types of alloys.

The alloy used for the wheel was based on the AlSi9 type, slightly modified, whilst for the winch a typical forging alloy, P-AlSi1MgMn or 6082 type, was used, moreover for a second series of the last component the AlSi4.5 MnMg alloy was also used.

Owing to the advantage of liquid forging all the finishing operations were avoided and the integrity of the parts or the presence of possible defects was previously verified by means of radiography, successively some of them were T6 heat treated.

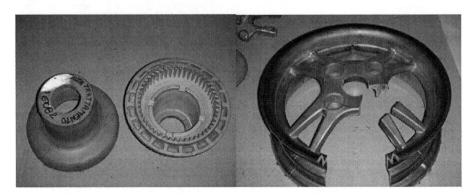

**FIGURE 2.** view of the winch and wheel produced by liquid forging.

Samples for metallographic analysis were obtained by cutting operations from the different produced parts and the observations were performed on the most important/critical sections, like feeding zones, surface and core. For example concerning the winch, it is well known that from a mechanical point of view the most critical zone are teeth and the observations were also focused there. All the examined samples always shown very compact microstructures with the absence of any significant porosity or of other type of defects.

The microstructure observed on the other prepared samples appeared as normal for the type of the alloy and never highlighted noticeable aspects. The observed results clearly demonstrate that the liquid forging process is fully applicable to all kind of aluminium alloys without any distinction between casting or alloys for plastic working.

The pictures in figure 3 are related to samples prepared from the wheel, they show that the alloy is constituted by a very small dendritic structure and that the interdendritic zones are filled with very fine and rounded eutectic phase. Never was observed gas porosity, as well as shrinkage cavities. However, even if the dendritic structure appears always very fine some differences arises between the core zones and some surface ones. In fact, comparing the microstructures on the left part in figure 3, it is evident that dendrites in the zone closer to the feeding system are finer than those of the core zones, evidently the cooling rates were different.

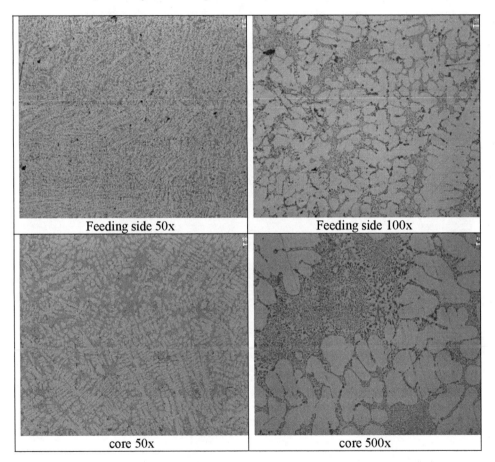

**FIGURE 3.** view of the microstructures of metallographic samples prepared from the wheel, in particular they are referred to feeding and core zones.

The observation of the microstructure at higher magnification highlights that the dendrites are well rounded and tend toward globular shape, in this manner they favour higher mechanical properties as well as toughness characteristics. Moreover, in the zone close to the feeding side some small porosity or defects can be observed, anyway they do not contrast the properties and performances of the part.

The evaluation of the mechanical properties was performed on suitable samples. The maximum attained hardness values have also been evaluated. Tensile strength higher than 440 MPa, with elongation up to 18% with hardness higher than 125 HB are easily attainable on Al alloys type A356.

## CONCLUSIONS

In this paper the liquid forging process was considered highlighting the very strong potential of this technology to produce important components characterised by high mechanical properties obtained thank to optimised microstructure characteristics with very low defect contents. It was demonstrated that all kind of aluminium alloys are prone for liquid forging without any distinction between casting or plastic working alloys.

The obtained results are very promising and indicate that liquid forging is a very reliable and powerful process. It allows a reduction of the costs trough a higher yield of the alloy because of the absence of risers and scraps from feeding systems, as well as a dramatic reduction of finishing operations of the produced parts is also attained.

## REFERENCES

1) ASM Metals Handbook, 10th ed., vol. 15, Casting (1990), ASM - Metals Park, Ohio.
2) J. Campbell and R. A. Harding, Casting Technology, in TALAT 2.0 cd-rom, EAA, Bruxelles (2000).
3) J. Campbell, Castings, Butterworth, Oxford (1991).
4) Perrone, F. Bonollo and V. Wagner, Alluminio Magazine, 4, (1998), p 32.
5) S. Schleg and D.P. Kamicki, Guide to casting and moulding processes, Engineered casting solutions, Technical articles (2000).
6) J.R. Brown, Non-ferrous foundry man's handbook, Butterworth, Oxford (1999).
7) P. Kaufmann, H. Wabusseg and P.J. Uggowitzer, Metallurgical and processing aspects of the NRC semi-solid casting technology.
8) M.R. Ghomashchi and A. Vikhrov, Journal of Materials Processing Technology, 101, (2000), p 1.
9) J.P. Gabathuler, H.J. Huber and J. Erling, La Metallurgia Italiana, 86, (1994), p 609.
10) B. Frulla, Es-Jotech s.r.l., Patent B22D 18/02 n. PCT/EP03/01016, 31.01.2003.
11) E. Romano, M. Rosso, "Caratteristiche microstrutturali e difettologia in getti di leghe di alluminio: confronto fra tecniche di colata tradizionali ed innovative", Proceedings of 29° Convegno Nazionale AIM, Cd-Rom Edition, 13-15 November 2002, Modena, Italy, Editor AIM, ISBN 88-85298-46-X.
12) M. Rosso, S. Guelfo, M. Leghissa, "Fatigue resistance of Al diecasting components", Proc. 2nd Int. Conf. High Tech Die Casting, Montichiari (Bs), Italy, 21-22 aprile 2004, AIM-EDIMET (2004), ISBN 88-86259-26-3, p. 207 - 214.
13) M. Rosso, S. Guelfo, "Investigation for the optimisation of T6 parameters of semisolid castings", Proc. of the "2nd International Conference & Exhibition on New Developments in Metallurgical Process Technology" – Riva del Garda 19-21 settembre 2004, AIM (2004).
14) M. Rosso, S. Guelfo, "Fatigue behaviour of rheocast parts", Proc. of the "8th International Conference on Semi Solid Processing of Alloys and Composites", CD rom, Limassol, Cipro, 21-23 September 2004, p. 1-8.

# Adaptive remeshing method in 2D based on refinement and coarsening techniques

L. Giraud-Moreau, H. Borouchaki and A. Cherouat

*Institut Charles Delaunay (FRE CNRS 2848)*
*University of Technology of Troyes*
*12 rue Marie Curie BP2060*
*10010 Troyes Cedex - France*
*e-mail: laurence.moreau@utt.fr*

**Abstract.** The analysis of mechanical structures using the Finite Element Method, in the framework of large elastoplastic strains, needs frequent remeshing of the deformed domain during computation. Remeshing is necessary for two main reasons, the large geometric distortion of finite elements and the adaptation of the mesh size to the physical behavior of the solution. This paper presents an adaptive remeshing method to remesh a mechanical structure in two dimensions subjected to large elastoplastic deformations with damage. The proposed remeshing technique includes adaptive refinement and coarsening procedures, based on geometrical and physical criteria. The proposed method has been integrated in a computational environment using the ABAQUS solver. Numerical examples show the efficiency of the proposed approach.

**Keywords:** adaptive remeshing, forming process in 2D, refinement and corsening..
**PACS:** Forming, 81.20.Hy

## INTRODUCTION

The analysis of mechanical structures using the Finite Element Method, in the framework of large elastoplastic strains, needs frequent remeshing of the deformed domain during computation [1-4]. Due to the imposition of large plastic strains, damage and friction, the finite element mesh representing the workpiece undergoes severe distortion and hence, necessitates remeshing or the generation of a new mesh for the deformed/evolved geometric representation of the computational domain [5]. It is therefore necessary to update the mesh in such a way that it conforms to the new deformed geometry and becomes dense enough in the critical region while remaining reasonably coarse in the rest of the domain.

This paper presents a remeshing method for the numerical simulation of forming processes in two dimensions. This method is based on geometrical and physical criteria. A geometrical criterion is considered to detect the change of the geometry and hence the need for remeshing in these area. The geometric curvature is estimated at each boundary vertex of the domain and is used to refine all elements sharing this boundary vertex. A physical criterion is used to refine the current mesh of the part with respect to one of the mechanical fields (for examples: the equivalent plastic

CP907, *10th ESAFORM Conference on Material Forming,* edited by E. Cueto and F. Chinesta
© 2007 American Institute of Physics 978-0-7354-0414-4/07/$23.00

deformation or the ductile damage). Based on this physical criterion, a physical size map is computed. It governs the adaptive remeshing of the domain.

The remeshing procedure is applied to the computational domain after each small displacement step of forming tools. It involves four steps:
- the kill of elements in totally-damaged area,
- the coarsening of the mesh with respect to the physical size map,
- the refinement of the boundary elements,
- the refinement of the mesh with respect to a given physical size map.

One of the major advantages of our approach is that the remeshing is applied without the knowledge of the forming tools under contact with the part. Some application examples are presented in order to show the pertinence of our approach.

## GENERAL REMESHING SCHEME

The simulation of the forming process is based on an iterative process. At first, a coarse initial mesh of the part is generated with triangular or quadrilateral elements. At each iteration, a finite element computation is realized in order to simulate numerically the forming process for a small displacement step of forming tools. Then, remeshing is applied after each deformation increment, if necessary, according to the following scheme:

*Coarsening :*
- killing procedure applied to elements in totally damaged area,
- coarsening procedure applied to elements with respect to the physical size map,
- iterative refinement to restore mesh conformity.

*Refinement:*
- refinement procedure applied to elements near the boundary of the domain (the refinement is applied in the vicinity of nodes for which the geometry of the boundary is modified and only if the minimal element size is not reached),
- refinement procedure applied to the whole part by using the physical criterion (with respect to the physical size map),
- iterative refinement to restore mesh conformity.

This process (simulation of the forming process for a small displacement step of forming tools, remeshing of the part) is repeated until the final tool displacement is reached.

## Geometrical And Physical Criteria

During the refinement procedure, a geometrical criterion is used to refine elements of the boundary. The geometric curvature is estimated at each boundary vertex of the domain. If this curvature has been modified during the deformation of the computational domain, all elements sharing this boundary vertex must be refined.

A physical criterion is then used to refine the current mesh of the whole part with respect to one of the mechanical fields. In this paper, the ductile damage has been chosen to define the physical size map. This mechanical field is quantified by a real value between 0 and 1. With a totally damaged element (value 1) is associated a minimal element size and, with a 0-damaged element, a maximal element size. A critical value Dc (for example: 0.9) can be defined from which the minimal size must be reached. For the other elements, a linear size variation can be used. A physical size ($h_D$) can thus be defined for each element of the part. These size specifications constitute a physical size map. For a given element, the physical criterion represents the ratio between the average size of its edges ($\bar{h}$) and its physical size ($h_D$). if this ratio ($\bar{h}/h_D$) is greater than a given threshold, the element must be refined. During the step of remeshing using the physical criterion, the mesh refinement is repeated as long as the physical criterion is violated. Totally damaged elements with the minimal element size are removed from the current mesh.

## Mesh Refinement And Coarsening Methods

The adaptive remeshing technique consists in improving the mesh in order to conform to the geometry and the mechanical fields of the current part surface during deformation. The mesh refinement and coarsening methods are now detailed.

The refinement technique consists in subdividing mesh elements. An element is refined if it is a boundary element which needs to be refined or if its size is greater than its physical size (physical criterion). There is only one element subdivision which allows to preserve the element shape quality: the uniform subdivision into four new elements. Figure 1 shows the triangular and quadrilateral element refinements.

**FIGURE 1.** Triangular and quadrilateral element refinements.

After each refinement procedure (geometrical criterion or physical criterion), an iterative refinement to restore mesh conformity is necessary. Indeed, after applying the subdivision according to the geometrical or physical criteria, adjacent elements to subdivided elements must be modified. As the edges of the subdivided elements are divided in two, there is a node in the middle of the edges common to the subdivided element and its adjacent elements. The mesh is then not conforming. To retrieve the mesh conformity, adjacent elements to subdivided elements must be also subdivided. This last subdivision can not be a homothetic subdivision in four elements because it would result in the systematic homothetic subdivision of all mesh elements.

There are three different configurations for adjacent elements which must be subdivided in order to ensure the mesh conformity:
- no edge is saturated (i.e. containing a new added node),
- only one edge is saturated,
- at least two edges are saturated.

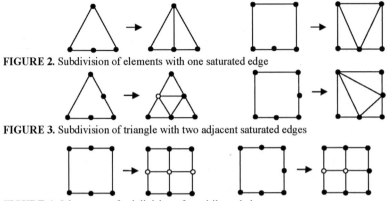

**FIGURE 2.** Subdivision of elements with one saturated edge

**FIGURE 3.** Subdivision of triangle with two adjacent saturated edges

**FIGURE 4.** Other cases of subdivision of quadrilateral element

Depending on the configuration, a subdivision is applied if necessary. In the first case (no saturated edge), the element is not subdivided and is not modified. In the second case (one saturated edge), a triangular element is subdivided in two triangles and a quadrilateral element in three triangles (see figure 2). This subdivision allows to stop the propagation of the homothetic subdivision. In the third case, if all the edges are saturated the element is subdivided in four homothetic elements. Otherwise, in the case of triangular elements (having two saturated edges), all possible subdivisions lead to the formation of poor shaped elements (stretched elements). It is then necessary to add a new node in order to subdivide also this element into four homothetic elements (see figure 3). In the quadrilateral case, when only two edges are saturated and are adjacent, the quadrilateral is subdivided in four triangles (see figure 3). This subdivision allows to stop the propagation of the homothetic subdivisions. In the other cases, the element is subdividing into four homothetic quadrilateral elements (see figure 4). This refinement procedure is iteratively applied until no new node is added.

**FIGURE 5**: Ordinary elements and extraordinary elements

From an algorithmic point of view, the mesh is composed of two types of element: ordinary and extraordinary. An ordinary element is a triangle or a quadrilateral without saturated edges (see figure 5). An extraordinary triangle is a triangle with one and only one saturated edge. An extraordinary quadrilateral is a quadrilateral with only one saturated edge or two adjacent saturated edges. Figure 5 shows ordinary and extraordinary elements. The remeshing algorithm must take into account these two element types. During the refinement operation, geometrical and physical criteria are applied to elements of both types. An ordinary or extraordinary element which is curved or whose size is greater than the physical size, is then subdivided into four ordinary elements. After this operation, all ordinary elements with at least two saturated edges, except the case of two adjacent edges for quadrilateral elements, are iteratively subdivided into four ordinary elements. Then, all the elements with at least one saturated edge are transformed to extraordinary elements and the other elements remain unchanged.

At the end of the refinement operations, for the mechanical computational purpose, the extraordinary elements of the resulting mesh are transformed : an extraordinary triangle is divided in two triangles, an extraordinary quadrilateral with one middle node is divided in three triangles and an extraordinary quadrilateral with two middle nodes is divided in four triangles.

The coarsening technique is the reciprocal operation of the refinement procedure. It can only be applied to a set of four ordinary elements, called associated elements, obtained during a homothetic element refinement. Thanks to the coarsening technique, the initial element is restored when the area in which this element belongs, becomes flat (see figure 6).

FIGURE 6. Triangular and quadrilateral elements coarsening

During the refinement procedure, the mechanical fields are simply associated to the new created elements from the subdivision. During the coarsening procedure, the mechanical fields associated to four associated elements are averaged and the result is associated to the new element.

## APPLICATION

Two application examples are now presented in order to show the pertinence of our approach. In these two cases, the solver ABAQUS/EXPLICIT has been used and the material model used is an isotropic elastoplastic von-Mises model with isotropic hardening.

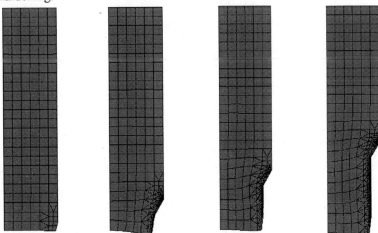

FIGURE 7. Deformed cylindrical bar for different extrusion steps

The proposed remeshing procedure is first applied to simulate 'virtually' the extrusion of a cylindrical bar. Figure 7 presents differents steps of the 2D axisymetric extrusion simulation. We can note that the mesh has been refined in regions where the geometry changes during the simulation and that some elements have been coarsened (third picture) by the remeshing method.

The second example presents a simulation of traction for a steel sheet. In this example, the mesh refinement is localized on areas where the damage is important. We can note that totally damaged elements have been killed by the remeshing procedure. Figure 7 presents different steps during the simulation of traction.

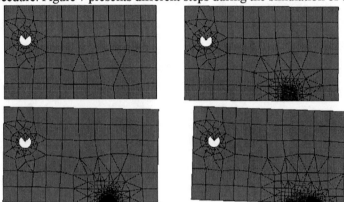

**FIGURE 8.** Simulation of traction in 2D

## CONCLUSION

The different steps necessary to the adaptive remeshing of the computation domain in large elastoplastic deformations with damage have been presented. Geometrical and physical criteria have been used. The proposed remeshing method has been implemented with in the ABAQUS code. Numerical simulations of forming process in two dimensions have validated the proposed approach and proved its efficiency. The extension in three dimensions based on an adaptive remeshing of the surface of the deformable piece is currently under progress.

## REFERENCES

1. K. Saanouni, A. Cherouat and Y. Hammi, "Optimization of hydroforming processes with respect to fracture", Esaform, Belguim, 1, 2001, pp. 361-364.
2. L. Fourment and J.-L. Chenot, "Adaptive remeshing and error control for forming processes", *Revue européenne des éléments finis 3*, 2, 1994, pp. 247-279.
3. H. Borouchaki, A. Cherouat, P. Laug and K. Saanouni, "Adaptive remeshing for ductile fracture prediction in metal forming", *C.R. Mecanique 330*, 2002, pp. 709-716.
4. L. Giraud-Moreau, H. Borouchaki, A. Cherouat, Sheet metal forming using adaptive remeshing, The 8th international Conference on Numerical Methods in Industrial Forming Process (Numiform 2004), Columbus, Etats-Unis, 2004, pp. 133 (book of abstracts).
5. P. Ladeveze et J.P. Pelle, La maîtrise du calcul en mécanique linéaire et non linéaire, *études en mécanique des matériaux et des structures*, Hermès, Paris, France, 2001.

# Plastic Behavior and Fracture of Aluminum and Copper in Torsion Tests

José Divo Bressan[†]

†Department of Mechanical Engineering  - UDESC Joinville – 89.223-100 Joinville – SC – Brazil.
www.joinville.udesc.br .  email: dem2jdb@joinville.udesc.br

**Abstract.** Present work investigates the plastic behavior, work hardening and the beginning of plastic instabilities, of cylindrical specimens deformed by high speed cold plastic torsion tests and at low speed tensile test. The tests were carried out in a laboratory torsion test equipment and an universal tensile test machine. The tensile tests were performed at room temperature in an universal testing machine at low strain rate of 0.034/s. Experimental torsion tests were carried out at constant angular speed that imposed a constant shear strain rate to the specimen. In the tests, the rotation speed were set to 62 rpm and 200 rpm which imposed high strain rates of about 2/s and 6.5/s respectively. The torsion tests performed at room temperature on annealed commercial pure copper and aluminum. Two types of torsion specimen for aluminum were used: solid and tubular. The solid aluminum specimen curves presented various points of maximum torque. The tubular copper specimens showed two points of maximum. Shear bands or shear strain localization at specimen were possibly the mechanism of maximum torque points formation. The work hardening coefficient n and the strain rate sensitivity parameter m were evaluated from the equivalent stress versus strain curve from tensile and torsion tests. The n-value remained constant whereas the m-value increased ten folds for aluminum specimens: from tensile test m= 0.027 and torsion test m= 0.27. However, the hardening curves were sigmoidal.

**Keywords: Torsion Test, Plasticity, Work hardening, Strain rate, Aluminum, Copper.**

## INTRODUCTION

Torsion testing of metal alloys have been used by researchers for investigating the plastic behavior of metals in high strain rates forming processes such as sheet rolling and upsetting. In these processes, strain rates values are commonly high and can attain values of 10/s. This level of strain velocity can be obtained in the torsion testing of solid or tubular specimens as related in previous work [1]. Both strain hardening and strain rate hardening effects are very important in the plastic behavior response of metals at high deformation velocities. In addition, torsion test can produce a larger hardening curve of stress vs strain than in the tensile test of metals.

The metal plastic behavior depends on the imposed strain rate. In torsion testing, the stress and strain state is simple shear, thus, shear strain is the main deformation mechanism: strain localization and shear bands can also be investigated as well as the redundant work produced in metal forming processes. On the other hand, in tensile tests, the stress state is uniaxial and the strain state is tri-dimensional of principal

CP907, *10th ESAFORM Conference on Material Forming,* edited by E. Cueto and F. Chinesta
© 2007 American Institute of Physics 978-0-7354-0414-4/07/$23.00

strains but mainly homogeneous deformation processes that occur up to the instability point.

In previous work by the author [1,2] on torsion tests of steel, brass, copper and aluminum at strain rates of about 2/s, the onset of specimen rupture were at the point of maximum torque or at the instability point. In the experimental test carried out with a solid aluminum specimen, the curve showed two points of maximum torque prior to rupture, possibly due to shear bands localization and softening mechanisms. This anomalous behavior of aluminum specimen in shear stress-strain curve will be investigated in the present work at a higher strain rate of 7.7/s.

The aims of the present work were to perform torsion tests in solid and tubular specimens of annealed aluminum and copper for obtaining the plastic behavior at strain rates similar to that occurs at some classes of industry processes of metal forming: high strain rates of about 2/s and 6.5/s. Secondly, to investigate the fracture and instability mechanisms in torsion specimens compared to tensile tests. Thirdly, to confirm the anomalous behavior of solid aluminum specimens in the hardening curve of shear stress versus shear strain.

## MATERIALS AND EXPERIMENTAL PROCEDURES

The aluminum and copper torsion and simple tension specimens were prepared from commercial bars of 19 mm in diameter. The specimens were cut at the appropriate length from the bar and machined to the specific shape. The geometry and dimensions for torsion specimens can be observed in Table 1 and are similar to the previous work [1]. The uniaxial tensile specimens were machined according to the ABNT Standard. Thus, the specimens for torsion tests were tubular and solid while for tensile tests were solid only.

All specimens were annealed at the apropriate temperature of 450 °C to yield large ductility. The mechanical properties and hardening equation obtained from tensile tests at strain rate of $0.17 \times 10^{-2}$/s and $3.34 \times 10^{-2}$/s are presented in Table 2. The main geometric characteristics of torsion specimen are shown in Table 3. All torsion specimens have the same useful length of 16 mm and various diameters.

TABLE 1. Conditions of Annealing Heat Treatment for Torsion and Tension Test Specimens.

| Material | Temperature (°C) | Heating Rate (°C/min) | Time (min) | Hardness (HV-15) | Reduction of Hardness After the Annealing Treatment (%) |
|---|---|---|---|---|---|
| Copper | 450 | 10 | 60 | 66 | 19 |
| Aluminum | 450 | 10 | 40 | 33 | 33 |

TABLE 2. Tensile test mechanical properties for strain rate of $0.17 \times 10^{-2}$/s and $3.34 \times 10^{-2}$/s .

| Material | Young Modulus E (GPa) | Yield Limit $\sigma_{esc}$ (MPa) | Strength Limit $\sigma_t$ (MPa) | Enlongation % | Hardening Equation $\sigma = K(\varepsilon_o + \varepsilon)^n \, \dot{\varepsilon}^m$ (MPa) |
|---|---|---|---|---|---|
| Copper | 82 | 182 | 257 | 21 | $\sigma = 450 \, (0.025 + \varepsilon)^{0.20}$ |
| Aluminum | 74 | 73 | 138 | 26 | $\sigma = 235 \, (0.01 + \varepsilon)^{0.21} \dot{\varepsilon}^{0.027}$ |

TABLE 3. Dimensions of the torsion test specimens.

| Material | Solid Specimen | | | |
|---|---|---|---|---|
| | $L_{useful}$ (mm) | $L_{total}$ (mm) | $D_{useful}$ (mm) | rpm |
| Aluminum - 1 | 16 | 50 | 10.04 | 62 |
| Aluminum - 2 | 16 | 50 | 9.50 | 62 |
| Aluminum - 3 | 16 | 50 | 12.05 | 200 |
| | Tubular Specimen | | | | | |
| | $L_{useful}$ (mm) | $L_{total}$ (mm) | $D_1$ (mm) | $D_2$ (mm) | Thickness (mm) | rpm |
| Aluminum - 4 | 16 | 50 | 8.0 | 12.15 | 2.07 | 61 |
| Aluminum - 5 | 16 | 50 | 8.0 | 12.15 | 2.07 | 61 |
| Aluminum - 6 | 16 | 50 | 8.0 | 12.04 | 2.02 | 200 |
| Copper - 6 | 16 | 50 | 7.96 | 12.07 | 2.05 | 200 |
| Copper - 7 | 16 | 50 | 8.0 | 12.07 | 2.035 | 200 |

The torsion tests were carried out in a laboratory torsion test machine, keeping constant the strain rate during the torsion test. The specimens were not allowed to deform in the axial direction. An acquisition data system registered the values of torque in "Nm" and the twist angle in "degree" simultaneously in a computer by a Labview program. Thus, it was possible to determine the metals experimental curves of shear stress versus shear strain and to calculate the materials constitutive hardening equations and its plasticity parameters for the selected strain rate and materials. The torsion tests were carried out at 61 and 200 rpm which correspond to strain rates of 2/s and 6.5/s respectively. The shear stress $\tau$ and strain $\gamma$ at the external surface of tubular specimen were calculated using the following equations [2],

$$\tau_a = \frac{M}{2\pi\left(a_2^3 - a_1^3\right)}(3+n+m) \tag{1}$$

$$\gamma_a = \frac{\theta}{L_u}\left(\frac{a_1+a_2}{2}\right) \tag{2}$$

where $M$ is the torque, $a_1$ is the internal radius, $a_2$ is the external radius, $n$ is the work-hardening coefficient, $m$ is the strain rate sensitivity coefficient, $\theta$ is the angle of twist in radians and $L_u$ is the useful length of the specimen. For solid specimens with radius r ($D_{useful} = 2$ r), the equations are,

$$\tau_a = \frac{M}{2\pi r^3}(3+n+m) \quad \text{and} \quad \gamma_a = \frac{r\,\theta}{L_u} \tag{3}$$

The equivalent von Mises equations for stress and strain used to calculate the plasticity constitutive equations were,

$$\overline{\sigma}=\sqrt{3}\,\tau_a \quad \text{and} \quad \overline{\varepsilon}=\frac{1}{\sqrt{3}}\gamma_a \tag{4}$$

523

# EXPERIMENTAL RESULTS

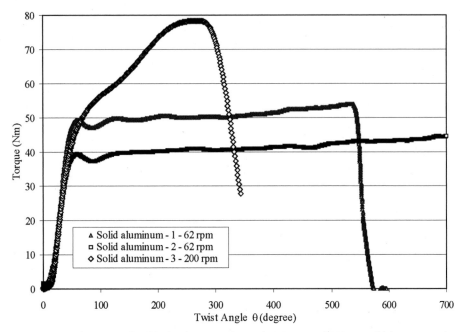

**FIGURE 1.** Torsion Tests of Solid Aluminum Specimens for 62 rpm and 200 rpm which correspond to Shear Strain Rate of 2/s and 7.7/s respectively.

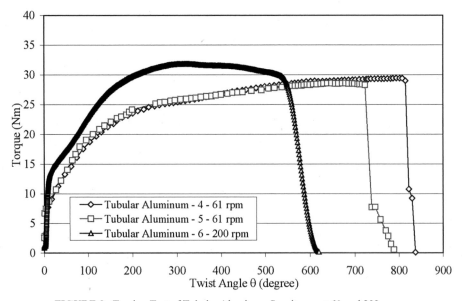

**FIGURE 2.** Torsion Test of Tubular Aluminum Specimens at 61 and 200 rpm.

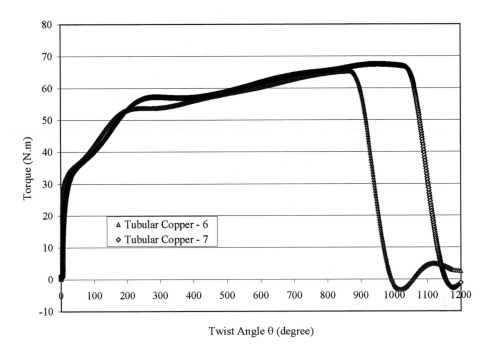

**FIGURE 3.** Torsion Tests of Tubular Copper Specimens at 200 rpm or Shear Strain Rate of 6.5/s.

Figure 1 and Fig.2 shows the torsion tests curves of solid and tubular aluminum specimens at rotation speed of 62 and 200 rpm, which corresponds to shear strain rates of 2/s and 7.7/s respectively. Note that specimen 3 was tested at 200 rpm and has diameter 12.05 mm which is 20% higher than specimens 1 and 2. This yielded a hardening curve for specimen 3 more inclined. In addition, the strain to rupture decreased approximately one half. In Fig.1, the presence of intermediate points of maximum torque is confirmed in the hardening curves for specimens 1 and 2 at 62 rpm, but only one maximum plus softening for specimen 3 at 200 rpm.

In Fig.2 the difference in the hardening curves for tubular specimens 4, 5 and 6 is lower. The specimens have quite similar diameters, thus, the difference in the hardening curves is due only to the shear strain rate of 2/s and 6.5/s. Again, the strain to rupture decreased about 20% for higher strain rate. Therefore, the aluminum specimens showed strain hardening and strain rate hardening mechanisms. The correspondent equivalent von Mises plasticity equation for tubular aluminum presented in Fig.4 was $\overline{\sigma}=120(0.035+\overline{\epsilon})^{0.295}\dot{\epsilon}^{0.27}$ MPa. Thus, the strain rate sensitivity coefficient obtained from the torsion tests was m = 0.27, ten times greater than the value calculated from tensile test of 0.027 as shown in Table 2. The work hardening coefficient n = 0.295 is fairly constant compared to the tensile test of n = 0.21.

In Fig.3 for tubular copper specimens at 200 rpm, the hardening curves showed two points of maximum torque or instability point. This was not reported in the previous work by the author [2]. The microstructure strain mechanism is possibly formation of shear bands or strain localization. Fracture occurred in the second instability point.

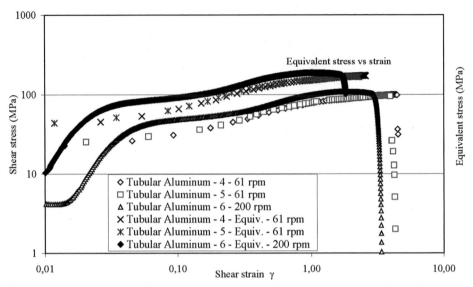

**FIGURE 4.** Shear Stress versus Shear Strain and Equivalent Stress-Strain Curves of Tubular Aluminum Specimens at 61 and 200 rpm from torsion tests.

## CONCLUDING REMARKS

From the present experimental results on torsion tests of solid and tubular specimens of annealed aluminum and copper the following conclusions can be drawn,

- The occurrence of more than one point of maximum torque in solid aluminum specimens have been confirmed for rotation speed of 62 rpm, but only one point of maximum torque at 200 rpm. Increase in strain rate decreased the fracture shear strain.

- Two points of maximum torque have been observed in tubular copper specimens at 200 rpm. Rupture occurred immediately after the second point of instability,

- In torsion test of tubular aluminum the strain rate sensitivity coefficient increased ten folds from m=0.027 in simple tension to m=0.27 in torsion at strain rate of 6.5/s.

- The constitute equation for plasticity has a sigmoidal shape in torsion test.

## ACKNOWLEDGMENTS

The authors would like to gratefully acknowledge the financial support received from the University of Santa Catarina State/Brazil, CNPq and CAPES.

## REFERENCES

1. J.D. Bressan and R.K. Unfer, "Plastic Instabilities and Fracture of Metals in Torsion and Tensile Tests" in Proc. ESAFORM – 2006, edited by Neal Juster & Andrzej Rosochowski, Glasgow, Scotland, 2006, pp.327-330.
2. J.D. Bressan and R.K. Unfer, Construction and validation tests of a torsion test machine. J. Mater. Proc. Technol., v.179, 2006, pp.23-29.

# Friction and wear in hot forging of steels

E. Daouben[(*)(x)], L. Dubar[(*)], M. Dubar[(*)], R. Deltombe[(*)], A. Dubois[(*)],
N. Truong-Dinh[(x)], L. Lazzarotto[(+)]

*(\*) LAMIH UMR CNRS 8530 – Université de Valenciennes*
*Le Mont Houy – 59313 Valenciennes Cedex 9 – France*
*(x) CONDAT Lubrifiants- Avenue Frédéric Mistral*
*38670 Chasses sur Rhônes – France*
*(+) CETIM – Etablissement de Saint Etienne*
*7, rue de la Presse – BP 802*
*42952 Saint Etienne  Cedex 9*

**Abstract.** In the field of hot forging of steels, the mastering of wear phenomena enables to save cost production, especially concerning tools. Surfaces of tools are protected thanks to graphite. The existing lubrication processes are not very well known: amount and quality of lubricant, lubrication techniques have to be strongly optimized to delay wear phenomena occurrence. This optimization is linked with hot forging processes, the lubricant layers must be tested according to representative friction conditions. This paper presents the first part of a global study focused on wear phenomena encountered in hot forging of steels. The goal is the identification of reliable parameters, in order to bring knowledge and models of wear. A prototype testing stand developed in the authors' laboratory is involved in this experimental analysis. This test is called Warm and Hot Upsetting Sliding Test (WHUST). The stand is composed of a heating induction system and a servo-hydraulic system. Workpieces taken from production can be heated until 1200°C. A nitrided contactor representing the tool is heated at 200°C. The contactor is then coated with graphite and rubs against the workpiece, leaving a residual track on it. Friction coefficient and surface parameters on the contactor and the workpiece are the most representative test results. The surface parameters are mainly the sliding length before defects occurrence, and the amplitude of surface profile of the contactor. The developed methodology will be first presented followed by the different parts of the experimental prototype. The results of experiment show clearly different levels of performance according to different lubricants.

**Keywords:** Hot Forging, friction, wear, graphite, solid contents.
**PACS: 81.20.Hy, 81.40.Pq**

## OUTLINES

The mastering of tool life is a major topic in the field of hot forging: tool life must be predicted to save cost production. The tool life depends on the process and the lubrication, including lubrication techniques and lubricants.  The goal of the main studies is the integration of wear models in the numerical simulation of processes, in order to predict wear occurrence. Unfortunately reliable wear models do not really exist. The existing lubrication processes are not very well known: amount and quality of lubricant, lubrication techniques have to be strongly optimized to delay wear phenomena occurrence.

Scientific studies have to bring knowledge on the behavior of lubricant at the tool workpiece interface, in order to establish reliable wear markers. That kind of studies can not be fully performed on the production plan because of schedule and cost. One way to face this problem is the development of a prototype testing stand to reproduce the key parameters and phenomena acting on wear occurrence in laboratory [1]. The warm and Hot Upsetting Sliding Test (W.H.U.S.T) has been developed at the LAMIH (UMR CNRS 8530) laboratory of the

CP907, *10th ESAFORM Conference on Material Forming*, edited by E. Cueto and F. Chinesta
© 2007 American Institute of Physics 978-0-7354-0414-4/07/$23.00

University of Valenciennes in the context of a partnership with CONDAT and CETIM. It reproduces contact and friction phenomena occurring at the tool work piece interface during hot forging. The new testing stand is first presented, describing its main parts : the lubrication device, the heating device and the mechanical part. The associated methodology is then summed up. At last, results are shown in the case of hot forging of a 3D part.

# EXPERIMENTAL STRATEGY

## Warm and Hot Upsetting Sliding Test

The identification of contact parameters is performed using the laboratory Warm and Hot Upsetting Sliding Test (WHUST). The WHUST involves a contactor which slightly penetrates a specimen and slides along its surface with a constant penetration. The WHUST parameters are the geometry of the contactor, the value of its penetration within the specimen, its sliding velocity and the contactor and specimen temperatures (Figure 1). A reliable identification of tribological data requires the simulation of the mechanical, physical and chemical parameters of the surfaces in contact since coefficients of friction and thermal contact resistances are strongly related to these parameters [2]. Obviously it is impossible for a laboratory test to simulate each parameters. Nonetheless, the WHUST uses a part of the industrial workpiece as specimen, a part of actual tools as contactor and lubricants taken from the process tanks. This peculiarity allows the WHUST to respect the chemical and physical characteristics of the contact (reactivity, adsorbed gas, material structure, hardness, roughness, wear particles, properties of surface coatings, etc.). The mechanical parameters of the contact (mainly contact pressure, strain rate and sliding velocity) are adjusted using the test parameters (penetration, geometry of the contactor).

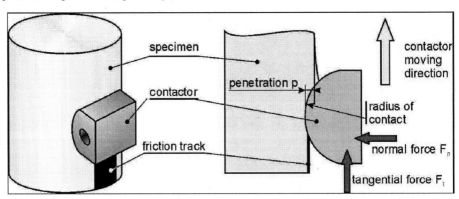

**FIGURE 1.** Overview of the W.H.U.S.T and main parameters.

A heating inductor is part of the testing stand. The workpiece can be heated up to 1200 °C, with a heating time less than one minute. The average oxide mass created on the specimen is less than 1% (Figure 1) [3]. The elapsed time for transfer between heating and testing never exceeds 15s.

# Methodology for friction and wear analysis

The methodology for friction and wear analysis has been led in four steps as follows :

***Process mechanical analysis***: the aim of this first step is the identification of the mechanical properties to be reproduced on the testing stand WHUST. Process parameters (temperatures, forces, surface roughness...) are first identified. The finite element simulation of the process is then performed respecting the identified parameters [4]. The friction coefficient has been taken equal to 0.15. Contact pressure at the tool/workpiece interface, strain and strain rate are the determined mechanical properties via the numerical simulation.

***Identification of optimum WHUST parameters***: contactor penetration and geometry are the parameters to identify in order to reproduce the contact pressure at the tool/workpiece interface, strain and strain rate determined in step 1. An inverse methodology of identification is used [5].

***Test performance***: tests are now carried out. Heating of the specimen is first performed via a heating inductor up to 1200°C. Temperature is controlled by three thermocouples. The contactor is then heated by a regulated heating cartridge up to 200°C [6]. Graphite lubricant is sprayed on the contactor respecting the industrial spraying conditions and so reproducing a precise graphite thickness. Finally, the penetration of the contactor is adjusted within the specimen. The contactor can now slide along the specimen. The sliding velocity is regulated by an hydraulic jack up to 0.5m/s. Normal and tangential forces on the contactor are recorded during the test. A sliding residual track is present on the specimen.

***Identification of a lubrication index and surface analysis***: A lubrication index $L_i$ is first calculated according to normal ($F_n$) and tangential ($F_t$) forces as follows: $L_i = \dfrac{F_t}{F_n}$. Secondly, the sliding length before defects occurrence $L_c$ is measured on the specimen by optical microscopy. Finally, 3D optical surface profile measurements are carried out on the contactor and the specimen to precisely analyze surface roughness evolution after friction tests.

# HOT FORGING OF A 3D PART : FRICTION AND WEAR ANALYSIS

## Hot Forging of a tripod

The studied process in this paper is the hot forging of a tripod. The constitutive material of the part is a C38 steel. Temperatures are respectively 1000°C for the part and 200°C for the nitrided dies . The press is a mechanical press with a maximum velocity of 1m/s (Figure 2). The analysis of friction and wear concerns the contact zone corresponding to the exit of the burr. In this zone, a critical point corresponding to the radius of the lower die is selected to analyze contact pressure and sliding velocities. The wear impact on this critical radius is very high, the lubrication strategy must be efficient to delay damage in this zone.

The advance of the burr from the critical radius to the exit is divided in 4 time steps 1,2,3 and 4. From step 1 to step 3, the increasing of the contact pressure is very low (figure 2). The value of the contact pressure at step 3 is 190 MPa. This value becomes the reference pressure for the setting of the testing stand. The corresponding local sliding velocity is found to be 60 mm/s.

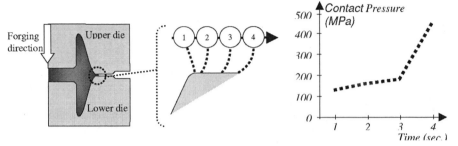

**FIGURE 2.** Hot forging of a tripod: 4 steps of the exit of the burr, contact pressure on critical radius versus time

## Experimental results

Friction tests are performed thanks to the WHUST testing stand presented above. The test is settled to reproduce the contact pressure identified above (190 MPa), it corresponds to a 0.4 mm penetration of the contactor into the workpiece. The workpiece is heated to 1100°C and the contactor is heated to 200°C. The lubricant is graphite suspension in water. Three types of tests are performed: without lubricant and with two different graphite lubricants L1 and L2. For each type of test, two friction passes are performed. The contactor is also lubricated between pass1 and pass2. All the tests simulate the same industrial contact conditions corresponding to the burr exit.

Figure 3 gives the values of the lubrication index according to three different lubrication conditions and two passes. The sensitivity of the test is found to be good. The lubrication index is lower when the contactor is coated with graphite. The graphite suspension sprayed onto the contactor leads to different lubrication index according to its composition. Both graphite coatings are 30 µm thick. Lubricant L1 leads to a 0.48 mean lubrication index in pass 1 and a 0.52 mean lubrication index in pass 2. Lubricant L2 leads to a 0.67 mean lubrication index in pass 1 and pass 2.

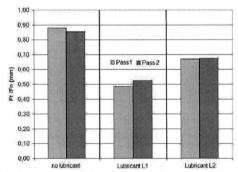

**FIGURE 3.** Hot forging of a tripod: lubrication index for pass 1 and pass 2

*Sliding Length before Defect*

The graphite layer deposited on the tool surface has to delay defect occurrence on the surface of the workpiece. After a given sliding length $L_c$, scratches appear on the friction track

530

as shown in figure 4. The sliding length is plotted in figure 4 according to the three different conditions of lubrication and the two passes.

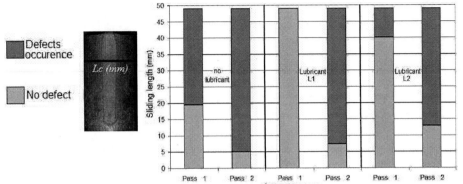

**FIGURE 4.** Hot forging of a tripod: sliding length for pass 1 and pass 2

We can first notice that Lc is greater with graphite. Secondly, L1 delays more the defect occurrence compared L2. Lubricant L1 also gives the best lubrication index. These interesting results are a first step towards the identification of wear markers and towards an optimum choice of lubricant.

*Surface profile measurements*

Figure 5 gives the evolution of the surface profile of the contactor after one pass.

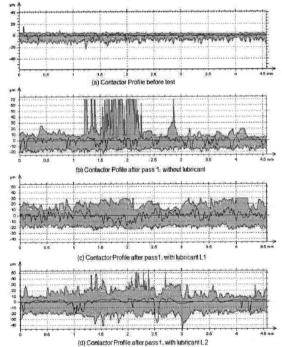

**FIGURE 5.** Surface profile measurements after pass 1 for the three different lubrication conditions

Figure 5-a represents the contactor before test and figures 5-b, c and d represent the contactor according to the different lubrication conditions. Each profile: a, b, c and d is made of 1024 profiles scanned on the contactor. The lower and upper bounds have been represented in grey and the mean profile is in black. The defect occurrence is clearly shown with increasing surface amplitude due to friction scratches and the presence of embedded oxide residues. We can notice that lubricant L1, which had already given interesting results, gives the most homogeneous profile and the smallest roughness amplitude. Surface profile measurements are interesting wear markers and can help for an optimum choice of lubricant.

## CONCLUSION

An original methodology focused on the analysis of wear phenomena in hot forging of steels has been presented. A prototype testing device called WHUST has been developed in the authors' laboratory. This stand, composed of a heating inductor, a spraying device and a servo-hydraulic system can precisely reproduce contact conditions at the tool/workpiece interface. Three reliable wear markers have been underlined after the friction test. The first one called the lubrication index represents a simple ratio between the normal and the tangential forces. The other two parameters come from the surface analysis: one is the sliding length before defects occurrence on the specimen and the second one is the roughness amplitude profile on the contactor. This methodology has been able to distinguish two graphite lubricants. These results are very promising and will help us to better understand wear mechanisms in a very near future. Actually, the second part of this wide study will be the understanding of the behaviour difference between lubricants mainly by means of surface microscopic observations to identify complementary wear markers on the tool.

## ACKNOWLEDGMENTS

The present research work has been supported by the European Community, the Délégation Régionale à la Recherche et à la Technologie, the Ministère de l'Education Nationale, de la Recherche et de la Technologie, the Région Nord-Pas de Calais, the Centre National de la Recherche Scientifique, CONDAT SA and CETIM, ASCOFORGE and FORGES de Courcelles. The authors gratefully acknowledge the support of these institutions and industrial partners.

## REFERENCES

1. M. Tercelj, P.Panjan, I. Urankar, P. Fajfar and R.Turk, *A newly designed laboratory hot forging test for evaluation of coated tool wear resistance.* Surface & Coatings Technology, 2006, Vol. 200, pp. 3594-3604..
2. E. Felder, *Mécanismes physiques et modélisation mécanique du frottement entre corps solide*, Mécanique & industries, 2000, Vol.1, 6, pp. 555-561.
3. B.Picqué, P.O. Bouchard, P. Montmitonnet, M. Picard, *Mechanical of iron oxide scale: Experimental and numerical study.* WEAR, 2005, Vol.260, pp. 231-242.
4. F. Hélénon, E.Vidal-Salle and J.C.Boyer, *Lubricant flow between rough surfaces during closed-die forming.* Journal of Material Processing Technology, 2004, Vol. 153/15, pp. 707 –713
5. A. Verleene, *Vers la maîtrise de la dégradation des outillages de mise en forme à froid.* Thesis, Université de Valenciennes et du Hainaut Cambrésis, 2000.
6. D. J. Jeong, D. J. Kim, J. H. Kim, B. M. Kim and T.A. Dean, *Effects of surface treatments and lubricants for warm forging die life.* Journal of Materials Processing Technology, 2001, Vol.113, pp. 544-550.

# 7 – EXTRUSION AND DRAWING

## (S. Støren and J. Huetink)

# Modelling of drawing and rolling of high carbon flat wires

## C. BOBADILLA[1], N. PERSEM[2], S. FOISSEY[3]

[1]MITTAL STEEL EUROPE R&D, BP 140 57360, Annéville, France
[2]CEMEF, 1, Rue Claude Daunesse, BP 06904 Sophia-Antipolis, France
[3]TREFILEUROPE, 25 bis, avenue de Lyon, BP 96 01000, Bourg-en-Bresse, France

**Abstract:** In order to meet customer requirements, it is necessary to develop new flat wires with a high tensile strength and a high width/thickness ratio. These products are manufactured from wire rod. The first step is to draw the wire until we have the required mechanical properties and required surface area of the section. After this, the wire is rolled from a round to a rectangular section. During the flat rolling process it can be reduced by more than 50%. Then the wire is exposed to a high level of stress during this process. Modelling allows us to predetermine this stress level, taking into account the final dimensions and the mechanical properties, thus optimising both rolling and drawing process. Forge2005 was used in order to simulate these processes. The aim of this study is to determine the value of residual stresses after drawing and so to optimise rolling. Indeed, the highest stress values are reached at this step of the process by changing the section of the wire from a round to a rectangular one. In order to evaluate the stress value accuracy for high strain levels, a behaviour law has been identified. This is a result of tensile tests carried out at each step of the drawing process. Finally, a multi-axial damage criterion was implemented using Forge2005. The optimisation of the rolling is directly linked to the minimisation of this criterion.

**Key words:** behaviour law, rolling and drawing, high carbon steel, damage criterion.

**PACS:** 81.20.Hy, 83.50.-v, 82.20.Wt, 02.30.Zz

## Introduction

Due to the high cost of industrial trials and the strong competition in the European market, drawers are obliged to use numerical models in order to develop new products. The manufacturing of high carbon flat wires including drawing and flat rolling is a complex process requiring the validation of numerous trials. Thanks to numerical modelling this development time will be reduced.

To validate the development of the existing process used to produce high carbon flat wires, it is necessary to adapt the parameters in order to achieve the required mechanical properties (tensile strength, ductility...). Managing both the reduction rate in drawing and parameters of the flat rolling process these aforementioned mechanical properties and the dimensions must be obtained without provoking a deterioration of the wire. Numerical simulation, i.e. Forge2005, allows us to predict stress level inside the wire to ensure deterioration will not occur.

CP907, 10th ESAFORM Conference on Material Forming, edited by E. Cueto and F. Chinesta

The accuracy of the stress level prediction depends on the calculation of the behaviour law and the precision between the parameters of the numerical model and the industrial process. Test trials and measurements on the process line are essential to define accurate modelling parameters.

This paper will present the 3 different stages of the methodology used to carry out the study i.e. :
- Identification of the behaviour law of the material
- Characterization of the process parameters and simulations
- Analysis of the simulations

## Identification of the behaviour law

*Selected trials*

The manufacturing process, used by the drawers, to produce     flat wire is divided into 3 parts. Each step of the process will be described in order to follow the evolution of the mechanical properties of the product thus allowing an adequate behaviour law to be determined.

First of all the wire rod is patented in order to improve its metallurgical properties before drawing. Secondly, the wire is drawn in order to achieve the stipulated mechanical properties. At this stage the required tensile strength of the final product has been achieved. The wire is exposed to high compression stresses on the surface and high tensile stresses in the core. Thirdly, the wire is flat rolled and the product goes from a round section to a rectangular one, corresponding to the required final dimensions of the flat. At this step, there is no reduction of the section. Therefore the mechanical properties remain constant. As in the drawing process, compression stresses are applied to the surface and tensile stresses to the core. However, at this stage flat rolling is a 3D process.

Since the objective of the study is to simulate both drawing and rolling processes, a behaviour law is required, which is valid for large strains. In the case of large strains, the behaviour identification as a result of only one tensile test cannot be applied. In this case the torsion test is the most commonly used. However solicitations on the wire, due to torsion tests, differ greatly to those resulting from the drawing and rolling process. Therefore an original identification method [1] of the behaviour law was chosen – combining drawing and tensile tests, without torsion.

The technique of identifying the behaviour law consists in plotting the strain – stress curve, which will be defined by carrying out tensile tests after patenting and after each pass of drawing. Only the first part of the tensile test after patenting (until the Ys value) is taken into account, and the maximum stress value of all other tensile tests. As we can see on the graph below, these values are combined with the equivalent strain value, obtained by calculation after each drawing pass, to define the stress – strain curve - all this work is based on the hypothesis that at each step of the process the material remains homogenous.

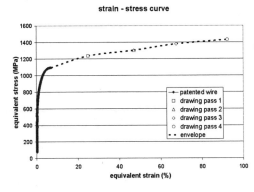

strain - stress curve

FIGURE 1. strain - stress curve obtained by
tensile tests

Thus, an envelope of the behaviour curve of the steel for large strains was defined. The most appropriate behaviour implemented in Forge2005 for this kind of simulation the elasto-plastic power law, is described below:

$$\sigma = \sqrt{3}K(1 + a\varepsilon^n) \qquad (1)$$

$\sigma$ equivalent stress, $\varepsilon$ equivalent strain, **K** constant term of the mechanical law, **a** strain hardening regularization, **n** sensitivity to strain hardening.

## Identification method

There are a lot of algorithms used in inverse analysis problems such as the problem of identifying the parameters (K, a and n) of the behaviour law. Some of the most well known are: the simplex algorithm, Newton's algorithms, and more recently the genetics algorithms. It was decided to use a Newton algorithm (easier to implement and giving good results in a small optimisation space: 3 dimension in our case). The inverse analysis method including a Newton optimisation can be summarised as follows:

FIGURE 2. Inverse Analysis diagram

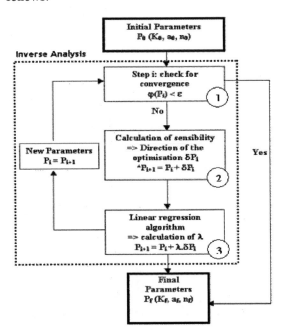

Pi is a vector representing the value of K, a and n at the iteration number i, $*P_{i+1}$ is an estimation of the value of $P_{i+1}$ before using a linear optimisation, $\varphi$ is the criterion function – in this case the least square function was used, $\varepsilon$ is the convergence threshold, $\lambda$ is the coefficient of the linear regression, $\delta Pi$ is the direction chosen at step number i to decrease the value of $\varphi$ in order to reach 0.

As a result of selecting the Newton algorithm, it is necessary to resolve the following equation at the second step of the solver:

$$0 = \varphi(*P_{i+1}) = \varphi(P_i) + \delta P_i \frac{d\varphi(P_i)}{dP_i} + o(\delta P_i) \Rightarrow \delta P_i = -\frac{\varphi(P_i)}{\dfrac{d\varphi(P_i)}{dP_i}} \qquad (2)$$

This equation gives us the direction of the optimisation at iteration number i. At this point the best solution in this direction of optimisation must be found. In order to minimise the risk of local solution, an approximation of the evolution of the criterion along the direction of decrease is made using a polynomial function whose degree can be determined before calculation.

Using this inverse analysis, the parameters of the behaviour law were determined allowing us the best approach of the envelope curve.

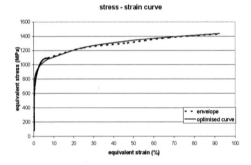

Results obtained using this method were very good. The results were validated by the modelling of the tensile tests after each simulated drawing pass. The tensile curves obtained were equivalent to the tensile curves produced by actual tensile test trials.

**FIGURE 3.** Comparison between curves obtained by tensile tests and by modeling

## Characterisation of the process parameters and simulations

Considerable work was carried out in collaboration with TréfilEurope Bourg-en-Bresse and CEMEF (Centre of Material Forming) in order to evaluate all the parameters of the process which are necessary to make the simulation.

Information regarding the shape of the dies and rolling cylinders, the speed of drawing, the rotation speed of the rolling cylinders and the reduction rate at each step of the process was provided by TréfilEurope Bourg-en-Bresse.

Other simulation parameters had to be identified. In the case of cold processes, the friction parameter for drawing and for rolling is the most sensitive. Since the process is supposed stationary, the Tresca model was selected to provide the friction between wire and dies. CEMEF took charge of the instrumentation of the dies in order to measure the drawing force. By applying this information at the third and fourth drawing pass, the drawing friction coefficient could thus be identified. This figure illustrates the identification of the friction coefficient and therefore the validation of the modelling for the drawing process.

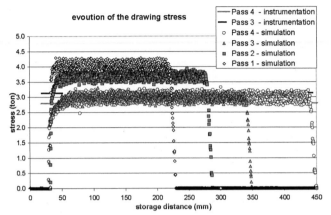

FIGURE 4. Measurements and simulation of the drawing effort

Using some samples from the rolling cylinders, bi-punching were performed at the CEMEF to define the friction coefficient for rolling. Checking final dimensions of the product allows us to validate the simulation of the process.

## Analysis of the simulations

*Prediction of stress level*

Now with this completed model for the forming of flat wires, we are able to predict the stress level inside the wire. Therefore minimising the stress level during the flat rolling process. After evaluating the residual stresses after drawing, all the process parameters, which can affect the stress level inside the wire, must be taken into consideration. As is illustrated on the following figure, maximum stresses are observed on the core of the wire and on its surface.

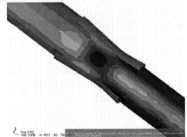

FIGURE 5. Longitudinal stresses during the drawing process (maximum stresses in black)

FIGURE 6. Von Mises stresses during the flat rolling process (maximum stresses in black)

As soon as the most sensitive parameters have been determined, the optimisation of the process can be completed. Using our model, we accomplished an improvement of the process.

*Introducing a damage criterion*

Because flat rolling creates tensile stresses in two directions, analysing only the maximum stress levels is insufficient. Therefore, a damage criterion was introduced in order to take into account the triaxiality of the stresses. So, we will have to define a maximum damage value and check that the process does not go beyond the critical value. In this case modelling will be used to modify the process.

The Mc Clintock criterion [2] was selected and implemented in Forge2005. The formula of this criterion is the following:

$$D = \int_0^\varepsilon C \left( \frac{2}{\sqrt{3}(1-n)} . \sinh\left( \frac{\sqrt{3}(1-n)}{2} \right) . \frac{\sigma_1 + \sigma_2}{\sigma_{eq}} \right) d\varepsilon \qquad (3)$$

$\sigma_1$, $\sigma_2$ and $\sigma_{eq}$ are respectively the first and the second main stresses and the equivalent Von Mises stress, $\varepsilon$ is the strain, n is the strain hardening coefficient, C is a constant to identify, D is the damage value. The interest of this formulation is to allow an elliptic deformation of the cavities contrary to the common law simulating the evolution of spherical cavities. Indeed, in our case, due to the process, elliptic cavities seem to be the best representation.

We observe that there is only one parameter to identify. Because tensile tests do not allow the simulation of multi-axial stresses these trials cannot be used for the identification of the parameter. Thus the identification of the criterion will be made from industrial trials.

As the main interest of this study is to check if the process might damage the material the damage is not coupled with the behaviour law. Indeed, as soon as the damage criterion reaches the critical value the process has to be modified. The purpose of this particular study does not involve the prediction of the propagation of cracks.

# Conclusion

In this study, we describe how to define a material law for large strains by combining drawing and tensile tests. Then we present an application with a high strain level process: drawing and flat rolling of high carbon steel grade. The trials necessary in order to identify all the parameters of this model are described in this paper. We propose a multi-axial criterion in order to evaluate damage evolution of the wire during both processes.

*References*

1. T. Balan: Parameter identification of steel behaviour laws through the simulation of wire drawing, Proc. of the International Conference on Drawing, Zakopane 2005
2. Y. Bao: A comparative study on various ductile crack formation criteria, Journal of Engineering Materials and Technology, July 2004, Vol. 126

# Research concerning the mechanical and structural properties of warm rolled construction carbon steels

## C. Medrea[1], G. Negrea[2], S. Domsa[2]

[1]*Technological Educational Institute of Piraeus, Depmartment of Physics, Chemistry and Materials Technology, 250 Thivon and P. Ralli Str, 12244, Aigaleo, Athens, Greece*
[2]*Technical University of Cluj-Napoca, Faculty of Materials Science and Engineering, ClujNapoca, Romania*

**Abstract.** Construction carbon steels represent an important steel class due to the large quantity in which it is produced. Generally, these steels are delivered in as-rolled or normalized condition heaving a ferrite-pearlite microstructure. For a given chemical composition, the mechanical characteristics of this microstructure are largely influenced by the grain size. Rolling is the deformation process which is most widely used for grain size refinement. Situated in the intermediate temperature range, warm-rolling presents certain advantages as compared to classical hot- or cold-working processes.
The paper presents a study on the microstructure and mechanical properties of Ck15 carbon steel samples warm-rolled. After deformation, the microstructure was investigated by light microscopy. Hardness measurements were made on the section parallel to the rolling direction. The mechanical properties of the steel after warm-rolling were assessed by tensile and impact tests. Additional information concerning the fracture behavior of warm-rolled samples was obtained by examining the fracture surface by scanning electron microscopy. The microstructure of the steel proved to have good mechanical properties. By considering the technologic and energy aspects, the paper shows that warm-rolling can lead to the improvement of mechanical properties of construction carbon steels.

**Keywords:** Construction carbon steels, warm rolling, microstructure, mechanical properties
**PACS:** 81.05.Bx,81.20.Hg,89.20.-a

## INTRODUCTION

Construction carbon steels are iron–carbon alloys with generally less than 0.25% C. They are used mainly for welded structures and, considering the large production scale, represent an important class of steels [1]. Construction carbon steels are supplied, generally, in as hot-rolled or normalized state. Due to their poor hardenability, the transformation during cooling from the rolling temperature or from the austenitization temperature for normalizing leads to a microstructure composed of ferrite and pearlite grains. Construction carbon steels have to fulfill a range of requirements regarding the mechanical characteristics. Among these, two appear as essential: the tensile strength and the brittle fracture strength [2]. For a given chemical composition, the mechanical characteristics depend on the microstructure. The mechanical properties of the ferrite – pearlite structure are strongly influenced by the size of the ferrite grains [3]. A range of methods to refine the ferrite – pearlite

CP907, *10th ESAFORM Conference on Material Forming,* edited by E. Cueto and F. Chinesta
© 2007 American Institute of Physics 978-0-7354-0414-4/07/$23.00

structure are applied in the industrial practice: altering of the chemical composition [4], normalizing heat treatment [5], plastic forming by controlled rolling [6] , rapid cooling [7]. The possibilities to alter the chemical composition are very restraint and the high cost of the rapid cooling equipment is limiting the industrial use of this method. The normalizing treatment gives good results only for fully deoxidized steels and increases the price of the products. Semi-deoxidized steels are suitable for controlled rolling but the restrictions it imposes are changing the manufacturing process into a very complex one. Therefore, rolling is the deformation process which is most widely used for grain size refinement The practice of rolling in the upper ferritic region instead of the austenitic region has been termed as warm rolling. Warm rolling causes significant changes on the final microstructure of the product. There have been several researches to study the process. The effect of warm rolling on the structure and properties of a low carbon steel has been investigated by Hawkins [8] and Haldar [9]. The effect of strain on the microstructure and mechanical properties of multi-pass warm caliber rolled low carbon steel have been conducted by Torizuka [10,11]. The modeling of warm rolling of a low carbon steel has been considered by Serajzadeh [12]. Ultra-fine ferrite microstructure in a warm rolled C-Mn steel has been obtained by Santos [13]. A submicron mild steel microstructure has been produced by one simple warm deformation by Liu [14]. The physical metallurgy involved during warm rolling is not yet fully understood and calls for further research activity on this process. The process is used for the manufacturing of small and middle sized parts and for a narrow range of steels. This paper presents the mechanical and structural properties of a warm rolled construction steel Ck15.

## EXPERIMENTAL DETAILS

Parallelepipedic samples with the dimensions of 20x20x200 mm were cut from normalized Ck15 carbon steel bars and subjected to full annealing (austenitizing followed by slow cooling) before rolling. The samples had the following chemical composition: 0.16%C, 0.32%Si, 0.56%Mn, 0.02%P, 0.01%S. Therefore, deformation of samples started from a coarse microstructure. The warm-rolling was performed by using a laboratory duo-reversible rolling mill under the following conditions: strain rate: 2.04 s$^{-1}$, deformation ratio: 36.4% and sample temperature: 700°C at the beginning and 550 °C, respectively, at the end of rolling[15].

After deformation, the microstructure was investigated by light microscopy on a section parallel to the rolling direction. The microstructure of warm-rolled specimens was compared with the microstructure of normalized samples, normalizing being frequently used for microstructure refinement of low carbon steels.

In order to evaluate the uniformity of deformation across the section of warm-rolled specimens, hardness measurements were made on the section parallel to the rolling direction. The results were compared with those determined on a separate group of samples, which were cold-rolled with the same deformation ratio.

The mechanical properties of the steel after warm-rolling and normalizing were assessed by tensile and impact tests. The tensile tests, performed according to standard procedure, allowed for determination of the yield strength, tensile strength, elongation and reduction in area. Additional information concerning the fracture behavior of

542

warm-rolled samples was obtained by examining the fracture surface by scanning electron microscopy. Toughness of warm-rolled and, for comparison, of normalized samples was determined by Charpy impact tests carried out at room temperature on U–notch specimens using a Charpy impact machine with a potential energy of 300 J.

## RESULTS AND DISCUSSION

Usually, the carbon steels are delivered in normalized condition which, for Ck15 consists of fine-grained ferrite–pearlite microstructure (fig. 1.a). The effects of previous plastic deformations are eliminated during normalizing, and the microstructure consists of equiaxed grains with a uniform distribution of pearlite. The samples subjected to warm-rolling display a specific microstructure with pearlite grains distributed in bands (fig. 1.b). The ferrite grains, which are more ductile, are strongly elongated along the rolling direction. The ferrite grains are subjected to strain hardening, and their grain boundaries are difficult to distinguish. The pearlite grains undergo two simultaneous processes during deformation. Firstly, the pearlite grains are divided into subgrain blocks embedded in ferrite and they appear in the microstructure as very fine pearlite grains distributed in bands parallel to the rolling direction. Secondly, inside these small pearlite grains, the cementite lamellae are broken into small fragments which will undergo a spheroidization process and take a globular shape before cooling. Thus, the pearlite grains are just slightly deformed and are broken into smaller fragments that are redistributed along the rolling bands [15].

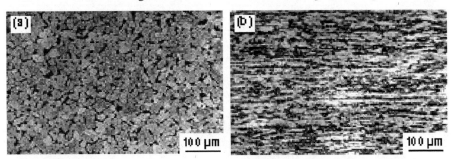

**FIGURE 1.** The microstructure of normalized (a) and warm rolled (b) samples.

Hardness measurements showed a very uniform hardness distribution in the cross–section of warm-rolled samples as opposed to the cold worked samples (Table 1).

**TABLE 1.** Hardness values of warm-rolled and cold-rolled samples: point A – in the middle of the section, point B – at half distance between A and C, point C – near the surface.

| Type of deformation | Hardness, HV20 | | |
|---|---|---|---|
| | A | B | C |
| Warm-rolling | 199 | 200 | 201 |
| Cold-rolling | 214 | 249 | 276 |

This is an indication of a uniform deformation across the section produced by warm-rolling. During warm-rolling, strain hardening and recrystallization processes take place simultaneously. Strain hardening is indicated by the vanishing of the ferrite grain boundaries and by the increase of the hardness from an initial value of 119 HV (in the annealed state) to 201 HV after warm-rolling. Strain hardening is more significant in the case of cold-rolling, from 119 HV to 276 HV. Therefore, the warm-rolled sample is partially strain hardened and partially recrystallized. The incomplete recrystallization consists in a first stage recovery (at lattice level only) of ferritic grains, a process that lowers the internal stresses induced during rolling. Due to this process, warm-rolling allows much higher deformations to be achieved as compared to cold-rolling, without the risk of cracking. If the hardness variation across the section is insignificant after warm-rolling, it reaches a value of 29 % after cold-rolling.

Table 2 shows the mechanical properties of warm-rolled and normalized samples, determined by tensile and impact tests. The results indicate that warm-rolled steel has a higher yield strength and tensile strength compared to normalized steel. However, due to partial strain hardening during warm-rolling, the elongation, reduction in area and toughness are slightly lower. Because the mechanical properties of the steel after warm-rolling and normalizing are comparable, it appears that warm-rolling can be used as an adequate process for achieving the required delivery state properties for construction carbon steels.

TABLE 2. Mechanical properties of samples after warm-rolling and in normalized condition.

| Condition | Yield strength, MPa | Tensile strength, MPa | Elongation, % | Reduction in area, % | Toughness, J/cm$^2$ |
|---|---|---|---|---|---|
| Warm-rolled | 359 | 493 | 24.7 | 59 | 162.5 |
| Normalized | 338 | 471 | 25.5 | 61.5 | 185 |

The examination of the fracture surfaces by scanning electron microscopy after tensile testing showed similar surface morphologies for both, warm-rolled and normalized samples as shown in figure 2.

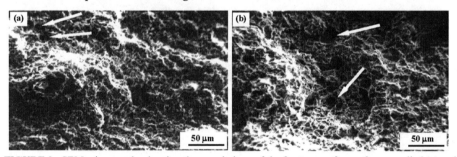

FIGURE 2. SEM micrographs showing the morphology of the fracture surfaces of warm-rolled (a) and normalized (b) samples subjected to tensile testing.

The samples display very rough intergranular fracture surfaces with the classical cup–and–cone appearance. The cup dimensions for warm-rolled specimens were considerably smaller, most probably due to strain hardening induced by warm-rolling (figure 2.a). Small cavities can be seen on the central area of the fracture surfaces (indicated by arrows). These cavities were found to be finer and more uniformly distributed on the warm-rolled sample compared to the normalized sample (figure 2.b). At the interface between the straining zone and the pulling-out one, inter-granular straining zones occurred (figure 3.a). According to the crystalline orientation of the grains, these zones show different aspects. In one grain, the straining strip appears in a "painting" shape (figure 3.b,d), while in another it looks like a wall in whose interior parallel straining strips, of variable thickness, occur (figure 3.c).

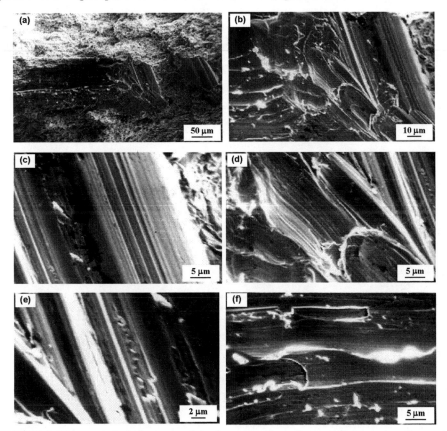

**FIGURE 3.** The morphological aspects of the fracture surface of warm-rolled samples after tensile testing.

For large magnifications, the wall shows fracture crests, as jig-saw teeth, which represents a front of edge dislocations (figure 3.e). The warm-rolled structure is reach in this kind of defects, due to a partial work hardening supported by the material during forming. The existence of the edge dislocations fronts is confirming the theory describing the spherulization of pearlite through the globulization of the cementite in its structure as a result of the motion of dislocations at the rolling temperature.

## CONCLUSIONS

Warm-rolling of Ck15 carbon steel generated a fine microstructure. The two constituents of the steel, ferrite and pearlite, behave differently during rolling. Ferrite grains, are plastically deformed and elongated along the rolling direction. The pearlite grains undergo only a slight plastic deformation. Instead, they are broken during rolling and then gain a globular appearance. After warm-rolling, pearlite grains are very small in size and are distributed in bands parallel to the rolling direction. After warm-rolling, the steel is partially strain hardened and partially recrystallized.

The mechanical properties after warm-rolling are similar to those obtained by normalizing. The warm-rolling process is of practical interest because it has certain economical advantages as compared to both, hot- and cold-working: reduction of material losses (there are no significant oxidation or decarburization processes), superior surface quality and dimensional control, and significant energy savings (lower deformation forces required and normalizing can be eliminated). In addition, warm-rolling can be applied to a broader range of steels, higher deformation can be achieved, the deformation is more uniform in the cross–section of the rolled product and the microstructure is less strain hardened.

# REFERENCES

1. I. Hrisulakis and D. Pandelis, "Hardening methods for metallic materials", in *Science and Technology of Metallic Materials*, edited by Papasotiriu Publishing House, Athens,1996, pp. 427.
2. S. Domsa and M. Bodea,C. Prica,"Design of Construction Steels" in *Design of Materials*, edited by Casa Cartii de Stiinta, Cluj-Napoca, 2005 pp.119-131.
3. A. J.DeAdro, C.L. Garcia and E.J. Palmiere," Heat Treating", in *ASM Handbook*, Vol. 4, 1991, pp. 237-255.
4. G. Vermesan, " Guide for Heat Treating" ,in *Thermochemical Treatments* , edited by Dacia Publishing House, Cluj-Napoca, 1987 , pp. 167-234.
5. T. Dulamita, et. al., "Normalizing", in *Technology of Heat Treatments*, edited by Editura Didactica si Pedagogica, Bucharest, 1982, pp. 108-115.
6. J. K. Mac Donald, "Developments in the production of notch ductile steels", *Journal of Australian Institute of Metals*, No. 5, 1965, pp. 52-58.
7. E. R. Morgan," Improved steels through hot strip mill controlled coding", *Journal of Metals*,August, 1965, pp. 829-835.
8. D. N. Hawkins and A. A. Shuttleworth ,"The effect of warm rolling on the structure and properties of a low carbon steel", *Journal of Mechanical Working Technology*, Vol.2,1979, pp. 333-345.
9. A. Hadar, R.K.Tay, "Microstructural and textural development in a extra low carbon steel during warm rolling".Materials Science and Engineering A, Vol. 391, 2005, pp. 402-407.
10. S. Torizuka, A. Ohmori, S. V. S. Narayama Murty and K. Nagai, "Effect of Strain on the Micrustructure and Mechanical Properties of Multi-pass Warm Caliber Rolled Low Carbon Steel", *Scripta Materialia*, Vol. 54, 2006, pp.563-568.
11. S. Torizuka, E. Muramatsu, S. V. S. Narayama Murty and K. Nagai, "Microstructure Evolution and Strenght-reduction in Area Balance of Ultrafine-grained Steels Produced by Warm Caliber Rolling", *Scripta Materialia*, Vol. 55, 2006, pp.751-754.
12. S. Serajzadeh, "Modeling the Warm Rolling of a Low Carbon Steel", *Materials Science and Engineering A*, Vol.371, 2004, pp.318-323.
13. D. B. Santos, R.K. Bruzszek, P.C.M.Rodriguez and E.V.Pereloma, " Formation of Ultra-fine Ferrite Microstructure in a Warm Rolled and Annealed C-Mn Steel", *Materials Science and Engineering*, Vol. 346,2003, pp.189-195.
14. M. Liu, Bi Shi, H.Cao, X. Cai and H. Song ,"A Submicron Mild Steel Produced By Simple Warm Deformation", *Materials Science and Engineering* , Vol. 360,2003, pp.101-106.
15. C. Medrea-Bichtas,I. Chicinas, S. Domsa, "Study on Warm Rolling of AISI1015 Carbon Steel", International Journal of Materials Research and Advanced Techniques,Vol. 6, 2002, pp554-449.

# Study Friction Distribution during the Cold Rolling of Material by Matroll Software

## H. Abdollahi*, K. Dehghani*

*Department of Mining and Metallurgical Engineering, Amirkabir University of Technology, Tehran, Iran

**Abstract.** Rolling process is one of the most important ways of metal forming. Since the results of this process are almost finished product, therefore controlling the parameters affecting this process is very important in order to have cold rolling products with high quality. Among the parameters knowing the coefficient of friction within the roll gap is known as the most significant one. That is because other rolling parameters such as rolling force, pressure in the roll gap, forward slip, surface quality of sheet, and the life of work rolls are directly influenced by friction. On the other hand, in rolling calculation due to lake of a true amount for coefficient of friction a supposed value is considered for it. In this study, a new software (Matroll), is introduced which can determine the coefficient of friction (COF) and plot the friction hills for an industrial mill. Besides, based on rolling equations, it offers about 30 rolling parameters as outputs. Having the rolling characteristics as inputs, the software is able to calculate the coefficient of friction. Many rolling passes were performed on real industrial aluminum mill. The coefficient of friction was obtained for all passes. The results are in good agreement with the findings of the other researchers.

**Keywords:** Matroll Software, Friction Hill, Cold Rolling, Coefficient of Friction

## INTRUDUCTION

Without the knowledge of the effectiveness of the coefficient of friction (COF), the satisfactory operation of commercial rolling facilities is very difficult to achieve. Moreover, it may become more difficult to use mathematical models of the cold rolling process for engineering purposes [1-4].

Friction itself influences rolling process or stability in other ways. Although, rolling without friction is impossible, the high value of COF is never recommended in the case of cold rolling of aluminum [1]. The development of a deep understanding of friction in cold rolling has been impaired most by the extreme difficulty of measuring frictional stresses in the roll gap. The only currently developed method of carrying out such measurements uses pressure-sensitive pins embedded in laboratory mill work rolls [5, 6]. Regardless of many correction factors that can lead to errors, further difficulties in interpreting pressure pin friction measurements arise from the fact that pressure pin tests are necessarily performed under conditions very different from those typical of production rolling. On the other hand, even using this technique, it is difficult to have any idea pertaining to the changes in friction or pressure distribution in the roll gap during real industrial rolling. Besides, one may agree that the conditions

CP907, *10th ESAFORM Conference on Material Forming*, edited by E. Cueto and F. Chinesta
© 2007 American Institute of Physics 978-0-7354-0414-4/07/$23.00

dominated in the laboratory simulations are completely different from those in industrial mill and cannot be exactly the same.

Considering the concerns (e.g. wear debris, oil contamination, roll pickup, roll coatings, roll marks, sheet stains, herring-bone defect, chattering marks, scratches, work roll life, undesirable surface quality, losing roll efficiency etc. [7,8] that directly or indirectly are affected by COF) arose from the unknown and consequently the uncontrolled friction between the work rolls and sheet, determination the amount of COF and, in turn, the friction hill are the major tasks in strip rolling industry.

The purpose of the present paper is therefore to introduce a new software (Matroll) that following determination the COF, it can offer a general assessment of the variations in friction across the arc of contact (i.e. plotting the friction hill) during the cold rolling of metal strips. This can be done in a simple way and more important for a real industrial mill. The Matroll software is established based on years of practical and experimental observations from cold rolling of metal strips especially aluminum [9].

## TECHNICAL PROCEDURE

Using a new software (Matroll), established by the authors [10], the COF values and friction hills were obtained for many passes carried out on industrial mill for aluminum alloys. The Matroll software was programmed using Delphi 6. Fig.1 shows one of the typical forms of the software out-puts offering the obtained COF (for a pass) along with some other rolling parameters. It should be mentioned that the Matroll software is consisted of 21 forms offering about 30 rolling parameters, including the plotted friction hills [10]. It considers various situations; thus, computations can be performed for different realistic or industrial ranges of rolling parameters. The results are then printed out which only COF is discussed here.

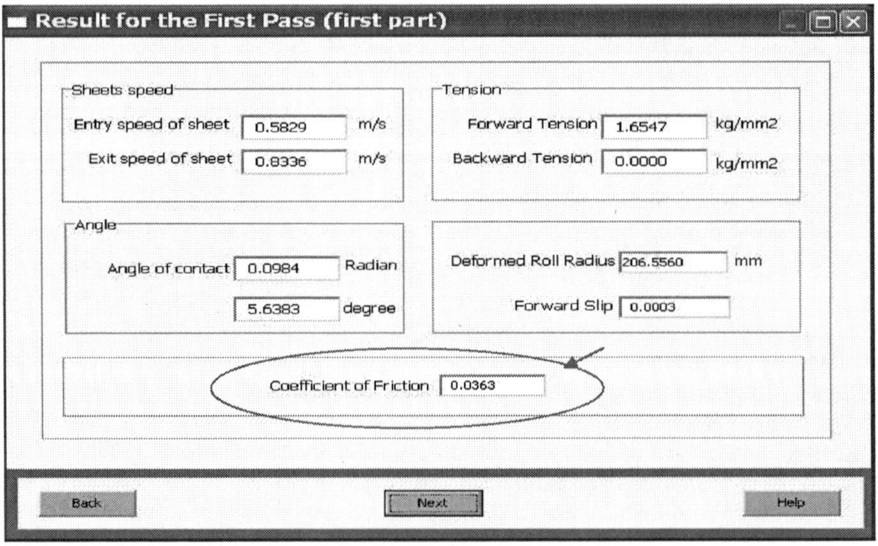

FIGURE 1.One of the typical forms of the software out-puts offering the obtained COF

As for obtaining the COF, Avitzur [11] suggests the following equation. Compared to other methods to calculate the COF [8, 12-16], this equation is believed that is more completed one, including more important parameters affecting the rolling process and COF. Avitzur's equation used in Matroll is as follows:

$$COF = \frac{\frac{1}{2}\sqrt{\frac{h_f}{R}}[\ln\left(\frac{h_f}{h_o}\right)+(\frac{1}{4}\sqrt{\frac{h_f}{R}}\sqrt{\frac{h_o}{h_f}-1})+\frac{t_o-t_f}{\frac{2}{\sqrt{3}}\sigma}]}{\left\{[(\ln\frac{h_o}{h_f}-1)\times\frac{(t_f-t_o)}{\frac{2}{\sqrt{3}}\sigma\sqrt{\frac{h_f}{h_o}-1}}]-\left[(\frac{1}{\frac{2}{\sqrt{3}}\sigma}(t_b-\frac{t_f-t_o}{\frac{h_f}{h_o}-1})-1).\tan^{-1}\sqrt{\frac{h_o}{h_f}-1}]\right]\right\}} \tag{1}$$

Where $h_o$ and $h_f$ are respectively the entry and exit thicknesses, $t_o$ and $t_f$ are also respectively the back and front tensions, R is the work roll radius and $2/\sqrt{3}\sigma$ is the flow stress of the strip being rolled.

Following obtaining the COF value by Matroll, for plotting the friction hill, the software divides the arc of contact into a lot of equal sectors in order to solve the rolling equations. The shorter the intervals or sectors, the more precise the outputs will be.

From the industrial production line, two coils (numbered 1 and 2 here) were selected. Each of them was subjected to four cold rolling passes. Various ranges of rolling parameters were employed. Both coils were aluminum alloy 3003 in a full-annealed condition. Then by using Matroll software the COF values and friction hills were obtained for eight passes carried out on industrial mills.

Using Matroll software established by the authors [12], the cold rolling process of aluminum was analyzed. The designed rolling schedules were performed on industrial rolling mills. The industrial aluminum mill had the following characteristics:

| | |
|---|---|
| Aluminum Mill: | Industrial four-high reversing mill |
| Back up diameter: | 1000 mm |
| Work roll diameter: | 400 mm |
| Lubricant: | light mineral oil with 4% additives |
| Entry and exit gauges: | 7 mm and 0.3 mm respectively |
| Rolling speed: | 0 to 400 m/min |

The cold rolling process of aluminum was then analyzed using Matroll software.

## RESULTS AND DISCUSSION

Some of the results obtained, after performing Matroll on the mentioned industrial mill, are summarized in Table1.

This is the paragraph spacing that occurs when you use the Enter key.

**TABLE 1.** Some of the results offered by Matroll software as outputs for aluminum alloy 3003

| | Coil No. 1 | | | | Coil No. 2 | | | |
|---|---|---|---|---|---|---|---|---|
| | Pass 1 | Pass 2 | Pass 3 | Pass 4 | Pass 1 | Pass 2 | Pass 3 | Pass 4 |
| Entry gauge (mm) | 6.65 | 4.65 | 3.2 | 2.3 | 6.62 | 4.8 | 3.5 | 2.35 |
| Exit gauge (mm) | 4.65 | 3.2 | 2.3 | 1.5 | 4.6 | 3.5 | 2.35 | 1.5 |
| Mill speed (m/min) | 50 | 90 | 82 | 82 | 100 | 100 | 100 | 100 |
| Back tension (kg/mm$^2$) | 0 | 1.03 | 1.25 | 2.48 | 0 | 0.90 | 0.24 | 1.82 |
| Front tension (kg/mm$^2$) | 1.67 | 0.52 | 2.41 | 1.38 | 0.93 | 0.37 | 1.92 | 1.01 |
| COF | 0.036 | 0.048 | 0.029 | 0.039 | 0.034 | 0.046 | 0.029 | 0.037 |
| Rolling load (ton) | 301.2 | 346.2 | 265.6 | 282.7 | 288.5 | 317.8 | 322.3 | 295.0 |
| Mill power (kW) | 670 | 532 | 595 | 405 | 670 | 553 | 675 | 396 |
| Exit flow stress (kg/mm$^2$) | 18.3 | 21.2 | 22.4 | 23.3 | 17.8 | 20.7 | 22.3 | 23.2 |
| Neutral point pressure(kg/mm$^2$) | 16.79 | 21.29 | 20.38 | 23.14 | 17.05 | 20.70 | 20.60 | 23.20 |
| Forward slip (%) | 0.034 | 0.122 | 0.109 | 0.34 | 0.041 | 0.101 | 0.036 | 0.270 |

Except the first three rows in Table1, the others values are the outputs of Matroll software. As already indicated, the results mentioned in Table I are only some of the software outputs.

As for the values obtained for COF in the present work, they are in good agreement with those reported by others. In case of aluminum cold rolling, Tselikov [17] found that the COF amounts fell between 0.029 and 0.069 when using different oils as lubricant. Jnuszkiewicz et al. [18] concluded that the COF amount of 0.036 for cold rolling of aluminum was a sufficiently high friction level to achieve a clean surface sheet for can body stock. R. Lazic et al. [15] obtained the COF between 0.02 and 0.06 for cold rolling of Al99.4. Depending on mill speed, Tsao et al. [19] pointed out that COF for cold rolling of aluminum could change between 0.02 and 0.07. For the same aluminum alloy, Sargent and Stawson [20] mentioned that the COF varied between 0.047 and 0.079. Kondo [8] obtained the COF values of 0.02 to 0.08 for cold rolling of aluminum.

In general, any judgment about load and pressure in rolling cannot be so valid without specifying a value for COF generated in roll throat. However, direct measurement of COF during a real industrial mill is very difficult, if not possible. In such a case, it is believed that a better procedure is to determine the COF value using a computerized program based on the well-known rolling equations, but solved by industrial inputs to verify it. This is the method applied here. While, the conventional procedure employed is to already choose a value of COF which is supposedly associated with the rolling conditions for which the load is required.

## CONCLUSION

A new software (Matroll) was introduced. The Matroll software is able to offer about 30 rolling parameters including COF and the curve of roll pressure distribution. To verify the software operation, it was performed on the real industrial mill. The COF values offered by Matroll are in good agreement with the finding of others (even better), indicating the high accuracy of software operation. Therefore, the accurate output values of about 30 important rolling parameters can be use to analyze the cold rolling process very precisely.

In addition, the software has the advantage of working both on-line and off-line, i.e. no stop in production line.

## REFERENCES

1.  Z. Y. Jiang and A.K. Tieu, "A 3-D Finite Element Method Analysis of Cold Rolling of Thin Strip with Friction Variation," Tribology International, Vol. 37, 2004, pp. 185-191.
2.  Z.Y. Jiang, A.K. Tieu, X.M. Zhang, C. Lu and W.H. Sun, "Finite Element Simulation of Cold Rolling of Thin Strip," Journal of Materials Processing Technology, Vol. 140, 2003, pp. 542-547.
3.  J. Larkiolda, P. Myllykoski, J. Nylander and A.S. Korhonen, "Prediction of Rolling Force in Cold Rolling by using Physical Models and Neural Computing," Journal of Materials Processing Technology, Vol. 60, 1996, pp. 381-386.
4.  H. Keife and C. Sjögren, "A Friction Model Applied in the Cold Rolling of Aluminum Strips," Wear, Vol. 179, 1994, pp. 137-142.
5.  Y.J. Liu, A.K. Tieu, D.D. Wang and W.Y.D. Yuen, "Friction Measurement in Cold Rolling," Journal of Materials Processing Technology, Vol. 111, 2001, pp. 142-145.
6.  D.A. Stephenson, "Friction in Cold Strip Rolling," Wear, Vol. 92, 1983, pp. 293-311.
7.  K. Komori, "Analysis of Herring-Bone Mechanism in Sheet Rolling," Journal of Materials Processing Technology, Vol. 60, 1996, pp. 377-380.
8.  S. Kondo, "Improving Rolling Stability during Cold Rolling of Aluminum," Light Metal Age, December, 1974, 5-7.
9.  K. Dehghani, "Measurement the Coefficient of Friction during the Cold Rolling of Industrial Aluminum Mill," Master Thesis, Tehran University, 1990 .
10. H. Abdollahi and K. Dehghani, "A new software (Matroll) to determine the coefficient of friction and to plot the friction hill during the cold rolling of aluminum," In the proceedings of COM 2005, Calgary, Canada, July 2005, pp. 319-332.
11. B. Avitzur, Metal Forming: Processes and Analysis, McGraw Hill Book Company, New York, NY, USA, 1979.
12. S. Jianlin and Z. Xinming, "Evaluation of Lubricants for Cold Rolling Aluminum Strips," Journal of Central South University of Technology, Vol. 4, 1997, pp. 65-68.
13. P.M. Lugt, A.W. Wemekamp, W.E.T. Napel, P. V. Liempt and J.B. Otten, "Lubrication in Cold Rolling", Wear, Vol. 166, 1993, pp. 203-214.
14. S. Zhang and J. G. Lenard, "The Effects of the Reduction, Speed and Lubricant Viscosity on Friction in Cold Rolling," Journal of Materials Processing Technology, Vol. 30, 1992, pp. 197-209.
15. R. Lazic, R. Dodok and V. Milenkovic, "Influence of Front and Back Tensions in Friction in the High Speed Cold Rolling of Aluminum Strip," Aluminum, Vol. 59, 1983, pp. 381-384.
16. L.B. Sargent, "Lubricants for Cold Rolling Aluminum Alloys- a Laboratory Appraisal," Light Metal Age, June, 1975, pp. 15-17.
17. A. Tselikove, Stress and Strain in Metal Rolling, MIR Publishers, Moscow, Russia, 1967, pp.60-80.
18. K.R. Januszkiewicz, G. Stratford and T. Ward, "Roll Bite Conditions Controlling Formation of Wear Debris during Cold Rolling of Aluminum," Journal of Materials Processing Technology, Vol. 45, 1994, pp. 117-12.
19. Y.H. Tsao and L.B. Sargent, "Friction and Slip in the Cold Rolling of Metals," ASLE Transactions, Vol. 21, 1978, pp. 20-24.
20. L.B. Sargent and C.J. Stawson, "Laboratory Evaluation of Lubricants for Cold Rolling Aluminum Alloys," Journal of Lubrication Technology, October, 1974, pp. 617-621.

# Evaluation of Friction Coefficient in Tube Drawing Processes

C. Bruni[1], A. Forcellese[1], F. Gabrielli[1], M. Simoncini[1] and L. Montelatici[2]

[1] Department of Mechanics, Università Politecnica delle Marche, Via Brecce Bianche, 60131 Ancona, Italy
[2] Dalmine SpA, Piazza Caduti 6 Luglio 1944 1, 24044 Dalmine (BG), Italy

**Abstract.** A methodology, based on a combined numerical–experimental approach, was developed to evaluate friction coefficient at the die-workpiece surface in tube drawing. In the experimental stage the upsetting sliding method, performed under contact conditions, in terms of contact normal stress and equivalent plastic strain, similar to those encountered in the tube drawing process under investigation, was used. The values of friction coefficient calculated were dependent on tube geometry.

**Keywords:** Cold Drawing, Tube, Friction Test, FE Analysis.
**PACS:** 81.40.Pq

## INTRODUCTION

The correct design of tube drawing operations requires the knowledge of the process conditions for the obtaining of tubes with the required quality, in terms of strength, dimensional and form tolerances. In order to meet these requirements, an effective and time saving approach, based on the development of FE (Finite Element) models able to simulate the cold drawing process, can be used.

The effectiveness of the FE models can be improved by developing accurate methodologies for the determination of frictional conditions at the tool-workpiece interfaces. Several methods were developed for the determination of friction in metal-forming operations, such as ring compression test, double cup extrusion test, etc [1-4]. Among them, a method very suitable in drawing processes is the Upsetting Sliding Test (UST) [5], consisting in sliding of an indenter on the specimen surface with a proper penetration. The UST is based on a combined numerical-experimental approach in order to simulate the contact conditions, in terms of contact normal stress (CNS) and equivalent plastic strain (EPS), at the tool-workpiece interfaces.

In this framework, the present investigation aims at the definition of a combined numerical–experimental methodology for the evaluation of friction coefficient at the die-workpiece surface in tube drawing using the UST method. Using an iterative procedure involving both the FE simulation and experimental UST, the amount of indenter penetration (p) to be used in the experiments was found and the resulting forces, parallel ($F_t$) and perpendicular ($F_n$) to the tube axis, were measured. The present approach is characterized by the implementation of a loop in which the

CP907, 10th ESAFORM Conference on Material Forming, edited by E. Cueto and F. Chinesta
© 2007 American Institute of Physics 978-0-7354-0414-4/07/$23.00

experimental conditions, in terms of the force perpendicular to the tube axis and of the width of the deformed zone (w), are compared to the numerical ones in order to guarantee the correct penetration of the indenter. From these data, using a proper analytical model, the friction coefficient was evaluated and the results discussed in detail.

## METHODOLOGY

The methodology used for the evaluation of Coulomb friction coefficient at the die-workpiece contact surface in tube drawing processes is shown in Figure 1.

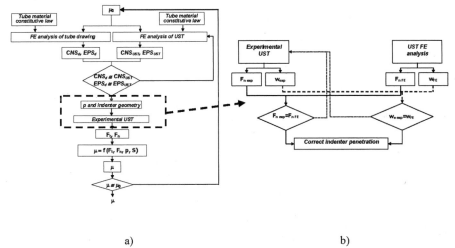

a)                                                              b)

**FIGURE 1.** Flow chart showing the methodology used for the evaluation of μ in tube drawing.

The procedure began with the FE simulation of the tube drawing process in order to identify the local CNS and EPS at the die-workpiece contact surface using an initial value of the friction coefficient ($\mu_0$) [1,5]. In this stage, a cylindrical indenter and an inclined front face indenter were considered in order to determine the geometry leading to local CNS and EPS values close to those predicted in the FE analysis of tube drawing processes. To this end, the geometry of the two indenters, in terms of the cylindrical radius (cylindrical indenter), front face angle and tool radius (inclined front face indenter), was varied. In particular, the use of inclined front face indenter (Figure 2) allowed the obtaining of lower CNS values, according to the literature [5,6], and closer those provided by the analysis of the drawing process. Then, once the choice of the front face angle and sizing length was made, the FE analysis of UST was performed using different values of the indenter penetration; the p value, to be used in the experimental UST was defined as the value whose contact conditions, in terms of CNS and EPS, match those predicted in the FE analysis of the tube drawing process (Figure 3).

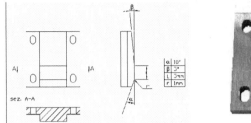

**FIGURE 2.** Inclined front face indenter geometry: α= front face angle; β=rear face angle; L=sizing length; r= tool radius.

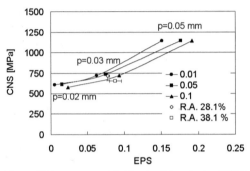

**FIGURE 3.** Typical CNS vs. EPS curves provided by the FE analysis of UST of tube B (p= 0.02, 0.03 and 0.05 mm; μ=0.01, 0.05 e 0.1) and related values obtained in the FE analysis of the tube drawing for each reduction of area (R.A. = 28.1 and 38.1 %).

The experimental UST was performed using the penetration values supplied by the FE analysis: the specimen was plastically deformed by the indenter and both the $F_t$ and $F_n$ forces were measured.

From these data the stress normal to the specimen-indenter contact surface ($\sigma_n$) and the one parallel to the same surface ($\sigma_t$) were determined as follows:

$$\sigma_n = \frac{F_t b + F_n a}{(a^2 + b^2)} \quad (1)$$

$$\sigma_t = \frac{F_t a - F_n b}{(a^2 + b^2)} \quad (2)$$

where $a$ and $b$ are given by the following relationships:

$$a = S_1 cos\alpha + S_2 \quad (3)$$

$$b = S_1 sin\alpha \quad (4)$$

with $S_1$ and $S_2$ indenter-specimen contact surfaces along the front face and along the sizing length parallel to the tube axis, respectively.

Therefore, using equations (1) and (2), the Coulomb friction coefficient was evaluated as follows:

$$\mu = \frac{\sigma_t}{\sigma_n} \quad (5)$$

By iterating the procedure using as initial the $\mu$ value provided by the first loop, until the absolute difference between the input and the output friction coefficient value was less than 10%, the Coulomb friction coefficient was obtained.

The proposed methodology, with the respect to one proposed in the literature [5,6], allows to perform a further loop (Figure 1b) in which the experimental values of the force normal to the tube axis ($F_n$) and of the width (Figure 4) of the deformed zone (w) are compared to those predicted by the FE model under the same conditions of indenter geometry and penetration, in order to impose the correct penetration in the experimental UST.

**FIGURE 4.** Typical plastically deformed zone of the specimen after UST.

## FE ANALYSIS AND EXPERIMENTAL TESTS

The drawing process of tubes, in desulphurized steel, lubricated using a zinc phosphate coating chemically combined with soap, with different geometries, under different conditions, were considered (Table 1). The constitutive equation was given by the tube manufacturer.

**TABLE 1.** Geometric characteristics of tubes used in the drawing process.

| Tube | Outer diameter (mm) | Thickness (mm) | Reduction of area (%) |
|------|---------------------|----------------|-----------------------|
| A    | 168.3               | 12.5           | 20.11, 25.30, 29.36   |
| B    | 121                 | 9              | 28.1, 38.1            |

The FE analysis of the drawing process was performed using the friction Coulomb model. The FE analysis of UST was carried out using the same friction coefficient; a quarter of cylinder (particular of YZ symmetry plane in figure 5) with an half of the indenter was considered. Along the sliding direction 100 8-nodes elements were used; 4 elements were distributed along the thickness and 14 along the circumferential direction, the element sizes being progressively smaller as the element approach to the indenter-specimen contact zone. The adaptive meshing option was used, in order to optimize the element-size in the deformed area.

The experimental USTs were performed using an indenter with the same die material (hardened and tempered steel) and surface preparation of the tube drawing process. To this end, a special device was designed (Figure 6). It was installed in a hydraulic press and it was characterized by: i) a part fixed along the direction parallel to the tube axis, in which the indenter is mounted, and ii) a part moving in the same direction in which the tube is positioned. The specimen was clamped in a vee-block.

Before the beginning of the test, a penetration of the indenter was imposed; then, the test started with the prescribed velocity. The $F_t$ and $F_n$ were measured using proper load cells.

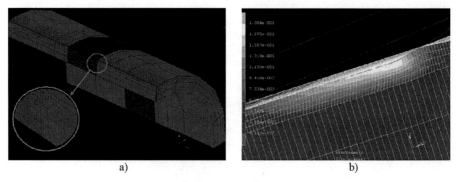

FIGURE 5. FE model for UST analysis (a) and particular of the deformed specimen (b).

FIGURE 6. Experimental equipment for UST.

## EVALUATION OF FRICTION COEFFICIENT

The optimization of the indenter geometry and the penetration were obviously done before the beginning of the experimental tests and, therefore, before identifying the friction coefficient. Therefore, the effect of friction coefficient on CNS and EPS was studied in depth as shown in Figure 3. In particular, a negligible effect of friction coefficient on CNS was obtained, whilst, as expected a more evident dependence of EPS on $\mu$ was provided at different penetrations. At constant $\mu$, an increase in CNS and EPS was observed with increasing penetration. Such results are in agreement with those obtained by other authors under similar conditions [5,6]. The friction coefficient values, computed using the equations from (1) to (5) and experimental $F_n$ and $F_t$ (Figure 7) are reported in Table 2. It is important to underline that for both the tube geometries the penetration of 0.03 mm allowed the obtaining of surface contact conditions, in terms of CNS and EPS, closer to the target values reported in Figure 3 than those provided using deeper penetrations. The calculated friction coefficients were used in the FE model of tube drawing in order to compare the numerical drawing force with that obtained during experimental tube drawing.

**TABLE 2.** Friction coefficients for different tube geometries.

| Tube geometry [mm] | Penetration [mm] | $\mu=\sigma_t/\sigma_n$ | Standard deviation |
|---|---|---|---|
| 168.3 x 12.5 (A) | 0.03 | 0.049 | 0.005 |
| 121 x 9 (B) | 0.03 | 0.032 | 0.006 |

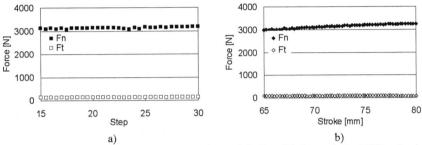

a)                                          b)

**FIGURE 7.** Typical numerical (a) and experimental (b) $F_t$ and $F_n$ forces during UST under the same conditions.

## CONCLUSION

A methodology, based on an experimental-numerical approach, was used to evaluate the friction coefficient at the die-workpiece contact surface in tube drawing. Typical values of friction coefficient obtained are equal to 0.049 with a standard deviation of 0.005 for the tube A and equal to 0.032 with standard deviation of 0.006 for the tube B.

## ACKNOWLEDGMENTS

Authors would like to thank Tenaris for funding this research. Dr. M. Pieralisi of the Università Politecnica delle Marche is also acknowledged for his help during experimental work.

## REFERENCES

1. K. Lange, "Tribology" in Handbook of Metal Forming, edited by Kurt Lange, Publisher McGraw-Hill Book Company, 1985, pp. 6.1-6.26.
2. O. Cuvalci, H.Sofuoglu, A. Ertas, *Tribology International* **36**, 757-764 (2003).
3. A.I. Obi, A.K. Oyinlola, *Wear* **194**, 30-37 (1996).
4. A. Forcellese, F. Gabrielli, A. Barcellona and F. Micari, *J. Mater. Process. Tech.* **45**, 619-624 (1994)
5. A. Dubois, J. Oudin and P. Picart, *Tribology International* **29-7**, 603-613 (1996).
6  A. Dubois, D. Patalier, P. Picart and J. Oudin, *J. Mater. Process. Tech.* **62**, 140-148 (1996).

# Effect Of Die Design On Strength And Deformability Of Hollow Extruded Profiles

## L. Donati[a], L. Tomesani[a]

[a] *Department of Mechanical Construction Engineering (D.I.E.M.), University of Bologna,*
*2 V.le Risorgimento, Bologna, 40136, Italy,*
*lorenzo.donati@mail.ing.unibo.it (corresponding author), luca.tomesani@mail.ing.unibo.it*

**Abstract.** In this paper, the behaviour of two different die design concepts in the production of the same industrial profile is described by both experimental tests and numerical FEM analysis. The investigated parameter is the shape of the leg for the extrusion of an AA6060 round tube 80mm in diameter and 1mm in thickness. Press loads and profile temperatures were monitored during the production. The processes are simulated by means of Deform 3D fully coupled simulations. and the results compared with the experimental ones. The profile are characterized by means of tensile tests in order to evaluate strength and deformability of the joints, and the results are related to the welding parameters as they were obtained by FEM. Finally, the die stress is investigated in order to explain tool life behaviour of the dies as it was observed during their whole life.

**Keywords:** Extrusion; AA 6060; Die design; Seam welds; Profile strength; Die life.
**PACS:** 81.20.Hy

## INTRODUCTION

In the last years, requirements for mechanical and aesthetical properties of aluminium extruded profiles have continuously increased: after early requirements for high strengths, precision and quality, extruded products must now guarantee high deformability, which is needed to prevent fracture occurrence in further processing operations such as bending or hydroforming [1]. The only way to join these requirements with ever lowering costs is to increase production rates without losing strength, deformability, precision and aesthetics.

This tendency is most clear in the extrusion of 6xxx aluminium series where complex shapes are produced at very high process rates. Such goal can be obtained only by optimizing the whole extrusion cycle, and in particular by carefully evaluating the early phase of die conception [2]. In fact, it is known that, at the shop floor level, hand-correction and process optimization (billet preheating temperature) can increase the process speed up to 15-20% with respect of the first trial. On the other hand, it has been shown [3] that with proper die design the extrusion speed can be raised up to three times those obtained in a typical configuration.

One of the major issue about die design is the effect of the leg design on the material flow in hollow profile production. In fact, the leg represents an obstacle to the

CP907, *10th ESAFORM Conference on Material Forming,* edited by E. Cueto and F. Chinesta
© 2007 American Institute of Physics 978-0-7354-0414-4/07/$23.00

material flow towards the exit, which locally modifies the distribution of strain and strain rate. This has consequences at different levels:

- in the maximum pressure in the welding chamber, which is related to the strength and deformability of the seam welds on the final product;
- in the exit speed difference of different points of the profile, which is related to tensile stresses at the die exit and the related occurrence of profile tearing when production speed are increased;
- in the tool life, inasmuch it depends on different failure mechanisms of the leg, principally related to their maximum stresses and creep resistance.

In order to understand how to trade between these effects, two different concepts of leg design for the same profile section are investigated by means of FEM simulations, the results being compared to the profile production results and to the mechanical testing campaign performed on the final product.

In the extrusion trials, AA6060 round tubes of 80mm diameter and 1mm thickness were extruded with the two dies at increasing production rates until the press limit was reached. The processes were then simulated by means of Deform 3D fully coupled simulations. Experimental extrusion loads and temperatures were then compared to the simulated ones.

The profile welds were characterized by means of tensile tests in order to evaluate strength and deformability of the joints, and the results were related to the welding parameters as they were obtained by FEM. Finally, the differences in die stress were investigated in order to explain the differences in tool life behaviour of the solutions.

## EXPERIMENTAL

In this paper the production of an AA6060 tube with φ 81 mm outer diameter, 1mm thickness, and a global extrusion ratio of 117 has been analyzed. The profile has been

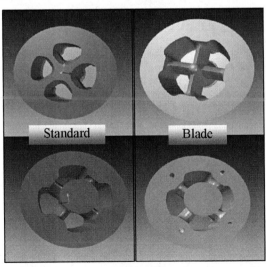

**FIGURE 1.** Standard design (left) and blade one (right); lower, back side of the die.

produced on an industrial 1600 ton. press, with two different die concepts: a standard design and a new one, named as 'blade' (fig. 1). The solutions are characterized by the same overall dimensions (welding chamber height, bearing length, etc) and materials (AISI H-13), but with different shapes of the legs. The 'blade' one (fig. 1 right) is characterized by a more gradual impact surface: this effect is obtained by shifting the legs towards the exit and, at the same time, by introducing a central splitter that contacts the deforming material upstream of the leg. Moreover, in the new solution, the leg has a sharper incoming edge and a bit thicker back side (fig. 1 bottom).

The two dies were pre-heated at 440°C and billets (152 mm in diameter, 425 mm in length) with a taper of 430-485°C were used; three billets for each die were extruded, by varying the profile speed between 35 and 60 mt/min with the blade die and between 35 and 72 mt/min with the standard one. No remarkable defects were found on the profiles in both cases, even at the maximum production speed.

During the tests the ram loads were 1250 [t] for the blade and 1293 [t] for the standard, both evaluated at 60 mt/min and full billet length; the blade solution thus evidencing a 3.5% load reduction with respect to the standard one.

The exit profile temperatures, monitored 2 mt after the die exit with K-type thermocouples were found to be 573°C for the blade and 560°C for the standard one, both approximately constant throughout the profile length at the maximum exit speed.

The extruded tubes, after sectioning, were tested by means of tensile tests (3 repetition for each condition): UTS (Ultimate Tensile Strength) and elongation at fracture were recorded (figure 3 left). UTS was found to be 180 MPa in the standard design against 170 MPa in the blade one, thus 5% smaller than with the standard design; both values were found to be almost independent from process speed.

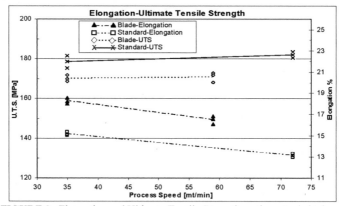

**FIGURE 2.** Elongation and Ultimate Tensile Strength on the extruded tubes.

The elongation at fracture evidenced a much more interesting behaviour: it was found to vary from 18% and 16% for the blade die in the 35-60 mt/min range and from 15% and 13% for the standard die between 35-70 mt/min. The increase in deformability of the blade solution is remarkable, being of about 20%, in particular if further processing is required on the final product. Also interesting is the sensitiveness of elongation to production rates: in both cases the increase in process speed of about

100% (from 35 mt min to 60-70 mt/min) determines a decrease of elongation of about 12% on the final product.

## FEM ANALYSIS

The complete thermo-mechanical FEM simulation of the two dies had been performed with the code Deform 3D; the flow stress curve of the deforming material was obtained by hot torsion tests [4] where the effect of strain, strain rate and temperature had been considered. Each simulation was made with approximately 250.000 elements, and at least three elements were placed inside the final profile thickness.

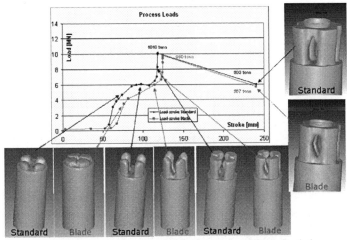

**FIGURE 3.** Extrusion loads predicted by means of FEM simulation.

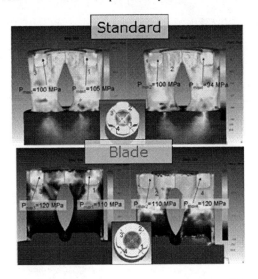

**FIGURE 4.** Welding pressures in the welding chambers of the 2 dies.

Friction was considered as sticking throughout the die and container, except on the bearing zones where a sliding model was used [5]. Heat transfer and heat generation by both friction and plastic strain were also considered.

Validation of the model was performed by extrusion loads and exit temperatures; maximum loads (fig. 3) were predicted in close agreement with experimental results and also the exit temperatures were found to be within 20°C to the experimental ones.

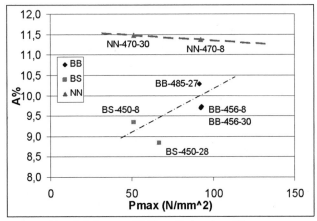

**FIGURE 5.** Seam Weld prediction chart [6].

The pressure in the welding chamber, which can be used as a useful parameter for assessing the quality of the seam weld, was around 100 MPa with the standard design, while it reached 120 MPa in the blade one (fig.4). Correspondingly, deformability increases from 15% to 18%, as explained. This result is in accordance to a previous work [6] where the increase in deformability of a different profile (of the same alloy but of different extrusion ratio) was obtained by increasing the pressure in the welding chamber from 50 to 100 MPa (fig.5).

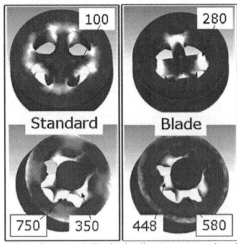

**FIGURE 6.** Von Mises die stress (MPa) in the dies (upper: top view, lower: bottom view).

Finally, the evaluation of the stress on the die was performed by means of Deform die stress module. In the standard solution a maximum equivalent stress of 750 MPa was found (fig.6) on the back side of the leg next to the central mandrel (due to the thinning of the leg); the mean stress throughout the leg was around 350 MPa, while the stress on the incoming edge was very low (about 100MPa). On the blade die a 23 % lower peak stress value is found (580MPa instead of 750MPa), but the mean stress is generally higher (448MPa versus 350MPa in the rear edge, 280MPa instead of 100MPa in the front). Correspondingly, the blade die life was 30% less than the standard one, probably due to creep-dependent failure mechanisms; this influence is at the moment under investigation.

## CONCLUSIONS

The effect of the leg shape in the extrusion process of AA6060 aluminium alloy is presented here. Two different die design concepts for the production of the same profile were tested: the "blade" shape and the standard one. The results were evaluated in term of process loads, profile resistance and die life and were discussed by means of FEM simulations. The "blade" solution showed a better elongation at fracture in the final product, as a consequence of the higher welding pressure. Die stresses in the "blade" shape gave 23% lower peak stress but an increase of 25% of the mean stress, thus increasing creep sensitivity of the die.

## ACKNOWLEDGMENTS

This work was carried out with the collaboration of Compes S.p.A. and Sepal -BS- (Italy), which are acknowledged.

## REFERENCES

1. Imaninejad M., Subhash G., Loukus A., "Influence of end-conditions during tube hydroforming of aluminum extrusions", Int. J. Mech. Science, 46 (2004), 1195-1212;
2. L. Donati, S. Andreoli "The die in extrusion: influence of the design choices on the performance of extruded profiles for automotive applications" Light Metal Age, April 2006, vol 64 n.2, pagg.28-33.
3. L. Donati, L. Tomesani, "3D FEM analysis of seam welds formation in aluminum extrusion and product characterization" Proceedings of 8th Esaform conference, Cluj Napocha –R- (2005) pagg. 561-564 vol. II.
4. Bariani P., Bruschi S. (Private communication);
5. I. Flitta, T. Sheppard, "Nature of friction in extrusion process and its effect on material flow", Int. J. of Materials Science and Technology, July 2003, vol. 19, pagg. 837-846.
6. L. Donati, L. Tomesani, "Extrusion welds in hollow AA 6060 profiles: FEM simulation and product characterization" Proceedings of 8th ICTP conference, Verona (2005), pag. 227-228.

# FE Simulation of Ultrasonic Back Extrusion

Malgorzata Rosochowska and Andrzej Rosochowski

*Department of Design, Manufacture and Engineering Management, University of Strathclyde,
Glasgow G1 1XJ, United Kingdom*

**Abstract.** The main benefit of using ultrasonic vibrations in metal forming arises from the reduction in the mean forming force. In order to examine mechanisms responsible for this effect FE simulations of ultrasonic back extrusion using ABAQUS/Explicit were carried out. In two analysed models, vibration of frequency of 20 kHz was imposed on the punch. In the first model, the die and the punch were defined as rigid bodies and in the second, the punch was modelled as an elastic body, this being the innovative feature of the research. The punch vibrated in a longitudinal mode. Simulations were performed for amplitude of vibrations of 8.5μm and different punch velocities for both friction and frictionless conditions. Results showed that the amplitude and the mean forming force depended on the process velocity. Further, the decrease in the mean forming force might be partly explained by the reduction in the friction force due to changes in the direction and magnitude of the frictional stress over the vibration period. A lower deflection of the elastic punch under oscillatory conditions was observed, which was an indirect evidence of the reduced forming force. It was also observed that amplitude of vibrations at the working surface of the elastic punch was smaller than the applied one.

**Keywords:** Back Extrusion, Ultrasonic Forming, FE Simulation.
**PACS:** 81.20.Hy, 43.25.Gf, 02.70.Dc.

## INTRODUCTION

The benefit of using ultrasonic vibrations in metal forming processes refers to the reduction in the forming force and to the better surface quality. The reduction in the forming force is attributed to the reduction in the flow stress of the work material and to the changes in friction at the interface between the vibrating tool and work material. Superposition of static and oscillatory stress, a mechanism proposed by Nevill and Brotzen [1], and material softening resulting from heat generated as suggested by Sigert and Ulmer [2] are responsible for the reduction in flow stress. Changes in friction may result from the changes in the coefficient of friction and/or changes in the magnitude and direction of the friction stress. The latter mechanism was theoretically investigated by Mitskevich [3] and subsequently verified experimentally by Sigert and Ulmer [4].

The influence of ultrasonic vibration on various metal forming processes has been experimentally investigated by several researchers over the last few decades. However, the extent to which each of the described mechanisms affects a particular process remains unclear; these mechanisms occur simultaneously and experimental techniques for isolating these influences have yet to be developed. Consequently, numerical analysis is a useful tool for acquiring a better insight into the phenomena occurring in ultrasonic metal forming. Akbari Mousavi at al. [5] used FE analyses to

CP907, *10th ESAFORM Conference on Material Forming*, edited by E. Cueto and F. Chinesta
© 2007 American Institute of Physics 978-0-7354-0414-4/07/$23.00

estimate influence of process parameters on the mean forming force in forward extrusion. A similar analysis was conducted for wire drawing by Hayashi at al. [6] and for wire flattening by Rosochowska [7]. Hung at al. [8] used FE simulations to investigate the influence of ultrasonic vibrations on upsetting of a model paste. In order to match the calculated mean force with that obtained experimentally they assumed, in the developed FE thermo-mechanical model, a quarter-scale coefficient of friction for the condition in which ultrasonic vibrations were applied. A similar approach has been adopted by Daud at al. [9] to interpret experimental results of the compression of aluminium specimens while being subjected to ultrasonic vibrations.

This paper reports on the influence of ultrasonic vibrations on back extrusion.

## FE MODELS

Two different 2D axisymmetric models were used to conduct FE simulations. In the simplified model, the die and the punch were defined as rigid bodies (Fig. 1a). The punch was subjected to vibrations of 20 kHz and amplitude of 8.0 μm by imposing a boundary constraint to the tool in the form of a sinusoidal displacement $A = A_o \sin(\omega t)$. In the second, more realistic model (Fig. 1b), the punch was regarded as an elastic body. The significant innovative feature of this model was that the punch was excited to vibrate in the longitudinal mode. To achieve this, the length of punch was designed to be of a half wavelength to sustain the resonant condition. The punch was forced to vibrate by applying sinusoidally changing pressure $p = p_o \sin(\omega t)$ to the surface shown in Fig.1b. In order to create vibration amplitude of a pre-specified magnitude, the damping properties of the punch material had to be specified.

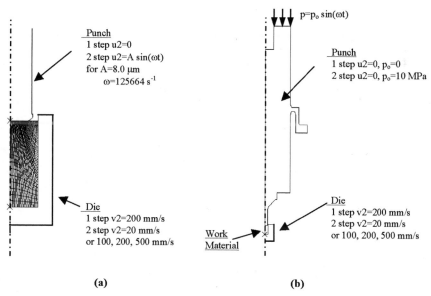

**FIGURE 1.** FE model for the rigid tools (a) and the elastic punch (b).

Simulations were performed in two stages. During the first stage, the work material was deformed without punch vibration by moving the die at a constant velocity of 200 mm/s; this continued until the steady-state phase of the process was established. Subsequently, in the second stage, the punch was excited at prescribed frequency and amplitude. Simulations were carried out for die velocities of 20, 100, 200 and 500 mm/s for both, the frictionless and frictional condition with friction coefficient of 0.1.

The aluminium test specimen of dimensions $\Phi = 5.0 \times 7.5$ mm was modelled as an elastic-plastic, isotropic, Huber-Mises material with Young's modulus 69000 MPa, Poisson's ratio 0.3 and a strain-hardening curve $\sigma = 159(0.02 + \varepsilon)^{0.27}$ MPa. The punch diameter was $\Phi = 4.0$ mm.

## RESULTS AND DISCUSSION

For a die velocity of 100 mm/s, the forming force derived using the model with rigid tools is shown in Fig. 2a. Figure 2b shows the average forming force for different die velocities. The curves were computed using an algorithm for moving average; ten successive data points were used to calculate the average. The relative forming force, defined as a ratio of the average force to the force under static conditions, was 0.42, 0.56, 0.67 and 0.87 for die velocities of 20, 100, 200 and 500 mm/s, respectively. Note

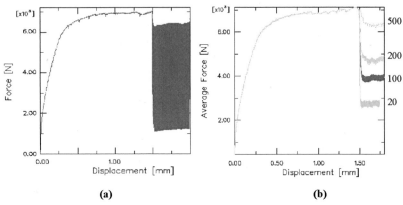

(a)                                    (b)

**FIGURE 2.** Variation of the forming force under frictional condition for a process velocity of 100 mm/s (a), the average forming force for a process velocity of 20, 100, 200 and 500 mm/s (b).

that the maximum value of the oscillating forming force is lower than the force under static condition. The difference between the forming force under static condition and the maximum forming force when the tool was subjected to vibrations, decreased with increasing die velocity. A possible explanation of this result is that the vibration reduced the friction force. In order to verify this hypothesis, FE simulations were conducted for the frictionless condition. The results for velocities of 200 and 20 mm/s are shown in Fig. 3. To enable comparison, this figure also contains the results obtained for simulations with friction. Under the frictionless condition, the maximum oscillating forming force was the same as the forming force under static conditions.

Therefore, it may be concluded that the superposition of vibrations reduces the friction force.

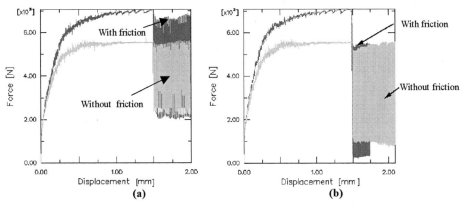

**FIGURE 3.** Force versus displacement for a friction coefficient 0 and 0.1 and a process velocity of 200 (a) and 20 mm/s (b).

Figure 4 shows variation of the forming force and the punch velocity with time when the die velocity was 20 mm/s. The force decreased when the difference between the punch and the die velocity was positive; this means that the punch moved away from the work material at a velocity higher than the velocity of the die. Consequently, the work material was temporarily unloaded. Figure 4b shows that the magnitude and direction of the shear stress (at instances 1, 2, 3, 4 and 5 in Fig. 4a) at the material/punch interface varied over the vibration cycle. Thus, in addition to the unloading effect, another mechanism responsible for the reduction in the forming force in back extrusion is the "friction vector effect".

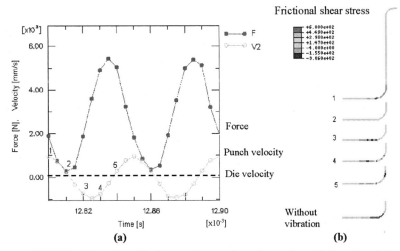

**FIGURE 4.** Variation of the forming force and punch velocity with time (a) and the frictional shear stress at the material/punch interface (b) for a process velocity of 20 mm/s.

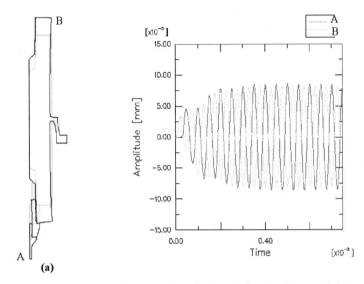

**(a)**

**FIGURE 5.** Mode of vibration (a) and amplitude at both ends of the punch (b).

A cyclic pressure of 10 MPa was used to excite the elastic punch to vibrate at an amplitude of 8.0 μm. Figure 5(a) shows the mode of vibration of the punch and Fig. 5(b) shows the amplitude of vibration of punch surfaces A and B when no load was applied. During extrusion, the amplitude of surface B remained unchanged while that of surface A, which was in contact with the work material, was reduced to 2.5 μm (see Fig.6a); further, there was not noticeable dependence of this amplitude on the die velocity. However, it depended on the load; for a steel work material this amplitude was 1μm. Figure 6b shows that vibration decreased deflection of the punch and, that this decrease depended on die velocity. Since deflection depends on the acting force, it was used to estimate the average forming force and its relative value with reference to

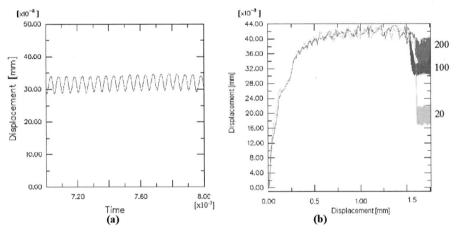

**FIGURE 6.** Amplitude of vibration of surface A for die velocity 100 mm/s (a) and deflection of the punch for die velocity 20, 100 and 200 mm/s (b).

568

the force under static condition. The latter was estimated to be 0.45, 0.78, 0.88 and 0.98 for the die velocity 20, 100, 200 and 500 mm/s. The relative forming forces for both models are shown in Fig. 7.

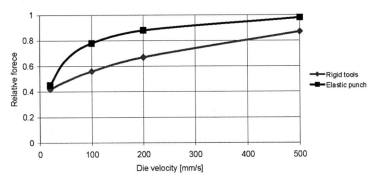

**FIGURE 7.** Relative force versus die velocity.

## CONCLUSIONS

The FE model based on an elastic punch provides a better representation of the actual forming conditions. Ultrasonic vibration was introduced by inducing the longitudinal standing wave in the punch. The amplitude of vibration was not constrained kinematically and depended on the process; the applied force reduced the amplitude from 8.0 μm to 2.5μm. The smaller deflection of the vibrating punch is an indirect evidence that vibrations reduced the forming force. It was shown, that in addition to the stress superposition, the reduction in the forming force in back extrusion can be explained by the changes in direction and magnitude of the friction stress.

## REFERENCES

1. G. I. Nevill, F. R. Brotzen, "The effect of vibration on the static yield strength of low-carbon steel", *Proc. Am. Soc. Test Material*, **57**, pp. 751-758 (1957).
2. K. Sigert, J. Ulmer, "Superimposing ultrasonic wave on tube and wire drawing", *Advanced Technology of Plasticity*, **III**, Proceedings of the 6th ICTP, pp. 1763-1774 (1999).
3. A. W. Mitskevich, "Motion of the body over a vibrating surface", *Soviet Physics - Acoustic*, **13/3**, pp. 348-351 (1968).
4. K. Sigert, J. Ulmer, "Reduction of sliding friction by ultrasonic wave", *Production Engineering*, **V/1**, pp. 9-12 (1998).
5. S. A. A. Akbari Mousavi, H. Feizi, "Simulation of extrusion process using the ultrasonic vibration", *Proceedings of the 9th International Conference on Materials Forming ESAFORM 2006*, Glasgow, UK, April 2006, pp. 499-500.
6. M. Hayashi, M. Jin, S. Thipprakmas, M. Murakawa, J. C. Hung, Y. C. Tsai, C. H. Hung, "Simulation of ultrasonic-vibration drawing using the finite element method (FEM), *Journal of Materials Processing Technology*, **140**, pp. 30-35 (2004).
7. M. Rosochowska, Influence of ultrasonic vibration on plastic flow of materials, PhD Thesis, 2003.
8. Z. Huang, M. Lucas, M. J. Adams, "Influence of ultrasonics on upsetting of a model paste", *Ultrasonics* **40**, pp. 43-48 (2002).
9. Y. Daud, M. Lucas, Z. Huang, "Superimposed ultrasonic oscillations in compression test of aluminum", *Available online at www.siencedirect.com*, *Ultrasonics*, **44**, Supplement 1, pp. e511-e515 (2006).

# On The Evaluation Of The Through Thickness Residual Stresses Distribution Of Cold Formed Profiles

Barbara Rossi[1], Anne-Marie Habraken[2] and Frederic Pascon[3]

[1]Research Fellow - National Foundation for Scientific Research - University of Liège – Division MS²F
of ArGenCo department - barbara.rossi@ulg.ac.be
[2]Research Director - NFSR - Ulg - MS²F - anne.habraken@ulg.ac.be
[3]Scientific Research Worker - NFSR - Ulg - MS²F - f.pascon@ulg.ac.be

**Abstract.** The aim of this research is to evaluate the through thickness residual stresses distribution in the walls and in the corners of a cold-formed open section made of a material presenting a non linear hardening behaviour. To get results as close as possible to the reality, the complete process is modeled, including coiling and uncoiling of the sheet before the cold bending of the corner itself. The elastic springback after flattening as well as after final shaping are also taken into account. In order to validate the model in predicting the residual stresses distribution, the presented results are confronted to experimental measurements and FE results collected from the literature.

**Keywords:** Residual stresses, Process, Nonlinear hardening behaviour, Cold-formed

## INTRODUCTION

In the design of structural cold-formed steel members, residual stresses count for a lot and may be decisive in the evaluation of the ultimate load in steel structures. The current paper presents theoretical equations aiming at evaluating the through thickness residual stresses distribution in cold-formed members due to coiling, uncoiling and cold braking of the corners. Through these equations, the effect of the radius of coiling and folding on the stresses and on the final radius after springback can be examined for carbon steel as well as for material presenting a non linear hardening behaviour as stainless steel. This is of great interest for the analysis of ultimate load in structural cold-formed members whilst compressive residual stresses cause a direct redistribution of the stresses during the loading and alters the behaviour in carrying.

CP907, 10ᵗʰ ESAFORM Conference on Material Forming, edited by E. Cueto and F. Chinesta
© 2007 American Institute of Physics 978-0-7354-0414-4/07/$23.00

# COMPUTATIONS OF RESIDUAL STRESSES

## Coiling, Flattening And Springback

In this section, the equations proposed by W.M. Quach et al. (4) are modified in order to add the non linear hardening behaviour. Due to the complexity of the theoretical solution, MATLAB is used to solve the equations numerically.

It is assumed that the flat sheet is free of stresses before its storage thanks to the annealing step. In this modified approach, the steel can exhibit an isotropic non-linear hardening behaviour. The longitudinal direction, denoted x, is the coiling direction, y corresponds to the thickness of the sheet and z is the transverse direction. As the width of the plate is assumed to be large enough to disregard the presence of transverse strains, the problem can be summarized as the *plane strain pure bending of a sheet in the x-y plane*.

**FIGURE 1.** Plane strain pure bending of a sheet in the x-y plane

### *Theoretical Model*

The procedure followed in these investigations is quickly described here. Hooke's law is chosen to characterise transversely isotropic material behaviour and one needs a longitudinal and transverse Young's modulus E and $E_t$ as well as Poisson's ratio to describe it. As soon as $\varepsilon_z$ and $\gamma_{yx}$ are equal to zero, the incremental part of the strain field $\underline{d\varepsilon}$ and the corresponding stress field $\underline{d\sigma}$ are given by:

$$\underline{d\varepsilon} = \begin{bmatrix} d\varepsilon_x \\ d\varepsilon_z \end{bmatrix} = \begin{bmatrix} d\varepsilon \\ 0 \end{bmatrix} = \begin{bmatrix} d\kappa.y \\ 0 \end{bmatrix} \tag{1}$$

$$\underline{d\sigma} = C_e^{-1} \underline{d\varepsilon} - C_e^{-1} \underline{d\varepsilon}^p \tag{2}$$

One assumes that $\underline{d\varepsilon} = \underline{d\varepsilon}^e + \underline{d\varepsilon}^p$ .and $\underline{C_e} = \begin{bmatrix} 1/E & -v_t/E_t \\ -v_t/E_t & 1/E_t \end{bmatrix}$ and $\underline{d\varepsilon}^p = \begin{bmatrix} d\varepsilon_x^p \\ d\varepsilon_z^p \end{bmatrix}$ is the unknown part of the problem.

Since plastic strains occur, Hill's quadratic flow surface $F_p(\underline{\sigma}, \underline{X}, \alpha_k, \underline{L})$ and the corresponding consistency equation have to be fulfilled.

$$F_p = \sigma_{equ} - \sigma_F = 0 \tag{3}$$

$$\frac{d\varepsilon^p_x}{d\varepsilon^p_z} = \frac{\dot{\lambda}\left(\frac{\partial F_p}{\partial \underline{\sigma}}\right)_x}{\dot{\lambda}\left(\frac{\partial F_p}{\partial \underline{\sigma}}\right)_z} = \frac{(\underline{L\sigma})_x}{(\underline{L\sigma})_z} \tag{4}$$

with $\sigma_{equ} = \sqrt{\frac{1}{2}(\sigma)^T \underline{\underline{L}} (\sigma)}$; $\underline{\underline{L}}$ is the tensor of the anisotropic parameters :

$$\underline{\underline{L}} = \begin{bmatrix} G+H & -H \\ -H & H+F \end{bmatrix}.$$

The numerical solution is thus summarised in a system of two equations with two unknowns ($d\varepsilon_x{}^p$ and $d\varepsilon_z{}^p$) and is solved in MATLAB. One divides the thickness into several layers; at each step of the numerical computation, the curvature increases and the system of equations is solved; $\sigma_F$ is described by Swift law $\sigma_F = K(\varepsilon_0 + \varepsilon_{equ}^p)^n$ (chosen for representing the non linear hardening behaviour where $\varepsilon^p{}_{equ}$ is the equivalent plastic strain), and is re-evaluated at each step (according to the increased $\varepsilon_{equ}^p$) and the flow surface $F_p$ grows up isotropically. Under reverse yielding, one supposes isotropic hardening behaviour in each case developed in this work. For further information, see (3) and (4).

## Finite Element Model

A finite element model of the bending of a steel sheet has also been conducted. The boundary conditions of the finite element model are depicted below.

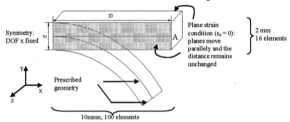

**FIGURE 2.** Boundary conditions for the FE model

The displacements are directed by two lines; *the neutral axis* of the sheet (x direction) is located at the half-width and fits an arc of circle (the radius of which being variable during the coiling); *the right edge* (y direction) remains all the time perpendicular to the neutral axis at the point A. The plans in the z direction are constrained to follow the displacement parallely with respect to the hypothesis of the problem. For symmetry reasons, only one half of the sheet is modelled. The present model has been implemented in the LAGAMINE finite element code, which has been developed at the University of Liège for more than 20 years. For more information see (2).

## Residual Stresses Distribution

In the current paper, the so-called DP1000 (dual phase) steel has been chosen. This steel presents a non-linear hardening behaviour and authors (7) give material parameters of Swift law in order to characterise the hardening behaviour.

## Table 1. Material parameters for DP 1000

| F | G | H | N | K | n | $\varepsilon_0$ |
|---|---|---|---|---|---|---|
| 1.051 | 1.036 | 0.925 | 3.182 | 1626 | 0.17 | 0.00487 |

$K, \varepsilon_0, n$ are material parameters determined by tensile test in the rolling direction z.

Moreover, such steel presents an anisotropic yield locus which can be described by means of Hill's 1948 quadratic equation (equ. (3)) using the material parameters given in Table 1.

MATLAB results and finite element results have been reported on the graph (**FIGURE 3**: left graph), where the left part and the right part represent respectively: *the stresses (obtained by numerical resolution) due to the coiling of the sheet; the stresses (obtained by numerical and finite element resolution) due to the coiling ($s\_c$) and the uncoiling ($s\_r$) of the shee.* In the core of the sheet (y<0.8mm), where the stresses remained elastic after coiling, the uncoiling process leaves no stresses. After the coiling and uncoiling process, the resulting moment, corresponding to the flexural stresses remaining, can be easily evaluated. Unloading via a reverse bending moment produces elastic strains and stresses corresponding to the springback of the sheet (**FIGURE 3**: right graph).

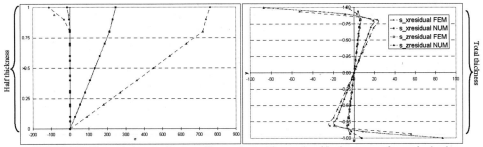

**FIGURE 3.** Flexural stresses due to coiling and uncoiling and residual stresses after springback (Rc=250mm)

## Residual Stresses Due To Press Braking

### *Theoretical Model*

In case of press braking, the sheet is set into a die and is shaped by means of a punch (press braking on a V press). First, one supposes that press braking can again be modelled as pure bending, thus the final stress distribution in the corner can easily be found out by means of the considerations developed previously. Such assumption is not far from reality if the length of the bent profile allows ignoring deformation through the longitudinal direction. Moreover, in reality, loads are applied via a punch and they could be more accurately modelled by means of a uniformly distributed internal pressure. In addition, the radius of the fold decreases a lot and the previous calculations can not be conducted anymore because of the presence of the radial (through thickness) stresses. Nevertheless, in the frame of the present research, one has chosen to keep the previous hypothesis disregarding the presence of the radial stresses.

The calculations are conducted with considering an initial state of stress (*sz_residualNUM* and *sx_residualNUM*) and plastic deformations corresponding to process preceding the press-braking. In that part of the problem, *x* indicates the longitudinal direction of the fold (initial state of stresses: *sz_residualNUM*) and *z* indicates the transverse direction (ISS: *sx_residualNUM*) as the fold is bent in a plane perpendicular to the plane of coiling-uncoiling.

## *Residual Stresses Distribution*

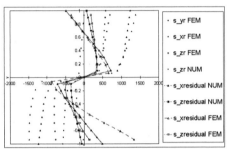

**FIGURE 4.** Residual stresses due to coiling - uncoiling -springback - bending - springback: R$_c$=250mm, R$_f$=6mm. results for a DP1000 steel.

As explained, the presence of radial stresses influences the distribution of stresses through the thickness during bending and after springback (**FIGURE 4**). Nevertheless, after springback, the distribution of stresses approached by the theoretical (MATLAB resolution) method is quite good in regards with the amplitude of stresses provided by the FE calculations.

In Hill's quadratic yield locus equation (equ. (3)) developed as equ. (5), one can notice that $\sigma_x$ and $\sigma_z$ are influenced by $\sigma_y$. Moreover $\sigma_y$ increases the equivalent plastic strain, which also modifies the value of $\sigma_F$. In the tensed part of the sheet, theoretical investigation recommends a stronger hardening while in the compressed part, the conclusion is the opposite.

$$\frac{d\varepsilon^p{}_x}{d\varepsilon^p{}_z} = \frac{\lambda\left(\dfrac{\partial F_p}{\partial \underline{\sigma}}\right)_x}{\lambda\left(\dfrac{\partial F_p}{\partial \underline{\sigma}}\right)_z} = \frac{(\underline{L\sigma})_x}{(\underline{L\sigma})_z} \tag{5}$$

Discrepancies between the calculations are also due to the distribution of the elastic stresses after springback (**FIGURE 5**). In the case of the bending of a curved sheet, this distribution is not linear as described in 8.

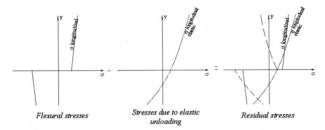

Flexural stresses | Stresses due to elastic unloading | Residual stresses

**FIGURE 5.** Residual stresses due to pure bending followed by springback: case of a small radius

## *Agreement With Experimental Measurements*

Finite element predictions given by Quach et al. (5) are confronted here with Weng and White's (6) measurements and put side by side with theoretical results in **FIGURE 6**. These results concern the cold-bending of a HY-80 steel (high yield strength, low carbon, low alloy steel with nickel, molybdenum and chromium). The HY8 steel behaviour is assimilated to a non linear one, the Swift law parameters have been fitted on a stress-strain curve corresponding to a tensile test in the longitudinal direction (see Table 2). The yield strength is equal to 593.3MPa.

Table 2. Material parameters for HY80

| K | n | $\varepsilon_0$ |
|---|---|---|
| 1035 | 0.1063 | 0.002 |

The finite element simulations model the complete press-braking process. They provide accurate results and overcome difficulties of measurement but such model requires a certain expertise and are time consuming. Additionally, in particular cases, they also suffer bad convergence and accurate results cannot be accomplished. Otherwise, theoretical results are in good agreement with the measurements in regard with the total amplitude of the residual stresses diagram.

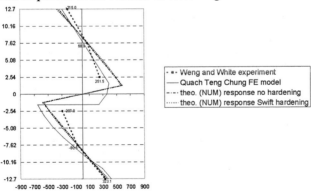

**FIGURE 6.** Residual stresses due to press-braking: Rf=139.7mm

# CONCLUSIONS

The theoretical calculations of residual stresses in cold-formed steel members including non-linear hardening behaviour are complex and require a certain number of assumptions. In this paper, theoretical (equations solved via MATLAB) and numerical (FE) calculations have been conducted to model the whole process: the coiling-uncoiling operations followed by natural springback and the press braking process also followed by elastic springback. Plane strain condition is assumed both in the coiling-uncoiling and in the press-braking process. In the theoretical calculations, no radial stresses are taken into account in the calculations. The material is supposed to be anisotropic and to obey to Hill's quadratic yield surface and the subsequent plastic flow rule. Non linear hardening behaviour is taken into account in the calculations.

The results confirm a complex distribution of flexural stresses and show good agreement with measurements collected in literature.

In cold formed column, residual stresses cause a direct redistribution of the stresses during the loading and alter the behaviour in carrying (6). Consequently, a good knowledge of the amplitude of the residual stresses and of their distribution is of great interest for the analysis of ultimate loads in structural cold-formed members. Pragmatic convention is to model residual stresses in cold formed profile as the sum of a membrane type and flexural type. This assumption is unreasonable while the calculations show more complex variation of the distribution through the thickness.

# ACKNOWLEDGEMENTS

The authors B. Rossi, A.-M. Habraken and F. Pascon, respectively Research Fellow, Research Director and Postdoctoral Researcher, are supported by a grant from the Belgian National Fund for Scientific Research (FNRS), which is gratefully acknowledged.

# REFERENCES

1. R. Hill, "A theory of the yielding and plastic flow of anisotropic materials", Proceedings Royal Society of London, A193, 281-297
2. Y.Y. Zhu, S. Cescotto, "Unified and mixed formulation of the 8-node hexahedral elements by assumed strain method", Computational Methods in Applied Mechanics and Engineering, 129 (1996), 177-209
3. A. Ragab, S. E. Bayoumi, "Engineering solid mechanics – fundamentals and applications", CRC Press, ISBN 0849316073 ,1998
4. W.M. Quach, J.G. Teng , K.F. Chung, "Residual stresses in steel sheets due to coiling and uncoiling : a closed-form analytical solution", Engineering structure 26 (2004) 1249-1259
5. W.M. Quach, J.G. Teng , K.F. Chung, "Finite element predictions of residual stresses in press-braked thin-walled steel sections ", Engineering structure, accepted 24 february 2006
6. C.C. Weng, R.N. White, "Residual stresses in cold-bent thick steel plates", Journal of structural engineering, Vol. 116, No. 1, January, 1990, ASCE, ISSN 0733-9445/90/0001-0024, Paper No. 24218

7.    P. Flores, L. Duchêne, T. Lelotte, C. Bouffioux, F. El Houdaigui, A. Van Bael, S. He, J. Duflou, A.H. Habraken ; "Model identification and FE simulations: Effect of different yield loci and hardening laws in sheet forming", Proceedings of the 6th International Conference and Workshop on Numerical Simulation of 3D Sheet Metal Forming Process, NUMISHEET 2005, Volume 778, pp. 371-381, August 2005

8.    S. Timoshenko, J.N. Goodier, "Théorie de l'élasticité", Librairie polytechnique CH. Béranger, 1961

# Fine Coining of Bulk Metal Formed Parts in Digital Environment

[a]T. Pepelnjak, [b]V. Krušič, [a]K. Kuzman

[a]Faculty of Mechanical Engineering Ljubljana, Aškerčeva 6, SI-1000 Ljubljana, Slovenia
[b]Iskra Avtoelektrika, Polje 15, SI-5290 Šempeter pri Gorici, Slovenia

**Abstract.** At present the production of bulk metal formed parts in the automotive industry must increasingly fulfil demands for narrow tolerance fields. The final goal of the million parts production series is oriented towards zero defect production. This is possible by achieving production tolerances which are even tighter than the prescribed ones. Different approaches are used to meet this demanding objective affected by many process parameters. Fine coining as a final forming operation is one of the processes which enables the production of good manufacturing tolerances and high process stability. The paper presents the analyses of the production of the inner race and a digital evaluation of manufacturing tolerances caused by different material parameters of the workpiece. Digital optimisation of the fine coining with FEM simulations was performed in two phases. Firstly, fine coining of the inner racer in a digital environment was comparatively analysed with the experimental work in order to verify the accuracy and reliability of digitally calculated data. Secondly, based on the geometrical data of a digitally fine coined part, tool redesign was proposed in order to tighten production tolerances and increase the process stability of the near-net-shaped cold formed part.

**Keywords:** Bulk Metal Forming, Fine Coining, FEM, Tool Redesign.
**PACS:** 81.20.Hy, 02.70.Dh

## 1. INTRODUCTION

Bulk metal forming of automotive components represents a demanding production process affected by numerous influential parameters. These parameters are ranging from heat generation due to the forming, elasto-plastic behaviour of the formed material, different friction conditions on all contact surfaces such as a tool-workpiece and a die-shrink ring, elastic deformations of the pre-stressed tool to the elastic deformations of the forming machine [1, 2, 3]. Therefore, the bulk metal forming process can be assumed as a complex thermo-tribological-elasto-plastic system where particular influential parameters have to be well known [4]. Furthermore, with time and cost pressure caused through market globalization, these processes used for the mass production of automotive components are increasingly designated as final manufacturing operations. Machining operations required to produce parts to their final shape in a prescribed tolerance field are to be omitted where this is possible. Therefore, forming operations have to be optimized and well controlled [5]. To assure high production accuracy and narrow tolerances also in the bulk metal forming, fine coining processes are to apply after the cold extrusion in order to assure narrow

CP907, 10th ESAFORM Conference on Material Forming, edited by E. Cueto and F. Chinesta
© 2007 American Institute of Physics 978-0-7354-0414-4/07/$23.00

578

tolerances on major active surfaces of the produced part. There are several development directions to achieve the accurate production of bulk formed parts ranging from the design and optimization of the entire forming system of forming machining, a forming tool, tribological conditions and a forming process in the CAE environment [3, 6] to the implementation of innovative concepts of the forming process such as local heating of the workpiece during the forming [7]. Through all these efforts, the production of the part's near-net-shape or even the final part shape can be achieved.

The research of near-net-shape technologies in forming bulk formed parts is an industrial challenge in all developed industrial countries which have to compete with the enormous price pressure of the produced automotive components from the Far East. The production of quality products at lower prices forces manufacturers to implement new and innovative development tools in combination with the process optimization in the digital environment.

## 2. NEAR-NET-SHAPE PRODUCTION OF AN AUTOMOTIVE PART

The Slovene Company Iskra Avtoelektrika d.d. was invited to produce the inner race of the outer constant velocity (CV) joint – Fig. 1 - in a near-net-shape technology to minimize the following machining operations. Production tolerances of the most important part dimensions of $d_{ref} \pm 0.02$ mm between ball bearings and the working radius of $R_{ref} \pm 0.045$ mm of the inner race represent a special challenge for the mass production with forming processes. Therefore, the fine coining process of the part was analyzed in a digital environment parallel to the production of the first parts in order to verify the reliability of the obtained FEM results in the first development phase. Furthermore, in the following development phase the tool optimization in a digital environment was performed to minimise possible geometrical deviations of parts according to the measured results of the test series.

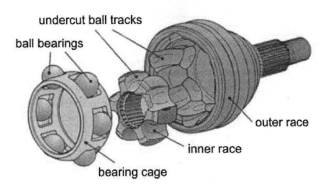

**FIGURE 1.** Assembly of the outer constant velocity (CV) joint.

Despite a great deal of experience in terms of near-net-shape production which the company has in the field of bulk metal formed parts the FEA approach in the optimization of a new part presented a challenge due to extremely narrow production tolerances of the inner race active surfaces which are in contact with ball bearings. The analyses of the fine coining of the inner race in a digital environment were therefore a matter of systematic research [8] in order to determine, evaluate and stabilize the forming process itself as well as to evaluate its influential parameters and their impact on the tolerance field.

To assure reliable data of the FEM simulation of the fine coining of the inner race it was necessary to consider the elasto-plastic behaviour of the entire system shrink ring-die-workpiece-punch – Fig. 2. FEM simulations were performed with the ABAQUS Standard program using quadrilateral continuum elements for the shrink ring and the die. The workpiece and the punch were meshed with tetrahedral elements due to their more complex shapes. To shorten simulations times, only one twelfth of the entire inner race was analysed considering part symmetry with appropriate boundary conditions. The workpiece was stress-release annealed and lubricated prior to the fine coining process. The formed material underwent the elasto-plastic Hollomon potential material law and the tool parts were assumed as elastic ones. The contacts between all surfaces were assumed with the Coulomb's friction law using the friction coefficient of $\mu$=0.06.

Detail : workpiece          FEM model of the fine coining          Detail: die

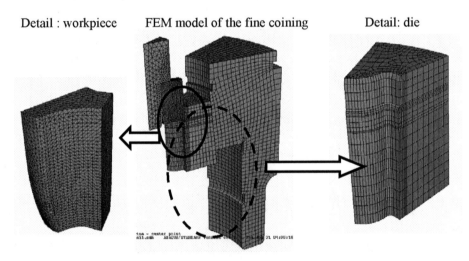

**FIGURE 2.** FEM model for the fine coining of the inner race.

The fine coining process was optimized according to the measured data of produced parts. Therefore, the geometry of the active area of the part was measured on five levels – Fig. 3 and compared with numerically obtained results [9]. It was ascertained that the scatter of measured values spread over a 39% of the prescribed tolerance field of ± 0.045 mm of the reference radius. Furthermore, the level of the distance between ball bearings was above the prescribed tolerance field for 0.012 to 0.031 mm – Fig. 4. Both levels were not acceptable to produce parts in a stable process window as far as mass production is concerned. The comparison of measured results with a numerically obtained geometry after the fine coining showed good correlation between the physical and digital approach – Fig. 4. Therefore, the optimisation phase, that is the tool redesign, was performed in a digital environment only in order to decrease scatter values of both reference dimensions.

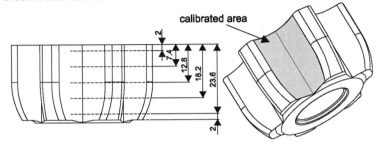

**FIGURE 3.** Analysed inner race and measured levels of active calibrated area [9].

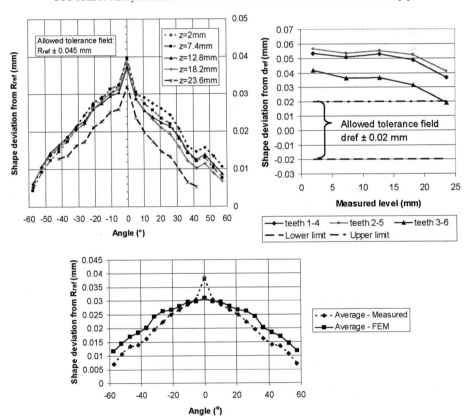

**FIGURE 4.** Level of shape deviations of the active radius (top - left) and a ball bearings diameter (top - right) as well as comparison of measured and FEM obtained average shape deviations of $R_{ref}$ (bottom).

Based on geometrical deviations of the simulated part from the standpoint of reference values as well as comparative average values of the measured calibrated components, the tool redesign was proposed [9]. The initial geometry of the workpiece remained unchanged. The performed simulations with the adjusted die geometry have shown improvements of both required reference values – $d_{ref}$ and $R_{ref}$. The scatter of the $R_{ref}$ was decreased to 28.9 % of the allowed tolerance field and moved towards its mean value. Furthermore, the distance between the ball bearings, which had prior been out of the prescribed range, was, after the tool redesign, almost entirely in the prescribed tolerance field – Fig. 5.

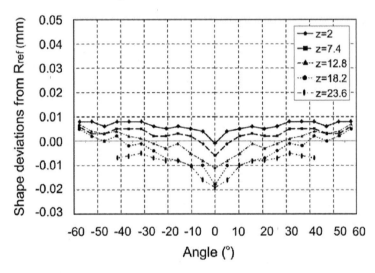

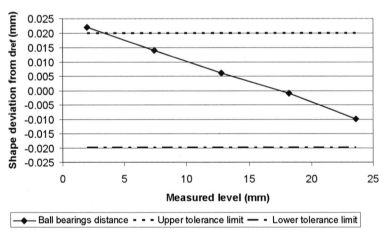

FIGURE 5. Level of shape deviations of the active radius (top) and ball bearings diameter (bottom) after tool redesign.

# 3. CONCLUSIONS

Demanding forming technologies force modern industry into the use of new concepts and technologies to shorten production chains and decrease production costs. The presented case study of the near-net-shape forming of the inner race of a CV joint with a fine coining process has shown that the FEA approach represents a reliable and fast tool for process redesign and optimization in order to minimise production scatter values caused by the forming of the workpiece. Production tolerances which were initially too high for the reference value $R_{ref}$ and out of the prescribed range for $d_{ref}$ could drastically be decreased by implementing the tool redesign in a digital environment. On the development level only the influence of tool redesign was analysed to evaluate the robustness of the forming process and to stabilize the production around the mean value of prescribed shape tolerances. Further progress of the presented work is focused on the analysis of the scatter of material properties and the workpiece geometry according to existing industrial conditions as well as the material selection of the die used in the presented fine coining process in order to increase its rigidity.

# ACKNOWLEDGMENTS

The work presented in this paper was partially performed in the frame of the joint project *Evaluation of the Material and Product Development in the Frame of the Polycentric Technological Centre* led by the Slovene Automotive Cluster (ACS). The project has partially been financed partially financed by the European Structural Fund. Their assistance is gratefully acknowledged.

# REFERENCES

1. H. Kudo, "Towards Net Shape Forming" in *J. Mater. Proc. Technol.*, edited by M.S. Hashmi, Elsevier, 22 (1990), pp. 307-342.
2. K. Kuzman, B. Štok, "Total Process Control – a Precondition for Net Shape Forming Implementation", Proc. 9th ICFC, Solihull, UK (1995), pp. 123-128.
3. V. Krušič, M. Arentoft, T. Rodič, "The impact of a mechanical press on the accuracy of products and the reliability of tools in cold forging" in Proc. of 5th ICIT Conf., Velenje, April 2005. pp. 35-41.
4. K. Kuzman, "Problems of accuracy control in cold forming" in *J. Mater. Process. Technol.*, edited by M.S. Hashmi, Elsevier, 113 (2001) No. 1/3, special issue "5th APCMP, Seoul, Korea", pp. 10-15.
5. B.-A. Behrens, E. Doege, et. all., "Precision forging processes for high-duty automotive components" in *J. Mater. Process. Technol.*, edited by M.S. Hashmi, Elsevier, in press 2006.
6. N. R. Chitkara and Y. J. Kim, "Near-net shape forging of a crown gear: some experimental results and an analysis" in *Int. J. of Mach. Tools and Manuf.*, Edited by T. Dean, Elsevier, 41 (2001) 3, pp. 325-346.
7. K.L. Schlemmer and F.H. Osman, "Differential heating forming of solid and bi-metallic hollow parts" in *J. Mater. Process. Technol.*, edited by M.S. Hashmi, Elsevier, (2005) pp. 564-569.
8. K. Kuzman, T. Pepelnjak, G. Gantar, D. Švetak, *"Evaluation of the Material and Product Development in the Frame of the Polycentric Technological Centre - Subproject: Stability Control of the Cold Forging System – Phase 2: First Phase Report"*, Faculty of Mechanical Engineering Ljubljana, Forming Laboratory, Ljubljana, 2006. 21 p. (in Slovene).
9. M. Jerič, *Control of fine coining in cold forming process*, University Diploma Work Nr. 5495, Faculty of Mechanical Engineering Ljubljana, Ljubljana, 2005. 71 p. (in Slovene).

# Manufacturing of Profiles for Lightweight Structures

Sami Chatti and Matthias Kleiner

*Institute of Forming Technology and Lightweight Construction (IUL)*
*University of Dortmund*
*Baroper Str. 301, D-44227 Dortmund, Germany*

**Abstract.** The paper shows some investigation results about the production of straight and curved lightweight profiles for lightweight structures and presents their benefits as well as their manufacturing potential for present and future lightweight construction. A strong emphasis is placed on the manufacturing of straight and bent profiles by means of sheet metal bending of innovative products, such as tailor rolled blanks and tailored tubes, and the manufacturing of straight and curved profiles by the innovative procedures curved profile extrusion and composite extrusion, developed at the Institute of Forming Technology and Lightweight Construction (IUL) of the University of Dortmund.

**Keywords:** Lightweight Construction, Sheet Metal Bending, Profile Bending, Curved Profile Extrusion, Composite Extrusion
**PACS:** 81.20.Hy

## INTRODUCTION

Lightweight construction is a multidisciplinary engineering science with great growth dynamics, which requires a holistic technological approach covering the whole system "design-material-manufacturing". Manufacturing technologies, and in particular forming technologies, are of major importance in this context. Forming technologies represent the most important procedure group for the production of multifunctional lightweight profiles involving new design and material aspects.

Lightweight construction is one of the central challenges for engineering scientific research as well as its implementation, for instance in traffic and civil engineering. Supporting structures, which are subdivided into light shells (sheet metal walls reinforced with profiles) and light space frames (profile frameworks covered with sheet metal), are of great importance in this field. For this reason, the demand for aluminium, magnesium steel, and composite profiles as important structural elements increased during the last years. These profiles are used in various fields of traffic engineering in a linear and curved contour. Curved profiles, especially three-dimensional curved profiles, allow the construction of structures with more advantages regarding e.g. space saving and aerodynamics. Tailored profiles which allow an adaptation of the properties and particularly of the component cross-sections to the local load state represent a further potential for weight reduction and product property

CP907, *10th ESAFORM Conference on Material Forming*, edited by E. Cueto and F. Chinesta
© 2007 American Institute of Physics 978-0-7354-0414-4/07/$23.00

optimization. A high manufacturing quality of profiles is an important condition for subsequent production steps like assembly and welding and can therefore reduce the costs of the entire production chain [1].

The paper shows the current investigations and the latest forming procedure developments in order to achieve more complex straight and curved profiles needed for complex lightweight structures.

# MANUFACTURING OF STRAIGHT PROFILES

Straight lightweight profiles are today used in almost all fields of technology. In addition, they are used as semi-finished products for subsequent processing operations (e.g. by means of bending and hydroforming) for the manufacturing of profiles with arbitrary contours of the longitudinal axis as important components of lightweight structures.

In principle, the profile manufacturing procedures can be subdivided according to the forming technology into two groups: cold forming procedures for the manufacturing of profiles usually from semi-finished sheet metal products (sheet metal profiles or cold profiles) and hot forming procedures for the manufacturing of "massive" profiles (hot profiles); bar extrusion is in this group the most important procedure for lightweight construction.

## Cold Profiles

Cold profiles are profiled semi-finished products with open or closed cross-section, characterized by the fact that they are manufactured at room temperature (thus in cold state) from flat rolled sheet material. The sheet semi-finished product can be a hot-rolled or a cold-rolled strip with untreated, refined, or coated surface. The profile is produced, thereby, through forming at unchangeable wall thickness dimensions.

The main advantage of sheet metal profiles compared to massive profiles is weight saving combined with increased strength. With the same material and weight a profiled sheet metal has a higher strength. This higher strength allows a lighter construction with the same statical characteristics. During bending processes when profiling, the originally soft material hardens and, thus, offers a higher stability. Further advantages of cold profiles are close cross-sectional tolerances and a high surface quality.

The manufacturing procedures of cold profiles depend on the demanded dimensions, the profile shapes, and the production quantities which are to be manufactured as profitably as possible. Following manufacturing processes are used: air and die bending on press brakes, folding on folding machines, roll forming on roll forming plants by progressive bending in driven shaped rolls, and drawing through several, non-driven shaped rolls or through rigid dies.

To support the manufacturing of load-adapted and weight-optimized lightweight profiles from innovative sheet metal semi-finished products of variable thickness over the longitudinal axis (tailor rolled blanks) the air bending process was further developed at the IUL.

The main problem in the bending of such tailor rolled sheet metal parts is the inhomogeneous springback due to the different amounts of thickness and strength in the sheet metal. To compensate such influences a certain tool strategy has been pursued within the scope of a research project at the IUL. The strategy is based on the modification of the air bending process by use of structural elements, enabling the setting of the effective punch displacement and/or die width according to the specific region. For this purpose tool sets have been manufactured, the lower tools of which consist of single die elements that can be individually handled (**figure 1**).

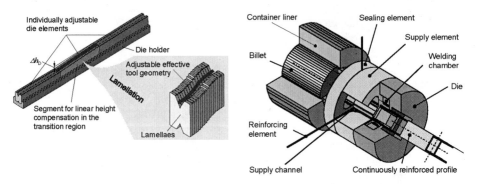

**FIGURE 1.** Tool design for air bending with variable die height in specific regions.

**FIGURE 2.** Principle of composite extrusion.

## Extruded Profiles

Another way to manufacture profiles is bulk forming. Here, the forming procedure bar extrusion offers by far the largest potential for the economic production of lightweight profiles. The variety of the profile cross-sectional shapes thereby producible is enormous. Through the use of extruded hollow profiles, particularly light and torsion-stiff supporting structures can be obtained. In contrast to roll formed and draw formed profiles, material increasing joints can be avoided and large wall thickness differences as well as undercuts can be produced in bar extrusion. The material can be distributed in the profile cross-section according to its function.

The manufacturing of straight extruded profiles has to be passed through the following three stations: extrusion, stretching and/or straightening, and thermal treatment. Afterwards, the strength-optimized profiles are further processed depending on their application state (straight or curved) by cutting and/or forming procedures.

A new process variant for bar extrusion was developed at the IUL, offering a great potential for lightweight construction. The new process allows the reinforcement of aluminum profiles with continuous elements such as high strength steel wires and steel wire ropes, which are embedded into the aluminum matrix during the extrusion process, using a modified porthole die (**figure 2**). The reinforcing elements are fed from the outside, led through supply channels, and are introduced into the material flow. Each element gets in contact with the aluminum in the welding chamber where the fusion takes place. The reinforcing material leaves the die along with the manufactured profile, pulled by the basic material. For optimal results the choice of reinforcing material and the design of the welding chamber to control the material flow is important. Composite extrusion allows the increase of the specific strength of profiles for certain applications.

# MANUFACTURING OF CURVED PROFILES

Curved profiles as structural parts imply highest quality requirements regarding the measurement and shape accuracy. The time and cost situation in mass production make such demands on a trouble-free assembly and automatic joining of curved profiles particularly necessary.

In some application fields, e.g. in the automotive industry, high tolerances of curved profiles are required which can not be achieved by conventional bending procedures or only at large expenditure. This forces the use of new and/or improved forming techniques, using process control systems or new tools, combined with calibration procedures. Only innovative procedures and/or procedure combinations can meet the requirements. As calibration procedure for hollow profiles the hydroforming process is often used. With this procedure complex three-dimensional curved profiles can be manufactured in a single manufacturing step under reproducibility of highest quality claims. The quality of the hydroformed workpiece and the reliability of the process is, however, defined by the material and geometrical properties of the initial parts, e.g. after bending.

## Profile Bending

A relatively new type of profiles supporting the lightweight construction is load-adapted and weight-optimized. These profiles are characterized by intentionally applied variations of material properties or cross-sections along the longitudinal profile axis. In general, such profiles are made of special sheets, like tailor welded blanks or tailor rolled blanks (TRB), or have different diameters across the profile length (tailored tubes). The bending of such rather complex profiles to desired shapes calls for adapted bending strategies which require increased process knowledge, the observation of different material behaviors, and the development of designated adaptive forming processes and tools.

### Bending of Profiles Made of Tailor Rolled Blanks

During the bending of the profiles made of TRB's (see section cold profiles) in a three-roll-bending process, it was obvious that the profiles used already showed certain deviations not only of the cross-section geometry, but also of the material properties (e. g. residual stresses) leading to inhomogeneous springback. The combined interaction of these influences causes a different forming behavior along the longitudinal axis of the semi-finished profile components which could be compensated with a finely tuned process control strategy during bending.

A control strategy was used which allows to bend profiles with thickness transitions to constant radii. By setting appropriate roll adjustment values for the constant profile areas and a non-linear roll adjustment in the area of the sheet metal thickness transitions the afore mentioned inhomogeneities could thus approximately be compensated and approximately constant bending radii over the total of the examined profile length could be achieved. The manufactured crash structures (**figure 3**) for the automotive industry based on tailor rolled blanks were not only lighter in weight by 13% than comparable structures of a constant initial material, but also showed a significantly improved structural behavior in static and dynamic load tests.

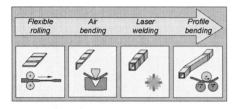

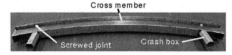

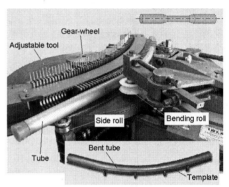

**FIGURE 3.** Process chain and manufactured crash structure from tailor rolled blanks.

**FIGURE 4.** Rapid tooling set-up for bending of tailored tubes to variable radii.

## *Bending of Tailored Tubes*

Tailored tubes are used particularly as flexible semi-finished products for a further forming by tube hydroforming. The manufacturing of curved tubular workpieces with complex geometry by hydroforming is a well suited process for lightweight construction. To improve the capability of forming in the hydroforming process, a new concept of using tailored semi-finished tubes with variable diameters has been developed at the IUL. The diameter of a tube is reduced here by metal spinning at required areas with respect to the geometry of the final product. Due to the fact that the pre-form usually has to be bent to assure that it fits into the hydroforming tool, a specialized tube bending process has been developed to realize the bending of tubes with variable diameter. The bending tool is a segmented tool based on the rapid tooling principle, permitting a locally independent adjustment of the bending radius under load to compensate springback differences in the different tube zones (**figure 4**).

## **Curved Profile Extrusion**

The conventional process chain for the production of curved extruded profiles involves a successive extrusion, stretching, and bending of the profiles. However, such semi-finished products feature disadvantages typical for bending which complicate the process design and can negatively influence the manufacture. Using the innovative extrusion process variant Curved Profile Extrusion (CPE), developed at the IUL, the exiting strand is being deflected by a guiding tool and, as a result, the profile exits the die directly in a curved shape (**figure 5**). This means that an additional bending process is not necessary. The curved extrusion is a result of a certain velocity profile of the material flow, caused by tensile and pressure stresses as well as a lateral force. Here, the contour radius of the curved profile is solely determined by the position of the guiding tool in relation to the die [2].

During CPE, the plasticity results from the extrusion process itself, not from the lateral force. Therefore, the properties of curved profiles manufactured by CPE are better than those of bent profiles. These are, among others, a high accuracy of shape (no springback), a minimal cross-section deformation, reduced residual stresses, and an unreduced formability. CPE also supports lightweight construction. It allows the curving of profiles with very small wall thicknesses. Furthermore, a design for bending of the profile cross-section, in most cases increasing the component weight, is not required for CPE. Finally, the formability of 2D and 3D-curved profiles (**figure 6**) is as good as of straight profiles, thus allowing a lighter shaping in further processing.

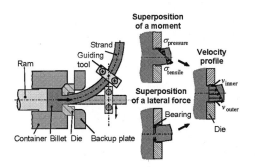

**FIGURE 5.** Process principle of curved profile extrusion (CPE).

**FIGURE 6.** Three-dimensional curved profile extrusion.

## SUMMARY

The manufacturing of straight and curved profiles is one of the most potent fields of the forming technology and supports the lightweight construction. New and innovative forming procedures have been developed using adapted process control strategies and adjustable tools enabling the manufacturing of complex lightweight profiles with arbitrary contours. Innovative semi-finished products, like e.g. tailor rolled blanks and tailored tubes, could be formed by bending to load-adapted and weight-optimized lightweight profiles. New forming procedures, like e.g. the curved profile extrusion and the composite extrusion, allow the manufacturing of high-strength thin-walled profiles with extended mechanical properties.

## REFERENCES

1. S. Chatti, *Production of Profiles for Lightweight Structures*, Habilitation thesis, University of Dortmund – University of Franche-Comté, Books on Demand GmbH, Germany, 2006.
2. A. Klaus, *Optimization of Process Accuracy and Reliability of Curved Profile Extrusion*, PhD Thesis (in German), University of Dortmund, 2002.

# Solid State Bonding Mechanics In Extrusion And FSW: Experimental Tests And Numerical Analyses

G. Buffa[1], L. Donati[2], L. Fratini[1] , L. Tomesani[2]

[1]*Dipartimento di Tecnologia Meccanica, Produzione e Ing. Gestionale, Università di Palermo, Italy*
[2]*Department of Mechanical Construction Engineering (D.I.E.M.), University of Bologna, Italy*

**Abstract.** In the paper the authors compare the different solid state bonding mechanics for both the processes of hollow profiles extrusion and Friction Stir Welding (FSW), through the results obtained from a wide experimental campaign on AA6082-T6 aluminum alloys. Microstructure evaluation, tensile tests and micro-hardness measurements realized on specimens extracted by samples of the two processes are discussed also by means of the results obtained from coupled FEM simulation of the processes.

**Keywords:** Extrusion, Friction Stir Welding, Bonding, FEM model
**PACS:** 81.20.Hy

## INTRODUCTION

Solid state bonding recurs in several manufacturing processes, in particular in extrusion of hollow profiles and in Friction Stir Welding (FSW). Such processes are nowadays of particular industrial interest because of the large diffusion of the produced parts (the former) and of the specific advantages with respect to the classic welding technologies (the latter). In the extrusion of hollow profiles, solid state welding occurs inside the die, the welding surface being characterized by a gap of the order of the interatomic distance at any location [1]. Initially, the deforming material divides into seams around the die legs supporting the mandrel; in a second step, the seams become into contact in the so called "welding chamber", where high hydrostatic pressures are reached for the bonding to be performed. In the third step, the material undergoes the final deforming stage between the outer and the inner part of the die (the bearing zones), thus assuming the final shape. At the beginning of the process, the material is in contact with air and the seams surfaces are covered by a layer of oxides; after the initial stage of die filling, the weld occurs without any formation of oxides. This behavior is clearly evidenced by the much lower strength (up to about 50%) of the joints in the first meters of the extrudate. Usually, a hollow profile has more welds, each one produced under different conditions of temperature, strain, strain rate and pressure. The welding mechanics is, moreover, quite complex, being produced by different paths each one with its own thermo-mechanical history. Past studies have shown that temperature weakly affects welds quality, while the increase of extrusion

CP907, *10th ESAFORM Conference on Material Forming*, edited by E. Cueto and F. Chinesta
© 2007 American Institute of Physics 978-0-7354-0414-4/07/$23.00

rate generally produces a modest decrease in the profile elongation [2]; furthermore, a great effect on resistance is obtained by controlling the shape of the welding chamber, especially if higher pressure is reached in combination with the increasing of contact time [3, 4]. In particular, it is generally established that pressure is a crucial factor in determining the mechanical resistance of the joint, and many criteria, mainly related to the pressure, have been proposed by different authors in order to help the die design and predict the effectiveness of the welding process.

On the other hand, FSW is obtained by inserting a specially designed rotating pin into the adjoining edges of the sheets to be welded and then moving it all along the joint [5]. During the process, the tool rotation speed (R) and feed rate ($V_f$), determining the specific thermal contribution conferred to the joint, are combined in a way that an asymmetric metal flow is obtained. In particular, an advancing side and a retreating side are observed: the former being characterized by the "positive" combination of the tool feed rate and of the peripheral tool velocity, the latter having velocity vectors of feed and rotation opposite to each other [6]. A material flow coming from the retreating side towards the advancing one is observed; such flow is particularly strong in the upper layers of the blanks due to the action of the tool shoulder. Furthermore, the material is forced downwards towards the bottom of the joint as the tool is inclined; a change of direction is observed for such material and then an ascendant laminar flow in the advancing side is found out. Finally, the material bonds with the undeformed one of the advancing side on an inclined plane. Actually it should be observed that the effectiveness of the obtained joint strongly depends on several operating parameters, both geometrical and technological [5]. Thus, in FSW the solid state bonding is obtained between an undeformed "cold" material already placed in the advancing side of the joint and the "hot" material flow incoming from the retreating side. Proper conditions of pressure, temperature, strain and strain rate occur in order to get the final effective bonding.

In the paper the authors compare the different solid state bonding mechanics for the two described processes, through the results obtained from a wide experimental campaign on AA6082-T6 aluminum alloys. The experimental results are finally compared with the numerical ones obtained by means of thermo-mechanical FEM simulations in order to discuss analogies between the two processes.

## BONDING MECHANICS IN FSW

The weld was developed on AA6082-T6 aluminum alloy in 3mm thick sheets, characterized by a yield stress of 280MPa, an ultimate tensile stress of 319MPa and a microhardness of 120HV. As far as the utilized tool is regarded, it was made in H13 steel; a cylindrical pin was used with a pin diameter equal to 4.00mm and pin height equal to 2.80mm. The shoulder diameter was equal to 12mm. The following process parameters were utilized: rotating speed (R) equal to 1000r.p.m., tool feed rate ($V_f$) ranging from 75 to 215mm/min, nuting angle $\theta=2°$ and finally tool sinking $\Delta h=2.90mm$. In Figure 1 the typical transverse section of the developed joint is shown. It should be observed that the actual bonding line, approximated in the picture by a straight line, differs from the welding one, i.e. from the trace of the edges of the starting blanks, and it is located in the advancing side of the weld. During the FSW, in fact, a material flow on the back of the tool from the retreating side towards the advancing one is observed [6].

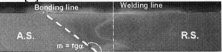

**FIGURE 1.** Transverse section of a FSW developed joint (R=1000rpm, Vf=100mm/min)

The tensile tests performed at the varying of the advancing velocity gave results, calculated as average value of the percentage of the joint UTS with respect to the base material UTS, ranging from about 60% up to about 75%, and confirmed the expected results in terms of fracture location, i.e. in the advancing side where the actual bonding line is (Figure 2). Micro hardness tests were also performed for the analyzed case studies, at 1.5mm from the bottom of the joint (Figure 3). In particular a significant softening is observed in the stirred zone, with a relative local maximum in the weld nugget. It should be observed that the minimum value is found in the advancing side in correspondence of the actual bonding line highlighted in Figure 1. What is more, the best performance is obtained with an advancing velocity $V_f$=100mm/min, in accordance with the tensile tests results previously shown.

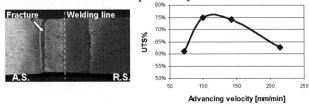

**FIGURE 2.** Fracture location and UTS of the joints at the varying of the tool feed rate.

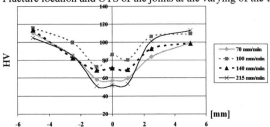

**FIGURE 3.** Microhardness distribution in FSW, in a transverse section, at a distance of 1.5mm from the bottom of the sheets, at the varying of the advancing velocity

In order to further highlight the process mechanics the FSW process was simulated utilizing the software DEFORM-3D™ through a non-linear fully coupled thermo-mechanical analysis [7]. The tool was modeled as rigid body while, as far as the modeling of the workpiece is regarded, a "single block" continuum model (sheet blank without a gap) is used in order to avoid contact instabilities due to the intermittent contact at the sheet-sheet and sheet-tool interfaces. The tool moves forward and welds a crack left behind the pin as it advances along the welding line. In Figure 4 the strain and strain-rate values of two material paths along the joint transverse section are considered: in particular two lines (Z1 and Z2) were chosen in the joint section at two different height of the section, namely 1mm and 2mm from the bottom, respectively. Examining the obtained values all along the two lines, both the material which undergoes the flux coming from the retreating side and the part of the joint which does not move are considered; it should be observed that the field variable values occurring in the transverse section containing the tool pin were plotted. In other words, for sake of simplicity, it has been assumed that the welding occurs in the transverse section of the joint containing the tool axis. In particular, the strain values reported are characterized by almost constant values for the two different considered layers with values in the range 5 to 6 in the welding zone, while the strain rate values are larger at the top of the joint because of the shoulder action, where a maximum of about 11 is observed in the welding zone. The bonding occurs at different distances from the welding line (see again Figure 1); in particular in the considered case study the material bonded at about 3mm (line 1) and 4mm (line 2) from the welding line. In this way ranges of the field variables have to be considered in order to highlight the welding conditions in FSW.

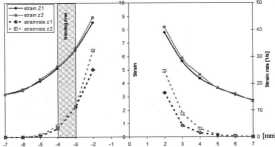

**FIGURE 4.** FSW: Strain and strain rate in the material paths (R=1000rpm and $V_f$=100mm/min).

## BONDING MECHANICS IN EXTRUSION

An I shape profile with a central wall thickness of 4 mm was realized by the extrusion of a cylindrical billet of 265 mm in diameter and 800mm in length. In order to create a weld in the middle section of the profile, a leg was inserted in the feeder part of the die assembly, thus creating a welding chamber of 23mm length (Figure 5). The same profile has been realized also by the use of a flat die, thus allowing the production of the profile without any weld. The billet (AA6082-O) was homogenized following industrial standards, then it was preheated at 435°C and extruded at 5.7 mm/s punch speed. The profile was quenched on the press immediately after the extrusion and then aged to a T6 condition. The profiles were tensile tested: in the welded profile an ultimate tensile strength of 350 MPa and an elongation of 8.7% were found with respect to 330MPa and 10.5% elongation of the profile without weld.

In figure 6 a portion of the transverse section of the profile is shown: the location of the welding line is reported and a micrograph of the zone is enclosed. In the extruded profile a perfect symmetry of the joint is usually evidenced; the welding line can be highlighted only by the use of HF etching; the microstructure is homogenous across the section except for the welding line where an higher distribution of intermetallic particles are usually found (white line in the micrograph of fig.6). Moreover, it's common to find grains crossing the welding line, this observation being congruent with the static recrystallization that occurs on the profiles after the die exit.

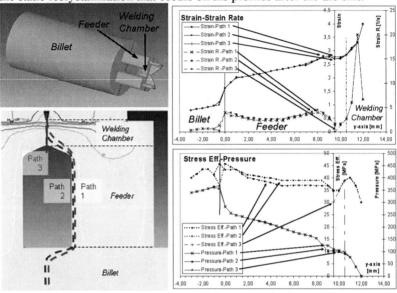

**FIGURE 5.** Extrusion: Process scheme and welding material paths (left), Distribution of strain, strain rate, effective stress and hydrostatic pressure in each path (right).

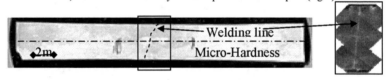

**FIGURE 6.** Transverse section of the extruded profile and seam welds magnification.

Micro-hardness measurements (figure 7) were then realized along the transverse section on the middle line of the I shape, where the axis origin represents the location of the welding line. In the profile without the weld a mean value of 95 HV was found, while in the welded profile a mean value of 110 HV was calculated regardless of the distance from the welding line. The overall mechanical proprieties of the joint are almost comparable with that of the profile without welds. In fact, in extrusion the bonding is a high pressure solid state joint with a following deforming stage that creates very strong recrystallised structure completely free of contamination such as oxides. Moreover, in welded profiles the strain path is more complex than in profiles without welds, thus creating a final higher strain and consequently a finer microstructure that improves the mechanical proprieties of the product.

A complete thermo-mechanical simulation of the process was realized by means of Deform 3D FEM software [8]. The distribution of strain, strain rate, effective stress and hydrostatic pressure for the welded joint are reported in figure 5: the simulated values are almost similar to the FSW ones (table 1) except for the welding pressures where about double values are found.

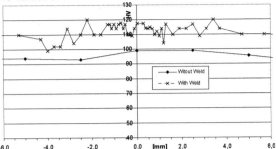

**FIGURE 7.** Microhardness distribution in extrusion in a transverse section (middle of the profile).

**TABLE 1.** Calculated field variable values.

|  | Strain | Strain rate [s⁻¹] | Effective stress [MPa] | Press. [MPa] | Temp. [°C] |
|---|---|---|---|---|---|
| Extr. | 4 | 20 | 38 | 77 | 510 |
| FSW | 5-6 | 4-11 | 22-27 | 39-45 | 536-567 |

## CONCLUSIONS

Looking at the welding conditions in the two processes, quite coherent values of the most relevant field variables are observed. Differences can be referred to the particular processing conditions considered in the two processes. Oxides and uniformity of the microstructure actually represent the major differences between the two processes. In extrusion, the welding occurs between two oxide-free seams in similar conditions of temperature, strain and strain rate. In FSW, on the other hand, the oxide- contaminated flowing material bonds on an undeformed surface.

## ACKNOWLEDGMENTS

This work was made using MIUR (Italian Ministry for University and Scientific Research) funds. Thanks also to Compes S.p.A. for founding the extrusion project.

## REFERENCES

1. R. Akeret, Proc. Fifth Int. Aluminum Extrusion Technology Seminar (ET 1992), Vol. 1, pp.319-336.
2. L. Donati, L. Tomesani, Proceedings of 8th ICTP conference, Verona (2005), pp. 227-228.
3. L. Donati, L. Tomesani, Journal of Materials Processing Technology 153-154 (2004), pp. 366-373.
4. L. Donati, L. Tomesani, Journal of Materials Processing Technology 164-165 (2005), pp.1025-1031.
5. A. Barcellona, G. Buffa, L. Fratini, VII Esaform 2004 Conference, (2004), pp. 371-374.
6. L. Fratini, G. Buffa, D. Palmeri, J. Hua, R. Shivpuri, Sci. and Tech. of Welding and Joining, Vol. 11, 4, (2006), pp.412-421.
7. G. Buffa, J. Hua, R. Shivpuri, L. Fratini, Mat. Science and Eng. A, 419/1-2, (2006), pp. 381-388.
8. L. Fratini, L. Tomesani, L. Donati, G. Buffa , Journal of Materials Processing Technology 177 (2006), pp.344-347.

# Optical Measurement Technology
# For Aluminium Extrusions

Per Thomas Moe, Arnfinn Willa-Hansen and Sigurd Støren

*Norwegian University of Science and Technology*
*Department of Engineering Design and Materials*
*Richard Birkelandsvei 2B, N-7491 Trondheim, Norway*

**Abstract.** Optical measurement techniques such as laser scanning, structured light scanning and photogrammetry can be used for accurate shape control for aluminum extrusion and downstream processes. The paper presents the fundamentals of optical shape measurement. Furthermore, it focuses on how full-field in- and off-line shape measurement during pure-bending of aluminum extrusions has been performed with stripe projection (structured light) using white light. Full field shape measurement is difficult to implement industrially, but is very useful as a laboratory tool. For example, it has been clearly shown how moderate internal air pressure (less than 5 bars) can significantly reduce undesirable cross-sectional shape distortions during pure bending, and how buckling of the compressive flange occurs at an early stage. Finally, a stretch-bending set-up with adaptive shape control using internal gas pressure and optical techniques is presented.

**Keywords:** Aluminium, Extrusion, Pure Bending, Optical Measurement Technology
**PACS:** 40, 80

## INTRODUCTION

Extruded and stretch-bent aluminum sections are commonly used in crash systems, engine cradles and other automotive structural elements. A real challenge encountered when joining and assembling aluminum extrusions is excessive shape variability. Downstream processes pose strict requirements on shape while there are limits to the capability of the extrusion process. Typical shape deviations may easily exceed 0.5 mm. Important steps have been made towards the development of a higher accuracy extrusion process [1,2], but industrial implementation of such a process is still some steps away. In the meantime, expensive shape calibration is needed to fine-tune part shapes before assembly. Furthermore, adaptive control of bending processes is often desirable to secure satisfactory shape. A key to successful implementation of a new extrusion process is the innovative combination of sensor and actuator technology.

In this paper we present some state-of-the-art optical shape measurement methods available today. Furthermore, we demonstrate how one technique, stripe projection with white light (structured light), can be used in- and off-line to measure undesirable cross-sectional shape deviations due to sagging and buckling during pure bending of hollow extrusions. Finally, we show how optical measurements can be integrated in a high-speed high-accuracy in-line system for shape control during industrial bending.

CP907, *10th ESAFORM Conference on Material Forming*, edited by E. Cueto and F. Chinesta
© 2007 American Institute of Physics 978-0-7354-0414-4/07/$23.00

# OPTICAL MEASUREMENT TECHNOLOGY

During the last 20 years we have witnessed a highly successful commercialization of optical shape measurement technologies. High resolution cameras, laser and light projectors and more computer power has made this possible. Today, optical techniques are used to establish full-field CAD and FEM models as well as for very advanced and robust process control with high-speed point or stripe measurements [2].

Most commercial measurement systems for metal forming applications are based on the same basic principle of distance measurement, triangulation (the cosine law). By observing an object from two different positions in space it is possible to estimate its position. The approach is for example used by the human eyes. Photogrammetry is a measurement method in which pictures of an object taken from different directions are used to calculate the coordinates of a set of characteristic points on the object. This is in principle the reversal of the process of photography, in which a fully 3D world is mapped onto a 2D plane. Photogrammetry is regarded as a passive optical technique since no external light sources are needed (although reflective markers are often used).

Triangulation is still possible if a light-emitting device is used instead of a camera. This may be a laser or light projector, projecting a characteristic pattern. In this case the technique is regarded as active. The locations of physical surface points are estimated in the same way as for the passive techniques. In the case of laser scanning, we need to know the exact locations and directions of the laser and camera. By turning the laser or a mirror reflecting the laser beam we can scan the surface of an object.

Instead of illuminating the object with only one point, we can project one or several stripes on the object. These may be either laser or white light stripes. The stripes are light intensity variations, and in the case of white light, variations may only be gradual (sinusoidal). Such full-field techniques are highly accurate and speed up the scanning process, but when using multiple stripes it may be difficult to distinguish the stripes close to steps in the surface. In this case coding of the projected light makes sense. By quickly projecting and recording sequences of light patterns a full-field model can be obtained. There are many ways to code the light. Very often a combination of Gray code black and white patterns is used in combination with patterns of sinusoidal varying light intensity. It is also possible to use colored light. Anyway, each single pixel of a CCD chip of a camera experiences a unique light sequence, which is used to establish a link between the coordinates of the camera and projector coordinates and a physical point in space (world coordinates). The system is calibrated simply by observing a reference cross or chess board from a different angles and distances.

The main advantage of optical techniques is the speed at which full fields can be measured. Modern stripe projection systems measure millions of points in a second. The systems can be used for a great range of object dimensions with the same relative accuracy and resolution. The spatial resolution is typically 1 ‰ of the measurement volume length. The accuracy may be better than 0.1 ‰ of the length. The accuracy of photogrammetry is often better than 0.01 ‰ of the measurement volume length. Thus, photogrammetry is often used in combination with other methods to secure accurate scanning of large objects. Laser and white light stripe projection are comparable with regard to performance. An important problem in relation to metal forming processes is reflections from very bright surfaces, and control of ambient light is often necessary.

# PURE BENDING EXPERIMENTS

The objectives of the experiments were to assess the effects of a moderate internal air pressure (0 - 5 bars) on the cross-sectional shape distortion of hollow aluminum extrusions during bending and to evaluate how to measure full-field section shape during and after bending with optical methods. Full-field shape measurement is a tool for very detailed studies of bending. The feasibility of performing in-line high-speed adaptive shape measurement was also an important issue.

During pure bending of single chambered sections the external flange is in tension while the internal flange is in compression. Tensile stresses contribute to pulling the outer flange towards the neutral plane (sagging). Compressive stresses cause the inner flange to buckle during experiments, and buckling can be prevented by stretching the section during bending. Sagging can be prevented by an internal tool or by applying an internal gas or fluid pressure. The last technique was patented by Everet and Miller [3] and demonstrated by Miller et al. for stretch-bending [4,5,6]. Moe et al studied pure bending of single chambered aluminium sections [7] with internal pressure.

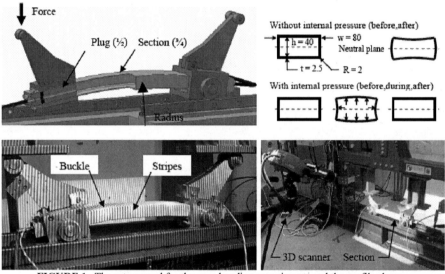

**FIGURE 1.** The set-up used for the pure bending experiments and the profile shape

Pure bending experiments were run with single chambered sections of external width 80 mm and height 40 mm. The wall thickness was 2.5 mm. An AA6060 alloy in the T4 condition was used. After extrusion the material was heated for an hour in an oven at 480 °C and consequently water-quenched and annealed for an hour at 100 °C. The result was a ductile material with a sufficiently matt surface for optical measuring.

During bending a constant bending moment was applied along the length of the section as shown in Figure 1. At the start of the experiments the distance between the tools was 360 mm, and the distance between the rotation points was 480 mm. The set-up has previously been used by Paulsen et al. who deduced analytic expressions for sagging and buckling [8,9]. The total length of the aluminum section was 660 mm.

New features that we introduced were inserts or plugs for sealing off the ends of the profiles and an air pressure control system (Norgren VP 51 actuator). The plug design is simple, but effective (Figure 1). A rubber plate is compressed between two metal cores. It expands and seals off the interior of the profile. After bending the system is unloaded and pulled out of the profile. The system can be implemented industrially. The metal cores prevent the end parts of the section from sagging or buckling.

During bending force, rotation and displacements were measured. In addition the full-field shape of the outer flange and a web was recorded by a 3D optical scanner (GOM ATOS III) during pauses for every 25 mm ram movement. It scans a 550 mm wide area with a resolution of 0.25 mm and accuracy of 0.05 mm. The measurement time is 8 seconds, and only one scan was necessary for capturing the upper side of the profile. In principle it is possible to scan also the inner flange of the profile.

During the experiments the maximum ram movement was either 25 or 40 mm. The movement corresponds to a tool tilt of 12.0 and 17.5° and an ideal bending radii of 900 and 600 mm. In reality, the inner flange buckles at a nominal bending radius of approx 1.3 m, after which the local curvature deviates from the ideal value. Experiments were run with a constant internal pressure of 0, 2.5 and 5.0 bars. 0 bar corresponds to the reference experiments of Paulsen. 5.0 bar is close to the maximum possible pressure delivered by the system. The numerical analysis predicted that a constant pressure of 2-3 bars suffice to prevent sagging (but not buckling). In all experiments we applied an accurate constant pressure although the analysis [6] strongly indicates that the best solution is a pressure that increases proportionally with the section curvature.

The experimental results were in good agreement with earlier results. During pure bending with no internal pressure, there was some initial sagging. Buckling occurs at a bending radius of approx 1.3 m. When an internal pressure is introduced, the profile bulges, and during bending sagging is reduced. Buckling still takes place, but the buckle shape differs. Due to the profile bulging, the first outward buckle close to the tool is most pronounced in the 5.0 bar pressure case. Figure 2 shows shape differences for sections bent with a pressure of 2.5 and 5.0 bar. The buckle height differs by 1.4 mm. When the inner flange starts to buckle, the webs and outer flanges are affected. The internal pressure reduces the effects of buckling on the web and outer flange.

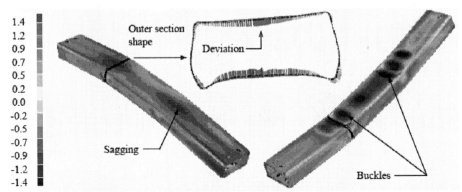

**FIGURE 2.** Typical results obtained from 3D scanning of the extrusions after bending to a tool rotation of 12.0° (nominal radius of 0.9 m). The plots show the difference in shapes for sections bent with an internal pressure of 2.5 and 5.0 bar. The sagging and buckling is less pronounced for a high pressure.

# DISCUSSION AND PROPOSAL

Pure bending is always an interesting case to study since the boundary conditions are relatively simple. Most importantly, there is no friction between the profile and a lower tool. This makes it easier to play with different analytical and numerical models. The buckling phenomenon is worth a much closer study, but also complicates matters significantly. An adaptive system for profile shape calibration by internal pressure and optical methods of measurement is more efficient and useful during stretch-bending of aluminum sections, which is a process of significant commercial importance.

Shape calibration by moderate internal air pressure is a method that has existed for a long time [5] although it is not very well known. The recent introduction of optical measurement techniques makes it possible to more closely control the outcome of the process. Shape measurement by fringe projection is still not practical during in-line adaptive control. Scanning is relatively slow and generates a large amount of data. The method used in this work is most useful for a close analysis of phenomena such as buckling or spring-back. However, there are a range of other optical measurement approaches that can be used for in-line shape control in combination with internal pressure. Laser (point or stripe) measurement is a well-known alternative, which has been used during high speed extrusion experiments [6]. Photogrammetry is commonly used for high-speed full-field strain measurements [7]. The technique can easily be used for measuring a small set of points at very high-speed during bending. One of the difficulties related to photogrammetric measurement is that characteristic points must somehow be detected and/or defined. The points may for example be sprayed on the surface of the extrusion before bending. A conceptual model of an adaptive stretch-bending process with adaptive internal pressure control is shown in Figure 3.

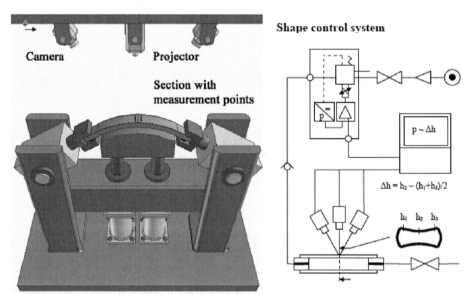

**FIGURE 3.** The conceptual design of a stretch-bending system with adaptive shape control.

# CONCLUSION

Experiments have shown that by applying moderate internal air pressure (2.5 to 5.0 bar) in single chamber hollow aluminum extrusions during pure bending distortions of the cross-sectional shape of profiles can be significantly reduced. Sagging due to the tensile stresses in the outer flange is almost completely prevented at the early stages of bending. However, during pure bending buckling occurs at a nominal bending radius of approx 1.3 m. Buckling cannot be prevented by an internal air pressure, but the air pressure affects the shape of the buckles and significantly reduces sagging of the upper flange connected to or caused by the buckling. By using full-field optical measurement techniques based on white light stripe projection, it is possible to closely study sagging, buckling and spring-back phenomena.

It is possible to use shape calibration with air pressure in combination with shape measurement with stripe projection to prevent distortion of the cross-sectional shape of hollow aluminum extrusions during pure bending. However, alternative optical techniques such as photogrammetry or laser scanning can more effectively used to measure only a few points instead of full fields. Data relating to shape distortions can be fed into a system that regulates pressure according to an adaptive scheme.

From a commercial perspective pure bending is a much less interesting process than stretch-bending. Furthermore, it is much more sensitive to flange buckling. Therefore, future experiments with adaptive shape control should focus on stretch-bending phenomena and on the development of a fully adaptive measurement system.

## ACKNOWLEDGMENTS

The financial support of the Norwegian University of Science and Technology, Hydro Aluminium and the Norwegian Research Council is gratefully acknowledged. Øystein Skotheim and Fred Couweleers of SINTEF ICT have provided most valuable support related to 3D scanning and optical systems design. We would like to thank colleagues Odd-Geir Lademo, Henry Ako Baringbing, Snorre Fjeldbo and Halvard Støwer for their contribution in discussions and planning of experiments.

## REFERENCES

1. S. Støren, *Int J. Mech Sci.* **35**, 1007-1020 (1993).
2. P. T. Moe, "Pressure and Strain Measurement during Hot Extrusion of Aluminium", PhD-thesis, Norwegian University of Science and Technology, 2005.
3. R. P. Everet and J. A. Miller, U.S. Patent No. 4,704,886 (10 November 1987)
4. J. E. Miller and S. Kyriakides and A.H. Bastard, *Int J. Mech Sci.* **43**, 1283-1317 (2001).
5. J. E. Miller and S. Kyriakides and E. Corona, *Int J. Mech Sci.* **43**, 1319-1338 (2001).
6. J. E. Miller, S. Kyriakides, *Int J. Mech Sci.* **45**, 115-140 (2003).
7. P. T. Moe, H. A. Baringbing, O.-G. Lademo, S. Støren and T. Welo, "A study of pure bending of hollow extrusions with internal pressure." in *Proceedings of the 9th ESAFORM Conference on Material Forming*, edited by N. Juster and A. Rosochowski, Glasgow, UK, 2005, pp. 423-426.
8. F. Paulsen and T. Welo, *Int. J. Mech. Sci.* **43**, 109-129 (2001).
9. F. Paulsen and T. Welo, *Int. J. Mech. Sci.* **43**, 131-152 (2001).

# A SUPG approach for determining frontlines in aluminium extrusion simulations and a comparison with experiments.

A.J. Koopman*, H.J.M. Geijselaers*, J. Huétink*, K.E. Nilsen[†] and P.T.G. Koenis[†]

*University of Twente, P.O. Box 217, 7500 AE Enschede, The Netherlands
[†]BOAL Beheer B.V., P.O. Box 75, 2678 ZH De Lier, The Netherlands

**Abstract.** In this paper we present a method to determine the frontlines inside the container and inside the extrusion die based on a steady state velocity field. Using this velocity field the convection equation is solved with a SUPG stabilized finite element method for a variable that represents the time it takes from the initial front to a certain point in the domain. When iso-lines in this field are plotted the development of fronts can be tracked. Extrusion experiments are performed with aluminium billets cut in slices. When extrusion is stopped the billet and extrudate are removed from the container and cut in half in the extrusion direction, copper foils between the slices show the frontlines. These lines show good agreement with the iso-lines from the numerical solution of convection equation.

**Keywords:** Aluminium, Extrusion, SUPG, Experiments, Frontlines
**PACS:** 81.20.Hy

## INTRODUCTION

In designing multi hole extrusion dies and porthole dies the velocity field and deformation of initially straight fronts are used as parameters for typical features as feeder holes and sink-ins. The ability to simulate an accurate front deformation leads to better design rules for the dies.

## MODELING

In this paper we focus on the solution of the convection equation. However in section "Determining the velocity field" we will briefly discuss the DiekA-FEM model used to determine the velocity field. Solving the convection equation is done as a post processing step.

### Governing equations

Consider the following convection equation over a domain $\Omega$ with boundary $\Gamma$.

$$\frac{Df}{Dt} = \frac{\partial f}{\partial t} + \mathbf{v} \cdot \nabla f \qquad (1)$$

CP907, 10th ESAFORM Conference on Material Forming, edited by E. Cueto and F. Chinesta
© 2007 American Institute of Physics 978-0-7354-0414-4/07/$23.00

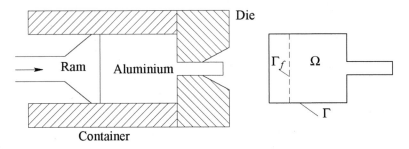

Die

Ram    Aluminium

Container

**FIGURE 1.**    Aluminium domain and boundaries

$f$ is the convected quantity and $\mathbf{v}$ the velocity field. When the velocity field is steady state the convected quantity is not time dependent. Now defining the total derivative of $f$ with respect to the time as 1, will yield a field $f = f(\mathbf{x})$ we call the "residence-time".

$$\frac{Df}{Dt} = \mathbf{v} \cdot \nabla f = 1 \tag{2}$$

The iso-lines of the $f$-field represent the location of a front at time $f = t$ that had the initial location on the boundary $\Gamma_f$ where $t = f = 0$. The essential boundary condition for this equation is:

$$f = 0 \quad on \quad \Gamma_f \tag{3}$$

$\Gamma_f$ can be a subset of $\Gamma$ but it can also be freely defined inside $\Omega$. $\Gamma_f$ is the location of the initial front that is convected through the domain, as can be seen in figure 1. The stabilized finite element formulation of (2) can be written as follows:

$$\int_{\Omega} (w + \alpha h \frac{\mathbf{v}}{\|\mathbf{v}\|} \nabla w)(\mathbf{v} \cdot \nabla f) d\Omega = \int_{\Omega} (w + \alpha h \frac{\mathbf{v}}{\|\mathbf{v}\|} \nabla w) \ d\Omega \tag{4}$$

With $\alpha$ the SUPG stabilizing parameter which is chosen constant for every element and $h$ the typical element length. We define $h$ as the length of the projection of an element on the velocity vector as in figure 2.

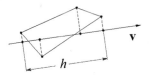

**FIGURE 2.**    Projection of element on $\mathbf{v}$

The magnitude of $\alpha$ is chosen based on the results as a compromise between the unstable solution with a small $\alpha$ and a highly diffusive solution with high $\alpha$ values.

# Determining the velocity field

The simulations are performed in two steps. First a velocity field will be determined with the use of an ALE / Eulerian formulation in the in-house code DiekA. The aluminium domain is chosen as half of the domain in figure 1, as determined by the axis of symmetry. The dimensions are chosen similar to the dimensions of the experiments, described in the next section. In figure 4 the boundary conditions on $\Gamma$ are shown. It is assumed that the aluminium sticks to the container wall and die, except for the bearing area where a frictionless slip situation is modeled. The normal condition is applied only on the corner node so that the velocity in x and y direction in the corner node is coupled with a direction that has a 45 degree angle with the symmetry axis. The aluminium domain is meshed with 4-node elements with bilinear interpolation function (Q4). After 100 to 500 time increments the velocity field reaches it steady state. These values are used in the second step of the simulations.

# EXPERIMENTS

The experiments were performed at BOAL. The billets were cut into cylindrical slices of approximately 15mm thick then joined together with copper foil between the slices. A total of 3 billets of 210mm were extruded and each was stopped at different remaining billet length (see table 1) The ram speed in the experiments was held constant at approxemately 1 mm/s and the initial temperature of the billet was approximately 723K. The exit temperature was not measured. However based on experience the estimated temperature rise due to the work done is about 100K. After the stop the billets were pressed out of the container. The deformation on the specimen, visible on the photos, is due to this extraction. Note that this deformation will work throughout the entire billet and also deform the position of the copper foil. The diameter of the billet and

TABLE 1.   Experiments

| Experiment | Remaining billet length | Extrusion time |
|---|---|---|
| Experiment 1 | 186 | 24 |
| Experiment 2 | 160 | 50 |
| Experiment 3 | 120 | 90 |

extrudate are 92mm and 20mm respectively. This is an extrusion ratio of approximately 21 which is slightly lower than usually used in every day practice. It is not known whether the discretisation of the billet with copper foil in between will cause a shear between the slices and therefore a different velocity field. To minimize this effect, not a continuous foil is used but a grid of copper. In between this grid the aluminium can weld together due to pressure and friction. Because of this the resulting lines in the billet are discontinuous, which is visible in figure 3a. In figure 3b the results are edited to improve the visual comparison.

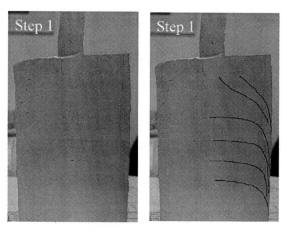

**FIGURE 3.** Experiment 1 after cutting

## ANALYSIS

In the postprocessor we use the steady state velocities and assemble the matrix and right-hand-side as described before and solve the system. We perform a number of analyses with $\Gamma_f$ at different equidistant locations. For every analysis the matrix and right hand side are identical, but since $\Gamma_f$ is different the system has to be solved for every analysis. From each analysis contour plots of $f$ are made. Next is a combination of plots from all analysis with only one contour line at $f = C$. These plots show the locations of different foils after extrusion time $t = C$. The combined plots are made at t=24, 50 and 90s (See table 1). A result is shown in figure 5b.

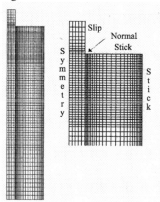

**FIGURE 4.** Mesh and boundary conditions

The axial symmetric mesh exists of approx. 5000 degrees of freedom with approx. 2500 rectangular elements. The time-iso-lines are plotted for initially straight fronts at y-positions from 0 up to 135 with slice thickness increments of 15 mm.

605

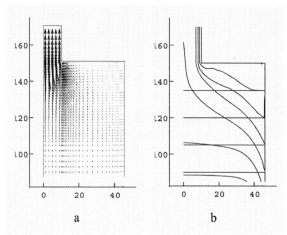

**FIGURE 5.** velocity field (a) and f-contour plot at 24s (b)

## Material Properties

The material is described rigid visco-plastic by the Sellars-Tegart law:

$$\sigma_y = s_m \text{arcsinh}\left(\left(\frac{\dot{\kappa}}{A}\exp(\frac{Q}{RT})\right)^{\frac{1}{m}}\right) \tag{5}$$

The material properties are based on the properties for AA6063 from Lof [2]. The properties for the material used in the experiments AA6061 are, except for $m$, chosen equal to AA6063. With parameter $m$ the results from the simulations are fitted on Experiment 1, resulting in a value $m = 10$.

**TABLE 2.** Material parameters for AA6063 alloy [2]

|  | Symbol | Used Value |
|---|---|---|
| Plastic properties (T=773K) | $s_m$ [MPa] | 25 |
|  | $m$ [-] | 5.4 |
|  | $A$ [1/s] | $6 \cdot 10^9$ |
|  | $Q$ [J/mol] | $1.4 \cdot 10^5$ |
|  | $R$ [J/molK] | 8.314 |

## COMPARISON

In this section we compare the results of the simulation with the experiments. In the simulations the deformation of the first ten front-lines are calculated. The deformations of the lines in the middle are used to fit the material properties so they correspond with the experimental results. Since the simulation is done with an Eulerian description the movement of the ram is not included. The difference between the experimental results in the lines around this fitted line can be attributed to several effects. Firstly the velocity field can be inaccurate. The difference can be caused by the fact that the determination of the velocity field is performed isothermally and in the experiment the billet will heat up locally due to deformation near the container wall, near and in the die. This will affect the yield stress and therefore the velocity field. Secondly the exact ram speed is not known from the experiment, so therefore deviations can occur. Also the Streamline Upwind diffusion in the postprocessing step can give deviations with the reality. Specially in regions where the velocity gradient is high a bigger error can occur. Furthermore, in the experiments the extraction of the billet has caused deformation. This deformation can be found in the non rectangular shape of the billet in figure 6. In this region the velocity

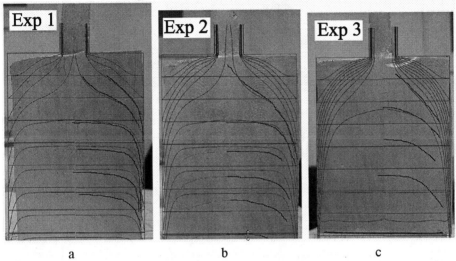

a                                    b                                    c

**FIGURE 6.**   Comparison between experiments and simulations at 24s (a), 50s (b) and 90s (c)

field is most smooth and easily convected, therefore the diffusive effect cannot cause the deviations and the velocity field is not correct.

## DISCUSSION

Based on the results we can conclude that the accuracy of the method is mainly determined by the accuracy of the velocity field. Extrusion is most of the times a 3D non isothermal process and the extension to 3D simulations with heat generation is the next step.

## REFERENCES

1.  J. ED. Akin, and T. E. Tezduyar, *Comput. Methods Appl. Engrg.*, Houston, 2004, pp. 1909–1922.
2.  Y. Lof, *PhD Thesis, University of Twente*, Enschede, 2000.

# Analysis of metal forming processes by using physical modeling and new plastic similarity condition

Z. Gronostajski, M. Hawryluk

*Institute of Mechanical Production and Automation Wroclaw University of Technology – ul. Lukasiewicza 3/5, 50-371 Wroclaw, Poland,*

*URL: www.pwr.wroc.pl*      *e-mail: zbigniew.gronostajski@pwr.wroc.pl*
*marek.hawryluk@pwr.wroc.pl*

**Abstract.** In recent years many advances have been made in numerical methods, for linear and non-linear problems. However the success of them depends very much on the correctness of the problem formulation and the availability of the input data. Validity of the theoretical results can be verified by an experiment using the real or soft materials. An essential reduction of time and costs of the experiment can be obtained by using soft materials, which behaves in a way analogous to that of real metal during deformation. The advantages of using of the soft materials are closely connected with flow stress 500 to 1000 times lower than real materials. The accuracy of physical modeling depend on the similarity conditions between physical model and real process. The most important similarity conditions are materials similarity in the range of plastic and elastic deformation, geometrical, frictional and thermal similarities. New original plastic similarity condition for physical modeling of metal forming processes is proposed in the paper. It bases on the mathematical description of similarity of the flow stress curves of soft materials and real ones.

**Keywords:** physical modeling, filia, similarity condition.
**PACS:** 62.20.Fe

## INTRODUCTION

In recent years the great progress has been reached in numerical methods, such as FEM or BEM for linear and non-linear 2D and 3D problems. The success of them depends very much on the correctness of the problem formulation and input data. The application of the real materials requires considerable investments in equipment and is time consuming. An essential reduction of time and costs can be obtained by using physical modeling. The main principle of such a modeling is to substitute the real material by the soft model material, which behaves analogously to the real metal during deformation.

The advantages of using the soft materials for study of the deformation of metals are closely connected with the fact that their flow stress is of 500 to 1000 times lower than that of real materials. Consequently there is possibility to use cheaper tool materials and simple and smaller presses. The proper results are depended on the similarity conditions between physical model and real process. The most important similarity conditions are

CP907, *10th ESAFORM Conference on Material Forming,* edited by E. Cueto and F. Chinesta
© 2007 American Institute of Physics 978-0-7354-0414-4/07/$23.00

plastic similarity in the range of plastic and elastic deformation, geometrical, frictional and thermal similarities. New original plastic similarity condition for physical modeling of metal forming processes was elaborated. It bases on the mathematical description of similarity of the flow stress curves of soft and real hard materials. The description of this plastic similarity condition for physical modeling is given in the paper.

## THE PLASTIC SIMILARITY CONDITION

One of the most crucial similarity conditions for metal forming analysis is the similarity of model and real materials behavior in the plastic range. The behavior of both materials during plastic is represented by their flow curves. Thus the plastic similarity condition can be simplified to selection of such a model material, which the shape of flow curve has the best matched to the flow curve of a real material.

The flow curse can be presented in a numerical or analytical form, very often as the complex functions.

One of the most simple and popular model describing flow curve is Alder-Phillips's model [1]

$$\sigma = C\varepsilon^{n}\dot{\varepsilon}^{m} \tag{1}$$

where: $C$ – material constant, $\dot{\varepsilon}$ – strain rate, $n$ – work hardening coefficient, $m$ – strain rate sensitivity.

It is commonly assumed that the plastic similarity condition is fulfilled when coefficients $n$ and $m$ of stress – strain curves for model and real material are very similar in equation (1). Studies conducted by numerous scientists, also by paper authors, have shown that the mathematical description of the plastic similarity condition presented above is not sufficient, especially for hot deformation condition then the increase of flow stress is observed at lower strain and softening for strain lager than the critical strain $\varepsilon_c$ (fig.1).

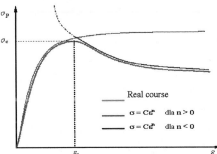

**FIGURE 1.** The approximation of the curve $\sigma_p$ - $\varepsilon$ by equation (1) for hot deformation conditions

In such a case the flow stress curve cannot be described in the whole strain range with only one coefficient $n$. Then the more precise description of the flow curve requires other description in the initial phase of deformation and in the larger strains. The acceptance of hardening coefficient as constant and greater from zero describes the monotonic growth of stress in whole range of deformations. Some authors, in order to describe the weakness of material, accept the negative values of coefficient n at large deformation. [2]. The differential courses of flow stress of the real materials is explained by recovery and recrystallisation processes. These phenomena do not occur during deformation of non - metallic model materials.

Some researchers for describing the course of flow stress use more complex function as Alder-Phillips's model. The essential difficulty in applying above mentioned models is great problem connected with estimation of many coefficients occurring in these models.

Due to many above presented disadvantages in applying model material the new description of the plastic similarity condition based on numerical description of flow curve of both model and real materials was elaborated [3-5]. It assumes the quantitative estimation of a matching of the flow stress curves of a model material and a real ones. Two parameters are used in the proposed description:

Scale coefficient

$$R = \frac{1}{k} \sum_{i=0}^{k} \frac{\sigma_i^r}{\sigma_i^p} \qquad (2)$$

and a similarity coefficient

$$t = \frac{\sum_{i=1}^{k} \dfrac{\left|\sigma_i^r - C\sigma_i^p\right|}{\sigma_i^r + C\sigma_i^p}}{2} \qquad (3)$$

Where: $\sigma_i^r$ – the flow stress of a real material at the point $i$, $\sigma_i^p$ – the flow stress of a model material at the point $i$, $k$ – the number of points on the flow curves of the model and real materials, for which the similarity coefficient is determined.

The coefficient $t$ is dimensionless and it allows to simple and quick determination of the matching of the flow curves of model and real materials. In the case of theoretically perfect matching of the both curves, the similarity coefficient is equal to zero. The graphic interpretation of the parameters appearing in the equations (2) and (3) are presented in the fig. 2.

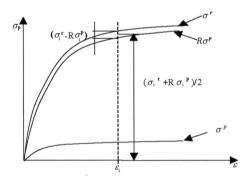

**FIGURE 2.** The graphic interpretation of parameters appearing in the equations (2) i (3)

Conducted research has shown that the accuracy of a physical model can be estimated on the basis of the coefficient $t$, which describes the similarity between a model material and a real material at conditions that other similarity conditions are correct. Assumptions are as follows:

- if values of $t$ belong to the range of 0 – 0,04: very good correlation is between model and real materials,

- if a values of $t$ belong to the range of 0,04 – 0,07 good accuracy of flow behavior in model and real materials exists but force parameters can be inaccurate
- if a value of $t$ is greater than 0,07, physical modeling does not describe a real process correctly.

Physical modeling with application of new description of plastic similarity condition has been used to analyze axisymmetric backward extrusion of lead and compare the two forging processes of CV joint body.

## Axisymmetric backward extrusion of lead

Schemes of axisymmetric backward extrusion is presented in Fig. 3. The researches were performed for four different wax mixtures having diversified matching factor of their flow curves to the flow curve of the real material, characterized by the parameter $t$. Chemical composition as well as values of the scale coefficient $R$ and similarity coefficient $t$ for particular mixtures are presented in the Table 1. The flow stress curves of the selected model materials and lead are shown in the Fig. 4.

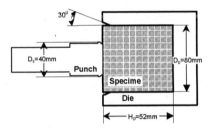

**FIGURE 3.** Schemes of the processes analyzed: a) axisymmetric backward extrusion: $D_p$ – specimen diameter, $H_p$ – specimen height, $D_s$ – punch diameter

**TABLE 1.** The t and R coefficient and chemical composition of the tested model materials

| Chemical composition | t | R |
|---|---|---|
| filia | 0,035 | 172,96 |
| filia+10% of lanolin | 0,158 | 306,85 |
| filia+8 %of kaolin | 0,043 | 113,1 |
| filia+16 % of kaolin | 0,049 | 62,35 |

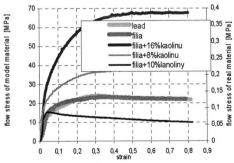

**FIGURE 4.** Flow stress vs. strain diagrams for the selected model materials and for the lead for the strain rate of $0,01 s^{-1}$

It was found on the basis of the similarity coefficient $t$ that the flow stress curve of filia the best matches the flow curve of the real material while the worst fitting was obtained for the mixture of filia and 10% of lanolin.

Deformation of grids placed on the model and real specimen surfaces were used for confirmation of chosen model material (Fig. 5).

Deformation of specimens made of the mixture of filia and 8 or 16% of kaolin is very similar to deformation of the pure filia and the real material. This is because these mixtures, similarly to filia, have low values of the similarity coefficient $t$. It testifies also to good matching of the flow curves of the model and real material. High values of the coefficient $t$ have been obtained for the mixture of filia and lanolin what proves wrong matching of the flow curves. The flow behavior of this material is also considerably different than the flow behavior of other mixtures.

**FIGURE 5.** Deformed coordination grids obtained for the lead and the different model materials

## Forging of the casing of CV joint body

The new plastic similarity condition was applied to to compare two forging processes of CV joint body. They differ both the shape of the die and initial perform dimensions. Unfortunately, it could not be done in industrial process because of the lack of results in real process.

In the first process the arched die and perform with diameter of 55 mm and height of 76.45 mm were used and in the second process the conical die and perform with diameter of 50 mm and height of 91.75 mm were applied. A schematic of the tools and preforms is shown in figure 6.

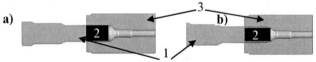

**FIGURE 6.** Schematic of tools: a) conical die, b) arched die, 1 – punches, 2 – preforms, 3 – dies.

In order to match model materials to the real material (steel UC1) different wax mixtures were tested and a *database of model materials* was used [6].

A preliminary qualitative analysis was made to select model materials best matching the real material flow curves. Because in the forging process the most operations were performed at 1000 °C and with strain rate of $10^{-1}$ therefore the stress – strain curve of model material should be matched to the curve of real material just at this conditions. The final choice was made using a new plastic similarity condition. The chemical composition, scale coefficient $R$ and similarity coefficient $t$ for the particular mixtures are shown in table 2. According to the similarity coefficient values given in table 2, the stress – strain curve of filia wax is best fitted to the UC1 flow curve.

**TABLE 2.** Coefficients $t$, $R$ and chemical composition of materials used for modelling forging of steel UC1

| Chemical composition | t | R |
|---|---|---|
| filia + 5% paraffin + 5 %lanolin | 0.06 | 269.3 |
| filia + 10 % paraffin + 10 % lanolin | 0.09 | 307 |
| filia + 5% Vaseline + 5 % lanolin | 0.035 | 344.6 |
| filia | 0.023 | 366.7 |

Figure 7 shows the deformed forgings in physical modelling with flow lines for a) arched dies and b) conical dies. Macroscopic examination revealed differences in the material flow for both tools. When the arched dies and a smaller diameter preforms are used, the material flow in the specimen's cross section is more uniform than in the case of the conical tools. It is confirmed by lower bending of the flow line for arched die than for conical die ($l_a < l_b$).

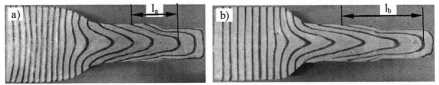

**FIGURE 7.** Comparison of model material flow in: a) arched tools, b) conical tools.

## CONCLUSIONS

The new approach to the description of plastic similarity condition, proposed by the authors, enables easy and quick qualitative and especially, quantitative matching of a model material to a real one. Under this condition physical modelling is very efficient way to analyse complex metal forming processes and can be used for verification of the designed processes by FEM.

The used of improvement of physical modeling together with FEM is very chip and economical method of metal forming design.

## REFERENCES

1. J. Alder and K.A. Phillips, The effect of strain–rate and temperature on the resistance of aluminum, copper and steel to compression, *J. Inst. Metals*, 83 (1954) 80–88.
2. R. Cacko, Aplication on wax material to simulation of deformation behaviour for structural component, *The Technical Universitet of Denmark Institute of Manufacturing Engineering, Internal report*, (1995), Chapter 4.
3. Z. Gronostajski, Physical and mathematical modelling of metal forming processes, In *Metal Technology Information*, Gliwice, 2003, 213-256.
4. Z. Gronostajski, M. Hawryluk, P. Karbowski, P. Bandoła: Physical modeling application for analysis of metal forming processes, *Journal of Machine Engineering*, 6 (2006) 124-133.
5. Z. Gronostajski, M. Hawryluk, Possibility determination of new mathematical description of plastic similarity condition in physical modeling of metal forming processes application, *Prace Naukowe-Politechnika Warszawska, Mechanika*, z.. 207 (2005) 115-120.
6. Hawryluk M.: Influence of plastic similarity condition on accuracy of physical modeling the processes of extrusion, Ph. D. Thesis, Wroclaw 2006.

# An upper bound solution for the spread extrusion of elliptical sections

K. Abrinia and M. Makaremi

*Mechanical Engineering Department, University College of Engineering, University of Tehran
Kargar Shomali St., PO Box 14395-515, Tehran, I.R.Iran
Email: Cabrinia@ut.ac.ir*

**Abstract:** The three dimensional problem of extrusion of elliptical sections with side material flow or spread has been formulated using the upper bound theory. The shape of the die for such a process is such that it could allow the material to flow sideways as well as in the forward direction. When flat faced dies are used a deforming region is developed with dead metal zones. Therefore this deforming region has been represented in the formulation based on the definitions of streamlines and stream surfaces. A generalized kinematically admissible velocity field was then derived for this formulation and strain rate components obtained for the upper bound solution. The general formulation for the deforming region and the velocity and strain rate fields allow for the optimization of the upper bound solution so that the nearest geometry of the deforming region and dead metal zone to the actual one was obtained.

Using this geometry a die with similar surfaces to those of the dead metal zone is designed having converging and diverging surfaces to lead the material flow. The analysis was also carried out for this die and results were obtained showing a reduction in the extrusion pressure compared to the flat faced die. Effects of reduction of area, shape complexity, spread ratio and friction on the extrusion process were also investigated.

**Keywords**: spread extrusion, elliptical section, analytical solution, upper bound, stream line and surfaces.
**PACS:**

## Introduction

Analytical solutions to the problem of forward extrusion have been considered difficult and sometimes impossible by many workers. However many workers have presented solutions for this problem. One of the early works done on this field was that of Chen and Ling [1] who presented an upper bound solution to axisymmetric extrusion problem. Extrusion of elliptical section from round billet was analyzed by Nagpal and Altan [2] who used dual stream functions, with appropriate numerical techniques, allowing determination of an optimum smooth die configuration. Yang and Lee [3] presented an analysis of three dimensional extrusions of sections through curved dies using conformal transformation technique. A new approach for the generalized three dimensional extrusions of sections from round billets by conformal transformation was later presented by Yang, Kim and Lee [4]. Hoshino and Gunasekara [5], [6] and [7] published their work on upper bound solutions to the extrusion of square, rectangle and polygonal sections from round bars through converging dies. Chitkara and Abrinia [8] gave a generalized upper-bound solution for three-dimensional extrusion of shaped sections using CAD/CAM bilinear surface dies. Abrinia and Bloorbar [9] presented their formulation for the solution to the three dimensional problem of the extrusion of shaped sections using a deforming region which unlike previous work did not

CP907, *10th ESAFORM Conference on Material Forming,* edited by E. Cueto and F. Chinesta
© 2007 American Institute of Physics 978-0-7354-0414-4/07/$23.00

considered the entry and exit to the deforming region as flat surfaces but rather curved surfaces.

In this paper the method of reference [14] has been developed further to take into consideration the particulars of the forward extrusion of shaped sections with larger dimensions than the initial billet.

# Theory

In order to develop the theory, a deforming region for the extrusion of shaped sections with larger dimensions than the initial billet is defined.

**Deforming region:** As shown in figure (1), the deforming region is defined and $OBB'O'$ is taken as a stream surface and $BB'$ as a streamline. Any point in the deforming region could be defined by the following vector:

$$\vec{R} = f(u, q, t)\vec{i} + g(u, q, t)\vec{j} + h(u, q, t)\vec{k}$$

Where f, g and h define the positions of x, y and z coordinates and u, q and t are parameters by changing of which all the points in the deforming region are defined:

$$t = L/z \qquad q = n\varphi/\pi \qquad u = OB/R$$

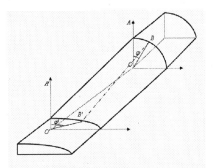

**Figure (1)**-Illustration of the deforming region for the spread extrusion of elliptical section

The surfaces of discontinuities on which the material flows could be defined by:

$$\vec{r} = \begin{bmatrix} 1 & t \end{bmatrix} \begin{bmatrix} -1 & 0 \\ -1 & 1 \end{bmatrix} \begin{bmatrix} \vec{r}_1 \\ \vec{r}_2 \end{bmatrix}$$

The entry and exit surfaces of discontinuities are defined as follows:

$$\vec{r}_1 = OB\sin\varphi\vec{i} + OB\cos\varphi\vec{j}$$

$$\vec{r}_2 = O'B'\sin\varphi'\vec{i} + O'B'\cos\varphi'\vec{j} + L\vec{k} = F_2\vec{i} + G_2\vec{j} + L\vec{k}$$

A kinematically admissible velocity field based on the above deforming region is given by:

$$v_x = \frac{f_t}{h_t}M, \ v_y = \frac{g_t}{h_t}M \text{ and } v_z = M(u,q,t)$$

Where M is a function which is obtained from the incompressibility conditions

615

## *Upper bound solution:*

The upper bound solution is given by:

$$J^* = W_f + W_i + W_e + W_x$$

Where $W_f$ is the power due to friction $W_i$ and is power due to internal deformation and $W_e$ and $W_x$ are the components of power due to velocity discontinuities at entry and exit.

# Results and Discussions

The analytical formulation presented in this paper was applied to the case of the forward extrusion of an elliptical section with larger dimensions than the initial billet and the following results were obtained.

In figure (1) the effect of the relative die length on the extrusion pressure and strain is shown for different ratios of $\dfrac{b}{2R}$ (ratio of the major half length of the ellipse to the diameter of the initial billet). It could be seen that as the ratio $\dfrac{b}{2R}$ increases the extrusion pressure also goes up. The optimum die length also increases form about 0.8 to 1.2 as the ratio $\dfrac{b}{2R}$ goes up. The strain is also very high for small die lengths and decreases as the die length increases. It could see that for optimum lengths of the die the strain value is small compared to smaller die lengths.

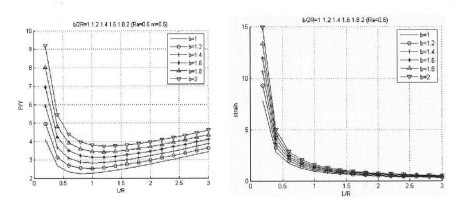

**Figure (2)**-Effect of the die shape and spread ratio on the (a) relative extrusion ratio (b) strain rate

The influence of the friction factor on the relative extrusion pressure could be seen in figure (3). Clearly as the value of friction factor increases, the pressure increases. The optimum die length also decreases as the friction goes up.

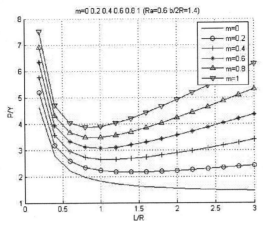

**Figure (3)**-The influence of the friction factor on the relative extrusion pressure

To study the effect of the reduction of area on the extrusion process figure (4)-(a) and on the strain figure (4)-(b) are illustrated. The more the reduction of area the more is the extrusion pressure. The optimum die length increases as the percentage of reduction of area increases. For the strain values however as the die length increases, they decrease and no matter how much the percentage of reduction of area increases, for larger die length no appreciable increase in strain is observed.

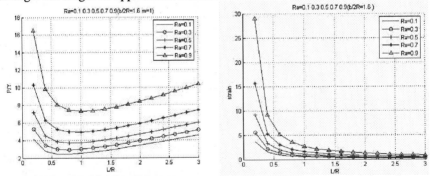

**Figure (4)**-The effect of reduction of area on the (a) extrusion pressure and (b) strain

In figures (4)-(a) and (4)-(b), the effect of aspect ratio of the elliptical cross section which expresses the complexity of the shape, on the extrusion pressure and strain values are shown respectively. Clearly as the aspect ratio increases and the shape gets more complex, the extrusion pressure increases and so does the strain. However it could be observed that for only very low aspect ratios (high shape complexities) the values of pressure and strain increase appreciably and otherwise small effects are observed. The optimum die length also increases as the shape of the extruded cross section becomes more complex.

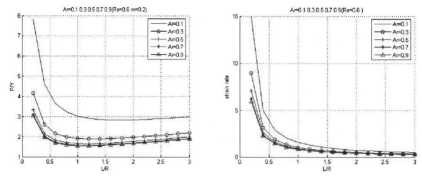

Figure (5)-The influence of aspect ratio on (a) relative extrusion pressure and (b) strain values.

# Conclusions

Considering the analysis carried out and the results presented the following conclusions could be drawn:

1- A new formulation based on upper bound was presented which was successfully applied to the three dimensional problem of extrusion of elliptical section with larger dimensions than the initial billet.

2- It was shown that optimum die lengths could be obtained for different process parameters which made the relative extrusion pressure minimum.

3- The larger the die length in general the smaller the strain values. Optimum die length usually delivers small strain value too.

4- It was concluded that although higher values of reduction of area make the relative extrusion pressure to go up but the strain values do not change appreciably provided the optimum die length is chosen.

5- The complexity of the final extruded cross section expressed by the aspect ratio which in this paper has been defined as the minor to major length of the ellipse only increases the extrusion pressure at very small values of the aspect ratio.

# References

1- Chen, C. T. and Ling, F. F. "Upper-bound solutions to axisymmetric extrusion problems" International Journal of Mechanical Sciences, Volume 10, Issue 11, November 1968, Pages,863-879 .

2- Nagpal V and Altan T. Analysis of the three dimensional metal flow in extrusion of shapes with the use of dual stream functions. 3rd. NAMRC Conference, Pa., Pittsburgh, May 1975.

3- Yang D.Y. and Lee C.H. "Analysis of three-dimensional extrusion of sections through curved dies by conformal transformation" International Journal of Mechanical Sciences 1978; 20(9):541-52.

4- Yang D.Y., Kim M.U. and Lee C.H. "A new approach for generalized three dimensional extrusion of sections from round billets by conformal transformation" IUTAM Symposium on Metal Forming Plasticity, Germany, 1979:204-21.

5- Hoshino S. and Gunasekera J.S. "An upper-bound solution for the extrusion of square section from round bar through converging dies", 21st. Machine Tool Design Research Conference 1980:97.

6- Gunasekera J.S. and Hoshino S. "Extrusion of non-circular sections through shaped dies", Annals of The CIRP 1980; 29(1): 141-5.

7- Gunasekera J.S, and Hoshino S. "Analysis of extrusion or drawing of polygonal sections through straightly converging dies", Transactions of the ASME Journal of Engineering for Industry 1982; 104:38.

8- Chitkara N.R., Abrinia K. "A generalized upper-bound solution for three-dimensional extrusion of shaped sections using CAD/CAM bilinear surface dies" 28th. International Matador Conference, 18-19 April 1990.

9- Abrinia K. and Bloorbar H., "A new improved upper bound solution for the extrusion of shaped sections using CAD techniques", COMPLAS VI Conference- Barcelona Spain- 2000.

# Experimental Investigation and Numerical Simulation During Backward Extrusion of a Semi-Solid Al–Si Hypoeutectic Alloy

Adriana NEAG[1,2], Véronique FAVIER[2], Regis BIGOT[3], Traian CANTA[1], Dan FRUNZA[1]

1.Technical University 103-105 Bwd.Muncii, 400641 Cluj-Napoca, Romania
2.LPMM, ENSAM, 4 rue Augustin Fresnel, 57078 Metz Cedex 3, France
3.LGIPM, ENSAM, 4 rue Augustin Fresnel, 57078 MetzCedex 3, France

**Abstract.** This work has been performed along two main directions. First of all we present the experimental results and effects obtained by backward extrusion tests on semi-solid aluminum alloy at three different forming temperatures and different holding times in isothermal conditions. The semi-solid billets were fabricated by the re-melting heat treatment method. Semi-solid extrusion tests were carried out to investigate the load-displacement curves and the deformation behaviour at different temperatures. The load level clearly decreases with increasing temperature and increasing holding time. Numerical simulations of semi-solid extrusion has been made too, using Forge 2005,. Experimental and simulated results are compared and discussed.

**Keywords:** semi-solid, backward extrusion, numerical simulation.

**PACS: 83.50.-v**

## INTRODUCTION

Semi-solid material forming (SSM) discovered in the early 1970s, is one of the near-net-shape forming processes which manufacture the final part by using the material at the temperature between liquidus and solidus. Spencer et al., discovered that the metallic alloys shows shear rate and time dependent flow behaviour in the semi-solid state. The solid fraction, the solid phase morphology and the thermomechanical history are the parameters which influence the semi-solid flow behaviour [1]. The semi-solid forming combines the advantages of conventional casting and forging. It benefits from the material capacity to behave like a solid at rest, that eases its transportation, and like a liquid under deformation, that eases its flow). Other major advantages include near-net-shape production of complex geometry, weight savings in components with less porosity than conventionally. In general, that components exhibits better mechanical properties than conventional casting or forging. The semi-solid processing had a significant impact in a number of industries including aerospace, automotive and electronic components [2].

CP907, *10th ESAFORM Conference on Material Forming*, edited by E. Cueto and F. Chinesta
© 2007 American Institute of Physics 978-0-7354-0414-4/07/$23.00

The purpose of the present study is to report the results of a mixed experimental-numerical method based on backward extrusion test and to examine the rheological behaviour of a semi-solid Al–Si hypoeutectic alloy.

## Materials and Experimental Procedures

### Material

The Al–Si alloy system is characterized by high specific strength, excellent corrosion resistance, good castability as well as good thermal and electrical conductivities. Normally hypoeutectic Al–Si alloys form coarse columnar and equiaxed αAl grains during solidification. The percentage of each depends on a multitude of parameters including cooling rate, pouring temperature and the temperature gradient in the liquid. The mould material can also influence thermal gradient established within the molten alloy.

The integrity of hypoeutectic Al–Si cast products is dependent on the fraction, size and morphology of primary αAl. The control of αAl formation is very important in foundry operations. The quality of the casting is improved by reducing the αAl grain size and manipulating its morphology, i.e. refinement process. Generally, as it was mentioned by Suéry in 2002 [3], there are three principal methods for achieving grain refinement: chemical modification [4] which produces fine fibrous silicon structure through the addition of several elements, such as sodium, antimony, potassium, calcium, strontium and barium; mechanical modification and thermal modification.

In our experimental procedures, we have utilized a hypoeutectic aluminum alloy with the chemical composition shows in Table 1.

**TABLE1.** Chemical compositions of aluminium alloy AlSi7Mg used for experiment.

| Si % | Fe % | Cu % | Mn % | Mg % | Al % |
|------|------|------|------|------|------|
| 7,86 | 0,42 | 0,25 | 0,16 | 0,27 | Bal. |

The aluminium alloy was melted at 750°C in a gas heated crucible furnace and formed in a permanent mould with a wall thickness of 9 mm. A covering flux (20% NaCl, 45% KCl, 35% NaF) was added at the beginning of melting. Argon was later purged through a stainless steel tube into the melt to degas hydrogen. The degassing time was 11 min. The molten alloy was then poured into the mould at a pouring temperature of 750°C, followed by water quenching. The ingot was 50 mm in diameter and 140 mm in height. A fine dendritic microstructure was observed.

### Partial Remelting

The cylindrical samples (39 mm diameter and 24 mm height) were machined in the axis of the billet. To reduce the temperature loss which occurs during the billet is transferred from the furnace to the press, the sample was heated together with the die (see Figure1a) in a cylindrical 1 kW electrical furnace. The whole sample-die system was heated at the investigation temperature. In our experiments, a 0.5 mm diameter (cromel-alumel) K-type thermocouple was inserted into the furnace, at the middle of the die. The theoretical liquidus temperature of AlSi7Mg alloy is 615°C and solidus temperature is 555°C. Generally, the solid fraction used is about 50%. Various temperatures were investigated in the range of 580–610°C. The isothermal holding times varied from 40 to 120 min. The holding time was measured from the moment when the system sample-die reached required temperature. The transfer time from electrical furnace to the press was 30s.

## Rheological testing - Backward extrusion

The deformation behaviour of the partially re-melted alloy was studied using backward extrusion. The semi-solid forming was carried out with the Heckert type hydraulic press, at the Technical University of Cluj-Napoca. The maximum capacity of the press is 200 kN. To limit the heat loss during the forming process, the inferior support of the press, where the die was located, was preheated at 550°C. In this experiment, the punch displacement, the time and the extrusion force are recorded simultaneously using a specific data acquisition system. The punch velocity in our experiment was about 1mm/s. An oil+graphite mixture was used for lubrication. A typical extrusion output is shown in Figure 1b.

As expected, the results prove that an increase in forming temperature and holding time lead to a decrease of the load needed to compress the sample. Higher temperatures increase the amount of liquid phase so that less load is needed for deformation. However, the load which is needed to deform the sample is still big [5] and we supposed that is due to the friction phenomenon, microstructure and disturbance problems in press.

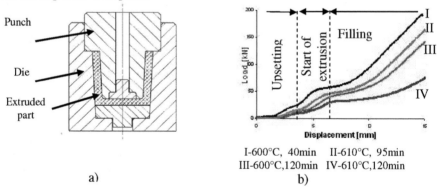

I-600°C, 40min    II-610°C, 95min
III-600°C,120min   IV-610°C,120min

a)                                        b)

**Figure 1**a) The system sample-tool; b) Experimental extrusion load as a function of punch displacement

## Analysis of experimental results

Aiming to elaborate a predicted behavior in terms of deformation load, temperature variation and material flow, a finite element simulation in FORGE 2005 code was made [1]. A good estimation can be done if good models are used as well as adequate constitutive law. The size and geometry of the sample and dies are the same as the experimental ones. Because of the symmetry, only one-half of the cross section needs to be considered in the analysis.

First, a cooling test for the sample-die system was numerically simulated with two sensors using the implicit solver. The simulation reveals that the experimental method of reheating allow us to begin the semi-solid forming at about 50% of solid fraction. Table 2 illustrates the temperature variation in the sample depending on the transfer time (between the furnace to the press).

**Table2** Temperature variation on the sample

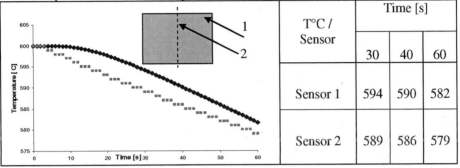

| T°C / Sensor | Time [s] | | |
|---|---|---|---|
| | 30 | 40 | 60 |
| Sensor 1 | 594 | 590 | 582 |
| Sensor 2 | 589 | 586 | 579 |

The constitutive equation for semi-solid material behavior is generally characterized by a power law, as show in several studies [6, 7]. A Norton-Hoff law was programmed:

$$\sigma = A * T^{m1} * \dot{\varepsilon}^{m2} \tag{1}$$

where: $\sigma$ is the stress tensor, A is the material consistency, $m_1$ is the sensitivity exponent of temperature T and $m_2$ is the sensitivity exponent of the strain rate $\dot{\varepsilon}$. Then, different simulation tests were performed, trying to identify the material parameters (see Table 3). A value of friction stress at the interface between the bodies in contact needs to be defined in the finite element analysis. In our case, the interface friction shear stress $\tau$ is:

$$\tau = \mu\sigma_n \qquad \text{if} \qquad \mu\sigma_n < m\frac{\sigma_0}{\sqrt{3}}$$

$$\tau = m\frac{\sigma_0}{\sqrt{3}} \quad \text{if} \qquad \mu\sigma_n > m\frac{\sigma_0}{\sqrt{3}} \tag{2}$$

where $\mu$ is the friction shear factor, m is the Tresca friction coefficient.

**TABLE3.** Values of modeling parameters identified on the experimental tests at 600°C

| Material consistency | $m_1$ | $m_2$ | Friction (Coulomb- Tresca) | | Thermal Exchange | |
|---|---|---|---|---|---|---|
| A | | | m | $\mu$ | Transf.Coeff | Emissivity |
| 40 | 0,01 | 0,5 | 0,3 | 0,15 | 2000 | 11763,62 |

The rheological behaviour in the Load-Displacement curves of simulation (see Figure 2a), can also be divided into three stages: first, when the material is upset (weak resistance of the semi-solid – low force), second, when initiate the flow and third, the filling of the die.

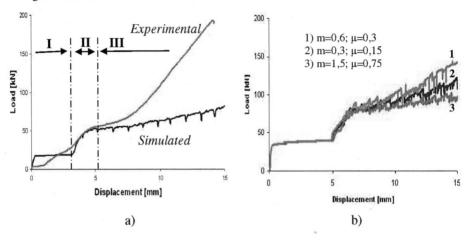

FIGURE 2 The comparison of experimental and simulated curves of backward extrusion force at 50% solid

In the first stage, it can be observed that the calculated values and measured values of the deformation force are different: the predicted curve displays a plateau whereas the experimental one exhibits a load increase with displacement. It is probably due to the viscoplastic constitutive equation that do not account for the history of the material deformation. In the second stage the deformation force presents an intensive growth when the material starts to flow in the tool cavity (Figure 3). This phase was selected to identify the constitutive equation parameters because it is the extrusion step. In this phase, and as wanted, the model shows a good approximation of the experimental data. In the third stage the material flows into the die. In this stage the experimental force grows intensively. It has been checked that this increase cannot be captured by the modeling whatever the friction coefficient value (see Figure 2b). Figure 4 reveals that the material temperature lowly decrease at the top of the die. Consequently, its consistency increases in those places. This effect could partly explain the strong load increase. However, though thermal exchanges and the temperature-sensitivity of the material consistency are accounted for in the modeling, the predicted load level is strongly underestimated. As it can be see, the applied rheological numerical model is considerably different respect to the experimental one. The strong load increase observed at the end of the forming process may be attributed to the evolution of both (i) solid fraction and (ii) semi-solid microstructure. The evolution of material microstructure due to deformation and temperature evolution is not taken into account into the constitutive equation and this could explain the discrepancy between experiments and simulations. These results demonstrate the difficulty in modeling the semi-solid behaviour with a viscoplastic law.

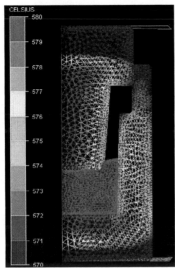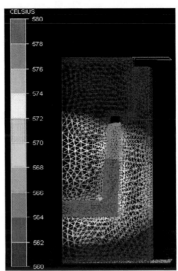

**FIGURE 3** Die filling at 1mm/s punch speed; second stage (left), final step at 65% $f_s$ (right)

## Conclusion

It is clear that the power law of Norton-Hoff appeared insufficient to describe some characteristics of semi-solid behavior such as the influence of strain rate, the temperature or the microstructure.

In our future works we will try to implement a micro-macro model based on the behavior of the liquid and the solid [8].

## REFERENCES

1. R.Bigot, V.Favier, C.Rooff, *J.Mat.Process.Technol* 160, 43-53 (2004)
2. H.V. Atkinson, *Prog.Mater.Sci.* 50, 341-412 (2005)
3. M. Suéry, «Obtention du matériau de base », in *Mise en forme des alliages métalliques à l'état semi-solide,* edited by Lavoisier, Paris 2002
4. K. Nogita, and A. K. Dahle, *Scripta Materialia* **48**, 307–313 (2003)
5. P.Kapranos., D.H.Kirkwood, H.V.Atkinson, Development of hypereutectic aluminium alloys for thixoforming based on the A390 composition, Proc.6th Int. Conf.Semi-Solid Proc of Alloy and Composites,Turin, 741-752, 2000
6. H.L.Yang, Z.L. Zhang, I. Ohnaka, J. *Mat. Process. Technol.* 151, 155–164 (2004)
7. S.Turenne, N.Legros, S.Laplante, F.Ajersch, *Met.and Mat.Trans.*vol.30A, 1138-1146 (1999)
8. V. Favier, C. Rouff, R.Bigot, , M.Berveiller, Micro-macro modeling of the isothermal steady-state behaviour of semi-solids, International Journal of Forming Processes, 7, 177-194, 2004

# 8 – MICROFORMING AND NANOSTRUCTURED MATERIALS

## (U. Engel and A. Rosochowski)

# Influences Of Size Effects
# On The Rolling Of Micro Strip

Koos van Putten, Reiner Kopp & Gerhard Hirt

*Metal Forming Institute, RWTH Aachen University, Intzestrasse 10, D-52056, Germany*

**Abstract.** Comparison between down-scaled flat rolling experiments of thin round wire and numerical simulation of those experiments have shown that the production process of manufacturing micro strip out of thin round wire is influenced by size effects. From plane strain compression tests, used as a physical simulation of the rolling process it is concluded that second order size effects of mechanical strength cause decreasing resistance to forming with decreasing wire diameters for rolling experiments with 25% and 50% reduction.

**Keywords:** rolling, plane strain compression, size effects, surface roughness, FEM simulation
**PACS:** 81.05-t, 81.20.Hy, 81.40.Ef, 81.40.Lm

## SIZE EFFECTS IN THE ROLLING OF MICRO STRIP

Comparison of physical experiments with numerical simulations, not including any size effects, have shown that the production of micro strips by the flat rolling of thin round wire is influenced by size effects [1]. Within the experiments the rolling process was scaled out of the macroscopic into the microscopic domain. Soft annealed OF-Cu rods of 1.5 m length cut out from wire with 4.0, 2.0, 1.0 and 0.5 mm in diameter ($d_{wire}$) were rolled flat with a reduction of 25%, 50% and 75% respectively.

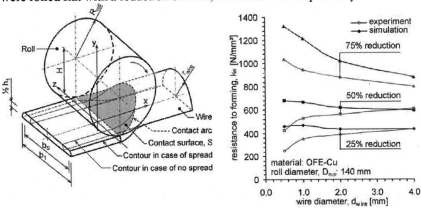

**FIGURE 1.** Left: Model of the flat rolling of a wire. Right: Resistance to forming for the flat rolling of pure copper (OF-Cu) wires with various wire diameters.

CP907, *10th ESAFORM Conference on Material Forming,* edited by E. Cueto and F. Chinesta
© 2007 American Institute of Physics 978-0-7354-0414-4/07/$23.00

The resistance to forming ($k_w$) of the rolling process is chosen as the variable for comparison between experiments as well as for comparison with finite element simulations. It is defined by the quotient of the rolling force and the contact area between the cylindrical roll and the flat rolled wire. The resistance to forming in dependency of the wire diameter for each combination of wire diameter and reduction is plotted in Figure 1. It shows that in the macroscopic domain, namely for 4 mm wire diameter, there exists a quite good agreement between experiment and simulation. However, with decreasing wire diameter the deviation between simulation and experiment increases. For 25% and 50% reduction there exists a completely different trend between simulation and experiment. For 75% reduction the trend between simulation and experiment is the same, but the absolute values of the resistance to forming differ. The differences between simulation and experiments indicate that the experiments are influenced by size effects.

## Analysis Of The Determined Size Effects

The different trends between simulation and experiment can hypothetically be explained by two size effects described in literature: First, the yield stress reduces with decreasing specimen size [2,3]. Second, the friction coefficient increases with decreasing specimen dimensions [3]. It is assumed that the effect of decreasing yield stress is dominant for the rolling experiments with 25% and 50% reduction and will cause decreasing resistance to forming with decreasing wire diameters. However for the rolling experiments with 75% reduction frictional effects are assumed to be of major importance and cause increasing resistance to forming with decreasing wire diameters.

## SIZE EFFECTS OBTAINED BY PLANE STRAIN COMPRESSION

To determine if wire flat rolling is truly affected by second order size effects of mechanical strength, plane strain compression (PSC) tests, used as a physical simulation of the rolling process, are scaled down with similarity [4]. The experiments are conducted with aid of a precision PSC device with exchangeable compression platens as described in [1].

## Specimen Preparation

All PSC specimens are made out of pure copper (OF-Cu) rolled sheet material. The length of the specimen is parallel to the rolling direction (RD), the specimen breadth is equal to the transversal rolling direction (TD). The specimen thickness ($h_0$) agrees to the normal direction (ND) of the rolled sheet.

Specimens of constant size and different mean grain size diameters are made out of 2.0, 1.0 and 0.5 mm thick sheet material. To obtain 90° specimen corners, the specimens with 2.0 and 1.0 mm thickness are milled and not cut. To overcome the negative effects of burrs and a poor quality of conventional cut faces and to obtain specimen with 90° corners, the specimen out of 0.5 mm sheet material are cut by wire

EDM. Varying the mean grain size is done by annealing the specimen at temperatures varying from 300 to 900 °C in a high vacuum oven for different times.

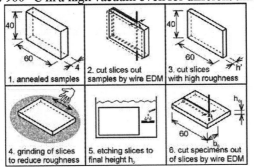

**FIGURE 2.** Scaled specimens maintaining geometrical similarity with the same mean grain size diameter and texture are all made out of the same 4 mm thick sheet material.

Scaled specimens maintaining geometrical similarity with the same mean grain size diameter and texture are all made out of the same 4 mm thick sheet material (Figure 2). First, the mean grain size is set by annealing the 4 mm sheet material at 700 °C for 30 minutes in a high vacuum oven. Then, slides of 2.1, 1.1 and 0.6 mm thickness (in ND) were cut out of the samples parallel to the RD-TD plane by wire EDM. The slides were grinded manually with 4000 grid grinding paper, to reduce the surface roughness. Following on, to overcome any effects of workhardened layers the slides were etched to their final thickness of 2.0, 1.0 and 0.5 mm. Finally the slides are cut in length direction to their accompanying breadth by wire EDM.

## Hall-Petch Effects In Scaled PSC experiments

In order to study the influence of the grain size diameter ($d_k$) at constant specimen dimension on the yield stress for decreasing specimen dimensions, PSC tests on sets of geometrical similar scaled specimens of 2.0, 1.0 and 0.5 mm thickness ($h_0$) with different grain sizes were performed.

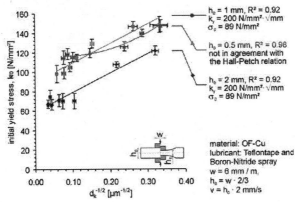

**FIGURE 3.** Influence of the grain size diameter at constant specimen dimension on the yield stress for decreasing specimen dimensions.

It is expected that the effect for each specimen size will be according to the Hall-Petch equation [5]. However, the Hall-Petch effect has never been obtained by PSC testing of miniaturized specimens. Figure 2 shows the relation between initial yield stresses with their accompanying standard deviations and the measured mean grain size diameters with their accompanying standard deviations.

The related pairs of standard deviations for each specimen size are connected by regression lines calculated by the method of least-squares best fit. The measurements of the 2 and 1 mm thick specimen can be, analog to the Hall-Petch relation, well fitted by a linear line. The Hall-Petch constant $k_y$ and $\sigma_0$ follow from the regression analysis. The measurements of the 0.5 mm thick specimens can only be suitably fitted by a parabolic line. This indicates that the behavior of small specimens characterized by a limited number of grains is not in agreement with the Hall-Petch relation.

## Size Effects In Surface Evolution

After the PSC of specimens with higher grain size diameter to specimen thickness ratio a change in surface roughness can be observed. From macroscopic metal forming processes it is known that the surface roughness of a specimen compressed by any tool reduces if the tool itself has a smooth surface [6]. Grains in the deformed zone of the specimen that are near the surface are unable to deform freely (as in e.g. tensile testing), they have to fulfill the boundary constraint imposed by the tool.

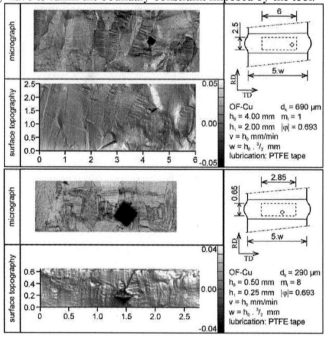

**FIGURE 4.** Correlation between surface topography and microstructure.

In spite of that, the surface roughness of the small PSC specimen with higher grain size diameter to specimen thickness ratio has increased. The $R_t$ and $R_z$ value of the compressed zone of a 4 mm thick specimen with $d_k/h_0$ of 0.17 are equal to 190 and 78 µm. For the compressed zone of a 0.5 mm thick specimen with $d_k/h_0$ of 0.58, $R_t$ and $R_z$ measure 123 and 47 µm. The surface roughness $R_a$, $R_t$ and $R_z$ of the compression platen measures 1.26, 11.7 and 8.8 µm respectively for all tools. The surface roughness of the compressed zone even exceeds the surface roughness of the tools by several orders of magnitude.

A comparison between the surface topography, measured by an opto-electronic 3D measurement system and a micrograph of the compressed specimen surface, shows a clear correlation between the surface topography (or roughness) and the grains in the specimen (Figure 4). The single grains from the micrograph can clearly be recognized in the surface topography. The influence of the Teflon lubricant is not completely clear yet. It might be, that the lubricant, as known from macroscopic forming with abundant lubrication, forms a viscous layer that enables to deform like a free surface. But preliminary PSC experiments with $MoS_2$ and without lubrication have also shown an increasing surface roughness for specimen with higher ratio of $d_k/h_0$.

## Free Surface Effects In Scaled PSC Experiments

Similarity scaled PSC tests were performed on specimens of 2.0, 1.0 and 0.5 mm thickness with constant grain sizes in order to study the effect of the ratio of grain size diameter to characteristic specimen dimension for decreasing specimen dimensions. Differences between the scaled processes maintaining geometric similarity are indicative of size effects.

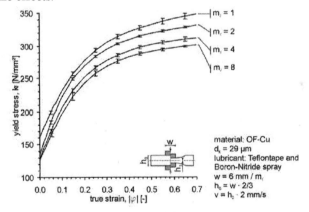

FIGURE 5. Effect of the ratio grain size diameter to characteristic specimen dimension for decreasing specimen dimensions scaled by different length scale parameters $m_l$ maintaining geometric similarity.

From tensile tests and cylindrical upsetting experiments, it is known that the yield stress reduces with ongoing miniaturization. This effect is attributed to the share of grains located at the specimen surface and the fact that the behavior of those grains differs from that of grains located within the specimen volume [3]. However, in tensile test all surfaces are free of any contact and in upsetting experiments the share of free surface is much bigger then the share of surface that contacts the tools. In contrast, in rolling the share of free surface is smaller then the share of tool surface. Until now, it is unknown if the effect of decreasing yield stress with ongoing miniaturization will be similar under rolling conditions or if the effect is diminished by the boundary conditions imposed by the roll or compression tool. The PSC experiments scaled with similarity show, despite a totally different ratio of free- to tool surface as in tensile and upsetting experiments, decreasing yield stresses with ongoing miniaturisation (Figure 5). From that it is concluded that decreasing yield stress with ongoing miniaturization occurs under rolling conditions as well.

## CONCLUSIONS

Down scaled PSC tests maintaining similarity, used as a physical simulation of the rolling process, have shown that second order size effects of mechanical strength also occur in circumstances quite similar to rolling. At small dimensions, the effect of the grain size diameter at constant specimen size on the yield stress is not according to the Hall-Petch relation. The surface roughness after the compression of the small PSC specimen is influenced by the grain size diameter to specimen thickness ratio and the topography of the roughness correlates with the micro structure. The effect of the ratio grain size diameter to characteristic specimen dimension is in agreement with the free surface model; decreasing yield stresses with ongoing miniaturization. From this it might be concluded that the assumption, that the effect of decreasing yield stress causes decreasing resistance to forming with decreasing wire diameters made for the rolling experiments with 25 and 50% reduction seems to be correct.

## ACKNOWLEDGMENTS

The authors gratefully acknowledge the financial support of the Deutsche Forschungsgemeinschaft (DFG) within the priority program (SPP) 1138: "Modelling of scaling effects on manufacturing processes". The authors also would like to express their thanks to Wieland-Werke AG for providing the copper wire and sheet material as well as Prymetall for the given possibility to use their precision rolling mill.

## REFERENCES

1. K. van Putten, R. Kopp and G. Hirt, "Size effects in the production of micro strip by the flat rolling of wire", edited by W. Menz et al., Proc. 2nd Int. Conf. Multi-Material Micro Manufacture, Grenoble-France, 2006, pp. 277-280.
2. R.T.A. Kals, "Fundamentals on the miniaturisation of sheet metal working processes", Ph.D. Thesis, Friedrich-Alexander University Erlangen-Nürnberg, Germany, 1999.
3. A Messner, „Kaltmassivumformung metallischer Kleinstteile", Ph.D. Thesis, Friedrich-Alexander University Erlangen-Nürnberg, Germany, 1997.
4. R. Hergemöller, „Anwendung der Ähnlichkeitstheorie auf Probleme der Umformtechnik", Ph.D. Thesis, RWTH Aachen University, Germany, 1982.
5. N.J. Petch, "The cleavage strength of polycrystals", J. Iron Steel Inst. 13, 1953, pp. 25-28.
6. Lange, K. Umformtechnik – Grundlagen (2nd ed.) Springer Verlag, Berlin, 2002.

# Processing Ultra Fine Grained Net-Shaped MEMS Parts Using Severe Plastic Deformation

Ruslan Z. Valiev*, Yuri Estrin†, Georgy I. Raab*, Milos Janecek¶, and Aikaterini Zi†

*Institute of Physics of Advanced Materials, Ufa State Aviation Technical University,
12 K. Marx Str., Ufa 450000, Russian Federation
†Institut für Werkstoffkunde und Werkstofftechnik, Technische Universität Clausthal, Agricolastrasse 6, D-38678 Clausthal-Zellerfeld, Germany
¶Department of Physics of Materials, Charles University, Ke Karlovu 5,
CZ-12116 Praha 2, Czech Republic

**Abstract.** This paper presents the results of investigation of the process of severe plastic deformation by means of ECAP method for producing the mini billets of pure aluminium. The strain state of the process and peculiarities of structure formation have been studied. The achieved results testify to the high efficiency of the process.
**Keywords:** SPD, ECAP, FE Simulation, MEMS.
**PACS:** 81.40.

## INTRODUCTION

Fabrication of miniaturized parts, particularly components for biomedical devices or microsystems such as MEMS, requires careful scaling of their internal structures. This calls for ultrafine grained (UFG) alloys as materials of choice. Indeed, UFG materials can fulfil two conditions at the same time: (i) that the average grain size be smaller than the smallest dimension of the structural component and (ii) that the grain size be small enough to ensure sufficiently high strength and fracture toughness. Condition (i) implies that reproducible properties of the component are warranted due to averaging over a large number of grains (e.g. over the thickness of a thin film, the cross-section of a wire or a tooth of a cog-wheel), while condition (ii) is an expression of the fact that strength and fracture toughness are generally improved on grain refinement.

Among the grain refinement techniques, severe plastic deformation (SPD) has advanced to the most promising one [1]. A significant body of literature has emerged over the past decade, cf. [2] and references therein. The arguably most popular SPD method, equal channel angular pressing (ECAP) [3], has been shown to produce sub micrometer scale grain structure in numerous metallic materials [4, 5]. Average grain sizes in the range of 50-300 nm are not uncommon in ECAP processed materials. In the quest for developing nanostructured materials for structural applications, numerous researchers are looking for the way to up-scale the known ECAP techniques and experimental rigs.

CP907, 10th ESAFORM Conference on Material Forming, edited by E. Cueto and F. Chinesta
© 2007 American Institute of Physics 978-0-7354-0414-4/07/$23.00

Rarely have there been attempts to *down-scale* the process with a view to manufacturing miniaturized parts. An interesting example of using an ECAP processed material for producing micro-gears was recently presented by Kim and Sa [6]. These authors have demonstrated the viability of micro-extrusion of ECAP pre-processed Mg alloy AZ31 using a millimetre-scale extrusion die as a way of manufacturing micro-gears. Zi et al. [7] have suggested to down-scale the ECAP process itself in order to produce severely deformed aluminium wires with a fine-scale grain structure across the wire diameter (2 mm). After this first demonstration that grain refinement can be achieved in a millimetre-scale ECAP process, further experiments with an improved die design were carried out. The results of these experiments are reported below. It is shown by transmission electron microscopy and hardness measurements that a substantial grain refinement can be achieved. It is further suggested to combine the ECAP pre-processing and the final extrusion step to produce the desired profile of an axisymmetric product in an integrated die design.

## EXPERIMENTAL

Pure aluminum (99.99 %) was used as a test material. Prior to the ECAP experiment, the material was heat treated for 30 minutes at 130 °C. The grain size determined after heat treatment by light microscopy was about 200μm.

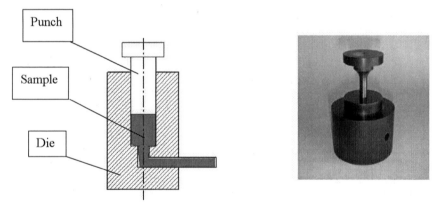

**FIGURE 1.** Schematics of the die design (left) and the actual die used in this work (right).

The miniaturized ECAP design used is shown in Fig. 1. The ECAP device consists of the following parts: the body where the die is placed, the die itself, a plunger and a cylinder, which is placed at the bottom of the device to support the die. A single channel die with a diameter of 3 mm is used. In order to facilitate the pressing, the feed material is introduced via a funnel-shaped channel, i.e. it is subjected to extrusion prior to the ECAP process. All parts were manufactured from tool steel 1.2344 (DIN X 40 Cr Mo V 5 1), except for the plunger, which was made from tool steel 1.2709 (DIN X 3 Ni Co Mo Ti 18 9 5). All parts were hardened to a maximum yield stress of 1900-2000 MPa.

The tests were carried out at room temperature on an INSTRON 5582 machine. The cross-head speed was 0.1 mm/min. A test was terminated when the punch covered the distance of 10 mm. The maximum load reached was about 37 kN. The microstructure of the ECAP processed material was determined using transmission electron microscopy. The first characterization of the mechanical properties was done by means of Vickers hardness measurements using a Struers Duramin hardness tester with a 50 g load applied for 10 sec.

## ECAP SIMULATION

A computer simulation was performed using Deform-3D, a program specially developed for solving forging tasks. The results of the simulation are presented in Fig. 2. The images show the process of strain accumulation at every stage of straining. The analysis of these data allows drawing several conclusions. The total average level of the accumulated strain at the die exit reaches the value of e=2.4. Also some inhomogeneity of the strain state in the cross-section of the billet is observed.

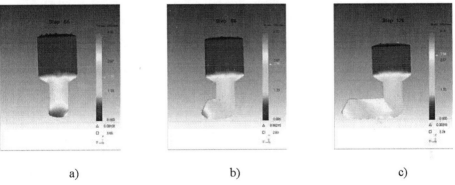

a)                                 b)                                 c)

**FIGURE 2.** The images of the stage-by-stage straining of the billet: a – extrusion, b – first stage of ECAP, c – stable stage of straining.

## EXPERIMENTAL RESULTS

In Figs. 3 and 4, the exemplary TEM micrographs in the transversal and longitudinal sections of the ECAP processed aluminium billet reveal a significant grain refinement achieved. The findings of the TEM investigation can be summarized as follows. The grain structure in the transversal sections (normal to the wire axis) is characterized by elongated grains 1 – 5 μm in length and 0.5 – 1 μm in width. In the longitudinal sections (along the wire axis), a rather equiaxed grain structure, with the average grain size of 1 – 2 μm, was observed.

**FIGURE 3.** Grain structure in the transversal cross-section after 'mini-ECAP' (left).
**FIGURE 4.** Grain structure in the longitudinal cross-section after 'mini-ECAP' (right).

Vickers hardness tests returned values of 42.5 HV 0.05 for the ECAP processed material – nearly double the value of 23.2 HV 0.05 exhibited by the unprocessed, heat treated Al. The observed extreme reduction of the grain size, from 200 μm to a few microns in a single ECAP step, is not uncommon in conventional ECAP processing [8], but a demonstration that this is also possible in a downscaled, miniaturized ECAP is a new and promising result. The concomitant increase in hardness is also an encouraging finding.

## FEASIBILITY OF AN INTERGRATED ECAP + EXTRUSION PROCESS

The tests performed provide a demonstration of the viability of miniaturized ECAP processing as a means of producing axisymmetric parts with millimetre-scale cross-sectional dimensions (Fig. 5a) and a grain structure in the micrometre-scale. With the aim of producing net-shape MEMS parts in mind, the micro ECAP process described above can be modified to allow for more than one ECAP pass within a die and also to include a subsequent extrusion step as part of an integrated process. The principal design of such a two-pass ECAP [9, 10] + extrusion die is presented in Fig. 5b.

To assess feasibility of the integrated two-pass ECAP + extrusion process, finite element (FE) simulations were conducted for 99.95% purity copper. As the axisymmetric part to be produced, a cylindrical part with the diameter of 1 mm was chosen. The input data for simulation were as follows. The temperature of processing: room temperature, the angle of intersection of pressing channels: 120°, the reduction ratio during pressing: ~10. The simulation was carried out using DEFORM 3D. The FE analysis of the proposed integrated process for manufacturing miniaturized parts showed a high level of cumulative strain (e=3.7 per pass, cf. Fig. 6a) that will likely result in the formation of ultrafine grained state in the processed Cu billets. However, this processing technique will pose a problem of dealing with the die working capacity because of a very high pressing force required for copper and the resulting die stress reaching 4 GPa, cf. Fig. 6b.

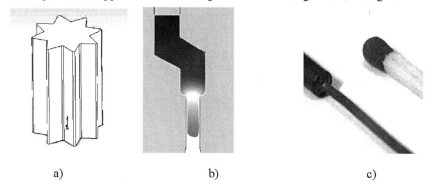

a)                              b)                              c)

**FIGURE 5.** Example of an axisymmetric article – a gear (a). Schematics of an integrated die design combining two ECAP passes (channel angle 120°) and a final extrusion step (b), Example of an extruded article (sample material) (c).

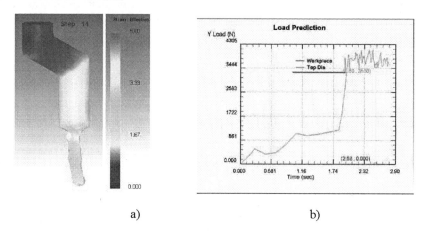

a)                              b)

**FIGURE 6.** The pattern of strain distribution in the bulk of the billet (a) and a diagram showing the pressing force (b) for integrated ECAP processing of Cu at room temperature.

Thus, manufacturing of miniaturized parts from high-strength metals using this method is not likely to be feasible without modifications. New approaches that would reduce specific pressing forces considerably need to be applied. The use of superplastic forming [11] should be considered as the most substantive step to be taken in this direction. A fortunate circumstance is that ECAP processing promotes superplasticity [5]. Integrated processing of aluminium that exhibits superplastic behaviour in the ultrafine grained state has verified this approach. Figure 5c shows a successfully processed part with the dimensions suitable for miniaturized gear-wheels manufacturing. The aim of the ongoing studies is to optimize the integrated processing technique suggested and to identify materials for which it can be used. The primary goal is to find suitable combinations of the material and the processing conditions for which it will be possible to capitalize on the superplastic properties developing during the ECAP stage of the processing pathway.

## CONCLUSIONS

The experiments performed on pure Al, as well as finite element simulations conducted, have demonstrated that a viable technology for manufacturing ultrafine grained MEMS parts can be developed on the basis of the processes proposed. In particular, an integrated two-step ECAP + extrusion process appears to be attractive, especially as the die profile at the exit end of the extrusion channel can be shaped to produce the desired profile of a MEMS part, for instance of a cog-wheel. Strains accumulated in such processes were shown to be at a level sufficient in principle for establishing conditions for superplastic forming, which may provide serious benefits in the manufacturing of MEMS parts. Further research will be directed at optimising the process design and producing prototype MEMS parts using the technology proposed.

## ACKNOWLEDGEMENTS

The preparation of this paper was made possible through partial support from the Russian Foundation for Basic Research under grant RFFI 06-08-00635-a as well as from the German Academic Exchange Service (DAAD) and the Academy of Sciences of the Czech Republic (AVCR) under grant D8-CZ5/06-07.

## REFERENCES

1. R. Z. Valiev, R. K. Islamgaliev and I. V. Alexandrov, *Prog. Mat. Sci.*, **45**, 103-189 (2000).
2. R. Z. Valiev, Y. Estrin, Z. Horita, T. G. Langdon, M .J. Zehetbauer and Y. T. Zhu, *JOM*, **58**, 33-39 (2006).
3. V. M. Segal, *Mater. Sci. Eng.*, **A197**, 157-164 (1995).
4. Z. Horita (ed.), Nanomaterials by Severe Plastic Deformation, Proc. of NanoSPD3 Conference (held at Fukuoka, Japan on September 22-26, 2005), *Materials Science Forum,* **503-504**, Trans Tech Publications Ltd, Switzerland (2006).
5. R. Z. Valiev and T. G. Langdon, *Prog. Mat. Sci.*, **51**, 881-981 (2006).
6. W. K. Kim and Y. W. Sa, *Scripta Mater.*, **54**, 1391-1395 (2006).
7. A. Zi, Y. Estrin, R. J. Hellmig, M. Kazakevich and E. Rabkin, *Sol. Stat. Phenom.*, **114**, 265-269 (2006).
8. S. C. Baik, R. J. Hellmig, Y. Estrin and H. S. Kim, *Z. Metallkunde*, **94**, 754-760 (2003).
9. G. I. Raab, R. Z. Valiev, G. V. Kulyasov, V. A. Polozovsky, Russian Patent No. 2181314 (2002).
10. G. I. Raab, *Mater. Sci. Eng.*, **A410-411**, 230-233 (2005).
11. Y. Saotome, H. Iwasaki, *J. Mater. Proc. Tech.*, **119**, 307-311 (2001).

# New Schemes of ECAP Processes for Producing Nanostructured Bulk Metallic Materials

Georgy I. Raab, Alexander V. Botkin, Arsentiy G. Raab, and Ruslan Z. Valiev

*Institute of Physics of Advanced Materials, Ufa State Aviation Technical University, K. Marx Str., 12, Ufa, 450000, Russia*

**Abstract.** During the last decade severe plastic deformation (SPD) has become a well established method of materials processing used for fabrication of ultrafine-grained (UFG) materials with advanced properties. Nowadays SPD processing is rapidly developing and is on the verge of a transition from lab-scale research to commercial production. This paper focuses on several new trends in the development of SPD techniques for effective grain refinement aiming to reduce the material waste and to obtain uniform UFG structure and properties in bulk billets.

**Keywords:** SPD, ECAP, FE Simulation.
**PACS:** 81.40.

## INTRODUCTION

High-pressure torsion (HPT) and equal-channel angular pressing (ECAP) are the SPD techniques that were first used to produce nanostructured metals and alloys possessing submicron- or even nano-sized grains [1, 2]. Since the time of the earliest experiments, processing regimes and routes have been established for many metallic materials, including some low-ductility and hard-to-deform materials [3-5]. High pressure torsion and ECAP die sets have also been essentially modernized. However, to date these techniques have been usually used for laboratory-scale research. The requirement of economically feasible production of ultrafine-grained metals and alloys that is necessary for successful commercialization raises several new problems in the SPD techniques development. The most topical tasks are to reduce the material waste, to obtain uniform microstructure and properties in bulk billets and products, and to increase the efficiency of SPD processing.

We solve these tasks by developing continuous ECAP and other ECAP's modifications [6,7] and multi-step combined SPD processing [8] for fabrication of long-sized rods in the framework of a joint IPP-DOE project with Los Alamos National Laboratory, MN, USA that aimed at setting up commercial production of nanostructured Ti materials for medical applications. Some new results of these works are presented below.

## CONTINUOUS ECA-PRESSING

So far, among all SPD techniques, equal-channel angular pressing (ECAP), also known as equal-channel angular extrusion (ECAE) [6], has attracted most attention,

CP907, *10th ESAFORM Conference on Material Forming,* edited by E. Cueto and F. Chinesta
© 2007 American Institute of Physics 978-0-7354-0414-4/07/$23.00

because it is very effective in producing UFG structures and can be used to produce UFG billets sufficiently large for various structural applications [7, 9].

However, the ECAP technique in its original design has some limitations, in particular, a relatively short length of the work piece that makes ECAP a discontinuous process with low production efficiency and high cost. In addition, the ends of a work piece usually contain nonuniform microstructure or macro-cracks and have to be thrown away, thus a significant portion of the work piece is wasted and the cost of the UFG materials produced by ECAP is further increased.

The key to wide commercialization of UFG materials is to lower their processing cost and waste through continuous processing. Several attempts have been made to this end. For example, repetitive corrugation and straightening (RCS) [10, 11] has been recently developed to process metal sheets and rods in a continuous manner. The co-shearing process [12] and the continuous constrained strip shearing (C2S2) process [13] were recently also reported for continuously processing thin strips and sheets to produce UFG structures. However, the question of further improvement of microstructure uniformity and properties remains topical in the development of these techniques.

In our studies we have worked on combining the Conform process with ECAP to continuously process UFG materials for large-scale commercial production. In this invention, the principle used to generate frictional force to push a work-piece through an ECAP die is similar to the Conform process [14], while a modified ECAP die design is used so that the work-piece can be repetitively processed to produce UFG structures.

We have designed and constructed an ECAP-Conform set-up which is schematically illustrated in Fig. 1. As shown in the figure, a rotating shaft in the center contains a groove, into which the work piece is fed. The work piece is driven forward by frictional forces on the three contact interfaces with the groove, which makes the work piece rotate with the shaft. The work piece is constrained in the groove by a stationary constraint die. This set-up effectively makes ECAP continuous. Other ECAP parameters (die angle, strain rate, etc.) can also be considered.

In our recent work [6] we used commercially pure (99.95%) coarse-grained long Al wire with a diameter of 3.4 mm and more than 1 m in length for processing at room temperature with 1-4 passes using ECAP route C, i.e. the sample was rotated 180° between ECAP passes. The starting Al wire had a grain size of 5-7 μm. Presently we are working on processing similar rods from CP Ti (Grade 2).

Figure 2(a) shows an Al work piece at each stage of the ECAP-Conform process, from the initial round feeding stock to rectangular Al rod after the first ECAP pass. As shown, the rectangular cross-section was formed shortly after the wire entered the groove. The change was driven by the frictional force between the groove wall and the Al work piece. The frictional force pushed the wire forward, deformed the wire to make it conform to the groove shape. After the wire cross-section changed to the square shape, the frictional force per unit of wire length became larger because of larger contact area between the groove and the wire. The total frictional force pushed the wire forward from the groove into the stationary die channel, which intersects the groove at a 90° angle.

TEM observations showed that ECAP-Conform led to microstructure evolution typical of the ECAP process [15, 16]. Figure 2(b) clearly indicates that the ECAP-Conform process can effectively refine grains and produce UFG structures.

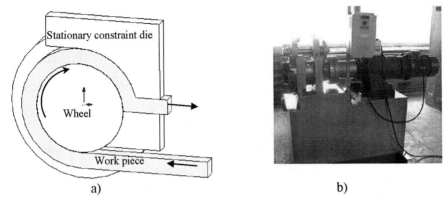

a)                                        b)

**FIGURE 1.** A schematic illustration of the ECAP-Conform set-up (a) and a machine (b).

The tensile mechanical properties of the as-processed Al samples after 1 to 4 passes are listed in Table 1. It is obvious that the ECAP-Conform process has significantly increased the yield strength ($\sigma_{0.2}$) and the ultimate tensile strength ($\sigma_u$), while preserving a high elongation to failure (ductility) of 12-14%. These results are consistent with those for Al processed by conventional ECAP [15]. We also found that, for CP Ti, there was more than a twofold strength increase after processing compared to the initial material; this fact is also consistent with Ti subjected to conventional ECAP.

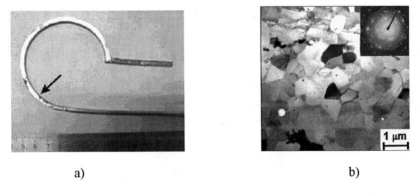

a)                                        b)

**FIGURE 2.** An Al work piece after processing by ECAP-Conform (a); a TEM micrograph of the longitudinal section of Al wire processed by four passes of ECAP-Conform (b).

Thus, the newly developed continuous SPD technique, ECAP-Conform can successfully produce UFG materials. The continuous nature of the process makes it promising for production of UFG materials on a large scale in efficient and cost effective manner. However, further study is needed to investigate its ability with respect to grain refinement and properties improvement of various UFG materials.

**TABLE 1.** Yield strength $\sigma_{0.2}$, ultimate tensile strength $\sigma_u$, elongation to failure $\delta$, and cross-section reduction (necking) $\psi$ of Al samples processed with 1 to 4 passes.

| Processing State | $\sigma_{0.2}$, MPa | $\sigma_u$, MPa | $\delta$, % | $\psi$, % |
|---|---|---|---|---|
| Initial Al Rod | 47 | 71 | 28 | 86 |
| After 1 Pass | 130 | 160 | 13 | 73 |
| After 2 Passes | 140 | 170 | 12 | 72 |
| After 3 Passes | 130 | 160 | 14 | 76 |
| After 4 Passes | 140 | 180 | 14 | 76 |

An FE simulation of the stress-strain state generated by the ECAP-Conform method was conducted using a dedicated "Deform 3D" program. The 6 mm diameter CP Ti wire was assumed as the initial material. The final dimensions of the cross-section of the processed wire were 5.3x5.3 mm. The process temperature was 200 °C. Friction coefficient was assumed to be within the range of 0.1 – 0.25. The angle of intersection between the channels was 120°. The character of friction and the stress-strain state were studied in the deformation region. The results of FE simulation are presented in the form of strain distribution (Fig. 3). Figure 3 shows that the front part of the billet, about two cross-sectional sizes long, is characterized by a non-homogeneous strain distribution; the remaining part is homogeneous, which means that the strain state satisfies the condition of stable flow. The averaged level of the accumulated strain for one pass is e=0.8, which corresponds quite well with the value e=0.75 calculated according to the formula e=1.18tg($\varphi$/2) [5] for the channel intersection angle of 120°. The investigation of strain distribution in the deformation zone shows that the most intense straining occurs over a distance of about 2.3 mm while the overall length of the strain growth region is 3.5 mm (Fig.4).

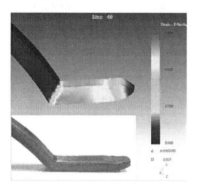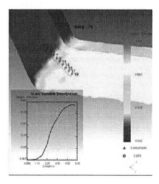

**FIGURE 3.** General view of the experimental billet and strain distribution in this billet subjected to ECAP-Conform at 200°C (3D simulation).
**FIGURE 4.** Zoom of the deformation region for ECAP-Conform performed at of 200°C.

# DEVELOPING ECAP WITH PARALLEL CHANNELS

Some early research made use of this approach [17,18] but the most recent development, combining a two-dimensional finite element method (2-D FEM) simulation and direct experiments, provides a very clear demonstration of the advantage of pressing material through two parallel channels [19-21]. The principle of this procedure is illustrated schematically in Fig. 5 [21] where $\Phi$ is the angle of intersection between the consecutive channels and $K$ is the distance between the channel axes.

A distinctive feature of ECAP with parallel channels is that, during a single processing pass, two distinct shearing events take place [20,21]. This means in practice that there is a considerable reduction in the number of passes required for the formation of an ultrafine-grained structure. The values for the channel distance, $K$, and the angle of channel intersection, $\Phi$, are the main parameters of the die geometry which influence both the flow pattern and the strain–stress state of the ECAP process.

The influence of the $\Phi$ and $K$ parameters on the flow pattern and strain homogeneity of a copper specimen was investigated using a 2-D FEM simulation [21]. It has been established that the optimal values of these parameters, leading to the largest strain homogeneity in the cross-section of the pressed billet, are $\Phi = 100°$ and $K \approx d_c$, where $d_c$ is the channel diameter or thickness. This means in practice that the optimum condition is achieved when the channel distance is approximately equal to the thickness of the channel. Under these conditions, the accumulated strain for one pass is approximately equal to 2.

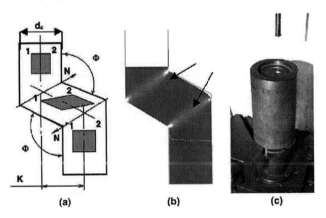

(a)          (b)          (c)

**FIGURE 5.** The principle of ECAP with parallel channels: (a) a schematic illustration where $N$ is in the shear direction, $K$ is the distance between the two channels, $\Phi$ is the angle of intersection between two consecutive channels and the internal shaded areas depicts the shearing as the sample traverses the shearing zone, (b) a view of the deformation zones obtained by 2-D FEM simulation for ECAP with parallel channels and (c) a general view of the experimental ECAP die-set where $\Phi = 100°$ and $K$ is equal to the channel diameter of 18 mm [21].

An important feature has been revealed concerning the nature of metal flow during ECAP in parallel channels. After one full pass, a mesh on the sample appears undistorted [21], thereby demonstrating that a uniform strain distribution is achieved including the tail-pieces. Thus, unlike in conventional ECAP, the sample shape after pressing remains identical to the initial sample.

The calculations results obtained for parallel channels were used for the design and fabrication of an ECAP die-set for operation at temperatures up to 500 °C, as shown in Fig. 5(c). Samples of Cu and Ti were produced by carrying out 4 passes of the process. The TEM observations showed that the structural refinement in these samples corresponded to the formation of an ultrafine grained structure observed after conventional ECAP when pressing through 8 passes. Furthermore, the UFG structure was rather homogeneous along the length of the bulk sample including the ends. Such a high microstructural uniformity is of much practical importance because of the increasing potential for utilizing the material after ECAP [19].

## CONCLUSIONS

Several new trends in SPD processing for fabrication of bulk nanostructured materials have been presented in this article, based on the recent results of our collaborative work on the development of commercial technology for the production of nanostructured Ti for medical and some other applications. We demonstrated how new tasks, connected with economically feasible production of UFG metals and alloys, can be solved by decreasing the material waste, obtaining homogeneous structure and properties in bulk billets and products using the modernization of ECAP processing.

## REFERENCES

1. R. Z. Valiev, N. A. Krasilnikov, N. K. Tsenev, *Mater. Sci. Eng.*, **A137**, 35-40 (1991).
2. R. Z. Valiev, A. V. Korznikov, R. R. Mulyukov, *Mater. Sci. Eng.*, **A168**, 141-148 (1993).
3. T. G. Langdon, M. Furukawa, M. Nemoto, Z. Horita, *JOM*, **52**, No. 4, 30-33 (2000).
4. T. C. Lowe and R. Z. Valiev, *JOM*, **56**, No 10, 64-68 (2004).
5. V. M. Segal, *Mater. Sci. Eng.*, **A197**, 157-164 (1995).
6. G. J. Raab, R. Z. Valiev, T. C. Lowe, Y. T. Zhu, *Mat. Sci. Eng.*, **A 382**, 30-34 (2004).
7. R.Z. Valiev, T.G. Langdon, *Prog. Mat. Sci.*, **51**, 881-981 (2006).
8. V. V. Latysh, I. P. Semenova, G. H. Salimgareeva, I. V. Kandarov, Y. T. Zhu, T. C. Lowe, R. Z. Valiev, *Mat. Sci. Forum.*, **503-504**, pp.763-768 in *Nanomaterials by Severe Plastic deformation*, edited by Z. Horita, Trans Tech Publications Ltd, Switzerland (2006).
9. R. Z. Valiev, R. K. Islamgaliev, I. V. Alexandrov, *Prog. Mater. Sci.*, **45**, 103-189 (2000).
10. J. Huang, Y. T. Zhu, H. Jiang, T. C. Lowe, *Acta Mater.*, **49**, 1497-1505 (2001).
11. Y. T. Zhu, H. Jiang, J. Huang, T. C. Lowe, *Metall. Mater. Trans.*, **32A**, 1559 (2001).
12. Y. Saito, H. Utsunomiya, H. Suzuki, T. Sakai, *Scripta Mater.*, **42**, 1139-1144 (2000).
13. J. C. Lee, H. K. Seok, J. Y. Suh, *Acta Mater.*, **50**, 4005-4019 (2002).
14. C. J. Etherington, *Eng. for Industry*, **No. 8**, 893 (1974).
15. Y. Iwahashi, Z. Horita, M. Nemoto, T. G. Langdon, *Acta Mater.*, **45**, 4733-4741 (1997).
16. M. Furukawa, Z. Horita, M. Nemoto, T. G. Langdon, *J. Mater. Sci.*, **36**, 2835-2843 (2001).
17. Z. Y. Liu, G. X. Liang, E. D. Wang, Z. R. Wang, *Mater. Sci. Eng.*, **A242**, 137-140 (1998).
18. Z. Liu, Z. Wang, *J. Mater. Proc. Tech.*, **94**, 193-196 (1999).
19. G. I. Raab, R. Z. Valiev, G. V. Kulyasov, V. A. Polozovsky, Russian Patent No. 2181314 (2002).
20. G. I. Raab, *Fizika i Tekhnika Vysokikh Davleniy* (Physics and Engineering of High Pressures), **14 (4)**, 83 (2004).
21. G. I. Raab, *Mater. Sci. Eng.*, **A410-411**, 230-233 (2005).

# Mesoscopic Model - Advanced Simulation of Microforming Processes

Stefan Geißdörfer, Ulf Engel and Manfred Geiger

*Chair of Manufacturing Technology, University of Erlangen-Nuremberg,*
*Egerlandstrasse 11, D-91058 Erlangen, Germany*

**Abstract.** Continued miniaturization in many fields of forming technology implies the need for a better understanding of the effects occurring while scaling down from conventional macroscopic scale to microscale. At microscale, the material can no longer be regarded as a homogeneous continuum because of the presence of only a few grains in the deformation zone. This leads to a change in the material behaviour resulting among others in a large scatter of forming results. A correlation between the integral flow stress of the workpiece and the scatter of the process factors on the one hand and the mean grain size and its standard deviation on the other hand has been observed in experiments. The conventional FE-simulation of scaled down processes is not able to consider the size-effects observed such as the actual reduction of the flow stress, the increasing scatter of the process factors and a local material flow being different to that obtained in the case of macroparts. For that reason, a new simulation model has been developed taking into account all the size-effects. The present paper deals with the theoretical background of the new mesoscopic model, its characteristics like synthetic grain structure generation and the calculation of micro material properties - based on conventional material properties. The verification of the simulation model is done by carrying out various experiments with different mean grain sizes and grain structures but the same geometrical dimensions of the workpiece.

**Keywords:** Simulation, Microforming, Synthetic grain structure, Simulation of process scatter.
**PACS:** 82.20.Wt, 87.53.Vb.

## INTRODUCTION

Microproduction technology is supposed to be one of the key technologies of the next years. As an evidence, the NEXUS market analysis of the microproduction technology [1] predicted an increasing market volume of this area of up to 68 billion US$ in 2005 with an annual growth rate of 20 per cent. One of the driving forces for this demand of microparts is the increasing sales figures of technical consumer products like digital cameras, camcorders, mobile phones and MP3 players. The metallic parts needed for these electro-mechanical systems, with characteristic geometrical dimensions in the range of few millimetres are up to now mostly manufactured by turning, milling or electrochemical processes. Putting in mind that these processes are predominately appropriate to small quantity production, other technologies like microforming are more likely to be used. If this technology, which is well suited for serial production at an industrial scale, is applied, some specifics issues have to be considered. When scaling down the geometrical dimensions from

CP907, *10th ESAFORM Conference on Material Forming,* edited by E. Cueto and F. Chinesta
© 2007 American Institute of Physics 978-0-7354-0414-4/07/$23.00

conventional scale to micro scale, so-called size-effects appear which are mainly caused by the material forming behaviour and a change in the friction conditions. The influence of the material on the forming process has been investigated for various microforming processes showing a significant dependency of the forming results on the initial state of the material structure. Further investigations have confirmed these influences to be the main reasons for the occurring size-effects [2, 3]. An approach to describe these effects, using simulation methods from a global point of view, was introduced in [4]. Based on the Ashby's theory, the influence of a free surface on the forming behaviour of microparts was considered in a simulation model by subdividing the material into three volume fractions with different material properties. This enabled the description of flow stress reduction in the open die processes when scaling down to a microscale.

Since the size-effect on the material flow is mainly controlled by the ratio of the mean grain size and specimen dimensions [4], there are, in general, two ways of investigating the influence of the material structure on the local forming behaviour. Firstly, the specimen's geometrical dimensions can be scaled down according to a similarity theorem [5]. However, in order to exclude the influence of friction, it is more favourable to keep the specimen's dimensions unchanged. This can be realized by applying various heat treatments to the specimens, in which the resulting fine grained and coarse grained material represents the micro and the macro case, respectively. This second approach is used in the present study, in which the cylindrical flat upsetting test of a micro specimen is considered.

## EXPERIMENTAL INVESTIGATIONS

As mentioned above, the flat upsetting test is well suited for the investigation of the influence of the material structure on the microforming results due to its robustness against varying friction conditions and the plain strain condition in the centre of the specimen required for 2D simulation. The experiments are carried out using a universal testing machine (UTS 5K with a Walter & Bai controller) which is equipped with a HBM 5 kN force measurement system. The specimens are made from CuZn15, a material frequently used for the production of microparts in electronic applications. The dimensions of the cylindrical workpieces are 0.5 mm in diameter and 3 mm in length. Even if the influence of friction on the experimental results is very small due to the chosen test, $MoS_2$ is used as solid lubricant to keep friction reproducible. The variation of material structures for different specimens is realized by applying different heat treatments. This leads to a variation of the mean linear grain size in a wide range, determined by standard methods (DIN 50601). 25 specimens per batch are chosen to get an adequately reliable statistical basis for the experimental investigations. The evaluation of the flat upsetting tests is done in terms of the mean load $\bar{F}$ as well as its scatter, characterized by standard deviation $s_F$ and a respective variation coefficient $s_F/\bar{F}$. In the case of fine grains, the total grain structure can be regarded as almost homogeneous yielding rather reproducible results. However, for coarse grains, the grain structure might be more different from specimen to specimen even if the nominal mean grain size is kept constant, causing the increasing scatter in material behaviour.

# SIMULATION APPROACH

The first attempt to describe the size-effect in the material forming behaviour has been made in [4]. This approach considers the influence of the different material forming behaviour on the integral flow stress while decreasing the size of the workpiece but it is not able to consider the influence of the material structure on the scatter of the forming results including shape evolution.

Thus the material definition of the simulation system has to be enhanced by describing the size-dependent local material forming behaviour. This has been realized by the development of a mesoscopic simulation model which generates, based on the Monte-Carlo-Potts algorithm, a synthetic material structure comparable to a real material structure. The material properties for each synthetic grain have been calculated individually based on its size and position within the specimen. As it is given by the theory of Hall-Petch and Ashby [6], the materials plastic response to external forces is mainly controlled by the grain size and the position of grain within a specimen. The first can be described by the correlation between grain size and flow stress as given by eq. (1) where $\tau$ is the shear stress, $\tau_0$ the critical shear stress, K the Hall-Petch slope and $d_G$ the mean grain size.

$$\tau = \tau_0 + \frac{K}{\sqrt{d_G}} \qquad (1)$$

If the forming behaviour over the scale of a few grains is to be considered, eq. (1) can be modified into eq. (2) taking into account the influence of neighbouring grains on the mechanical properties of a single grain.

$$\tau = \tau_0 + \frac{m_{1,2} \cdot \tau_i \cdot \sqrt{\delta}}{\sqrt{d_G}} \qquad (2)$$

In eq. (2), $m_{1,2}$ represents a transformation matrix considering the different sliding systems in adjoining grains, $\tau_0$ is the critical shear stress in the considered grain and $\delta$ represents the distance between a pile-up source and the dislocation. In this case, dislocations can be assumed to pile-up until the stress concentration at the grain boundary exceeds the stress $\tau_i$. From this point on, dislocation sources will be activated in the neighbouring grains.

Following the fact that eq. (2) is valid for a single grain only, the Hall-Petch factor $K_i$ (valid for a single grain) is given by eq. (3) and the correlation between the integral Hall-Petch factor K (cf. eq. (1)) and $K_i$ by eq. (4):

$$K_i = m_{1,2} \cdot \tau_i \cdot \sqrt{\delta} \quad \text{and} \quad K_i = K \cdot \xi_i \qquad (3)(4)$$

Thus, in case of a single grain, eq. (2) can be rewritten as:

$$\tau = \tau_0 + \frac{K \cdot \xi_i}{\sqrt{d_G}} \tag{5}$$

The newly introduced factor $\xi_i$ describes the amount of dislocation pile-up at the grain boundary region. Considering the influence of the free surface on the forming behaviour of the specimen and thus the non-capability of dislocations to pile up in this region, a more detailed view on the influence of the grain boundary on the properties is given in eq. (6).

$$\xi_i = \frac{\overline{k}_{nG}}{k_G} = \frac{1}{k_G} \frac{1}{\alpha_G} \sum \alpha_c \cdot k_{nG} \tag{6}$$

In eq. (6), $\overline{k}_{nG}$ represents the mean value of yield stresses $k_{nG}$ of the adjoining grains related to the real contact area fraction $\alpha_c/\alpha_G$ of the grains in the case of 3D studies. In the case of a free surface grain boundary it can easily be seen by eq. (6) that there $k_{nG}*\alpha_c$ is zero and thus decreasing the Hall-Petch slope $K_i$ for the considered grain. Furthermore, this approach even reflects the influence of the free surface on grains being located not directly at the surface due to the consideration of the ratio between $k_G$ and $k_{nG}$. Applying this model to macroscopic dimensions, there is a large number of smallest grains within the considered specimen and thus the influence of the free surface on the overall material properties is rather low.

A further enhancement of the mesoscopic model is done based on the theory of Meyers and Ashworth [7] considering the dislocation pile-up in the region of a grain boundary in a more detailed way. This is achieved by a subdivision of each single grain into two main regions, actually the grain boundary volume fraction $\alpha_{GB}$ yielding a higher flow stress $k_{f,GB}$ due to the occurring dislocation pile-up compared to the rather low flow stress $k_{f,I}$ within the material volume fraction $\alpha_I$ yielding no pile-up. The integral material behaviour of a single grain is expressed by linear superposition of the yield stresses of these two volume fractions.

$$k_f = \alpha_I \cdot k_{f,I} + \alpha_{GB} \cdot k_{f,GB} \tag{7}$$

In order to simplify the analytical model, the shape of a single grain is considered to be spherical. Thus, the above mentioned two volume fractions $\alpha_I$ and $\alpha_{GB}$ can be expressed by:

$$\alpha_{GB} = 2\left[ 3\left(\frac{t}{d_G}\right) - 6\left(\frac{t}{d_G}\right)^2 + 4\left(\frac{t}{d_G}\right)^3 \right] \text{ and } \alpha_I = 1 - \alpha_{GB} \tag{8)(9}$$

with a thickness of the grain boundary layer t and grain size $d_G$. Inserting eq. (8) and eq. (9) into eq. (7) leads to

$$k_f = k_{f,I} + 6(k_{f,GB} - k_{f,I}) \cdot t d_G^{-1} - 12(k_{f,GB} - k_{f,I}) \cdot t^2 d_G^{-2} + 8(k_{f,GB} - k_{f,I}) \cdot t^3 d_G^{-3} \qquad (10)$$

Obviously, different cross sections of a single grain will produce different area fractions of $\alpha_G$ and $\alpha_{GB}$. Hence Meyers and Ashworth recommended using the mean values of t and $d_G$, $\bar{t}$ and $\bar{d}_a$, which can be approximated by

$$\bar{t} = 1.57 \cdot t, \bar{d}_G = \frac{\pi}{4} d_G \qquad (11)$$

Considering the variation of the thickness t of the work hardening layer to be

$$t = (k_1 k_2 d_G)^{1/2} = k_{MA} d_G^{1/2} \qquad (12)$$

it has to describe two effects: the fluctuation of the stress field varies with $d_G$ if the grain size is decreased, leading to a dependency $t = k_1 d_G$ and the dislocation spacing to be unchanged and the dislocation interactions will dictate a constancy in t, thus $t = k_2 d_{G0}$. Assuming the term $\bar{t} d_a^{-1}$ to be approximately equal to $2 t d_G^{-1}$, eq. (10) can be written as

$$k_f = k_{f,I} + 12 k_{MA}(k_{f,GB} - k_{f,I}) \cdot d_G^{-1/2} - 24 k_{MA}^2$$
$$(k_{f,GB} - k_{f,I}) \cdot d_G^{-1} + 16 k_{MA}^3 (k_{f,GB} - k_{f,I}) \cdot d_G^{-3/2} \qquad (13)$$

In the case of grains being in the micrometer range, the $d_G^{-1/2}$ term dominates and thus the Hall-Petch relation is obtained where the Hall-Petch slope K is equal to

$$K = 12 k_{MA}(k_{f,GB} - k_{f,I}) \qquad (14)$$

Finally, inserting eq. (3) and eq. (5) into eq. (7), the flow stress of the boundary region $k_{f,GB}$ and the inner region $k_{f,I}$ of the grain can be calculated by

$$k_{f,GB} = k_{f,0} + \frac{K \cdot \xi_i}{\alpha_{GB} \cdot \sqrt{d_G}} \text{ and } k_{f,I} = k_{f,0} \qquad (15)(16)$$

The thickness of the grain boundary region t can be calculated by inserting eq. (14) into eq. (12). For the investigations in this paper, t was assumed to be

$$t = 0.133 \cdot d_G^{0.7} \qquad (17)$$

# VERIFICATION AND DISCUSSION

For the purpose of verification of the advanced mesoscopic model, the simulation results of the previously described flat upsetting test are compared with the results obtained experimentally. The parameters of interest for this comparison are the mean forming force and its standard deviation. As it is shown in Fig. 1a, the simulation model describes well the material forming behaviour in a scaled down process.

a)                                              b)

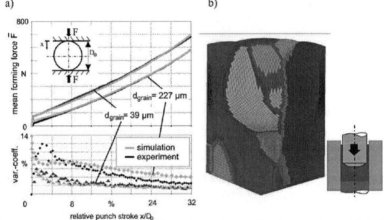

**FIGURE 1.** a) comparison between simulation and experiment using flat upsetting test b) influence of the material structure on the shape evolution of can in a backward extrusion process.

Since the shape evolution at microscale is mainly influenced by the grain orientation, the mesoscopic model presented in this paper gives only the first idea about the influence of the material structure on the local forming behaviour (Fig. 1b). Thus this model will have to be enhanced if a more detailed representation of shape evolution is required.

# ACKNOWLEDGMENTS

The authors would like to thank the German Research Foundation for the financial support. Also, this work was carried out within the EC Network of Excellence 4M.

# REFERENCES

1. R. Wechsun et al., Nexus Market Analysis for microsystems, 2000-2005. Nexus Task Force Market Analysis. Wicht Technologie Consulting. München, 2002.
2. M. Geiger, M. Kleiner, R. Eckstein, N. Tiesler, U. Engel, Microforming, Keynote Paper, Annals of the CIRP, 50 (2) (2001) 445-462.
3. U. Engel, R. Eckstein, Microforming - from basic research to its realization, J. Mater. Process. Technol. 125-126 (2002) 35-44.
4. U. Engel, A. Meßner, M. Geiger, Advanced Concept für the FE-Simulation of Metal Forming Processes for the Production of Microparts. In: Altan, T. (Edtr.): Advanced Technology of Plasticity 1996, Vol. II, p. 903 - 907
5. O. Pawelski, Beitrag zur Ähnlichkeitstheorie, Archiv für das Eisenhüttenwesen 35 (1964) 1
6. M.F. Ashby, The Deformation of Plastically Non-homogeneous Materials, Phil. Mag. 21, (1970) 254-255.
7. M. A. Meyers, E. Ashworth, Phil. Mag. A, 1982, 46, 737.

# FEM Simulation of Incremental Shear

Andrzej Rosochowski* and Lech Olejnik[†]

*Design, Manufacture and Engineering Management, University of Strathclyde, 75 Montrose Street, Glasgow G1 1XJ, United Kingdom
[†]Institute of Materials Processing, Warsaw University of Technology, Narbutta 85, 02-524 Warsaw, Poland

**Abstract.** A popular way of producing ultrafine grained metals on a laboratory scale is severe plastic deformation. This paper introduces a new severe plastic deformation process of incremental shear. A finite element method simulation is carried out for various tool geometries and process kinematics. It has been established that for the successful realisation of the process the inner radius of the channel as well as the feeding increment should be approximately 30% of the billet thickness. The angle at which the reciprocating die works the material can be 30°. When compared to equal channel angular pressing, incremental shear shows basic similarities in the mode of material flow and a few technological advantages which make it an attractive alternative to the known severe plastic deformation processes. The most promising characteristic of incremental shear is the possibility of processing very long billets in a continuous way which makes the process more industrially relevant.

**Keywords:** Incremental Shear, Severe Plastic Deformation, Ultrafine Grained Metals, FEM Simulation.
**PACS:** 81.07.Bc; 81.20.Hy; 81.40.Ef; 81.40.Lm.

## INTRODUCTION

Bulk ultrafine grained (UFG) metals draw substantial attention due to their unique mechanical and physical properties. For example, at cryogenic temperatures, an ultrafine grain size (<1 μm) doubles toughness of the material [1] while at high temperatures it leads to superplastic behaviour at a strain rate which is one order higher than for traditional superplastic materials [2]. At room temperature, UFG metals possess the yield strength which is 2-4 times higher than for their coarse grained equivalents. The ultimate tensile strength is usually doubled while elongation is reduced [3].

The preferable method of producing bulk UFG metals, which avoids the technical, economical and health issues associated with nanopowders, is severe plastic deformation (SPD). In this method, shear banding associated with large plastic deformation (true strain of approximately 3) causes the creation of sub-micron dislocation cells in coarse grains of bulk metals. As a result of further plastic deformation, these cells evolve into ultrafine grains with high angle boundaries. SPD processes are different from traditional metal forming processes because they retain the shape of the billet.

There are two groups of SPD processes – batch and continuous. Batch processes deal with discrete billets which are usually used for laboratory purposes to produce

CP907, 10th ESAFORM Conference on Material Forming, edited by E. Cueto and F. Chinesta
© 2007 American Institute of Physics 978-0-7354-0414-4/07/$23.00

samples for metallurgical and mechanical tests. The most popular batch process is equal channel angular pressing (ECAP), also known as equal channel angular extrusion [4]. In ECAP, a square or cylindrical billet is pushed from one section of a constant profile channel to the next section orientated at an angle $\geq 90°$ to the previous one. Plastic deformation of the material is caused by simple shear in a thin layer along the crossing plane of the channel sections. The length of the leading channel limits the length of the billet processed to about 6 lateral dimensions. It must not be too long to avoid an excessive force caused by friction and the associated tool design problems. This, together with end effects, causes poor utilisation of the material, often not exceeding 50% [5].

To address this problem, continuous SPD processes are being developed. One group of these processes is based on ECAP with different forms of the continuous feeding of a billet [6-9]. The main feature of all these processes is that feeding of billets is based on friction which, despite relatively low force required for ECAP, proves to be problematic. Another approach is accumulative roll bonding (ARB) [10] in which the sheet is hot rolled to 50% of its initial gauge, cut, cleaned, stacked and hot rolled again. This sequence can be repeated several times until a desired strain is achieved. Due to many intermediate operations involved and the manageable sheet size, ARB is not a true continuous process. Its success depends critically on the quality of the bond, which may be difficult to achieve. Metals subjected to ARB are made up of multiple layers of elongated grains. Yet another proposition is repetitive corrugation and straightening (RCS) [11]. RCS comprises bending of a straight plate/bar between corrugated rolls and then restoring the straight shape with smooth rolls. The process does not use simple shear, which is known to be the most appropriate mode of SPD. It has also problems with strain and structure uniformity.

This paper introduces an alternative approach based on the concept of incremental shear. Finite element method (FEM) modelling was instrumental in developing this new process. It provided vital clues regarding tool geometry and process parameters.

## CONCEPT OF INREMENTAL SHEAR

Figure 1a presents the schematics of a new process of incremental shear (IS). The process uses three simple dies. Dies A and B define the input channel while A and C the output channel for a billet. Dies A and B can be fixed or movable to provide billet clamping. Die C is a working die, which moves in a reciprocating manner at an appropriate angle to the billet. The movement of the working die is synchronised with the feeding movement of the billet. The sequence of operations is as follows:
(1) Die C is moved away from the billet by a distance of at least "a".
(2) This enables the billet I to be moved forward by a distance "a" so that it sticks out by the same distance beyond die B. Without contact with die C, feeding of the billet requires only a small force.
(3) The billet is fixed in position for the next stroke, for example, by supporting it by a punch located between dies A and B.
(4) The working die C moves in a predetermined way towards the billet causing plastic deformation of the billet in the narrow zone marked by the dashed lines. The billet assumes the form II.

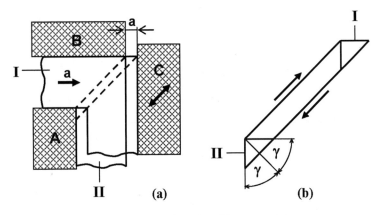

**FIGURE 1.** Schematics of incremental shear.

(5) The whole sequence described above is repeated as many times as necessary to process the whole length of the billet. Due to its incremental nature, IS can continuously process infinite billets.

The geometrical analysis of the material flow in the "dashed" zone in Fig. 1a leads to the conclusion that the mode of deformation is that of simple shear. For the purpose of process analysis it is split into two steps as shown in Fig. 1b. Shearing the initial parallelepiped I by the angle $\gamma$ produces a rectangle, for which the shear strain is tg$\gamma$ and the equivalent von Mises strain is $\varepsilon=\text{tg}\gamma/3^{0.5}$. Continuing this shearing by another angle $\gamma$ converts the rectangle to the final parallelepiped II and doubles the equivalent strain to $\varepsilon=2\text{tg}\gamma/3^{0.5}$. For $\gamma=45°$, the total equivalent strain is $\varepsilon=1.155$, which is the value known from the classical ECAP with the channel angle of 90°. Thus, in terms of the type and value of the strain produced, the proposed process is equivalent to ECAP.

# FEM SIMULATION

## FEM Model

A commercial FEM program Abaqus/Explicit was used to simulate the elastic/plastic flow of the material in IS. The tools were assumed to be rigid. The die channel was 10 mm wide. There was a billet feeding/supporting punch in the input channel and a reciprocating die at the channel exit. The feeding pattern of the punch followed a prescribed ramp while the reciprocating die followed a sine wave (Fig. 2). The synchronisation of these two movements allowed feeding to be carried out without the billet contacting the reciprocating die. The billet was divided into 900 plane-strain, bilinear, quadrilateral elements with reduced integration. Despite large strains, no re-meshing was attempted so the material flow was easier to observe and interpret.

The material used in the simulation was CP aluminium. It was modelled as an elastic-plastic, isotropic, Huber/Mises material with a strain-hardening curve described by $\sigma=159(0.02+\varepsilon)^{0.27}$ MPa. Friction was assumed to follow Coulomb's law with friction coefficient μ=0.1. The process reference velocity was increased to 1 m/s to reduce the computation time. Calculations were carried out with double precision.

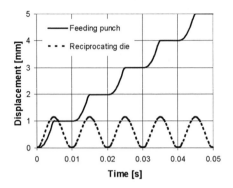

## Material Flow

According to the analysis above, the billet in IS undergoes simple shear like in ECAP. This has been confirmed by FEM results of the classical ECAP process and an IS process for which the tool geometry, billet properties and friction conditions were the same. The only difference was that the ECAP channel had an outer radius of 2 mm while the IS channel had no outer radius specified; the inner radius was 2 mm in both cases. The reciprocating die in IS was moving at an angle of 37.5° to the axis of the input channel. The feeding stroke was 1 mm. Figure 3 presents the equivalent strain results for both cases.

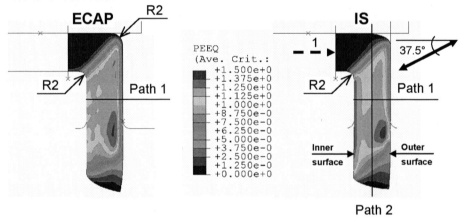

**FIGURE 3**. Distribution of equivalent plastic strain in ECAP and IS.

Some differences in strain distribution are better visible when observed across the billet. Figure 4a displays the strain distribution along Path 1 defined in Fig. 3. The strain resulting from ECAP reaches 1.15 at the inner surface, then climbs slowly to about 1.3 towards the outer surface and drops to about 0.6 on the outer surface. The highest value of strain produced by IS is also 1.3, however, it occurs on the inner surface. Then there is a strain plateau at 1.15, a week local maximum and a similar drop to 0.6 on the outer surface as in the case of ECAP. IS is relatively insensitive to the angle at with the reciprocating die works the billet. For the given channel geometry and friction coefficient of 0.1, the angle of 30°seems to be a good choice.

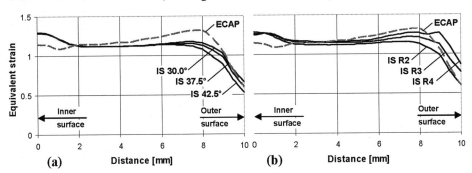

**FIGURE 4**. Distribution of equivalent plastic strain along Path 1 in ECAP and IS performed at 30°, 37.5° and 42.5° with the inner radius of 2 mm (a) and at 37.5° with the inner radius of 2, 3 and 4 mm (b).

As far as the channel geometry is concerned, the main parameter is the inner radius. Its effect on strain distribution was investigated by using three values equal 2, 3 and 4 mm. As illustrated in Fig. 4b, increasing the inner radius does not affect strain distribution on the inner surface. Instead, it slightly increases the plateau level and more visibly the value of strain closer to the outer surface.

Strain uniformity depends also on the feeding stroke (Fig. 5). The stroke of 1 mm used above gives pretty uniform strain distribution along Path 2 defined in Fig. 3. However, the stroke of 5 mm leads to strain undulations. The stroke of 3 mm, that is 30% of the billet thickness, seems to be acceptable.

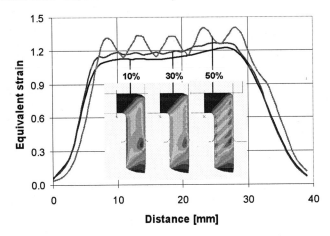

**FIGURE 5**. Distribution of equivalent plastic strain along Path 2 in IS performed at 37.5° for feeding strokes of 10%, 30% and 50% of the billet thickness.

# CONCLUSIONS

The new SPD process of incremental shear (IS) is similar to the well know process of equal channel angular pressing (ECAP). This is due to the same level and mode of deformation which is simple shear. Small differences are revealed by the FEM analysis of material flow. The details of strain distribution in IS depend on the inner radius of the die channel, the angle at which the reciprocating die works the material and the feeding stroke of the billet. The preliminary analysis carried out for friction coefficient of 0.1 suggests that the reasonable values for the above three factors are all equal to 30, either 30% of the billet thickness as in the case of radius and stroke or 30° for the working angle. More detailed analysis, revealing possible interactions between all the factors, should enable the optimum factor levels to be identified.

Despite material flow similarities with ECAP, IS is fundamentally different. Instead of a fixed (or movable for friction reduction) ECAP die, IS uses a reciprocating die which plastically deforms a fixed billet. Feeding of the billet takes place when there is no contact between the billet and the reciprocating die. Thus feeding and deformation have been separated. This should help processing much longer billets and, provided an appropriate billet feeding mechanism is used, also infinite billets. In result, the main problem of short billets and poor utilisation of the material in many SPD processes can be solved. This should increase industrial interest in UFG metals produced by SPD and lead to new applications.

# ACKNOWLEDGMENTS

The financial support of the Scottish Enterprise Proof of Concept Fund is gratefully acknowledged.

# REFERENCES

1. Y. Wang, E. Ma, R. Z. Valiev and Y. Zhu, *Advanced Materials* **16/4**, 328-331 (2004).
2. S. Komura, M. Furukawa, Z. Horita, M. Nemoto and T. G. Langdon, *Mater. Sci. Eng.* **A297**, 111-118 (2001).
3. A. Rosochowski, L. Olejnik and M. Richert, "3D-ECAP of square aluminium billets" in *Proc. of the 8th Int. Esaform Conference on Material Forming*, Cluj-Napoca, Romania, April 27-29, 2005, pp. 637-640.
4. V. M. Segal, V. I. Reznikov, A. E. Drobyshevskiy and V. I. Kopylov, *Russ. Metall.* (Engl. Transl.) **1**, 99-105 (1981).
5. R. E. Barber, T. Dudo, P. B. Yasskin and K. T. Hartwig, "Product yield of ECAE processed material" in *Ultrafine Grained Materials III, 2004 TMS Annual Meeting*, Charlotte, North Carolina, U.S.A., March 14-18, 2004, pp. 667-672.
6. Y. Saito, H. Utsunomiya and H. Suzuki, "Conshearing" in *Advanced Technology of Plasticity*, edited by M. Geiger, Springer, 1999, pp. 2459-2464.
7. J. C. Lee, H. K. Seok, J. H. Han and Y. H. Chung, *Mater. Res. Bull.* **36**, 997-1004 (2001).
8. G. J. Raab, R. Z. Valiev, T. C. Lowe and Y. T. Zhu, *Mater. Sci. Eng.* **A328**, 30-34 (2004).
9. Srinivasan R., Chaudhury P. K., Cherukuri B., Han Q., Swenson D. and Gros P., "Continuous Plastic Deformation Processing of Aluminum Alloys, Final Technical Report" (2006), http://www.osti.gov/bridge/servlets/purl/885079-37CRhi/885079.pdf, last accessed January 5, 2007.
10. Y. Saito, N. Tsuji, H. Utsunomiya, T. Sakai and R. G. Hong, *Scripta Mater.* **39/9**, 1221-1227 (1998).
11. J. Y. Huang, Y. T. Zhu, H. Jiang and T. C. Lowe, *Acta Mater.* **49**, 1497-1505 (2001).

# Rheological Model and Mechanical Properties of Hard Nanocoatings in Numerical Nanoindentation Test

## M. Kopernik, M. Pietrzyk

*Department of Modelling and Information Technology, Faculty of Metals Engineering and Industrial Computer Science, Akademia Górniczo-Hutnicza, Mickiewicza 30, 30 – 59 Krakow, Poland*

**Abstract.** The objective of the work is development of the method, which is based on inverse analysis principles, and which allows determination of material model for lower material layer in the hard nanocoatings system. Hardness of such a system is additionally calculated in the paper. Design of block diagram of inverse analysis for the identification of mechanical properties of hard nanocoating is the main part of the solution. The method for the identification of mechanical properties of the layers is described in detail. Results of calculations are presented and discussed.

**Keywords:** inverse analysis, finite element method (FEM), nanocoatings, nanoindenatation test, mechanical properties, rheological material models
**PACS:** 46.35.+z, 91.60.Dc

## INTRODUCTION

Tribological hard nanocoatings of the third generation, eg. (Ti, Zr)N, (Ti, Cr)N, (Ti, Al, V)N, (Ti, Al, Si)N are usually investigated in experimental nanoindentation tests. Analytical methods of the analysis of these tests, which lead to evaluation of mechanical properties, were developed by Oliver and Pharr [1]. These solutions are, however, dedicated to monolayer materials only. Development of an alternative method, which is based on the inverse analysis principles and allows determination of material model for lower material layer in hard nanocoatings system, is the objective of the present work.

## THE NANOINDENTATION TEST

### Specification of Experimental Nanoindentation Test

Numerical nanoindentation test is based on the experimental one, which is performed in load-controlled mode using a Nano Test System [2]. Indentations are 20-cycle load-controlled load-partial unload experiments from 1 mN to 20 mN maximum load. Diamond (Young modulus $E = 1141$ GPa, Poisson ratio $\nu = 0.07$), Berkovich shape, pyramid, deformable indenter tip (tip radius $R = 150$ nm, pyramid angle $\alpha = 65.03°$) penetrates into specimen. 3D model of this indenter is shown in FIGURE 1. 2D approximation (conical shape) is used in simulation in this work.

All the experimental results used in the present work for identification of material models were obtained by the Authors of paper [2]. 50 experimental tests were carried out, each of them had 20 stages and the final sum of indents was equal to 1000.

CP907, *10th ESAFORM Conference on Material Forming,* edited by E. Cueto and F. Chinesta
© 2007 American Institute of Physics 978-0-7354-0414-4/07/$23.00

Multistage process of deformation is required in the case of testing multilayer material to eliminate the effect of scatter in results and to create a possibility to achieve the response of bottom layers during long-term process. Deformed specimen is a material system of 11 hard nanocoatings. This technical system [3], deposited by physical vapor deposition (PVD) technique, is composed of mixed elastic and elastoplastic material layers, which have two different thicknesses. The two nanocoatings are deposited periodically, respectively coating 1 (elastic) is repeated six times and coating 2 (elastoplastic) is repeated five times. Indentation test supplies force versus indentation depth data.

a)

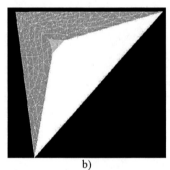
b)

**FIGURE 1.** a) Real view of Berkovich indenter, b) 3D model of Berkovich indenter [4].

## Numerical Model

Finite element code FORGE 2 is used in the simulations. The following flow stress function is introduced in the plastic part of the elastoplastic constitutive law:

$$\sigma_p = K\varepsilon^n \tag{1}$$

where, $\sigma_p$ – flow stress, $\varepsilon$ - effective strain, $K$, $n$ – material parameters.

Number of indentations steps is fixed to 20 (20 stages of deformation process), as it was in experiments. Deformed mesh from previous computing step is set into the next step as an initial mesh. Remeshing procedure is performed when elements are too distorted. Numerical model of specimen has 18062 nodes and 35564 triangular elements. Frictionless contact is assumed in the simulations, what is justified by the sensitivity analysis performed by the Authors in [4], see FIGURE 2. Sensitivity of loads with respect to friction coefficient is negligible comparing to that with respect to the shape of the tip.

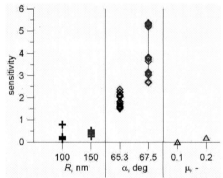

**FIGURE 2.** Sensitivity of loads with respect to tip radius $R$, tip angle $\alpha$ and friction coefficient $\mu$ [4].

280 simulations were made assuming 2D axisymmetric model of the indenter with the round shape of the tip. Local mesh refinement in the FEM program leads to fine mesh in the contact region, what gives good results. These refinements were done in the die-specimen contact region and in the thinnest nanocoatings.

## IDENTIFICATION OF MATERIAL PROPERTIES

### Methodology

Analytical methods in experimental nanoindentation test lead to evaluation of mechanical properties and they produce desired, calculated results, but only for monolayer materials. For example, there is no possibility to evaluate rheological parameters of lower layers in such material system.

The loads measured for the multilayer system are the main output from the experiment, which is indirectly used in the objective function of the inverse analysis. Martens hardness $HM_C$ of coating 1 is defined as:

$$HM_C = \frac{F}{A_s(d)} = \frac{F}{26.43d^2} \tag{2}$$

where: $A_s(d) = \frac{3\sqrt{3}tg\alpha}{\cos\alpha}d^2 = 26,43d^2$ – indent area of Berkovich indenter, $d$ – depth of indent, $\alpha$ – Berkovich tip angle, $F$ – force. Depth of the indent is measured either on FEM mesh in simulations or using AFM (atomic force microscope) in experiment.

Martens hardness $HM_S$ of specimen composed of 11 nanocoatings can be calculated using the following equation:

$$HM_S = \frac{1}{m^2 A_s(h)/h^2} = \frac{1}{26.43m^2} \tag{3}$$

where: $h$ – displacement of indenter, $A_S(h) = 26.43h^2$, $m$ – slope of force-displacement curve, $\alpha$ – Berkovich tip angle.

Force-displacement experimental data for 11 hard nanocoatings are used in this work as input for the FEM model. Numerical test is depth controlled. Knowing velocity $v = 10$ nm/s and twenty tool displacements (96.1; 50.8; 50.9; 55.3; 59.5; 62.8; 72.6; 74.7; 75; 76.3; 80.4; 82.9; 84.4; 87.6; 84.1; 89.8; 93.5; 90; 96.8; 96.8 nm), force versus displacement and force versus depth are calculated and measured. Pressure, equivalent strain and stress distributions are computed, as well.

Flow chart of the identification of properties of nanocoating based on simulations of tests is proposed in FIGURE 3. FEM simulation of the nanoindentation test for the mentioned system was performed. Elastic material model is set for directly deformed, first nanocoating and its properties are obtained directly from the experiment. Optimization variables ($E$, $K$, $n$ for the coating 2) are determined by searching for the minimum of the goal function defined for the Martens hardness of the first coating:

$$\phi = \sqrt{\frac{1}{N}\sum_{i=1}^{N}\left(\frac{HM_{NUM(i)} - HM_{EXP(i)}}{HM_{EXP(i)}}\right)^2} \tag{4}$$

where: $HM_{NUM(i)}$, $HM_{EXP(i)}$ – calculated and measured Martens hardness, respectively, $N$ – number of sampling points

Minimum of function (4) is searched with respect to parameters of the elastoplastic material model for coating 2. This layer has no contact with the indenter, and its parameters are obtained in numerical test using optimization procedures. Stop criterion is achieved, when the predicted value of Martens hardness for the coating 1 is close enough to the experimental one. Hardness is determined from the calculated force-depth data using equation (2). It has been noticed that the first step (first deformation) in the multistep deformation process, which mainly specifies the specimen response, has the greatest influence on character of deformation of the investigated material. This observation helps to reduce time of reaching correct material model, because only a few simulations have to be made, instead of tens. Proposed numerical procedure in the nanoindentation test is shown schematically in FIGURE 3.

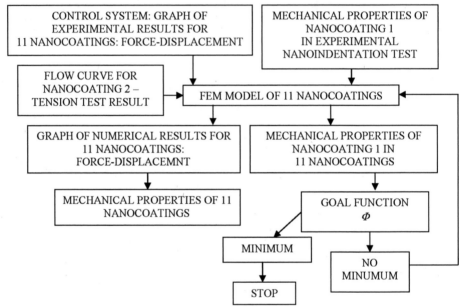

**FIGURE 3.** Block diagram of the identification procedure of mechanical properties of hard nanocoating and mechanical properties of multicoating system.

## Results

Experimental tension tests for coating 2 were performed in the project [5]. Material model for coating 2 based on equation (1) was determined as $K = 446.55$ and MPa, $n = 0.1456$, and used as starting point in the analysis. Such a model is not adequate for the material system investigated in the present work. It should be modified, because of multimaterial character of examined system and other deposition technique.

Experimental nanoindentation test gives Martens hardness $HM_{EXP} = 30$ GPa for coating 1, Young modulus $E = 368$ GPa and Poisson ratio $\nu = 0.177$. 120 simulations were made with constant value of Young modulus $E = 439$ GPa and Poisson ratio

$v = 0.25$ for coating 2, as well as with material model parameters $n$ and $K$, which were set as shown in FIGURE 4a. Selected results of these simulations are presented in TABLE 1 and they are numbered as simulations 1-4 and 6,7, respectively. Results presented in FIGURE 4 and in TABLE 1 show that parameters $K$ and $n$ do not have large influence on material model in the investigated range of values. Young modulus of coating 2 seems to be more important and it is considered in further analysis. Therefore, different values of this modulus were considered and they were chosen from data available in literature (physical range, which determines the area of search).

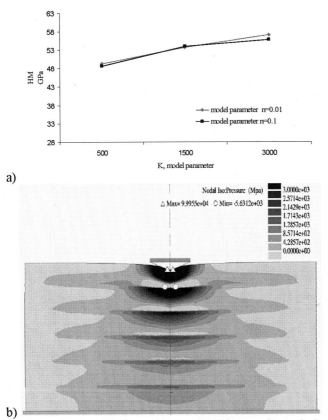

a)

b)

FIGURE 4. a) Martens hardness $HM$ for coating 1 versus $K$ for two values of $n$ in material model for coating 2 (TABLE 1, simulations numbered as 1-4 and 6,7); b) average stresses in the sample at the final stage of the test calculated by the model with the optimized materials parameters.

Selected results of this analysis are presented in TABLE 1 in rows 5 and 8-14. Finally, material model of coating 2 giving minimum of the goal function (4) is $K = 50$ MPa, $n = 0.1$, $E = 25$ GPa, computed Martens hardness $HM_{NUM} = 31$ GPa and final goal function $\Phi = 0.01$ (last line with bold face numbers in TABLE 1).

This identified model was used in the following simulations and Martens hardness of multilayer specimen was calculated from equation (3) as 26 GPa and the experimental value of specimen hardness is equal to 23 GPa. Selected results showing distribution of average stresses in the sample at the final stage of the test, calculated by the model with the optimized materials parameters, are shown in FIGURE 4b.

**TABLE 1.** Parameters of material model for coating 2 and results of inverse analysis for coating 1 ($HM_{EXP}$ = 30GPa, $m$ = 20 stages).

| SIMULATION NUMBER (each of 20 stages $m$) | PARAMETERS OF MATERIAL MODEL OF COATING 2 ($v$ = 0.25) | | | INVERSE ANALYSIS – RESULTS FOR COATING 1 | |
|---|---|---|---|---|---|
| | $K$, MPa | $n$ | $E$, GPa | $HM_{NUM}$, GPa | Objective function $\phi$ |
| 1 | 3000 | 0.01 | 439 | 57 | 0.201 |
| 2 | 3000 | 0.1 | 439 | 55.8 | 0.192 |
| 3 | 1500 | 0.1 | 439 | 54 | 0.179 |
| 4 | 1500 | 0.01 | 439 | 53.7 | 0.177 |
| 5 | 3000 | 0.01 | 352 | 53.3 | 0.174 |
| 6 | 500 | 0.01 | 439 | 49.3 | 0.144 |
| 7 | 500 | 0.1 | 439 | 48.6 | 0.139 |
| 8 | 100 | 0.1 | 616 | 41 | 0.082 |
| 9 | 100 | 0.1 | 232 | 40.2 | 0.076 |
| 10 | 100 | 0.2 | 232 | 39.7 | 0.072 |
| 11 | 100 | 0.1 | 100 | 38.8 | 0.066 |
| 12 | 50 | 0.1 | 50 | 35.6 | 0.042 |
| 13 | 100 | 0.1 | 25 | 32.2 | 0.016 |
| **14** | **50** | **0.1** | **25** | **31** | **0.010** |

# CONCLUSIONS

Methods suggested in the paper, which are based on the inverse analysis principles and observations from earlier simulations of tests [3], allow to obtain more information about examined material from the experiments, especially regarding models of materials located in the lower layers of the system. Proposed sequence of simulations of deformation in numerical test can be used in getting more realistic material models and hardness of hard nanocoatings system and its inner coatings.

*Acknowledgements: Financial assistance of MNiSzW, project no. 11.11.110.643, is acknowledged.*

# REFERNECES

1. C. Oliver, G.M. Pharr, An improved technique for determining hardness and elastic modulus using load and displacement sensing indentation experiment, *J. Mater. Res.*, **7**, 1564-1583 (1992).
2. B.D. Beake, S.P. Lau, Nanotribological and nanomechanical properties of 5–80 nm tetrahedral amorphous carbon films on silicon, *Diamond and Related Materials*, **14**, 535 – 1542 (2005).
3. M. Kopernik, M. Pietrzyk, 2D numerical simulation of elasto – plastic deformation of thin hard coating systems in deep nanoindentation test with sharp indenter, *Arch. Metall. Mater.* (2007), in press.
4. M. Kopernik, D. Szeliga, Modelling of nanomaterials – sensitivity analysis to determine the nanoindentation test parameters, *Computer Methods in Materials Science*, **7**, 255-261 (2007).
5. Y.L. Su, S.H. Yao, C.S. Wei, C.T. Wu, Tension and fatigue behavior of a PVD TiN-coated material, *Thin Solid Films* **315**, 153–158 (1998).

# Bulk Forming of Industrial Micro Components in Conventional Metals and Bulk Metallic Glasses

M. Arentoft[1], N.A. Paldan[1], R.S. Eriksen[2], T. Gastaldi[2],
J.A. Wert[3], M. Eldrup[3]

[1])IPU, Produktionstorvet, Building 425, DK-2800 Kgs. Lyngby, Denmark, www.ipu.dk
[2])Dept. of Manuf. Eng. and Managem., Techn. Univ. of Denmark, DK-2800 Kgs. Lyngby,Denmark
[3])Risø National Laboratory, Frederiksborgvej 399, P.O. 49, DK-4000 Roskilde, Denmark

**Abstract.** For production of micro components in large numbers, forging is an interesting and challenging process. The conventional metals like silver, steel and aluminum often require multi-step processes, but high productivity and increased strength justify the investment. As an alternative, bulk metallic glasses will at elevated temperatures behave like a highly viscous liquid, which can easily form even complicated geometries in 1 step. The strengths and limitations of forming the 2 materials are analyzed for a micro 3D component in a silver alloy and an Mg-Cu-Y BMG.

**Keywords:** Micro forming, BMG, forging
**PACS:** 81.20.Hy, 61.43.Fs, 71.23.Cq, 07.10.Cm

## INTRODUCTION

The growing market for compact products with high functionality is increasing demand for precision metallic micro components. Manufacturing such components by traditional machining operations or by chemical milling is costly and wasteful of material. The latter factor can be significant, even for micro components, in cases where precious metals such as palladium are used to confer the required degree of corrosion resistance in the final product. An alternative is to use metal forming processes, but adaptation of macroscale forming processes to the microscale is challenging. A flexible microforming system is designed and tested [1,2]. This concept has been used to form a number of industrial micro components in aluminum, silver, brass and titanium. Since all of the components included in the evaluation have intricate 3D shapes, multiple forming steps are needed. However, multistep processes can be difficult to implement because of the complexity of micro component transfer and gripping devices, in addition to the cost of developing and manufacturing multiple tools. As an alternative, forming of a bulk metallic glass (BMG) has been investigated.

### Bulk metallic glasses

BMGs are a new material class whose discovery was announced in 1989 [3,4]. These are solid materials in which the atomic scale structure is little different from that of a liquid and which only require modest cooling rates to retain the amorphous atomic structure. Below the glass transition temperature ($T_g$), BMG's are hard solid materials, however when BMGs are heated above $T_g$, the hard glassy state transforms

CP907, 10th ESAFORM Conference on Material Forming, edited by E. Cueto and F. Chinesta
© 2007 American Institute of Physics 978-0-7354-0414-4/07/$23.00

to a supercooled liquid state in which the material has a flow response characteristic of extremely viscous liquids, such as warm asphalt. In this temperature regime, the flow stress is strongly strain rate dependent and can be extremely small for low strain rates. There is no work hardening mechanism in BMGs, so umlimited shape change is possible in the supercooled liquid state. The main limitation is the metastability of the amorphous structure in the supercooled liquid regime above $T_g$. Holding in this temperature range results eventually in crystallization of the material. When a BMG is cooled below $T_g$, it spontaneously transforms to a glass, automatically restoring the high hardness associated with the glassy state.

From the description of the response of BMGs to thermomechanical conditions, it is apparent that BMGs have advantages for some shaping operations. But the brittleness of BMGs in general loading conditions prevent their application as large scale components where damage tolerance properties are normally mandatory. However, small-scale components which are used extensively in modern electronic and medical devices, for example, escape from damage tolerances requirements to some degree. Metallic components may be used in such applications to take advantage of their electromagnetic shielding, acoustic, eleastic or wear properties, while being subjected to moderate loads. The excellent formability of BMGs could be exploited in such cases, and the limited ductility would not be a serious drawback [5,6].

As a result of the excellent formability of BMGs, an investigation has been undertaken to explore forming properties for these novel materials for micro component applications. The work described in this article involves comparison of processing methods for manufacturing a millimeter scale component from silver alloy and from a Mg-Cu-Y BMG.

## GEOMETRY AND MATERIAL FOR FORMED COMPONENT

A number of micro components formed by cold forging processes have already been reported [7,8]. Most are industrial components designed for machining and due to this they are axi-symmetric with sharp corners and are made of metallic materials. The components are normally slightly modified as regards material and geometry to enable cold forging. To prove the concept of micro forming and at the same time compare processing of BMG's and conventional metals, it is chosen to design a component, which is difficult to produce by machining, involves very small details and gaps, and needs at least 2 steps if produced in conventional metals. Figure 1 shows the component geometry.

The material chosen for conventional micro forming is a heavily work hardened silver alloy (97%Ag3%Cu) with a flow stress at 455 MPa. The material is delivered as Ø2 mm rods, which in 2 steps are drawn down to Ø1.87 mm. This rod is then cropped by the cropping device invented by IPU into 1.12 mm long cylindrical billets [9].

The $Mg_{60}Cu_{30}Y_{10}$ BMG material is fabricated at Risø in following manner: 22.4 g Cu and 10.4 g Y are alloyed by melting in Ar environment in an arc melting furnace. The roughly spherical ingots were turned over and remelted 5 times to promote chemical homogeneity. The ingots produced by this method are placed in a stainless steel crucible together with 17.2 g Mg in an induction casting furnace. This furnace is evacuated and filled with Ar. The alloy is heated to 850°C and gravity cast into a bulk

Cu mould with a 2 mm opening. In this way, a slightly irregular plate with dimensions approximately 20 x 30 x 2 mm is obtained. Cylindrical billets with a diameter of 1.4 mm are trepanned from the plate by electrodischarge machining (EDM) using a tubular stainless steel tool with an inner diameter of 1.5 mm. The reason for the modified billet geometry compared to the conventional silver alloy is to avoid length reduction of the BMG material, since BMG's are impossible to form in cold conditions. The diameter is therefore reduced to keep volume constancy. This causes increased deformation in the forming process, but since BMG's do not deformation harden, it has no influence on the success of the final process. No evidence of crystallinity can be detected by X-ray methods after billet preparation. The mechanical properties of the BMG are described in [10].

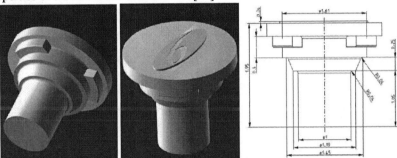

**FIGURE 1.** Geometry of formed component.

## FORMING OF SILVER AND BMG COMPONENT

A die set is designed and manufactured for mounting of dies manufactured in agreement with the tool standard used by IPU. Punch, die and ejector are positioned inside the die set and the package is fixed on a press table.

The tools for the process are produced by a micro-electrodischarge machining (EDM) process in Vanadis 23 hardened to 62 HRC using an electrode of 60 μm. The tool design and manufacturing methods are described in [7,8].

The forming of the silver component is carried out at low punch speed on a 100kN hydraulic press. The billet is lubricated with Molycote[TM] and re-lubricated after step 1. The $1^{st}$ and $2^{nd}$ steps are performed with loads of 4.5 kN and 15.5 kN, respectively. The high load for the second step caused damage to the punch. Figure 2 shows the formed silver components.

The BMG component is formed in a press with a heating chamber located at Risø [11]. The die set and dies are heated in the furnace for 1 hour at 170 °C ($T_g$=145 °C) after which the BMG billet is positioned in the die cavity. To reduce friction and ease the ejection, the billet is sprayed with Molykote[TM]. The pressing is carried out with a 30 second loading ramp to 4 kN, which is held for 30 seconds followed by removing of die set from the furnace and cooling. The punch can easily be removed from the die, in which the component will stick. By a 1 mm pin, the component is carefully ejected. In no cases fracture occurred, even though the BMG is brittle and therefore sensitive to impact loading. Figure 3 shows the formed BMG component.

a)                                  b)                                  c)

**FIGURE 2.** The formed silver component a) step 1, b) step 2, c) detail showing the imprint on top.

a)                                  b)                                  c)

**FIGURE 3.** The formed BMG component.

When comparing the silver and BMG components in figures 2 and 3, the 3 grooves on the back-side of the head and the cylindrical part at the middle are incompletely filled for the silver part, whereas the BMG shows a geometrical almost perfect component. Furthermore some scratching at the extruded surfaces due to insufficient lubrication can be identified for the silver part. Looking at figure 2c, a backward extruded flash at the rim of the component is seen. This is caused by the high pressure which expands the die and thereby allows the material to extrude into the gap between punch and die.

The surface of the BMG component looks porous. A closer study indicates that the surface has replicated the surface of the die, which is generated by the EDM process. This is underlined by the scratches found next to the logo at figure 3c, which originate from grinding of the punch.

The load for producing the incomplete silver component is 4 times the load for the completely filled BMG. Contrary, the forming time for the silver part is just a few seconds compared to 1 minute for the BMG process. As regards the mechanical properties of the formed components, the strength and ductility of the BMG component are limited.

To increase the productivity and thereby the industrial relevance of the BMG forming process, a study of the minimum forming time is carried out. For this purpose, the experimental work is supported by numerical analyses using the commercial FE-program Deform 3D. The load ramp is kept constant at 133 N/sec. The lubrication by

Molycote is expected to cause a friction factor at 0.12. The material is described by a constitutive model relating stress to temperature and strain rate [11].

Four steps from the loading cycle are chosen for further analysis, corresponding to 8, 13, 17 and 30 seconds. Figure 4 shows the load procedure and corresponding FE-predictions. As seen, a complete filling is expected to be obtained after 17 seconds (situation no. 3), which corresponds to a load of 2260 N

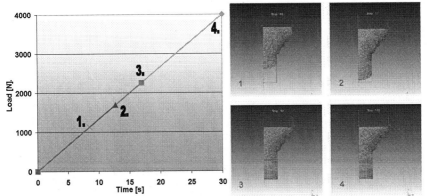

**FIGURE 4.** Finite Element analysis of loading of BMG. Left: 4 steps of the loading cycle chosen for further analysis. Right: FE predicted formfilling corresponding to the 4 loading steps.

**FIGURE 5.** Partially formed BMG components corresponding to points 1-4 in figure 4.

The FE-predictions from figure 4 are investigated experimentally by forming of 4 BMG components. Figure 5 shows the partially formed components. When comparing the experimental results to the FE predictions, it is clear that the FE-simulations overestimate the form filling, and that the complete geometry only is obtained for a 30 second loading ramp. This may indicate that friction for forming BMG components is much higher than for conventional metals. Taking into account the replication of the tool surface identified at figure 3, it is tempting to assume that the higher friction is caused by the ability of the BMG material to flow into all gaps and thereby behave more sticking than conventional metals. Supplementary simulations have shown agreement between FE-simulations and experimental work for a friction factor at 0.8. Alternatively, the discrepancy between FE simulations and experimental observations may indicate that refinement of the material model used for the BMG material may be required.

## CONCLUSION

A complex 3D micro component is produced in a silver alloy and an Mg-based BMG. In both cases, a successful forming is obtained. For the silver component, some minor under-fillings can be identified and a 2 step process is needed, whereas the BMG component replicates the geometry in an almost perfect way in 1 step. From an industrial point of view, a forming time of 30 seconds and a brittle component is a disadvantage for the BMG material. Risø has successfully produced Zr-based BMGs, with improved strength, but that requires forming temperatures in the range of 500°C.

## ACKNOWLEDGMENTS

The authors wish to thank the European Commission and the Danish Ministry for Science Technology and Innovation for supporting the work by the EU project MASMICRO, no. 500095-2 and the Innovation Consortium MIKROMETAL, no. 66334.

## REFERENCES

1  M. Arentoft, N.A. Paldan: "Production equipment for manufacturing of micro metal components", 9th ESAFORM 2006, Glasgow, UK, pp 579-582
2  C.P. Withen, J.R. Marstrand, M. Arentoft, N.A. Paldan: "Flexible tool system for cold forging of micro components", First International Conference on Multi-Material Micro Manufacture, 29June-1 July 2005, Forschungszentrum Karlsruhe, Germany, Elsevier, pp. 143-146
3  A. Inoue, "Stabilization of metallic Suipercooled Liquid and Bulk Amorphous Alloys", Acta Materialia, 48;2000:279-306.
4  W.L. Johnson, "Bulk Glass-Forming Metallic Alloys Science and Technology" MRS Bulletin, 24 (10);1999:42-56.
5  Y. Saotome, H. Iwazaki, "Superplastic Backward Microextrusion of Microparts of Micro-Electro-Mechanical Systems", J. Materials Processing Technology 119;2001:307-311.
6  J.A. Wert, N. Pryds, Materialenyt 1;2003:1-25.
7  M. Arentoft, N. Paldan, R.S. Eriksen: "Cold forging of industrial micro components", Proceedings of the 39th ICFG plenary meeting, Changwon, Korea, pp. 83-86, 2006
8  M. Arentoft, N.A. Paldan: "Press design, transfer mechanism and tooling for micro forming", Journal of Materials Processing Technology, special issue from ESAFORM 2006, to be published.
9  N.A. Paldan, M. Arentoft, R.S. Eriksen: "Production equipment and Processes for Bulk Formed Micro Components", Proceedings of Esaform 2007, April 2007.
10 J.A. Wert, N. Pryds, E. Zhang "Rheological Properties of of Mg60Cu30Y10 Alloy in the Supercooled Liquid State", in Proc. 22nd Risø International Symposium on Materials Science, Risø National Laboratory, Roskilde, Denmark, 2001, pp. 423-428.
11 J. Wert, C. Thomsen, R. Debel: "Forming of a Bulk Metallic Glass Microcomponent", Proc. of Danish Metallurgical Society, Dept. of Materials Research, Risø National Laboratory, January 2007.

# Injection moulding of micro-parts: applications to micro-gears

T. Barriere, J.C. Gelin, G. Michel, M. Sahli, C. Quinard

*FEMTO-ST Institute / Applied Mechanics Department, ENSMM, 24 chemin de l'épitaphe, 25000 Besançon, France*

**Abstract.** The paper is concerned with the development of the injection molding process for micro-parts with an application to injection moulding of micro-gears with the volume less than 0.2 mm$^3$. First the paper describes the equipment that is used consisting in a micro-injection equipment that that is used consisting in a micro-injection that permits simultaneously bi-polymer injection. A polypropylene polymer is then used and characterized. The mould cavity corresponding to a micro-gear and related equipment is related. Finally the micro-injections are realized and the geometry of the micro-gears is reported in using an optical measurement system.

**Keywords:** Micro-moulding, micro-components.
**PACS:** 81.20.Hy

## INTRODUCTION

Since five years, the micro-injection domain knows an important growth and interesting development perspectives are emphasized. A lot of different injection moulding equipment systems have been developed devoted to micro-injection: servo hydraulic and electric equipments, micro injection moulding systems [1], optical measurement units to check the quality of the components.

Some research activities has been started in micro-PIM with conventional machine or specific equipments such as Battenfeld Microsystem$^©$ or Ferromatik Milakron$^©$ ones equipments [2-3]. Micro-components with an injected volume inferior to 1 mm$^3$ have been realized with Battenfeld Microsystem$^©$. Actually, there are more than hundred micro-injection moulding systems arround the world.

In our lab., a lot of research developments carried out concerning PIM domains [4-6]. A two-injection micro-injection Battenfeld Microsystem$^©$ has been specially developed with Battenfeld in order to begin some research activities in micro-PIM and micro-co-injection and micro-bi-injection in biomedical, electronic domains or automotive fields.

CP907, *10$^{th}$ ESAFORM Conference on Material Forming,* edited by E. Cueto and F. Chinesta
© 2007 American Institute of Physics 978-0-7354-0414-4/07/$23.00

# THE MICRO-INJECTION MOULDING EQUIPMENT

Two injections Battenfeld Microsystem 50 equipment has been used in this study. This fully electric production cell was specially designed for parts weights as low as 0.1 milligrams. This equipment consists of the following modules: clamping, injection, swivel, removal and handling, quality monitoring and an external vacuum pump. Typical components realized with the Microsystem 50 equipment are related in Figure 1 b. In our case, the equipment is only used with one injection module.

The injection unit, described in Figure 1 below, is built-up in three stages specially for highly accurate processing with the minimum injection moulding volume: 14 mm extruder screw, 5 mm piston for predosing with an accuracy of 0.001 cm$^3$ and 5 mm piston injection, the maximum injection speed is equal to 760 mm/s$^{-1}$, the maximum injection pressure is equal to 250 Mpa and the theoretical injection volume varies from 0.025 cm$^3$ to 1.1 cm$^3$. The maximum clamping force is 50 kN and the ejection force is 1.2kN.

Some micro-components with different polymers (PP, POM, and PEEK) are given in [7].

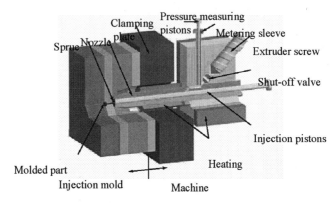

**FIGURE 1.** BATTENFELD Microsystem50 injection module.

# MOULD MICRO-CAVITIES FOR INJECTION MOULDING OF MICRO-GEAR

Thanks to the swivel module two movable moulds parts were used composed of a classical die holder unit and a core with the four toothed wheel form cavities. Four ejector pins are used with a diameter equal to 0.4 mm at the extremity. The fixed mould part is composed of a fixed plate with runners, the sprue bushing channel and a sprue-puller pin combined with a sprue picker (Figures 2a). A special mould with 4 micro toothed wheels was manufactured (Figure 2c). CAD shape is shown Figure 2b. The die cavity dimensions were measured with a Werth Benchtop ScopeCheck$^{©}$ 200 CNC using a image processing video sensor with fixed telecentric objective.

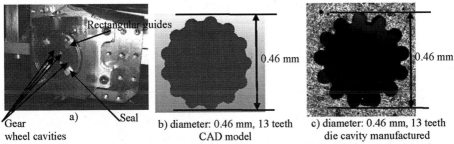

| a) Gear wheel cavities | b) diameter: 0.46 mm, 13 teeth CAD model | c) diameter: 0.46 mm, 13 teeth die cavity manufactured |

**FIGURE 2.** a) Injection Mould with 4 gear mould cavities; b) and c) CAD model gear wheel specimen compared to the manufactured.

## TEST MATERIALS AND EXPERIMENTAL PROCEDURE

### Rheology

The polymer used in the experiments was injection grade Polyethylene (PP, EP548N) produced by Basel company consisting of 3 mm granulate average size. Its density is 0.9 g/cm$^3$. Table 1 summarizes the polypropylene characteristics, relevant to the micro-injection process.

**TABLE 1.** Material properties of polypropylene

| Properties | Values |
|---|---|
| Melt density $\rho$ | 0.73026 g/m$^3$ |
| Specific heat $C_p$ | 2960 J/kgK |
| Thermal conductivity K | 0.16 W/mK |
| Melt flow index MFI | 16 g/10min |

The rheological characteristics for high shear rate values were characterized using a capillary rheometer. The polypropylene was extruded through a die in tungsten carbide. These experiments were conducted with 1 mm diameter and 16 mm length die cavity. All the experiments were performed at different temperatures under a shear rate ranging from 1 to $10^5 \, s^{-1}$.

Figure 3 shows the measured viscosity at five different temperatures. It has to be noticed that a temperature dependency of the viscosity is clearly visible. The viscosity is approximately are half between 220 to 260°C.

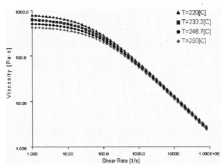

**FIGURE 3.** Shear viscosity evolution for polypropylene polymer obtained by capillary rheometry at different temperatures.

## MOLDFLOW ANALYSIS

Moldflow© software is used to simulate the filling of the die cavity. The analysis was carried out in order to simulate the polymer flow and the thermal behaviours in mould. A 3D simulation has been performed (the gravity and inertia have been neglected during the solving of Stoke equation). Figures 4a, b and c show the mould filling at different filling rate 74%, 90% and 100%.

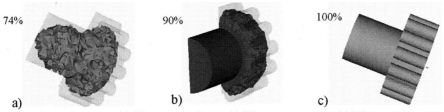

**FIGURE 4.** a) Mould filling to 74%; b) Mould filling to 90%; c) Mould filling to 100%.

## EXPERIMENTAL PROCEDURE

Injection moulding experiments were carried out on a BATTENFELD machine from ENSMM. A Polypropylene was injected at 240°C into the mould (Fig. 5). The mold temperature is controlled through the use of a heating cartridge.

Initially, the die cavity filling characteristics were estimated with Moldflow© software. When the injected material enters in contact with the metallic mould, it is quickly cooled making it difficult to fill the cavity fill properly. In order to facilitate injection, the pressure and temperature were gradually increased until to reach a final pressure of 100 MPa. In order these processing conditions, the material is easily injected and parts are obtained.

The injection process parameters were listed in Table 2.

**TABLE 2.** Injection Parameters

| Injection Velocity | Melt temperature | Mold temperature | Holding time |
|---|---|---|---|
| 20-350 mm$^3$.s$^{-1}$ | 240-250°C | 50-60 °C | 10 s |

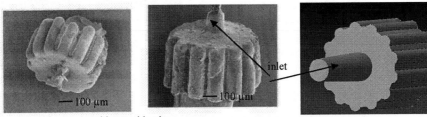

Gear wheel in upper and lower side view
**FIGURE 5.** A gear wheel specimen compared to the CAD model.

# INJECTION VOLUME VARIATION

Injection moulding has been realized by increasing step by step of injection volume from 58 mm$^3$ to 128 mm$^3$. For injection volume lower than 100 mm$^3$, there are only sprue and runners filled. The polymer fills the microcavities for injection volume corresponding to 128 mm$^3$ (Table 3 and Fig. 6).

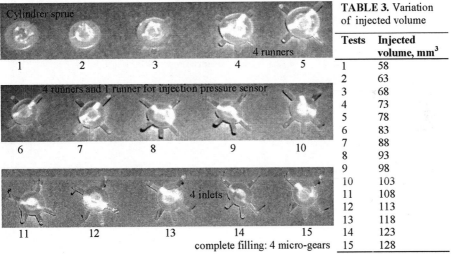

**TABLE 3.** Variation of injected volume

| Tests | Injected volume, mm$^3$ |
|---|---|
| 1 | 58 |
| 2 | 63 |
| 3 | 68 |
| 4 | 73 |
| 5 | 78 |
| 6 | 83 |
| 7 | 88 |
| 8 | 93 |
| 9 | 98 |
| 10 | 103 |
| 11 | 108 |
| 12 | 113 |
| 13 | 118 |
| 14 | 123 |
| 15 | 128 |

**FIGURE 6.** Comparison of injected volume from short shot to complete filling.

# CONCLUSION

The paper relates investigations carried out in polymer injection moulding for micro-composants with volume less than 0.2 mm$^3$. In a first stage, one has investigated the micro-injection equipment that can be used for such a purpose, and one focus on a injected moulding equipment that lead to render as low as possible the injection sprue. This different polymer have considered and characterized for their fluidity and their possibility to fill small cavities with very low volume, typically in the range 0.05 to a 2 mm$^3$, and a polypropylene was chosen. Then a die cavity consisting in micro-gear with 0.46 mm diameter was chosen. The adjustment of the injection moulding parameters permit to perform the required shape that is obtain with a good accuracy. So, one has demonstrated the possibility to get micro-parts, by a proper characterisation of polymer characterisation rheology and optimisation of processing conditions.

# REFERENCES

1. C.G. Kukla, Micro injection Molding, Int. J. of Forming Processes **4**, n° 3-4, 253-269 (2001).
2. T. Osada, K. Nishiyabu, Y. Karasaki, S. Tanaka and H. Miura, Investigations on the variation of feedstock properties in Micro MIM Products, PIM 2003, Powder injection molding, Penn State College, Ed. by R.M. German, Pennsylvannia, USA, pp. 1-11.
3. A.C. Rota, F. Petzoldt and P. Imgrund, Micromolding of advanced material combinations, PIM 2003, Powder injection molding, Penn State College, Ed. by R.M. German, Pennsylvannia, USA, pp. 1-18.
4. J.C. Gelin., T. Barriere and M. Dutilly, Experiments and computational modelling of metal injection molding for forming small parts, Annals of the CIRP **48**, n° 1, 179-182 (1999).
5. J.C. Gelin, T. Barriere and B. Liu, Mould design methods by experiment and numerical simulation in metal injection molding, J. of Engineering Manufacture **126**, Part B, 1533-1547 (2002).
6. T. Barriere et J.C Gelin, Nouvelles technologies de réalisation d'empreintes et de gravures pour le moulage par injection de poudres et optimisation des pièces frittées, New advances in die design and manufacturing for metal injection moulding and optimization of powder sintering processes, Bulletin du Cercle d'Etudes des Métaux, INNOVMECA, Congres international sur les procédés innovants, International congres on innovative process, 8-10 mars 2005, Lyon, 2005, pp. 1-10.
7. C. G. Kukla, Micro-injection Moulding, Int. J. of Forming Process, vol. 4, n° 3-4, 2001, pp. 253-267.

# Flexible Manufacturing System For Vibration Assisted Microforming

Wojciech Presz

*Institute of Materials Processing, Warsaw University of Technology, Narbutta 85, 02-524 Warsaw, Poland*

**Abstract.** Miniaturisation generates necessity of micro-parts production. Micro-scale means closer tolerances and better surface roughness. These requirements can be achieved with metal forming processes but under high pressure and sufficient relative sliding distance between tool and workpiece surface. It makes such process galling prove. This tendency increases with diminishing of component dimensions. Literature search and previous investigations shows that implementation of vibrations might be a solution for limitation of galling tendency in microforming. The group of so called "reference micro-parts" has been chosen as a representation of micro-products. The tooling system for vibration-assisted microforming of referenced parts has been designed. Vibrations are performed with vibrators based on stacked ceramic multilayer technology, which provides accurate frequency and amplitude control. Proposed reference micro-components and designed laboratory system can be used for investigations of technological parameters for utilisations of microforming. After laboratory investigations it is intended to design industrial system working on same principles.

**Keywords:** Plastic Forming, Microforming, Vibrations, Flexible tooling.
**PACS:** 81.20.Hy

## INTRODUCTION

Miniaturization of products in the electronics industry results needs smaller and smaller metallic parts. Diminishing sizes requires improvement of dimensional accuracy and surface smoothness. Microforming processes carried out at sufficiently high contact pressure and large relative movement at the material-tool interface assure these requirements: narrow dimensional tolerances and a very low surface roughness [1], especially using ultra fine-grained [2] or nano-materials. However, these conditions are perfect for the occurrence of a very dangerous surface phenomenon - galling. Unfortunately, diminishing the objects' size to a range of microforming increases the risk of galling because of worse lubrication conditions [3]. It means that retarding undesirable surface phenomena which specially regars to galling becomes a critical factor in microforming processes deciding about successful implementation of this technology. Based on the literature reports, it appears that adding vibrations to metal forming processes could be a promising solution for surface phenomena problems.

CP907, *10ᵗʰ ESAFORM Conference on Material Forming,* edited by E. Cueto and F. Chinesta
© 2007 American Institute of Physics 978-0-7354-0414-4/07/$23.00

# VIBRATIONS IN METAL FORMING

The first application of vibrations during deformation of metals was reported by Garskij and Efromov [4] in 1953. Two years later Blaha and Langekert [5] used ultrasonic vibrations in tensile tests of a Zink monocristal. Since that time in laboratories all over the world many investigations of plastic forming with a help of vibrations have been performed. Although there is no uniform opinion about mechanisms of mainly advantageous phenomena observed during processes with a help of vibrations. Industrial applications so far mainly regard to wires and tubes drawn with ultrasonic waves of tools [6,7,8]. Significant drop of process force and surface quality improvement was observed. Only few works refer to other metal forming processes and the low frequency vibrations - under 1kH, but also in this area some positive effects were observed [9,10].

Although long history of laboratory investigations which shown profitable influence of vibrations on course of plastic deformations there were hardly reported any spectacular industrial uses. The reason of such limited industrial applications seems to be: large energetic expense and high costs of setting in vibrations of heavy tools which usually used in metal forming. These troubles occur because of small masses of products in principle do not refer to microforming. A preferable opinion is that the influence of vibrations on metal forming processes refer to two phenomena. First, the so-called *volume effect* [5] links lowering of yield stress with the influence of vibrations on the dislocation movement. Second, the so-called *surface effect* [11, 12] explains lowering of the effective coefficient of friction by a periodic reduction in the contact area and/or periodic changes of direction of the friction force vector. This effect (surface effect) seems to be quite good theoretically supported and has been chosen as a dominant and goal phenomena within this project.

As it was already pointed the contact problems are very critical in metal forming because of a very high pressure and increasing of contact surface and temperature that favors adhesions and galling. With scaling down dimensions of products the contribution of surface phenomena increases because surface to volume factor increases. It makes application of surface effect of vibrations especially suitable to micro-forming processes. In the so far utilizations diminishing of positive effect of vibrations was observed with lowering of vibrations amplitude [13]. On the other hand amplitude increasing results with increasing of technical complication and might lead to serious technical problems. A sort of compromise must be than achieved during system design. Within this project the relative surface movement has been assumed within a range of 25-50 μm [14] This assumption has a background coming out of a typical industrial tool surface constitution because precise grinding to the surface roughness about Ra = 0.32 μm results distance between asperities about 25-50 μm.

# TYPES OF VIBRATIONS

Surface effect can be observed only if changes in relative movement of areas being in contact during metal forming process occur. Practically it means reversing of sliding direction of material on a tool surface or temporary loosing of contact between them. It can be obtained by vibration in two main ways. First, standing waves in tools

[6] that cause complicated usually 3D periodic deformation of tool, resulting local surface movement – points of tool are moving on along different paths. Second, periodic movement of tool without its deformation – points of tool are moving along the same paths [7,9].

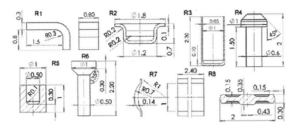

**FIGURE 1.** Dimensions of reference micro-parts: R1-R8.

First method is based on propagation waves inside tool in resonant conditions to obtain possible high amplitude. Since eigenfrequency of tool is usually very high obviously problems with relative movement amplitudeoccur. Additionally along the work piece deformation self-frequency of vibration system, which consists of tool and work piece is changing. This change draws correction of excitation frequency to keep in resonant range. It makes system rather complicated and certainly not flexible.

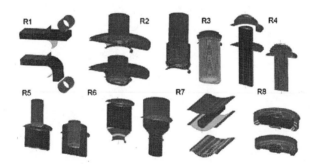

**FIGURE 2.** FEM simulations of reference micro-processes.

The second type of vibration or excitation is usually called *direct vibration or excitation*. In this project vibrations are utilized in the meaning of direct excitation of tools that limits vibration frequency to relative low values – less than 1kHz.

In general, metal forming manufacturing system consists of three main parts: 1 moving tool – mainly punch that is in "forming contact" with work piece, 2 stable tool – die, 3 blank holder or ejector - only in some cases. It is possible to apply direct vibrations to each part in 6 ways: 3 translations and 3 rotations. It makes quite a big amount of theoretically possible design variants. Most probably all of them would be able to successfully support forming processes with vibrations, but not all of them are at the same level of construction difficulties. In this project only two longitudinal translations have been chosen: one for moving tool and one for stable tool.

# REFERENCE MICRO-PARTS

There was a specified a group of micro-parts to manufacturing of which system is dedicated. These parts called "reference micro-parts" cover main groups of metal forming components and can be used for analysis technological and design problems in microforming

**TABLE 1.** Forces of Micro-processes

| Reference Process | Force [N] | | | |
|---|---|---|---|---|
| | Aluminum | | Steel | |
| | Stable | Top | Stable | Top |
| R1 Bending | 12 | 12 | 58 | 58 |
| R2 Deep drawing | 8 | 7 | 50 | 50 |
| R3 Ironing | 27 | 27 | 195 | 198 |
| R4 Heading (closed) | 118 | 180 | 874 | 1160 |
| R5 Backward cup extrusion | 75 | 75 | 570 | 570 |
| R6 Forward bar extrusion | 76 | 76 | 572 | 572 |
| R7 Closed-die forging | 21 | 83 | 231 | 693 |
| R8 Closed-die forging (disc) | 250 | 350 | 1500 | 2300 |

Technical drawings of these parts are shown in Fig.1. In Table 1 are collected some basic information about reference parts and their manufacturing processes also treated as "reference processes".

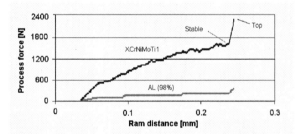

**FIGURE 3.** Process force for reference process R8 (FEM)..

All reference microforming processes were simulated with FEM MSC-Marc code under 2d conditions, Fig.2. Rigid tools and coulomb friction coefficient $\mu = 0.1$ was assumed. Two very different materials were taken: soft Al 98 % and pretty hard alloy steel XCrNiMoTi18. At this stage of analysis the main goal of simulation was to find process forces because of they strong impact on vibrators design. Contact pressure distribution and relative material movement were this way also obtained. Comparison of process forces is shown in Table 1. In case of some forging processes forces increase rapidly in the final stage of deformation. For a better description there are two forces expressed in the table 1: so called "stable" force – process force just before rapid increase and the top force. It is explained in Fig.3. In case of component R8 very high force is obtained because of three main reasons: 1. relatively big cross section area of this component, 2. not optimal material flow in the final stage of process, 3. high flatness that increases the negative role of friction.

# DESIGN OF MANUFACTURING SYSTEM

The flexible manufacturing system is shown in Fig.4. System consists of two piezoelectric vibrators: moving vibrator *1* and stable one *2* mounted in the frame *3*. Vibrating heads *4* and *5* are exciting to longitudinal vibrations with stacked ceramic multilayer actuators *6* and *7*.

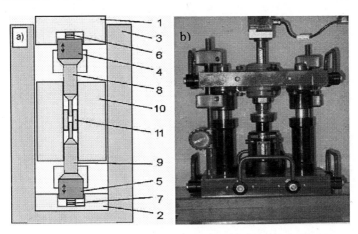

**FIGURE 4.** System principle (a) and overview (b).

Heads are connected with upper vibrating tool *8* and lower vibrating tool/ejector *9*. Shanks of vibrating tools have standard ends for stable joining with vibrating heads. For convenient manufacturing and maintaining they are relative big - 3 mm in diameter. Vibrating tools ends are much smaller and have shapes depending on micro product geometry. Both vibrating tools are guided in the tool block *10*. Vibrators are based on the multi-layer PZT ceramics [15]. This technology can be used for relative long distance movements [16]. Each piezo-stack is placed inside housing and have ball joints on the both sides to prevent bending load. Stacks of 6x6x50 mm dimensions have been used. They are pre-stressed with disk springs to balance tensile inertia forces acting on them during vibrations.

## FUTURE WORK

System is going to be used for testing of influence of low frequency vibrations on two parameters: 1st - surface quality and 2nd tool load. First seems to be important for softening and prone to adhesion materials like aluminum. Second is critical for extrusion processes of higher strength materials like steel. On base of experimental program it is planed to state new values (author believes that higher) of limiting factors for referenced microforming processes.

# CONCLUSIONS

- A group of reference micro-parts for microforming have been proposed for experimental and FEM analysis of these type of technology.

- Principle of designed *flexible system* is assisting of microforming operations with low frequency (under 1kH) longitudinal and direct vibrations of tools.

- Designed system consists of two vibrators acting on the work piece from the top and the bottom.

- According to theoretical analysis stacked ceramic multilayer actuators with dimensions of 6x6x50 mm are suitable for designed system.

# ACKNOWLEDGMENTS

The Polish State Committee for Scientific Research partially supported this research under contract No. 1508/T07/2004/27

# REFERENCES

1. W. Presz, A. Rosochowski, "The influence of grain size on surface quality of microformed components", Proceedings ESAFORM 2006 Conference, Glasgow, pp. 587-590.
2. A. Rosochowski, W.Presz, L. Olejnik and M. Richert, "Micro-extrusion of ultra-fine grain aluminium", Multi-Material Micro Manufacture 4M2005, Oxford 2005, pp. 161-164.
3. N. Tiesler, "Microforming, Size effects in friction and their influence on extrusion process", Wire 52 (02) pp. 34-38.
4. F.K. Garskii, V.I. Efromov, " Effect on ultrasound on the decomposition of solid solutions", Izviestia Akademii Nauk Belorousk SSR, 3, 1953, (in Russian).
5. F. Blaha, B. Langenecker, "Strain of zinc monocrystals under the effect of ultrasonic vibration", Naturwissenschaften, 42, 556, 1955, (in German).
6. L. Lang, X. Li, "Wire drawing with ultrasonic vibration", Wire Industry, January 1994.
7. K. Siegert, A. Mock, "Wire drawing with ultrasonically oscilating dies", Journal of Materials Processing Technology 60, 1996, pp. 657-660.
8. B. Langenecker, S. Illiewich, O. Vodep, "Basic and applied research on metal deformation in macrosonic fields at PVL Austria", Proceedings of Ultrasonics International 1973, pp. 34-37.
9. G.R. Dawson, C.E. Winsper, D. H. Snasome, "Application of high- and low-frequency oscillation to the plastic deformation of metals", Metal Forming, April, 1970, pp. 158-162.
10. D. Godfrey, "Vibration reduces metal to metal contact and causes an apparent reduction in friction", ASLE 10, 1967, pp. 183-192.
11. K. Sieger, J. Ulmer, "Influencing the Friction in Metal Forming Process by Superimposing Ultrasonic Waves", Annals of the CIRP, 50/1,2001, pp. 81-84.
12. W. Presz, B.G. Ravn, J. Sha, B. Andersen, T. Wanheim, " Application of piezoelectric actuators in pressing process with vibrating tools", Proceedings of KomPlasTech 2005 Conference, pp. 155-162.
13. W. Presz, "Micro-punching with vibrating punch", Proceedings of ICIT 2005 Conference, pp.93-95.
14. W. Presz, "The influence of punch vibration on surface phenomena in micro-extrusion", Proceedings of ESAFORM 2006 Conference, Glasgow, 583-586.
15. Noliac A/S, Hejreskovvej 18, Kvistgaard, www.noliac.com
16. W. Presz B. Andersen, T. Wanheim, "Piezoelectric driven Micro-press for microforming", Journal of Achievements in Materials and Manufacturing Engineering 18, 2006, pp. 411-414.

# 9 – MACHINING AND CUTTING

## *(F. Micari and Ph. Lorong)*

# High Speed Machining: A New Approach To Friction Analysis At Tool-Chip Interface

J. Brocail[a], M. Watremez[a], L. Dubar[a], and B. Bourouga[b]

a: Laboratoire d'Automatique, de Mécanique et d'informatique Industrielles et Humaines, UMR 8530,
Université de Valenciennes et du Hainaut Cambrésis, F-59313 Valenciennes Cedex 9, France
b: Laboratoire de Thermocinétique de Nantes, Polytech' Nantes, UMR 6607,
Rue Christian Pauc, BP 50609, F-44306 Nantes Cedex 3, France

**Abstract.** Numerical approaches of high-speed machining process are necessary to increase productivity and to optimisize tool wear and worked piece residual stresses. In this purpose, rheological behaviour and friction model have to be correctly determined. Actual numerical approaches with current friction models don't lead to good correlations of process variables like cutting forces or chip-tool contact length. This paper proposes a new approach to characterize the friction behaviour at the chip-tool interface. First, a mechanical analysis enables to determine contact pressure, plastic strain, velocity and temperatures in this chip-tool interface. This study concerns the interface zone near the cutting edge. A low sliding velocity combined with a contact pressure higher than 1 GPa is characteristic of this zone. An experimental device has been designed to simulate the friction behaviour at chip-tool interface. During this upsetting-sliding test, the indenter contacts the specimen and rubs against it with a constant velocity and generates a residual friction track. The forces, in both normal and tangential directions are measured in order to calculate the contact pressure $\sigma_n$ and the friction stress $\sigma_t$. These variables are then used for identification of the friction data depending on the interface temperature and sliding velocity. Finally, perspectives are given for implementation of data in FEA machining models.

**Keywords:** Friction, chip-tool interface
**PACS:** 46.55.+d

## INTRODUCTION

Numerical approaches of high-speed machining process are necessary to increase productivity and to optimisize tool wear and worked piece residual stresses. In this purpose, rheological behaviour and friction model have to be correctly determined. A Coulomb's law is used in most numerical approaches to describe contact at the tool-chip interface. Other nonlinear relations take into account of the shear flow stress k [1-3] in order to correlate the experimental profiles. As shown by two recent studies [4,5], good correlations of process variables are not obtained with current friction models. An average error of 50% appears for the chip-tool contact length, and an error of 25% is generally observed for the cutting forces.

This paper proposes a new approach to characterize the friction behaviour at the tool-chip interface. The extreme contact conditions of process require a mechanical analysis of the high speed machining. Contact pressure, plastic strain, sliding velocity

CP907, 10th ESAFORM Conference on Material Forming, edited by E. Cueto and F. Chinesta
© 2007 American Institute of Physics 978-0-7354-0414-4/07/$23.00

and temperatures at tool-chip interface have to be determined. The first part of this study deals with the mechanical analysis of the interface zone near the cutting edge. In a second part, an experimental device designed to simulate the friction behaviour at chip-tool interface called Upsetting-Sliding Test (UST) is presented.

## STATE OF THE ART

According to numerical approaches of orthogonal cutting [6-9], very intense plastic strains (from 1 to 5) and high strain rates (until $10^5$ $s^{-1}$) are representative of high speed machining process. Experimental results obtained by the split tool method [10] show a pressure on the tool tip higher than 1 GPa. This contact pressure value has been recently confirmed by a thermomechanical model [11] and numerical approaches [5,9]. The interfacial temperature is mainly generated by the plastic strains undergone in the primary shearing zone and by the friction at the tool-chip interface [12]. A temperature of 800 °C is observed in the tool tip against 1100 °C in a hot zone located at some tenth of millimetres from the tool tip [13]. The contact conditions in the tool-chip interface are characterized by seizure and/or sliding phenomena [14-18] and a sliding velocity reduced in the tool tip. A low sliding velocity combined with a contact pressure higher than 1 GPa are characteristic of the interface zone near the cutting edge, which is the subject of this paper.

## GENERAL STRATEGY

### Upsetting Sliding Test

An upsetting sliding test is used to simulate the specific contact conditions of the interface zone. This friction test involves the specimen and the contactor, which respectively represent the chip and the tool (Figure 1). The contactor and the specimen are respectively machined in the tool material and in the workpiece.

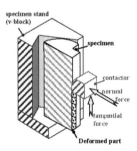

**FIGURE 1.** Design of upsetting sliding test

Before the test, a relative penetration of the contactor into the specimen is imposed. During the test, the contactor contacts the specimen and rubs against it with a constant velocity and generates a residual friction track. The relative penetration, the geometry and the displacement velocity of the contactor are the test parameters. Forces in both normal and tangential directions are measured during the test. The contact pressure $\sigma_n$ and the friction stress $\sigma_t$ are calculated with an analytical model. These variables lead to the identification of the friction data as a function of the interfacial temperature and of the sliding velocity.

## Mechanical Analysis

The mean friction coefficient is determined by the ratio of the mean friction stress and the mean contact pressure. The real surface of contact during the test depends on the front bulge $\omega$ which is compared with the equivalent surface of contact (Figure 2).

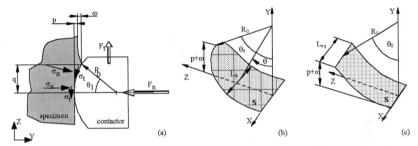

**FIGURE 2.** (a) mechanical analysis; (b) real surface of contact; (c) equivalent surface of contact

The mechanical equilibrium of the contactor during the stationary phase leads to the determination of the mean friction stress $\sigma_t$ and the mean contact pressure $\sigma_n$ in accordance with the following equations:

$$\sigma_t = \frac{F_t q - F_n (p+\omega)}{L_{eq}[q^2 + (p+\omega)^2]}, \quad \sigma_n = \frac{F_n q + F_t (p+\omega)}{L_{eq}[q^2 + (p+\omega)^2]}. \tag{1}$$

where Leq and q are respectively the equivalent seating width and the seating length. Finally, the mean Coulomb's friction coefficient $\mu=\sigma_t/\sigma_n$ can be expressed by:

$$\mu = \frac{-(p+\omega) + q\dfrac{F_t}{F_n}}{q + (p+\omega)\dfrac{F_t}{F_n}}. \tag{2}$$

## EXPERIMENTS

### Working process

The specimen was dimensioned in collaboration with the LTN (Thermocinetic Laboratory of Nantes) to thermomechanically study the tool-chip interface. The specimen presents a prismatic zone of friction allowing a 2D thermal study (Figure 3a). The contactor has a radius of 14 mm and a seating of 3 mm (Figure 3b).

For these first tests, the experiment is carried out with an AISI 1035 specimen steel and an AISI M2 tool steel. The tool is heated at 140 °C using a heating cartridge. The specimen is heated in the induction furnace up to 630 °C before to be positioned in the v-block using an arm manipulator. The contactor moves up with a constant velocity equal to 200 mm.s$^{-1}$.

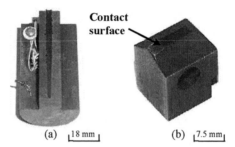

(a) |18 mm|     (b) |7.5 mm|

**FIGURE 3.** UST antagonists: specimen (a) and contactor (b)

## Results

The test is performed three times. After a fast rising in efforts, a stationary zone is observed. Then, the efforts decrease when the contactor reaches the end of specimen. The mean normal and mean tangential forces are equal to 4455 N and 2660 N in the stationary zone (Figure 4).

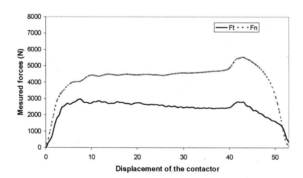

**FIGURE 4.** Mean normal and mean tangential forces during the upsetting-sliding test.

The mean Coulomb's friction coefficient is calculated with equation 2. This relation is based on the measured forces $F_t$ and $F_n$, the penetration p, and the front bulge $\omega$. The parameter $\omega$ is determined by a numerical model of the upsetting-sliding test. The specimen is discretized with CPE4R elements, and the contactor is dimensionally stable. The contactor is modelled with R2D2 type rigid elements. The numerical model does not usually correlate the measured forces because of lack of rheological behaviour data. However, previous works [19] show the relation of the front bulge $\omega$ with on the penetration p, the indenter radius $R_0$ and the shape of the specimen. The numerical value of the front bulge is then supposed correct and equal to 157 µm. The depth of trace, measured by profilometry is equal to 70 µm.

From equation 2, a mean Coulomb's friction coefficient $\mu=0.48$ is obtained for a mean contact pressure $\sigma_n=787$ MPa and an interface temperature of 750 °C [20]. These variables are then used for the identification of the friction data depending on the interface temperature and sliding velocity. Indeed, an equivalent Coulomb's friction coefficient $\mu_{eq}$ is defined for a pressure value. The tests will be performed for other contact pressures in order to determine a set of points. Thus, the friction stress $\sigma_t$ appears as a function of the contact pressure $\sigma_n$ and interfacial temperature $T_{int}$ (Figure 5). Finally, these friction data could be implemented in FEA machining models.

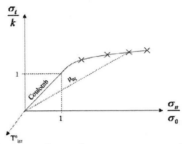

**FIGURE 5.** Friction data depending on the contact pressure and interfacial temperature

## CONCLUSION

The contact at the tool-chip interface has to be master in order to accurate numerical approaches of high-speed machining process. This new approach to friction analysis at tool-chip interface enables to define the contact near the cutting edge. However, the mean contact pressure $\sigma_n$ is less than characteristic value of the interface zone near the cutting edge which is about 1 GPa. Higher values would be reach if the indenter radius $R_0$ is decreased.

This new approach allows the identification of the friction data depending on the interface temperature and sliding velocity. Future work will be carried out with other materials in need of machining process optimization as stainless steel AISI 304L. The mean Coulomb's friction coefficient and the mean contact pressure will be more precisely determined with a thermomechanical numerical model. The rheological behaviour data have to be defined in the range of plastic strains, strain rates, and temperature of the test.

Lastly, higher sliding velocity zones can't be studied with the UST and an other experimental device is actually in developpement.

## ACKNOWLEDGMENTS

The authors are grateful to the fondation de France / CETIM for supporting this work. The present research work has been also supported by the European Community, the Délégation Régionale à la Recherche et à la Technologie, the Ministère de l'Education Nationale, de la Recherche et de la Technologie, the Région Nord-Pas de Calais, the Centre National de la Recherche Scientifique.

# REFERENCES

1. E. Usui, T. Shirakashi, *Mechanics of machining - from descriptive to predictive theory*, ASME Publications, PED Sciences 39, 1997, pp. 369-389.
2. T.H. Childs, K. Maekawa, T. Obikawa, Y. Yamane, *Metal Machining: Theory and Applications*, Elsevier, Amsterdam, 2000, pp. 408.
3. M.H. Dirikolu, T.H. Childs, K. Maekawa, *Finite element simulation of chip flow in metal machining*, International Journal of Mechanical Sciences 43, 2001, pp. 2699-2713.
4. L. Filice, F. Micari, S. Rizzuti, D. Umbrello, *A critical analysis on the friction modelling in orthogonal machining*, International Journal of Machine Tools & Manufacture, 2006.
5. Tugrul Özel, *The influence of friction models on finite element simulations of machining*, International Journal of Machine Tools & Manufacture 46, 2006, pp. 518-530.
6. Tugrul Özel, Taylan Altan, *Determination of workpiece flow stress and friction at the chip–tool contact for high-speed cutting*, International Journal of Machine Tools & Manufacture 40, 2000, pp. 133-152.
7. Mahmoud Shatla, Christian Kerk, Taylan Altan, *Process modeling in machining. Part I: determination of flow stress data*, International Journal of Machine Tools & Manufacture 41, 2001, pp. 1511-1534.
8. Franck Girot, Daniel Géhin, *Perçage à sec des alliages d'aluminium aéronautique*, Mécanique & Industries 3, 2002, pp. 301-313.
9. K.C. Ee, O.W. Dillon Jr., I.S. Jawahir, *Finite element modelling of residual stresses in machining induced by cutting using a tool with finite edge radius*, International Journal of Mechanical Sciences 47, 2005, pp. 1611-1628.
10. L.C. Lee, X. D. Liu, K.Y. Lam, *Determination of Stress Distribution on the Tool Rake Face using a Composite Tool*, International Journal of Machine Tools and Manufacture 35, 1995, pp. 373-382.
11. A. Moufki, A. Molinari, D. Dudzinski, *Modelling of orthogonal cutting with a temperature dependent friction law*, Journal of Mechanical Physics of Solids 46, 1998, pp. 2103-2138.
12. J.-L. Battaglia, H. Elmoussami, L. Puigsegur, *Modélisation du comportement thermique d'un outil de fraisage : approche par identification de système non entier*, C. R. Mecanique 330, 2002, pp. 857-864.
13. J.C. Outeiro, A.M. Dias, J.L. Lebrun, *Experimental assessment of temperature distribution in three dimensional cutting process*, Mach. Sci. and Tech. 8, 2004, pp. 357-376.
14. E.M. Trent, Metal Cutting and the Tribology of Seizure (I-III), Wear 128, 1988, 29-81.
15. H.S. Qi, B. Mills, *Formation of a transfer layer at the tool-chip interface during machining*, Wear 245, 2000, pp. 136-147.
16. V.R. Marinov, *Hybrid analytical-numerical solution for the shear angle in orthogonal metal cutting Part I: theoretical foundation*, International Journal of Mechanical Sciences 43, 2001, pp. 399-414.
17. V. Madhavan, S. Chandrasekar, T.N. Farris, *Direct Observations of the Chip-Tool Interface in the Low Speed Cutting of Pure Metals*, Transactions of the ASME Journal of Tribology 124, 2002, 617-626.
18. H.S. Qi, B. Mills, *Modelling of the dynamic tool–chip interface in metal cutting*, Journal of Materials Processing Technology 138, 2003, pp. 201-207.
19. L. Lazzarotto, " Maîtrise des conditions de contact et de frottement en mise en forme par frappe à froid ", Ph.D. Thesis, Université de Valenciennes et du Hainaut Cambrésis, 1998.
20. E. Guillot, B. Bourouga, B. Garnier, L. Dubar, *Experimental study of thermal sliding contact with friction: application to high speed machining of metallic materials*, ESAFORM congress, Zaragoza, 2007.

# A study of the influence of the metallurgical state on shear band and white layer generation in 100Cr6 steel: application to machining

Malek HABAK, Jean-Lou LEBRUN and Anne MOREL

*LPMI/MSPTF- EA 1427. ENSAM. 02, Bd du Ronceray. BP 93525- 49035 Angers Cedex 01. France.*
*E-mail: malek.habak@angers.ensam.fr*

**Abstract.** The aim of this paper is to better understand the material behaviour involved in machining operations. During machining, the workpiece experiences large strains, high strain rate, high temperatures, complex loading histories, and recovery. To reproduce these loadings and to understand the behaviour of 100Cr6 bearing steel, quasi-static and dynamics mechanical shearing tests were carried out. These tests made it possible to reproduce the primary shear zone observed on the chips after cutting using specimens with special geometries "hat-shaped specimens". The geometry of these specimens results in a localised shearing zone when loaded in compression. Two metallurgical states of the material were investigated (with and without carbides). For each state, three material hardnesses are used (46, 51 and 55HRc). The tests parameters investigated were the strain rate and temperature. For all tests, the microstructures of the shear zones were examined. Results show that the presence of carbides has the tendency to increase the material resistance. The micrographic observations of the sheared zones highlighted the effect of the microstructure and the link between the thermo-mechanical effects and the characteristics of the white zones. It is possible to produce a white layer, similar to those obtained in machining, by quasi-static and dynamic shearing tests. The presence of carbides has a strong effect on the generation of the shear bands and the white layers. Increasing the test temperature and strain rate tends to increase the width of shear band and white layers. A comparison between the white layers obtained by the dynamic tests and those observed on the chip in hard turning are carried out. The results show good agreement.

**Keywords**: microstructure, carbides, material hardness, shear band, white layer, Hopkinson bars.
**PACS:** 81.20.Wk

## INTRODUCTION

In turning, the formation of chips is accompanied by generation of Shear Bands and white layers (Adiabatic shear Band: ASB), which affects the mechanical, metallurgical and physical properties of the workpiece surface and the cutting tool. Hence, it is necessary to understand these ASB. In order to comprehend the phenomena occurring during the formation and propagation of shear bands, many studies have been carried out using different materials [1-4]. However, in spite of the great number of studies relating to this topic, none of them treat the influence of microstructure on the on ASB formation during deformation of bearing steel.

The objectives of this study are to better understand the material behaviour involved in machining operations, to reproduce the primary shear zone (shear band

CP907, *10ʰ ESAFORM Conference on Material Forming,* edited by E. Cueto and F. Chinesta
© 2007 American Institute of Physics 978-0-7354-0414-4/07/$23.00

and the white layer) observed on the chips after cutting using a specimen with special geometries "Hat-shaped shear specimen" and to understand the effect of the microstructure and the temperature on the ASB generation.

## EXPERIMENTAL PROCEDURE

In this study, we reproduce in a controlled laboratory environment the shear band typically formed in machining operations. These shear bands are obtained at low strain rate (quasi-static test) and high strain rates (dynamic test) using "Hat-shaped shear specimen". These specimens have a particular geometry that gives them the possibility of developing a localised shear zone when loaded in compression (figure 1). They were introduced by Hartmann [5] and have since often been used to identify material behaviour laws in machining [4, 6, 7]. Using this type of specimen it is possible to produce a plastic instability close to that observed in the primary shear zone created during the process of chip formation in machining operations.

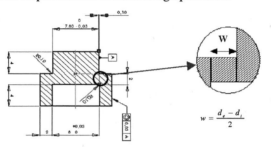

**FIGURE 1.** Geometry of the hat-shaped shear specimens.

The material used in this study was 100Cr6 (AISI 52100) bearing steel of average composition; 1.04% C, 1.48% Cr, 0.24% Si, 0.033% Cu, 0.018% S and 0.01% Mo. In order to study the influence of the microstructure on the formation of shear band, two metallurgical states of the material were investigated. These states were obtained by different heat treatments, which result in two microstructures, one with non-dissolved carbides dispersed in the martensitic matrix and one without carbide (bainitic matrix). For each state, three hardnesses (46, 51 and 55 HRc) were also obtained via quenching and different tempering times. In this work the term carbides does not concern the fine carbides created by tempering. To understand the behaviour of the 100Cr6 bearing steel and ASB formation, quasi-static and dynamic shear tests were carried out. A Split Hopkinson Pressure Bar system was used for the dynamic tests.

## RESULTS AND DISCUSSIONS

### Quasi-static shear tests

The quasi-static shear tests were carried out using an INSTRON servo-hydraulic testing machine, with a maximum capacity of 100 kN. Tests were done at room temperature, at a constant test rate of 1 mm/s and for the two metallurgical states with various hardnesses. For each case, the forces-displacement curves were obtained. An example of these curves is presented in figure 2 (The error in the measurement is approximately ± 3 %).

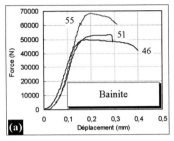 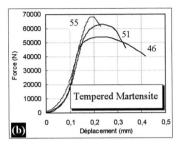

**FIGURE 2.** Force-displacement curves obtained for the quasi-static tests for various hardnesses (a) without carbide and (b) with carbides.

Figure 2 present the effect of the hardness on the material behaviour for the two metallurgical states. For both states, it is shown that the material with a hardness of 55 HRc has the highest resistance. Indeed, for 100Cr6 with carbides, the maximum load supported by the specimens is 68 kN for a hardness of 55 HRc, whereas it is only 54 kN for the 46 HRc. With a hardness of 55 HRc, it was noted that the presence of carbides decreases slightly the ductility of material. A comparison between the different metallurgical states shows that the presence of carbides increases slightly the maximum material resistance.

## Dynamic shear tests

Dynamic shear tests were carried out using a direct impact Split Hopkinson Pressure Bar (SHPB). This technique can be used to study the dynamic plastic flow stress of metals that undergo large strains at strain rates between $10^2$ to $10^5$ s$^{-1}$. The same hat-shaped specimen geometry, as discussed above, was used in these tests.

Tests were carried out at there different temperatures: room temperature 25°C, 250°C and 350°C. The two different metallurgical states of 100Cr6 (with and without carbides) with different hardnesses (46, 51 and 55 HRc) and different temperature (25, 200 and 260 °C) at test rate of $12.5 \times 10^3$ mm/s (strain rate > $10^4$ s$^{-1}$) were investigated. A stopper ring was used to control the maximum displacement of the specimen ~1mm.

Figure 3 shows an example of the force-displacement curves obtained from these tests. The comparison between the different metallurgical states shows that for the same displacement, the presence of carbides results in a higher maximum load. These observations show the same trend as those observed in the quasi-static tests.

The difference observed between the two microstructures can be explained due to the fact that for the microstructure with carbides, the increase in the compressive stress generates slip in which the dislocations within the grains are blocked due to the presence of the carbides. The dislocation density increases and their movement becomes increasingly difficult. It is thus necessary to increase the plastic yield stress in order to further plastically deform the material. Thus the maximum load is higher.

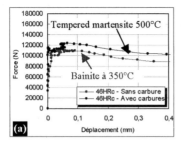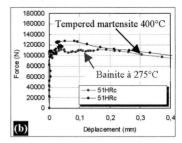

**FIGURE 3.** Effect of the microstructure on the force-displacement curves for the dynamic tests (a) for a hardness of 46 HRc (b) for a hardness of 51 HRC ($V_{impactor}$ = 12.5x10$^3$ mm/s at room temperature).

These observations are made on the basis of macroscopic test results. A characterization of the mechanisms of dislocation propagation was not made. It would be judicious to carry out an analysis under a TEM microscope in order to support the observations discussed above.

## Shear band in the "Hat-shaped shear specimens"

In the following, the microstructures in the shear zone of the hat-shaped specimens, before and after quasi-static and dynamic tests, are analysed. The specimens were sectioned, polished and etched (using Nital). Microstructural evaluations were carried out using an optical microscope. A comparison between the microstructures obtained after testing at room temperature dynamically and quasi-statically is presented in figure 4.

(a). Homogenous white zone (microstructure without carbide)

(b). Localised white zone (microstructure with carbides)

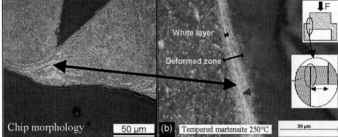

**FIGURE 4.** Comparison between the white zone obtained on chips and those obtained in the shear zone of Hat-shaped shear specimens (55 HRc, $V_{impactor}$= 12.5x10$^3$ mm/s, 25°C).

694

On this figure, the direction in which the charge was applied can be observed. It was noted that the characteristics of the shear bands are completely different when the microstructure changes. A broad shear zone appears at slow strain rate. On the other hand at high speed, the zone is very localised. In the case with carbides a relatively wide shear band can be seen (~10 μm) and at the centre of this band there is a white layer approximately 1 μm wide. This is in contrast to the case without carbide in which the width of the white zone is equal to the width of the shear band.

For the quasi-static tests, at very low deformation rates (tests rate ~ 1 mm/s), no white band is observed (figure 5a). It only starts to appear for test conducted at rate greater than approximately 10 mm/s (figure 5b).

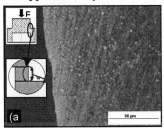 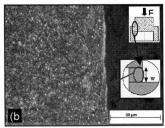

**FIGURE 5.** Examples of sheared zones obtained at room temperature for the case with carbides (46 HRc) for a low strain rate tests (a) 0.1 mm/s and (b) 10 mm/s.

For the dynamic tests, the presence of the white zone is systematic (figure 4). It is more intensely white for tests conducted at higher temperatures (figure 6) and higher deformation rates. Also, the higher these are, the wider the zone is. As shown in figure 6, the width of the white zone is approximately 15 μm at a temperature of 200°C and 30 μm for 260°C. Thus, this confirms that a rise in the cutting speed causes an increase in the width of the white layer. In effect, the characterisation of the shear band and primary shear zone (observed on the chip) with respect to the test parameters has shown that the strain rate and high temperature favour the formation of the white layer and this is the cause of a phase transformation [4]. Also, it was observed that the carbides increase the width of shear zone and localise the white layer in this zone.

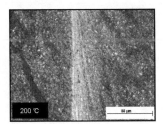 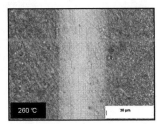

**FIGURE 6.** Shear zones in hat-shaped specimens tested using a SHPB setup at various temperatures (100Cr6 with carbides, 51 HRc, $V_{impactor}$ = 12.5x10³ mm/s).

From a comparison between the two metallurgical states it is possible to conclude that the shearing is more localised for the metallurgical state without carbide and that for the case with carbides the deformed band is more is larger. The ASB are formed by a local plastic deformation, causing strain hardening, at a local defect. The plastic strain energy is converted into thermal energy, leading to a rise of the temperature and causing adiabatic heating. The material properties in the local band are changes and the ASB is formed [10].

## CONCLUSION

The following conclusions can be made from this work:
- Specimen geometry contribute to the development of shear band and white layer similar to those observed on the chip in hard turning;
- Higher strain-rates and temperatures result in greater formation and growth of ASB.
- The origin of the white layer is thermo-mechanical. It is intensified by the increase in strain rate and temperature. They are generated initially by a mechanical effect involving a very localised shear force. This high strain results in a localised temperature rise which in turn generates a thermal effect that aides the formation of the whites bands.

This study will make it possible to predict the behaviour of the 100Cr6 steel in hard turning and to understand the influence of the microstructure on the machinability and the surface integrity.

## ACKNOWLEDGMENTS

This study has been carried out in the framework of a *Contrat Plan Etat Région Pays de la Loire* gathering EC Nantes, ENSAM Angers and CETIM. Angers Loire Métropole and the local collectivise has financed it.

## REFERENCES

1. S. Larbi, "Contribution à l'Étude de l'Usinage à Grandes Vitesses de Matériaux Métalliques par Simulation sur un Banc d'Essai à Base de Barres de Hopkinson". PhD thesis, Univ. Nantes 1990.
2. M. N. Bassim, *J. Mater. Proc. Tech.* (2001) Vol. 119 Issue 1-3, pp. 234-236.
3. G. Poulachon, "*Aspects phénoménologiques, mécaniques et métallurgiques en tournage c-BN des aciers durcis. Application : usinabilité de l'acier 100Cr6*". PhD thesis, ENSAM, 1999-14.
4. M. Habak, "*Etude de l'influence de la microstructure et des paramètres de coupe sur le comportement en tournage dur de l'acier à roulement 100Cr6*". PhD thesis, ENSAM, 2006-57.
5. K.H. Hartmann, H.D. Kunze and L.W. Meyer, "Metallurgical effects on impact loaded materials" *in Shock-Wave and High-Strain-Rate Phenomena in Materials*, M.A. Meyers & L.E. Murr, eds, Marcel Dekker, NY, 1981, p. 325 (1981).
6. B. Lesourd, "Etude et modélisation des mécanismes de formation de bandes de cisaillement intense en coupe des métaux : Application au tournage assisté laser et de l'alliage de titane TA6V", Ph.D. thesis, EC Nantes, ED 82-174 (1996).
7. B. Changeux, "Loi de comportement pour l'usinage. Localisation de la déformation et aspects microstructuraux", Ph.D. thesis, ENSAM, 2001-12.
8. G. Poulachon, A.L. Moisan, M. Dessoly, *Mécanique & Industries* 3, (2002) pp. 291–299.
9. M. Habak, J-L. Lebrun, B. Huneau, G. Germain, P.Robert, "Effect of carbides and cutting parameters on the chip morphology and cutting temperature during orthogonal hard turning of 100Cr6 bearing steel with a cutting tool", in 9th CIRP. Bled, Slovenia, pp. 517-524 (2006).
10. H. Feng, M.N. Bassim, Mater. Sci. Eng. (1999) A 266, pp. 255–260.

# Advanced Prediction of Tool Wear by Taking the Load History into Consideration

## K. Ersoy*, G. Nuernberg*, G. Herrmann*, H. Hoffmann*

*Institute of Metal Forming and Casting, Technische Universität München, Garching, Germany*

**Abstract.** A disadvantage of the conventional methods of simulating the wear occurring in deep drawing processes is that the wear coefficient, and thus wear too, is considered to be constant along loading duration, which, in case of deep drawing, corresponds to sliding distance and number of punch strokes. However, in reality, it is a known fact that wear development is not constant over time. In former studies, the authors presented a method, which makes it possible to consider the number of punch strokes in the simulation of wear. Another enhancement of this method is introduced in this paper. It is proposed to consider wear as a function of wear work instead of the number of punch strokes. Using this approach, the wear coefficients are implemented as a function of wear work and fully take into account the load history of the respective node. This enhancement makes it possible to apply the variable wear coefficients to completely different geometries, where one punch stroke involves different sliding distance or pressure values than the experiments with which the wear coefficients were determined. In this study, deep drawing experiments with a cylindrical cup geometry were carried out, in which the characteristic wear coefficient values as well as their gradients along the life cycle were determined. In this case, the die was produced via rapid tooling techniques. The prediction of tool wear is carried out with REDSY, a wear simulation software which was developed at the Institute of Metal Forming and Casting, TU-Muenchen. The wear predictions made by this software are based on the results of a conventional deep drawing simulation. For the wear modelling a modified Archard model was used.

**Keywords:** Deep drawing, wear simulation, tool wear, modified Archard model
**PACS:** 46.55+d, 81.20.Hy, 07.05.Tp

## INTRODUCTION

Wear is the progressive damage and material loss, which occurs on the surface of a component as a result of its motion relative to the adjacent working parts. Along with the advent of many high-strength steels in the market and the increasing usage of rapid tooling (RT) techniques for tool production, the abrasive wear of the forming tools becomes an important problem, and the need to predict tool wear is increasing.

The key for making a prediction of the development of wear in metal forming tools is numerical simulation. In forming applications, wear is commonly described using models based on contact mechanics, the most important one being the Archard wear equation. Several studies [1,2,3] were carried out coupling the Archard wear equation to the finite element simulation. In these studies it was shown that the wear simulation using the Archard model is capable of determining wear sensitive regions in the model

CP907, *10th ESAFORM Conference on Material Forming,* edited by E. Cueto and F. Chinesta
© 2007 American Institute of Physics 978-0-7354-0414-4/07/$23.00

with excellent qualitative agreement with experiments. In [4], a simulation scheme was suggested which makes it possible to obtain quantitatively reasonable results.

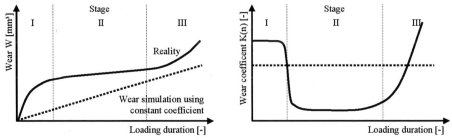

**FIGURE 1.** Wear vs. loading duration, the deviation of wear prediction using constant wear coefficients.

It is a known fact that wear is a function of loading duration, i.e. production time or sliding distance [5,6] (FIGURE 1). Yet, the Archard's equation does not include a term for this characteristic of wear. The authors therefore suggested a modified Archard's model by introducing the wear coefficient as a function of the number of punch strokes accumulated by the tool. It was shown that with this modelling it was possible to obtain wear simulation results, which were in a good qualitative and quantitative agreement with the experiments [4].

However, the punch stroke dependent wear coefficients, determined using a certain experimental geometry, may fail to yield accurate results for another geometry, if in this case a punch stroke involves completely different loading conditions (pressure values) and sliding distances. In this case, modelling the wear coefficients as a function of number of punch strokes can lead to over- or underestimation of the wear for another geometry and the coefficients determined are bound to be geometry dependent. Moreover, in the same geometry different locations can be under different loading conditions, in which case using a single coefficient for the whole geometry may lead to over- or underestimation of the results.

In this study a modified version of the Archard's model is proposed, where the wear coefficients are implemented to be path dependent. The wear coefficients, and therefore the wear, are expressed in terms of total wear work done instead of the number of punch strokes. This approach allows taking the load history of the each node into account and is therefore independent from the tool geometry.

## THEORY AND ALGORITHM

### Archard's Wear Equation

The principles of the Archard wear equation are straightforward [6]: It defines the wear volume W as directly proportional to the dimensionless wear coefficient K, the normal force $F_N$, sliding distance s and inversely proportional to the surface hardness H of the wear-affected tool material.

$$W = K \frac{F_N}{H} s \qquad (1)$$

In order to transform the common wear equation into the elemental equation, finite element discretization is necessary. The wear volume change per unit area at a certain time t is defined in equation (2)

$$\dot{w} = K \frac{p(t) \cdot v_{slid}(t)}{H} \qquad (2)$$

where $\dot{w}$ is the wear depth change, p is the normal pressure and $v_{slid}$ is the relative sliding velocity in the tangential direction. The wear volume of a certain area is obtained by integrating equation (2) over time

$$w = \frac{K}{H} \int_t p(t) v_{slid}(t) dt \qquad (3)$$

where w is the wear depth.

$$Z = \int_t p(t) v_{slid}(t) dt \qquad (4)$$

By introducing wear work Z as integral $p \cdot v_{slid}$ over time (equation (4)), w can be written in the form of equation (5).

$$w = \frac{K}{H} \cdot Z \qquad (5)$$

## Variable Wear Coefficient Taking the Loading History into Account

It is a known fact that wear has three characteristic stages, beginning with running-in wear, normal wear and accelerated wear as illustrated in Figure 1. An optimization of Archard's wear model would make it possible to obtain reasonable quantitative wear prediction. Therefore, a wear coefficient $K(Z_{total})$, which depends on total wear work done at each location of the tool, was introduced into equation (1) to obtain the modified Archard wear equation (6):

$$W = K(Z_{total}) \frac{F_N}{H} s \qquad (6)$$

In this case the modified Archard equation can be expressed as follows at the elemental level at location i and stroke number n, where t corresponds to the time of the respective punch stroke:

$$w_i(n) = \frac{K_i(Z_{total})}{H} \int_t p_i(n,t) \cdot v_{slid,i}(n,t) \cdot dt \qquad (7)$$

Based on the wear equation for finite elements, the wear simulation is carried out (Figure 2). The geometry and all necessary nodal variables are imported from the forming simulation. With this input REDSY calculates the nodal wear work done at this punch stroke and based on this the nodal wear. The procedure of the geometric update scheme GUS published in [3] is shown in figure 2 (left).

The new algorithm however, for the first time, allows the wear coefficient K to be updated for each number of punch strokes for all nodes. With the updated algorithm, it is not only possible to update the tool geometry at each step, but also to choose the wear coefficient correspondingly in order to achieve an accurate estimation. The implementation of the variable wear coefficient into the wear simulation tool REDSY is shown in figure 2 (right).

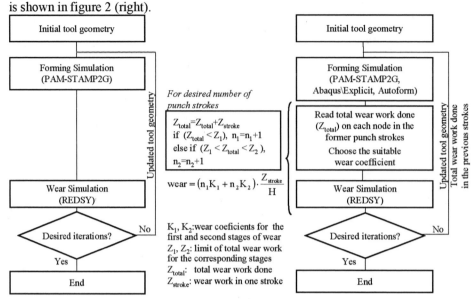

**FIGURE 2.** Original Geometric-Update-Scheme (GUS) algorithm (left) and updated algorithm (right)

## Determination of the Variable Wear Coefficients

In this study, wear depth w and hardness H are determined with experiments and both pressure p and sliding velocity $v_{slid}$ are computed by FE simulation. Wear coefficient K is determined using these values.

A simple axially symmetric cup deep drawing geometry is used to determine the wear coefficients. Only die wear is examined, as most abrasive wear is found there.

The experiments are carried out on a follow-on composite tool, run in a precision automatic punching press at 60 strokes per minute. In the first stages, the small sheet for deep drawing with a diameter of 60 mm is blanked out of the 10 mm wide and 0.8 mm thick strip. Then the deep drawing process takes place. Finally, the strip is chopped.

In the described experiments, the blank sheet material is a mild steel DC05 and the die material is LaserForm A6 Steel. The tool was produced by Indirect Metal Laser Sintering (IMLS), which is a commonly used rapid tooling method. The process parameters were: draw depth 16.5 mm and blankholder force 22.4 kN. The hardness of the material was measured to be 120 MPa. Wear measurements were performed after certain punch stroke intervals to determine the wear coefficient of this material pair over the lifetime of the die. These measurements were conducted by MahrSurf XCR20, Mahr GmbH, Göttingen, Germany, which is a tactile measurement system, working with an accuracy about 0.005 mm. The measurements are done at four locations as shown in figure 3 (left, top), where the top view of the die is depicted.

In parallel to the experiments, PAM-STAMP 2G forming simulations were conducted. A quarter cylindrical cup geometry with appropriate symmetry boundary conditions was used for conducting these forming simulations. The simulation variables were modelled to be identical with the real process described above. For the forming simulations the digitalised die surface was used instead of the CAD Data in order to increase the accuracy of the forming simulation.

## RESULTS

Wear in the die radius affects the part quality significantly. Therefore, surface measurements were conducted at four specified locations on the beginning of the draw die radius, where the largest amount of wear is expected. The measured wear depth of these locations over the number of punch strokes are shown in figure 3 (left). Experiments indicate that the first 1000 punch strokes cause a chief amount of wear in comparison to the second stage. This region can therefore be regarded as the first stage, i.e. the running-in wear, whereas the next interval is part of the second stage, i.e. the normal wear.

Figure 3 (left) indicates that in both stages the wear depth can be modelled to be linearly proportional with number of punch strokes. In this case, due to small wear values, the change in radius geometry and therefore in the contact area, pressure and wear work is negligibly small. Thus, it can be assumed that wear work is directly proportional to the number of punch strokes (Figure 3 (mid)). The two stages are distinguished by having different slopes, which correspond to two different wear coefficients (Figure 3 (right)). Here, the limit wear work ($Z_1$) corresponds to 1100 GPa.mm, which is the occurring wear work after 1000 punch strokes at the beginning of the die radius. This value is driven from the smoothed average of the wear work value for a single punch stroke at that location.

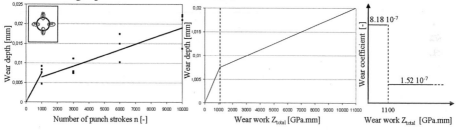

**FIGURE 3.** Wear depth values measured at the beginning of the die radius at four locations with certain punch strokes (left), corresponding wear depth vs. wear work diagram and the computed wear coefficients

Figure 4 depicts the measured and the computed wear depth on die radius. The results obtained by using the suggested algorithm with two wear work dependent coefficients are in a good quantitative agreement with the experiments. Here, as the wear occurring is so small that the change in contact pressure is negligible, the geometry update can be skipped and the computation is done with a single forming simulation. It can be seen that very good results can be obtained even with one step.

701

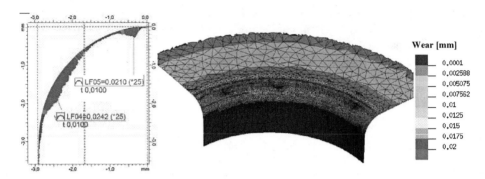

**FIGURE 4.** The measured wear on radius profile and the computed wear with the suggested algorithm.

# CONCLUSION

This study offers an enhancement of the previous wear simulation methods. This is achieved by using variable wear coefficients which take the wear work accumulated by each tool node into account. The algorithm enabling the consideration of the loading history is shown. That way it is possible to make the variable coefficients independent of the geometry. It also has great advantages for the wear calculations, if there are huge differences in loading conditions on different locations within a tool geometry. The first results are in a good agreement with the experimental data. Still, the suggested model should be verified by experiments involving different geometries, where different sliding distances and mechanical loading conditions appear.

# ACKNOWLEDGMENTS

The author is grateful to Bayrische Forschungsstiftung, which made this work possible within ForWerkzeug program. The die was produced at iwb Anwenderzentrum, Augsburg.

# REFERENCES

1. H. Hoffmann, C. Hwang, K. Ersoy, "Advanced wear simulation in sheet metal forming" in: CIRP Annals 54, No. 1, pp. 217-220, 2005
2. K. Ersoy, G. Nuernberg, G. Herrmann, H. Hoffmann, "Wear Simulation of Deep Drawing Tools With Abaqus/Explicit", in Proceedings 18. Abaqus Benutzerkonferenz Erfurt, 2006, Part 2.7
3. G. Nuernberg, K. Ersoy, G. Herrmann, H. Hoffmann, "Prediction of Wear in Deep Drawing Tools Using REDSY and Pamstamp 2G", in Proceedings Europam 2006, Toulouse
4. K. Ersoy, G. Herrmann, G. Nuernberg, M. Golle, "Enhancing The Wear Simulation of Sheet Metal Forming Process with Variable Wear Coefficients", in Proceedings iddrg pp. 635-642, 2006
5. K. H. Habig, V*erschleiß und Härte von Werkstoffen*, Hanser, 1980, pp. 49-54
6. L. A. Sosnovskiy, *Tribo Fatigue*, Springer, 2005, pp. 65
7. J. F. Archard, "Contact and rubbing of flat surfaces", Journal of Applied Physics, Vol. 24, 1953, pp. 981 – 988

# A New Modelling of Dynamic Recrystallization – Application to Blanking Process of Thin Sheet in Copper Alloy

Sébastien Thibaud, Abdelhamid Touache, Jérôme Chambert, and Philippe Picart

*Université de Franche-Comté, Institut FEMTO-ST – UMR 6174,*
*Laboratoire de Mécanique Appliquée R. Chaléat, 24 rue de l'Epitaphe, 25000 Besançon, France*

**Abstract.** Blanking process is widely used by electronic and micromechanical industries to produce small and thin components in large quantities. To take into consideration the influence of strain rate and temperature on precision blanking of thin sheet in copper alloy, a thermo-elasto-visco-plastic modelling has been developed in [1]. Furthermore the blanking of thin sheet in Cua1 copper presents dynamic recrystallization. A new modelling of dynamic recrystallization based on the thermodynamics of irreversible processes is proposed. Blanking simulations of Cua1 copper sheet are performed to analyze the softening effect induced by dynamic recrystallization.

**Keywords:** blanking, dynamic recrystallization, copper alloys, thermo-elasto-visco-plasticity.
**PACS:** 81.20.Hy, 81.10.Jt, 81.40.Ef, 46.15.-x, 81.05.Bx, 46.35.+z, 81.40.Lm, 81.70.Bt.

## INTRODUCTION

In precision blanking operations for very thin sheet about 0.25 mm, an accurate prediction of the maximum cutting force and cut edge profile are essential for designers and manufacturers. A finite element software untitled Blankform [2] has been developed in our laboratory to simulate the complete blanking operation from elastic deformation to the complete rupture of the sheet. In blanking operation like in many forming processes, the material undergoes large strains, high strain rates and considerable variations in temperature. In order to take strain rate and thermal effects into account, a specific thermo-elasto-viscoplastic modelling has been developed for Cua1 copper alloy, which is used in the manufacturing of electronic components by high precision blanking. The mechanical characterization and the modelling of Cua1 behaviour has been performed in the same way as in [1].

By using microhardness and fractography techniques along the cut edge profile, Gréban [3] shows that Cua1 copper alloy exhibits dynamic recrystallization (DRX) during the blanking process. The main purpose of this paper is to propose a new modelling of DRX within the framework of the thermodynamics of irreversible processes [4] in the context of blanking process. The new thermo-elasto-viscoplastic law with DRX has been implemented in the finite element code Blankform by

CP907, *10th ESAFORM Conference on Material Forming*, edited by E. Cueto and F. Chinesta

Touache [5]. Then, blanking simulations are carried out with or without DRX of Cua1 copper alloy to demonstrate the interest of such an approach.

## EXPERIMENTAL PROCEDURE

### Materials

Table 1 provides the chemical composition of two copper alloys which are used in blanking of precision parts for electronic components industry. Vickers microhardness tests (see Fig. 1), performed by Gréban [3], along the cut edge profile have highlighted a softening effect for Cua1 copper alloy (99.9% of Cu) contrary to other copper alloys such as CuNiP, for which it is observed a rise in microhardness. At a depth of about 150 μm in the shear zone, a minimum value of 65 Hv (respectively a maximum value of 170 Hv) is obtained for Cua1 (respectively CuNiP) copper alloy.

**TABLE 1.** Chemical composition of two copper alloys in weight percent.

| Material | P | Fe | Ni | Pb | Zn | Sn |
|----------|--------|--------|--------|--------|-------|------|
| CuNiP | 0.0175 | 0.02 | 0.2711 | 0.014 | 0.03 | 0.02 |
| Cua1 | 0.004 | 0.0008 | 0.0003 | 0.0016 | 0.001 | |

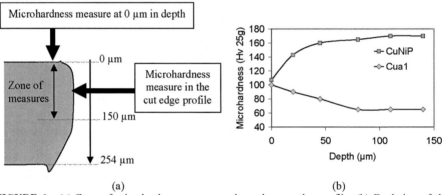

(a)                         (b)

**FIGURE 1.** (a) Zone of microhardness measures along the cut edge profile; (b) Evolution of the Vickers microhardness in terms of the depth in the cut edge profile for two copper alloys [3].

According to Gréban [3], the softening observed for the Cua1 results from the phenomenon of DRX. Indeed, scanning electron micrographs in the shear zone of the cut edge profile show that the average grain size is 22 μm before blanking and 3 μm after blanking. After tempering without hardening of 30 minutes, the temperature of recrystallization $T_{drx}$ is 250°C for Cua1 and approximately 450°C for CuNiP [3]. This is another reason why DRX occurs during the blanking of thin sheet in Cua1.

## Tensile Tests

Similarly as [1], the thermo-elasto-viscoplastic behaviour of Cua1 copper alloy has been characterized by carried out several tensile tests. Five temperatures have been considered: $T = 20°C$, $100°C$, $200°C$, $300°C$ and $400°C$. Three strain rates have been imposed: $\dot{\varepsilon} = 0.67e\text{-}3/s$, $27.02e\text{-}3/s$ and $67.65e\text{-}3/s$. If the imposed temperature is lower than $200°C$, the displacement is measured by using an extensometer. Otherwise, the displacement of the movable crosshead is used. Figures 2a and 2b illustrate respectively the influences of temperature and strain rate on the flow curves. It can be pointed out that, not only flow stress decreases with the rise in temperature, but also flow stress increases with strain rate.

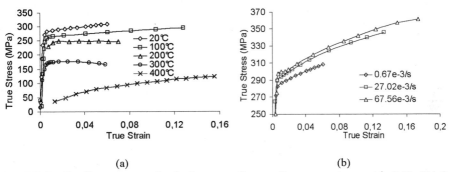

(a)                          (b)

**FIGURE 2.** Tensile test curves for Cua1 copper alloy at: Constant strain rate ($\dot{\varepsilon} = 0.67e\text{-}3/s$) for different temperatures (a); Room temperature ($T = 20°C$) for different strain rates (b).

None of these experimental flow curves for Cua1 present DRX even if the imposed temperature is higher than the temperature of recrystallization. The main reason is that the deformation at rupture obtained by tensile tests never exceeds 18% whatever temperature and strain rate.

## MODELLING

### Thermo-Mechanical Modelling

In a first approximation, the blanking process is assumed to be performed in an adiabatic condition so that we consider a weak thermo-mechanical coupling. By following the same procedure as in [1] for CuNiP, the sensitivity to temperature and strain rate on the flow stress of Cua1 has been taken into account:

$$\sigma_0 = \left(\sigma_Y + Q\left(1 - \exp(-b\bar{\varepsilon}^p)\right) + \frac{K_0}{T}\bar{\varepsilon}^p\right)\left[1 + a_v \ln\left(1 + \frac{\dot{\bar{\varepsilon}}^p}{\dot{\varepsilon}_0}\right)\right]\exp\left(\frac{\alpha}{T}\right) \quad (1)$$

where $\sigma_Y, Q, b, K_0, a_v, \dot{\varepsilon}_0$ and $\alpha$ are material parameters. These parameters are identified by the way of a genetic algorithm (see Table 2). From the flow curves, the Young modulus $E$ is 99500 MPa. From literature, the Poisson ratio $v$ is 0.31.

| $\sigma_Y$ (MPA) | $Q$ (MPa) | $b$ | $K_0$ (MPa.°C) | $a_v$ | $\dot\varepsilon_0$ (/s) | $\alpha$ (°C) |
|---|---|---|---|---|---|---|
| 21.71 | 276.2 | 402.5 | 279.7 | 1.436e4 | 9758 | -732.5 |

TABLE 2. Identified material parameters of the strain-hardening law for Cua1.

## Dynamic Recrystallization Modelling

The major part of DRX models proposed in literature are based on either physical laws (dislocation theory, etc.), or empirical rules. In this paper, we propose a new phenomenological model by using the thermodynamics of irreversible processes [5].

We assume the Helmholtz specific free energy depends on the following states variables: elastic strain tensor, isotropic hardening, recrystallized volume fraction and temperature. The first and second principles of thermodynamics are used to derive the state laws. By using the framework of "generalized standard materials", we assume the existence of a dissipation potential. The normal dissipative laws are then expressed. The potential function is divided into two parts: a viscoplastic one and a recrystallization one.

For the sake of simplicity, we assume that the critical strain for the onset of DRX $\varepsilon_{drx}$ is equal to the peak strain corresponding to the maximum stress. We propose the following expression:

$$\varepsilon_{drx} = \zeta \langle T - T_{drx} \rangle \ln\left(1 + \frac{\dot{\bar\varepsilon}^p}{\varepsilon_{drx}^0}\right) \qquad (2)$$

where $\zeta$ and $\varepsilon_{drx}^0$ are material parameters and $\zeta$ is a function of temperature. If the temperature is lower than the temperature of recrystallization $T_{drx}$, then $\zeta$ tends to infinity and DRX is not activated. The operator $\langle \cdot \rangle$ is defined as: if $a>0$ then $\langle a \rangle = a$, or else $\langle a \rangle = 0$.

The internal variable $X$ denotes the recrystallized volume fraction: $0 \le X \le 1$ ($X=0$ if no DRX and $X=1$ if completion of DRX). We assume that the thermodynamic force $\pi$, associated with $X$, takes the following form:

$$\pi = -\chi \, d_0^q \, Z^m \left\{1 - \exp\left(-k\langle \bar\varepsilon^p - \varepsilon_{drx}\rangle\right)\right\} \qquad (3)$$

where $\chi$, $q$, $m$ and $k$ are material parameters, $d_0$ the initial grain size, $Z$ the Zener-Hollomon parameter.

The evolution equation of recrystallized volume fraction reads:

$$\dot{X} = -\pi(1 - X)\dot{\bar\varepsilon}^p \qquad (4)$$

To take account of the influence of DRX on the flow stress, we propose the following isotropic hardening law:

$$\sigma_0 = \sigma_0(\bar\varepsilon^p, \dot{\bar\varepsilon}^p, T) \qquad \text{if } \bar\varepsilon^p < \varepsilon_{drx}$$
$$\sigma_0 = \sigma_0(\varepsilon_{drx}) - X \Delta\sigma_{drx} \quad \text{if } \bar\varepsilon^p \ge \varepsilon_{drx} \text{ and } T \ge T_{drx} \qquad (5)$$

where $\sigma_0(\bar\varepsilon^p, \dot{\bar\varepsilon}^p, T)$ is given by Eq. (1) and $\Delta\sigma_{drx}$ is the variation between the maximum stress and the steady state stress at large strains.

706

The numerical resolution of the preceding equations allows us to obtain the typical shape for DRX: strain hardening to a peak stress followed by a decrease in stress to a steady state level at large strains.

## NUMERICAL SIMULATION

### Blanking Test Description

The blanking test geometry is described on Fig. 3a. The sheet thickness is 0.254 mm, the punch width is 1.7 mm and the punch-die clearance is 8% of thickness. The punch and the die cutting edge radii are equal to 0.025 mm. The material parameters of the strain-hardening law for Cua1 are listed in Table 2. The chosen values of DRX parameters are given in Table 3. Let us note that these values are purely numerical and they do not correspond to any material.

TABLE 3. Material parameters of DRX.

| $\zeta$ | $\varepsilon_{drx}^0$ (/s) | $T_{drx}$ (°C) | $m$ | $k$ | $\chi d_0^q$ | $\Delta\sigma_{drx}$ (MPa) |
|---|---|---|---|---|---|---|
| 0.1 | 1 | 250 | -0.5 | 0.1 | 100 | 30 |

The punch speed is 1 mm/s. Sliding contact conditions are assumed between tools and specimen. The simulation has been performed in 2D plane strains conditions. The initial mesh is made up of 502 triangular 3 nodes-elements with 1 integration point. To reduce the influence of the mesh distortion, a global periodic remeshing of the sheet is generated every 10% of punch penetration.

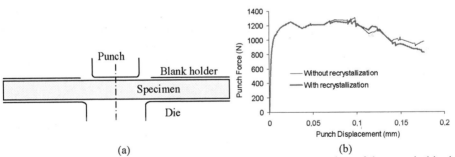

(a)  (b)

**FIGURE 3.** Schematic representation of the blanking process (a). Comparison of the numerical load-displacement curves computed by the model with and without DRX (b).

### Results and Discussion

Figure 3b shows that, as soon as DRX occurs, the numerical load-displacement curve obtained by the thermo-elasto-viscoplastic model with DRX is below the one performed without DRX.

Figure 4 presents the distributions of recrystallized volume fraction $X$ in the shearing zone for three values of the punch penetration. DRX initiates near the punch corner for a punch penetration of 35 % of the sheet thickness. Then this phenomenon grows quickly in this zone: $X$ is about 90% for a punch penetration of 55%. From this penetration, DRX appears in the die corner region. On the distribution corresponding to 69% of punch penetration, volume fraction $X$ is localized in a narrow band between the punch and die radii. We notice also that the growth of DRX leads to a decrease in equivalent stress inside this small band.

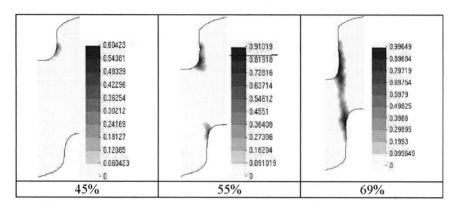

**FIGURE 4.** Evolution of recrystallized volume fraction in terms of the punch penetration for Cua1.

# CONCLUSION

In this paper, a thermo-elasto-viscoplastic model with DRX associated to Cua1 alloy has been proposed and implemented in the FE code Blankform. The thermo-mechanical parameters of this model have been identified and the parameters related to DRX should be identified by direct or inverse methods. The numerical simulation of blanking process enables us to check that the proposed model of DRX is qualitatively in agreement with phenomena observed experimentally in blanking.

# REFERENCES

1. S. Thibaud, A. Touache, P. Picart and J. Chambert, "Mechanical characterization and modelling of copper alloys behaviour in precision blanking by a thermo-elasto-viscoplastic approach" in *9th ESAFORM International Conference on Material Forming*, European Scientific Association for Material Forming (ESAFORM), edited by N. Juster and A. Rosochowski, Glasgow, U.K., 2006, pp. 635-638.
2. V. Lemiale, "Contribution à la modélisation et à la simulation numérique du découpage des métaux", Ph.D. Thesis, Université de Franche-Comté, France, June 2004 [in French].
3. F. Gréban, "Découpabilité du cuivre et des alliages cuivreux", Ph.D. Thesis, Université de Franche-Comté, France, February 2006 [in French].
4. J. Lemaitre and J.-L. Chaboche, *Mechanics of Solid Materials*, Cambridge: Cambridge University Press, 1990.
5. A. Touache, "Contribution à la caractérisation et à la modélisation de l'influence de la vitesse et de la température sur le comportement en découpage de tôles minces", Ph.D. Thesis, Université de Franche-Comté, France, December 2006 [in French].

# Evaluation of Fracture Initiation in the Mannesmann Piercing Process

S. Fanini[1,*], A. Ghiotti[1], S. Bruschi[2]

1 DIMEG, University of Padova, Via Venezia 1 – 35131 Padova, Italy
2 DIMS, University of Trento, Via Mesiano 77 – 38050 Trento, Italy
*corresponding author: silvio.fanini@unipd.it

**Abstract**. One of the challenging objectives in studying the Mannesmann piercing process is to predict the fracture initiation, known as "Mannesmann effect", in order to design and optimize the working parameters of the piercing process. The objective of the paper is to investigate the workability of a tube steel tested in the same conditions of the Mannesman piercing process. The stress and strain states as well as temperature fields arising during the process are identified through numerical simulations. The hot tensile test is chosen for fundamental studies on fracture initiation, as a tensile state of stress in the centre of the billet in the first stages of the piercing process before the plug arrival seems to be one of the main causes of the crack initiation. The material constants of energy-based models implemented in FEM codes are calculated and numerical results are compared with non-plug piercing tests carried out on the industrial plant.

**Keywords**: tube piercing, workability, hot tensile test, continuous casting defects
**PACS**: 81.05.Bx

## INTRODUCTION

Seamless tubes are generally used in applications where safety plays a decisive role like sea or land oil-gas lines, pipe lines, oil rigs, and structural elements in mechanical and automotive industry. The industrial manufacturing of seamless tubes consists of two different phases: (i) the rotary piercing process to obtain the tube from the initial cylindrical steel billet, and (ii) the multi-pass rolling for the final tube calibration in length and thickness. This sequence of forming operations allows improving the dimensional and mechanical characteristics of the tubes compared to the welded technology, as they are characterised by an elevated resistance to high pressures, obtained thanks to the fine and uniform microstructure granted by the process.

The cylindrical billet, generally obtained by continuous casting, enters the Mannesmann piercing mill at the initial temperature of 1250-1300°C. It passes through the gap of two skew conical rolls that cause the axial fracture along the billet longitudinal axis, thanks to the stress-strain state arising at the billet centre (Figure 1). Once the fracture is initiated, the billet is rolled between the external rolls and the plug that, acting as a mandrel, enlarges the initial cavity. After piercing, the hollow cylinder is moved on other rolling mills where the final diameters and thickness of the tube are reached. The fracture initiation in the piercing process is usually known as the

CP907, 10th ESAFORM Conference on Material Forming, edited by E. Cueto and F. Chinesta
© 2007 American Institute of Physics 978-0-7354-0414-4/07/$23.00

"Mannesmann effect" [1-10]: an example is shown on the right of Figure 1, where a steel billet pierced without the action of the plug is shown.

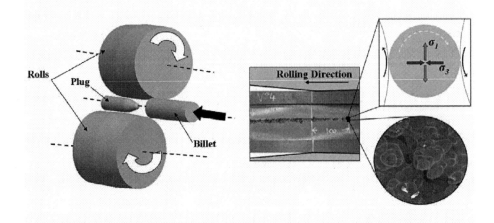

**FIGURE 1.** Scheme of the piercing process (on the left) and fracture in the centre of the billet in a piercing industrial trial without the use of the plug (on the right).

The position of the fracture initiation has a great importance in the service life of the mandrel: if the fracture in the central part of the billet is not initiated when the plug pierces the material, the wear is excessive. Otherwise, in case of large distance of the plug form the billet, the internal surface of the crack will experience an excessive oxidation leading to defects in the final tube. Due to this reason, the optimization of the Mannesmann piercing process passes through the accurate knowledge of the fracture initiation, in order to set more accurately the plug position and in this way to minimize plug wear and tube defects.

The objective of the paper is to investigate the workability of a tube steel in the same condition of the industrial process. In the first part of the paper, the numerical model of the piercing tests carried out on an industrial plant is described in order to identify the evolution of the thermal and mechanical parameters relevant to the crack initiation. The second part of the paper presents the hot tensile laboratory tests, carried out on samples machined from continuous casting billets in different positions, and the identification of the constants of energy-based damage models [11-13]. Finally, the values of damage corresponding to fracture in the industrial trials are compared to the ones obtained from the experiments.

## NUMERICAL SIMULATION OF THE MANNESMANN PIERCING PROCESS

The three-dimensional numerical simulation of the non-plug Mannesmann piercing process was carried out with the commercial finite element code *Forge2005$^{TM}$*.

Geometrical and kinematic parameters were set to replicate those of an industrial trial. In the implemented FE model of the tube piercing process, the plug action was not modelled, in order to compute the stress-strain condition in the billet centre leading to the Mannesmann effect only due the action of the skewed rolls. The initial temperature of the billet at the beginning of the process was set equal to 1250 °C.

The material used in this study was a low alloyed steel for tube production; its rheological behaviour was obtained by hot compression tests performed on a *Gleeble*[TM] thermo-mechanical simulator under conditions that were representative of the process; material constants of the Hansel-Spittel law were calculated through non-linear regression method and implemented into the code [14].

Lagrangian virtual sensors were placed along the billet axis in the FE-model in order to follow the evolution of the stress and strain states during the process simulation. Since the shear components can be considered negligible, the principal stresses were taken equal to the three normal components of the stress tensor calculated on the billet axis. The evolutions of the principal stresses, of the hydrostatic pressure and of the triaxiality factor (here defined as the ratio between the mean stress and the Von Mises equivalent stress) are reported in Figure 2 and confirm the presence of a stress condition that is favourable of voids enlargement and therefore of crack formation.

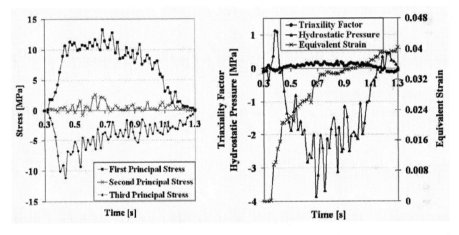

**FIGURE 2.** Evolution of principal stress components, hydrostatic pressure, triaxiality factor and equivalent strain in the centre of the cylindrical billet during non-plug piercing (resulting from FE simulations).

## EXPERIMENTS

The tube steel used for this study is obtained through continuous casting, that can cause a random distribution of porosities along the radial direction and hence variations in the strength behaviour of the billet material.

The hot tensile test was chosen to investigate the material workability, being the secondary tensile stress arising in the billet axis (Figure 2) the most relevant one; moreover, the tensile test is characterised by values of the triaxiality factor and amount of strain comparable to those encountered in the process (Figure 2).

Tensile specimens were machined oriented both parallel or normal to the billet axis, and extracted both in the centre and in the periphery of the billet; in this way, a random distribution of porosities can be taken into account (Figure 3 on the right).

The hot tensile tests were carried out on a $Gleeble^{TM}$ machine equipped with a dilatometer to monitor the instantaneous section of the sample after necking (a way to obtain the instantaneous value of the strain), the temperature of the gauge length of each specimen was set equal to 1250°C and the displacement profile of the movable punch was chosen to have an average constant strain rate in the gauge length (Figure 3 on the left). Strain rate was varied from $5 \cdot 10^{-4}$ s$^{-1}$ to 0.5 s$^{-1}$ to study the steel sensitivity to strain rate: notice that, from numerical simulation results, the average value of the strain rate at the Mannesmann billet axis during the process before the plug arrival was in the order of $10^{-3}$-$10^{-4}$ s$^{-1}$.

Each experimental condition was repeated at least 5 times; however, the random voids distribution even in specimens machined from the same position in the cast billet can assure repeatibility of results (in terms of stress and strain at fracture) only in the range of ±20%.

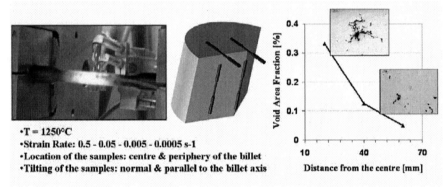

·T = 1250°C
·Strain Rate: 0.5 - 0.05 - 0.005 - 0.0005 s-1
·Location of the samples: centre & periphery of the billet
·Tilting of the samples: normal & parallel to the billet axis

**FIGURE 3.** Hot tensile test set-up and voids percentage as function of the distance from the billet axis.

The graph on the left of Figure 4 describes how the flow stress is affected by the machining position of the samples all deformed under the same processing conditions ( T=1250°C and $\dot{\varepsilon}$ =0.5 s$^{-1}$): in general a specimen belonging to the billet centre zone is less resistant than a peripheral one (in terms of maximum stress and maximum strain at fracture) and that a normal to axis specimen reaches an higher maximum stress value but a lower strain at fracture than a parallel to axis sample. These differences on the material behaviour are due to the voids distribution in each specimen; porosities and voids due to cooling are larger and more numerous in the centre of the billet, and additionally they are characterised by a stretched shape in the casting direction (that corresponds to the rolling direction during piercing). The diagram on the right of Figure 4 underlines the material sensitivity to strain rate; in particular it reports the results of tensile tests on specimens normal to the billet axis and machined from the peripheral zone of the billet; for each curve it is also reported the value of the damage calculated according to the Cockcroft&Latham criterion. As it can be expected, this value is strongly affected by the strain rate at fixed temperature.

From numerical simulation results, the tensile tests carried out at $5 \cdot 10^{-4}$ s$^{-1}$ give a better description of the mechanical conditions on the billet centre when Mannesmann fracture arises during piercing. In fact, the strain rate of the test is of the same order of magnitude of the one on the billet axis during the first part of the process, and the maximum tensile stress (see Figure 4 – peripheral normal to the axis specimen) is comparable to the values of the first principal component of the stress tensor. This seems to confirm that the most important cause of the cavity formation is the secondary tensile stress normal both to the rolling direction and to the compressive load direction due to the rolls narrowing. Furthermore, from the numerical simulation of non-plug piercing industrial trials, the fracture occurs when the damage according to Cockcroft&Latham criterion reached values around 0.30-0.35, that were approximately of the same order of magnitude of the value calculated from the tensile test, that is 0.76.

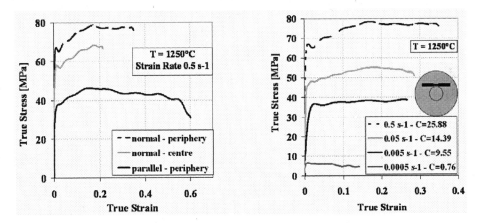

**FIGURE 4.** Hot tensile test results: influence of the position of the sample in the initial billet on the tensile behaviour (left) and material sensitivity to strain rate with damage calculations (right).

## CONCLUDING REMARKS

A three-dimensional FEM model of Mannesmann piercing process before the plug arrival was developed in order to identify the strain and stress states at the billet axis. These data were then utilised to set up the experimental campaign. As the cavity initiation for Mannesmann effect seemed to be mainly induced by the secondary tensile stress (first principal component of stress), and the contributions of shear stress components as well as low cycle fatigue could be considered negligible, the hot tensile test was chosen to evaluate material workability. Tests carried out on specimens machined in different locations at different orientation from cast billets showed a significant scatter in results and underlined the need to pay particular attention to the specimen extraction. The workability behaviour resulting from normal to axis samples was strongly influenced by strain rate; the related damage parameters calculated according to Cockroft&Latham criterion seemed in quite good agreement with the results from industrial trials, when considering the actual low strain rate encountered in the process.

To reach a closer agreement with pierced billets, multi-axial loading in the experiments should be considered, in order to include the effect of the third principal component of stress (compressive loading, normal to the tensile one and to the rolling direction) and the effect of the rotation (dynamically, the tensile load is applied to different crystallographic planes). Moreover, being the scatter in voids distribution too high and variable, data about material workability should be modelled following a stochastic approach related to initial voids distribution.

Finally, investigations concerning the typology of the Mannesmann fracture are in progress: in fact, the very low strain rate condition leads to a fracture in the tensile specimen that cannot be considered a pure ductile fracture, but more similar to a kind of low strain rate intergranular fracture. Some confirmation on this sense is given by SEM observations of the fracture surface of the billet coming from industrial trials.

# REFERENCES

1. G. Capoferri, E. Ceretti, C. Giardini, A. Attanasio, F. Brisotto, "FEM Analysis of Rotary Tube Piercing Process" in *Tube & Pipe Technology* (May-June 2002), pp.55-58, 2002.
2. E. Giardini, C. Giardini, F. Brisotto, "Development of a Simulation Model of the Tube Piercing Process and FEM Application to Improve the Quality of Seamless Tubes" in *Advanced Technology of Plasticity 2005*, edited by P. F. Bariani, Proceedings of the 8th ICTP, Verona, Italy, 2005.
3. K. Mori, K. Osakada, "Finite Element Simulation of Three-Dimensional Deformation in Shape Rolling" in *International Journal of Numerical Methods in Engineering*, 30-8, pp. 1431-1440, 1990.
4. S. Urbanski and J. Kazanecki, "Assessment of the Strain Distribution in the Rotary Piercing Process by the Finite Element Method" in *Journal of Materials Processing Technology* 45, pp. 335-340, 1994.
5. J. Yang, G. Li, W. T. Wu, K. Sawamiphakdi and D. Jin, "Process Modeling for Rotary Tube Piercing Application" in *Materials Science & Technology 2004* Vol. 2, pp. 137-148, New Orleans LA, 2004.
6. K. Mori, H. Yoshimura, K. Osakada, "Simplified Three-dimensional Simulation of Rotary Piercing of Seamless Pipe by Rigid-Plastic Finite-Element Method" in *Journal of Materials Processing Technology* 80-81, pp. 700-706, 1998.
7. K. Komori, "Simulation of Mannesmann Piercing Process by the Three-dimensional Rigid-Plastic Finite-Element Method" in *International Journal of Mechanical Sciences* 47, pp. 1838-1853, 2005.
8. F. Piedrahita, L. Garcia Aranda, Y. Chastel, "Prediction of Internal Defects in Cross-Wedge Rolling of Bars" in *The 9th International Conference on Material Forming ESAFORM 2006, Glasgow UK*, edited by N. Juster and A. Rosochowski, pp. 459-462, 2006.
9. Q. Li, M. R. Lovell, W. Slaughter, K. Tagavi, "Investigation of the Morphology of Internal Defects in Cross Wedge Rolling" in *Journal of Materials Processing Technology* 125-126, pp.248-257, 2002.
10. Y. Dong, K. A. Tagavi, M. R. Lovell, Z. Deng, "Analysis of Stress in Cross Wedge Rolling with Application to Failure" in *International Journal of Mechanical Sciences* 42, pp.1233-1253, 2000.
11. M. G. Cockcroft and D. J. Latham, "Ductility and the Workability of Metals" in *Journal of the Institute of Metals*, Vol. 96, pp. 33-39, 1968.
12. D. C. J. Farrugia, "Prediction and Avoidance of High Temperature Damage in Long Product Hot Rolling" in *Journal of Materials Processing Technology* 177, pp. 486-492, 2006.
13. B.P.P.A. Gouveia, J.M.C. Rodrigues, P.A.F. Martins, "Fracture Predicting in Bulk Metal Forming" in *International Journal of Mechanical Sciences* 38-4, pp.361-372, 1996.
14. Fanini S., D.I.M.E.G. Internal Report, September 2006.

# Numerical modelling of tool wear in turning with cemented carbide cutting tools

P. Franco, M. Estrems and F. Faura

*Departamento de Ingeniería de Materiales y Fabricación, Universidad Politécnica de Cartagena,
Campus Muralla del Mar, C/ Doctor Fleming s/n, 30202 Cartagena (Spain)*

**Abstract.** A numerical model is proposed for analysing the flank and crater wear resulting from the loss of material on cutting tool surface in turning processes due to wear mechanisms of adhesion, abrasion and fracture. By means of this model, the material loss along cutting tool surface can be analysed, and the worn surface shape during the workpiece machining can be determined. The proposed model analyses the gradual degradation of cutting tool during turning operation, and tool wear can be estimated as a function of cutting time. Wear-land width ($VB$) and crater depth ($KT$) can be obtained for description of material loss on cutting tool surface, and the effects of the distinct wear mechanisms on surface shape can be studied. The parameters required for the tool wear model are obtained from bibliography and experimental observation for AISI 4340 steel turning with WC-Co cutting tools.

**Keywords:** Metal cutting; Numerical modelling; Tool wear; Adhesion; Abrasion; Fracture.
**PACS:** 81.20.Wk; 07.05.Tp; 81.40.Pq; 46.55.+d.

## 1. INTRODUCTION

In metal cutting, tool wear is one of the most critical aspects due to its influence on surface finish, dimensional accuracy and final cost of machined part. Many authors have analyzed tool wear as a function of factors involved in machining, such as cutting conditions, cutting tool geometry and workpiece material [1-4]. Lo Casto et al. [13] analysed the influence of thermal stress on cracks formation for ceramic tools, and Evan and Marshall [12] provided a fracture toughness threshold for wear rate. Dearnley [9], Ruppi et al. [10] and Brandt and Mikus [11] studied wear resistance of coated WC-Co inserts, while Kwon [15] developed models for abrasion wear on flank face for TiN, TiCN and Al$_2$O$_3$ coated inserts. In this work, a numerical model for tool wear analysis including not only flank face but also rake face is proposed. This model is not limited to abrasion wear, but also considers adhesion and fracture, as well as initial and fast wear. The tool wear model is used to estimate worn surface shape of cutting tool and wear parameters such as wear-land width ($VB$) and crater depth ($KT$).

CP907, *10th ESAFORM Conference on Material Forming,* edited by E. Cueto and F. Chinesta
© 2007 American Institute of Physics 978-0-7354-0414-4/07/$23.00

# 2. NUMERICAL MODELLING OF TOOL WEAR

## 2.1. Normal stress at tool surface

For tool wear simulation, normal stress variation along chip-tool interface can be described by Zorev's model [9], and a uniform distribution can be considered on flank face. According to Equation (1), normal stress on rake face can be obtained from maximum normal stress $\sigma_{fd,max}$, surface position $x$ and chip-tool contact length $l_f$, with maximum normal stress given by average friction coefficient, average shear strength and ratio between sticking zone and chip-tool contact length [9]. The expression provided by C. Arcona and T.A. Dow [10] for a similar workpiece material can be applied to estimate normal stress on flank face from hardness $H_{mi}$ and elastic modulus $E_{mi}$ of workpiece material, as shown in Equation (2).

$$\sigma_{fd} = \sigma_{fd,max}\left(\frac{x-l_f}{l_f}\right)^n \tag{1}$$

$$\sigma_{fi} = kH_{mi}\sqrt{\frac{H_{mi}}{E_{mi}}} \tag{2}$$

## 2.2. Wear mechanisms

Wear mechanisms of adhesion, abrasion and fracture can be involved in machining operations, and initial and fast wear must be also identified for tool wear modelling. From Archard's equation for adhesion, Rabinowicz's model for abrasion and Evan and Marshall's model for fracture, the following expression can be deduced for resultant material loss in cutting tool surface in terms of Archard's equation [11], where $k_{adh}$, $k_{abr}$, $k_f$, are wear coefficients for adhesion, abrasion and fracture, $k_{ini}$ and $k_r$ correspond to initial and fast wear, $F_n$ is normal force on tool surface, $K_{c,h}$ is fracture toughness of workpiece material, and $\delta$ is initial deformation due to contact between part and tool surfaces:

$$V = \frac{k_{adh}F_nl}{H_h} + \frac{k_{abr}F_nl}{H_h} + \frac{k_fN(E_h/H_h)F_n^{9/8}l}{K_{c,h}^{1/2}H_h^{5/8}} + \frac{k_{ini}F_n\delta}{H_h}\left(1-e^{-n_{ini}l}\right) + \frac{k_rF_n}{H_h}\left(e^{n_rl}-1\right) \tag{3}$$

From material loss contribution of each wear mechanism, if adhesion, abrasion and fracture are represented by a same wear coefficient $k$, the worn depth of cutting tool in normal direction to fresh tool can be defined by the following expression, as a function of surface position $x$ and cutting time $t_c$, where the first and last terms represent initial and fast wear, respectively:

$$dh(x,t_c) = \sigma_f(x,t_c)\left\{k_{ini}(x,t_c)\,d\!\left[e^{-n_{ini}\,v\,t_c}\right] + k(x,t_c)\,v\,dt_c + k_r(x,t_c)\,d\!\left[e^{n_r\,v\,t_c}\right]\right\} \qquad (4)$$

## 2.3. Tool wear model

For numerical calculation of tool surface degradation caused by mechanical contact with machined surface and removed chip, the finite-difference method can be applied due to its benefits in model simplicity and computational time, according to the mathematical procedure illustrated in Figure 1a. In this model, tool wear is obtained separately on flank and rake surfaces of cutting tool. On flank face, uniform and constant normal stress $\sigma_{fi}$ and wear coefficients $k_{s,i}$ are considered, and material loss can be deduced analytically from cutting time. Nevertheless, rake face normal stress $\sigma_{fd}$ and wear coefficients $k_{s,d}$ change along chip-tool interface, and different values are achieved during turning process, since cutting zone slides along rake face due to flank wear (Figures 1b and c). For this reason, cutting time and surface position are discretized according to cutting time step $dt_c$ and position step $dx$ in this surface.

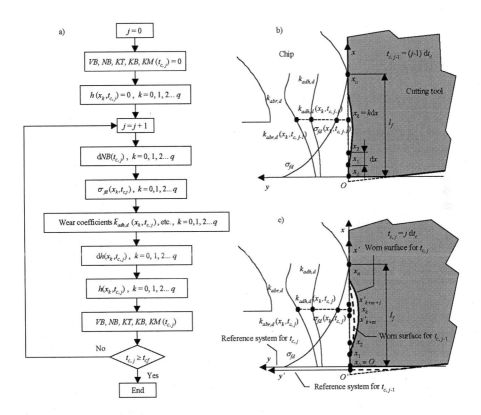

**FIGURE 1.** Tool wear modelling: a) calculation equations; b) diagram for $t_{c,j-1}$; and c) diagram for $t_{c,j}$.

Wear normal to land $dNB_j = dNB(t_{c,j})$ for cutting time $t_{c,j}$ (with $j = 0, 1, 2...$) can be deduced from Equation (5), and then rake face normal stress and wear coefficients can be determined for different nodes of chip-tool interface. From normal stress and wear coefficients for $t_{c,j}$, the increase in worn depth at chip-cutting tool interface $dh_{k,j} = dh(x_k, t_{c,j})$ for surface position $x_k$ and cutting time $t_{c,j}$ wil be given by Equation (6):

$$dNB_j = \begin{cases} 0 & , j = 0 \\ \\ k_{ini,i}\left(e^{-n_{ini,i} v t_{c,j-1}} - e^{-n_{ini,i} v t_{c,j}}\right) + k_i \, v \, dt_c + k_{ri}\left(e^{n_{ri} v t_{c,j}} - e^{n_{ri} v t_{c,j-1}}\right) & , j > 0 \end{cases} \tag{5}$$

$j = 0, 1, 2...$

$$dh_{k,j} = \begin{cases} 0 & , j = 0 \\ \\ (\overline{\sigma}_{fd})_{k,j}\left[(\overline{k}_{ini,d})_{k,j}\left(e^{-n_{ini,d} v t_{c,j-1}} - e^{-n_{ini,d} v t_{c,j}}\right) + (\overline{k}_d)_{k,j} \, v \, dt_c + \right. \\ \left. + (\overline{k}_{rd})_{k,j}\left(e^{n_{rd} v t_{c,j}} - e^{n_{rd} v t_{c,j-1}}\right)\right] & , j > 0 \end{cases} \tag{6}$$

$k = 0, 1, 2... q$
$j = 0, 1, 2...$

where $q$ is the number of nodes along chip-tool interface, $(\overline{\sigma}_{fd})_{k,j} = \overline{\sigma}_{fd}(x_k, t_{c,j})$ is the average value of normal stress for surface position $x_k$ during time interval $t_{c,j} - t_{c,j-1}$, and wear coefficients $(\overline{k}_{ini,d})_{k,j} = \overline{k}_{ini,d}(x_k, t_{c,j})$, $(\overline{k}_d)_{k,j} = \overline{k}_d(x_k, t_{c,j})$ and $(\overline{k}_{rd})_{k,j} = \overline{k}_{rd}(x_k, t_{c,j})$ represent the average values for initial, linear and fast wear that correspond to surface position $x_k$ and cutting time $t_{c,j}$. The worn depth $h_{k,j} = h(x_k, t_{c,j})$ and worn surface shape on rake face for cutting time $t_{c,j}$ are given by numerical calculations based on finite-difference method. Tool wear parameters such as wear-land width $VB_j$ and crater depth $KT_j$ can be easily deduced from the worn surface shape of cutting tool as a function of wear normal to land $NB_j$ and rake face worn depth $h_{k,j}$.

## 3. RESULTS AND DISCUSSION

The tool wear model is applied to workpiece machining with $a_p = 1.27$ mm, $v = 240$ m/min and $f = 0.254, 0.285, 0.330$ and $0.406$ mm/rev from experimental data provided by Cho and Komvopoulos [1], and cutting tool material loss is analysed according to mechanical efforts on contact surface during turning process. Numerical and experimental results for wear-land width ($VB$) and crater depth ($KT$) are depicted in Figure 2 when $f = 0.285$ mm/rev and wear mechanisms of adhesion, abrasion and fracture are considered. Additional information about variation of these wear parameters can be obtained from numerical predictions for intermediate cutting times.

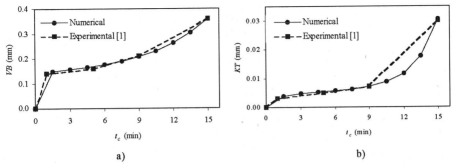

**FIGURE 2.** Wear-land width (*VB*) and crater depth (*KT*) as a function of cutting time when $f$ = 0.285 mm/rev by adhesion, abrasion and fracture.

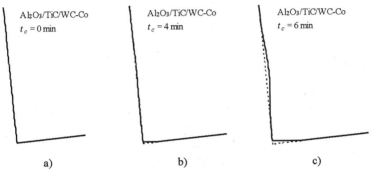

**FIGURE 3.** Worn surface shape for $f$= 0.406 mm/rev by adhesion, abrasion and fracture: a) $t_c$ = 0 min; b) $t_c$ = 4 min; and c) $t_c$ = 6 min.

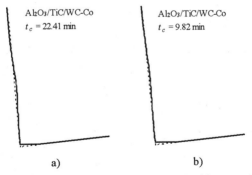

**FIGURE 4.** Worn surface shape for tool life by adhesion, abrasion and fracture: a) $f$= 0.254 mm/rev; and b) $f$= 0.330 mm/rev.

Figure 3 illustrates the worn tool surface shape obtained by adhesion, abrasion and fracture wear mechanisms when $f$ = 0.406 mm/rev and cutting time $t_c$ = 0, 4 and 6 min. The formation of a crater-shape cavity in the intermediate region of chip-tool interface, and a wear-land in machined surface direction can be observed in this figure. The numerical model can be used to deduce the limit of tool life for cutting conditions applied in machining operation. From the tool wear criterion of $VB_0 = 0.3$ mm and $KT_0 = 0.06 + 0.3\,f$ proposed by ISO, a tool life $t = 22.41$ min is deduced for $f = 0.254$ mm/rev by material loss in flank face, as can be seen in Figure 4a. If $f = 0.330$ mm/rev, a tool life $t = 9.82$ min is also given by flank wear, as shown in Figure 4b. The accepted criterion for rake wear is not achieved until cutting time $t_c$ exceeds 25 min tool life when $f = 0.254$ mm/rev. A smaller crater depth is obtained when $f = 0.330$ mm/rev, even for $t = 9.82$ min.

# 4. CONCLUSIONS

In this work a numerical model is proposed to study the material loss on cutting tool surface as a consequence of adhesion, abrasion and fracture wear mechanisms. By means of this model, the worn surface shape of cutting tool can be estimated as a function of process parameters and cutting time, and parameters such as wear-land width ($VB$) and crater depth ($KT$) can also be determined. The proposed model is probed to be useful to describe the effects of wear mechanisms such as adhesion, abrasion and fracture during turning processes. This wear model could be coupled to other numerical algorithms in order to analyze metal cutting problems like prediction of cutting forces, chatter, dimensional quality or surface finish.

# ACKNOWLEDGMENTS

The authors thank Comisión Interministerial de Ciencia y Tecnología of Spain (CYCIT) and European Comission, by financial support received from research project with reference DPI2005-04969.

# REFERENCES

1. S.-S. Cho and K. Komvopoulos, J. Trib. **119**, 8-17 (1997).
2. S. Lo Casto, E. Lo Valvo, E. Lucchini, S. Maschio and V.F. Ruisi, Wear **208**, 67-72 (1997).
3. S.-S. Cho and K. Komvopoulos, J. Trib. **120**, 75-81 (1998).
4. T. Kitagawa, A. Kubo and K. Maekawa, Wear **202**, 142-148 (1997).
5. S. Lo Casto, E. Lo Valvo, E. Luchini, S. Maschio, M. Piacentini, V.F. Ruisi, *Key Engineering Materials* **114**, 105-134 (1996).
6. A.G. Evans and D.B. Marshall, "Wear Mechanisms in Ceramics", in *Fundamentals of Friction and Wear of Materials. AIP Conference Proceedings Vol. 620*, edited by D.A. Rigney, Metals Park, Ohio: ASM, 1980, pp. 439-452.
7. P.A. Dearnley, J. Eng. Mater. & Tech. **107**, 68-82 (1985).
8. S. Ruppi, B. Högrelious and M. Huhtiranta, Int. J. Refract. Met. & Hard Mater. **16**, 353-368 (1998).
9. G. Brandt and M. Mikus, Wear **115**, 243-263 (1987).
10. P. Kwon, J. Trib. **122**, 340-347 (2000).
11. G. Boothroyd and W.A. Knight, *Fundamentals of Machining and Machine Tools*, New York: Marcel Dekker, 1989.
12. C. Arcona and T.A. Dow, *J. Manuf. Sc.Eng.* **120**, 700-707 (1998).
13. B. Bhushan, *Principles and Applications of Tribology*, New York: John Wiley & Sons, 1999.

# Adaptiv modeling and simulation of shear banding and high speed cutting

## Christian Hortig and Bob Svendsen

*Chair of Mechanics, University of Dortmund, 44227 Dortmund, Germany*

**Abstract.** The purpose of this short work is the thermomechanical modeling of shear band and chip formation during high-speed cutting. Shear bands develop in areas of maximal mechanical dissipation in which temperature-dependent softening dominates strain- and strain-rate-dependent hardening. In the simulations, the well-known problem of the mesh-dependence of the shear-band development is addressed, involving both mesh size and mesh orientation. An example simulation is presented.

**Keywords:** high-speed cutting, adiabatic shear-banding, finite-element analysis, mesh dependence

## INTRODUCTION

High-speed cutting is a process of great interest in modern production engineering. In order to take advantage of its potential, a knowledge of the material and structural behavior in combination with the technological conditions is essential. To this end, investigations based on the modeling and simulation of the process are necessary. Initial such investigations were analytical in nature and focused on the process of machining (*e.g.*, [11, 12]). For the significantly more complex processes and geometries of today, approaches based on numerical and in particular finite-element simulation represent the state of the art [1, 22, 3, 4, 10, 13, 26, 17]. In order to account for the effects of high strain-rates and temperature on the material behavior, most of these approaches are based on thermoviscoplastic material modeling. For example, the Johnson-Cook model [25] is used in [4, 26, 17] and in the current work.

Experimental results [23, 7, 17, 29] show that shear banding represents the main mechanism of chip formation and results in reduced cutting forces. In the context of a finite-element analysis, such shear banding can be modeled using thermo-viscoplastic material models including in particular the effect of thermal softening (and in general damage as well: *e.g.*, [17]). As is well-known, this results in a loss of solution uniqueness, resulting in so-called pathological mesh-dependence of the simulation results. Usually, this dependence is expressed in terms of the size of the elements used, *i.e.*, the element edge-length. However, it is not restricted to this property of the elements. Indeed, other properties, *e.g.*, element orientation, or interpolation order, are just as, if not more, influential in this regard. All such element properties are relevant in the context of, *e.g.*, the use of adaptive remeshing techniques [1, 22, 3, 13, 26] to deal with large element distortion, resulting almost invariably in unstructured meshes. In the literature, remeshing techniques using structured meshes can also be found. In [1, 3], for example, an arbitrary Lagrangian-Eulerian-like approach is used to rearrange and refine a structured mesh. As will be shown in the current work, the influence of the mesh orientation

CP907, *10th ESAFORM Conference on Material Forming*, edited by E. Cueto and F. Chinesta
© 2007 American Institute of Physics 978-0-7354-0414-4/07/$23.00

becomes significant in the context of adiabatic shear banding, especially in connection with structured meshes.

The use of remeshing techniques may lead to a reduction of mesh-dependence, but of course cannot eliminate it. This can be achieved only by working with models based on additional criteria (*e.g.*, penalization of "vanishingly thin" shear-bands via regularization).

## MATERIAL MODELING

As is well-known, metal cutting is influenced by a number of competing physical processes in the material, in particular heat conduction and mechanical dissipation. Consider for example the cutting of the material X20Cr12 at different cutting speeds as shown in Fig. 1.

**FIGURE 1.** Metal cutting experiments with X20Cr13 at cutting speeds $v_c$ of 8 m/min (left) and 200 m/min (right) (courtesy of S. Hesterberg, Department of Machining Technology, University of Dortmund).

At lower cutting speeds (left) and so strain-rates, heat conduction is fast enough to prevent the temperature increase due to mechanical dissipation, resulting in thermal softening. At higher speeds (right) and so strain-rates, however, heat conduction is too slow to prevent the temperature from increasing to the point where thermal softening occurs, leading to shear banding and chip formation.

The strong dependence of this process on strain-rate and temperature implies that the material behavior of the metallic work piece is fundamentally thermoelastic, viscoplastic in nature. For simplicity, isotropic material behavior is assumed here. Restricting attention then to metals and to small elastic strain, the model for the stress can be based on the thermoelastic Hooke form

$$K = \{\lambda_0 (I \cdot \ln V_E) - (3\lambda_0 + 2\mu_0)\alpha_0 \theta\} I + 2\mu_0 \ln V_E \qquad (1)$$

for the Kirchhoff stress $K$ in terms of the elastic left logarthmic stretch $\ln V_E$ and temperature $\theta$. Here, $\lambda_0$ and $\mu_0$ represent the elastic longitudinal and shear moduli, and $\alpha_0$ the thermal expansion, all at a reference temperature $\theta_0$. The evolution of $\ln V_E$ is given by the objective associated flow rule

$$- \ln \overset{*}{V}_E = \partial_K \phi_P \qquad (2)$$

as based on an inelastic flow potential $\phi_P$ given by

$$\phi_P = \dot{\varepsilon}_{P0} C_0 \sigma_{Yd} \exp\{\langle \sigma_P - \sigma_{Yd}\rangle / (C_0 \sigma_{Yd})\} \quad , \tag{3}$$

with

$$\sigma_{Yd} = (A_0 + B_0 \varepsilon_P^{n_0})\{1 - [\langle \theta - \theta_0\rangle / (\theta_{M0} - \theta_0)]^{m_0}\} \quad , \tag{4}$$

the quasi-static yield stress. In the context of the corresponding evolution relation

$$\dot{\varepsilon}_P = \partial_{\sigma_P} \phi_P \tag{5}$$

for the equivalent inelastic strain rate $\dot{\varepsilon}_P$, these are consistent with the Johnson-Cook model [25]. Here, $\dot{\varepsilon}_{P0}$ represents a characteristic inelastic strain-rate. Further, $A_0$ represents the initial quasi-static yield stress, while $B_0$ and $n_0$ govern quasi-static isotropic hardening. In addition, $\theta_{M0}$ represents the melting temperature, and $m_0$ the thermal softening exponent. Further, $C_0$ influences the dynamic hardening behavior and strain-rate dependence, and $\langle x \rangle = \frac{1}{2}(x + |x|)$. The thermodynamic force driving the evolution of $\varepsilon_P$ is given by

$$\sigma_P = \sigma_{vM}(K) - \partial_{\varepsilon_P} \psi_P \quad , \tag{6}$$

representing the effective flow stress in terms of the von Mises effective stress $\sigma_{vM}(K)$ determined by the Kirchhoff stress $K$ and inelastic part $\psi_P$ of the free energy density. Together with $\dot{\varepsilon}_P$, $\sigma_{vM}(K)$ determines the rate of inelastic mechanical dissipation

$$\varpi_P = \{\sigma_{vM}(K) - \partial_{\varepsilon_P} \varepsilon_P\} \dot{\varepsilon}_P \tag{7}$$

of inelastic heating. Lacking a model for the cold-work term $\partial_{\varepsilon_P} \varepsilon_P$ (e.g., [27]), one often works with the alternative form

$$\varpi_P = \beta \, \sigma_{vM}(K) \, \dot{\varepsilon}_P \tag{8}$$

of this last relation in terms of the Taylor-Quinney coefficient $\beta$. In [27], it has been shown that $\beta$ is in fact not a constant but rather depends on strain and strain rate to varying degrees. In the following, this coefficient will be treated as constant as there is no experimental data concerning $\beta$ for the material (Inconel 718) considered in this study.

Give such parameter values, one can investigate the model behavior. In particular, when the metal is deformed plastically, the part of the inelastic mechanical dissipation transformed into heat (as determined by $\beta$) results in a temperature rise. In particular, this temperature increase is shown as a function of equivalent strain in Fig. 2 (left). In contrast to accumulated inelastic strain, an increase of temperature results in softening. At points of maximal mechanical dissipation in the material, softening effects may dominate hardening (Fig. 2, right), resulting in material instability, deformation localization and shear-band formation.

This completes the summary of the model.

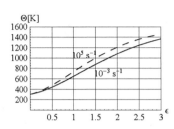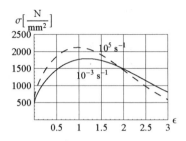

**FIGURE 2.** Temperature (left) and yield stress (right) as a function of equivalent strain in dynamic uniaxial tension at two different strain rates as based on the Johnson-Cook model.

## ADAPTIVE SCHEME

In the following, a finite element model of the workpiece-tool system is considered. Here, the workpiece is discretized with 2D-triangular elements and the adaptive procedure is performed with the help of a python script, using ABAQUS as the FE-solver. The main idea is, to separate FE-solver and mesh generator from the rest of the adaptive scheme to achieve best possible flexibility. Starting with a coarse mesh, the calculation is stopped after fixed time increments and the result data is analyzed to generate the mesh parameter. The mesh parameter provides the mesh generator with information on the requested mesh density. Here it is connected to the equivalent plastic deformation. Thus the procedure becomes best controllable. After mapping information from the old to the new mesh, the next calculation step is started.

## RESULTS

In the literature, the mesh is generally used as such to adjust the simulation results into agreement with experimental investigation [24, 28]. Such models are then used to investigate the influence of parameters like cutting speed, tool angle, and friction between tool and chip. **Fig. 3** shows the results for a simulation without adaptive remeshing, using an adjusted mesh and the results for using adaptive remeshing, respectively. Both results coincide quite well, where in the case of using adaptivity the influence of mesh orientation is eliminated. Mesh dependence is a hint that the local model is inappropriate to model deformation localization in a physically-reasonable way. To alleviate this, a non-local extension of Johnson-Cook is now being implemented and applied.

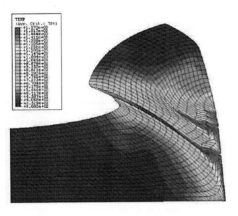
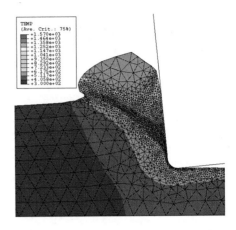

**FIGURE 3.** application of adaptive remeshing technique eliminates influence of mesh orientation [24]

# REFERENCES

1. M. Bäker, J. Rösler, C. Siemers, A finite element model of high speed metal cutting with adiabatic shearing, Comput. Struct. 80 (2002) 495-513.
22. M. Bäker, An investigation of the chip segmentation process using finite elements, Tech. Mech. 23 (2003) 1-9.
3. M. Bäker, Finite element simulation of high speed cutting forces, J. Mater. Process. Technol. 176 (2006) 117-126.
4. A. Behrens, B. Westhoff, K. Kalisch, Application of the finite element method at the chip forming process under high speed cutting conditions, in: H. K. Tönshoff, F. Hollmann (Eds.), Hochgeschwindigkeitsspanen, Wiley-vch, ISBN 3-527-31256-0, 2005, pp. 112-134.
5. C. Comi, U. Perego, Criteria for mesh refinement in nonlocal damage finite element analyses, European Journal of Mechanics A/Solids 23 (2004) 615-632.
23. E. El-Magd, C. Treppmann, Mechanical behaviour of materials at high strain rates, in: H. Schulz (Ed.), Scientific Fundamentals of High-Speed Cutting, Hanser, ISBN 3-446-21799-1, 2001, pp. 113-122.
7. T.I. El-Wardany, M.A. Elbestawi, Effect of material models on the accuracy of high-speed machining simulation, in: H. Schulz (Ed.), Scientific Fundamentals of High-Speed Cutting, Hanser, ISBN 3-446-21799-1, 2001, pp.77-91.
8. D.P. Flanagan, T. Belytschko, A Uniform Strain Hexahedron and Quadrilateral with Orthogonal Hourglass Control, International Journal for Numerical Methods in Engineering, 17 (1981), pp. 679-706.
25. G. R. Johnson, W. H. Cook, A constitutive model and data for metals subjected to large strain, high strain-rates and high temperatures, in: Proceedings of the 7th International Symposium on Ballistics, The Hague, The Netherlands, 1983, pp. 541-547.
10. T. Mabrouki, J.-F. Rigal, A contribution to a qualitative understanding of thermo-mechanical effects during chip formation in hard turning , J. Mater. Process. Technol. 176 (2006) 214-221.
11. M. E. Merchant, Mechanics of the metal cutting process, I. Orthogonal cutting and a type 2 chip, J. Appl. Phys. 16 (1945) 267-275.

12. E. H. Lee and B. W. Shaffer, The theory of plasticity applied to a problem of machining, J. Appl. Phys. 18 (1951) 405-413.

13. T. Özel, T. Altan, Process simulation using finite element method - prediction of cutting forces, tool stresses and temperatures in high speed flat end milling, J. Mach. Tools Manuf. 40 (2000) 713-783.

26. T. Özel, E. Zeren, Determination of work material flow stress and friction for FEA of machining using orthogonal cutting tests, J. Mater. Process. Technol. 153-154 (2004) 1019-1025.

15. F. Reusch, B. Svendsen, D. Klingbeil, Local and non local gurson based ductile damage and failure modelling at large deformation, Europ. J. Mech. A/Solid 22 (2003) 779-792.

27. P. Rosakis, A. J. Rosakis, G. Ravichandran and J. Hodowany, A thermodynamic internal variable model for the partition of plastic work into heat and stored energy in metals, J. Mech. Phys. Solids 48 (2000) 581-607.

17. R. Sievert, A. Hamann, H. - D. Noack, P. Löwe, K. N. Singh, G. Künecke, R. Clos, U. Schreppel, P. Veit, E. Uhlmann, R. Zettier, Simulation of chip formation with damage during high-speed cutting (in german), Tech. Mech. 23 (2003) 216–233.

28. R. Sievert, A. Hamann, H. - D. Noack, P. Löwe, K. N. Singh, G. Künecke, Simulation of thermal softening, damage and chip segmentation in a nickel super-alloy (in german), in: H. K. Tönshoff, F. Hollmann (Eds.), Hochgeschwindigkeitsspanen, Wiley-vch, ISBN 3-527-31256-0, 2005, pp. 446-469.

19. B. Svendsen, A. Flatten, D. Klingbeil, Continuumm thermodynamic multiscale modeling and simulation of dynamic adiabatic shear banding in metals, in preparation (2006).

29. H. K. Tönshoff, B. Denkena, R. Ben Amor, A. Ostendorf, J. Stein, C. Hollmann, A. Kuhlmann , Chip formation and temperature development at high cutting speeds (in german), in: H. K. Tönshoff, F. Hollmann (Eds.), Hochgeschwindigkeitsspanen, Wiley-vch, ISBN 3-527-31256-0, 2005, pp. 1-40.

21. Q. Yang, A. Mota, M. Ortiz, A class of variational strain-localization finite elements, International Journal for Numerical Methods in Engineering, 62 (2005), pp. 1013-1037.

22. M. Bäker, An investigation of the chip segmentation process using finite elements, Tech. Mech. 23 (2003) 1-9.

23. E. El-Magd, C. Treppmann, Mechanical behaviour of materials at high strain rates, Scientific Fundamentals of HSC (2001), H. Schulz, (ed.), Hanser, ISBN 3-446-21799-1, 113-122.

24. Hortig,C., Simulation of chip formation during high speed cutting, submitted to J. Mater. Process. Technol.

25. G. R. Johnson, W. H. Cook, A constitutive model and data for metals subjected to large strain, high strain-rates and high temperatures, in: Proceedings of the 7th International Symposium on Ballistics, The Hague, The Netherlands,(1983), 541-547.

26. T. Özel, E. Zeren, Determination of work material flow stress and friction for FEA of machining using orthogonal cutting tests, J. Mater. Process. Technol. 153-154 (2004) 1019-1025.

27. P. Rosakis, A. J. Rosakis, G. Ravichandran and J. Hodowany, A thermodynamic internal variable model for the partition of plastic work into heat and stored energy in metals, J. Mech. Phys. Solids 48 (2000) 581–607.

28. R. Sievert et al., Simulation der Spansegmentierung einer Nickelbasislegierung unter Berücksichtigung thermischer Entfestigung und duktiler Schädigung, Hochgeschwindigkeitsspanen (2005), H. K. Tönshoff, F. Hollmann (ed.), Wiley-vch, ISBN 3-527-31256-0, 446-469.

29. H. K. Tönshoff et al., Spanbildung und Temperaturen beim Spanen mit hohen Schnittgeschwindigkeiten, Hochgeschwindigkeitsspanen (2005), H. K. Tönshoff, F. Hollmann (ed.), Wiley-vch, ISBN 3-527-31256-0, 1-40.

# Simulation of Ball End Milling Process with Cutter Axis Inclination

Takashi Matsumura*, Takahiro Shirakashi*, Eiji Usui*

*Department of Mechanical Engineering, Tokyo Denki University
2-2 Kanda Nishiki-cho, Chiyoda-ku, Tokyo, 101-8457, JAPAN

**Abstract.** A force model based on the minimum cutting energy is applied to the simulation of the milling processes with the cutter axis inclination. The cutting model during the cutter rotation is made with coordinate transformation rotated at the tilt angle and the lead angle. Based on the orthogonal cutting data, the cutting force in the ball end milling process can be simulated with the chip flow model. The comparison of the analyzed forces with the measured ones verifies the simulation based on the presented force model.

**Keywords:** Milling, Ball End Mill, Multi-axis Machine Tool, Cutting Force, Chip Flow
**PACS:** 02.60.Cb, 02.70.-c, 07.05.Tp

## INTRODUCTION

Recent manufacturing industries have machined complex parts on multi-axis controlled machine tools. Many CAM systems have been developed to make the operation on the multi-axis machine tools easy with generating the cutter paths. However, most of machine shops have not yet taken full advantage of the high-end machine tools because the cutting processes such as the cutting forces cannot be estimated with the cutter inclinations. The cutting simulators, therefore, have been required for the multi-axis controlled machining.

Many studies have been made on the force model in the milling process with the cutter axis inclination so far. Most of them simulate the cutting force based on the cutting coefficient with calculating the cutting thickness [1][2][3]. The approaches, however, require many cutting tests to get the cutting coefficients because the coefficients change with the tool geometries and the cutter axis inclinations. Therefore the other models that are not based on the cutting coefficients have been required to reduce the cutting tests in the machine shops. This paper applies a force model based on the minimum cutting energy to the simulation of the ball end milling processes with the cutter axis inclinations. Because the model requires orthogonal cutting data that do not depend on the tool geometry and inclination [4], the database can be prepared in a small amount of cutting tests. The model, then, makes the chip flow with piling up orthogonal cuttings to predict the cutting force and the chip flow direction. The simulation is verified by comparison of the analyzed forces with the measured ones. The cutting force, then, is simulated to discuss the effect of the cutter axis inclination with changing the tilt angle.

CP907, 10th ESAFORM Conference on Material Forming, edited by E. Cueto and F. Chinesta

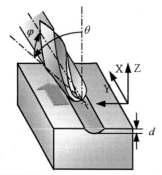
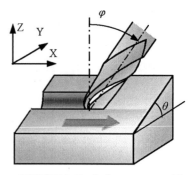

FIGURE 1. Cutting process with an inclined tool          FIGURE 2. Equivalent process to Fig. 1

# FORCE MODEL OF BALL END MILLING PORCESS BASED ON THE MINIMUM CUTTING ENERGY

This study discusses the cutting processes with the inclined ball end mills as shown in Fig. 1, where the cutter axis is inclined at the tilted angle $\theta$ and the lead angle $\varphi$. Three components of cutting force are calculated in the direction of each axis. In the simulation, the cutting process can be regarded as the equivalent process in Fig. 2, where the ball end mill inclined at the angle $\varphi$ to the cutter feed machines the workpiece inclined at the angle $\theta$ in the vertical plane with respect to the cutter feed. The cutting process with the inclined tool at the angle $\varphi$ is simulated with coordinate transformation. The coordinate system X-Y-Z in Fig. 3(a) is transformed to the coordinate system $X^*$-$Y^*$-$Z^*$ which inclines Z-axis at the tilt angle $\varphi$:

$$\begin{cases} x^* = (x - ft)\cos\varphi - \{z - \rho(1 - \cos\varphi)\}\sin\varphi \\ y^* = y \\ z^* = (x - ft)\sin\varphi + \{z - \rho(1 - \cos\varphi)\}\cos\varphi \end{cases} \tag{1}$$

where $\rho$ is the ball nose radius. Figure 3(b) shows the cutter locus of a point P viewed from $Z^*$ on $X^*$-$Y^*$-$Z^*$. The coordinate system $X^*$-$Y^*$-$Z^*$ are transformed to X'-Y'-Z' of the bottom of the cutter rotating at the angular velocity $\omega$ :

$$\begin{cases} x' = x^* \sin(\omega t - \gamma) + y^* \cos(\omega t - \gamma) \\ y' = -x^* \cos(\omega t - \gamma) + y^* \sin(\omega t - \gamma) \\ z' = z^* \end{cases} \tag{2}$$

where $\gamma$ is the delay angle of the point P referred to the bottom of the cutter. The coordinate system X'-Y'-Z', then, is transformed to the coordinate system X"-Y"-Z" where Y"-axis is associated with the direction of the cutting velocity by Eq. (3):

$$\begin{cases} x'' = x'\cos\Theta + y'\sin\Theta \\ y'' = (-x'\sin\Theta + y'\cos\Theta)\cos\Theta_z + z'\sin\Theta_z \\ z'' = -(-x'\sin\Theta + y'\cos\Theta)\sin\Theta_z + z'\cos\Theta_z \end{cases} \tag{3}$$

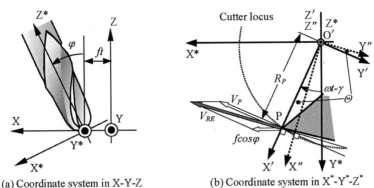

(a) Coordinate system in X-Y-Z      (b) Coordinate system in X*-Y*-Z*

**FIGURE 3.** Coordinate systems

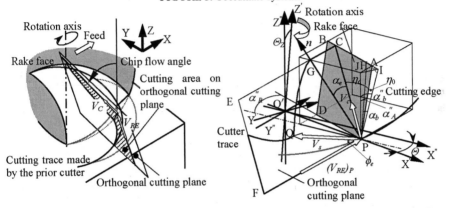

**FIGURE 4.** Chip flow model of a ball end mill     **FIGURE 5.** Orthogonal cutting at Point P

where $\Theta$ and $\Theta_Z$ are the wedge angles between Y'-axis and Y"-axis and between Z'-axis and Z"-axis, which can be given with the rotational velocity $V_P$ and the feed rate $f$ at the point P as follows:

$$\begin{cases} \Theta = \tan^{-1}(v_X / v_Y) \\ \Theta_Z = \tan^{-1}(v_Z / \sqrt{v_X^2 + v_Y^2}) \end{cases} \qquad (4)$$

where $v_x$, $v_y$ and $v_z$ are the following velocity components:

$$\begin{cases} v_X = f\cos\varphi\sin(\omega t - \gamma) \\ v_Y = V_P + f\cos\varphi\cos(\omega t - \gamma) \\ v_Z = -f\sin\varphi \end{cases} \qquad (5)$$

The force model interprets the chip flow as a piling of orthogonal cuttings in the planes containing the cutting velocities $V_{RE}$ and the chip flow velocities $V_c$ as shown in Fig. 4, where the chip is assumed to flow without plastic deformation between each plane. The chip flow direction is determined to minimize the cutting energy. In the simulation, the cutting edges are divided into small segments to consider the change of the edge shape. Figure 5 shows an orthogonal cutting at the point P in the coordinate system X"-Y"-Z". The plane PACBD shows the rake face of the cutter inclined at the radial rake angle $\alpha_R$" and the axial rake angle $\alpha_A$", which can be given with $\Theta$ and $\Theta_Z$. PA is the intersection on the rake face cut by the plane containing the cutting velocity. PI is the projection of Z-axis onto the rake face. The effective rake angle $\alpha_e$ can be given as follows:

$$\alpha_e = \sin^{-1}\left\{\sin(\eta_0 + \eta_c)\sin\alpha_R'' + \cos(\eta_0 + \eta_c)\tan\alpha_A''\cos\alpha_b''\cos^2\alpha_R''\right\} \qquad (6)$$

where $\eta_0$ is the wedge angle between PI and PA. The orthogonal cutting model at the center of the cutting area as shown in the plane PCGEF can be made by the following equation, which should be prepared for the simulation in the orthogonal cutting tests:

$$\left.\begin{array}{l} \phi = f(\alpha_e, V, t_1) \\ \tau_s = g(\alpha_e, V, t_1) \\ \beta = h(\alpha_e, V, t_1) \end{array}\right\} \qquad (7)$$

where $\phi$, $\tau_s$ and $\beta$ are the shear angle, the shear stress on the shear plane and the friction angle. $\alpha_e$, $V$ and $t_1$ are the rake angle, the cutting velocity and the undeformed chip thickness. The chip flow velocity at the center of the cutting area is given by:

$$V_c = \frac{\sin\phi_e}{\cos(\phi_e - \alpha_e)}V \qquad (8)$$

where $\phi_e$ is given with considering the inclination of the pre-machined surface [5]. The chip flow velocity at other areas of the cutting edge can be calculated geometrically to keep the angular velocity of the chip curl constant without the plastic deformation in the chip. Then the shear angles can be given by the chip velocities. As a result, the orthogonal cutting models can be made in all cutting areas on the edge.

The shear energy on the divided cutting edge $d\dot{U}_s$ can be calculated as follows:

$$d\dot{U}_s = \tau_s l_s dL_s \frac{\cos\alpha_e}{\cos(\phi_e - \alpha_e)}V \qquad (9)$$

where $l_s$ and $dL_s$ are the length and the width of the shear plane on the divided cutting edge. The friction energy $d\dot{U}_f$ can be calculated as follows:

$$d\dot{U}_f = dF_t V_c \qquad (10)$$

where $dF_t$ and $V_c$ are the friction force and the chip flow velocity on the divided cutting edge. $dF_t$ can be given by the following equation:

$$dF_t = \tau_s t_1 \frac{\sin\beta}{\cos(\phi_e + \beta - \alpha_e)\sin\phi_e}dL_f \qquad (11)$$

where $dL_f$ is the width of the tool chip contact area. The cutting energy $\dot{U}$, then, can be given by the integration over the height $[h_{min}, h_{max}]$ of the cutting area as follows:

$$\dot{U} = \int_{h\min}^{h\max}(d\dot{U}_s + d\dot{U}_f)dh \qquad (12)$$

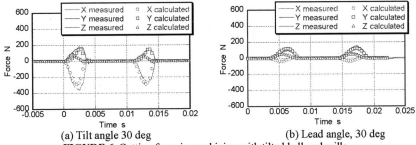

(a) Tilt angle 30 deg            (b) Lead angle, 30 deg

**FIGURE 6.** Cutting force in machining with tilted ball end mills

Cutting conditions: material cut, 0.5% Carbon steel; tool material, Carbide; tool diameter, 12mm; helix angle, 30deg; revolution rate, 2750rpm; axial depth of cut, 0.5mm; lubrication, dry.

The chip flow angle $\eta_c$ can be determined to minimize $\dot{U}$. The cutting force, then, can be predicted with the chip flow model having the minimum cutting energy. The tangential cutting force on the divided cutting edge $dF_H$ can be calculated as:

$$dF_H = d\dot{U}/V \qquad (13)$$

where $d\dot{U}$ is the cutting energy on the divided cutting edge. The normal force on the rake face $dF_n$ can be given as follows:

$$dF_n = \frac{dF_H - dF_t \sin\alpha_e}{\cos\alpha_b \cos\alpha_R''} \qquad (14)$$

where $dF_t$ is the friction force in Eq. (11). $\alpha_b$ is the inclination angle of the rake face from Z-axis direction. The radial component $dF_T$ and the axial one $dF_V$ are:

$$\left. \begin{array}{l} dF_T = -dF_t \cos\alpha_e \sin\eta_c' + dF_n \cos\alpha_b \sin\alpha_R'' \\ dF_V = dF_t \cos\alpha_e \cos\eta_c' - dF_n \sin\alpha_b \end{array} \right\} \qquad (15)$$

$\eta_c'$ is the projected angle of the chip flow direction on the vertical plane. $dF_H$, $dF_T$, and $dF_V$ can be converted into the components of X-, Y-, and Z- axis. The total cutting force, then, can be calculated as the sum of the forces loaded on divided areas of the cutting edges.

## CASE STUDY AND DISCUSSION

Figure 6 shows examples of the analysis results in machining of 0.5% carbon steel with carbide tool, where the components of the cutting force is shown in the directions as shown in Fig. 1. The analysis results designated by the symbols agree with the measured cutting forces as shown in the solid line. Figure 7 shows the cutting forces with changing the tilt angle. When the cutter having two flutes machines a groove without the cutter axis inclination, an edge starts cutting at another edge's end of cut. The cutter usually removes the material without the noncutting time. The actual cutting time during a rotation of the cutter reduces with increasing the tilt angle. The noncutting processes, during which neither of the edges cut the workpiece, are observed at tilt angles of 30 deg and 45 deg. From the point of view of thermal effect,

731

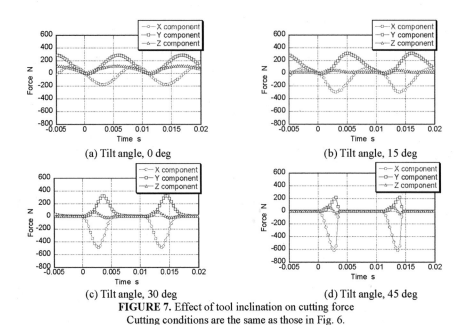

(a) Tilt angle, 0 deg

(b) Tilt angle, 15 deg

(c) Tilt angle, 30 deg

(d) Tilt angle, 45 deg

**FIGURE 7.** Effect of tool inclination on cutting force
Cutting conditions are the same as those in Fig. 6.

the short cutting time reduces the time for heat generation. Because heat conduction requires the heating time, the temperature rise is small in a short cutting time even if the cutting speed increases with the tool inclination. On the other hand, noncutting cools the cutting edges well. As a result, the tool wear can be suppressed at a low cutting temperature in cutting with the inclined milling tool.

## CONCUSION

The force model based on the minimum cutting energy is applied to the simulation of the cutting process with the inclined ball end mill. The model makes the chip flow with piling up the orthogonal cuttings in the plane containing cutting velocities and the chip flow velocities, in which the chip flow direction is determined to minimize the cutting energy. The model simulates the actual cutting process well. It is shown that the noncutting time appears at large tilt angles in the presented simulation.

## REFERENCES

1. I. Lazoglu and S. Y. Liang, "Modeling of Ball-End Milling Forces With Cutter Axis Inclination" in ASME J. of Manufact. Sci. and Eng., 122, 2000, pp. 3-11.
2. R. Zhu, S. G. Kapoor and R. E. DeVor, "Mechanistic Modeling Of The Ball End Milling Process For Multi-Axis Machining Of Free-FormSurfaces," in ASME J. of Manufact. Sci. and Eng., 123, 2001, pp. 360-379.
3. Yucesan, G., and Altintas, Y., "Prediction of Ball End Milling Force," in ASME J. of Eng. for Ind., 118, 1996, pp.95-103.
4. E. Usui, A. Hirota and M. Masuko, "Analytical Prediction of Three Dimensional Cutting Process Part1 Basic Cutting Model and Energy Approach," in ASME J.of Eng. for Ind., 100, 1978, pp.222-228.
5. T. Matsumura, T. Furuki, T. Shirakashi and Usui, E., "On the Development of A Milling Process Simulator Based on Energy Approach," in Proc. of the 7th Int. ESAFORM Conf., 2004, pp. 725-728.

# Finite Element Analysis Of The Dynamic Cutting Force Distribution In Peripheral Milling

G. Casalino, S.L. Campanelli, A.D. Ludovico and I. Launi

*Department of Mechanical and Operational Engineering (DIMeG), Politechnique of Bari, Viale Iapigia,182, 70126 Bari, Italy.*

**Abstract.** Among the machining errors generated in the peripheral milling the dimensional accuracy and surface roughness play a fundamental role in the quality of the machined component in the automotive and aerospace industry. The inadequacy of the surface roughness and the imprecision of the dimensional accuracy brings to the rejection of the part that causes financial loss and productivity slowdown.

In this paper an analysis of the machining errors due to the cutting forces in peripheral milling was performed by means of a numerical model based on the finite element method. The dynamic cutting forces distribution were evaluated for several cutting conditions that varied with the effective rake angle, the axial depth of cut, the radial depth of cut and the feed per tooth per revolution.

The model was tested with the cutting forces measured by Yucesan and Altintas [1] in field trials on a titanium alloy. An analytical model furnished by Liu et al. [2] and calibrated with the Yucesan's results gave a benchmark to the discussion of the results of this paper.

**Keywords:** Peripheral Milling, Cutting Force Distribution, Finite Element Analysis.
**PACS:** 02.60.–x, 81.20.–n.

## INTRODUCTION

Among the machining errors generated in the peripheral milling the dimensional accuracy and the surface roughness play a fundamental role in the quality of machined components in the automotive and aerospace industry. The inadequacy of the surface roughness and the imprecision of the dimensional accuracy bring to the rejection of the part that causes financial loss and productivity slowdown. Therefore decreasing the machining errors in peripheral milling is a target of the manufacturers.

Several sources of error have been identified in peripheral milling such as tool wear, friction, tool run-out, chatter vibration, and cutting forces. Errors due to cutting force originate from tool and workpiece deflections.

CP907, *10th ESAFORM Conference on Material Forming,* edited by E. Cueto and F. Chinesta
© 2007 American Institute of Physics 978-0-7354-0414-4/07/$23.00

A correct prediction of those forces can avoid the onset of the errors and the following flaws in the dimensional accuracy and surface roughness of the machined parts.

Smith and Tlusty [3] made a review of several models based on theoretical assumptions and experimental up to year 1991. A bunch of enhanced models followed that review. Many of those models did not take into consideration the dynamics in the tool/workpiece system [4, 5]. Moreover, in many works there was a lack of theoretical dynamics model that includes the size effect of undeformed chip thickness and the influence of the effective rake face angle [6, 7].

This kind of approximation is particularly ineffective when more than one tooth is engaged in cutting. In that case even if a tooth is not directly engaged in cutting, the forces acting on the other teeth that are engaged in cutting cause a deflection of the cutter [3].

Liu et al. proposed a theoretical model to get an ideal cutting force distribution so as to improve the dimensional accuracy and surface roughness while keeping a high productivity. The model was verified in the thin walled aerospace milling of titanium alloy. Those data were furnished by Yucesan and Altintas [1].

Based on the prediction of the cutting force distribution they proposed a careful selection of the cutter for the half-fine peripheral milling that ensure an ideal cutting force distribution and kept high feed rate [2].

The same investigation path and target proposed by Liu et al. were followed in this paper. So ideal cutting force distributions were evaluated for several cutting conditions that varied with the effective rake angle, the axial depth of cut, the radial depth of cut and the feed per tooth per revolution.

Li et al. proposed a different theoretical model for the simulation of the milling force. In the model, the action of a milling cutter was considered as the simultaneous work of a number of single point cutting tools, and milling forces were predicted from milling data of the workpiece material properties, the cutter parameters and tools and tooth geometry, the cutting conditions, and the kind of milling [8].

The use of the finite element method (FEM) for building a numerical model for the peripheral milling gave some advantages over the theoretical approach. In fact it was also possible to calculate the stress, the strain and the cutting pressure of the process. Graphical evidences of the correctness were also collected through the graphical check of the chip geometry and chip flow direction.

The results of the simulation permit to determine practical suggestions for reasonable selection of cutter and cutting parameters that minimize the machining error due to the cutting force and keep the productivity high.

One very recent paper stands for the effectiveness of the FEM approach. It was given by Lee & Choo who developed a reference cutting force model for rough milling federate scheduling using FEM analysis [9].

In this paper a two-dimensions FEM model for peripheral milling was built using the AdvantEdge software. The model was tested with the cutting forces measured by Yucesan and Altintas [1] during in-field trials on a titanium alloy. The analytical model furnished by Liu et al. [2], which itself was calibrated with the Yucesan's results [1], gave a benchmark to the discussion of the results of this paper.

# THE FEM MODEL

The design, the research as well as the optimization have benefited from the large diffusion of general purpose as well as targeted softwares among scientists, engineers and technicians operating in every field of the engineering sciences.

Increasing attention has been devoted to the application of numerical methods to the investigation of the machining processes over the last decade.

As a machining-dedicated software AdvantEdge provides an explicit dynamic, thermo-mechanically coupled finite element modeling package [10].

The software approaches the problem of chip flow using a Lagrangian system of coordinates in which the mesh follows the material. This approach can be used to simulate initial and steady state condition of the cutting process. In this way a number of cumbersome problems have been successfully dealt with.

The element topology used is a six-node quadratic triangle element with three corner and three midsize nodes.

Continuous mesh adaptation is used to update the spatial discretisation of the piece. In particular AdvantEdge employs the combination of r and h-adaptivity. R-adaptation is based on relocation of the nodes without altering the topology of the mesh. Otherwise h-adaptation changes the size of the mesh, which means that the number of elements and the connectivity of the nodes are changed from one discretisation to the successive one [2].

By mesh adaptation at regular interval the problem of element distortion due to high deformations is corrected.

The chip formation can be assumed to be due to plastic flow. As an alternative a separation criterion can defined. Continuous chip is automatically generated by continuously mesh refinement in the separation zone by either increasing the number of elements or relocation of the nodes [11].

A two-dimensional model of the end milling was performed by means of the software AdvantEdge.

**TABLE 1.** Layout of the simulations.

| Simulations | effective rake angle (°) | axial depth of cut (mm) | radial depth of cut (mm) | feed per tooth per revolution |
|---|---|---|---|---|
| 0 | 12 | 7.62 | 19.06 | 0.2030 |
| 1 | 12 | 7.62 | 19.06 | 0.1020 |
| 2 | 12 | 7.62 | 19.06 | 0.0508 |
| 3 | 12 | 7.62 | 19.06 | 0.0254 |
| 4 | 12 | 7.62 | 19.06 | 0.0127 |
| 5 | 12 | 5.08 | 9.52 | 0.2030 |
| 6 | 0 | 5.08 | 9.52 | 0.2030 |
| 7 | 20 | 5.08 | 9.52 | 0.2030 |
| 8 | 20 | 5.08 | 9.52 | 0.1020 |
| 9 | 20 | 5.08 | 9.52 | 0.0508 |
| 10 | 20 | 5.08 | 8.50 | 0.1020 |
| 11 | 20 | 5.08 | 7.50 | 0.1020 |
| 12 | 20 | 5.08 | 6.80 | 0.1020 |

All simulations were performed considering a carbide end mill composed by 90 % of WC, 10 % of Co, having hardness of 92 Rockwell and Young's Modulus of 630

GPa. A 5 mm workpiece of height, which was equal to the diameter of the mill and an endless length were considered. The workpiece was made of the titanium alloy whose chemical composition was 6% Al, 4% V. The Young Modulus was 110 GPa, Poisson ratio 0.34 and tensile strength 900 MPa.

The cutting parameters common to all simulations were: cutting edge radius, which was set to 0.02 mm, the relief angle $\gamma$ set to 10°, the spindle speed 500 rev/min, the cutter diameter 19.02 mm.

Some other cutting parameters were changed in order to find out the best combination that led to a cutting force reduction. They have been the feed per tooth $f$, the radial width of cut $d$, the axial depth of cut $b_a$ and the rake angle $\alpha_r$. In table 1 the levels for each numerical simulation are given. All the parameters were introduced in the software by means of windows as shown in Figure 1.

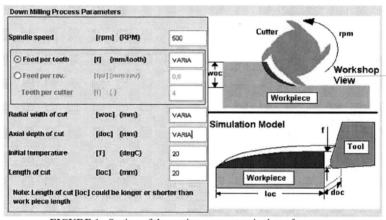

**FIGURE 1.** Setting of the cutting parameters in the software.

## RESULTS AND DISCUSSION

The first five simulations (Table 1) were used to test the reliability of the used software. These simulations were performed using the same parameters chosen by Yucesan and Altintas [1] for the experimental measurement of the cutting forces. In figure 2 the results of the five simulations in terms of cutting force F were compared with the results obtained experimentally by Yucesan and Altintas [1].

The first three simulations present results similar to the experiments: specifically for simulation 0 the error is 2.26%, for simulation 1 is 2.41%, for simulation 2 is 3.55% (figure 2). The last two simulations show 87% of error for the simulation 3 and 74 % for simulation 4; for these parameters combinations the software is not able to evaluate the cutting force during the milling operation. The same results were found by Smith e Tlusty [3] applying their numerical model. According to them it can be concluded that if the feed rate is lower than the cutting edge radius the feed force becomes more important than the cutting force. In this range the FEM model is not able to give results consisted with the experimental ones.

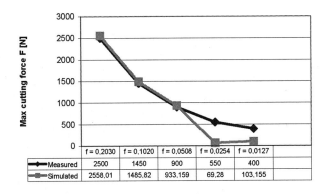

**FIGURE 2.** Chart of the measured and simulated max cutting forces.

| | f = 0,2030 | f = 0,1020 | f = 0,0508 | f = 0,0254 | f = 0,0127 | |
|---|---|---|---|---|---|---|
| Measured | 2500 | 1450 | 900 | 550 | 400 | |
| Simulated | 2558,01 | 1485,82 | 933,159 | 69,28 | 103,155 | |

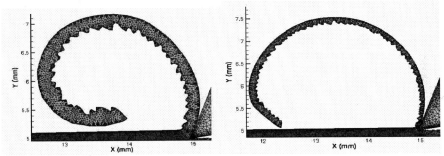

**FIGURE 3.** Chip generating for $f$ = 0.2030 mm/tooth a) and f = 0.1020 mm/tooth b), (7 and 8).

In the 5, 6, and 7 simulations the effective rake angle $\alpha_r$ was changed from 20° to 0°, that produced lower values for the cutting force and a reduction of the temperature at the tool-workpiece interface.

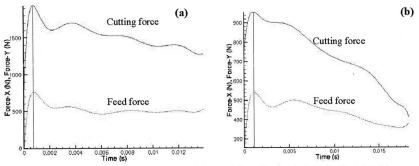

**FIGURE 4.** Trend of the cutting and the feed force in simulation 12 (a) and 7 (b).

The decreasing in temperature allows a longer tool life and a smaller wear rate leading to a better quality of the worked surface. In the next two simulations (8 and 9) only the feed rate was changed; it was found that decreasing $f$ from 0.1020 to 0.0508, the cutting force can be reduced about 69%. It has to observed the chip geometry and its flow direction. Figure 3 shows chip geometry for f = 0.2030 mm/tooth (part a) and f = 0.1020 mm/tooth (part b), (simulation 7 and simulation 8 respectively).

In number 10, 11, and 12 simulations only the radial depth of cut *d* was changed from 8.50 mm to 6.80 mm. The results show that the decreasing of *d* leads to a further reduction of the cutting force. The maximum value of F can be calculated for each simulation as in figure 4a; for simulation 12 this value is F = 955.8 N, which is further below the 1900 N of the cutting force in simulation 7 (figure 4 (b)).

Figure 5 shows the temperature distribution for the simulation number 7, which had a 20° rake face angle. In the proximity of the tool tip the temperature rose up to 658 °C. When the rake face angle was set to 0° the temperature in the same area was as high as 607 °C, which is 8% lower than the previous case. The temperature reduction during the operation allows the tool a longer life and better cutting performance and consequently the quality of the worked surface is higher.

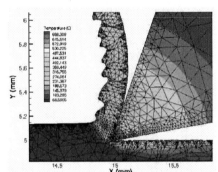

**FIGURE 5.** Temperature distribution at the tool tip (7).

## CONCLUSION

A two dimensional finite element method (FEM) code was used to the peripheral milling process. This methodology allowed to analyzing the influence of the main milling parameters on the cutting force distribution. Moreover, the FEM code resulted effective in calculate the stress, the strain and the cutting pressure of the process. Evidences of the effectiveness of this approach were also collected through the graphical output of the chip geometry and chip flow direction.

Based on the prediction of the cutting force distribution a targeted selection of the cutter for peripheral milling is possible.

## REFERENCES

1  G. Yucesan and, Y. Altintas, *Int. J. Mach. Tools Manufact.* **34**, 463-487 (1994).
2  X.-W. Liu, K. Cheng, D. Webb, X.-C. Luo. *Int. J. Mach. Tools Manufact.* **42**, 791-800 (2002).
3  S. Smith, J. Tustly, *Trans. of the ASME: J. of Eng. For Ind.*,113, 169-175 (1991).
4  W. A. Kline, R. E. DeVor I, A, Shareef, *Trans. of the ASME: J. of Eng. For Ind.*,104, 272-278 (1982).
5  J. W. Sutherland, R. E. Devor, *Trans. of the ASME: J. of Eng. For Ind.*,108, 269-279 (1986).
6  G. M. Zhang, S. G. Kapoor, *Trans. of the ASME: J. of Eng. For Ind.*,113, 137-144 (1991).
7  G. M. Zhang, S. G. Kapoor, *Trans. of the ASME: J. of Eng. For Ind.*,113, 145-153 (1991).
8  X. P. Li, A. Y. C. Nee, Y. S. Whong, H. Q. Zheng, J. Mat. Proc. Thec., **89-90**, 266-272, (1999).
9  Ann-Ul Lee, Dong-Woo Cho, *Int. J. Mach. Tools Manufact.* 347, 158-167 (2007).
10 AdvantEdge Software, Theoretical Manual, Version 3.5. Third way system, 2001.
11 M. Ortiz, Quigley, Compt. *Mech. Appl. Mech. Eng.* **90**, 1099-1114 (1991),

# Inverse Method for Identification of Material Parameters Directly from Milling Experiments

A. Maurel, G. Michel, S. Thibaud, M. Fontaine, J.C. Gelin

*FEMTO-ST Institute / Applied Mechanics Laboratory, ENSMM, 26 rue de l'Epitaphe, 25000 Besançon. France*

**Abstract.** An identification procedure for the determination of material parameters that are used for the FEM simulation of High Speed Machining processes is proposed. This procedure is based on the coupling of a numerical identification procedure and FEM simulations of milling operations. The experimental data result directly from measurements performed during milling experiments. A special device has been instrumented and calibrated to perform force and torque measures, directly during machining experiments in using a piezoelectric dynamometer and a high frequency charge amplifier. The forces and torques are stored and low pass filtered if necessary, and these data provide the main basis for the identification procedure which is based on coupling 3D FEM simulations of milling and optimization/identification algorithms. The identification approach is mainly based on the Surfaces Response Method in the material parameters space, coupled to a sensitivity analysis. A Moving Least Square Approximation method is used to accelerate the identification process. The material behaviour is described from Johnson-Cook law. A fracture model is also added to consider chip formation and separation. The FEM simulations of milling are performed using explicit ALE based FEM code. The inverse method of identification is here applied on a 304L stainless steel and the first results are presented.

**Keywords:** Material parameters, Identification, Milling, FEM, Cutting forces measure.
**PACS:** 81.20.Wk; 02.30.Zz;02.70.Dc;

## INTRODUCTION

In the last ten years, High Speed Machining has become one of the most used processes in macro and micro mechanics manufacturing with applications in aeronautic, automotive or mechanical engineering industries. The understanding of the physical phenomena and the identification of material behaviour in machining are the main issue for the industrial process optimization. In fact the workpiece is machined under severe mechanical solicitations (very high strain rate, vibrations, failure) as well as thermal ones (thermal transfer, friction). Some analytical models were previously proposed by Merchant [1] which were developed by Oxley [2] and Molinari & Dudzinski [3], these last works were adapted to milling [4]. But also a complete semi-empirical approach was developed by Altintas [5]. More recently, calibrated FEM models appear and seem to progress in the way of the real machining process simulation, Pantalé [6]. Nevertheless, the high thermo-mechanical coupling and the nonlinear aspects involved in these models give only a partial interpretation of the process's phenomena. To improve the models predictive capacities, one needs to characterize the material behaviour which is involved in the machining process.

CP907, *10th ESAFORM Conference on Material Forming,* edited by E. Cueto and F. Chinesta
© 2007 American Institute of Physics 978-0-7354-0414-4/07/$23.00

Therefore it is emphasized in the present paper to proceed by inverse method for the identification of material parameters from machining experiments.

Then this paper shows the developments carried out in our laboratory related to the identification methodologies [7][8][9], through three main tasks: The first one consists to set up experimental equipment allowing the measure of cutting forces during machining, then to develop a finite element model of machining process, and finally to lead on an identification procedure.

## EXPERIMENTAL INVESTIGATIONS

In the first step, a complete instrumentation including sensors and on line measuring system has been set up to obtain repetitive and reliable data. The sensor system is based on a Kistler dynamometer using the piezoelectric accelerometers technology, in order to give, through direct measurement, the cutting forces during the machining process. Then the 9272A dynamometer has been preferred for its robustness and its wide range of use in force and frequency as related in Figure 1. This device is based on four piezoelectric sensors, each of them delivering an electric signal function related to applied loads on each referenced axis.

| Natural frequency | $f_n (x,y)$ | kHz | ≈3,1 |
|---|---|---|---|
| (mounted on rigid base) | $f_n (z)$ | kHz | ≈6,3 |
| | $f_n (M_z)$ | kHz | ≈4,2 |
| Measuring range | $F_x, F_y$ | kN | −5 ... 5 [1)] |
| | $F_z$ | kN | −5 ... 20 [2)] |
| | $M_z$ | N·m | −200 ... 200 |

**FIGURE 1.** 9272A Kistler dynamometer's axis and characteristics.

The measuring channel has been completed by an high frequency Kistler charge amplifier and the software TPCut© for the treatment of the measured data. The software is also able to give the cutting areas in order to realize the tests in industrial conditions. The experiments were carried out on a KERN micro milling machine as shown in Figure 2. The dynamometer was fixed on the table of this 3 axes CNC machine, and then it was linked to the charge amplifier which was connected to a PC through an input/output acquisition card. The equipments chosen for these milling experiments are as following.

The type of the selected milling tools are two teeth end mills to have a better description of the cyclical teeth motion and to avoid the perturbation on the measured curves due to a poor balancing of end mills with three or four cutting edges. These tools are uncoated tungsten carbide mills with 30° nominal helix angle and 12° normal rake angle, and the tested diameters are in the range 2-6 mm. All the test specimens have been machined in the same 304L austenitic stainless steel bar and in the same dimensions: 80 mm x 40 mm x 10 mm.

**FIGURE 2.** Experimental acquisition system used in milling experiments.

All milling experiments have been carried out with a micro lubrication sprayed on the tooling area. And the used cutting velocities are defined between 50 and 400 m/min to reach a maximum spindle speeds at 40 000 rpm. Figure 3 relates a measured curve example: a forces rise and an instability at the input and the output of the test specimen must be noted. This effect is due to a very fast shift of the tool engagement conditions and of the initial surface integrity of the test specimen (hardening, residual stresses, and surface roughness). The central zone corresponds to the stabilized signal where sinusoidal like and repetitive curves are obtained.

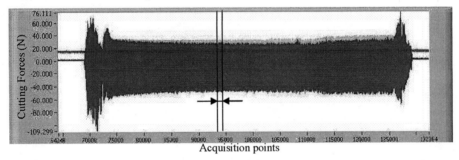

**FIGURE 3.** $F_x$ cutting forces measured curve with two tooth end mill Ø 4 mm, $V_c$ = 200 m/min, $f_t$ = 0.01 mm/tooth, $d_a$ = 0.5 mm (sampling: 20000 Hz).

In Figure 4 is related a zoom of the cutting forces stabilized signal in milling for the same curve than in Figure 3, where the cutting conditions and machining parameters correspond to:

    2 teeth end mill Ø 4 mm in slotting tests (full radial immersion);
    Spindle speed: $\Omega$ = 15920 rpm; Feed: $f_t$ = 0.01 mm/tooth;
    Cutting speed: $V_c$ = 200 m/min; Axial depth of cut: $d_a$ = 0.5 mm.

In these conditions, the end mill perform one rotation in 3 ms. In sampling at a 20 000 Hz frequency, the curve is composed with sixty points by cutter rotation. Figure 4 shows that is possible to get a good and accurate variation of the milling forces on the 3 machine axes. The Figure 4 shows cyclic and sinusoidal curves, where each cycle corresponds to a milling cutter revolution and, each cycle peak corresponds to a tooth path. A discrepancy in the forces level appears between the two peaks, it is mainly due to the tool run-out (position and/or angle offset between the cutter axis and the spindle axis).

741

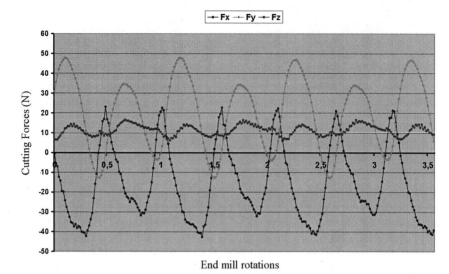

FIGURE 4. Analyzed curve of cutting forces $F_x$, $F_y$, $F_z$ function of end mill rotations.

A signal spectral analysis was conducted on forces components in order to check the stability of measured data and to recover the equivalent forces for a rigid case after appropriate low filtering. These accurate and reliable measures allow the understanding of a part of the end mill behaviour during the milling process and then the setting up of the identification process.

## NUMERICAL SIMULATION OF MILLING

In order to perform numerical simulation of milling process, a finite element model was implemented in LS-Dyna© FEM code. As machining is an high speed metal forming process characterized by large dynamic stresses and localized high strains, an explicit code solution scheme was adopted, resulting in the choice of LS-Dyna© FEM as the simulation one.

Accordingly with the experiments presented before, a Ø 4 mm end mill and a 304L stainless steel rectangular specimen have been designed with SolidWorks©, and furthermore the same cutting conditions than in experiments have been used in the numerical FEM model. The figure 5 relates the first simulation results. The test specimen is meshed with 160000 elements. The corresponding material behaviour has been firstly modelled using the classical Johnson-Cook law [10] with the 304L stainless steel material parameters provided by the CETIM foundation.

$$\sigma = \left[A + B\left(\varepsilon^p\right)^n\right]\left[1 + C\ln\left(\frac{\dot{\varepsilon}^p}{\dot{\varepsilon}_0}\right)\right]\left[1 - \left(\frac{T - T_f}{T_{seuil} - T_f}\right)^m\right] \tag{1}$$

The end mill is meshed with about 210000 elements and considered as a rigid body. This modelling uses an element failure criterion based on a limiting plastic strain critical value.

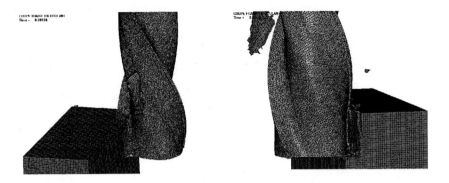

**FIGURE 5.** Simulation results of a milling process numerical model with LS-Dyna software.

## INVERSE METHOD OF IDENTIFICATION

The final aim of this work consists to lead an inverse method of identification process for determination of 304L stainless steel parameters. The identification procedure methodology needs an objective function to minimize, a set of parameters to analyze, a finite element method solver and a procedure for up dating the material parameters. For the milling experiments identification, the parameters system **p** (A, B, n, C, $T_{seuil}$, m) has been chosen in the Johnson-Cook constitutive law (1). Then an objective function dependent of the measured experimental criteria (cutting forces) has been defined in (2).

$$R(\mathbf{p}) = \frac{1}{2} \left( \sum_{i=1}^{Np} \left[ F_x^{exp} - F_x^{num}(\mathbf{p}) \right]^2 + \sum_{i=1}^{Np} \left[ F_y^{exp} - F_y^{num}(\mathbf{p}) \right]^2 + \sum_{i=1}^{Np} \left[ F_z^{exp} - F_z^{num}(\mathbf{p}) \right]^2 \right) \quad (2)$$

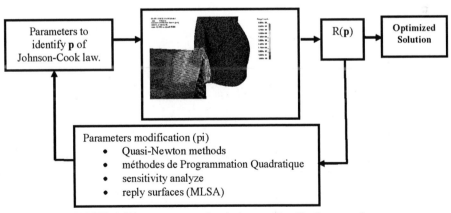

**FIGURE 6.** Diagram representing the inverse identification procedure.

The previous function measures difference between the cutting forces curves obtained with the numerical simulation software and with the experimental device. Figure 6 schematically relates the identification loop, the procedure is based on a set of parameters obtain through simulation and then compared with experimental results. After this estimate, whether the objective function is equal to zero and the optimized solution appeared, or the function is different to zero and the set of parameters has to be modified and the loop begin one more time. Several methods exist for the identification strategy, in this work the Moving Least Square Approximation (MLSA) have been preferred.

## CONCLUSIONS

These milling experiments allowed us to analyze precisely the measured cutting forces curves and to obtain reliable and useful measured data for the identification procedure. Then, a first 3D simulation of shoulder milling process has been set up in the experiments conditions. The numerical model is based on a dynamic transient explicit scheme and on lagrangian formulation. At the end, this paper exposed the identification procedure which will be set up soon. Moreover, the further developments in progress concern the numerical model improvements and the choice of a new material behaviour model more adapted to a high speed metal forming process in order to test the method in High Speed Milling conditions.

## ACKNOWLEDGMENTS

This work has been carried out with the financial support from the CETIM foundation in the frame of PGV1 national project.

## REFERENCES

1. E. Merchant, *J. of Applied Mechanics*, **66**, 168-175(1944).
2. P.L.B. Oxley, *ASME*, 50-60(1963).
3. A. Molinari, D. Dudzinski, *C.R. Acad. Sciences Paris*, **315 (II)**, 399-405(1992).
4. M. Fontaine, A. Devillez, A. Moufki and D. Dudzinski, *Int. J. Machine Tools and Manufacture*, **46**, 367-380(2006).
5. Y. Altintas, *J. Machining Science and Technology*, **3-4**, 445-478(2000).
6. O. Pantalé, "Modélisation et simulation tridimensionnelle de la coupe des métaux", Ph.D. Thesis, Bordeaux I University, 1996.
7. J.C. Gelin, C. Labergere, *Int. J. Forming Processes*, **7**, 141-158(2004).
8. O. Ghouati, J.C. Gelin, *J. Material Processing Technology*, **80-81**, 560-564(1998).
9. O. Ghouati, J.C. Gelin, *Computational Materials Science*, **21**, 57-68(2001).
10 G.R. Johnson, W.H. Cook, *Proc. 7th Symp. On Ballistics*, Netherlands(1983).

# A Critical Analysis on the Friction Modeling in Orthogonal Cutting of Steel

L. Filice[1], F. Micari[2], S. Rizzuti[1] and D. Umbrello[1]

[1]Dept. of Mechanical Engineering, The University of Calabria, 87036 Rende (CS) - Italy
[2]Dept. of Manufacturing and Management Eng., The University of Palermo, 90128 Palermo - Italy

**Abstract.** Numerical simulation of cutting process is today moving towards two different directions. The former concerns the development of high performance codes able to approach the 3D phenomena, the latter is already focused on the study of some fundamental aspects whose full understanding may be strategic for the knowledge enhancing in this very complex field. In the paper this second way was pursued and a wide analysis on the numerical robustness of the 2D orthogonal cutting process is presented. In particular, the role played by the friction modeling is discussed taking into account a wide integrated numerical and experimental campaign.

**Keywords:** Orthogonal cutting, Friction, Temperature distribution.
**PACS:** 81.20.Wk

## INTRODUCTION

It is well known that simulation of machining still represents a very hard task due to the geometric complexity of the real chip-tool systems and the cutting speed regime that requires very long simulation times even if high performance finite element codes have been recently introduced [1]. Having a look at the technical literature, it seems clear the researchers are focusing their attention on two different approaches. The former tends to allow the simulation of 3D phenomena which represent, of course, the higher industrial interest. In this field some preliminary investigations were proposed but the computational complexity, using the tradition FE method based on the updated-Lagrangian formulation, allows only the simulation of shorter process time or proper designed applications [2]. On the other hand, other scientist are investigating some fundamental aspects of the cutting simulation in order to assess the process modeling, increasing the process simulation reliability. Of course, the common target of the above scientists is the possibility to accurately simulate every cutting processes, obtaining reliable predictions of the process variables.

The fundamental studies on cutting processes are focused on the role of material behavior and modeling [3], chip formation [4], friction modeling [5,6]and, finally, tool and workpiece heating and heat transfer [7].

In the present paper a systematic investigation on the role played by the implemented friction model, within a 2D simulation of orthogonal cutting of five couples of tool-workpiece materials, was carried out. The main simulation results were compared with experimental measurements in order to verify if it is possible to

CP907, 10th ESAFORM Conference on Material Forming, edited by E. Cueto and F. Chinesta
© 2007 American Institute of Physics 978-0-7354-0414-4/07/$23.00

identify the best friction model between those proposed in literature for all the tool-workpiece materials. Once the comparison with mechanical variables was completed, a subsequent study on temperature prediction utilizing the above friction models was executed as well.

How it will be shown, today it is possible to well model the friction behavior if mechanical variables have to be calculated. Thermal simulation requires further analysis and, probably, up to now the Arbitrarian Lagrangian-Eularian (ALE) formulation constitutes the most promising strategy.

## EXPERIMENTAL TESTS

Cutting experiments were carried out in a lathe-turning, reproducing orthogonal cutting conditions (Figure 1). The workpiece-tool materials utilized were:

AISI 1020 - uncoated carbide ISO P25 tool (Case 1),

AISI 1045 - uncoated carbide ISO P30 (Case 2),

AISI 1045 - uncoated carbide ISO P40 (Case 3),

AISI 1045 - carbide ISO P40 with a TiN coating (Case 4) and

AISI 1045 - carbide ISO P40 with a TiAlN coating (Case 5).

All the turning tests were carried out using a feed of 0.1 mm/rev and a cutting speeds of 100 m/min. The depth of cut was generally 3 mm but it was fixed equal to 2.5 mm in the Case 1. The rake angle was equal to 0°C in the Case 1 and 2, while it was equal to +10° in the other ones.

**FIGURE 1.** Experimental setup

Cutting and thrust forces were measured by using a piezoelectric dynamometer while an optical microscope was used for estimating both the contact length and chip thickness. The temperatures were estimated starting form the measure executed using a single wire thermocouple that was directly embedded in the tool, at a know distance from the rake face.

Table 1 shows the experimental results in terms of cutting and thrust forces, contact length, chip thickness and measured temperature, obtained as average value of five experiments.

**TABLE 1.** Experimental Results.

| Measured parameters | CASE 1 | CASE 2 | CASE 3 | CASE 4 | CASE 5 |
|---|---|---|---|---|---|
| Cutting force $F_c$ [N] | 920 | 745 | 820 | 741 | 710 |
| Thrust force $F_t$ [N] | 463 | 600 | 620 | 450 | 386 |
| Contact length $l_c$ [mm] | 0.529 | 0.5 | 0.36 | 0.19 | 0.28 |
| Chip thickness t [mm] | 0.42 | 0.29 | 0.28 | 0.18 | 0.21 |
| Temperature [°C] | 483 | 542 | 514 | 427 | 417 |

# NUMERICAL PREDICTION

## Numerical Set-up

In this investigation, the finite element code DEFORM-2D, which is based on an updated Lagrangian formulation and an implicit integration method, was used to simulate the cutting process. For all the Cases the workpiece was modelled as elastic-plastic and the tool as rigid; 2D plane strain conditions were imposed in the simulation. How it is well-known, an accurate and reliable flow stress model has to be considered to represent work material constitutive behaviour in order to successfully run FE simulations. Furthermore, it is widely recognized that work material flow stress properties are mostly influenced by temperature, strain and strain rate factors.

Taking into account these considerations, a consistent model proposed by Oxley [8] was used in DEFORM-2D, for both AISI 1020 and AISI 1045 materials.

In addiction, a realistic characterisation of the friction interaction between the chip and the tool is important at least as flow stress characterisation of the work material.

In the present study the three most common friction models were taken into account. The first model (Model I) assumes a constant frictional stress ($\tau$) on rake face, equal to a fixed percentage of the shear flow stress of the working material k:

$$\tau = m \cdot k \qquad (1)$$

The second model (Model II) is based on the simple Coulomb's law, with a constant coefficient of friction $\mu$ on the whole contact zone:

$$\tau = \mu \cdot \sigma_n \qquad (2)$$

where $\tau$ is the frictional stress and $\sigma_n$ is the normal one.

Finally, the last model taken into account was the sticking-sliding one [9]. According to the stress distribution proposed by Zorev, the existence of two distinct regions on the rake face is modelled: the sticking zone, where the frictional stress is assumed to be equal to the shear flow stress of the material being machined and the sliding region, where, according to the Coulomb's theory, a constant coefficient of friction is utilized (3).

747

$$\tau = \mu \cdot \sigma_n \ when \ \tau < k$$
$$\tau = k \qquad when \ \tau \geq k$$
(3)

Tables 2-6 summarize the best numerical results for all the investigated Cases concerning cutting and thrust forces, contact length and chip thickness. The Tables also report the relative errors for each parameter and the average error $\bar{e}$.

**TABLE 2.** Best numerical results for each friction model (Case 1).

| Model | $F_c$ [N] | $e_{Fc}$% | $F_t$ [N] | $e_{Ft}$% | $l_c$ [mm] | $e_{lc}$% | t [mm] | $e_t$% | $\bar{e}$ % |
|---|---|---|---|---|---|---|---|---|---|
| I (m=0,84) | 853 | 7.3 | 530 | 14.5 | 0.43 | 18.7 | 0.31 | 26.2 | 16.68 |
| I (m=0.85) | 864 | 6.1 | 545 | 17.7 | 0.55 | 4.0 | 0.32 | 23.8 | 12.90 |
| II ($\mu$=0.59) | 821 | 10.8 | 467 | 0.8 | 0.34 | 35.7 | 0.28 | 33.3 | 20.15 |
| II ($\mu$=0.9) | 839 | 8.8 | 499 | 7.8 | 0.38 | 28.2 | 0.30 | 28.6 | 18.35 |
| III ($\mu$=0.6) | 815 | 11.4 | 466 | 0.6 | 0.34 | 35.7 | 0.28 | 33.3 | 20.25 |
| III ($\mu$=0.9) | 843 | 8.4 | 501 | 8.2 | 0.34 | 35.7 | 0.31 | 26.2 | 19.63 |

**TABLE 3.** Best numerical results for each friction model (Case 2).

| Model | $F_c$ [N] | $e_{Fc}$% | $F_t$ [N] | $e_{Ft}$% | $l_c$ [mm] | $e_{lc}$% | t [mm] | $e_t$% | $\bar{e}$ % |
|---|---|---|---|---|---|---|---|---|---|
| I (m=0.82) | 780 | 4.7 | 462 | 23.0 | 0.25 | 50.0 | 0.20 | 31.0 | 27.93 |
| I (m=0.9) | 810 | 8.7 | 507 | 15.5 | 0.28 | 44.0 | 0.21 | 27.6 | 23.95 |
| II ($\mu$=0.4) | 761 | 2.1 | 430 | 28.3 | 0.26 | 48.0 | 0.20 | 31.0 | 27.35 |
| II ($\mu$=0.8) | 779 | 4.6 | 466 | 22.3 | 0.23 | 54.0 | 0.20 | 31.0 | 27.98 |
| III ($\mu$=0.2) | 731 | 1.9 | 387 | 35.5 | 0.26 | 48.0 | 0.20 | 31.0 | 29.10 |
| III ($\mu$=0.4) | 766 | 2.8 | 432 | 28.0 | 0.23 | 54.0 | 0.20 | 31.0 | 28.95 |

**TABLE 4.** Best numerical results for each friction model (Case 3).

| Model | $F_c$ [N] | $e_{Fc}$% | $F_t$ [N] | $e_{Ft}$% | $l_c$ [mm] | $e_{lc}$% | t [mm] | $e_t$% | $\bar{e}$ % |
|---|---|---|---|---|---|---|---|---|---|
| I (m=0.97) | 740 | 9.8 | 345 | 44.4 | 0.27 | 25.0 | 0.21 | 25.0 | 26.05 |
| I (m=0.98) | 747 | 8.9 | 361 | 41.8 | 0.26 | 27.8 | 0.21 | 25.0 | 25.88 |
| II ($\mu$=0.85) | 654 | 20.2 | 240 | 61.3 | 0.18 | 50.0 | 0.18 | 35.7 | 41.8 |
| II ($\mu$=0.9) | 663 | 19.1 | 245 | 60.5 | 0.17 | 52.8 | 0.18 | 35.7 | 41.85 |
| III ($\mu$=0.8) | 667 | 18.7 | 257 | 58.5 | 0.16 | 55.6 | 0.18 | 35.7 | 42.13 |
| III ($\mu$=0.9) | 664 | 19.0 | 246 | 60.3 | 0.16 | 55.6 | 0.18 | 35.7 | 42.65 |

**TABLE 5.** Best numerical results for each friction model (Case 4).

| Model | $F_c$ [N] | $e_{Fc}$% | $F_t$ [N] | $e_{Ft}$% | $l_c$ [mm] | $e_{lc}$% | t [mm] | $e_t$% | $\bar{e}$ % |
|---|---|---|---|---|---|---|---|---|---|
| I (m=0.75) | 729 | 1.6 | 398 | 11.6 | 0.19 | 0.0 | 0.18 | 0.0 | 3.30 |
| I (m=0.79) | 743 | 0.3 | 421 | 6.4 | 0.18 | 5.3 | 0.18 | 0.0 | 3.00 |
| II ($\mu$=0.8) | 749 | 1.1 | 430 | 4.4 | 0.19 | 0.0 | 0.18 | 0.0 | 1.38 |
| II ($\mu$=0.92) | 752 | 1.5 | 437 | 2.9 | 0.19 | 0.0 | 0.19 | 5.6 | 2.50 |
| III ($\mu$=0.5) | 745 | 0.5 | 431 | 4.2 | 0.19 | 0.0 | 0.18 | 0.0 | 1.18 |
| III ($\mu$=0.8) | 753 | 1.6 | 442 | 1.8 | 0.19 | 0.0 | 0.18 | 0.0 | 0.85 |

**TABLE 6.** Best numerical results for each friction model (Case 5).

| Model | $F_c$ [N] | $e_{Fc}$% | $F_t$ [N] | $e_{Ft}$% | $l_c$ [mm] | $e_{lc}$% | t [mm] | $e_t$% | $\bar{e}$ % |
|-------|-----------|-----------|-----------|-----------|------------|-----------|--------|--------|-------------|
| I (m=0.8) | 762 | 7.3 | 417 | 8.0 | 0.20 | 28.6 | 0.18 | 14.3 | 14.55 |
| I (m=0.9) | 814 | 14.6 | 487 | 26.2 | 0.24 | 14.3 | 0.20 | 4.8 | 14.98 |
| II (μ=0.8) | 760 | 7.0 | 433 | 12.2 | 0.26 | 7.1 | 0.19 | 9.5 | 8.95 |
| II (μ=0.81) | 755 | 6.3 | 441 | 14.2 | 0.26 | 7.1 | 0.19 | 9.5 | 9.28 |
| III (μ=0.89) | 747 | 5.2 | 431 | 11.7 | 0.24 | 14.3 | 0.19 | 9.5 | 10.18 |
| III (μ=0.9) | 747 | 5.2 | 429 | 11.1 | 0.22 | 21.4 | 0.19 | 9.5 | 11.80 |

Analyzing each couple of materials it is possible to state that the different formulations of friction do not provide general relevant differences as far as the mechanical parameters are concerned.

On the other hand, taking into account all the Cases it is possible to highlight that a proper friction model choice is strictly related to the specific workpiece-tool system.

More in detail, as it is shown in Table 7, the constant shear model (Model I) seems to be more suitable to describe friction phenomena when uncoated tools are utilised. However, when compositions of steel and P-grade carbide change, different friction factors allow the best performance.

As far as coated tools are concerned, the sticking-sliding model (Model III) supplied the best results for the couple AISI 1045 - carbide ISO P40 with a TiN coating (Case 4). On the other hand, Coulomb model allowed to obtain the minimum average error when TiAlN coated tool was used.

**TABLE 7.** Best Results for each couple of materials.

| Couple | Model | $F_c$ [N] | $e_{Fc}$% | $F_t$ [N] | $e_{Ft}$% | $l_c$ [mm] | $e_{lc}$% | t [mm] | $e_t$% | $\bar{e}$ % |
|--------|-------|-----------|-----------|-----------|-----------|------------|-----------|--------|--------|-------------|
| 1 | I (m=0.85) | 864 | 6.1 | 545 | 17.7 | 0.55 | 4.0 | 0.32 | 23.8 | 12.90 |
| 2 | I (m=0.9) | 810 | 8.7 | 507 | 15.5 | 0.28 | 44.0 | 0.21 | 27.6 | 23.95 |
| 3 | I (m=0.98) | 747 | 8.9 | 361 | 41.8 | 0.26 | 27.8 | 0.21 | 25.0 | 25.88 |
| 4 | III (μ=0.8) | 753 | 1.6 | 442 | 1.8 | 0.19 | 0.0 | 0.18 | 0.0 | 0.85 |
| 5 | II (μ=0.8) | 760 | 7.0 | 433 | 12.2 | 0.26 | 7.1 | 0.19 | 9.5 | 8.95 |

## Temperature Prediction

How it is well known, the current finite element codes allow to simulates only a time of few milliseconds of the machining process. In these conditions the thermal steady state is not reached and different issues occur when some related aspects are investigated. Different approaches were proposed to overcome this problem. A common practice is based on an artificial relevant increasing of the global heat transfer coefficient between the tool and workpiece. This strategy is not completely physically consistent but permits to obtain a quicker reaching of the thermal steady state. In particular, some researchers [10] found that a value of h close to 1000 kW/m$^2$K permits a satisfactory agreement between the numerical data and the experimental evidences, for a specific couple workpiece-tool (AISI 1045 – ISO P20). This tentative value was utilised also in the present analysis.

More in detail, the influence of the friction model on temperatures was investigated for the Cases 1 and 2. The obtained results are summarized in Table 8.

**TABLE 8.** Predicted Temperature in the Tool.

| CASE 1 MODELS | $T_{FEM}$ [°C] | e% | CASE 2 MODELS | $T_{FEM}$ [°C] | e% |
|---|---|---|---|---|---|
| I (m=0.85) | 535 | 10.33% | I (m=0.82) | 560 | 3.32% |
| II (μ=0.9) | 739 | 53.00% | II (μ=0.4) | 555 | 2.40% |
| III (μ=0.6) | 515 | 6.63% | III (μ=0.2) | 425 | 21.59% |
| III (μ=0.9) | 670 | 38.72% | III (μ=0.4) | 565 | 4.24% |

It can be easily observed that the numerical temperatures are strictly dependent on the friction model and, within this, on the selected constants. It seem clear that, while the mechanical variables can be accurately estimated properly setting the friction model, further efforts are required as far as the thermal aspect are regarded.

# CONCLUSION

Friction modeling is one of the most critical aspects to be properly defined in FE simulation of machining. In this paper some common models were applied to different couples of tools and workpiece materials. It was assessed that a good prediction of the mechanical variables is possible via a proper choice and calibration of the model.

On the contrary, thermal simulation currently represents a point of weakness since a generally applicable model is not yet available. For this reason, further efforts in this direction are surely encouraged.

# REFERENCES

1. T.H.C. Childs, K. Maekawa, T. Obikawa, Y. Yamane, "Metal Cutting – Theory and Applications", Elsevier Ltd (2000).
2. J.C. Aurich, H. Bil, *Annals of the CIRP* **55/1**, 47-50 (2006).
3. D. Umbrello, J. Hua, R. Shivpuri, *Material Science and Engineering – A,* 374 90 – 100 (2004).
4. A.G. Mamalis, M. Horvath, A.S. Branis, D.E. Manolakos, *J. Mat. Proc. Tech.* **110**, 19-27 (2001).
5. T. Ozel, *Int. J. Mach. Tools & Manuf.* **46**, 518-530 (2006).
6. T.H.C. Childs, *Wear* **260**, 339-344 (2006).
7. Y.C. Yen, A. Jain, T. Altan, *J. Mat. Proc. Tech.* **146**, 72-81 (2004).
8. L.B. Oxley, Mechanics of Machining, an Analytical Approach to Assessing Machinability, Halsted Pr., 1989.
9. N.N. Zorev, Inter-Relationship between shear processes occurring along the tool face and shear plane in metal cutting, *International Research in Production Engineering, ASME*, New York (1963), 42-49.
10. L. Filice, F. Micari, L. Settineri and D. Umbrello, *Proceedings of the 8$^{th}$ ESAFORM Conference* **2**, 729-732 (2005).

# New Materials Design Through Friction Stir Processing Techniques

G. Buffa[1], L. Fratini[1], R. Shivpuri[2]

[1]Dipartimento di Tecnologia Meccanica, Produzione e Ingegneria Gestionale
Università di Palermo, Viale delle Scienze 90128 Palermo, Italy
[2]The Ohio State University, Department of Industrial, Welding and Systems Engineering
1971 Neil Avenue, 210 Baker Systems, Columbus, Ohio 43210, USA

**Abstract.** Friction Stir Welding (FSW) has reached a large interest in the scientific community and in the last years also in the industrial environment, due to the advantages of such solid state welding process with respect to the classic ones. The complex material flow occurring during the process plays a fundamental role in such solid state welding process, since it determines dramatic changes in the material microstructure of the so called weld nugget, which affects the effectiveness of the joints. What is more, Friction Stir Processing (FSP) is mainly being considered for producing high-strain-rate-superplastic (HSRS) microstructure in commercial aluminum alloys. The aim of the present research is the development of a locally composite material through the Friction Stir Processing (FSP) of two AA7075-T6 blanks and a different material insert. The results of a preliminary experimental campaign, carried out at the varying of the additional material placed at the sheets interface under different conditions, are presented. Micro and macro observation of the such obtained joints permitted to investigate the effects of such process on the overall joint performance.

**Keywords:** FSP, material flow, microstructure.
**PACS:** 81.20.Vj

## INTRODUCTION

Friction stir welding (FSW) is an energy efficient and environmentally "friendly" (no fumes, noise, or sparks) welding process, during which the parts are welded together in a solid-state joining at a temperature below the melting point of the workpiece material under a combination of extruding and forging mechanics. This relatively new process, patented by TWI in 1991, has been successfully used to join materials that are difficult-to-weld or unweldeable by fusion welding methods. FSW of butt joints is obtained by inserting a specially designed rotating pin into the adjoining edges of the sheets to be welded and then moving it all along the joint [1-3]. During the process, on the basis of the tool geometry, the tool rotation speed (R) and feed rate (Vf) determine the Specific Thermal Contribution (STC), i.e.. the ratio between the conferred power, resulting from the energy due to the friction forces dissipated into heat and from the material distortion energy, and are combined in a way that an asymmetric metal flow is obtained. In particular, an advancing side and a retreating side are observed: the former being characterized by the "positive"

CP907, 10th ESAFORM Conference on Material Forming, edited by E. Cueto and F. Chinesta
© 2007 American Institute of Physics 978-0-7354-0414-4/07/$23.00

combination of the tool feed rate and of the peripheral tool velocity while the latter having velocity vectors of feed and rotation opposite to each other. A detailed observation of the material microstructure in the joint section by Nelson et al. [4] indicated that there exists an area located at the core of the welding, called "nugget", where the original grain and subgrain boundaries appear to be replaced with fine, equiaxed recrystallized grains characterized by a nominal dimension of a few $\mu$m [1-6]. What is more, it should be observed that in FSW the actual bonding line differs from the original welding line, i.e. from the edges of the blanks to be welded. In former papers [7-8] the authors have described the material flow occurring in FSW and have outlined the actual bonding line which is located in the advancing side of the transverse section of the joint.

Based on the growing knowledge of the process, and due to the microstructual modifications occurring in FSW, Friction Stir Processing (FSP) has been demonstrated as an effective grain refinement technique and successfully used to achieve high strain rate superplasticity (HSRS) conditions in aluminum alloys [9-10]. At the moment no further application of FSP is known by the authors, though its features make it feasible for using in cladding and repairing of die cast alloys.

In the paper the authors investigate the feasibility of a new material design technique: the aim of the present research is the development, in the weld nugget, of a locally composite material through the Friction Stir Processing (FSP) of two AA7075-T6 blanks and a different material insert. The results of a preliminary experimental campaign, carried out at the varying of the additional material placed at the sheets interface under different conditions, are presented. Micro and macro observation of the such obtained joints permitted to investigate the effects of such process on the overall joint performance.

## EXPERIMENTAL PROCEDURE AND MATERIAL FLOW

Four different materials were tested at the interface between the aluminum sheets to be welded. In particular, commercially pure copper (99.95%), brass and Teflon, i.e. $C_2F_4$ polytetrafluoroethylene (PTFE), a polymer of fluorinated ethylene, 0.1mm thick foils (3x100mm) were placed between the adjoining edges of the blanks (Figure 1).

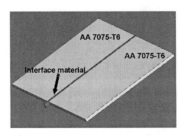

**FIGURE 1.** Schematic view of the experimental setup used for copper, brass and Teflon foils.

Finally titanium dioxide $TiO_2$ dust was added with water to produce a paste to be spread along the adjoining edges of the blanks.

As far as the experimental activity is regarded, a 12mm shoulder FSW tool, with a conical pin characterized by a major diameter equal to 5.00mm, a minor diameter of 1.90mm and a pin height equal to 2.70mm (cone angle 30°), was utilized. The tool was made in H13 steel quenched at 1020 °C, characterized by a 52 HRc hardness. The utilized base material was an AA7075-T6 aluminum alloy in 3mm thick sheets, characterized by a yield stress of 490MPa and an ultimate tensile stress (UTSb) of 530MPa. The AA7075-T6 sheets were reduced in specimens of 100x70mm edges. The edges of the blanks were properly prepared, i.e. milled, to avoid any uncertain contact at the interface between the sheet and the additional material.

All the test were carried out utilizing a nuting angle of 2° and a tool sinking of 2.9mm; as far as the conferred specific thermal contribution (STC) is regarded a tool rotational speed of 1000 rpm and a tool feed rate of 100mm/min was adopted on the basis of preliminary analysis on the joint resistance.

Each test was repeated five times and the transverse sections taken from the specimens were analyzed by both micro and macro observations. In order to obtain such results the specimens were embedded by hot compression mounting, polished and finally etched with Keller reagent for 45s and observed by a light microscope. Furthermore, for each welding, a tensile specimen of 140x20 mm edges, was taken transversally to the welding line and tensile tests have been performed.

A former research [7-8] developed by the authors on the material flow occurring during the FSW process highlighted, through a copper insert, a material flow on the back of the tool from the retreating side towards the advancing one. In Figure 2 a very interesting rebuilt 60X image of the top left corner of the nugget area in a transverse section is reported in which a 250X enlargement is also added.

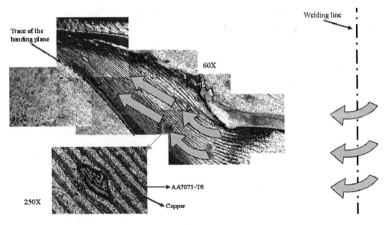

**FIGURE 2.** 60X rebuilt image of the top left corner of the nugget area in a transverse section with 250X enlargement.

First of all a highlighted line is immediately distinguished which separates the core of the welding from the portion of the transverse section in which no copper is found out any more; this line indicates the border of the material in the advancing side which undergoes to the metal flow due to the FSW process, and it is the trace of the real bonding surface occurring during the process. It has to be noticed that such a line is far from the welding line, also highlighted in the figure. Furthermore, alternated flow lines of aluminum and copper are clearly discernible, denoting that the occurring laminar material flow directed towards the top of the joint is a laminar one (see the arrows in Figure 2). It should be observed that the distance between two subsequent flow lines increases from the periphery towards the center of the joint. This peculiar aspect is proper of the so called onion rings [11] usually discernable in the transverse section of effective FSW joints. In the considered case the onion rings are not "closed" since the observed downwards laminar flow close to the middle of the joint section separates in two different directions. It arises that in FSW processes the actual bonding between the two adjoining edges occurs in the advancing side of the joints rather than along the welding line, and the enhancement of the weld nugget mechanical properties, from the beginning of the TMAZ in the retreating side, to the bonding line in the advancing side, is thus likely to improve the overall performances.

## RESULTS AND DISCUSSION

As mentioned before, tensile tests have been performed on the specimen from the developed welds and compared to a "traditional" weld obtained with the same operative parameters and no additional material. Figure 3 shows the average value of the obtained ultimate tensile stress (UTS), as a percentage of the UTS of the base material.

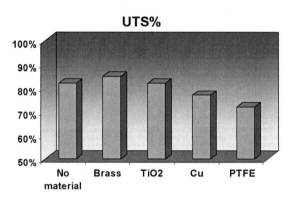

**FIGURE 3.** UTS of the developed welds as a percentage of the UTS of the base material.

Considering as reference testing condition the result obtained with no additional material, that is 82% of the base material UTS, a slight improvement is observed for the brass insert case study, due to the presence of the Zinc. It should be noticed that such an element is already present in the utilized AA7075 aluminum alloy and is responsible for the quite large UTS values of such alloy with respect to others aluminum alloys. Figure 4 shows the transverse section of the joint together with a micro observation of the bonding line area: a quite homogeneous microstructure is observed which indicates a good dispersion of the insert into the weld nugget.

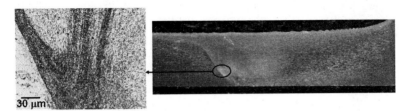

**FIGURE 4.** Macro observation of a transverse section for the brass insert case study and enlargement (625X) of the bonding line area.

The performance of the joint for the $TiO_2$ case study is very close to the reference testing condition one; such a result can be explained with a competition between two counteracting phenomena: the good mechanical properties of Ti are balanced by the detrimental effect of the paste used between the sheets, which remains "embedded" in big clusters in some areas of the nugget zone (see darker areas in Figure 5). Worse performances are observed for the Cu insert (77%) and for the PTFE insert (72%) case studies.

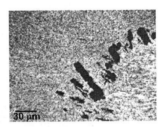

**FIGURE 5.** Micro observation of the weld nugget for the $TiO_2$ insert case study; 625X enlargement.

Finally, Figure 6 shows the elongation at rupture for the considered welds. All the analyzed case studies show worse performances with respect to the reference testing condition.

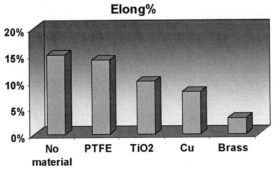

**FIGURE 6.** Elongation (%) for the considered case studies.

The largest elongation value, very close to the one obtained with no additional material, is observed for the PTFE case study, having such material no detrimental effect on the welding area ductility; on the contrary, the weld obtained with the brass insert is the less ductile, once again due to the presence of the Zinc which sensibly increases the brittleness of the joint.

## CONCLUSIONS

The feasibility of a new materials design technique through Friction Stir Processing (FSP) of two AA7075-T6 blanks in FSW of butt joints has been investigated. In particular, a preliminary experimental campaign, carried out at the varying of a few different inserts placed at the sheets interface under different conditions, permitted to highlight the possibility to enhance Friction Stir Welded joints mechanical performances through the use of additional materials. Future work will be focused on the study of the most effective type of insert which has to be used, for each different material, with the aim to improve the joint performance maintaining an as much homogeneous as possible weld nugget microstructure integrity.

## ACKNOWLEDGMENTS

This work was made using MIUR (Italian Ministry for University and Scientific Research) funds.

## REFERENCES

1. Liu, H. J., Fujii, H., Maeda, M., Nogi, K., *J. of Mat. Proc. Tech.* **142,** 692–69 (2003)
2. Rhodes, C. G., Mahoney, M.W., Bingel, W.H., Spurling, R.A., Bampton, C.C., *Scripta Materialia* **36/1,** 69-75 (1987)
3. Guerra, M., Schmidt, C., McClure, L.C., Murr, L.E., Nunes, A.C., *Materials characterization* **49** 95-101 (2003)
4. Su, J. Q., Nelson, T.W., Mishra, R., Mahoney, M. W., *Acta Materialia* **51** 713-729 (2003)
5. Shigematsu, I., Kwon, Y. J., Suzuki, K., Imai, T., Saito, N., *J. of Mat. Science Letters* **22** 343-356 (2003)
6. Fratini, L., Buffa, G., *Int. J. Machine Tools & Manufacture* **45** 1188-1194 (2005).
7. Fratini, L., Buffa, G., Palmeri, D., Hua, J., Shivpuri, R., *ASME Journal of Engineering Materials and Technology* **128/3** 428-435 (2006).
8. Fratini, L., Buffa, G., Palmeri, D., Hua, J., Shivpuri, R., *Science and Technology of Welding and Joining* **11/4,** 412-421 (2006).
9. Mishra, R.S., Charit, I., *Maert. Science and Eng. A* **359,** 290-296 (2003).
10. Mishra, R.S., Mahoney, M.W., *Mater. Science Forum* **507,** 357-359 (2001).
11. John, R., Jata K.V., Sadananda K., *International Journal of Fatigue* **25(9-11),** 939-948 (2003).

# Analytical and Finite Element Approaches for the Drilling Modelling

Mohamad JRAD, Arnaud DEVILLEZ and Daniel DUDZINSKI

*Laboratoire de Physique et Mécanique des Matériaux, UMR CNRS 7554, ISGMP, Université Paul Verlaine – Metz, Ile du Saulcy, 57044 Metz, France, jrad@lpmm.univ-metz.fr*

**Abstract.** Perform drill point design is one of the major problems for the drill manufacturers. To enhance drill performance they have to elaborate prototypes and carry out many tests to progress step by step to an optimised geometry. Model and simulate drilling operations is a very interesting way to obtain useful information for the drill manufacturing process. In this work, a geometrical and thermomechanical analytical model and a finite element approach of drilling were used. While the first gives very quickly some global information, the second gives more details but after a long calculation time. It is shown that the two approaches are complementary and that they may be used with advantage for the drill design.

**Keywords:** Drilling, model, thermomechanical analytical model, finite element model.
**PACS:** 81.20.Wk

## INTRODUCTION

Drilling is one of the more complex metal cutting operations. The drill geometry leads to a variation in cutting angles and cutting velocity along the cutting edges. The cutting conditions are severe leading in particular to high values of temperature. In addition, the cutting zone is inside the workpiece that makes impossible the observation of the process and of the chip flow. Few workers have studied this process; some of them were interested by the drill geometry [1, 2], others proposed models based on mechanistic and experimental approaches [3-9] to predict the torque and thrust. More recently, finite element method was used to simulate the drilling process [10-12]. However, in the majority of these works, the geometrical description is limited to classical twist drill or bevel ground drill. High speed drilling imposed new generation of drills with a more sophisticated geometry presenting for example curved cutting lips and thinned chisel edge. In the following, we used such a drill for modeling, simulation and experimental tests.

First, an analytical description of the drill is presented; it is based on a mathematical definition of tool faces, cutting lips and cutting angles. In order to calculate the thrust and torque, the cutting lips were divided into a series of linear elementary cutting edges. Each cutting element defined by an inclination angle and a normal rake angle works in oblique conditions. The thermomechanical analytical cutting model [13] was then applied on each element and the torque and thrust were calculated by summation. In a second step, a 3D thermo mechanically coupled finite

CP907, *10th ESAFORM Conference on Material Forming*, edited by E. Cueto and F. Chinesta
© 2007 American Institute of Physics 978-0-7354-0414-4/07/$23.00

element model of the drilling process was developed using commercial implicit FEM code Deform-3D$^{TM}$. The drill was specified in this model by its CAD definition. Finally, experimental tests were performed to validate the drilling models.

## ANALYTICAL APPROACH

The drill geometry is defined by two major surfaces: the flute and the flank surfaces. The intersection between the flank surfaces around the drill tip defines the chisel edge, and the cutting lips result from the intersection between the flute and flank surfaces. Then, the drill performance depends strongly of the flank surfaces definition. Depending on grinding parameters, various flank shapes are possible like conical (standard drills), ellipsoidal, hyperboloidal, planar, or multifacet. Tsai and Wu [1] developed a mathematical tool to accurately measure complex drill geometries and reveal how these geometries may be generated through their relationship to the grinding parameters. Using this approach, Paul et al. [14] proposed a methodology to obtain the optimum shape for the chisel edge and the cutting lips with arbitrary point geometries. Hsieh [2] presented a comprehensive and straightforward method for design and analysis of helical drills. The mathematical models for the flute and the flank surfaces were integrated, the cutting edges and chisel edges could be obtained by numerical calculation and the drill characteristic angles were calculated. An alternative mathematical model was proposed recently by Jrad et al. [15] to define the drill geometry and allow the application of a cutting force model to determine the drill performance. This approach is reminded and used in the following.

The tool used for this study was a TiN-coated drill with curved primary cutting edges and thinned chisel edge, the diameter was 8 mm. This drill was developed for high speed drilling applications on steels. A geometrical model of the drill was obtained from the CAD definition of the tool; it allowed the determination of the cutting angles along the cutting lips. To perform the force calculation, the cutting edges were decomposed into series of linear elementary cutting edges, Figure 1, corresponding to the radial increment *dr*. In the same way, the cutting faces in the vicinity of the cutting edges were decomposed into series of planar elementary rake faces. Each elementary cutting edge and its associated elementary rake facet form a cutting element which works in oblique condition during the drilling process.

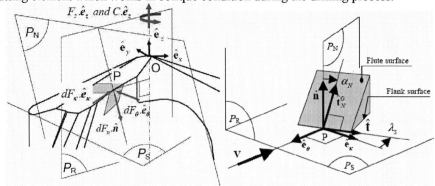

**FIGURE 1.** Elementary cutting edge and chip flow

758

The thermomechanical model of oblique cutting [13] was applied for each cutting element but through a modified version. During the drilling process the elementary chip flow is not free; it is constrained by the adjacent chip elements. The local chip flow direction is imposed by the global chip motion which depends on the flute geometry. More precisely, the chip curls and forms a spiral shape, the angle and diameter of the spiral chips can be calculated from the geometrical parameters of the drill.

The worked material was supposed to be isotropic and rigid and the thermomechanical response was described by a Johnson-Cook law. For each elementary oblique cutting process, the shear stress $\tau$, the shear strain $\gamma$, the shear strain rate $\dot{\gamma}$ and temperature $T$ in the primary cutting zone were calculated from the basic principles of continuum mechanics:

$$\tau = \tau(\gamma, \tau_0) = \rho(V \cos \lambda_s \sin \phi_n)^2 \gamma + \tau_0 ;$$

$$T = T(\gamma, \tau_0) = T_w + \frac{\beta}{\rho c}\left( \rho(V \cos \lambda_s \sin \phi_n)^2 \frac{\gamma^2}{2} + \tau_0 \gamma \right); \quad \frac{d\gamma}{dz_s} = \frac{\dot{\gamma}(\gamma, \tau_0)}{V \cos \lambda_s \sin \phi_n} \tag{1}$$

$$\dot{\gamma} = \dot{\gamma}(\gamma, \tau_0) = \dot{\gamma}_0 \exp\left( \frac{\tau\sqrt{3}}{m\, g_1(\gamma) g_2(\gamma)} - \frac{1}{m} \right); \quad g_1(\gamma) = \left[ A + B\left(\frac{\gamma}{\sqrt{3}}\right)^n \right], g_2(T) = \left[ 1 - \left( \frac{T - T_r}{T_m - T_r} \right)^v \right]$$

The friction at the tool-chip interface was defined by a Coulomb law.

The results obtained with this approach were compared to experimental results, Figure 2. Three feed rates and cutting velocities were tested, the deviations of the predicted global thrust and torque from the experimental results of these components were less then 10%. To evaluate the effect of drill geometries, the distributions of the axial force and torque were studied. Figures 2 c shows the predicted and experimental axial distributions.

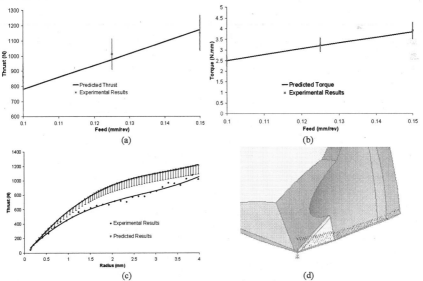

**FIGURE 2.** Thrust and torque obtained with an 8 mm diameter drill, 42CRMo4 workpiece a 2000 rpm spindle speed and a feed rate of 0.15 mm/rev.

The tool-chip interactions calculated with the proposed approach are presented in Figure 2 d where a homogeneous and global chip flow was imposed.

## FINITE ELEMENT MODEL FOR DRILLING PROCESS

An analytical finite element technique was developed by Strenkowski et al. [10] for predicting the thrust force and torque in drilling with twist drills. The approach was based on representing the cutting forces along the cutting lips as a series of oblique sections. Analytical force model was used to determine the cutting forces along the cutting lips. Cutting within the chisel edge region was treated as equivalent orthogonal cutting slices with large negative rake angles and a 2D Eulerian approach was applied to calculate cutting forces distribution in this region.

Bono and Ni [11] proposed also to use a finite element model associated to analytical equations to analyze the temperature profile along the cutting edges of a drill. The predicted results were consistent with experimental observations; in particular they obtained a temperature near the chisel edge larger than on the primary cutting edges.

More recently, a 3D numerical model based on the finite element method was used by Klocke et al. [12] for the prediction of feed force torque and temperature during the drilling process. The 3D FE analysis is more realistic because it gives a good idea about the chip flow and chip morphology and it determines the stress and temperature fields on the tool and the workpiece. These results depend on the introduced material behavior and friction conditions at the interfaces.

In the presented study, the commercial Code Deform 3D$^{TM}$ based on the Finite Element Method was employed to simulate drilling process. As previously in the analytical approach, the drill used was with curved cutting lips and thinned chisel edge, its diameter was 8 mm; the CAD definition of the drill was imported to the FE model. Workpiece was a cylinder of 10 mm diameter and the worked material was the 42CrMo4 steel.

Deform 3D$^{TM}$ provided an implicit 3D coupled thermo-mechanical numerical model of the drilling process with continuous remeshing of the workpiece. In this simulation, the material behavior was described by the Johnson-cook law:

$$\bar{\sigma} = \left(A + B\bar{\varepsilon}^n\right)\left(1 + C\ln\left(\frac{\dot{\bar{\varepsilon}}}{\dot{\bar{\varepsilon}}_0}\right)\right)\left(\frac{\dot{\bar{\varepsilon}}}{\dot{\bar{\varepsilon}}_0}\right)^\alpha \left(D - ET^{*m}\right),$$

$$where \ T^* = \frac{(T - T_{room})}{T_{melt} - T_{room}} \quad D = D_0 \exp\left[K\left(T - T_b\right)^\beta\right]$$

(2)

The material behavior constants were obtained from the code library. The physical parameters like Young's modulus, heat capacity and heat conduction coefficients were supposed temperature dependent. The tool was modeled as a rigid body with a TiN coating of 5µm thickness. The friction at the tool-chip and tool-workpiece interfaces was supposed given by a shear type law : $\tau = \bar{m}k$, where $k$ is the worked material shear flow stress and $\bar{m}$ a chosen constant and equal to 0.6.

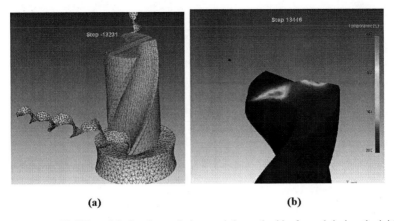

(a)                                                      (b)

**FIGURE 3.** (a) 3D FE models for the workpiece and the tool, chip formed during the initial drill penetration. (b) Computed temperature contour on the drill; the speed and the feed were fixed at 2000 rpm and 0.15 mm/rev.

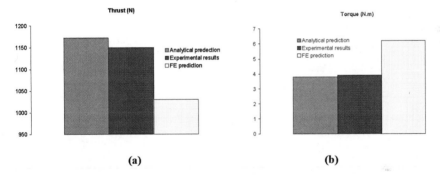

(a)                                                      (b)

**FIGURE 4.**Comparison between analytical and numerical results.

Figure 3a presents the 3-D FE models of the workpiece and of the tool and a chip formed during a initial drill penetration. A constant remeshing process was used to ensure that large deformations do not cause unacceptable element distortions. Zones of high mesh density are always situated in the cutting and contact zones are not visible in the figure. The obtained chip morphology is very similar to the one observed during the first stage of the experimental drilling tests.

The temperature distribution on the drill calculated at the beginning of the drilling process is shown in Figure 3b. The obtained values are of the same order than those measured by Bagci and Ozcelik [16] during dry drilling of steel. In opposition with experimental observations obtained by Bono and Ni [11], temperatures on the primary cutting edges are larger than near the chisel edge, however the results presented in figure 3b concern the first stage of the drilling process.

The thrust and the torque predicted by the analytical model and by the finite element computation were compared to the experimental measures, Figure 4. It appears that the analytical results are better than the numerical ones; however, the FE model needs more investigation and the predictions may surely be enhanced.

761

# CONCLUSION

In this paper, analytical and 3D finite element numerical models were developed for the drilling process simulation. The first one gives quickly results about the thrust and torque and the cutting forces distribution along the cutting edges. The obtained values depend on the drill geometry and on the cutting conditions; they are consistent with the experimental measures. The second model provides interesting information on the chip formation and on the temperature and stress distributions but the calculations are time consuming. The two methods may be used as complementary approaches to optimize cutting conditions and drill geometry.

# REFERENCES

1. W. D. Tsai, S. M. Wu, Computer Analysis of Drill Point Geometry, Int. J. Mach. Tool. Des. Res., Vol. 19, 1979, pp. 95-108.
2. Jung-Fa Hsieh, Mathematical Model for Helical Drill Point, International Journal of Machine Tools & Manufacture, Vol. 45, 2005, pp. 967-977.
3. C. J. Oxford, Rochester, Mich, On the Drilling of Metals 1-Basic Mechanics of the Process, Trans. ASME, 1955, pp. 103-114..
4. M. C. Shaw, C. J. Oxford, On the Drilling of Metals 2-The Torque and Thrust in Drilling, Trans. ASME, 1957, pp. 139-148.
5. R. A. Williams, A Study of the Drilling Process, Trans. ASME, Journal of Engineering for Industry, 1974, pp. 1207-1215.
6. S. Wiriyacosol, Thrust Torque Prediction in Drilling from a Cutting Mechanics Approach, Annals of the CIRP, Vol. 28, 1979, pp. 87-91..
7. E. J. A. Armarego, C. Y. Cheng, Drilling with Flat Rake Face and Conventional Twist Drills-1. Theoretical Investigation, Int. J. Mach. Tool Des. Res., Vol. 12, 1972, pp. 17-35..
8. A. R. Watson, Geometry of Drill Elements, Int. J. Tool. Des. Res., Vol. 25, 1985, 209-227.
9. A .R. Watson, Drilling Model for Cutting Lip and Chisel Edge and Comparison of Experimental and Predicted Results. Part1 and Part 2, Int. J. Tool. Des. Res., Vol. 25, 1985, pp. 347-365, pp. 367-376.
10. J. S. Strenkowski, C. C. Hsieh, A. J. Shih, An analytical finite element technique for predicting thrust force and torque in drilling, International Journal of Machine tool and Manufacturing, 2004, 44,pp. 1413-1421.
11. M. Bono, J. Ni, The location of the maximum temperature on the cutting edges of a drill, International Journal of Machine Tools & Manufacture, 46, 2006,pp. 901–907.
12. F. Klocke, D. Lung, K. Gerschwiler, M. Abouridouane, K. Risse, 3D Modeling and scaling effects in drilling, 9th CIRP International Workshop on Modeling of Machining Operations May 11-12, 2006, Bled, Slovenia.
13. A. Moufki, A.Devillez, D.Dudzinski, A. Molinari, Thermomechanical modeling of oblique cutting and experimental validation, International Journal of Machine Tools & Manufacture, Vol. 44, 2004, pp. 971-989.
14. Anish Paul, Shiv G. Kapoor, Richard E. DeVor, Chisel edge and cutting lip shape optimization for improved twist drill point design, Int. J. Mach. Tools& Manuf, 2005, 45, 421-431.
15. M.Jrad, A.Devillez, D.Dudzinski, "Thermomechanical Approach of Drilling Based on A CAD Definition", Proceedings of the 9th CIRP International Workshop on Modeling of Machining Operations, May 2006, pp. 247-253.
16. Eyup Bağci and Babur Ozcelik, Investigation of the effect of **drilling** conditions on the twist drill temperature during step-by-step and continuous dry drilling, *Materials & Design, Volume 27, Issue 6, 2006, Pages 446-454.*

# Proposal and Use of a Void Model for the Simulation of Shearing

Kazutake Komori

*Department of Mechanical Engineering, Daido Institute of Technology*
*10-3 Takiharu-town, Minami-ward, Nagoya-city, Aichi-prefecture 457-8530, Japan*

**Abstract.** In this study, a ductile fracture model derived from microscopic considerations is proposed and is demonstrated to be effective for simulating shearing. First, a method for predicting ductile fracture is proposed using a model obtained from microscopic considerations. The relationship between the void volume fraction and the strain to fracture calculated using this model agrees well with the one obtained using Thomason's model for internal necking. Next, a computer program was developed for analyzing the behavior of crack growth after ductile fracture has occurred; it was based on the proposed ductile fracture model and employs the finite-element method. The condition for a material to break into two upon shearing was simulated using this program. Finally, a simulation and an experiment into shearing of copper samples were performed, and the validity of our proposed model was demonstrated by comparing the analytical results with the experimental results.

**Keywords:** Shearing, Thomason's Model, Multi-Scale Simulation.
**PACS:** 46.35.+z, 46.50.+a, 62.20.Fe.

## INTRODUCTION

In previous studies, we have developed a computer program that employs the finite-element method to analyze the behavior of crack propagation after ductile fracture. Simulations of inner fracture defects produced in drawing [1] and of shearing [2] have been performed, and the validity of the computer program has been demonstrated by comparing the simulation results with experimental results. In these simulations, the criterion for fracture is when the void volume fraction of the material exceeds a certain critical value. However, such a fracture criterion does not necessarily have a definite physical meaning on the microscopic level.

Microscopically, ductile fracture occurs through the nucleation, growth and coalescence of voids. Hence, many numerical and experimental studies have investigated these behaviors of voids. Thomason proposed a model of void coalescence based on internal necking of the intervoid matrix ligaments [3]. This model was derived using the upper-bound method, in which the material is assumed to fracture when the energy required to coalesce voids by internal necking is less than that required to deform the material homogeneously. This model of void coalescence is useful since it has a definite physical meaning.

In the Thomason model, the following assumptions are made.

CP907, *10th ESAFORM Conference on Material Forming,* edited by E. Cueto and F. Chinesta
© 2007 American Institute of Physics 978-0-7354-0414-4/07/$23.00

* The voids are rectangular in shape.
* The longitudinal axis of a void is aligned with the direction of maximum principal stress.

In other words, the model assumes that the direction of principal strain does not change during plastic deformation. However, this is assumption does not hold during shearing and thus the model cannot be utilized to analyze shearing.

In previous papers by us [4, 5, 6] we proposed a new model based on the Thomason model, which can be utilized to analyze shearing. In our model, the following assumptions are made.

* The voids are parallelograms.
* The longitudinal axis of a void is not necessarily aligned with the direction of maximum principal stress.

In other words, in our model the direction of principal strain is assumed to vary during plastic deformation. Our model was incorporated into a computer program that uses the finite-element method, which we had previously developed to analyze the behavior of crack propagation after ductile fracture [1, 2]. In this paper, ductile fracture behavior in shearing was simulated, and the validity of our proposed model was confirmed by comparing the numerical results with experimental results.

## ANALYSIS METHOD

### Overview of Macroscopic Analysis

The deformation of a material during shearing was analyzed using the conventional axisymmetric rigid-plastic finite-element method [7]. The approximate yield function $\Phi'$, proposed by Gurson [8] and modified by Tomita [9], is adopted:

$$\Phi' = \frac{3}{2} \cdot \frac{\sigma'_{ij} \sigma'_{ij}}{\sigma_M^2} + \frac{f}{4} \cdot \left( \frac{\sigma_{kk}^2}{\sigma_M^2} \right) - (1-f)^2 = 0, \tag{1}$$

where $\sigma_M$ is the tensile yield stress of the matrix and $f$ is the void volume fraction of the material. In addition, the following time-evolution equation, which expresses the variation in the void volume fraction with time, is assumed:

$$\dot{f} = (1-f)\dot{\varepsilon}_{kk} + A \left\langle \frac{\sigma_{kk}}{3\bar{\sigma}} - B \right\rangle \dot{\bar{\varepsilon}}, \tag{2}$$

where $\bar{\sigma}$ is the equivalent stress, $\dot{\bar{\varepsilon}}$ is the equivalent strain rate and $A$ and $B$ are material constants. Here, the angular brackets ($<>$) that appear in Equation (2) denote Macauley's bracket; in other words, $\langle x \rangle$ equals $x$ when $x$ is positive, while $\langle x \rangle$ equals zero when $x$ is negative.

# Overview of Microscopic Analysis

An overview of the microscopic analysis is given below:

(1) The void volume fraction $f$ and the average deformation gradient $\overline{\partial x/\partial X}$ are calculated from the void volume fraction rate $\dot{f}$ and the deformation gradient rate $\partial \dot{x}/\partial X$, which are calculated using macroscopic rigid-plastic finite-element analysis.

(2) The void configuration and the void shape are calculated.

(3) The ratio of the energy-dissipation rate of internal necking to that of homogeneous deformation $E$ is calculated.

(4) Whether or not the material fractures is determined.

# Void Configuration and Void Shape

The following assumptions about the void shape are made.

* The void shape before forming is a square.
* The square, which is embedded in the material before forming, becomes a parallelogram after forming. The void shape after forming is identical to the parallelogram after forming.
* The void shape at void nucleation is a square.
* The square, which is embedded in the material at void nucleation, becomes a parallelogram after void nucleation. The void shape after void nucleation is identical to the parallelogram after void nucleation.

The following assumptions about the void configuration are made.

* The void configuration before forming is identical to the configuration of the intersection point of a hexagonal grid.
* The hexagonal grid, which is embedded in the material before forming, deforms after forming. The void configuration after forming is identical to the configuration of the intersection point of the deformed hexagonal grid after forming.
* The void configuration at void nucleation is identical to the configuration of the intersection point of a hexagonal grid.
* The hexagonal grid, which is embedded in the material at void nucleation, deforms after void nucleation. The void configuration after void nucleation is identical to the configuration of the intersection point of the deformed hexagonal grid after void nucleation.

# Estimation of Material Fracture

Figure 1 shows two neighboring voids and the velocity field. Figure 1(a) shows the two neighboring voids. The velocity discontinuity surfaces, represented by the broken lines in Fig. 1, are assumed. In this figure, $v$ denotes the material velocity in the direction of the maximum principal stress; $\theta_1$ and $\theta_2$ denote the angles between the direction of the maximum principal stress and the direction of the surface of the velocity discontinuity; $2l_1$ and $2l_2$ denote the lengths of the velocity discontinuity lines and $L$ denotes the distance between two neighboring voids. Figure 1(b) shows the velocity field represented as a hodograph. Here, $\Delta v_1$ and $\Delta v_2$ denote the values of the velocity discontinuities.

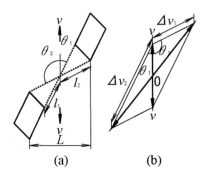

(a)　　　　　(b)

**FIGURE 1.** Two neighboring voids and velocity field. (a) Two neighboring voids (b) Velocity field

First, the energy dissipation rate on the velocity discontinuity surface when two neighboring voids coalesce by internal necking, i.e., the energy dissipation rate of internal necking, is expressed as

$$k \cdot \Delta v_1 \cdot 2l_2 + k \cdot \Delta v_2 \cdot 2l_1 = 4kv \frac{l_1 \sin\theta_2 + l_2 \sin\theta_1}{\sin(\theta_1 + \theta_2)}. \tag{3}$$

Next, the energy dissipation rate of homogeneous plastic deformation, i.e., the energy dissipation rate of homogeneous deformation, is given as

$$L \cdot 2k \cdot 2v = 4kvL. \tag{4}$$

Since void coalescence is assumed to occur when the energy dissipation rate of internal necking is less than that of homogeneous deformation, the criterion for void coalescence is expressed as

$$L \geq \frac{l_1 \sin\theta_2 + l_2 \sin\theta_1}{\sin(\theta_1 + \theta_2)}. \tag{5}$$

The ratio of the energy dissipation rate of internal necking to that of homogeneous deformation $E$ is defined as

$$E = \frac{l_1 \sin\theta_2 + l_2 \sin\theta_1}{L \sin(\theta_1 + \theta_2)}. \tag{6}$$

The criterion for void coalescence is satisfied when $E$ is less than unity and the material is assumed to fracture when this criterion is satisfied.

## ANALYTICAL RESULTS

A tensile test experiment was performed using tough-pitch copper. The following stress-strain relationship was obtained from the experimental results and used in the analysis:

$$\sigma_M = -310 + 780 \cdot \left(\varepsilon_M + 0.02\right)^{0.21} (\text{MPa}), \tag{7}$$

where $\varepsilon_M$ is the strain of the matrix.

The values for the material constants $A$ and $B$ that appear in Equation (2) were assigned as follows; first, $B$ was assumed to be zero to simplify the analysis, while $A$ was assumed to have a value that ensured that the contraction of the area calculated in the analysis was the same as that measured in the experiment. In this way, the material constants $A$ and $B$ were assigned to be:

$$A = 0.04, B = 0. \tag{8}$$

Figure 2 shows the finite-element meshes after the tensile test and the relationship between the nominal strain and the nominal stress in the tensile test. Figures 3, 4 and 5 show the finite-element meshes of chips and holes and the relationship between the punch displacement and the punch force for various punch-die clearances.

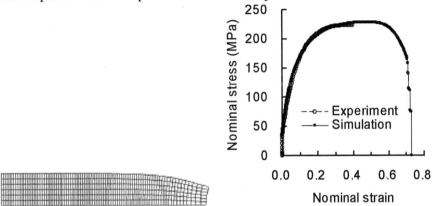

**FIGURE 2.** Finite-element meshes after tensile test (left) and relationship between nominal strain and nominal stress in tensile test (right).

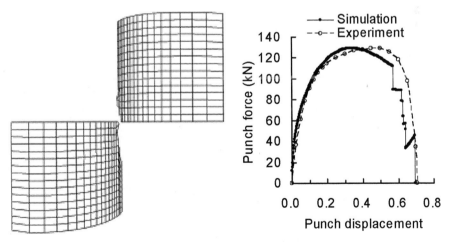

**FIGURE 3.** Finite-element meshes of chips and holes (left) and relationship between punch displacement and punch force (right). (Non-dimensional punch-die clearance of 0.03)

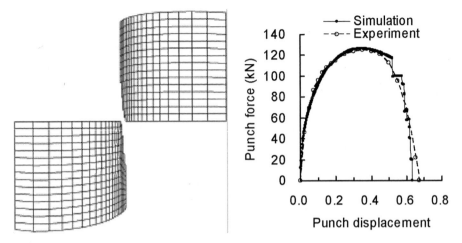

**FIGURE 4.** Finite-element meshes of chips and holes (left) and relationship between punch displacement and punch force (right). (Non-dimensional punch-die clearance of 0.05)

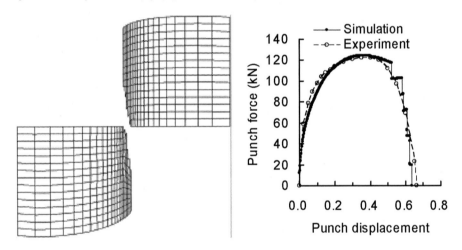

**FIGURE 5.** Finite-element meshes of chips and holes (left) and relationship between punch displacement and punch force (right). (Non-dimensional punch-die clearance of 0.07)

# REFERENCES

1. K. Komori, *Int. J. Mech. Sci.*, **45**, 141-160 (2003).
2. K. Komori, *Theor. Appl. Frac. Mech.* **43**, 101-114 (2005).
3. P. F. Thomason, *J. Inst. Metals,* **96**, 360-365 (1968).
4. K. Komori, *Acta Mater.*, **47**, 3069-3077 (1999).
5. K. Komori, *Mater. Sci. Eng.: A*, **421**, 226-237 (2006).
6. K. Komori, *Acta Mater.*, **54**, 4351-4364 (2006).
7. S. Kobayashi, S. I. Oh, T. Altan, *Metal Forming and the Finite-Element Method,* Oxford University Press, New York, 1989.
8. A. L. Gurson, *Trans. ASME J. Engng. Mater. Technol.*, **99**, 2-15 (1977).
9. Y. Tomita, *Numerical Elasticity and Plasticity,* Youkendou, Tokyo, 1990.

# Improving Quality in Machined Automotive Parts with the Finite Element Method

Unai Segurajauregui [1], Luc Masset [2], Pedro José Arrazola [1]

[1]*Manufacturing Department, Faculty of Engineering-Mondragon University, Mondragon, Spain*
[2]*Samtech SA, Liège, Belgium*

**Abstract.** One of the current trends in machining is the reduction or elimination of cutting fluids in order to reduce economical costs. As a consequence, part heating is increased, leading to greater thermal distortions. In this article we present a finite element approach of the thermal behavior of the part during drilling operations and we demonstrate how numerical simulation may improve the part quality. First, an aluminum sample is studied, for which numerous experimental data have been collected: thermocouples and infrared methods for temperatures and inductive sensors for displacements. Then, a thermo-mechanical finite element approach gives the temperature fields and the thermal distortions of the sample. The experimental data are used to validate the inputs of the numerical model, mainly the thermal fluxes entering the sample. Finally, we use the same approach to study the behavior of an aluminum casting gear box during drilling operations. At each time, we are able to predict the displacements of the holes. Thus, we can modify their position in the NC program in order to compensate the thermo-mechanical distortions and to reduce their position errors. This demonstrates how manufacturing industries may increase their production rate and quality through machining simulation.

**Keywords:** FEM, thermo-mechanical distortion, drilling, aluminum workpiece, MQL
**PACS:** 89.20.Kk

## INTRODUCTION

The pressure of increasing productive costs, mostly because of the irruption of countries with cheaper handwork in occidental markets, and the change in the environmental awareness have led manufacturers to reduce or eliminate cutting fluids. Traditional machining processes using lubricants are replaced by more ecological ones: dry cutting or MQL (Minimum Quantity of Lubricant). The MQL technique consists in the application of pressured air with small oil particles in the cutting zone. The oil has a lubrication function and the air a cooling function, meanwhile the high pressure helps the chip evacuation [1].

Without cutting fluids, heat dissipation is reduced, leading to increased thermo-mechanical distortions [2]. Nearly all the mechanical energy developed in the cutting process is transformed into heat [3]. A part of this heat is absorbed by the workpiece that tends to dilate. After the machining operation and the removal of the clamping devices, the part returns to its initial thermal state, which causes various dimensional defects such as position errors and form errors [4].

CP907, *10th ESAFORM Conference on Material Forming,* edited by E. Cueto and F. Chinesta
© 2007 American Institute of Physics 978-0-7354-0414-4/07/$23.00

The finite element method is well suited for predicting thermo-mechanical distortions during machining [5]. In this work, we do not study the heating at a local scale [6] but the behaviour of the whole part. Few researches on thermo-mechanical aspects at macroscopic scale are found in the specialized literature. Let us cite Weinert et al [7] for drilling processes, Kakade and Chow [8] in boring operations and Stephenson et al [4] who studied thermo-mechanical inaccuracies in turning.

In this article we use the finite element method to study the thermo-mechanical behaviour of aluminium parts during drilling operations. Drilling has been chosen because thermal aspects are critical and this operation is widely used for automotive parts such as aluminium gear boxes. For validation purpose, we have started with the study of a simple aluminium sample for which numerous experiments were carried out using different drills and cutting conditions. These tests allowed us to setup the inputs of the numerical approach, mainly the thermal fluxes to apply to the machined part. Then, we have studied a real industrial part (an aluminium casting gear box) and demonstrated how it is possible to reduce dimensional errors by compensating thermal expansions.

## FINITE ELEMENT MODELING

The SAMCEF finite element software [9] is used for all the numerical calculations. The sample is a beam with a rectangular section (figure 1). The machining process consists of 10 consecutive holes of Ø10 mm. The sample is constrained on one end and kept free at the other. The sample expansion is measured at its free end thanks to an inductive sensor so that we can evaluate for example the position error between the first and the last holes (figure 1).

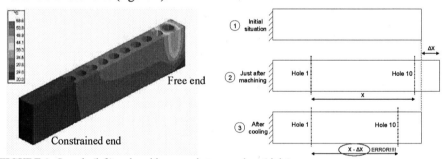

**FIGURE 1.** Sample (left) and position error between edges (right)

The numerical approach is uncoupled. First, a pure thermal numerical calculation is performed using the measured thermal flux generated during the drilling operation. Then, the obtained temperature fields are used as an input for the mechanical calculation during which we compute the displacement fields in function of time. Thermal fluxes are applied to all the internal surface of the holes, from the first one to the last one.

Calculations are performed for two different aluminium alloys: Al2017 T451 and AlSi9Cu3, which is the material of the gear box. Measures were made for three cutting speeds ($V_{c1}$= 30 m/min; $V_{c2}$=40 m/min; $V_{c3}$=300 m/min) and three feed rates ($f_1$=0,10 mm/rev; $f_2$=0,15 mm/rev; $f_3$=0,50 mm/rev). $V_{c3}$ and $f_3$ are the cutting conditions employed when machining the gearbox.

The most important input data to define for the numerical model is the quantity of the thermal energy that flows into the sample. In this work, it is employed an inverse methodology for calculating this parameter. Thus, we first consider that all the thermal energy is absorbed by the sample. Then we set the fraction of energy in order to match the temperature evolution measured with thermocouples at different zones of the sample. The total heat flux developed during drilling is obtained from the multiplication between the angular speed of the drill and the drilling torque, which is measured with a KISTLER piezoelectric sensor.

## EXPERIMENTAL MEASUREMENTS

Experimental measurements of temperature and displacement are performed to compare with numerical results. The temperature is measured with K type superficial thermocouples and the displacement is measured with a non contact inductive sensor. Figure 2 shows the experimental setup for temperature and displacement measurements. It can be seen that the distortion is measured on the free end and the thermocouples are placed in the external lateral surface of the sample.

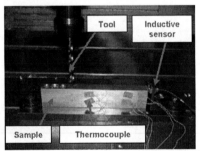

**FIGURE 2.** Experimental measurement of temperature and displacement

For the experimental measurements of temperature three thermocouples are considered, located at 25 mm deep of the first, fifth and tenth holes (T1, T5, T10). The measurement of the displacement is made by a non contact sensor located on the top of the free side of the sample. All the data obtained for each experiment is summarized in two graphs, one for temperatures and the other for displacements (figure 3).

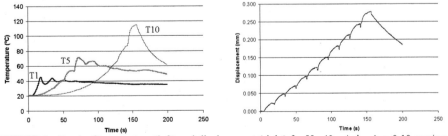

**FIGURE 3.** Graph of temperature (left) and displacement (right) for $V_c$=40 m/min, $A_v$= 0,10 mm/rev, Al2017 T451, dry drilling

Infrared techniques are also employed for temperature measurement of the sample, although the correct temperature measurement of aluminium alloys with this technique is very difficult because of the low emissivity of the material and the complicated definition of this parameter (it varies with temperature…) [10]. Thus, the results obtained with this technique are not as good as expected and it is considered only the thermocouple data for the temperature measurements.

# RESULTS

In order to compare the experimental and numerical results of temperature and displacements, the simulation results of these outputs are obtained for the same points of the experimental measurements. Thus, we are looking for the exact percentage of heat that goes into the workpiece, with the aim of applying this data in the analysis of the aluminium gearbox. For this, the same drilling conditions as for the gearbox are considered in the sample (figure 4: $V_{c3}$=300 m/min, $f_3$=0,50 mm/rev, AlSi9Cu3, MQL). By the other hand, the possibility of the totally dry drilling with the same cutting conditions is also studied. It can be seen that, considering the displacement of the top of the free side of the sample, the heat flux to the workpiece in a machining with MQL is around the 10% of the total heat generated during the drilling, whereas for the dry drilling case this percentage is augmented until 40%.

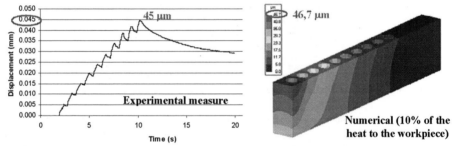

**FIGURE 4.** Analysis of the % of heat that goes into the workpiece for MQL drilling

Other interesting observations are also obtained at this point:
- In high cutting conditions the temperature and distortion of the workpiece is considerably lower than in low cutting conditions.
- There are differences according to the machined material. Al 2017 T451 reaches higher temperature and distortions than AlSi9Cu3.

# INDUSTRIAL APPLICATION

As an industrial application, we study the drilling operations of an aluminium gearbox. The aim of the simulation is to predict the temperature and expansion of the part during the drilling of 25 holes in the top face of the gearbox. A position tolerance of 80 μm is imposed between holes 24 and 25 distant from 425 mm. For the calculation, we have considered the cases of MQL and dry drilling. As it has been mentioned before, for the first case (figure 5) the 10% of the thermal energy goes into the workpiece, whereas for the second case the introduced heat is the 40%.

FIGURE 5. Simulation of the thermal distortion for the drilling operations of the carter with MQL

It is important to remark that the holes 24 and 25 are the latest ones. It means that all the other holes are already done and that the piece is heavily heated and distorted when machining holes 24 and 25. Also some machining operations (boring, milling) are performed before the drilling ones so that the part is already heated before starting the drilling of the 25 holes. These operations have not been considered here. We assume that drilling has a more significant impact on the final distortion of the gearbox.

With the aim of compensate the thermal distortions, the position of the holes in the NC programme can be modified. In figure 6, it can be seen the modifications proposed for the cases of MQL and dry drilling. We have to mention again that these compensations would be greater if all the machining operations of the gearbox were considered.

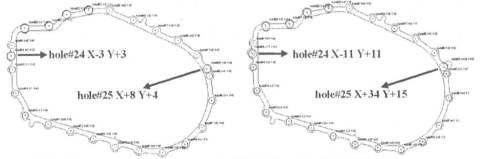

FIGURE 6. NC compensation in micrometers for MQL (left) and dry (right) drilling

## CONCLUSIONS

Thermo-mechanical simulations offer a good insight on machining issues, especially for aluminium parts. Industries may efficiently adapt and improve their process setup by selecting the correct machining strategy or the most appropriate clamping scheme. Moreover, the ability to adjust the NC program to compensate the thermal expansion of the part should ease the change from traditional machining (cutting fluids) to ecological machining (MQL or dry).

It has been seen that for dry drilling the displacements are higher than for MQL drilling. The distortions obtained by the numerical analysis in both cases are not big enough for keeping the critical dimensions of the gearbox (distance between holes 24 and 25) out of tolerance, but they are quite important to influence the process capacity index. Thus, for a correct analysis, there should be considered all the machining operations of the gearbox like the borings and millings that are performed before the drillings.

By the other hand, it has been said that the percentage of heat that goes into the workpiece is the most important and difficult to define input data. As this parameter is critical for the success of the numerical analysis, it would be studied more in detail (analytically, numerically and experimentally) for future works.

## ACKNOWLEDGMENTS

The authors would like to thank the Basque Government for its financial support of the projects EKOLAN (IT-2005/0000110) and MARGUNE (IE03-107). Special thanks are addressed to Ideko S.Coop Research Centre for the help in the experimental tests carried out and to Fagor Ederlan for the support given about technical information dealing with workpieces.

## REFERENCES

1. K. Weinert, I. Inasaki, J. W. Sutherland, T. M. Wakabayashi, *Dry Machining and Minimum Quantity Lubrication*, CIRP Annals - Manufacturing Technology, v 53, n 2, 2004, p 511-537.
2. R. P. Zeilmann, W. L. Weingaertner, *Analysis of temperature during drilling of Ti6Al4V with minimal quantity of lubricant*, Journal of Materials Processing Technology, v 179, n 1-3, 2006, p 124-127.
3. M. C. Shaw, *Metal Cutting Principes*, Oxford University Press, 1984.
4. D. A. Stephenson, M. R. Barone, F. M. Dargush, *Thermal Expansion of the Workpiece in Turning*, Journal of Engineering for Industry, Transactions of the ASME, v 117, n 4, 1995, p 542-550.
5. V.A. Sukaylo, A. Kaldos, G. Krukovsky, F. Lierath, T. Emmer, H. J. Pieper, J. Kundrak, V. Bana, *Development and verification of a computer model for thermal distortions in hard turning*, Journal of Materials Processing Technology, v 155-156, n 1-3, 2004, p 1821-1827.
6. P. J. Arrazola, *Modélisation Numérique de la Coupe: Étude de Sensibilité des Paramètres d'Entrée et Identification du Frottement entre Outil-Copeau*, Phd. Thesis, E.C. Nantes, 2003.
7. K. Weinert, M. Kersting, M. Schulte, C. D. Peters, *Finite Element Analysis of Thermal Stresses During the Drilling Process of Thin-Walled Profiles*, Production Engineering, v XII/1, 2005, p 101-104.
8. N. N. Kakade and J. G. Chow, *Finite Element Analysis of Engine Bore Distortions during Boring Operation*, Journal of Engineering for Industry, Transactions of the ASME, v 115, n 4, 1993, p 379-384.
9. www.samcef.com
10. C.-D. Wen, I. Mudawar, *Emissivity characteristics of roughtened aluminum alloy surfaces and assessment of multispectral radiation thermometry (MRT) emissivity models*, International Journal of Heat and Mass Transfer, v 47, n 17-18, 2004, p 3591-3605.

# Modeling of the flow stress for AISI H13 Tool Steel during Hard Machining Processes

Domenico Umbrello*, Stefania Rizzuti*, José C. Outeiro[†] and Rajiv Shivpuri[¶]

*Department of Mechanical Engineering, University of Calabria 87036 Rende (Italy)
[†]Portuguese Catholic University, 3080-024 Figueira da Foz (Portugal)
[¶] Industrial, Welding and Systems Engineering – Ohio State University 43210 Columbus (USA)

**Abstract.** In general, the flow stress models used in computer simulation of machining processes are a function of effective strain, effective strain rate and temperature developed during the cutting process. However, these models do not adequately describe the material behavior in hard machining, where a range of material hardness between 45 and 60 HRC are used. Thus, depending on the specific material hardness different material models must be used in modeling the cutting process. This paper describes the development of a hardness-based flow stress and fracture models for the AISI H13 tool steel, which can be applied for range of material hardness mentioned above. These models were implemented in a non-isothermal viscoplastic numerical model to simulate the machining process for AISI H13 with various hardness values and applying different cutting regime parameters. Predicted results are validated by comparing them with experimental results found in the literature. They are found to predict reasonably well the cutting forces as well as the change in chip morphology from continuous to segmented chip as the material hardness change.

**Keywords:** Material behaviour; Finite element analysis; Hard machining.
**PACS:** 83.60.-a; 02.70.Dc; 81.20.Wk.

## INTRODUCTION

Machining operations involve a substantial portion of the world's manufacturing sector. They create about 15% of the value of all mechanical components manufactured worldwide [1]. Because of its great economic and technical importance, a large number of researches have been carried out in order to optimise the machining process in terms of improving quality, increasing productivity and lowering cost. However, according to a working paper, conducted by Armarego et al. [2], was reported that in USA, the correct cutting tool is selected less that 50% of the time, the tool is used at the rated cutting speed only 58% of the time, and only 38% of the tools are used up to their full tool-life capability. This situation urges the need for developing more scientific approaches for improving the technological machining performance.

Several improvements in term of better understanding of machining process have been done in the last decade thank to the heavy use of Finite Element methodology. In fact, numerous researchers involved in this field have used FEM to predict the effect

CP907, 10th ESAFORM Conference on Material Forming, edited by E. Cueto and F. Chinesta
© 2007 American Institute of Physics 978-0-7354-0414-4/07/$23.00

of several variables, such as cutting forces, chip morphology, surface integrity, etc. These were well recognized by a reviewer paper proposed by Mackerle [3]. Furthermore, in the last couple of years great efforts were spent in understanding and simulating the hard machining process [4]. This process allows manufacturers to machine hardened materials to their finish part quality without the aid of grinding, increasing the efficiency and decreasing the cost and processing time for post finishing. Recently, several steel alloys, such as AISI 52100 bearing steel and AISI H13 tool steel are machined by using this technology. The latter, AISI H13 tool steel possesses good resistance to thermal softening and heat checking, high strength, high toughness and high hardenability (the hardness of AISI H13 varies with its application for different type of dies [5]).

Unfortunately, the known flow stress models published in literature for these materials are mainly based only on the effective strain, effective strain – rate and temperature. This may be acceptable in conventional machining or for a selected level of hardness but is unacceptable in hard machining where the flow stress varies with different heat treatment of the material. Therefore, it is very important to include the effect of the hardness in the flow stress model to reflect the influence of the different heat treatment on the selected material. In fact, successful modeling and analysis of any thermo-mechanical process is strictly related to the material model being utilized.

In this paper, a detail approach to develop a hardness based material model for the hard turning of AISI H13 tool steel is presented. Moreover, the proposed hardness based flow stress model takes also into account the increase of the hardness due to the high temperature and the severe cooling reached in a machining process so that the white layer is generated. In fact, such layer is associated with workpiece material that has melted and then cooled very rapidly to produce an amorphous structure [6]. In addition, both the critical damage value (Cockcroft & Latham's criterion) and the shear factor to model friction are proposed as a function of the initial hardness. Furthermore, an iterative procedure is utilized for determining the global heat transfer coefficient at the tool-chip and tool-workpiece interface. Finally, the proposed model for AISI H13 is implemented in the Finite Element (FE) basis by a proper user defined routine, and it is validated by comparing the predicted results, such as cutting forces, chip morphology, temperature, etc., with the experimental evidences found in published literature.

## FLOW STRESS OF AISI H13 TOOL STEEL AT DIFFERENT HARDNESS

The construction of the material model to account for the influence of the workpiece hardness is considered as follows. First, a reference flow stress curve at certain workpiece hardness is chosen. Particularly, a Johnson-Cook (JC) material model for a workpiece hardness of 46 HRC, proposed by Shatla et al. [7] was selected as a reference curve and given by Eq. (1).

$$\sigma_{\mathrm{Re}f} = \left(674.8 + 239.2\varepsilon^{0.28}\right)\left(1 + 0.027 \cdot \ln\left(\frac{\dot{\varepsilon}}{\dot{\varepsilon}_0}\right)\right)\left(1.16 - 0.88 \cdot \left(\frac{T - 20}{1487 - 20}\right)^{1.3}\right) \tag{1}$$

The variables included in this model were: plastic strain, $\varepsilon$, strain-rate, $\dot{\varepsilon}$, and workpiece temperature, $T$. Then, an additional component of stress is included to take into account the variation of the workpiece hardness on flow stress. Thus, the overall material flow stress model is presented by coupling these two parts as follows:

$$\sigma = (\varepsilon, \dot{\varepsilon}, T, HRC) = f\left(\sigma_{ref}(\varepsilon, \dot{\varepsilon}, T), \Delta\sigma(HRC)\right) \qquad (2)$$

where $\sigma_{ref}(\varepsilon, \dot{\varepsilon}, T)$ represents the flow stress curve at 46HRC while $\Delta\sigma(HRC)$ denotes the additional component of stress, reflecting the influence of workpiece hardness.

## Determination of the additional component of stress

For a given material, the hardness varies with different heat treatment, resulting in different material strength. This can be caused by designed heat treatment process or due to the temperature and short cooling time generated during the machining process. Consequently, it can be assumed that hardness is dependent of the high temperature reached along the primary and secondary shear zones. Therefore, in this study the hardness of the workpiece is incorporated in the flow stress using the following procedure:

1. Take yield stress and tensile strength as the start and the end points for a specific flow stress curve. If the hardness is higher, then both the yield stress and tensile strength are increased. The points within this range are obtained by assuming a linear behaviour which is added to the reference value;
2. Take the initial workpiece hardness as the reference one. If the temperature is higher than 980°C [8], then the new hardness is updated as follows:
   2.1 $HRC_{Ref} = HRC_{Initial}$
   2.2 IF (T > 980.0) THEN $\Delta_{HRC} = ((0.047059)*(T-980.0))$
      ELSE $\Delta_{HRC} = 0.0$
   2.3 $HRC_{updated} = HRC_{Initial} + \Delta_{HRC}$
   2.4 $HRC_{Ref} = USRE2(2)$ (Read from FEM at the current step)
   2.5 IF ($HRC_{updated} > HRC_{Ref}$) THEN HRC = $HRC_{updated}$ (New value to be used at the next step)
      ELSE HRC = $HRC_{Ref}$ (New value to be used at the next step)
3. Finally, for the given material the Young's modulus is assumed to be independent of hardness. This is often the case for most material.

This procedure was applied to the values reported by Shivpuri [5] and by Semiatin [8]. Some of percentage values used to increase or decrease from 46HRC (hardness of reference curve) to the other hardness values are shown in Figure 1.

Two 3$^{rd}$ order polynomial functions, namely $F$ and $G$, which taking into account the hardness are defined by using a regression analysis:

$$F(HRC) = -0.00283 \cdot HRC^3 + 0.49998 \cdot HRC^2 - 4.79458 \cdot HRC - 570.104 \qquad (3)$$

$$G(HRC) = -0.0001 \cdot HRC^3 + 0.1875 \cdot HRC^2 - 10.6558 \cdot HRC + 120.7277 \qquad (4)$$

In particular, the former, $F$, modifies the initial yield stress and the latter, $G$, the strain hardening curve. Following Eq. (2) and combining with Eqs. (3) and (4), the

777

flow stress model for AISI H13 tool steel at the different initial hardness can be expressed as follows:

$$\sigma_{eq} = \left(A + F + G \cdot \varepsilon + B\varepsilon^n\right)\left(1 + C \cdot \ln\left(\frac{\dot{\varepsilon}}{\dot{\varepsilon}_0}\right)\right)\left(D - E \cdot \left(\frac{T - T_{room}}{T_{melt} - T_{room}}\right)^m\right) \qquad (5)$$

Where $A$, $B$, $C$, $D$, $E$, $n$ and $m$ are the material constants found by Shatla et al. [7], $\dot{\varepsilon}_0$ is the reference strain-rate (1s$^{-1}$), $T$ the workpiece temperature, and $T_{melt}$ and $T_{room}$ are the material melting temperature and the room temperature, respectively.

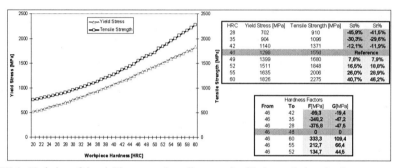

**FIGURE 1.** Variation of the yield stress and tensile strength with the workpiece hardness.

## MODEL CALIBRATION

The calibration of the proposed flow stress model was carried out by FE analysis for several cutting speeds and workpiece hardness by comparing the predicted results with those found experimentally. In particular, the aim of this calibration phase was to determine the critical damage value (CDV) and shear factor ($m$) as function of the initial hardness. Furthermore, the same iterative procedure was also utilized for determining the global heat transfer coefficient ($h_{int}$) at the tool-chip and tool-workpiece interfaces (Figure 2.a).

The FEM simulations were conducted in the same experimental conditions as those used by Ng et al. [9, 10]. In particular, a PCBN cutting tool was utilized with a chamfer geometry of 20°x0.2mm, a tool rake angle and a clearance angle of –5° and 5°, respectively. The cutting speed was varied between 75m/min and 200m/min, while the uncut chip thickness was kept constant and equal to 0.25mm. In addition, several workpiece hardnesses were considered (28HRC, 42HRC, 49HRC, 52HRC). The physical properties of the cutting tools and workpiece were taken from existing data [11]. The variation of the CDV with the workpiece hardness and different cutting speeds is shown in Figure 2.b. It can be noted that different CDV are found for different hardness, because the change of hardness for the same material causes the change of material flow stress, thus segmentation of the chip. This curve confirms that as hardness increases, the fracture toughness (or the critical damage value) decreases. Moreover, the shear factor value changes with the hardness according the following equation:

$$m(HRC) = 0.000276 \cdot HRC^2 - 0.016051 \cdot HRC + 0.933951 \qquad (6)$$

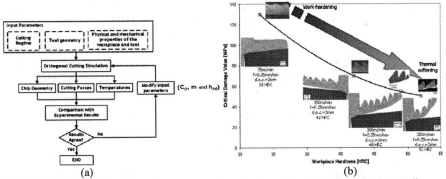

|                | (a)                                 | (b)                      |

**FIGURE 2.** (a) The utilized iterative procedure for calibration phase. (b) The obtained CDV diagram.

Using the proposed iterative procedure the global heat transfer coefficient, $h_{int}$, at the tool-chip and tool-workpiece was found, being equal to $100kW/m^2K$. This value permits to predict, with reasonable accuracy, the temperature distribution at the tool-chip interface.

## MODEL VALIDATION

The material model was then validated by comparing numerical predictions from FE simulations with the experimental results found in references [10, 12], concerning to the cutting forces, chip morphology and temperatures for different cutting speeds and material hardness.

In particular, a cutting speed of 200m/min and an uncut chip thickness of 0.25mm were used when AISI H13 with 42HRC (case 1) was machined using the PCBN cutting tool above mentioned, while a cutting speed of 75m/min and an uncut chip thickness of 0.25 mm were utilized when AISI H13 with 52HRC (case 2) was machined using the same PCBN cutting tool as case 1. The comparison between the predicted and the experimental results are reported in Table 1.

**TABLE 1.** Comparison between the experimental observations and the predicted results.

| | | Mechanical and Thermal Variables | | | Chip Morphology | | |
|---|---|---|---|---|---|---|---|
| | | Principal Force | Thrust Force | $T_{intAVE}$ | Pitch$_{AVE}$ | Pick$_{AVE}$ | Valley$_{AVE}$ |
| Case 1 | Experimental Results | 1290 N | 968 N | - | 312 μm | 278 μm | 83 μm |
| | Numerical Results | 1108 N | 913 N | - | 343 μm | 308 μm | 3104 μm |
| | Absolute Error | 14.1% | 16% | - | 9.9% | 10.8% | 25.3% |
| Case 2 | Experimental Results | 1499 | 1089 | 310°C | 385 μm | 314 μm | 186 μm |
| | Numerical Results | 1280 | 947 | 302°C | 322 μm | 291 μm | 157 μm |
| | Absolute Error | 14.6% | 13.0% | 2.06% | 16.4% | 7.3% | 15.6% |

Furthemore, the similarity between predicted and experimental chip morphology for cases 1 and 2 is shown in Figure 3 while the strain localization is depicted in Figure 4.

779

(a)                                    (b)

**FIGURE 3.** Experimental [10, 12] and predicted chip geometry: (a) 42HRC; (b) 52HRC.

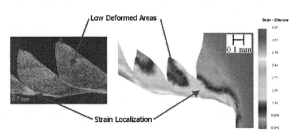

**FIGURE 4.** Experimentally [10] and numerically obtained strain localization.

# CONCLUSIONS

In this paper, a hardness based flow stress model for AISI H13 was developed and validated. This model takes into account the structural transformations generated during the machining process. Moreover, both the critical damage value and friction factor value were as a function of the material hardness. Based on the comparison between predicted and experimental results it can be concluded that the FE model incorporating the developed flow stress model can be successfully applied to simulate the hard machining of AISI H13 tool steel.

# REFERENCES

1. T. H. Chu and J. Wallbank, *J. Manuf. Sc. & Eng.* **121**, 259-263 (1998).
2. E. J. A. Armarego, I. S. Jawahir, V. A. Ostafiev, P. K. Venuvinod, "Modeling of Machining Operations", CIRP Working Group Paper, STC-C Paris, France, 1996.
3. J. Mackerle, *Int. J. Mach. Tools & Manuf.* **43**, 103-114 (2003).
4. H. K. Tonshoff, C. Arendt, R. Ben Amor, *Annals of the CIRP* **49/2**, 547-566 (2000).
5. R. Shivpuri, "Dies and Die Materials for Hot Forging", in *Dies and Die Materials*, edited by ASM Handbook, 2005, pp. 47-61.
6. M. C. Shaw and A. Vyas, *Annals of the CIRP* **42**, 29-33 (1993).
7. M. Shatla, C. Kerk, T. Altan, *Int. J. Mach. Tools Manuf.* **41**, 1511-1534 (2001).
8. S. L. Semiatin, "Introduction to Forming and Forging Processes", in *Forming and Forging*, edited by ASM Handbook, 2004, pp. 53-56.
9. E. G. Ng and D. K. Aspinwall, *J. Mat. Proc. Tech.* **127**, 222-229 (2002).
10. E. G. Ng and D. K. Aspinwall, *J. Manuf. Sc. & Eng.* **124**, 588-594 (2002).
11. Deform 2D® V9.0, User Manual, Columbus (OH), USA, 2006.
12. E. G. Ng, D. K. Aspinwall, D. Brazil, J. Monaghan, *Int. J. Mach. Tools Manuf.* **39**, 885-903 (2001).

# Deterioration Process of Sintered Material by Impact Repetition

## SHIRAKASHI Takahiro

*Department of Precision Machinery Engineering, Tokyo Denki University*
*2-2 Kandanishiki-cho, Chiyoda-ku, Tokyo, Japan*
*e-mail:sirakasi@cck.dendai.ac.jp*

**Abstract.** For prediction of time dependent tool breakage of sintered carbide tool in interrupted turning operation, the special impact stressing set-up is prepared. A change of fracture stress—
—deterioration process—of a sintered carbide tool material with both tensile and compressive impact stressing repetition is discussed and the process is evaluated through the fracture stress criterion superposed by Weibull's distribution. The reliability of fracture stress is decreased with the repetition, the maximum fracture stress, however, is not decreased. The equivalency between compressive and tensile stresses on the process is also discussed and the process is shown as change of probabilistic fracture locus with impact repetition times. Finally a deterioration state of sintered carbide tool under interrupted turning operation with the so called parallel entry and a very soft exit condition is estimated based on the deterioration process and the probability map of breakage occurrence on tool surface is shown under given cutting condition. The tool life based on breakage occurrence is also shown by fracture probability change with impact repetition and evaluated by experiments.

**Keywords:** Sintered carbide, impact stressing, fracture stress, Weibull's distribution, fracture locus, interrupted turning , tool breakage , tool life.
**PACS: 81.05.Je, 81.70.Bt.**

## 1. INTRODUCTION

It is very important to estimate a tool life in a practical machining process. Many theoretical and experimental reports for estimation of the life based on wear, but very few one based on tool chipping or breakage. It is clear that chipping of cutting edge take place when impact stress in a portion of the edge in interrupted cutting exceeds a fracture stress of tool material. The reason why a chipping occurrence probability increases with cutting time—number of interruption— may be thus attributed a decreasing of fracture stress with impact repetition. In other words, it will be possible to estimate analytically the time dependent occurrence of cutting edge chipping, when the decreasing of fracture stress with impact repetition can be evaluated beforehand from any impact test and transient distribution of the impact stress in cutting edge is also known.

CP907, *10th ESAFORM Conference on Material Forming,* edited by E. Cueto and F. Chinesta
© 2007 American Institute of Physics 978-0-7354-0414-4/07/$23.00

# 2. CHANGE OF FRACTURE STRESS
## WITH IMPACT REPETITION

## 2.1. Impact Fatigue Testing Set-up

Both impact compressive and tensile fatigue testing set-up are designed as shown in Fig.1. These are vertical Hopkinson bar type(1), where the rising time and the stressing duration can be adjusted. In the experiments loading rate 104 GPa/sec and loading duration 200 $\mu$ sec are mainly used as representative in actual interrupted cutting, and all test are conducted in room temperature.

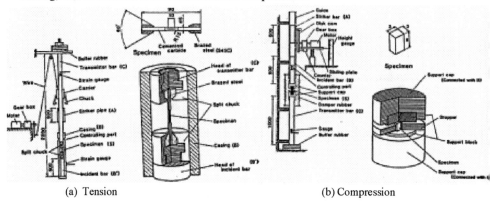

(a) Tension          (b) Compression

**FIGURE 1.** Impact Fatigue Testing Set-up

## 2.1. Change of Fracture Stress with Impact Repetition

It is well known that a fracture stress of brittle materials, such as a sintered carbide, is not determined uniquely, but some scatter of the stress is evitable. The Weibull's probability distribution has been recognized to fit successfully the stress of brittle materials. The following equation may be introduced,

$$G(\sigma^*, v) = 1 - \exp\{-v((\sigma^* - \sigma_u)^m / \sigma_0)\}, \qquad \sigma > \sigma_u$$

$$= 0 \qquad\qquad\qquad\qquad \sigma^* < \sigma_u \qquad (1)$$

Where $G(\sigma^*, v)$ is probability that specimen volume v has fractured when stress $\sigma^*$ is reached, and $\sigma_u$, $\sigma_0$, m are the Weibull's parameters. In Fig.2 are shown the Weibull's distribution of both tensile and compressive fracture stresses after each of tensile and compressive impact fatigue test. It is clearly seen that fracture stress of sintered carbide shows remarkable reduction and scatter with impact repetition, whereas the maximum fracture stress (99% probability) does not change.

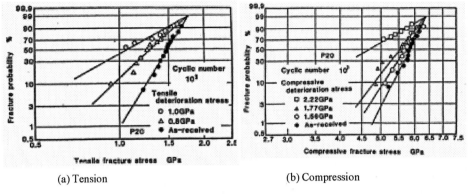

(a) Tension                (b) Compression

**FIGURE 2.** Impact Fracture Stress after Impact Fatigue

Since the impact stress within cutting edge is no uniform, but generally triaxial, a defferent extent of deterioration at each location will be expected. It is apparently impossible to prepare all the test data for various situations of the deteriorations, hence some methods of equivalent evaluation and conversion must be introduced. An important idea for the conversion may be obtained from Fig.3 which indicates that a same deteriorated state can be obtained either by tensile and compressive impact repetition, when the magnitude of the impact deterioration stresses is chosen appropriately. In the figure either tensile impact stress 0.8GPa or compressive one 1.77GPa yields a same distribution of both tensile and compressive fracture stresses. In Fig.4 is shown the equivalency between impact tensile and compressive deterioration stresses for giving the same deteriorated state. The two stresses are seen to be plotted on a unique straight line regardless of fracture manner and probability. The result indicates obviously the "identity of deterioration route" though the rate of deterioration is different correspondingly to the type and the magnitude of impact stress. Metallographic investigation also showed no appreciable crack of particular direction, but a uniform void structure in Co phase after deterioration.

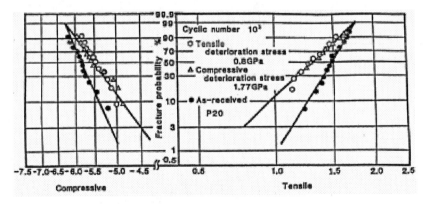

**FIGURE 3.** Equivalency of Deterioration Stress in Fracture Stress

Based on the results a conversion rule is proposed that the deterioration is solely governed by the sum of tensile principal strains, but not by compressive one. From this hypothesis, the relation between tensile and compressive deterioration stresses may be shown with Hooke's law as follows,

$$\epsilon_i = \{\sigma_i - \nu(\sigma_j + \sigma_k)\}/E \tag{2}$$

Where $\epsilon_i$, $\sigma_i$, ($\sigma_j$, $\sigma_k$) are principal strain and stress respectively, E is Young's modulus and $\nu$ is Poisson's ratio. Taking uniaxial deterioration, for example, effective tensile strain $\epsilon_e$ for particular deterioration state is

$$\epsilon_e = (\sigma_d)_t/E = 2\nu(\sigma_d)_c /E$$

Since Poisson' ratio of Sintered carbide (P20) is 0.22, the inclination of the line in Fig.4 (0.44) can be justified.

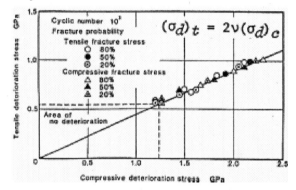

**FIGURE 4.** Equivalency between Tensile and Compressive Deterioration Stresses

## 2.2. Fracture Stress Criterion

Since the impact stressing cutting edge is generally triaxial, some fracture stress criterion under multi-axial stress state is required. One of the stress criterion is proposed by B. Paul[2] as follows,

$$\sigma_3/S_t < 1 - N_1^2: ((\sigma_1 - \sigma_3)/S_t)^2 + 2N_1^2((\sigma_1 + \sigma_3)/S_t)^2 + N_1^2(N_1^2 - 4) = 0$$

$$\sigma_3/S_t > 1 - N_1^2 : \sigma_1 /S_t = 1 \tag{3}$$

where 1 and 3 are algebraically maximum and minimum principal stresses acting on a specimen respectively, $S_t$ is uniaxial tensile fracture stress of the specimen, and $N_1$ is material constant. The Weibull's probability distribution is superposed upon the fracture criterion equation as shown in Fig.5. And an example of change of fracture stress criterion superposed by Weibull's distribution is shown in Fig.6.

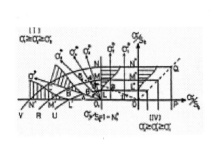

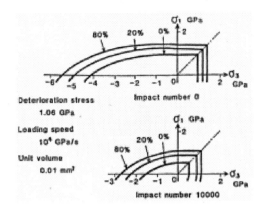

Deterioration stress
1.06 GPa

Loading speed
$10^4$ GPa/s

Unit volume
0.01 mm³

FIGURE 5. Fracture Locus     FIGURE 6. Change of Fracture Locus with Fatugue

## 3. FRACTURE PROBABILITY WITHIN CUTTING EDGE

A progress of tool edge fracture probability under interrupted turning with the so-called parallel entry and very soft exit condition is estimated and a change of fracture probability in tool with impact repetition is also estimated.

As the progress of deterioration directly depends on the sum of tensile principal strain, the deterioration map may be presumed from distribution of the sum as shown in Fig.7. The change of fracture probability distribution with impact repetition can be estimated based on Fig.7 as shown in Fig.8. The deterioration could be remarkably advanced along the nose edge and at a region just behind the cutting area shown by the chained line These results will lead the cumulative fracture probability that the fracture may occur in some where within cutting edge by the prescribed number of impact, which will be called as total fracture probability as shown in Fig.9, where very small edge chipping in around edge nose is also included.

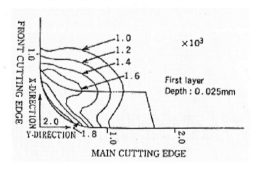

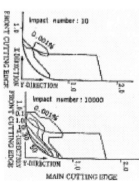

FIGURE 7. Deterioration Map in Tool     FIGURE 8. Change of Fracture Probability

785

In practical machining process the small edge chippings around nose edge is not so detrimental to continue the machining and their accumulation makes the cutting edge rather safe with formation of the edge roundness. The probability of large scale failure from behind contact area within impact repetition is shown in Fig.10. It is apparent that the predicted line is in good agreement with the experimental results shown by open circles.

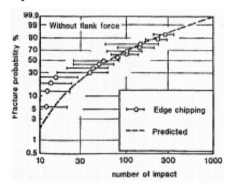 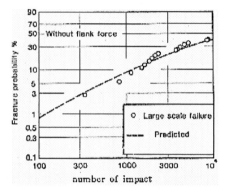

**FIGURE 9.** Total Fracture Probability  **FIGURE 10.** Fracture Probability without Small Chipping

## 4. CONCLUDING REMARKS

A cutting edge failure is most troublesome problems in practical machining station. In the paper an estimation method for the failure is proposed. For the estimation a change of mechanical characteristics of tool material is discussed through impact experiment. And combination of the change and the mechanical characteristic in machining, which will be obtained through numerical simulation on cutting process or experiment, will give the information of fracture occurrence with impact repetition in probability manner.

In order to obtain any practical information in machining process, the exact material characteristics are basic information. When characteristics are virtual ones, the obtained results are also virtual one. This type of discussion or research is only an exercise of computer.

REFERENCES

1. F. E .Hauser, Technics for Measuring Stress Strain Rate Relation at High Strain Rate , Experimental Machanics, 6, (1966)p. 395-401.
1. B. Paul, L. Mirandy, An Improved Fracture Criterion for Three Dimensional Stress State, Trans. ASME, 98, 2, (1976) p.159-165.

A4

Vanden Eynde, X., 1233
Vandepitte, D., 999
Van Hoof, T., 82
Van Houtte, P., 88, 309
Vanoverberghe, L., 1221
van Putten, K., 629
van Ravenswaaij, R., 9
Vegter, H., 187
Velasco, R., 405
Velay, V., 33
Veldman, E., 3
Vergnes, B., 1402
Verleye, B., 945
Verpoest, I., 945, 999, 1058
Viana, J. C., 795, 884
Vidal-Sallé, E., 1023
Villa Rodríguez, B. H., 1074
Villon, P., 1384, 1424
Vincent, C., 1205
Vladimirov, I. N., 139
Vos, M., 368
Vrh, M., 239
Vu Duong, A., 1033

# W

Wagner, S., 417
Wang, H., 337
Wang, J., 1092, 1098
Watremez, M., 685
Wauthier, A., 112
Welo, T., 127

Wen, S. W., 303
Wenner, M. L., 424
Wenzelburger, M., 1118, 1124
Wert, J. A., 665
Weyler, R., 1484
Wieczorek, A., 986
Wiggers, J., 1011
Willa-Hansen, A., 596
Willems, A., 999, 1058
Wiśniewski, B., 445

# Y

Yamamoto, T., 1506
Yang, J., 481
Yi, Y., 481
Youk, J. H., 853
Yu, W.-R., 853
Yvonnet, J., 1418

# Z

Zadpoor, A. A., 258
Zago, A., 509
Zagórski, R., 1086
Zamora, R., 1430
Zheng, C., 1472
Zi, A., 635
Zimmermann, C., 1124
Zineb, T. B., 47